Euler Identities

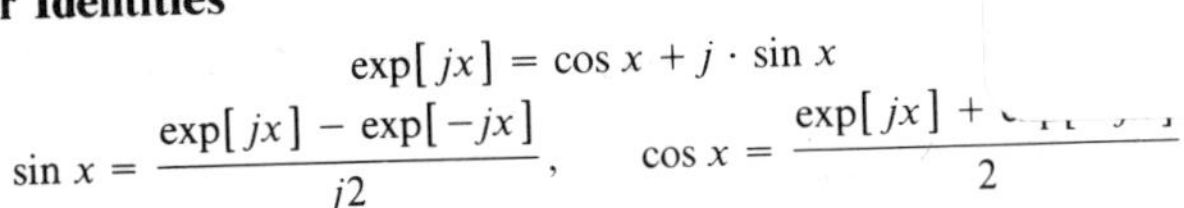

$$\exp[jx] = \cos x + j \cdot \sin x$$

$$\sin x = \frac{\exp[jx] - \exp[-jx]}{j2}, \qquad \cos x = \frac{\exp[jx] + \exp[-jx]}{2}$$

Derivatives

$$\frac{d}{dx} ax = a, \qquad \frac{d}{dx} x^n = nx^{n-1}$$

$$\frac{d}{dx}[f(x) + g(x)] = \frac{d}{dx} f(x) + \frac{d}{dx} g(x)$$

$$\frac{d}{dx}[f(x) \cdot g(x)] = f(x) \frac{d}{dx} g(x) + g(x) \frac{d}{dx} f(x)$$

$$\frac{d}{dx} \frac{f(x)}{g(x)} = \frac{1}{g(x)} \frac{d}{dx} f(x) - \frac{f(x)}{g(x)^2} \frac{d}{dx} g(x)$$

$$\frac{d}{dt}[f(t) \cdot u(t)] = u(t) \frac{d}{dt} f(t) + f(t) \cdot \delta(t)$$

$$\frac{d}{dx} \exp[ax] = a \cdot \exp[ax]$$

$$\frac{d}{dx} \sin ax = a \cdot \cos ax, \qquad \frac{d}{dx} \cos ax = -a \cdot \sin ax$$

$$\frac{d}{dx}\{f(x) \cdot \exp[ax]\} = \exp[ax] \cdot \frac{d}{dx} f(x) + a \cdot f(x) \cdot \exp[ax]$$

Indefinite Integrals

$$\int a \cdot f(x)\, dx = a \int f(x)\, dx$$

$$\int x^n\, dx = \frac{x^{n+1}}{n+1}, \qquad n \neq -1, \qquad \int x^{-1}\, dx = \ln x$$

$$\int \exp[ax]\, dx = \frac{1}{a} \cdot \exp[ax]$$

$$\int \cos[ax]\, dx = \frac{1}{a} \cdot \sin ax, \qquad \int \sin ax\, dx = -\frac{1}{a} \cdot \cos ax$$

$$\int \sin px \cdot \cos qx\, dx = -\frac{\cos[(p-q)x]}{2(p-q)} - \frac{\cos[(p+q)x]}{2(p+q)} \qquad \text{for} \quad p^2 \neq q^2$$

$$\int \frac{1}{a^2 + x^2}\, dx = \frac{1}{a} \cdot \tan^{-1} \frac{x}{a}$$

Definite Integrals

$$\int_a^b f(x)\, dx = \int_a^c f(x)\, dx + \int_c^b f(x)\, dx,$$

$$\text{where} \quad a < c < b$$

$$\int_0^{2\pi/a} \cos ax\, dx = \int_0^{2\pi/a} \sin ax\, dx = 0$$

$$\int_0^{2\pi/a} \cos[p(ax)] \cdot \cos[q(ax)]\, dx = 0 \qquad \text{for} \quad p^2 \neq q^2$$

$$\int_0^{\infty} \exp[-ax]\, dx = \frac{1}{a}$$

Transform Methods in Circuit Analysis

Cecil A. Harrison
University of Southern Mississippi

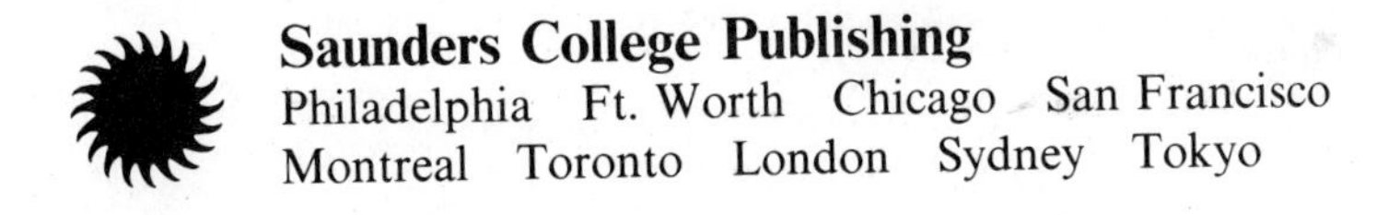
Saunders College Publishing
Philadelphia Ft. Worth Chicago San Francisco
Montreal Toronto London Sydney Tokyo

Text Typeface: 10/12 Times Roman
Compositor: Science Typographers, Inc.
Acquisitions Editor: Barbara Gingery
Production: Rachel Hockett, Spectrum Publisher Services
Manager of Art and Design: Carol Bleistine
Art and Design Coordinator: Doris Bruey
Text Designer: Hunter Graphics
Cover Designer: Lawrence R. Didona
Text Artwork: J & R Technical Services

Printed in the United States of America

Transform Methods in Circuit Analysis

0-03-20724-x (ISBN)

Library of Congress Cataloging-in-Publication Data

Harrison, Cecil A.
Transform methods in circuit analysis/by Cecil A. Harrison.
p. cm.
ISBN 0-03-20724-X
1. Electric circuit analysis. 2. Transformations (Mathematics)
I. Title.
TK454.H36 1990
621.319'2--dc20
89-24081
CIP

0123 0015987654321

To Iris

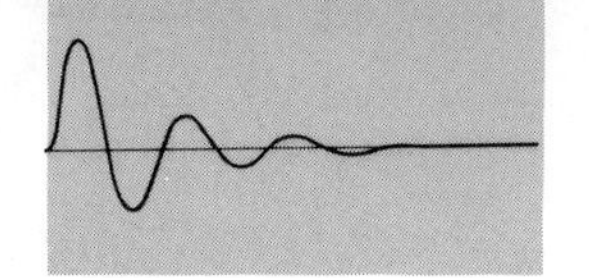

Preface

Course Application

In preparing the manuscript for this text, I have assumed that it will be used in an advanced circuit analysis course in a baccalaureate program in electrical or electronics engineering technology (EET), or in an intermediate circuit analysis course in an electrical engineering (EE) program. The manuscript began as a set of lecture notes for a one-semester transform circuit analysis course for third- or fourth-year students in an EET curriculum. The original material now resides in Chapters 2 through 7, and in Chapter 9.

Student Preparation

Student preparation for use of this text should consist of basic circuit analysis, college physics, and differential and integral calculus of one variable. The basic circuit analysis course and the calculus survey course in the typical EET program are adequate preparation, as are the core-curriculum calculus and circuit analysis courses in most engineering programs.

The experience level of students with regard to mathematics applications has been a major consideration in the design of this text. The mathematics preparation described above may be adequate in scope for linear circuit analysis, but it does not usually provide sufficient experience in applying mathematics to physical problems. I have sought to develop the student's ability to apply mathematics to circuit analysis by:

1. Placing mathematics **FLASHBACKS** within the text as a reminder of the mathematics principle being applied.
2. Employing careful step-by-step development in all example problems.

I use **FLASHBACKS** extensively in the early chapters, but I gradually reduce their frequency in later chapters.

Organization and Content

Chapter 1—Review of dc Circuit Analysis.

A compact review of dc circuit analysis which compensates for uneven backgrounds in circuit analysis and makes this text a single-source reference on linear circuit analysis techniques. In my classes, I assign Chapter 1 for outside study and proceed directly with Chapter 2.

Chapter 2—Waveform Analysis and Synthesis.

A presentation of the mathematical and graphical representation of piecewise-linear, piecewise-exponential, and sinusoidal waveforms.

Chapter 3—Time Domain Analysis.

A detailed presentation of time-domain analysis of first-order circuits, with emphasis on the operating characteristics of capacitors and inductors. I have emphasized understanding the origins of circuit transients.

Chapters 4 through 8—Laplace Transform Methods.

Chapter 4: A presentation of the properties of the Laplace transform and development of inverse transformation techniques.

Chapter 5: Introduction of the circuit transformation concept; application of Laplace transform methods to the same type of first-order circuits analyzed by time-domain methods in Chapter 3. I believe that this correlation between Chapters 3 and 5 builds confidence in transform analysis and reinforces understanding of time-domain analysis.

Chapter 6: A bridge between the fundamental techniques of Laplace transform analysis, and the more advanced applications which follow in Chapters 7 and 8. I explore the nature (e.g., overdamped, critically damped, or oscillatory) of higher-order transient responses, separate responses into transient and steady-state components, and relate response to circuit configuration.

Chapter 7: A compendium of most frequently used network analysis techniques—source conversion, node and mesh analysis, superposition, and Thevenin's and Norton's theorems—as they apply to transformed circuits.

Chapter 8: Application of Laplace transform methods to two-port networks with specific examples for OPAMP networks and loaded, linear transformers.

Chapter 9 through 11—Frequency Domain Analysis.

Chapter 9: A presentation of the essentials of frequency response analysis applied to the immitance function and the driving point response of one-port networks. I demonstrate that phasor (ac) analysis is a specialization of Laplace transform analysis. In consideration of uneven backgrounds in circuit analysis, and to further the usefulness of the text as a circuit analysis reference, I provide a compact review of ac analysis.

Chapter 10: Frequency response analysis of two-port networks with emphasis on the piecewise-linear Bode plot as a display technique. Under the premise that a filter is the preeminent example of a two-port network in which frequency response is the determinant characteristic, I have included a section on passive filters and modern filter design. I have also included a section on frequency response analysis of both ideal and nonideal transformers.
Chapter 11: A presentation of spectral analysis—both discrete (Fourier series) and continuous (Fourier transform).

Chapter 12—z-Transforms.
A brief treatment of the z-transform, and its applications to circuit analysis. I have emphasized the role of the sample-hold amplifier (SHA) and the analysis of coupling networks which contain a SHA.

Computer Applications

An appendix on the use of SPICE is included to permit the student to gain experience in using a readily available circuit analysis program. I have placed emphasis on the use of SPICE in the analysis of passive linear networks. The use of SPICE is illustrated in Chapters 1 through 3, and 5 through 11.

In Chapter 4, I have provided illustrations of the use of one of several personal computer programs available for inverse Laplace transformation.

Examples, Exercises, and Problems

My goal has been to provide at least one detailed example of every circuit analysis technique mentioned in this book; in many cases, I have provided more than one. I have also placed key exercises, with answers, at the end of nearly every section of text. My scheme for organizing end-of-chapter problems has been to group problems by type under a particular number (e.g., 7-3), and then provide several problems of that type [e.g., 7-3(a), 7-3(b), etc.]. I have provided answers for slightly more than one-half of the end-of-chapter problems. Examples, exercises, and problems which may be solved with SPICE, as well as solved analytically, are marked SPICE; end-of-chapter problems for which answers are provided at the end of the text are preceded by an asterisk (*).

Acknowledgments

From prospectus to release, the publication of this text has taken more than three years. The administration at the University of Southern Mississippi has been most supportive throughout this long period, and for this I thank them sincerely. I particularly thank Nancy Blessé, Senior Secretary of the School of Engineering Technology, who was most helpful in manuscript preparation.

Iraj Hajjar, Instructor of Computer Engineering Technology at the University of Southern Mississippi, checked all examples and exercises, and collaborated in preparing the solutions manual. I am very grateful for his patience throughout this tedious job.

Many reviewers participated at various stages of manuscript preparation, and I am indebted to each and every one for their constructive criticism and guidance: Albert McHenry (Arizona State), James Everly (Cincinnati), Doyle Ellerbruch (Manhattan State), Harold Broberg (Purdue), Russ Puckett (Texas A & M), Paul Wojnowiak (Southern Tech), John Polus (Purdue), Joseph Farren (Dayton), Frank Reave (Purdue), Nikola Sorak (Purdue), Lee Dickson (Weber State), and Bill Studyvin (Pittsburg State).

The staff at Saunders has been more than helpful to me, and I especially acknowledge the contribution of Barbara Gingery (Acquisitions Editor) and Charlene Squibb and Rachel Hockett (Production Managers).

My last acknowledgment is to Eric Harrison, a mathematician—and my son—who worked with me throughout the project, largely during the time he was pursuing a Master of Science degree in Mathematics. He has proofread and commented upon every page; if at any point a mathematical application or explanation appears especially lucid, this is probably thanks to Eric and his penchant for conciseness and clarity. He is also largely responsible for Appendices A and B.

Cecil A. Harrison

Contents

1 Review of Basic Circuit Analysis

1.1 INTRODUCTION

Basic circuit analysis courses and textbooks are commonly divided into three major sections:

1. **Dc Analysis.** Analysis of networks which contain only resistors and constant (dc) voltage or current sources.
2. **Ac Analysis or Phasor Analysis.** Analysis of networks which contain resistors, capacitors, inductors, and sinusoidal steady-state (ac) voltage or current sources.
3. **Transient Analysis.** Analysis of circuits with resistors, capacitors, and inductors, and which include switches (either mechanical or electronic) and/or sources which are neither constant (dc) nor sinusoidal steady state (ac).

This chapter will review the fundamental considerations of circuit analysis, the techniques of dc analysis, and some important network theorems. The elements of transient analysis will be reviewed and expanded upon in Chapter 3, and a review of phasors and ac analysis techniques is included in Chapter 9.

It should be emphasized here that the transform method presented in Chapters 5, 6, 7, and 8 is a general method for analysis of linear circuits which unifies dc analysis, transient analysis, and ac analysis. It will become clear in these chapters that dc analysis is an almost trivial specialization of the transform method. In Chapter 9, the phasor technique will be approached as a special case of the transform method.

1.2 Fundamental Considerations

We will begin our review of basic circuit analysis with the symbols for circuit elements and quantities and the units of measurement. Use of a consistent system of symbols is essential to the efficient analysis and design of electric circuits because it minimizes errors and provides for concise specification of parameters. An understanding and proper use of the units of measurement is essential because it serves to emphasize physical relationships, and it provides for continuous error checking through **dimensional analysis**. Dimensional analysis, the cancellation of units in formulas, is emphasized in basic science, technology, and engineering courses, and will be used in example problems in this text.

Symbols and Units

The symbols for circuit elements and quantities in Table 1.1 are consistent with common usage by technical writers and editors. The following scheme is used throughout the book when representing charge, energy, current, voltage, and power:

Lowercase symbols for instantaneous values (q, w, i, v, p).

Uppercase symbols for constant or steady-state values (Q, W, I, V, P).

Lowercase symbols with functional notation to emphasize time dependence [$q(t)$, $w(t)$, $i(t)$, $v(t)$, $p(t)$].

The units of measurement in Table 1.1 are consistent with the International System of Units (**le Système International d'Unités**, or SI) which also specifies the abbreviation to be used for each unit. Table 1.2 lists and defines the prefixes and their abbreviations for use with the SI units.

Note that the symbol for **capacitance** is the same as the abbreviation for **coulomb**, the symbol for **energy** is the same as the abbreviation for **watt**, and the symbol for constant **voltage** is the same as the abbreviation for **volt**. Care should be taken to distinguish the *symbol* for a quantity or element from the

Table 1.1
Symbols and Units

Quantity / Element	Symbol	SI Unit	Abbreviation
Charge	Q, q, $q(t)$	coulomb	C
Time	t	second	s (sec)[1]
Current/current source	I, i, $i(t)$	ampere (C/sec)	A
Energy, work	W, w, $w(t)$	joule	J
Voltage/voltage source	V, v, $v(t)$	volt (J/C)	V
Power	P, p, $p(t)$	watt (J/sec)	W
Resistance/resistor	R	ohm (V/A)	Ω
Conductance	G	siemens (A/V)	S
Capacitance/capacitor	C	farad [A/(V/sec)]	F
Inductance/inductor	L	henry [V/(A/sec)]	H

[1] The abbreviation for second, the SI unit of time, is "s," which leads to confusion between the unit of time and the symbol for the Laplace transform variable, which is "s" by convention. The abbreviation "sec" will be used for "second" in this book.

Table 1.2
Prefixes for SI Units

Prefix	Abbreviation	Multiplier
femto-	f	10^{-15}
pico-	p	10^{-12}
nano-	n	10^{-9}
micro-	μ	10^{-6}
milli-	m	10^{-3}
kilo-	k	10^{3}
mega-	M	10^{6}
giga-	G	10^{9}
tera-	T	10^{12}

abbreviation for a unit of measurement. (In print, the former is italicized, and the latter is spaced off.)

Terms, Definitions, and Conventions

Elements. An element is the most basic part of a circuit. Elements are used alone or in combination with other elements to **model**, or represent, practical electrical or electronic devices or components. Elements possess two **terminals** [Figure 1.1(a)], or points into which or from which current can flow. Between the two terminals, a **potential difference**, or voltage, can exist.

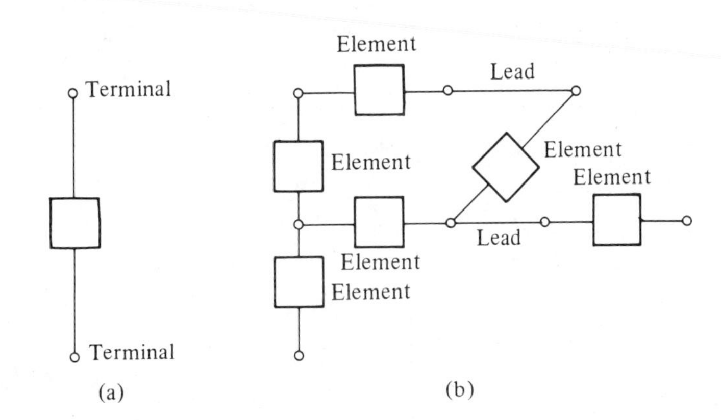

Figure 1.1
Elements and networks.
(a) Element.
(b) Network.

Networks. A **network** [Figure 1.1(b)] is formed when the terminals of two or more elements are connected either directly or by **leads**. A network consists of

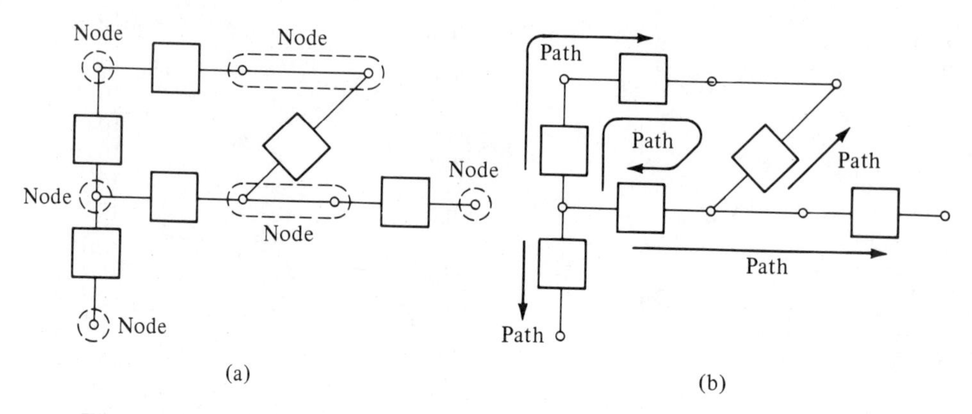

Figure 1.2 Nodes and paths in a network. (a) Nodes. (b) Paths (not all paths are indicated).

nodes and **paths**. A node is the point of connection for the terminals of two or more elements, and normally includes the leads [Figure 1.2(a)], while a path is a sequence of nodes and elements which always begins at a node and ends at a node [Figure 1.2(b)].

Circuits. A **closed path**, also called a **loop**, is a path which returns to the node from which it began [Figure 1.3(a)]. A loop which contains no other loops is called a **mesh**. The term **circuit** is commonly applied to any network which contains one or more loops. If there are no paths leading away from a given path, then the elements in the given path form a **branch** [Figure 1.3(b)].

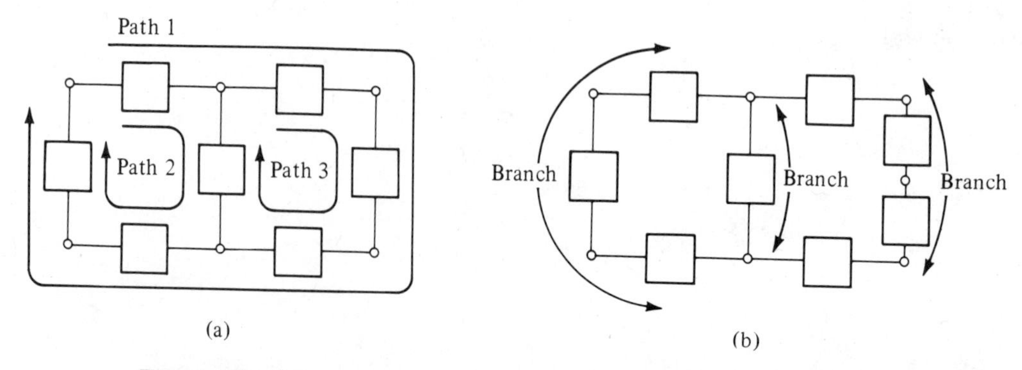

Figure 1.3 Loops, meshes, and branches. (a) All paths indicated are loops; only path 2 and path 3 are meshes. (b) Branches.

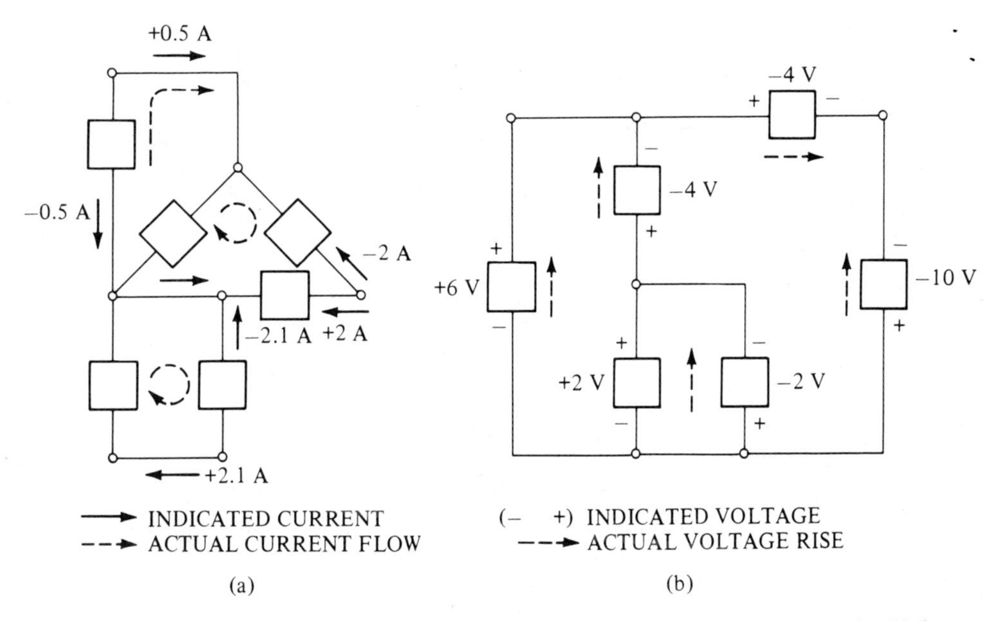

Figure 1.4 Current and voltage conventions. (a) Current flows. (b) Voltage rises and drops.

Current and Voltage Conventions. Arrows are used at terminals or parallel to leads to indicate current; the algebraic sign of the current quantity determines whether **current flow** is with or against the arrow [Figure 1.4(a)]. Plus (+) and minus (−) symbols are used at terminals or nodes to indicate potential difference, or voltage; the algebraic sign of the voltage quantity determines whether there is a **voltage rise** or a **voltage drop** from the − node to the + node [Figure 1.4(b)]. The terms **current direction** and **voltage polarity** will also be used with obvious meaning in the remainder of the text.

Energy and Power Conventions. When positive current flows *from* the terminal of highest potential, the element is **supplying** energy to the network [Figure 1.5(a)]. Such an element is said to be **active** and an **energy source**. When positive current flows *into* the terminal of highest potential, the element is **absorbing** energy from the network [Figure 1.5(b)]. Such an element is said to be **passive** and an **energy sink**. By convention, energy absorbed is given a positive algebraic sign, while energy supplied is given a negative sign. Since power is the rate of energy absorbed or supplied, the same sign convention applies to power.

Series and Parallel Connections. When elements are connected in such a manner that there is only one path for current flow [Figure 1.6(a)], and that

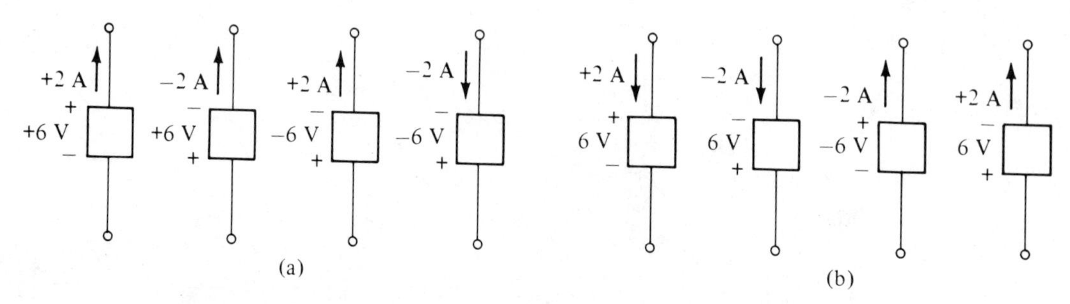

Figure 1.5 Energy conventions. (a) Active or energy-source elements; all elements supply energy. (b) Passive or energy-sink elements; all elements absorb energy.

current flows through every element, the elements are said to be **in series**. When elements are connected in such a manner that there is only one potential difference [Figure 1.6(b)], and that potential difference exists across the terminals of every element, the elements are said to be **in parallel**.

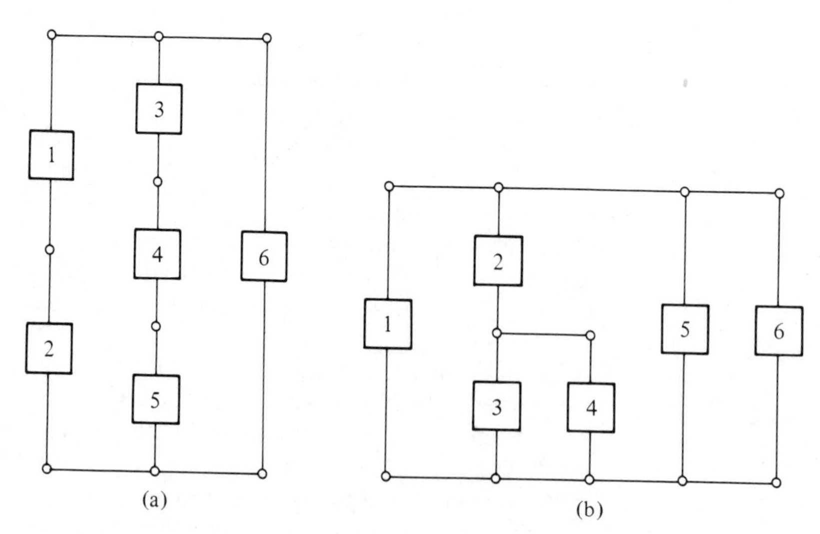

Figure 1.6
Series and parallel connections.
(a) Elements 1 and 2 in series; elements 3, 4, and 5 in series.
(b) Elements 1, 5, and 6 in parallel; elements 3 and 4 in parallel.

Short Circuits and Open Circuits. A short circuit is formed by placing a lead, rather than an element, between two nodes. The essential characteristic of a short circuit is that the potential difference between two nodes connected by a short circuit is 0 V. An open circuit is formed by removing any element or lead between two nodes. The essential characteristic of an open circuit is that the current flow in it is 0 A.

Source Elements

The two types of active elements to be considered here are **ideal independent sources**, hereafter referred to as **independent sources**, and **ideal linearly dependent sources**, hereafter referred to as **controlled sources**.

Independent Sources. An independent **voltage source** [Figure 1.7(a)] supplies a specified terminal voltage regardless of the amount of current which it is required to supply; hence, the source voltage is *independent* of any conditions which exist in the remainder of the network. Similarly, an independent **current source** [Figure 1.7(b)] supplies a specified amount of current regardless of its terminal voltage. The **terminal characteristic**, a means of describing the instan-

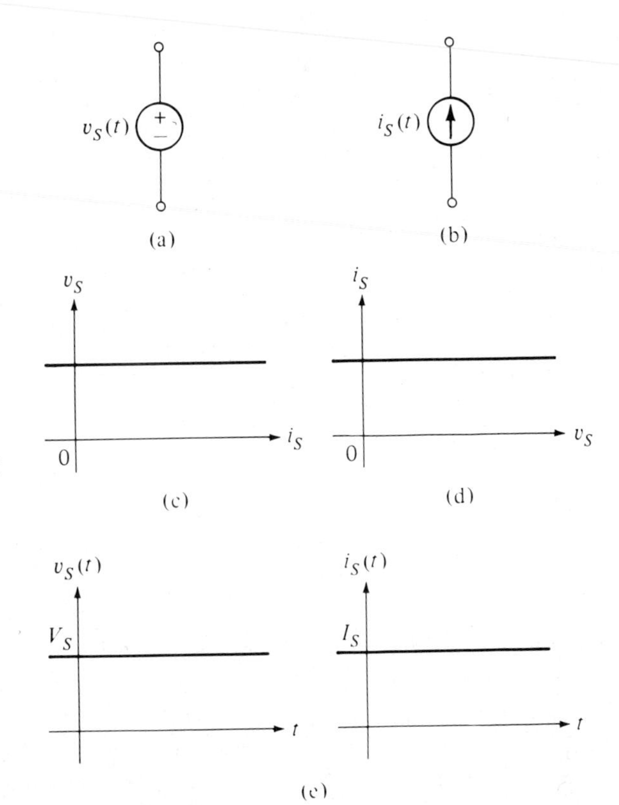

Figure 1.7 Independent sources. (a) Independent voltage source. (b) Independent current source. (c) Terminal characteristic of an independent voltage source. (d) Terminal characteristic of an independent current source. (e) Time dependence of *dc* sources.

taneous relationship between voltage and current at the terminals of an element, is shown in Figure 1.7(c) for the independent voltage source and in Figure 1.7(d) for the independent current source. In general, the voltage supplied by an independent voltage source or the current supplied by an independent current source may vary with time. Only those sources which do not display a time variation (commonly called **dc sources**) are considered in this chapter [Figure 1.7(e)]. Note that Figure 1.7(c) and (d) show the relationship between current and voltage, while Figure 1.7(e) shows the relationship between voltage and time and between current and time.

Controlled Sources. A **controlled voltage source** [Figure 1.8(a) or (c)] supplies a terminal voltage which *depends* upon a voltage $v_x(t)$ or a current $i_x(t)$ which exists elsewhere in the network. The voltage $v_x(t)$ is the **control voltage**, and $i_x(t)$ is the **control current**. The source voltage is related to the control voltage or current by a constant; hence, the controlled source is **linearly dependent** on the control voltage or current. Linearly dependent means that the source voltage depends upon the *first power* of the control voltage or current multiplied by a constant. For example, a controlled source $v_S(t) = 8v_x(t)$ is linearly dependent, as is $v_S(t) = 30i_x(t)$, but $v_S(t) = 2v_x(t)^2$ is *not* a linearly dependent source. Controlled voltage sources are identified as **voltage-controlled voltage sources** (VCVS) or **current-controlled voltage sources** (CCVS). Similarly, **voltage-controlled current sources** (VCCS) [Figure 1.8(b)] and **current-controlled current sources** (CCCS) [Figure 1.8(d)] supply current which is linearly dependent upon a control voltage or current existing elsewhere in the network. Examples of the use of controlled sources in modeling electronic devices are given in Figure 1.9.

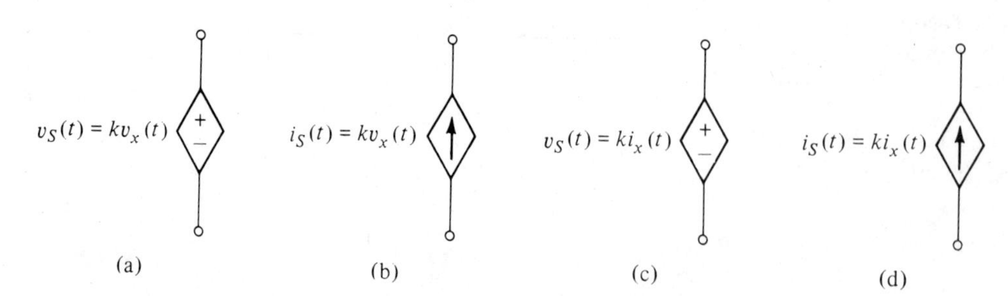

Figure 1.8 Linearly dependent, or controlled, sources. (a) Voltage-controlled voltage source (*VCVS*); k is dimensionless. (b) Voltage-controlled current source (*VCCS*); k is in siemens. (c) Current-controlled voltage source (*CCVS*); k is in ohms. (d) Current-controlled current source (*CCCS*); k is dimensionless.

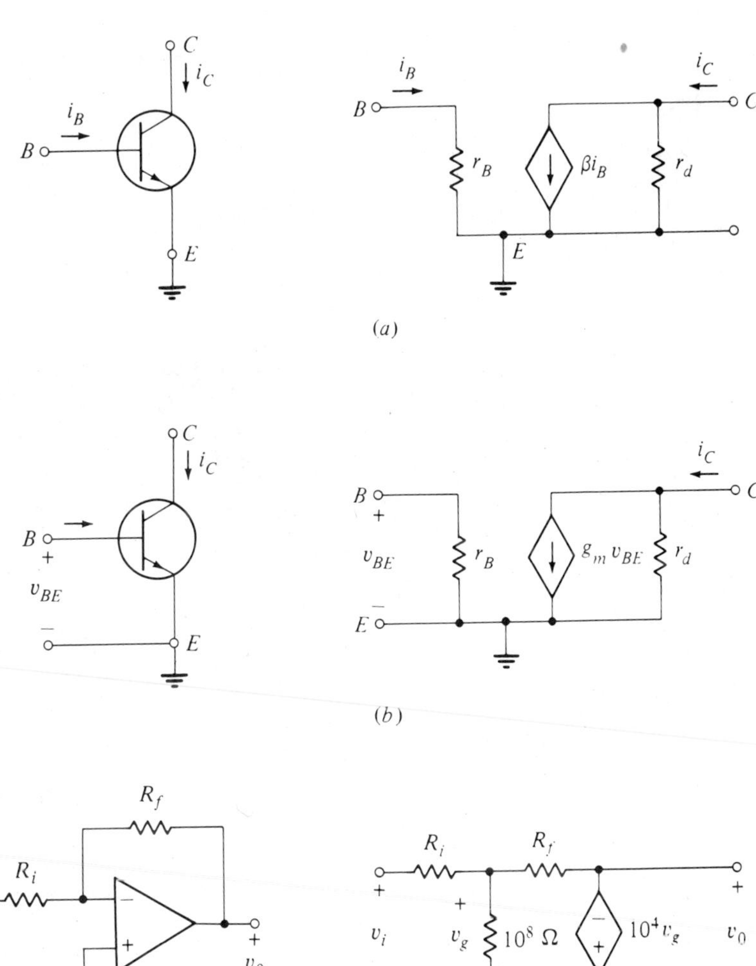

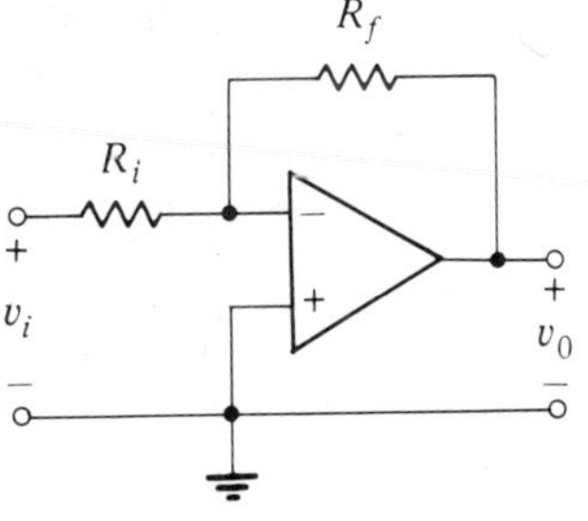

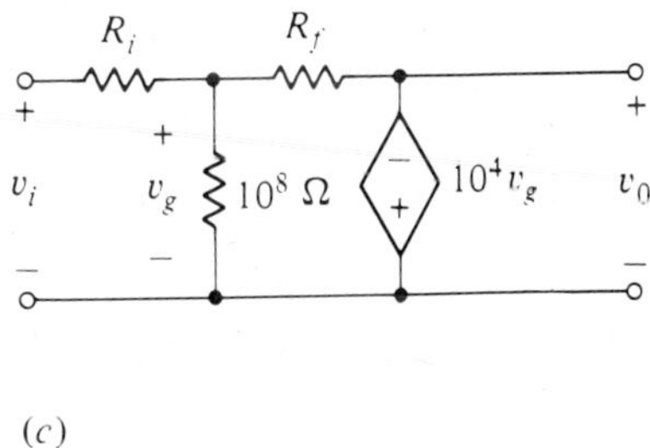

Figure 1.9
Examples of controlled-source applications.
(a) Hybrid π circuit model for an *NPN* transistor; β is dimensionless.
(b) Transconductance circuit model for an *NPN* transistor; g_m is in siemens.
(c) Circuit model for an OPAMP scaling circuit.

Ohm's Law: The Terminal Characteristic of a Resistor

The resistor is the only other circuit element which will be considered in this chapter. Resistors are passive elements, or energy sinks, which **dissipate** the absorbed energy as heat. A **linear resistor** has a **linear operating region**, symmetric about zero, for which current through the resistor is in direct proportion to the voltage across the resistor terminals. The observation that many materials possess a linear operating region is called **Ohm's law**, which is expressed as

$$v = Ri \tag{1.1}$$

where R is the **resistance** (in **ohms**), or

$$i = Gv \tag{1.2}$$

where G is the **conductance** (in **siemens**). Clearly, R and G are reciprocals.

The terminal characteristic of the resistor in Figure 1.10 is a graphic expression of Ohm's law. Operation of resistors within the linear region is implicit in linear circuit analysis.

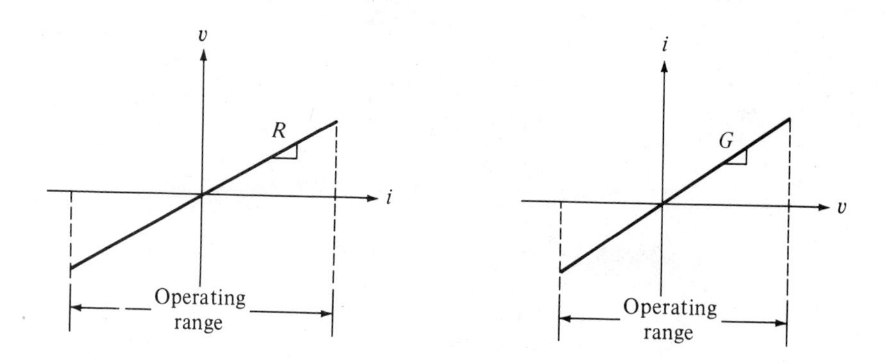

Figure 1.10 Terminal characteristic of a resistor.

The Kirchhoff Circuit Laws

The Kirchhoff circuit laws provide the basis for circuit analysis. The immediate consequences of the Kirchhoff laws are techniques for simplifying circuits and for determining the currents and voltages for nodes and branches throughout a circuit.

Kirchhoff's Current Law (KCL). KCL arises from the observation that charge cannot be stored at a node; hence, any current which enters the node must also leave the node. By using the current conventions given above, all current can be assumed to *enter* the node if the correct algebraic sign is used for the quantity of current in each path leading to the node. KCL then states that the algebraic sum of currents entering any node is zero, or

$$\sum_{k=1}^{N} i_k = 0 \tag{1.3}$$

where N is the number of paths leading to the node. Similarly, all currents can be assumed to *leave* the node, which again leads to Equation (1.3), this time in terms of current leaving a node.

Kirchhoff's Voltage Law (KVL). KVL arises from the observation that only one potential can exist at a node. Since a closed path (loop) which includes a particular node must begin and end at that node, any voltage rise in the loop must be offset by a corresponding drop in the same loop. The voltage conventions given above permit the assumption that all differences in potential are **drops** if the correct algebraic sign is used for the quantity of voltage between the terminals of each element in the loop. KVL then states that the algebraic sum of voltage drops in any loop is zero, or

$$\sum_{k=1}^{N} v_k = 0 \tag{1.4}$$

where N is the number of elements in the loop. Similarly, all differences in potential can be assumed to be **rises**, which again leads to Equation (1.4), this time in terms of voltage rises.

Duality. Observe the similarity between KCL and KVL. Replacing the words "current" and "node" in KCL with "voltage" and "loop" results in a statement of KVL, and replacing i_k in Equation (1.3) with v_k results in Equation (1.4). This similarity is the result of a powerful concept in circuit analysis called **duality**. Duality recognizes the mathematical similarity between related, but not identical, equations and physical quantities. Sufficient motivation for considering duality at this point is that duality will reduce by half the number of circuit analysis formulas to be remembered. Nodes and loops are dual features of a network, and current and voltage are dual electrical quantities. Other instances of duality will be pointed out as they occur.

1.3 Circuit Analysis Techniques

The *techniques* of circuit analysis are derived from application of the Kirchhoff circuit laws and the terminal characteristics of the circuit elements. In this section we will develop such basic techniques as circuit simplification through element combination, current and voltage division, source conversion, and assumed solutions.

Combination of Elements

Current and Voltage Sources. As a consequence of KCL, current sources connected in *parallel* may be combined and replaced with one equivalent source. If all sources are independent, the equivalent source is also indepen-

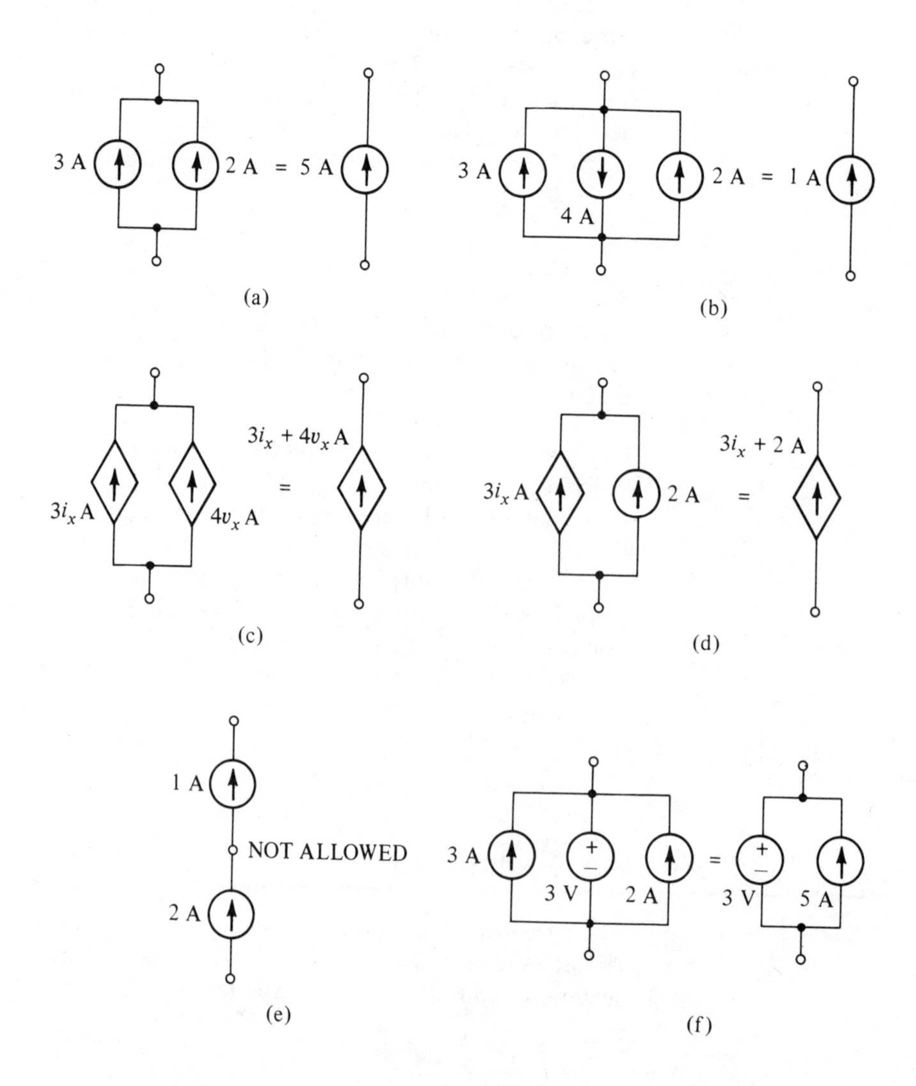

Figure 1.11 Combining current sources.

Figure 1.12 Combining voltage sources.

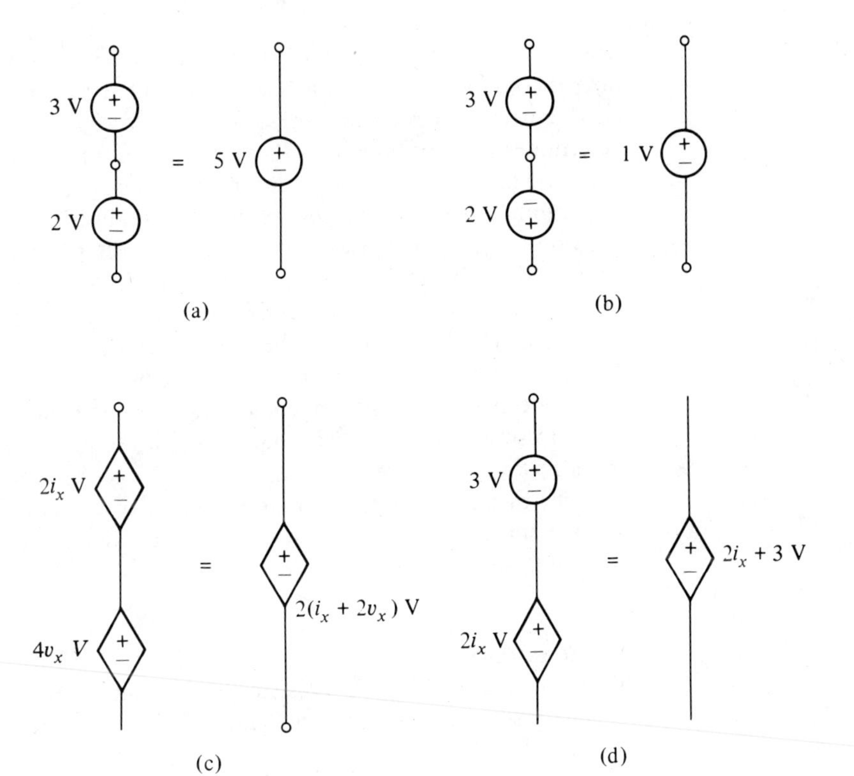

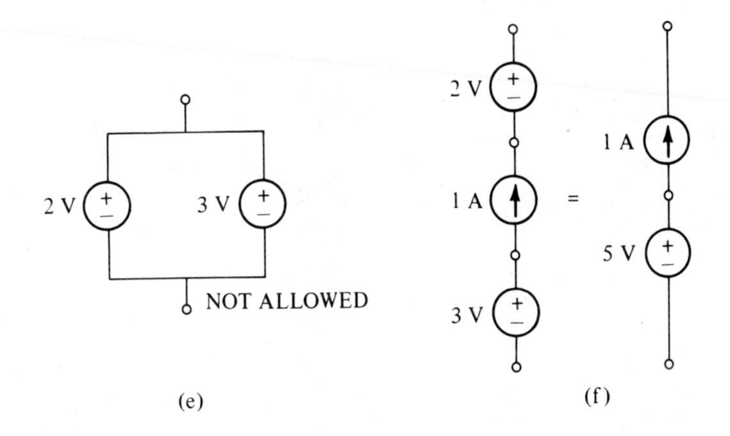

dent; however, if any of the parallel sources is a controlled current source, then the equivalent source is also a controlled current source. Connection of *ideal* current sources *in series* is not allowed, because a current source determines the current within its branch; if more than one current source is in the same branch, then the current in that branch is ambiguous. Some examples of combining current sources are given in Figure 1.11.

Similarly, as a consequence of KVL, voltage sources in *series* may be combined, with similar restrictions regarding controlled voltage sources. Connection of *ideal* voltage sources *in parallel* is not allowed, because a voltage source determines the voltage between its terminals; if more than one voltage source is in parallel, then the voltage between the terminals is ambiguous. There are no provisions for combining current sources with voltage sources either in series or in parallel. Some examples of combining voltage sources are given in Figure 1.12.

The observations regarding connection of current sources in parallel and voltage sources in series point out the fact that a parallel connection is the dual of a series connection.

Resistors in Series. For the circuit of Figure 1.13(b) with v and i equal to those of Figure 1.13(a), the resistance R_T is **equivalent** to the resistance of the N resistors in series. The equivalent resistance of any number of resistors

Figure 1.13
Combining resistors in series.
(a) Resistors in series.
(b) Equivalent resistance.

+
i
R_1
R_2
v
R_N
−
(a)

i
+
R_T
v
−
(b)

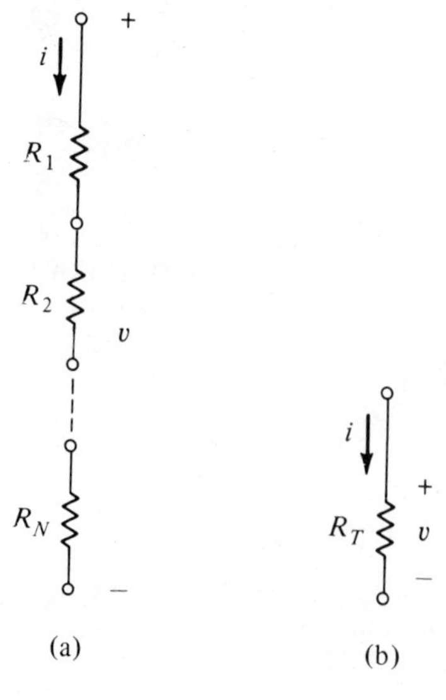

connected in series is found by adding the resistance values of the individual resistors, or

$$R_T = \sum_{k=1}^{N} R_k \tag{1.5}$$

Resistors in Parallel. For the circuit of Figure 1.14(b) with i and v equal to those of Figure 1.14(a), the conductance G_T is **equivalent** to the conductance of the N resistors in parallel. The equivalent conductance of any number of resistors connected in parallel is found by adding the conductance values of the individual resistors:

$$G_T = \sum_{k=1}^{N} G_k \tag{1.6}$$

If the equivalent resistance is required, then the reciprocal of the equivalent conductance can be calculated.

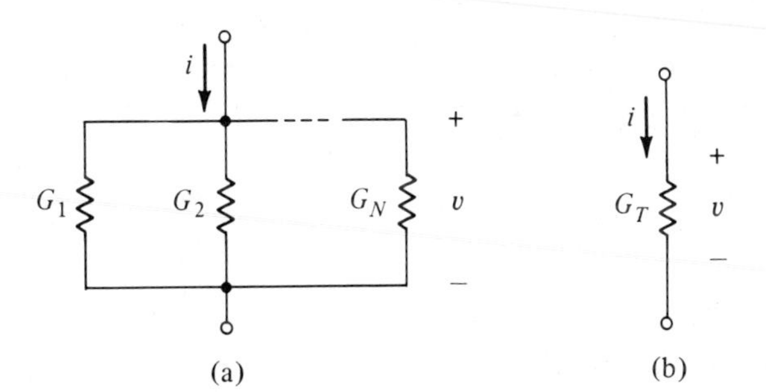

Figure 1.14
Combining resistors in parallel.
(a) Resistors in parallel.
(b) Equivalent conductance.

Equations (1.5) and (1.6) indicate that resistance is the dual of conductance, and that the formula for **series resistance** is the dual of the formula for **parallel conductance**. The proofs of Equations (1.5) and (1.6) follow directly from KVL and KCL, respectively, and are left as an exercise (Problem 1-2).

Voltage and Current Division Rules

Voltage Division. The **voltage division rule** (VDR) permits calculation of the voltage across any given resistor (R_x) in a branch consisting of any number

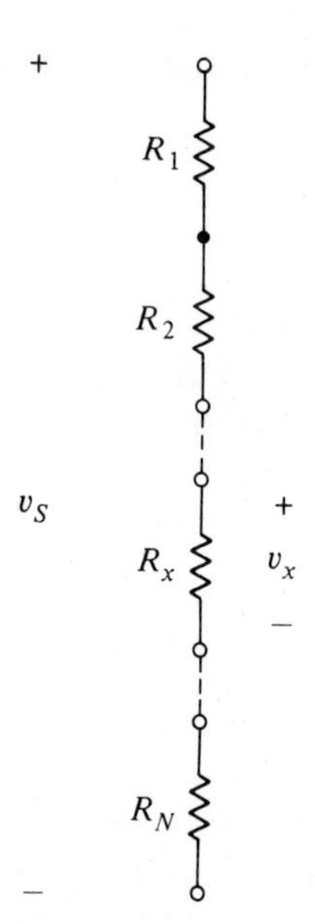

Figure 1.15
Voltage division rule.

(N) of resistors, provided that the branch voltage is known. For the circuit shown in Figure 1.15,

$$v_x = v_S \frac{R_x}{R_T} \tag{1.7}$$

Current Division. The **current division rule** (CDR) permits calculation of the current through any given resistor (R_x) in a parallel network of any number (N) of resistors, provided that the node current is known. For the circuit shown in Figure 1.16,

$$i_x = i_S \frac{G_x}{G_T} \tag{1.8}$$

Since voltage and current are duals, and resistance and conductance are duals, Equation (1.8) is the dual of Equation (1.7). Equation (1.7) follows from KVL and Equation (1.5), and Equation (1.8) follows from KCL and Equation (1.6); the proofs are left as an exercise (Problem 1-8). Examples 1.1 and 1.2 which follow illustrate the application of VDR and the CDR, respectively.

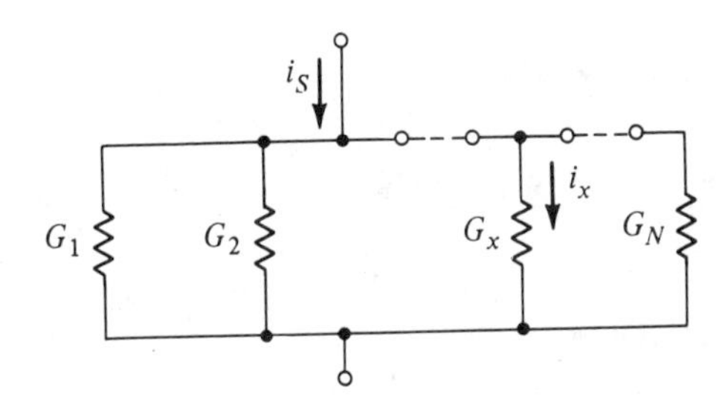

Figure 1.16
Current division rule.

Example 1.1

Problem:

Calculate the voltage v across the 30-Ω resistor in the series network shown in Figure 1.17.

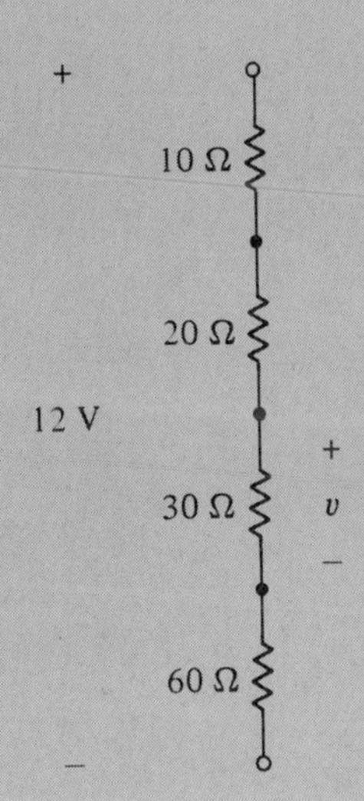

Figure 1.17
Example 1.1.

Solution:

From Equation (1.7),

$$v = (12\text{ V})\left(\frac{30\ \Omega}{10\ \Omega + 20\ \Omega + 30\ \Omega + 60\ \Omega}\right) = 3\text{ V}$$

Example 1.2

SPICE

Problem:

Calculate the current i through the 25-Ω resistor in the parallel network shown in Figure 1.18.

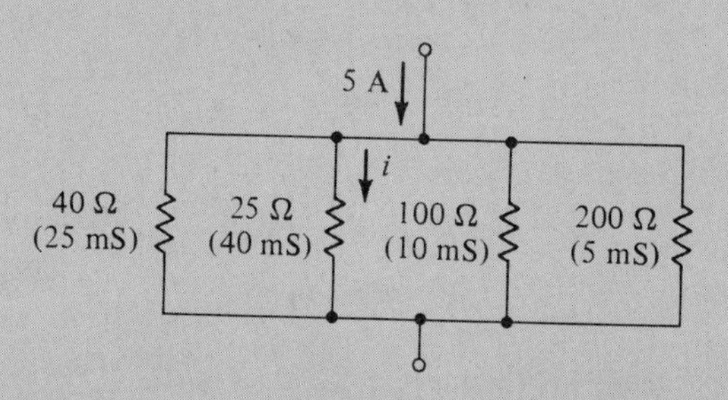

Figure 1.18
Example 1.2.

Solution:

Since the resistor specifications were given in ohms, as is usually the case, first calculate the conductance of each resistor (shown in parentheses in Figure 1.18). From Equation (1.8),

$$i = (5\text{ A})\left(\frac{40\text{ mS}}{25\text{ mS} + 40\text{ mS} + 10\text{ mS} + 5\text{ mS}}\right) = 2.5\text{ A}$$

Practical Sources and Source Conversions

The sources described in Section 1.2 are ideal elements because they maintain specified terminal conditions regardless of the **load**, that is, regardless of the amount of voltage dropped across the terminals of the current source or the amount of current flow from the voltage source. In practice, terminal conditions for the types of devices which provide electrical energy to a circuit—batteries, generators, etc.—vary with the load placed on the device.

In Figure 1.19(a), a battery is viewed as a voltage source. The voltage and current are observed as the load R_L is varied from $R_L \to \infty$ (open circuit) to $R_L \to 0$ (short circuit) and plotted in Figure 1.19(b). The linear relationship

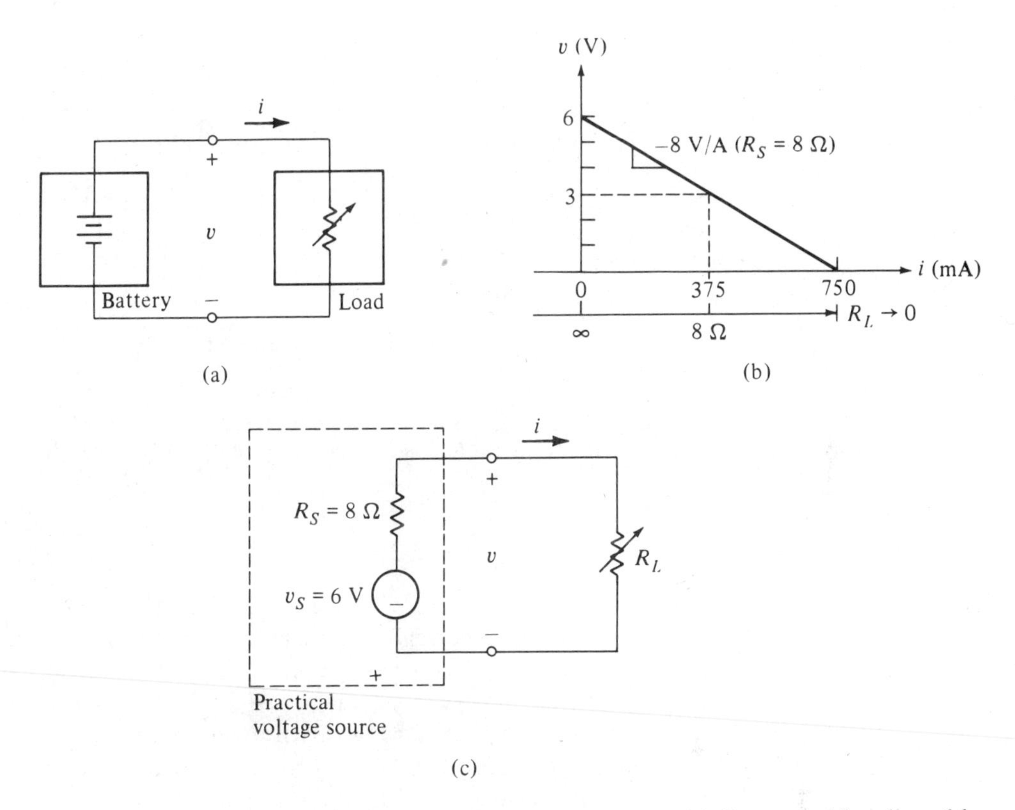

Figure 1.19 Model of a practical voltage source. (a) Battery with adjustable load. (b) v–i relationship for battery with adjustable load. (c) Model for battery with adjustable load.

between v and i suggests resistance within the battery. This **internal resistance** R_S, which can be determined from the negative of the slope of the v–i plot, leads to the model for a practical voltage source shown in Figure 1.19(c). In the model, the ideal voltage source has a value equal to the **open-circuit** voltage and is in series with the internal resistance. As a practical matter, internal resistance can also be determined directly by noting that the terminal voltage of the practical voltage source is one-half its open-circuit value when the load resistance is equal to the source resistance.

In Figure 1.20(a), a battery is viewed as a current source rather than as a voltage source, and the voltage and current are observed and plotted as the load is varied from $R_L \to 0$ $(G_L \to \infty)$ to $R_L \to \infty$ $(G_L \to 0)$. The linear relationship between i and v [Figure 1.20(b)] now suggests conductance within the battery. This **internal conductance** (G_S), which can be determined experimentally as the negative of the slope of the i–v plot or by noting the value of

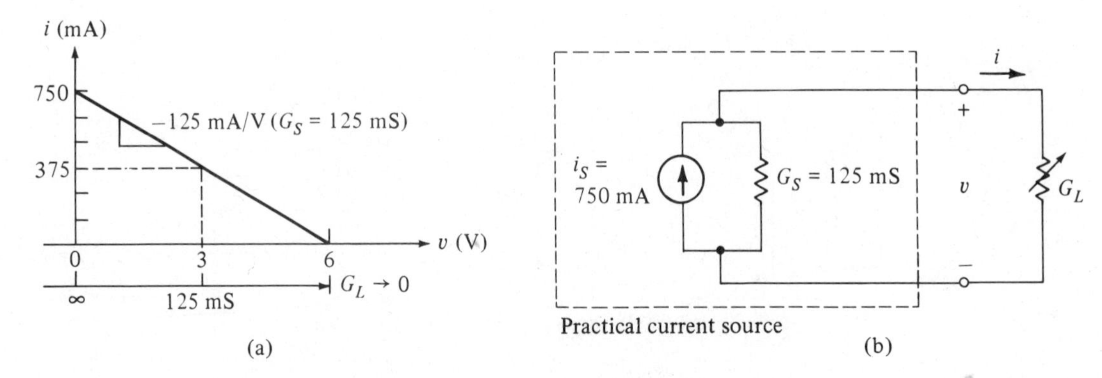

Figure 1.20 Model of a practical current source. (a) i–v relationship for battery with adjustable load (from Figure 1.19). (b) Model for battery with adjustable load.

G_L when the terminal current is one-half its short-circuit value, leads to the model for a practical current source shown in Figure 1.20(c). In this model, the ideal current source has a value equal to the short-circuit current and is in parallel with the internal conductance.

A relationship can now be seen between the practical-voltage-source model and the practical-current-source model. First, the internal *conductance* of the current-source model is the reciprocal of the internal *resistance* of the voltage-source model. Second, the ideal voltage source has a value equal to the short-circuit current multiplied by internal resistance,

$$v_S = i_{sc} R_S$$

or conversely, the ideal current source has a value equal to the open-circuit voltage multiplied by internal conductance,

$$i_S = v_{oc} G_S$$

These observations suggest that it is possible to *convert* from a practical-voltage-source model to a practical-current-source model, or vice versa. The usefulness of **source conversion** will become apparent in employing the circuit analysis techniques and network theorems to be reviewed later in this chapter. Verification of the equivalence of the two models for a given energy source is left as an exercise (Problem 1-9). It should be evident that the practical voltage source is the dual of the practical current source, and further, that the open circuit is the dual of the short circuit in a network.

Example 1.3

Problem:

The nominal **no-load** (open-circuit) voltage of a dry-cell battery pack is 9.2 V. According to the manufacturer's specifications, the voltage will be 8.7 V at its maximum rated current of 25 mA.

a. Assuming that the v–i relationship is linear between no-load and maximum rated load, what is the internal resistance (R_S) of the battery pack?

b. Sketch and label a practical voltage source model of the battery pack.

c. Use source conversion to find the practical current source model of the battery pack.

Solution:

a. Since the internal resistance is the negative of the slope of the v–i curve [Figure 1.21(a)],

$$R_S = -\frac{9.2\text{ V} - 8.7\text{ V}}{0\text{ mA} - 25\text{ mA}}$$

$$R_S = (20\text{ V/A})\left(\frac{1\ \Omega}{1\text{ V/A}}\right) = 20\ \Omega$$

b. See Figure 1.21(b).

c. For the practical-current-source model [Figure 1.21(c)],

$$G_S = 1/R_S$$

$$G_S = \left(\frac{1}{20\ \Omega}\right)\left(\frac{1\text{ S}}{1/(1\ \Omega)}\right) = 50\text{ mS}$$

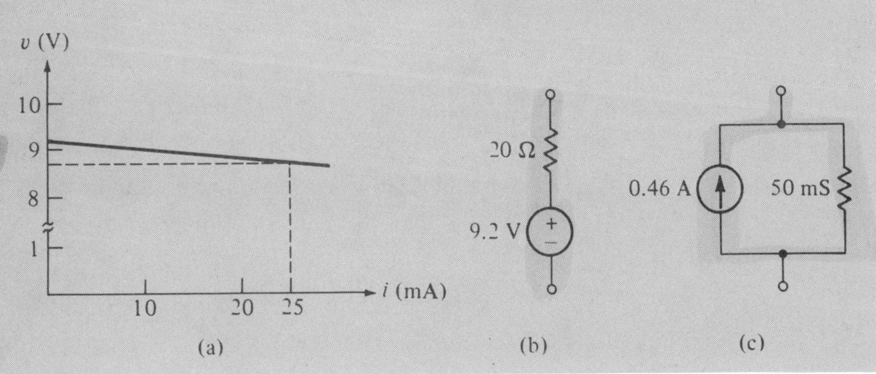

Figure 1.21 Example 1.3. (a) i–v relationship. (b) Practical-voltage-source model. (c) Practical-current source model.

and

$$i_S = v_{oc} G_S$$

$$i_S = (9.2 \text{ V})(50 \text{ mS})\left(\frac{1 \text{ A/V}}{1 \text{ S}}\right) = 0.46 \text{ A}$$

Example 1.4

Problem:

The short-circuit terminal current of a dc electronic power supply is measured as 0.3 A. A variable resistor, set to its minimum value, is connected across the terminals [Figure 1.22(b)], and the resistor current is

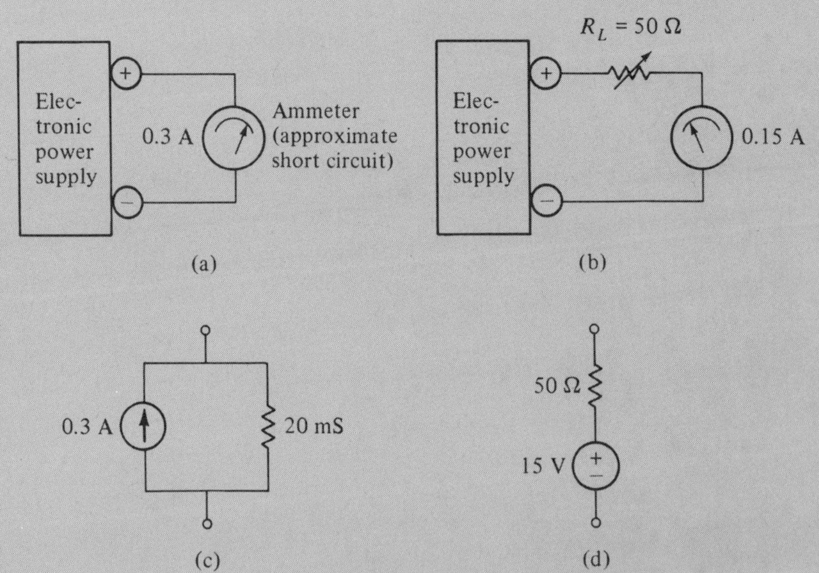

Figure 1.22
Example 1.4.
(a) Connection for short-circuit current measurement.
(b) Connection for load current measurement.
(c) Practical-current-source model.
(d) Practical-voltage-source model.

observed as the resistance is increased. When the current reaches 0.15 A, the resistor is disconnected and its resistance is measured as 50 Ω.

a. Sketch and label a practical-current-source model for the power supply.

b. Assuming that the power supply output is linear, what is maximum voltage available?

c. Use source conversion to find the practical-voltage-source model for the power supply.

Solution:

a. The value of the ideal current source in the practical-current-source model is equal to the short-circuit current, or 0.3 A. The internal conductance is equal to the load conductance required to reduce the load current to one-half its short-circuit value, or

$$G_S = \left(\frac{1}{50\ \Omega}\right)\left(\frac{1\ \text{S}}{1/(1\ \Omega)}\right) = 20\ \text{mS}$$

b. The maximum voltage will occur under open-circuit conditions and will be the voltage developed by the ideal current source across the internal conductance,

$$v_{oc} = i_S/G_S$$

$$v_{oc} = \left(\frac{0.3\ \text{A}}{20\ \text{mS}}\right)\left(\frac{S}{(1\ \text{A})/(1\ \text{V})}\right) = 15\ \text{V}$$

c. For the practical-voltage-source model,

$$v_S = v_{oc} = 15\ \text{V}$$

and

$$R_S = 1/G_S$$

$$R_S = \left(\frac{1}{20\ \text{mS}}\right)\left(\frac{1\ \Omega}{1/(1\ \text{S})}\right) = 50\ \Omega$$

Assumed Solutions

Perhaps the most fundamental technique for analyzing an electric circuit is the one most frequently overlooked—the method of **assumed solutions**. This technique is best explained by an illustration such as that given in Example 1.5.

Example 1.5

Problem:

SPICE

Determine all currents and voltages in the circuit shown in Figure 1.23(a).

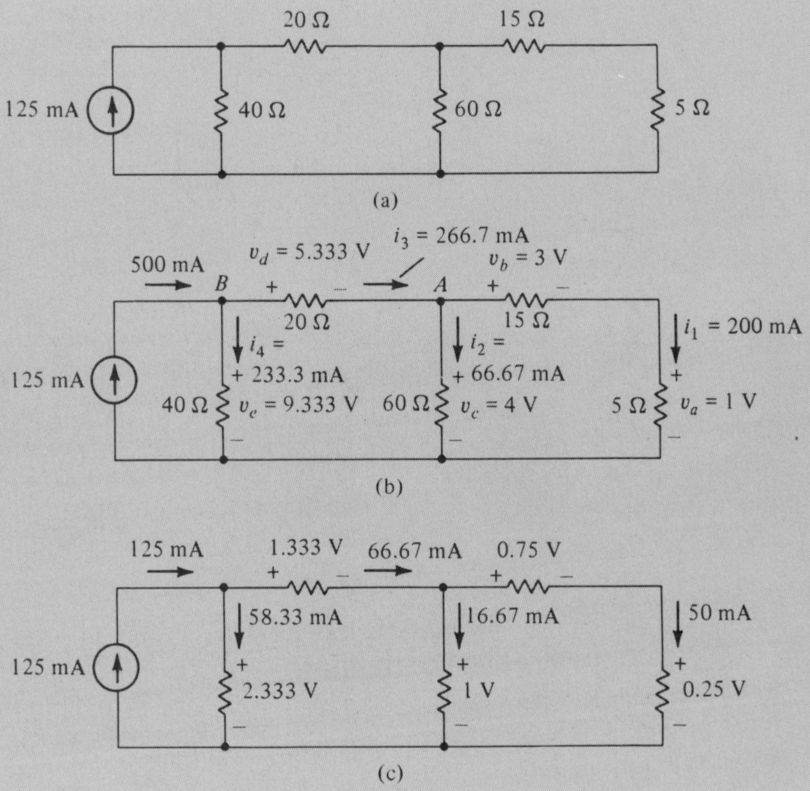

Figure 1.23
Example 1.5.
(a) Circuit to be analyzed for all voltages and currents.
(b) Assumed voltages and currents.
(c) Corrected voltages and currents.

Solution:

1. Begin with the element farthest from the source, the 5-Ω resistor, and assume that the voltage across that element is $v_a = 1$ V [Figure 1.23(b)]. According to Ohm's law, the current through the 5-Ω resistor is

$$i_1 = \left(\frac{1\text{ V}}{5\ \Omega}\right)\left(\frac{1\ \Omega}{1\text{ V/A}}\right) = 0.2\text{ A} = 200\text{ mA}$$

2. The same current flows through the 15-Ω resistor, so according to Ohm's law the voltage across the 15-Ω resistor is

$$v_b = (0.2\text{ A})(15\ \Omega)\left(\frac{1\text{ V/A}}{1\ \Omega}\right) = 3\text{ V}$$

3. According to KVL, the voltage across the branch containing the 5-Ω and 15-Ω resistors is

$$v_c = v_a + v_b = 1\text{ V} + 3\text{ V} = 4\text{ V}$$

4. Since this branch is in parallel with the branch containing only the 60-Ω resistor, the voltage across both branches is 4 V, and the current in the branch containing the 60-Ω resistor is

$$i_2 = \frac{v_c}{60\ \Omega} = \left(\frac{4\text{ V}}{60\ \Omega}\right)\left(\frac{1\ \Omega}{1\text{ V/A}}\right) = 66.67\text{ mA}$$

5. According to KCL, the current entering node *A* is

$$i_3 = i_1 + i_2 = 200\text{ mA} + 66.67\text{ mA} = 266.7\text{ mA}$$

6. The voltage drop across the 20-Ω resistor is

$$v_d = (i_3)(20\ \Omega) = (266.7\text{ mA})(20\ \Omega)\left(\frac{1\text{ V/A}}{1\ \Omega}\right) = 5.333\text{ V}$$

7. According to KVL, the voltage across the 40-Ω resistor is

$$v_e = v_c + v_d = 4\text{ V} + 5.333\text{ V} = 9.333\text{ V}$$

8. The current through the 40-Ω resistor is

$$i_4 = \frac{v_e}{40\ \Omega} = \left(\frac{9.333\text{ V}}{40\ \Omega}\right)\left(\frac{1\ \Omega}{1\text{ V/A}}\right) = 233.3\text{ mA}$$

9. KCL requires that the current entering node *B* must equal the current leaving node *B*. According to the assumed solution, the total current leaving node *B* is

$$233.3\text{ mA} + 266.7\text{ mA} = 500\text{ mA}$$

10. This contradicts the fact that the only current entering node *B* is the fixed source current, 125 mA. Since KCL must be satisfied at every node, the current in the assumed solution must be reduced by a factor of 4. In a circuit such as this which contains only *linear* elements, any change in current or voltage anywhere in the circuit will result in a *proportional* change in current or voltage everywhere in the circuit. When all voltages and currents are reduced by a factor of 4, the correct solution is realized [Figure 1.23(c)].

In summary, the assumed-solution technique involves beginning at some element, usually the farthest from the source, and assuming a current or voltage for that element. Ohm's law is then applied at each element, KVL at each loop, and KCL at each node until a contradiction is reached. The multiplication factor required to resolve the contradiction is used throughout the circuit to correct the results of the initial assumed value of voltage or current. If no contradiction is reached, the original assumption was a fortunate one.

The assumed-solution technique is especially useful in analyzing **ladder networks**. Example 1.6 illustrates analysis of a R–$2R$ ladder used in digital-to-analog converters.

Example 1.6

Problem:

SPICE

Let $R = 1\ \text{k}\Omega$, and determine the current i_1 in the R–$2R$ ladder in Figure 1.24(a).

Solution:

1. Assume $i_1 = 1$ mA; then in Figure 1.24(b),

$$
\begin{aligned}
v_a &= (2\ \text{k}\Omega)(i_1) = 2\ \text{V} \\
v_b &= (1\ \text{k}\Omega)(i_1) = 1\ \text{V} \\
v_c &= v_a + v_b = 3\ \text{V} \\
i_2 &= v_c/(2\ \text{k}\Omega) = 1.5\ \text{mA} \\
i_3 &= i_1 + i_2 = 2.5\ \text{mA} \\
v_d &= (1\ \text{k}\Omega)(i_3) = 2.5\ \text{V} \\
v_e &= v_c + v_d = 5.5\ \text{V} \\
i_4 &= v_e/(2\ \text{k}\Omega) = 2.75\ \text{mA} \\
i_5 &= i_3 + i_4 = 5.25\ \text{mA} \\
v_f &= (1\ \text{k}\Omega)(i_5) = 5.25\ \text{V} \\
v_g &= v_e + v_f = 10.75\ \text{V} \\
i_6 &= v_g/2\ \text{k}\Omega = 5.375\ \text{mA} \\
i_7 &= i_5 + i_6 = 10.625\ \text{mA} \\
v_h &= (2\ \text{k}\Omega)(i_7) = 21.25\ \text{V}
\end{aligned}
$$

2. KVL requires that the voltage drops must equal the voltage rises in the rightmost loop. According to the assumed solution, the total voltage

Figure 1.24
Example 1.6.
(a) Circuit to be analyzed.
(b) Assumed voltages and currents.
(c) Corrected voltages and currents.

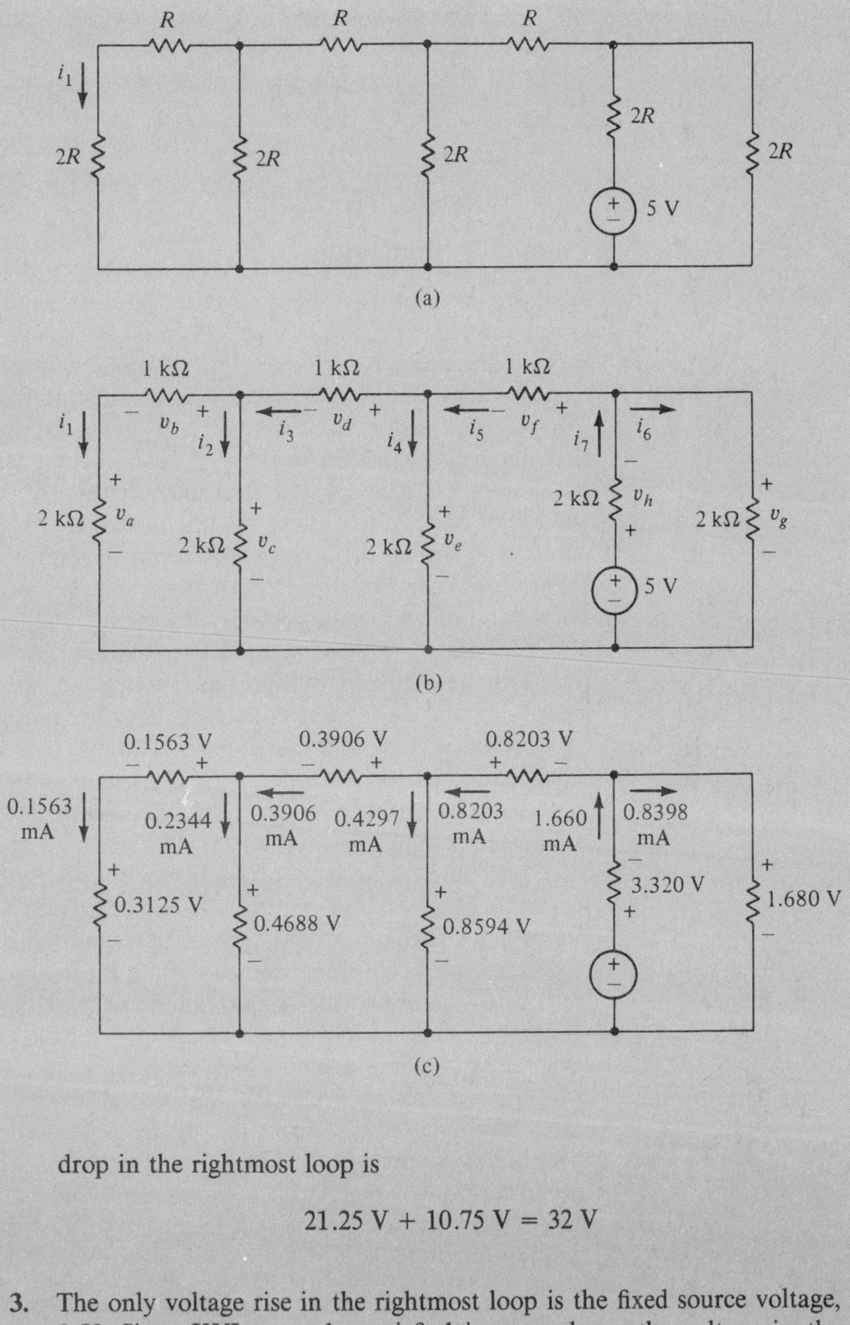

drop in the rightmost loop is

$$21.25 \text{ V} + 10.75 \text{ V} = 32 \text{ V}$$

3. The only voltage rise in the rightmost loop is the fixed source voltage, 5 V. Since KVL must be satisfied in every loop, the voltage in the assumed solution must be reduced by a factor of 5/32 (or 0.15625).

When the assumed value of i_1 (1 mA) is multiplied by 0.15625, the correct solution (0.15625 mA) is obtained. Any other electrical quantity in the circuit may also be obtained by multiplying the assumed quantity by 0.15625 [Figure 1.24(c)].

Computer Simulation

Digital computers have been used for circuit analysis for more than twenty years; however, efficient, economical, and easy-to-use circuit analysis programs have been readily available for less than half that period. The appearance of the simulation program called SPICE (for Simulation Program with Integrated Circuit Emphasis) in the public domain has made it possible for the users of virtually any mainframe computer or minicomputer to conduct computer simulations of electrical and electronic circuits. Within the past few years, a number of organizations have modified SPICE for use with personal microcomputers, and several of these organizations have provided SPICE versions at no charge for classroom use. Other circuit analysis programs are available, but SPICE in all its versions is certainly the most widely used.

All versions of SPICE require that the user provide a description of the circuit to be analyzed by specifying a unique number for each node and identifying the element(s) connected between nodes. The method of input depends upon the SPICE version and varies from noninteractive data files (assumed in this text) to fully interactive graphics displays. SPICE is capable of several modes of analysis, such as dc, ac, transient, etc., and provides tabular and graphic output displays, all of which must be specified by the user.

The reader who is not familiar with the SPICE should now turn to Appendix D for a brief tutorial on the details of node numbering, element description, selection of analysis mode, and selection of output display. Appendix D is *not* a comprehensive guide for the use of SPICE; however, most versions of SPICE provide a user's guide as part of the software, and many versions provide on-line user information (HELP). In addition, a number of comprehensive SPICE manuals have been published and are available either in bookstores or from the publishers.

A computer simulation[1] of the circuit in Example 1.6 was conducted with the result shown in SPICE 1.1. Note that the SPICE output **echoes** the input data for verification of the circuit configuration. Only the node voltages and the currents *entering* the voltage sources were printed by SPICE. A technique for

[1] This simulation and the others in this text were conducted with a Honeywell DPS-90 Vector Processor using SPICE Version 2G6.

causing SPICE to print currents in other branches will be illustrated in later simulations.

```
*******09/17/88 ********  SPICE 2G.6    3/15/83 ********11:56:38*****
SPICE 1.1
****     INPUT LISTING                           TEMPERATURE =   27.000 DEG C
*******************************************************************************
 R11 1 2 1K
 R12 2 3 1K
 R13 3 4 1K
 R21 1 0 2K
 R22 2 0 2K
 R23 3 0 2K
 R24 4 5 2K
 R25 4 0 2K
 V 5 0 DC 5
 .END
*******09/17/88 ********  SPICE 2G.6    3/15/83 ********11:56:38*****
SPICE 1.1
****     SMALL SIGNAL BIAS SOLUTION              TEMPERATURE =   27.000 DEG C
*******************************************************************************
  NODE   VOLTAGE      NODE   VOLTAGE      NODE   VOLTAGE      NODE   VOLTAGE
 (  1)     .3125     (  2)     .4687     (  3)     .8594     (  4)    1.6797
 (  5)    5.0000
     VOLTAGE SOURCE CURRENTS
     NAME        CURRENT
     V        -1.660D-03
     TOTAL POWER DISSIPATION   8.30D-C3  WATTS
         JOB CONCLUDED
```

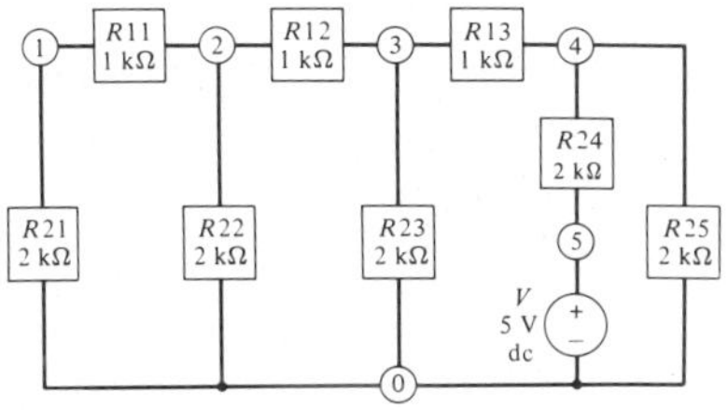

1.4 Node Analysis

Two circuit analysis techniques which follow directly from the Kirchhoff circuit laws are **node analysis** and **mesh analysis**. The two are duals, because node analysis is based on applying KCL at every node, and mesh analysis is based on applying KVL in every mesh. Both techniques result in a system of linear algebraic equations, which must be solved for the unknown quantities. In the system of node-analysis equations, the unknown quantities are the node voltages relative to a designated reference node. In the system of mesh-analysis equations, the unknown quantities are the mesh currents in a designated direction. Of the two techniques, node analysis is the more general, because mesh analysis is limited to **planar** circuits. Planar circuits are those circuits which can be drawn in a plane without crossover of elements or leads. The difference between a planar and a nonplanar circuit is illustrated in Figure

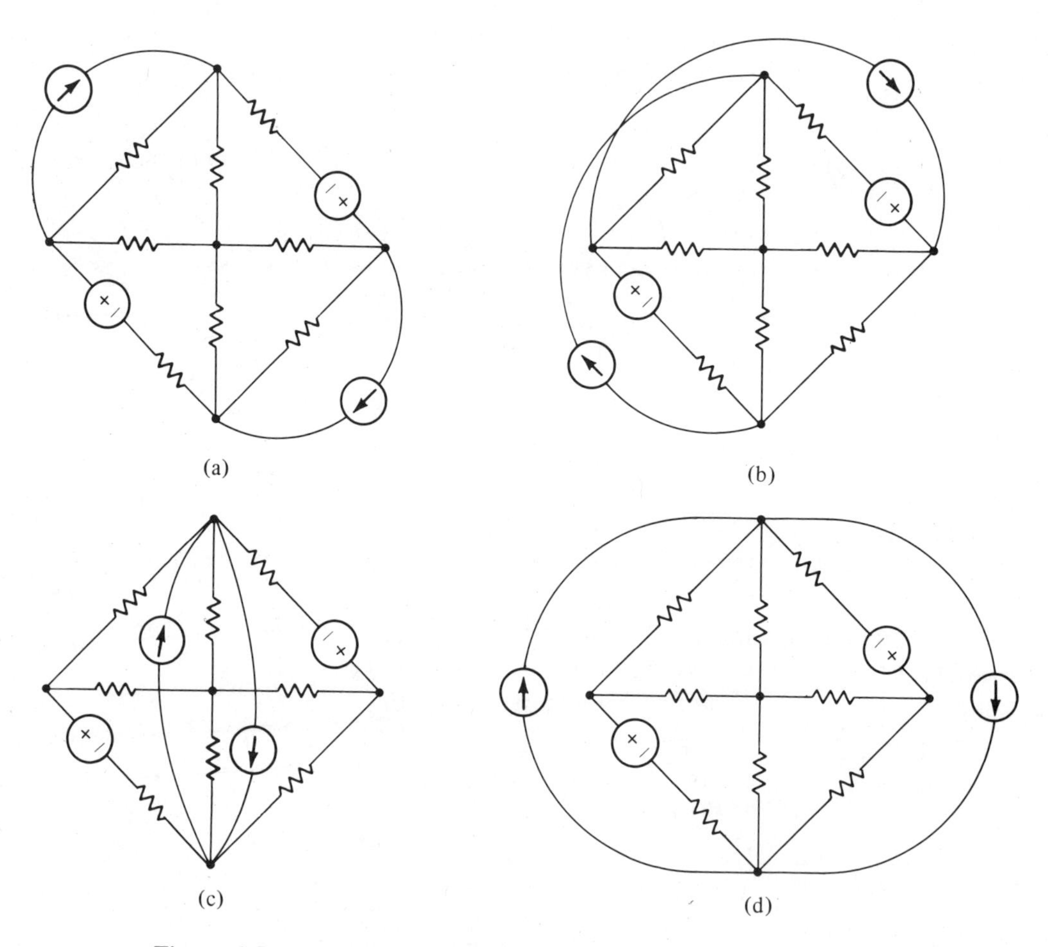

Figure 1.25 Planar and nonplanar networks. (a) Planar network. (b) Nonplanar network. (c) Network which appears to be nonplanar. (d) Network of (c) redrawn as planar network.

1.25(a) and (b). Some networks which appear at first to be nonplanar [Figure 1.25(c)] can be redrawn as planar networks [Figure 1.25(d)].

If KCL is applied at every node of a circuit except the ground, or reference, node, then the result is a system of *linear algebraic equations* for which the unknowns are the node voltages, the coefficients depend on the conductance between nodes, and the right-hand terms depend on the circuit current sources. Since there is one independent equation in the system for each node, the number of equations is equal to the number of unknown node voltages, which guarantees that the node voltages are uniquely determined. Node analysis is a general technique, applicable to any circuit.

Node Analysis of Circuits with Current Sources Only

Example 1.7 is a simple illustration of the node analysis technique applied to a circuit in which all sources are current sources. Note that the reference node is specified. If the reference node is not specified, the obvious choice is the **ground**, or **common**, node; otherwise, choose the node to which the most elements are connected as reference node for the simplest system of KCL equations. The nodes other than the reference node are called **independent nodes**.

Example 1.7

SPICE

Problem:

Find the voltage at node A and at node B, relative to the reference node, in the circuit shown in Figure 1.26(a).

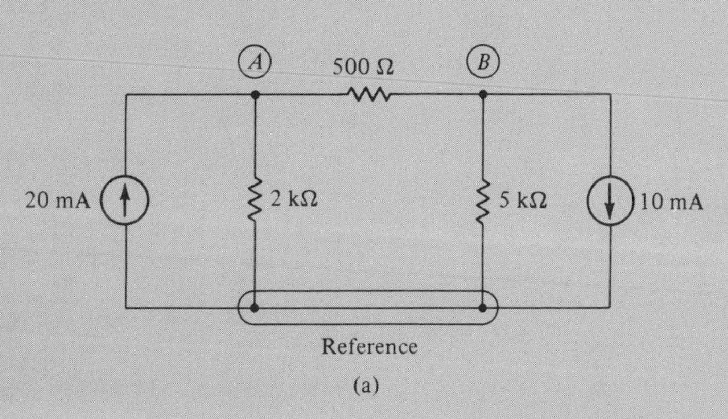

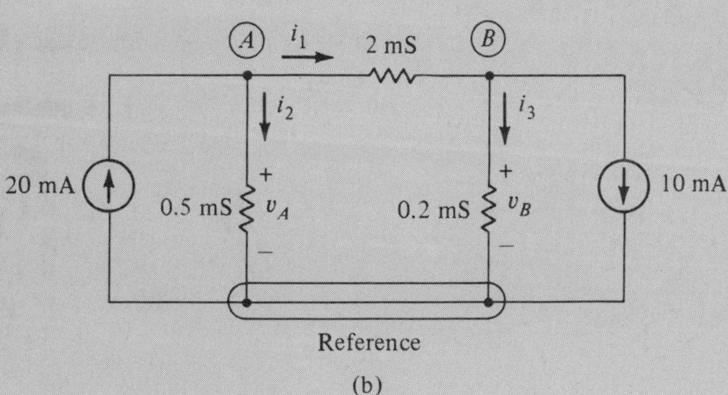

Figure 1.26 Example 1.7. (a) Circuit with current sources only. (b) Circuit labeled for node analysis.

Solution:

1. Assume branch currents as shown in Figure 1.26(b). Note that all resistance has been changed to conductance for the purpose of node analysis. Apply KCL at node A:

$$i_1 + i_2 - 20 \text{ mA} = 0$$

2. Apply KCL at node B:

$$-i_1 + i_3 + 10 \text{ mA} = 0$$

3. Use Ohm's law to express the assumed currents i_1, i_2, and i_3 in terms of node voltage and conductance:

$$i_1 \text{ (mA)} = 2(v_A - v_B)$$
$$i_2 \text{ (mA)} = 0.5v_A$$
$$i_3 \text{ (mA)} = 0.2v_B$$

4. Substitute the expressions for the assumed currents into the KCL equations and rearrange terms.

$$2(v_A - v_B) + 0.5v_A = 20$$
$$-2(v_A - v_B) + 0.2v_B = -10$$

or

$$2.5v_A - 2\ v_B = 20$$
$$-2\ v_A + 2.2v_B = -10$$

5. Multiply the first equation by 1.1, then add the two equations to eliminate v_B:

$$\begin{aligned} 2.75v_A - 2.2v_B &= 22 \\ -2\ v_A + 2.2v_B &= -10 \\ \hline 0.75v_A &= 12 \end{aligned}$$

or

$$v_A = 16 \text{ V}$$

6. Back-substitute v_A into either equation and solve for v_B:

$$v_B = 10 \text{ V}$$

Node Analysis of Circuits with Voltage Sources

If there is a voltage source in the circuit, the procedure for node analysis depends on whether the source is a practical voltage source (resistance in the branch containing the voltage source) or an ideal voltage source (no resistance in the branch containing the voltage source). If the voltage source is a practical source, it should be converted to a practical current source (Example 1.8); if the source is ideal, then an auxiliary equation must be introduced (Example 1.9).

Example 1.8

Problem:

SPICE

Find the voltage at node A and at node B for the circuit shown in Figure 1.27(a).

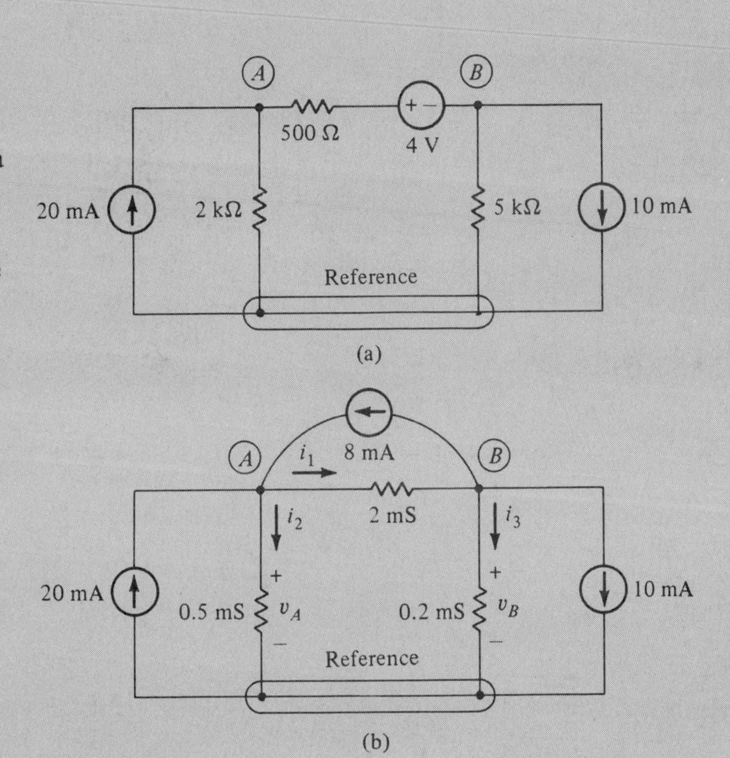

Figure 1.27 Example 1.8. (a) Circuit with current sources and a practical voltage source. (b) Circuit prepared and labeled for node analysis.

Solution:

1. Perform a source conversion between nodes A and B [Figure 1.27(b)], then apply KCL at node A and at node B:

$$i_1 + i_2 - 20\text{ mA} - 8\text{ mA} = 0$$
$$-i_1 + i_3 + 10\text{ mA} + 8\text{ mA} = 0$$
$$2(v_A - v_B) + 0.5v_A = 28$$
$$-2(v_A - v_B) + 0.2v_B = -18$$

or

$$2.5v_A - 2\ v_B = 28$$
$$-2\ v_A + 2.2v_B = -18$$

2. The solution of this system of equations is

$$v_A = 17.07\text{ V}$$
$$v_B = 7.333\text{ V}$$

Example 1.9

SPICE

Problem:

Find the voltage at node A and at node B for the circuit shown in Figure 1.28(a).

Solution:

1. Apply KCL at node A and at node B [Figure 1.28(b)]:

$$i_1 + i_2 - 20\text{ mA} = 0$$
$$-i_1 + i_3 + 10\text{ mA} = 0$$

2. The current i_1 cannot be determined by Ohm's law as in Examples 1.7 and 1.8, since the conductance of the branch is not known; therefore, the KCL equations must be written

$$i_1 + 0.5v_A = 20$$
$$-i_1 + 0.2v_B = -10$$

3. These two equations contain three unknowns, so a unique solution cannot be obtained unless a third equation involving v_A, v_B, or i_1 can be found. The presence of the ideal voltage source between nodes A and B provides such an equation, since the voltage source fixes the

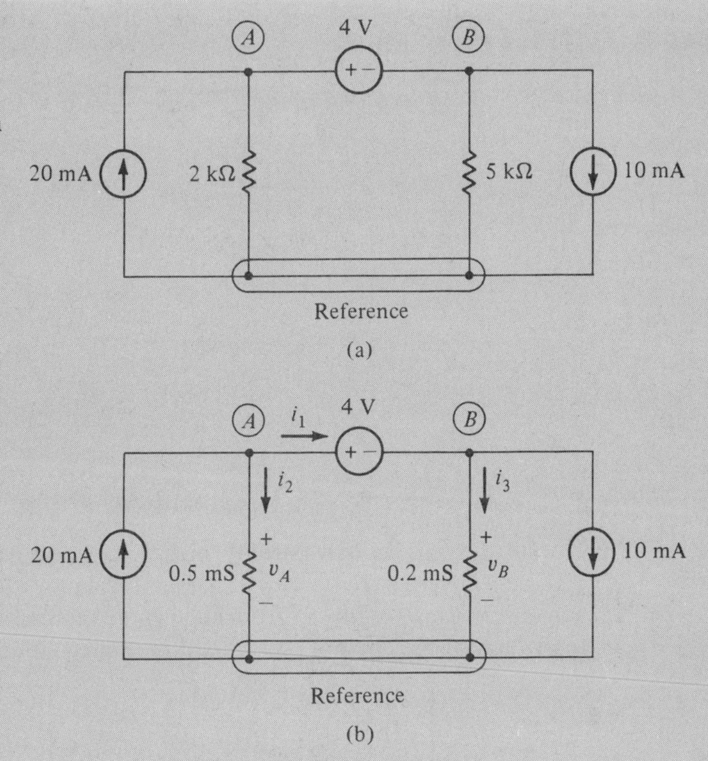

Figure 1.28
Example 1.9.
(a) Circuit with current sources and an ideal voltage source.
(b) Circuit labeled for node analysis.

potential difference between nodes A and B at 4 V. Thus, the system of equations to be solved for v_A and v_B is

$$
\begin{aligned}
i_1 + 0.5v_A \qquad\qquad &= 20 \\
-i_1 + \qquad\qquad 0.2v_B &= -10 \\
v_A - \quad v_B &= 4
\end{aligned}
$$

4. The system of equations may be solved by the method outlined in Appendix A, or by the procedure below.

a. Add the first two equations to eliminate i_1 and obtain a reduced system of equations:

$$
\begin{aligned}
0.5v_A + 0.2v_B &= 10 \\
v_A - \quad v_B &= 4
\end{aligned}
$$

or

$$0.5v_A + 0.2v_B = 10$$
$$0.2v_A - 0.2v_B = 0.8$$

b. Add these two equations to eliminate v_B:

$$0.7v_A = 10.8$$
$$v_A = 15.43 \text{ V}$$

c. Back-substitute v_A into either equation:

$$v_B = 11.43 \text{ V}$$

General Matrix Method for Node Analysis

If a circuit contains only current sources (Example 1.7), or the voltage sources can be converted to current sources (Example 1.8), then the system of equations in matrix form, or the augmented matrix, can be written directly by inspection of the circuit. The general matrix equation is

$$\begin{bmatrix} G_{11} & -G_{12} & \cdots & -G_{1N} \\ -G_{21} & G_{22} & \cdots & -G_{2N} \\ \vdots & \vdots & & \vdots \\ -G_{N1} & -G_{N2} & \cdots & G_{NN} \end{bmatrix} \begin{bmatrix} v_1 \\ v_2 \\ \vdots \\ v_N \end{bmatrix} = \begin{bmatrix} i_{S1} \\ i_{S2} \\ \vdots \\ i_{SN} \end{bmatrix}$$

where N is the number of independent nodes.

G_{jj}, sometimes called the **self-conductance**, is the total conductance of all resistors connected to node j, and G_{jk}, sometimes called the **mutual conductance**, is the total conductance between node j and node k. Collectively, the conductance entries form the matrix of coefficients, which is also known appropriately as the **conductance matrix**. Notice that the entries G_{jj} all have a plus sign, and the entries G_{jk} $(j \neq k)$ all have a minus sign. Clearly, the conductance matrix is a **square matrix** and is **symmetric**, that is, $G_{jk} = G_{kj}$.

The column of node voltages is also known as the **vector of unknowns**, and the column of source currents on the right-hand side of the equation is also known as the **source vector**. For the source-vector entries, i_{Sj} is the sum of all source currents *entering* node j.

Example 1.10

SPICE

Problem:

Write the matrix equation and the augmented matrix for node analysis of the circuit shown in Figure 1.29(a).

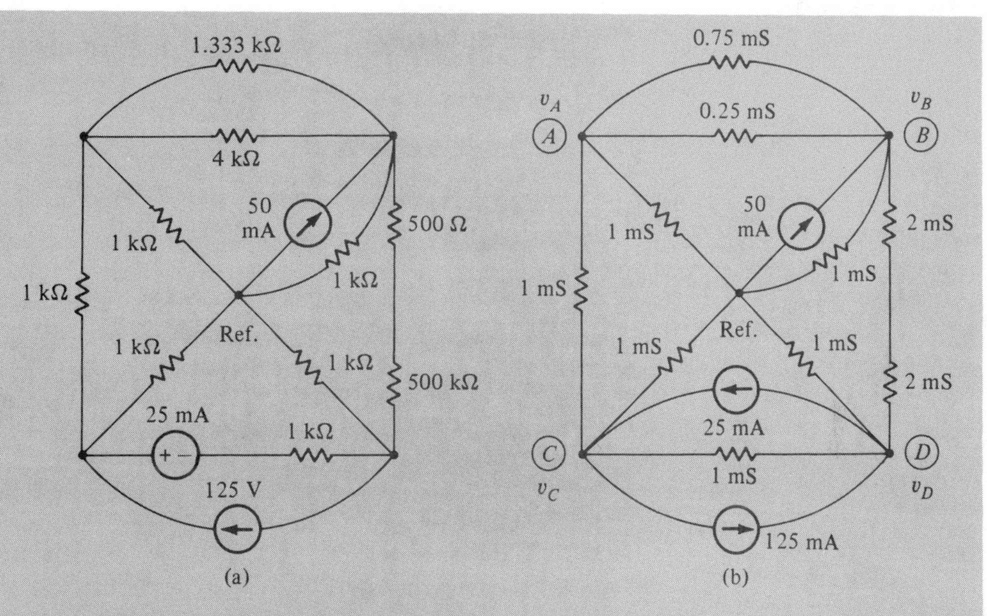

Figure 1.29 Example 1.10. (a) Circuit susceptible to general node analysis (b) Circuit prepared and labeled for general node analysis

Solution:

1. The matrix equation is

$$\begin{bmatrix} G_{AA} & -G_{AB} & -G_{AC} & -G_{AD} \\ -G_{BA} & G_{BB} & -G_{BC} & -G_{BD} \\ -G_{CA} & -G_{CB} & G_{CC} & -G_{CD} \\ -G_{DA} & -G_{DB} & -G_{DC} & G_{DD} \end{bmatrix} \begin{bmatrix} v_A \\ v_B \\ v_C \\ v_D \end{bmatrix} = \begin{bmatrix} i_{SA} \\ i_{SB} \\ i_{SC} \\ i_{SD} \end{bmatrix}$$

2. The conductance-matrix entries are

$$G_{AA} = 0.25 \text{ mS} + 0.75 \text{ mS} + 1 \text{ mS} + 1 \text{ mS} = 3 \text{ mS}$$
$$G_{BB} = 0.25 \text{ mS} + 0.75 \text{ mS} + 1 \text{ mS} + 1 \text{ mS} = 3 \text{ mS}$$
$$G_{CC} = 1 \text{ mS} + 1 \text{ mS} + 1 \text{ mS} = 3 \text{ mS}$$
$$G_{DD} = 1 \text{ mS} + 1 \text{ mS} + 1 \text{ mS} = 3 \text{ mS}$$
$$G_{AB} = G_{BA} = 0.25 \text{ mS} + 0.75 \text{ mS} = 1 \text{ mS}$$
$$G_{AC} = G_{CA} = 1 \text{ mS}$$
$$G_{AD} = G_{DA} = 0$$
$$G_{BC} = G_{CB} = 0$$
$$G_{BD} = G_{DB} = 1 \text{ mS}$$
$$G_{CD} = G_{DC} = 1 \text{ mS}$$

3. The source-vector entries are

$$i_{SA} = 0$$
$$i_{SB} = 50 \text{ mA}$$
$$i_{SC} = 25 \text{ mA} - 125 \text{ mA} = -100 \text{ mA}$$
$$i_{SD} = 125 \text{ mA} - 25 \text{ mA} = 100 \text{ mA}$$

4. The matrix equation is

$$\begin{bmatrix} 3 & -1 & -1 & -0 \\ -1 & 3 & 0 & -1 \\ -1 & 0 & 3 & -1 \\ 0 & -1 & -1 & 3 \end{bmatrix} \begin{bmatrix} v_A \\ v_B \\ v_C \\ v_D \end{bmatrix} = \begin{bmatrix} 0 \\ 50 \\ -100 \\ 100 \end{bmatrix}$$

Notice that each entry in the conductance matrix on one side of the equation is in millisiemens (mS), and each entry in the source vector on the other side of the equation is in milliamperes (mA). Specifying node voltages in volts makes the equation dimensionally correct, and allows units to be entirely suppressed from the matrix equation. The system of equations is solved as an example in Appendix A.

Computer Simulation

A computer simulation of the circuit of Example 1.10 is given in SPICE 1.2.

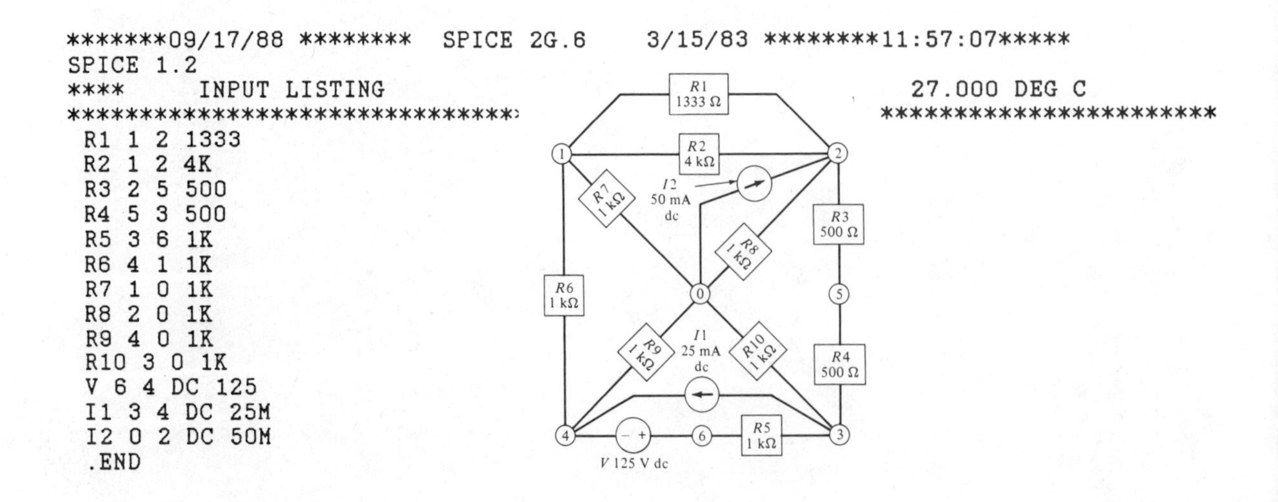

```
*******09/17/88 ********  SPICE 2G.6    3/15/83 ********11:57:07*****
SPICE 1.2
****      INPUT LISTING                                27.000 DEG C
*******************************                ***********************
 R1 1 2 1333
 R2 1 2 4K
 R3 2 5 500
 R4 5 3 500
 R5 3 6 1K
 R6 4 1 1K
 R7 1 0 1K
 R8 2 0 1K
 R9 4 0 1K
 R10 3 0 1K
 V 6 4 DC 125
 I1 3 4 DC 25M
 I2 0 2 DC 50M
 .END
```

```
*******09/17/88 ********  SPICE 2G.6    3/15/83 ********11:57:08*****
SPICE 1.2
****     SMALL SIGNAL BIAS SOLUTION        TEMPERATURE =    27.000 DEG C
***************************************************************************
 NODE    VOLTAGE      NODE    VOLTAGE      NODE    VOLTAGE      NODE    VOLTAGE
(  1)     3.3347     (  2)    29.9987     (  3)    36.6663     (  4)   -19.9997
(  5)    33.3325     (  6)   105.0003
    VOLTAGE SOURCE CURRENTS
    NAME        CURRENT
    V          -6.833D-02
    TOTAL POWER DISSIPATION   8.63D+00  WATTS

       JOB CONCLUDED
```

1.5 Mesh Analysis

As was indicated in the previous section, mesh analysis is limited to planar circuits (see Figure 1.25). If KVL is applied in every mesh of a circuit, the result is a system of linear algebraic equations for which the unknowns are the mesh currents, the coefficients depend on the resistance within the meshes, and the right-hand terms depend on the circuit voltage sources. Since there is one independent equation in the system for each mesh, the number of equations is equal to the number of unknown mesh currents, which guarantees that the mesh currents are uniquely determined.

Mesh Analysis of Circuits with Voltage Sources Only

Example 1.11 is a simple illustration of the mesh analysis technique applied to a circuit in which all sources are voltage sources. Note that all mesh currents are specified in the clockwise direction. The direction of mesh currents is arbitrary, and different directions can be specified for different meshes in the same circuit; however, to minimize errors in applying KVL, all mesh currents should be specified consistently in the same direction.

Example 1.11 **Problem:**

SPICE

Find the current in mesh 1 and in mesh 2, in the direction indicated, in the circuit shown in Figure 1.30(*a*).

Solution:

1. Assume element voltages as shown in Figure 1.30(b).
2. Apply KVL in mesh 1:

$$v_A + v_B + 10\ \text{V} = 0$$

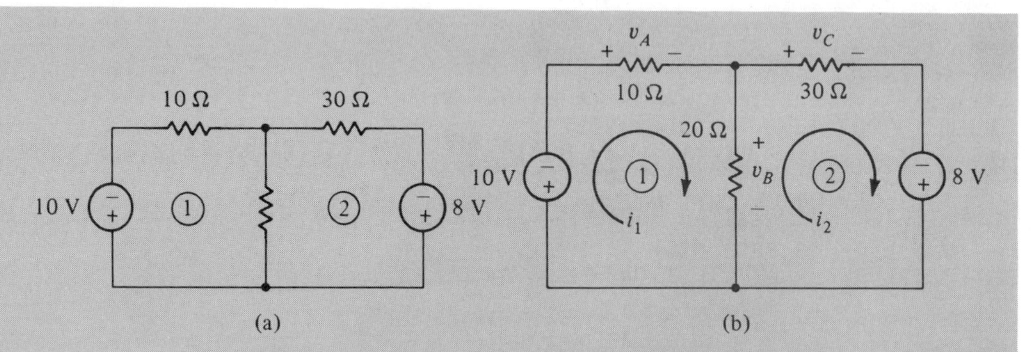

Figure 1.30 Example 1.11. (a) Circuit with voltage sources only. (b) Circuit labeled for mesh analysis.

3. Apply KVL in mesh 2:

$$v_C - v_B - 8\text{ V} = 0$$

4. Use Ohm's law to express the assumed voltages, v_A, v_B, and v_C, in terms of mesh currents and resistance:

$$v_A = 10i_1$$
$$v_B = 20(i_1 - i_2)$$
$$v_C = 30i_2$$

5. Substitute the expressions for voltage into the KVL equations and rearrange terms:

$$30i_1 - 20i_2 = -10$$
$$-20i_1 + 50i_2 = \quad 8$$

6. The solution for the system of KVL equations is

$$i_1 = -0.3091\text{ A}$$
$$i_2 = 0.03636\text{ A}$$

Mesh Analysis of Circuits with Current Sources

If there is a current source in the circuit, the procedure for mesh analysis depends on whether the source is a practical current source (conductance in parallel with the current source) or an ideal current source (no parallel conductance). If the current source is a practical source, it should be converted to a practical voltage source (Example 1.12); if the source is ideal, then an auxiliary equation must be introduced (Example 1.13).

Example 1.12

SPICE

Problem:

Find the current in mesh 1 and in mesh 2 for the circuit shown in Figure 1.31(a).

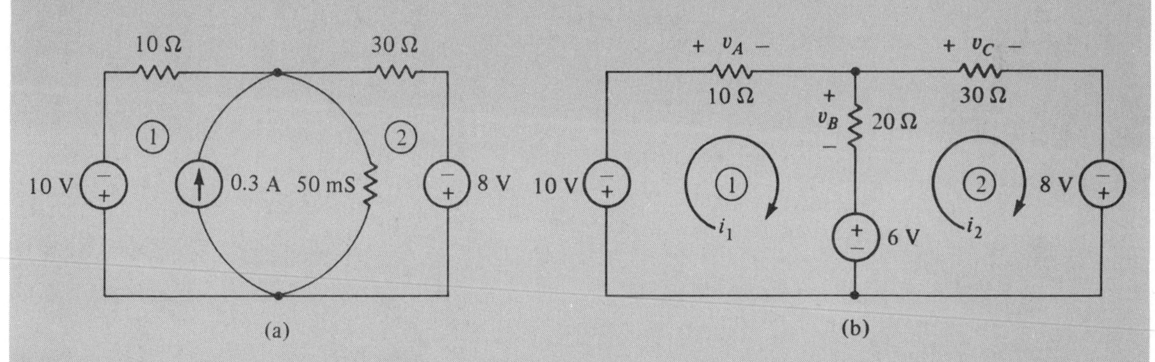

Figure 1.31 Example 1.12. (a) Circuit with voltage sources and a practical current source. (b) Circuit prepared and labeled for mesh analysis.

Solution:

1. Perform a source conversion [Figure 1.31(b)], then apply KVL in mesh 1 and in mesh 2:

$$v_A + v_B + 6\text{ V} + 10\text{ V} = 0$$

$$v_C - v_B - 8\text{ V} - 6\text{ V} = 0$$

$$10i_1 + 20(i_1 - i_2) = -16$$

$$30i_2 - 20(i_1 - i_2) = 14$$

or

$$30i_1 - 20i_2 = -16$$

$$-20i_1 + 50i_2 = 14$$

2. The solution of this system of equations is

$$i_1 = -0.4727 \text{ A}$$
$$i_2 = 0.09091 \text{ A}$$

Example 1.13. SPICE

Problem:

Find the current in mesh 1 and in mesh 2 for the circuit shown in Figure 1.32(a).

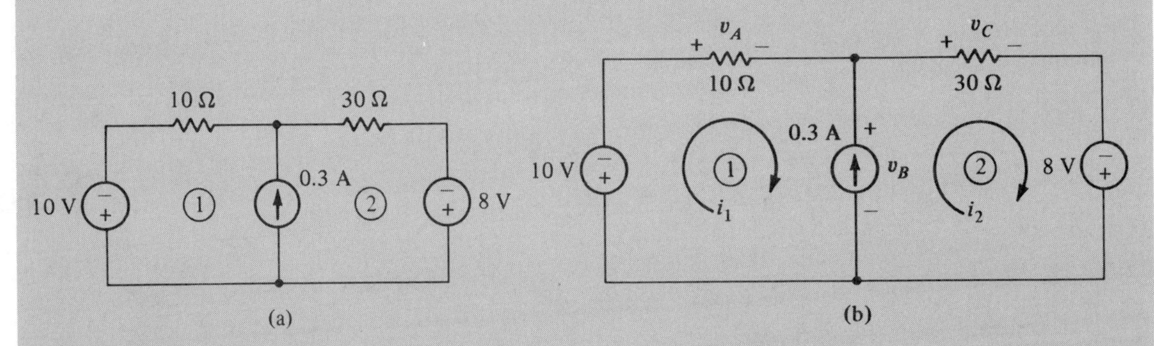

Figure 1.32 Example 1.13. (a) Circuit with voltage sources and an ideal current source. (b) Circuit labeled for mesh analysis.

Solution:

1. Apply KVL in mesh 1 and in mesh 2:

$$v_A + v_B + 10 \text{ V} = 0$$
$$v_C - v_B - 8 \text{ V} = 0$$

2. The voltage v_B cannot be determined by Ohm's law as in Examples 1.11 and 1.12, since the resistance of the branch is not known; therefore, the KVL equations are

$$10i_1 + v_B = -10$$
$$30i_2 - v_B = 8$$

3. These two equations contain three unknowns, so a unique solution cannot be obtained unless a third equation involving i_1, i_2, or v_B can be found. The presence of the ideal current source in the branch common to mesh 1 and mesh 2 provides such an equation, since the source fixes

the current in that branch at 0.3 A. Thus, the system of equations to be solved for i_1 and i_2 is

$$\begin{aligned} v_B + 10i_1 \qquad\quad &= -10 \\ -v_B + \qquad\quad 30i_2 &= 8 \\ i_1 \qquad - \quad i_2 &= 0.3 \end{aligned}$$

for which the solution is

$$i_1 = -0.275 \text{ A}$$
$$i_2 = 0.025 \text{ A}$$

General Matrix Method for Mesh Analysis

If a circuit contains only voltage sources (Example 1.11), or the current sources can be converted to voltage sources (Example 1.12), then the system of equations in matrix form, or the augmented matrix, can be written directly by inspection of the circuit. The general matrix equation is

$$\begin{bmatrix} R_{11} & -R_{12} & \cdots & -R_{1N} \\ -R_{21} & R_{22} & \cdots & -R_{2N} \\ \vdots & \vdots & & \vdots \\ -R_{N1} & -R_{N2} & \cdots & R_{NN} \end{bmatrix} \begin{bmatrix} i_1 \\ i_2 \\ \vdots \\ i_N \end{bmatrix} = \begin{bmatrix} v_{S1} \\ v_{S2} \\ \vdots \\ v_{SN} \end{bmatrix}$$

where N is the number of meshes.

R_{jj}, sometimes called the **self-resistance**, is the total resistance of all resistors in mesh j, and R_{jk}, sometimes called the **mutual resistance**, is the total resistance common to mesh j and mesh k. Collectively, the resistance entries form the matrix of coefficients, which is also known appropriately as the **resistance matrix**. Notice that the entries R_{jj} all have a plus sign, and the entries R_{jk} $(j \neq k)$ all have a minus sign. Clearly, the resistance matrix is a square matrix and is symmetric, that is, $R_{jk} = R_{kj}$.

The column of mesh currents is also known as the **vector of unknowns**, and the column of source voltages on the right-hand side of the equation is also known as the **source vector**. For the source vector entries, v_{Sj} is the sum of all source voltage rises *in the direction of mesh current* in mesh j.

Example 1.14

SPICE

Problem:

Write the matrix equation and the augmented matrix for mesh analysis of the circuit shown in Figure 1.33(a).

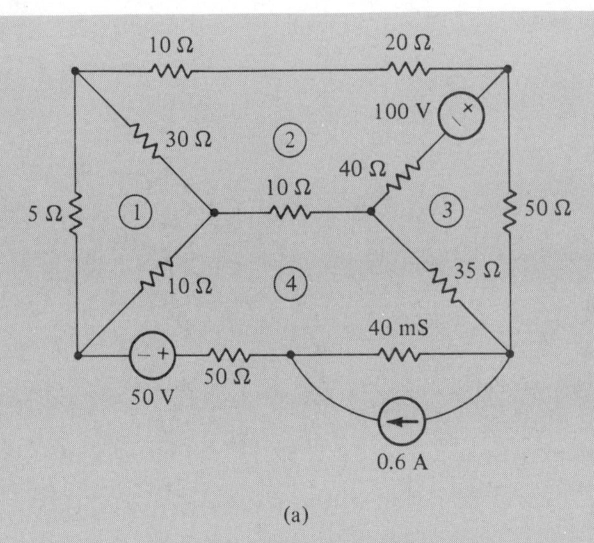

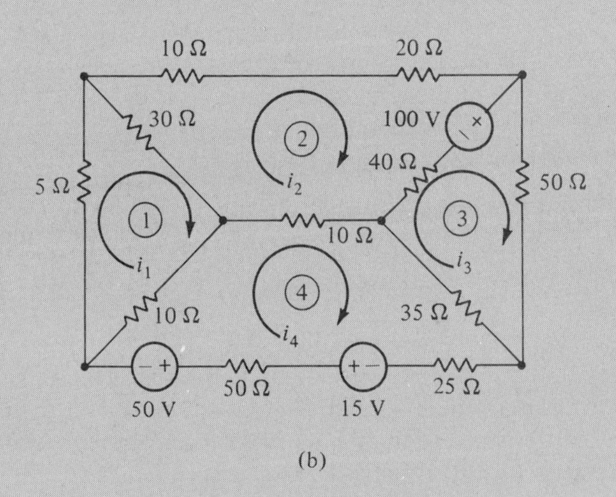

Figure 1.33 Example 1.14. (a) Circuit susceptible to general mesh analysis. (b) Circuit prepared and labeled for general mesh analysis.

Solution:

1. The matrix equation is

$$\begin{bmatrix} R_{11} & -R_{12} & -R_{13} & -R_{14} \\ -R_{21} & R_{22} & -R_{23} & -R_{24} \\ -R_{31} & -R_{32} & R_{33} & -R_{34} \\ -R_{41} & -R_{42} & -R_{43} & R_{44} \end{bmatrix} \begin{bmatrix} i_1 \\ i_2 \\ i_3 \\ i_4 \end{bmatrix} = \begin{bmatrix} v_{S1} \\ v_{S2} \\ v_{S3} \\ v_{S4} \end{bmatrix}$$

2. The resistance matrix entries are

$$R_{11} = 5\ \Omega + 30\ \Omega + 10\ \Omega = 45\ \Omega$$
$$R_{22} = 30\ \Omega + 10\ \Omega + 20\ \Omega + 40\ \Omega + 10\ \Omega = 110\ \Omega$$
$$R_{33} = 40\ \Omega + 50\ \Omega + 35\ \Omega = 125\ \Omega$$
$$R_{44} = 10\ \Omega + 10\ \Omega + 35\ \Omega + 25\ \Omega + 50\ \Omega = 130\ \Omega$$
$$R_{12} = R_{21} = 30\ \Omega$$
$$R_{13} = R_{31} = 0$$
$$R_{14} = R_{41} = 10\ \Omega$$
$$R_{23} = R_{32} = 40\ \Omega$$
$$R_{24} = R_{42} = 10\ \Omega$$
$$R_{34} = R_{43} = 35\ \Omega$$

3. The source vector entries are

$$v_{S1} = 0$$
$$v_{S2} = -100\ \text{V}$$
$$v_{S3} = 100\ \text{V}$$
$$v_{S4} = 15\ \text{V} - 50\ \text{V} = -35\ \text{V}$$

4. The matrix equation is

$$\begin{bmatrix} 45 & -30 & 0 & -10 \\ -30 & 110 & -40 & -10 \\ 0 & -40 & 125 & -35 \\ -10 & -10 & -35 & 130 \end{bmatrix} \begin{bmatrix} i_1 \\ i_2 \\ i_3 \\ i_4 \end{bmatrix} = \begin{bmatrix} 0 \\ -100 \\ 100 \\ -35 \end{bmatrix}$$

Notice that each entry in the resistance matrix on one side of the equation is in ohms (Ω), and each entry in the source vector on the other side of the equation is in volts (V). Specifying mesh currents in amperes makes the equation dimensionally correct, and allows units to be entirely suppressed from the matrix equation. Solution of the system is assigned as an exercise [Problem 1-20(a)] at the end of the chapter.

Computer Simulation

A computer simulation of the circuit in Example 1.14 is given in SPICE 1.3. Note the inclusion of the three additional voltage sources (VAM1, VAM2, and VAM3), each of which is entered in the input file without description. By default, these voltage sources will each have a value of 0 V, and thus have no effect on circuit performance; their purpose is to serve as **SPICE ammeters**. SPICE will only provide as a direct output those currents which *enter* voltage sources, so in order to obtain i_1, for example, there must be a voltage source present in a branch not shared by mesh 2, 3, or 4; VAM1 satisfies this requirement, and the current entering VAM1 is i_1. Similar observations may be made for i_2 and VAM2, and for i_3 and VAM3. In mesh 4, the 50-V source is already in position to serve as the SPICE ammeter, so an additional 0-V source is not required.

```
*******09/17/88 ********  SPICE 2G.6     3/15/83 ********13:03:49*****
SPICE 1.3
****      INPUT LISTING                       TEMPERATURE =   27.000 DEG C
******************************************************************************
 R1 2 3 5
 R2 4 5 10
 R3 6 7 20
 R4 8 9 50
 R5 9 0 25
 R6 1 0 50
 R7 4 11 30
 R8 2 11 10
 R9 9 10 35
 R10 10 11 10
 R11 10 12 40
 I 9 0 DC .6
 VAM1 3 4
 VAM2 5 6
 VAM3 7 8
 V1 1 2 DC 50
 V2 7 12 DC 100
 .END
*******09/17/88 ********  SPICE 2G.6     3/15/83 ********13:03:49*****
SPICE 1.3
****      SMALL SIGNAL BIAS SOLUTION          TEMPERATURE =   27.000 DEG C
******************************************************************************
  NODE    VOLTAGE      NODE   VOLTAGE      NODE   VOLTAGE      NODE   VOLTAGE
 (  1)    14.6155     (  2)  -35.3845     (  3)  -31.7719     (  4)  -31.7719
 (  5)   -21.9085     (  6)  -21.9085     (  7)   -2.1816     (  8)   -2.1816
 (  9)   -22.3078     ( 10)  -46.6269     ( 11)  -39.6866     ( 12) -102.1816
     VOLTAGE SOURCE CURRENTS
     NAME          CURRENT
     VAM1       -7.225D-01
     VAM2       -9.863D-01
     VAM3        4.025D-01
     V1         -2.923D-01
     V2         -1.389D+00
     TOTAL POWER DISSIPATION    1.67D+02  WATTS

         JOB CONCLUDED
```

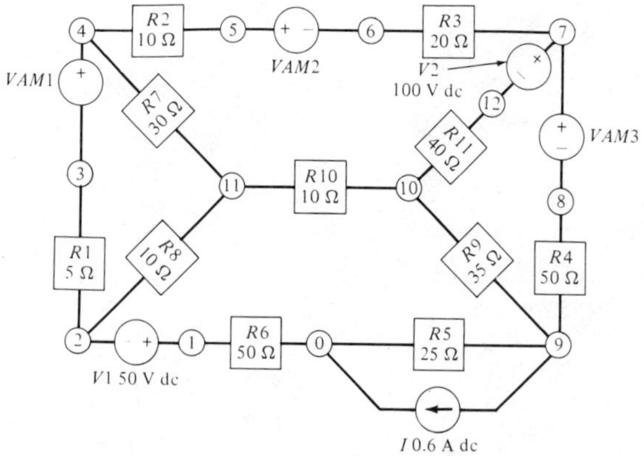

1.6 Superposition Principle

A number of network theorems have been established, among the more useful and important of which are the **superposition principle** (or **theorem**), **Thevenin's theorem, Norton's theorem**, and the **maximum power transfer theorem**. In the remaining sections of this chapter, we will state each theorem, discuss its immediate consequences, and provide several examples of its application to circuit analysis and design.

The superposition principle, a direct consequence of linearity, greatly simplifies the analysis of networks with multiple independent sources.

Superposition Principle

> In a **linear**, **bilateral** network with multiple **independent** sources, the voltage at any node or the current in any branch is the algebraic sum of the node voltages or the branch currents produced by each independent source acting alone.

A linear, bilateral network is one which consists entirely of independent sources, linearly dependent sources, and linear, bilateral passive elements. Linearly dependent sources and linear, bilateral resistors were described in Section 1.2. Other linear, bilateral elements will be described in later chapters.

In applying the superposition principle, the multiple-source circuit is decomposed into simpler circuits each containing exactly one of the independent sources of the original circuit (Figure 1.34). Independent sources are removed from the original circuit by specifying a zero output value. When a voltage source has an output of 0 V, it may be replaced with a short circuit, and when a current source has an output of 0 A, it may be replaced with an open circuit. Note that dependent sources are *not* removed in the decomposition process; however, the control variable becomes that of the simpler circuit. In Figures 1.34(b), (c), and (d), the superscripts (1), (2), and (3) are used to identify control variables and other unknowns pertaining only to the simpler circuit. For example, $i_1^{(1)}$ is the control variable for the CCCS in the circuit of Figure 1.34(b) only, $i_1^{(2)}$ is the control variable in Figure 1.34(c), and $i_1^{(3)}$ is the control variable in Figure 1.34(d).

Each of the simpler circuits must be analyzed separately for the required voltage or current. For example, if v in the circuit of Figure 1.34(a) is required, then $v^{(1)}$ must be obtained from the circuit of Figure 1.34(b), $v^{(2)}$ must be obtained from the circuit of Figure 1.34(c), and $v^{(3)}$ must be obtained from the circuit of Figure 1.34(d). According to the superposition principle, the values of $v^{(1)}$, $v^{(2)}$, and $v^{(3)}$ are the *linear components* of v, so

$$v = v^{(1)} + v^{(2)} + v^{(3)}$$

The analysis of three circuits is **superposed** to obtain an analysis of a more

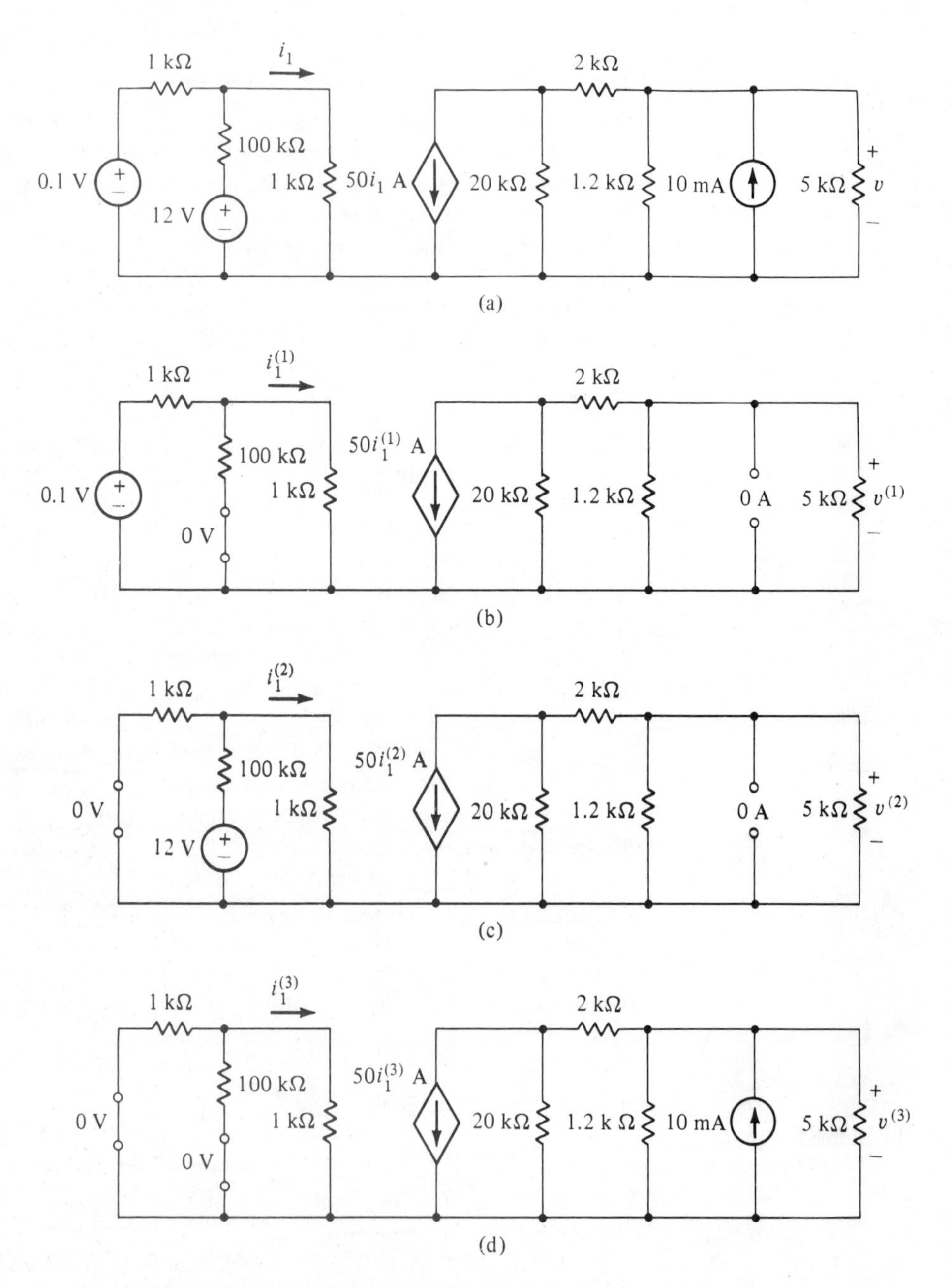

Figure 1.34 Decomposition of multisource circuit for application of superposition principle. (a) Multisource circuit. (b) Circuit with only 0.1-V independent voltage source. (c) Circuit with only 12-V independent voltage source. (d) Circuit with only 10-mA independent current source.

complex circuit from which the three circuits were derived; thus the term **superposition**. The actual analysis of the three circuits is presented in Example 1.16.

An important point regarding superposition requires special emphasis. The superposition principle applies to voltage and current calculations, but *does not* apply to power calculations. Power, which depends on the *product* of voltage and current (Gv^2 or Ri^2), is a **nonlinear response**, and is not amenable to superposition. For example, it is *incorrect* to calculate the power absorbed by a resistor from one source acting alone, then calculate the power absorbed in the same resistor from a second source acting alone, and add the two results.

Example 1.15

Problem:

Use the superposition principle to find i in the multisource circuit shown in Figure 1.35(a).

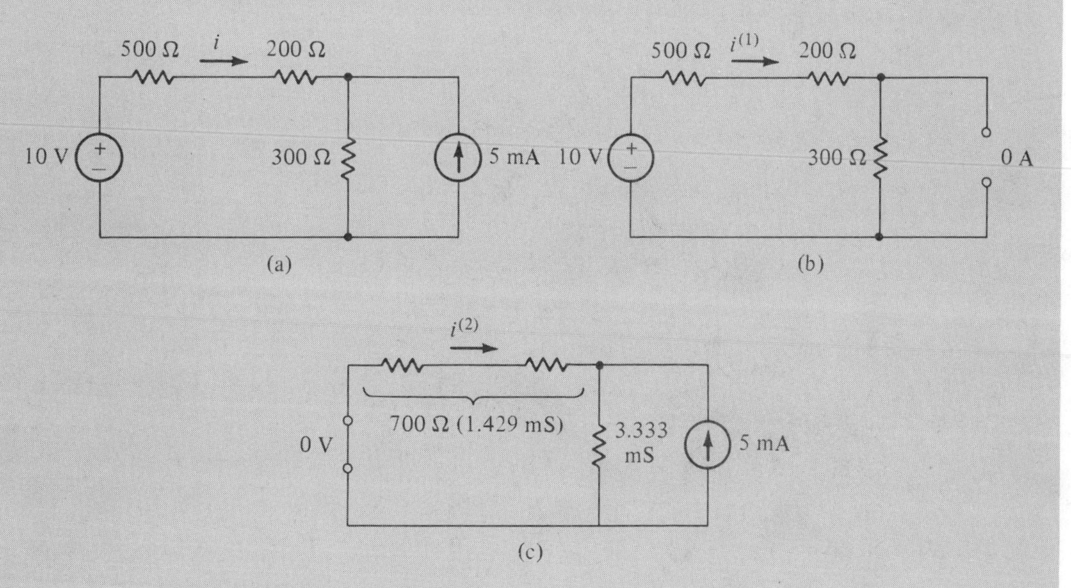

Figure 1.35 Example 1.15. (a) Multisource circuit. (b) Circuit with only 10-V independent voltage source. (c) Circuit with only 5-mA independent current source.

Solution:

1. Obtain the circuit of Figure 1.35(b) by removing the 5-mA source and replacing it with an open circuit.

2. Solve for $i^{(1)}$ by Ohm's law:

$$i^{(1)} = \frac{10}{500 + 200 + 300} = 1 \text{ mA}$$

3. Obtain the circuit of Figure 1.35(c) by removing the 10-V source and replacing it with a short circuit.
4. Solve for $i^{(2)}$ by the current division rule:

$$i^{(2)} = -5 \times \frac{1.429}{1.429 + 3.333} = -1.5 \text{ mA}$$

5. Solve for i by the superposition principle:

$$i = i^{(1)} + i^{(2)} = 1 \text{ mA} - 1.5 \text{ mA} = -0.5 \text{ mA}$$

Example 1.16

Problem:

Use the superposition principle to find v in the multisource circuit shown in Figure 1.34(a).

Solution:

1. Obtain the circuit of Figure 1.34(b) by removing the 12-V independent voltage source and the 10-mA independent current source. Replace the 12-V source by a short circuit, and the 10-mA source by an open circuit.
2. Solve for $v^{(1)}$, as follows:
   a. Perform a source conversion on the CCCS [Figure 1.36(a)], combine the 1.2-kΩ and 5-kΩ parallel network, and solve for $v^{(1)}$ in terms of $i_1^{(1)}$. Apply the voltage division rule:

$$v^{(1)} = -10^6 i_1^{(1)} \frac{0.9677}{0.9677 + 20 + 2} \simeq -42.13 \times 10^3 i_1^{(1)} \text{ V}$$

   b. Perform a source conversion on the 0.1-V source [Figure 1.36(a)], and solve for $i_1^{(1)}$. Apply the current division rule:

$$i_1^{(1)} = 0.1 \times \frac{1}{1 + 0.01 + 1} \simeq 0.04975 \text{ mA}$$

   c. Substitute the value of $i_1^{(1)}$ into the expression for $v^{(1)}$:

$$v^{(1)} \simeq -42.13 \times 10^3 (0.04975 \times 10^{-3}) = -2.096 \text{ V}$$

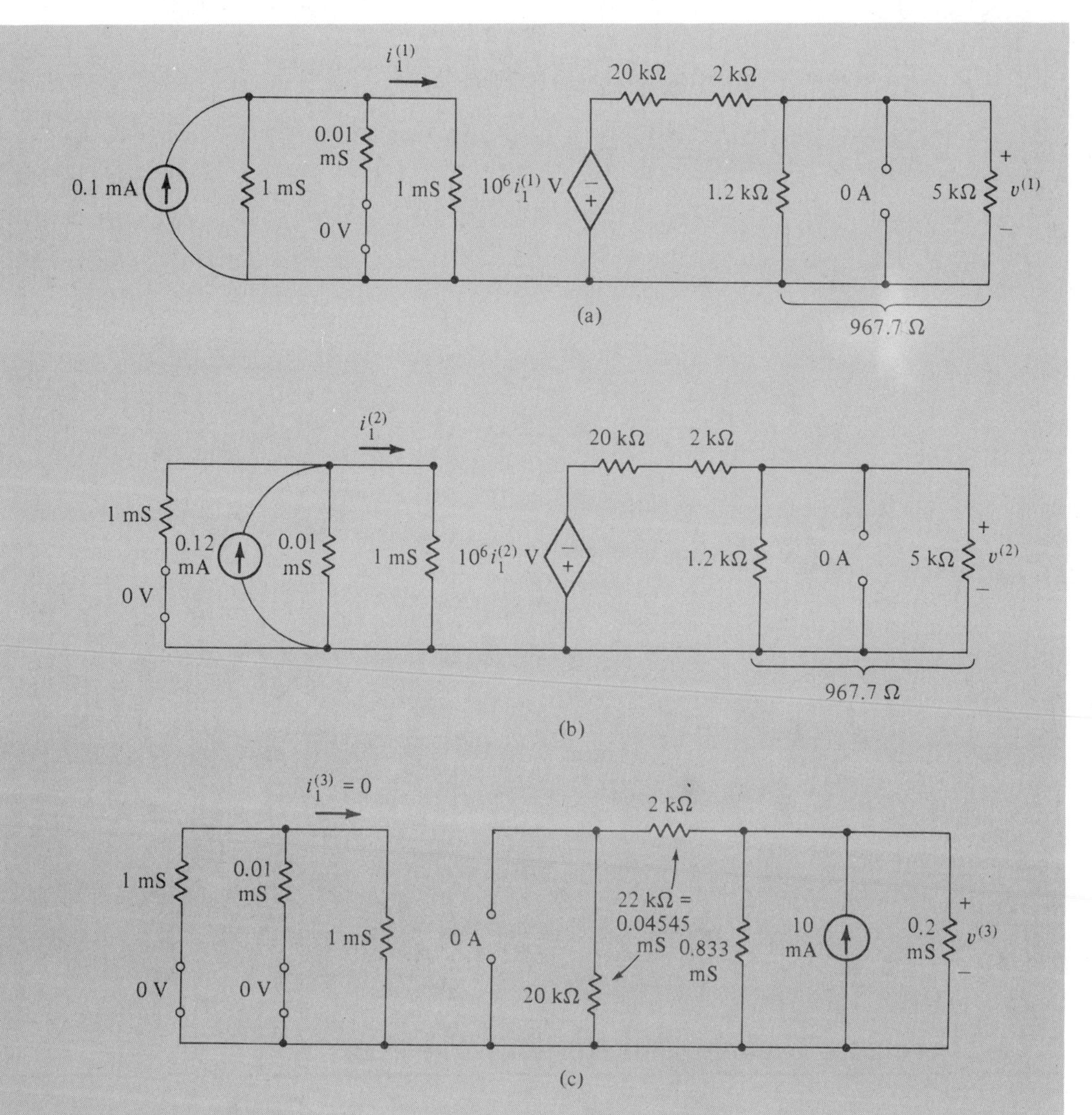

Figure 1.36 Example 1.16. (a) Circuit of Figure 1.34(b) after conversion of 0.1-V source and dependent source. (b) Circuit of Figure 1.34(c) after conversion of 12-V source and dependent source. (c) Circuit of Figure 1.34(d) with dependent source removed.

3. Solve for $v_B^{(2)}$ in a similar manner.
   a. The right-hand portion of the circuit in Figure 1.36(b) is the same as in Figure 1.36(a); thus, the expression for $v^{(2)}$ is the same as for $v^{(1)}$:

$$v^{(2)} \simeq -42.13 \times 10^3 i_1^{(2)} \text{ V}$$

   b. Perform a source conversion on the 12-V source [Figure 1.36(b)], and solve for $i_1^{(2)}$. Apply the current division rule:

$$i_1^{(2)} = 0.12 \times \frac{1}{1 + 0.01 + 1} \simeq 0.05970 \text{ mA}$$

   c. Substitute the value of $i_1^{(2)}$ into the expression for $v^{(2)}$:

$$v^{(2)} \simeq -42.13 \times 10^3 (0.05970 \times 10^{-3}) \simeq -2.516 \text{ V}$$

4. Solve for $v^{(3)}$ as follows:
   a. From Figure 1.34(d), $i_1^{(3)}$ is clearly zero; therefore, the CCCS can be replaced with an open circuit [Figure 1.36(c)].
   b. Combine the conductance and use Ohm's law to solve for $v^{(3)}$:

$$v^{(3)} \simeq \frac{10}{0.04545 + 0.8333 + 0.2} = 9.270 \text{ V}$$

5. Solve for v by the superposition principle:

$$v = v^{(1)} + v^{(2)} + v^{(3)}$$
$$v \simeq -2.096 \text{ V} - 2.516 \text{ V} + 9.270 \text{ V} = 4.658 \text{ V}$$

1.7 Thevenin's Theorem and Norton's Theorem

Thevenin's Theorem

Certainly the most useful of the network theorems is Thevenin's theorem, which allows the circuit analyst or designer to simplify less interesting portions of a circuit and concentrate on the portion which is of greater relevance to the problem at hand. We will rely heavily on Thevenin's theorem in Chapter 3, for establishing simplified circuits to facilitate the investigation of inductors and capacitors in electric circuits.

Thevenin's Theorem

> Any electric circuit may be separated into two networks, one called the **source network** and one called the **load network**, which are connected by a single pair of terminals called the **load terminals**. Once separated, the source network may be replaced by an equivalent network, consisting only of an ideal independent voltage source in series with a linear resistance, which will deliver the same voltage and current to the load terminals as the original source network.

Figure 1.37 illustrates this statement of Thevenin's theorem. For the circuits to be equivalent, v_L and i_L must be the same in Figure 1.37(b) as in Figure 1.37(a). The equivalent source network is called **Thevenin's equivalent source**, and together with the load network, **Thevenin's equivalent circuit**. The source in the equivalent source network is called **Thevenin's voltage** (v_{Th}), and the resistance is called **Thevenin's resistance** (R_{Th}).

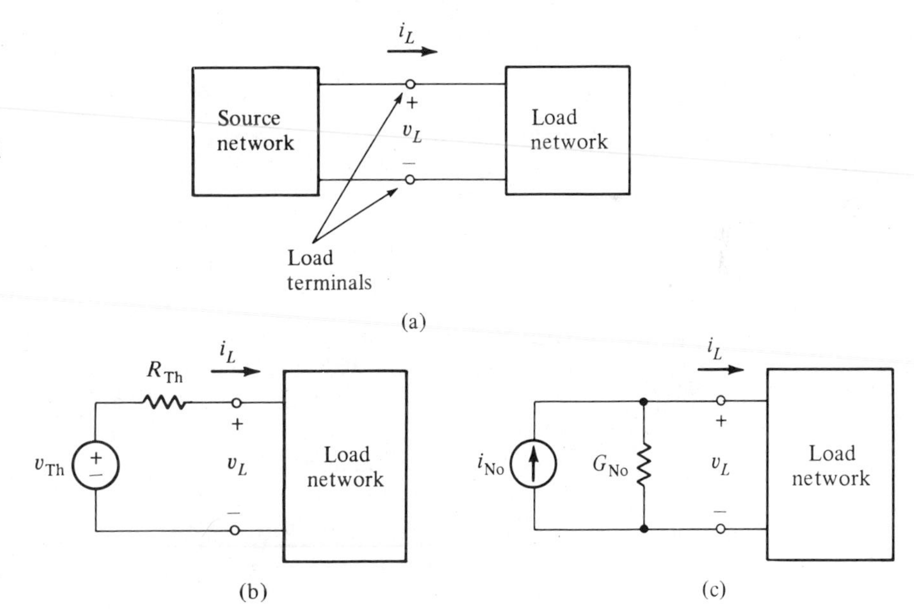

Figure 1.37 Thevenin's and Norton's equivalent circuits. (a) Circuit partitioned for application of Thevenin's theorem. (b) Thevenin's equivalent circuit. (c) Norton's equivalent circuit.

The reader familiar with electronics technology should recognize at this point that Thevenin's resistance is the same as the **output resistance** frequently specified for electronic devices such as linear amplifiers and drivers. All readers should recognize that Thevenin's equivalent source network is a **practical voltage source**, as described in Section 1.3, and that R_{Th} is identical to **source resistance** (R_S).

The most important consequence of Thevenin's theorem is that once the equivalent source network has been determined, the load network may be changed experimentally without any effect on the source network. The analyst and designer is then free to concentrate on the load network—where the action is, so to speak—without concern for the source, which may be a complex but uninteresting part of the circuit.

Norton's Theorem

Norton's theorem, which was established much later than Thevenin's theorem, is simply the dual of Thevenin's theorem. A statement of Norton's theorem can be obtained by substituting current for voltage, voltage for current, parallel for series, and conductance for resistance in the statement of Thevenin's theorem. The terms **Norton's equivalent source**, **Norton's equivalent circuit**, **Norton's current** (i_{No}), and **Norton's conductance** (G_{No}) [Figure 1.37(c)] are commonly used. Norton's equivalent source is a *practical current source*, with **source conductance** (G_S) equal to **Norton's conductance**. For a given source network, Norton's source is the *source conversion* of Thevenin's source, and vice versa.

Restrictions on Circuit Partitioning

There are some restrictions on the original circuit and how the original circuit must be partitioned for the application of Thevenin's and Norton's theorems:

1. The original circuit must have at least one *independent* source, which may reside in either the source network or the load network.
2. The original circuit must be partitioned so that *dependent* sources are in the same network with their control variables; that is, a dependent source cannot be separated from its control variable when the circuit is partitioned. This ensures that once the circuit has been partitioned, the load network can be changed without affecting the source network, and hence its Thevenin's or Norton's equivalent.
3. The source network must be linear, but the load network may be either linear or nonlinear, which is a useful feature is analyzing or designing circuits containing nonlinear electronic devices.

Obtaining Equivalent Sources

The circuit models in Figure 1.38 reveal that the equivalent source networks are obtained in a surprisingly straightforward manner. In Figure 1.38(a), the load has been removed from the load terminals, so the voltage (v_{oc}) measured (experimental) or calculated (analytical) at the load terminals is equal to v_{Th}. * In Figure 1.38(b), the load has been removed and replaced by a short circuit,

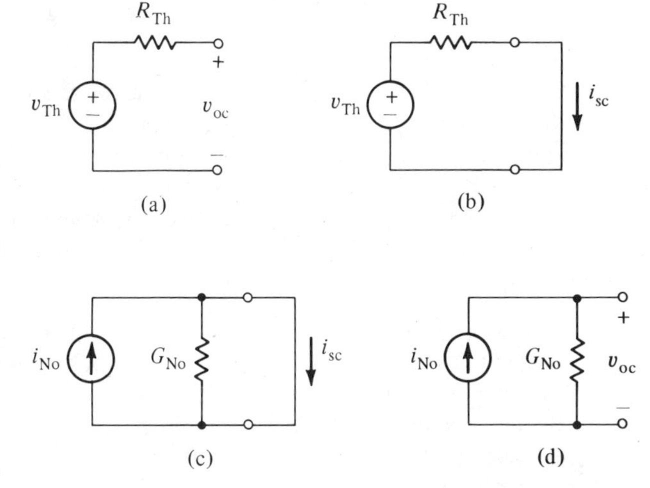

Figure 1.38
Obtaining Thevenin's and Norton's equivalent circuits.
(a) Thevenin's equivalent circuit with load removed.
(b) Thevenin's equivalent circuit with load replaced by short circuit.
(c) Norton's equivalent circuit with load replaced by short circuit.
(d) Norton's equivalent circuit with load removed.

so the current (i_{sc}) measured or calculated in the short circuit depends only upon v_{Th} (previously determined) and R_{Th}.

R_{Th} can be calculated:

$$R_{Th} = \frac{v_{Th} \text{ (or } v_{oc})}{i_{sc}}$$

In Figure 1.38(c), the load has been removed and replaced with a short circuit, so $i_{sc} = i_{No}$. In Figure 1.38(d), the load has been removed, so v_{oc} depends only upon i_{No} and G_{No}, and G_{No} can be calculated:

$$G_{No} = \frac{i_{No} \text{ (or } i_{sc})}{v_{oc}}$$

The necessary procedures for determining equivalent source networks are now clear—measure or determine analytically v_{oc} and i_{sc}. The value of v_{oc} is Thevenin's voltage (v_{Th}), and the ratio of v_{oc} to i_{sc} is Thevenin's resistance. On the other hand, i_{sc} is Norton's current, and the ratio of i_{sc} to v_{oc} is Norton's conductance. For a given circuit, determining v_{oc} and i_{sc} establishes both Thevenin's equivalent source and Norton's equivalent source. The preceding observations provide a general experimental, as well as analytical, technique for obtaining equivalent source networks; specialized analytical methods, which are sometimes more efficient, are available for use with particular types of source networks. In one specific type of source network, both v_{oc} and i_{sc} are zero, so the ratio is indeterminate, and another approach must be used to obtain R_{Th} or G_{No}.

The various analytical methods for obtaining Thevenin's and Norton's equivalent sources may be categorized as follows:

Case I Resistors and Independent Sources (Examples 1.17 and 1.18)
For Thevenin's equivalent source, determine v_{oc} only; for Norton's equivalent source, determine i_{sc} only. To determine R_{Th} or G_{No}, remove all sources from the source network; replace voltage sources by short circuits and current sources by open circuits. The **deactivated** source network now consists only of resistors, and R_{Th} or G_{No} can be measured or calculated at the load terminals.

Case II Resistors and Practical Independent Sources (Example 1.19)
Thevenin's or Norton's equivalent source can be determined analytically by repeated source conversions, and combination of series and parallel elements, without directly determining v_{oc} or i_{sc}. The equivalent source (Thevenin's or Norton's) will be apparent when all possible source conversions and element combinations have been accomplished.

Case III Resistors and Dependent and Independent Sources (Example 1.20)
The best approach to use in this case is to determine both v_{oc} and i_{sc} as previously described in the general case.

Case IV Resistors and Dependent Sources (Example 1.21)
In this case v_{oc} and i_{sc} are both zero, which is logical, because without independent sources in the source network, all control variables for the dependent sources are zero. The value of R_{Th} (or G_{No}), which depends on the ratio of v_{oc} to i_{sc} (0/0), is indeterminate. The situation can be resolved by placing a unit (1-V or 1-A) test source at the load terminals. If a 1-A source is used, the voltage rise *in the direction of the test source current* at the load terminals is numerically equal to R_{Th}; if a 1-V source is used, the current *from the source* into the more positive load terminal is numerically equal to G_{No}.

In Cases I and II, R_{Th} will be positive or possibly zero (short circuit), and G_{No} will be positive or possibly infinite (open circuit). In Cases III and IV, R_{Th} can be positive, negative, or zero, depending on the source-network configuration, and G_{No} will have the same sign, or will be infinite when R_{Th} is zero.

Example 1.17

SPICE

Problem:

Determine Thevenin's equivalent for the circuit shown in Figure 1.39(a).

Solution:

1. The equivalent source network shown in Figure 1.39(b) has been derived for the purpose of determining v_{oc} by node analysis, so the

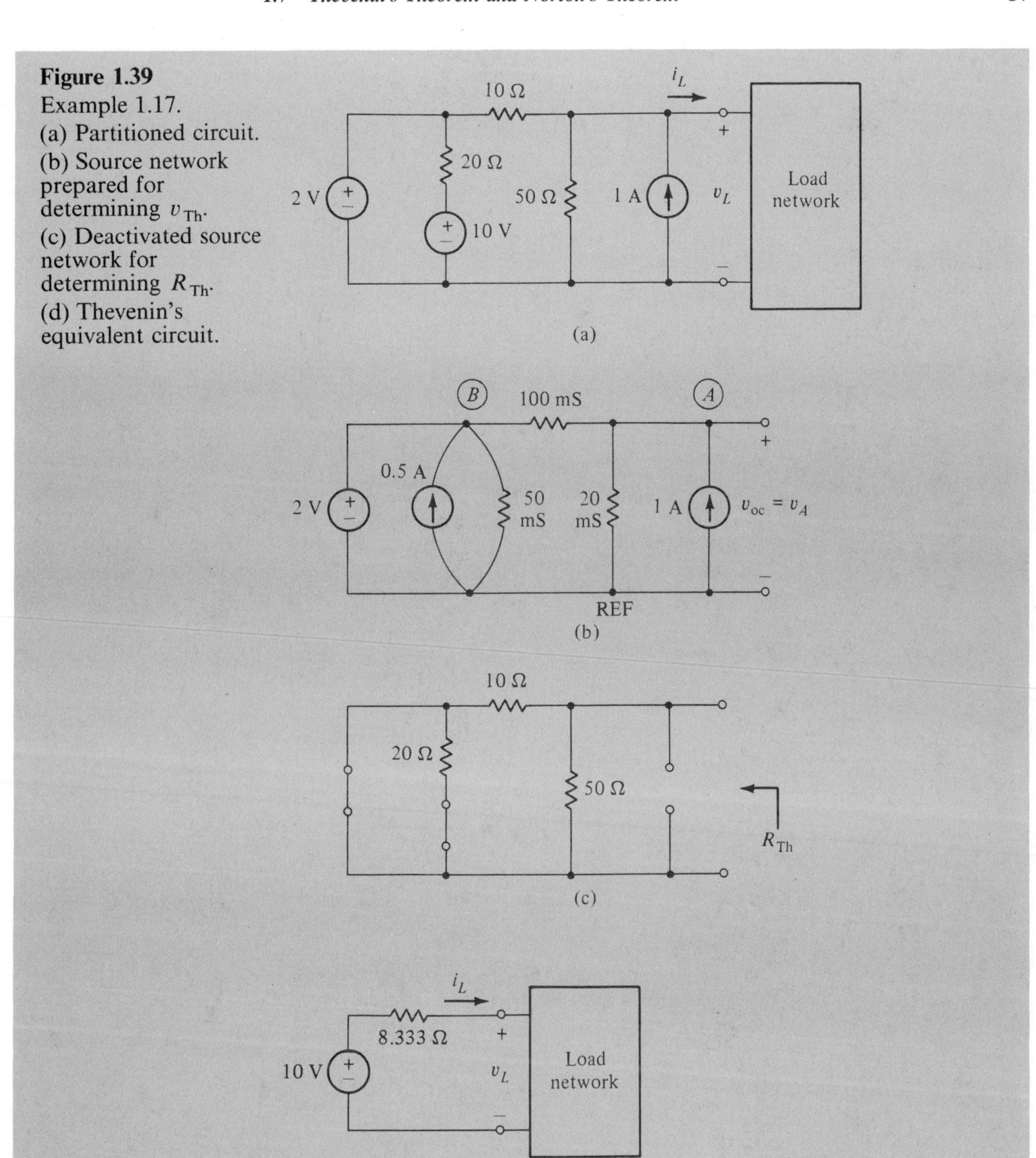

Figure 1.39
Example 1.17.
(a) Partitioned circuit.
(b) Source network prepared for determining v_{Th}.
(c) Deactivated source network for determining R_{Th}.
(d) Thevenin's equivalent circuit.

practical voltage source has been converted to a current source. With the reference node selected as indicated as in Figure 1.39(b), v_{oc} is identical to the node voltage at node A, v_A. Apply KCL at node A:

$$0.020 v_A + 0.100(v_A - v_B) = 1 \text{ A}$$

It is not necessary to apply KCL at node B, because the 2-V ideal source fixes the value of v_B at 2 V. Substitute v_B into the node-A equation:

$$0.020 v_A + 0.100(v_A - 2) = 1 \text{ A} \quad \Rightarrow \quad v_A = 10 \text{ V}$$

2. Since there are no dependent sources present, the source network can be deactivated by replacing the voltage sources with short circuits, and the current source with an open circuit [Figure 1.39(c)]. The deactivated source network now consists only of resistors, and Thevenin's resistance can be measured or calculated at the load terminals. In this case, replacing the 2-V ideal source with a short circuit has shorted the 20-Ω resistor and placed the 10-Ω resistor in parallel with the 50-Ω resistor, so

$$R_{Th} = \frac{1}{100 \text{ mS} + 20 \text{ mS}} = 8.333 \ \Omega$$

It must be emphasized that this method only applies to source networks which *do not* contain dependent sources.

Example 1.18

SPICE

Problem:

Obtain Norton's equivalent circuit for the circuit shown in Figure 1.40(a).

Solution:

1. When the load is removed and replaced by a short circuit, i_{sc} can be determined by mesh analysis. With the mesh current direction selected as indicated as in Figure 1.40(b), i_{sc} is identical to the current in mesh 1, i_1. Apply KVL in each mesh:

$$\begin{aligned} &\text{Mesh 1:} \quad 12 i_1 + 18(i_1 - i_2) = -v_x \\ &\text{Mesh 2:} \quad 12 i_2 + 18(i_2 - i_1) = v_x - 15 \end{aligned}$$

2. Add the KVL equations to eliminate v_x:

$$12 i_1 + 12 i_2 = -15 \quad \Rightarrow \quad 4 i_1 + 4 i_2 = -5$$

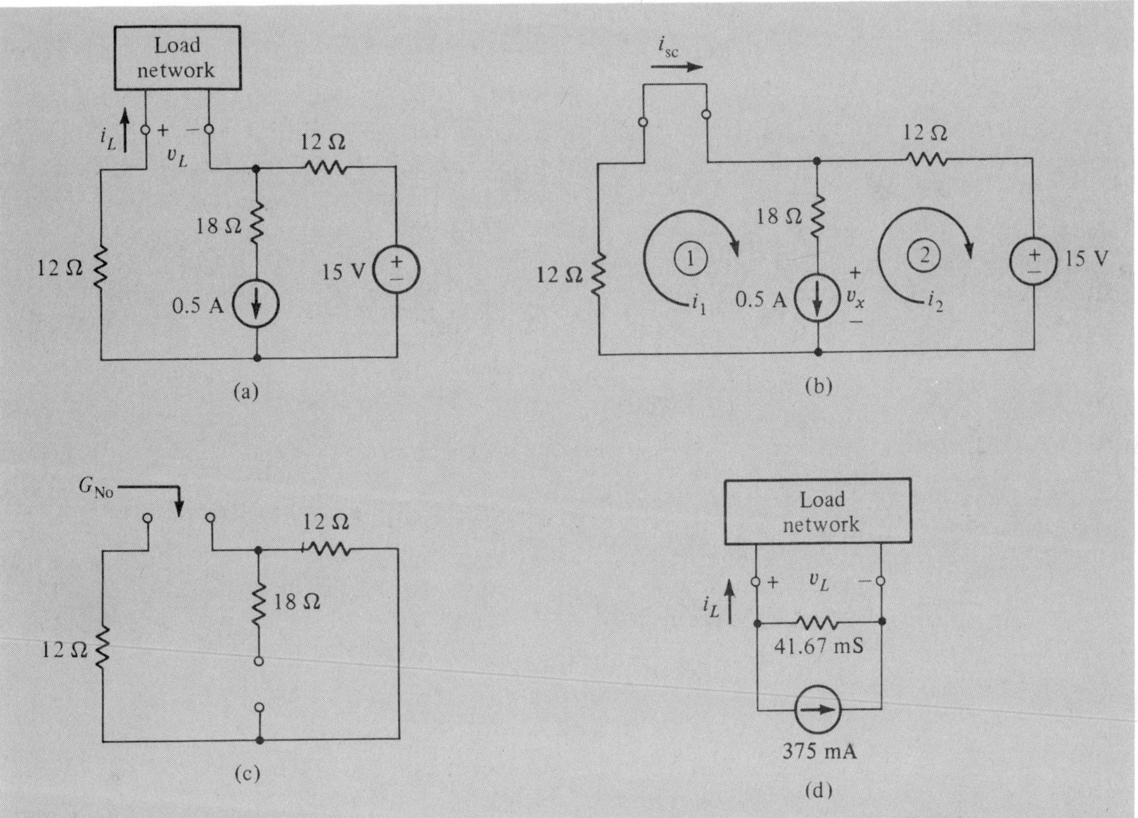

Figure 1.40 Example 1.18. (a) Partitioned circuit. (b) Source network prepared for determining i_{No}. (c) Deactivated source network for determining G_{No}. (d) Norton's equivalent circuit.

3. There is no unique solution for one equation with two unknowns, so write an auxiliary equation based on the 0.5-A ideal current source:

$$i_1 - i_2 = 0.5 \text{ A}$$

4. Now the two equations can be solved. Multiply the second equation by 4, and add the two equations:

$$\begin{aligned} 4i_1 + 4i_2 &= -5 \\ 4(\ i_1 - \ i_2) &= \ 0.5 \end{aligned} \quad \Rightarrow \quad \begin{aligned} 4i_1 + 4i_2 &= -5 \\ 4i_1 - 4i_2 &= \ \ 2 \\ \hline 8i_1 \qquad &= -3 \end{aligned}$$

$$i_1 = -0.375 \text{ A}$$

5. Since there are no dependent sources present, the source network can be deactivated by replacing the voltage source with a short circuit and the current source with an open circuit [Figure 1.40(c)]. The deactivated source network now consists only of resistors, and Norton's conductance can be measured or calculated at the load terminals. In this case, replacing the 0.5-A ideal source with an open circuit has left the 18-Ω resistor **dangling** and placed the two 12-Ω resistors in series, so

$$G_{No} = \frac{1}{12\ \Omega + 12\ \Omega} = 41.67\ \text{mS}$$

Once again, this method only applies to source networks which do not contain dependent sources.

Example 1.19

SPICE

Problem:

Determine Thevenin's and Norton's equivalent circuits for the circuit shown in Figure 1.41(a).

Figure 1.41
Example 1.19.
(a) Partitioned circuit.
(b) Circuit from (a) after source conversions.
(c) Circuit from (b) after conversion of 666.7-mA source.
(d) Thevenin's equivalent circuit.
(e) Norton's equivalent circuit.

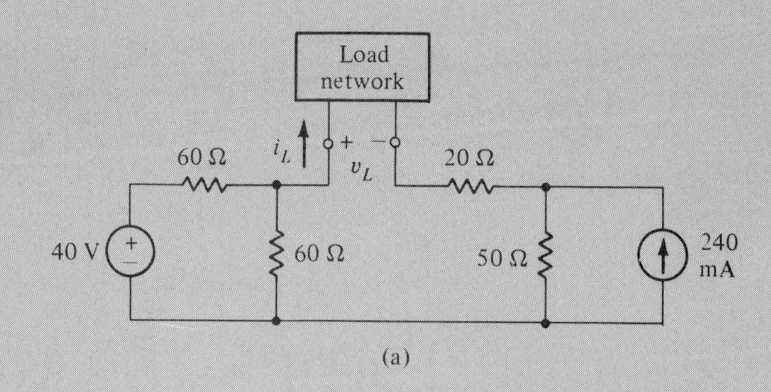

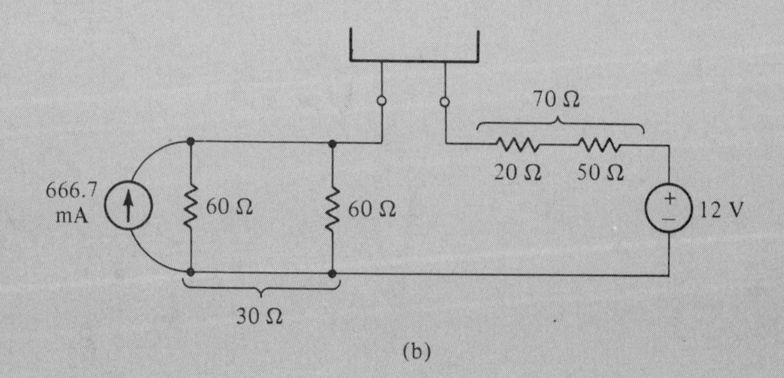

Figure 1.41
Continued

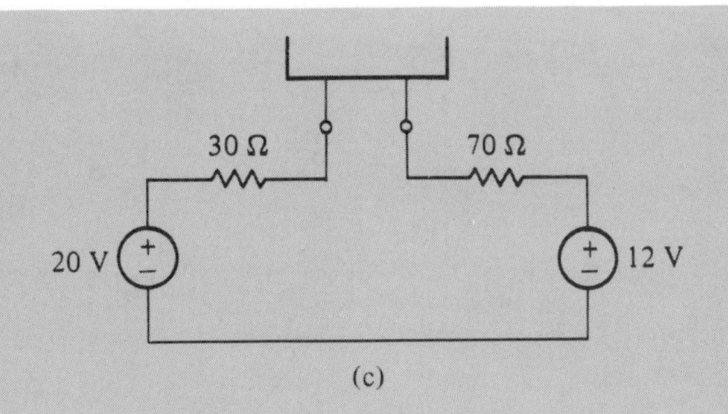

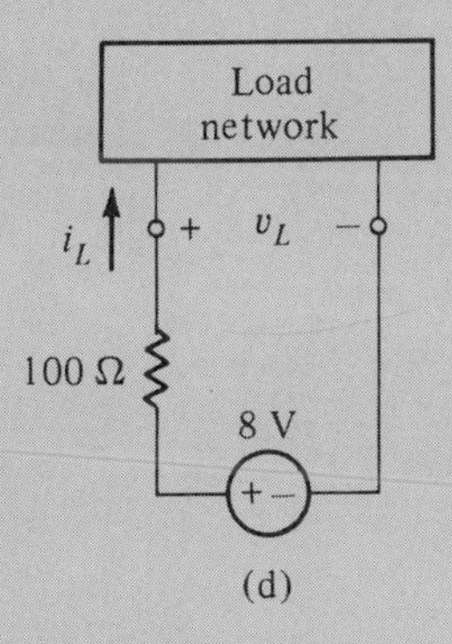

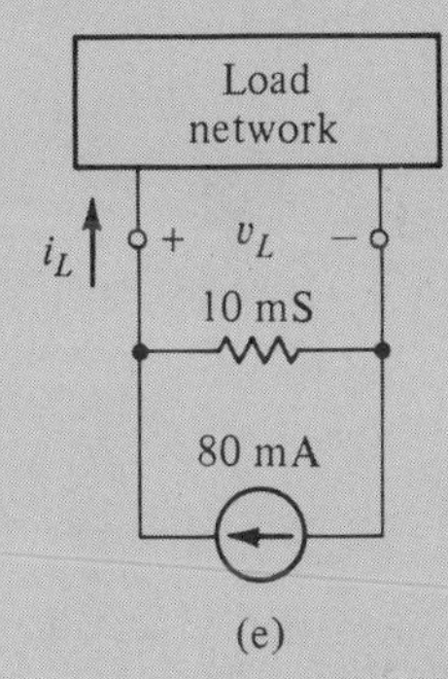

Solution:

1. Perform source conversions on each of the practical sources in Figure 1.41(a) to obtain the source network shown in Figure 1.41(b).
2. Combine the parallel 60-Ω resistors and the 20-Ω and 50-Ω series resistors, and perform a source conversion on the 666.7-mA source in Figure 1.41(b) to obtain the source network shown in Figure 1.41(c).
3. Combine the series voltage sources and series resistors in Figure 1.41(c) to obtain Thevenin's equivalent circuit [Figure 1.41(d)].
4. Perform a source conversion on Thevenin's equivalent source to obtain Norton's equivalent circuit [Figure 1.41(e)].

Example 1.20

SPICE

Problem:

Find Thevenin's equivalent circuit and Norton's equivalent circuit for the circuit shown in Figure 1.42(a).

Figure 1.42
Example 1.20.
(a) Partitioned circuit
(b) Source network, with load removed, prepared for node analysis.
(c) Source network, with load shorted, prepared for mesh analysis.
(d) Thevenin's equivalent circuit. (e) Norton's equivalent circuit.

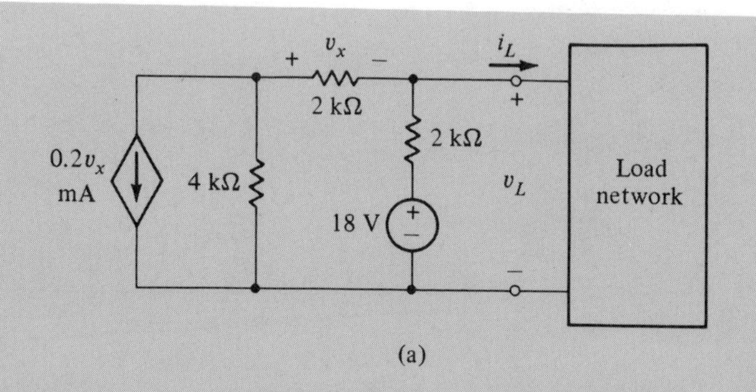

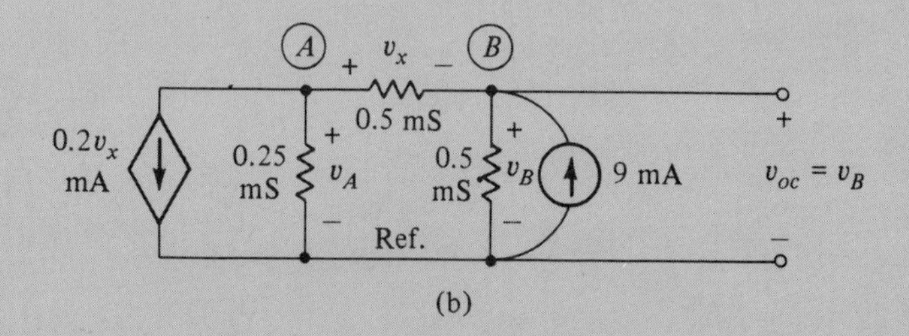

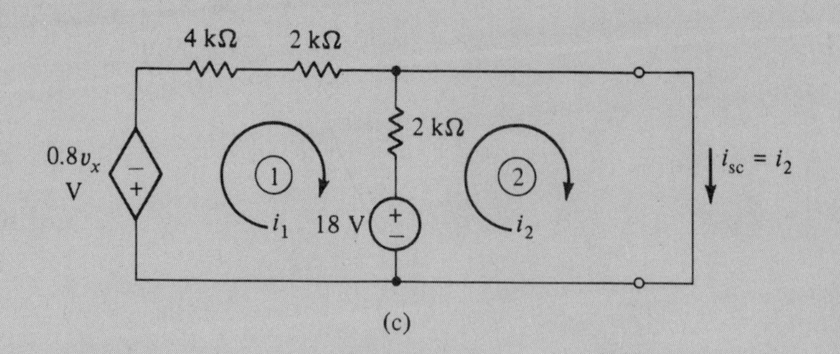

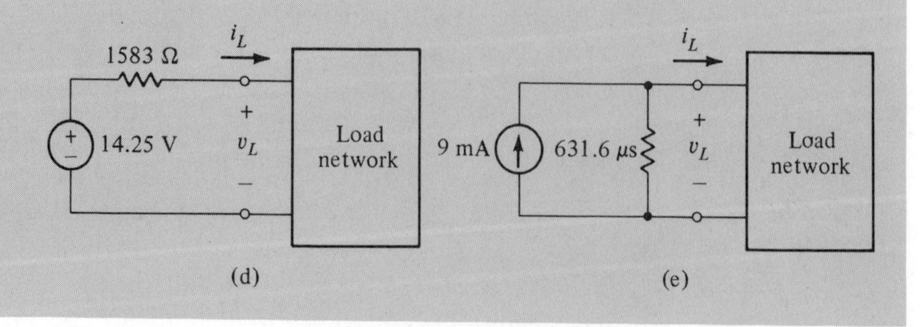

Solution:

1. Remove the load network, convert the independent voltage source to a current source, and determine v_B (= v_{oc}) by node analysis. Note that conductance is given in millisiemens (mS) and current in milliamperes (mA), so the equations are dimensionally correct as written:

$$\begin{aligned} \text{Node A:} \quad & 0.75v_A - 0.5v_B = -0.2v_x \\ \text{Node B:} \quad & -0.5\,v_A + v_B = 9 \\ & v_x = v_A - v_B \end{aligned}$$

2. Substitute for v_x in the node A equation, rearrange terms, and solve simultaneously with the node B equation for v_B:

$$0.75v_A - 0.5v_B = -0.2(v_A - v_B)$$

or

$$0.95v_A - 0.7v_B = 0$$

$$\begin{aligned} 0.95v_A - 0.7v_B &= 0 \\ 1.9(-0.5\,v_A + v_B &= 9) \end{aligned} \quad \Rightarrow \quad \begin{aligned} 0.95v_A - 0.7v_B &= 0 \\ -0.95v_A + 1.9v_B &= 17.1 \\ \hline 1.2v_B &= 17.1 \\ v_B &= 14.25\text{ V} \end{aligned}$$

3. Replace the load network with a short circuit, convert the dependent current source to a voltage source, and determine i_2 (= i_{sc}) by mesh analysis:

$$\begin{aligned} \text{Mesh 1:} \quad & 8000i_1 - 2000i_2 = -0.8v_x - 18 \\ \text{Mesh 2:} \quad & -2000i_1 + 2000i_2 = 18 \\ & v_x = 2000i_1 \end{aligned}$$

4. Substitute v_x into the mesh 1 equation, rearrange terms, and solve simultaneously with the mesh 2 equation for i_2:

$$8000i_1 - 2000i_2 = -0.8(2000i_1) - 18$$

$$9600i_1 - 2000i_2 = -18$$

or

$$
\begin{aligned}
9600i_1 - 2000i_2 &= -18 \\
-2000i_1 + 2000i_2 &= 18 \\
\hline
7600i_1 &= 0 \\
i_1 &= 0 \quad \Rightarrow \quad i_2 = 9 \text{ mA}
\end{aligned}
$$

5. Calculate R_{Th} and G_{No}:

$$R_{\text{Th}} = v_{oc}/i_{sc} \simeq 1583\ \Omega$$
$$G_{\text{No}} = i_{sc}/v_{oc} \simeq 631.6\ \mu\text{S}$$

Example 1.21

Problem:

SPICE

Find Thevenin's equivalent and Norton's equivalent for the circuit shown in Figure 1.43(a).

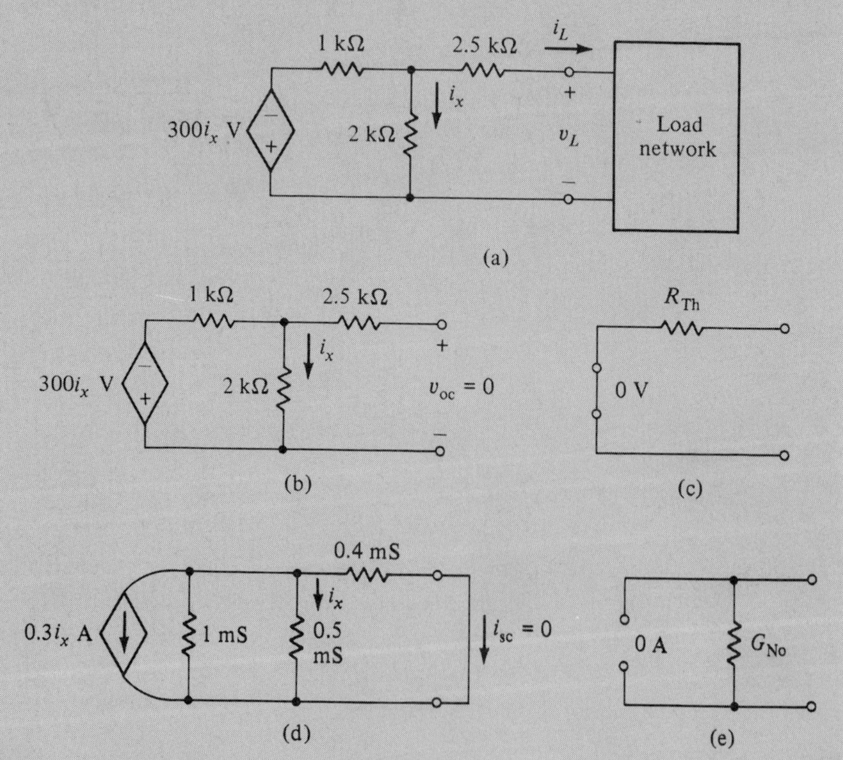

Figure 1.43 Example 1.21. (a) Partitioned circuit with dependent source only in source network. (b) Source network with load removed. (c) Thevenin's equivalent source network. (d) Source network with load replaced by short circuit. (e) Norton's equivalent source network. (f) Procedure for determining R_{Th}. (g) Source network prepared for determining R_{Th}. (h) Procedure for determining G_{No}. (i) Source network prepared for determining G_{No}.

Figure 1.43 Continued

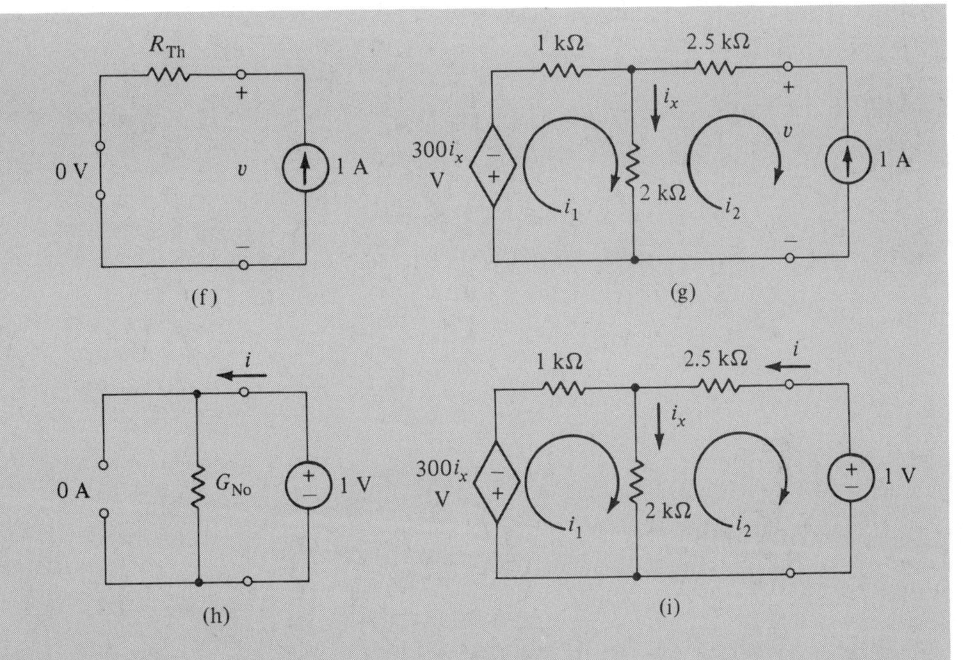

Solution:

1. In Figure 1.43(b), v_{oc} is zero, since there is no independent source in the source network. This can be verified from the KVL equation for the closed loop:

$$3000i_x + 300i_x = 0 \quad \Rightarrow \quad 3300i_x = 0 \quad \Rightarrow \quad i_x = 0$$

If $i_x = 0$, then $v_{oc} = 0$, which means that the Thevenin's equivalent source consists only of resistance, R_{Th} [Figure 1.43(c)].

2. That i_{sc} is also zero can be verified by the current division rule, once the dependent voltage source has been converted to a current source [Figure 1.43(d)]:

$$i_x = -0.3i_x\left(\frac{0.5}{1 + 0.5 + 0.4}\right) \simeq -0.07895i_x$$

$$i_x + 0.07895i_x = 0 \quad \Rightarrow \quad i_x = 0$$

If $i_x = 0$, then $i_{sc} = 0$, which means that the Norton's equivalent source consists only of conductance, G_{No} [Figure 1.43(e)].

3. R_{Th} cannot be calculated from the ratio v_{oc}/i_{sc}, since the result (0/0) is indeterminate. To determine R_{Th}, place a 1-A test current source at the load terminals [Figure 1.43(f)]; the response voltage v at the load

terminals will be numerically equal to R_{Th}:

$$v = (1\text{ A})R_{Th} \quad \Rightarrow \quad v = R_{Th}$$

4. Write the mesh equations for the circuit shown in Figure 1.43(g), with a 1-A source at the load terminals:

$$3000i_1 - 2000i_2 = -300i_x$$
$$-2000i_1 + 4500i_2 = -v$$

5. Write the auxiliary equations:

$$i_x = i_1 - i_2$$
$$i_2 = -1\text{ A}$$

6. Substitute the auxiliary equations into the mesh equations and rearrange terms:

$$3300i_1 \quad = -2300$$
$$-2000i_1 + v = \quad 4500$$

7. Solve these equations simultaneously for v:

$$v \simeq 3106\text{ V}$$

which means that

$$R_{Th} \simeq 3106\ \Omega \quad \text{and} \quad G_{No} = 1/R_{Th} \simeq 322.0\ \mu\text{S}$$

8. G_{No} could be obtained directly by placing a 1-V test source at the load terminals, in which case the response current into the upper load terminal would be numerically equal to G_{No} [Figure 1.43(h) and (i)]. The reader should verify this procedure (Problem 1-27).

Computer Simulations

The computer simulations in SPICE 1.4 (Example 1.19) and SPICE 1.5 (Example 1.20) demonstrate a technique for determining Thevenin's equivalent source network and Norton's equivalent source network. In SPICE 1.4, a simulation is conducted with the load replaced by a large resistance (ROPENCKT = 100 MΩ) to simulate an open circuit; the voltage at node 1 is the open-circuit voltage, and hence v_{Th}. A second simulation is conducted with the load replaced by a

```
*******09/17/88 ********  SPICE 2G.6    3/15/83 ********12:50:27*****
SPICE 1.4 (V-THEVENIN)
****     INPUT LISTING                   TEMPERATURE =   27.000 DEG C
*******************************************************************************
 R1 1 2 60
 R2 1 3 60
 R3 4 0 20
 R4 4 3 50
 ROPENCKT 1 0 100MEG
 V 2 3 DC 40
 I 3 4 DC 240M
 .END
*******09/17/88 ********  SPICE 2G.6    3/15/83 ********12:50:27*****
SPICE 1.4 (V-THEVENIN)
****     SMALL SIGNAL BIAS SOLUTION      TEMPERATURE =   27.000 DEG C
*******************************************************************************
  NODE   VOLTAGE     NODE   VOLTAGE      NODE   VOLTAGE      NODE   VOLTAGE
 (  1)    8.0000    (  2)   28.0000     (  3)  -12.0000     (  4)      .0000
    VOLTAGE SOURCE CURRENTS
    NAME        CURRENT
    V         -3.333D-01
    TOTAL POWER DISSIPATION   1.62D+01  WATTS

        JOB CONCLUDED
```

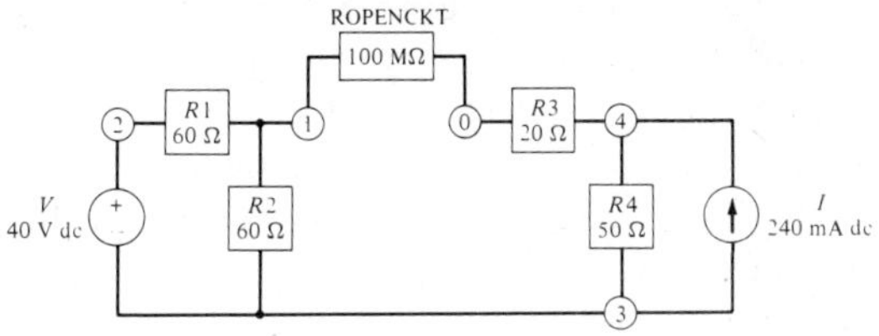

ROPENCKT 100 MΩ; R1 60 Ω; R2 60 Ω; R3 20 Ω; R4 50 Ω; V 40 V dc; I 240 mA dc

(a) Simulation to determine V_{Th}.

```
*******09/17/88 ********  SPICE 2G.6    3/15/83 ********12:58:04*****
SPICE 1.4 (R-THEVENIN)
****     INPUT LISTING                   TEMPERATURE =   27.000 DEG C
*******************************************************************************
 R1 1 3 60
 R2 1 3 60
 R3 4 0 20
 R4 4 3 50
 ITEST 0 1 DC 1
 .END
*******09/17/88 ********  SPICE 2G.6    3/15/83 ********12:58:04*****
SPICE 1.4 (R-THEVENIN)
****     SMALL SIGNAL BIAS SOLUTION      TEMPERATURE =   27.000 DEG C
*******************************************************************************
  NODE   VOLTAGE     NODE   VOLTAGE      NODE   VOLTAGE
 (  1)  100.0000    (  3)   70.0000     (  4)   20.0000
    TOTAL POWER DISSIPATION   1.00D+02  WATTS

        JOB CONCLUDED
```

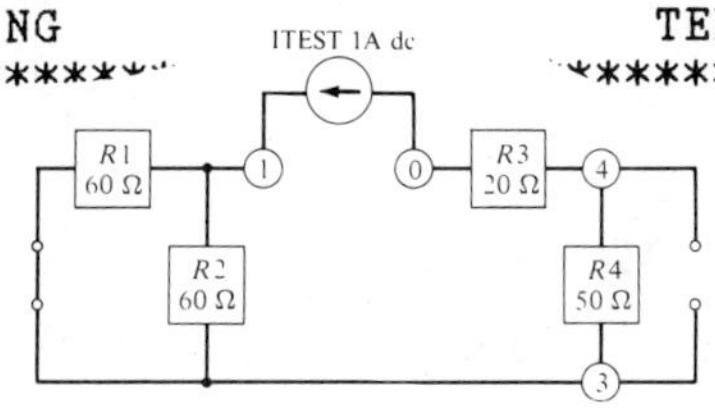

ITEST 1A dc; R1 60 Ω; R2 60 Ω; R3 20 Ω; R4 50 Ω

(b) Simulation to determine R_{Th}.

1-A test source (ITEST) and the independent sources deactivated; the voltage at node 1 is numerically equivalent to R_{Th}.

In SPICE 1.5, the first simulation is conducted with the load replaced by a SPICE ammeter (VSHRTCKT = 0 V), which effectively creates a short-circuit condition at the load terminals, so the current into VSHRTCKT is i_{No}. A second simulation is conducted with the load replaced by a 1-V test source (VTEST), and the *independent* source deactivated. The current into the more positive load terminal is numerically equivalent to G_{No}; hence, the current into VTEST is the negative of G_{No}.

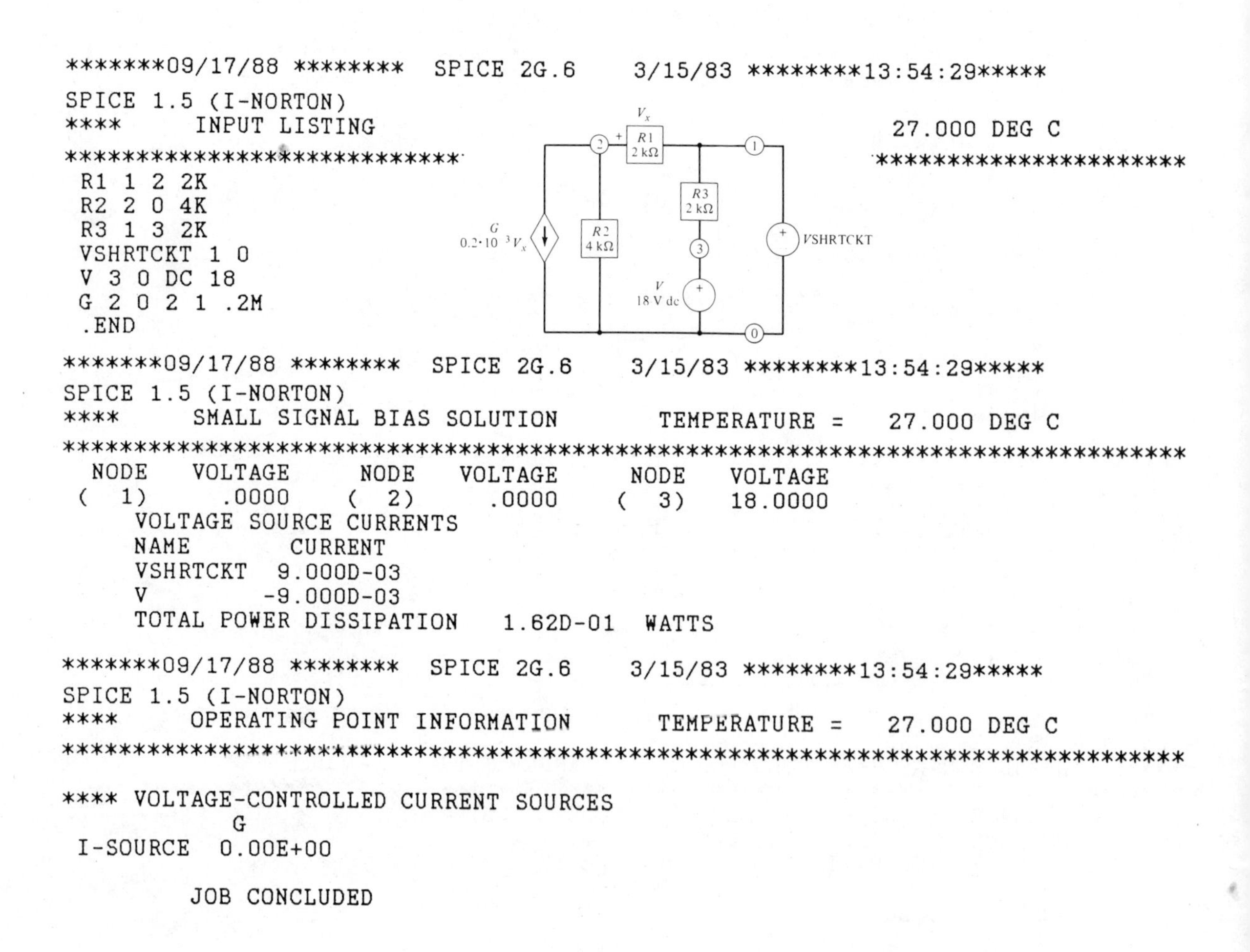


```
*******09/17/88 ********  SPICE 2G.6    3/15/83 ********13:54:29*****
SPICE 1.5 (I-NORTON)
****      INPUT LISTING                                   27.000 DEG C
****************************                    ***********************
 R1 1 2 2K
 R2 2 0 4K
 R3 1 3 2K
 VSHRTCKT 1 0
 V 3 0 DC 18
 G 2 0 2 1 .2M
 .END

*******09/17/88 ********  SPICE 2G.6    3/15/83 ********13:54:29*****
SPICE 1.5 (I-NORTON)
****      SMALL SIGNAL BIAS SOLUTION        TEMPERATURE =   27.000 DEG C
***********************************************************************
  NODE    VOLTAGE      NODE    VOLTAGE      NODE    VOLTAGE
 (  1)      .0000     (  2)      .0000     (  3)    18.0000
     VOLTAGE SOURCE CURRENTS
     NAME         CURRENT
     VSHRTCKT   9.000D-03
     V         -9.000D-03
     TOTAL POWER DISSIPATION    1.62D-01  WATTS

*******09/17/88 ********  SPICE 2G.6    3/15/83 ********13:54:29*****
SPICE 1.5 (I-NORTON)
****      OPERATING POINT INFORMATION       TEMPERATURE =   27.000 DEG C
***********************************************************************

**** VOLTAGE-CONTROLLED CURRENT SOURCES
              G
 I-SOURCE  0.00E+00

         JOB CONCLUDED
```

(a) Simulation to determine I_{No}.

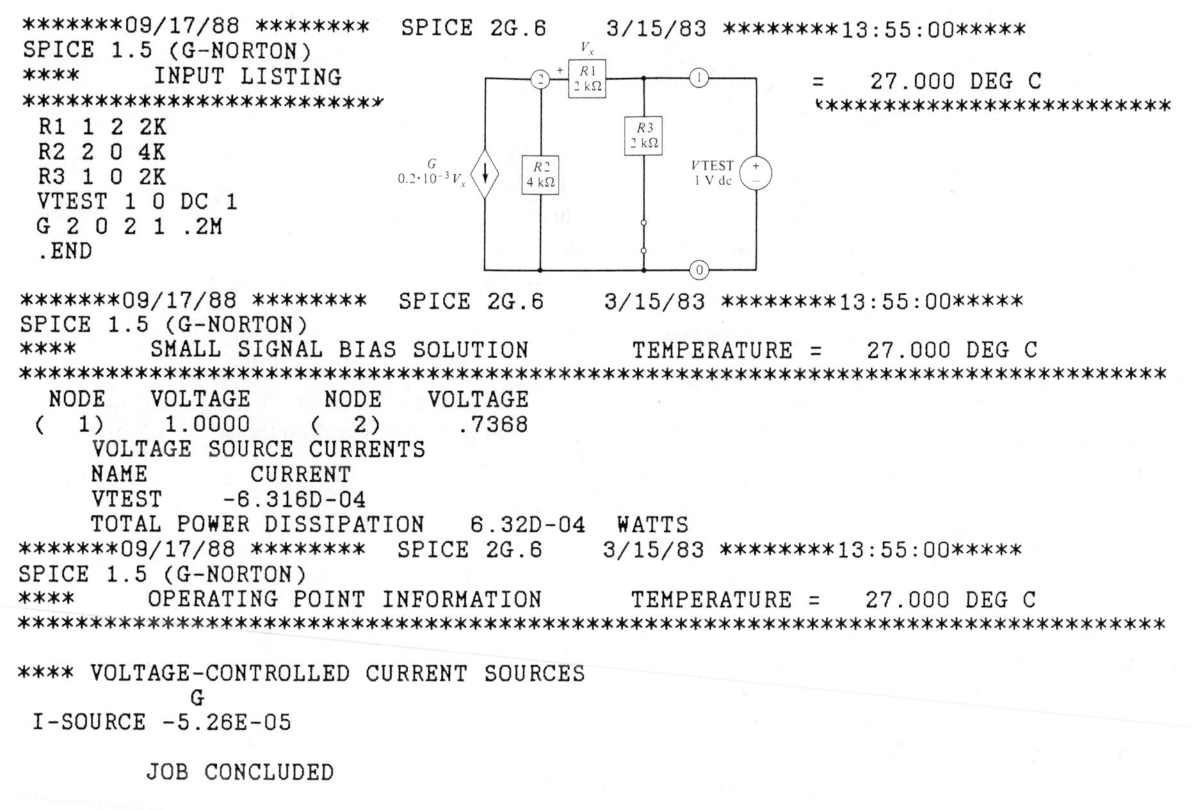


```
*******09/17/88 ********  SPICE 2G.6    3/15/83 ********13:55:00*****
SPICE 1.5 (G-NORTON)
****      INPUT LISTING                                   =    27.000 DEG C
************************                                  ***********************
 R1 1 2 2K
 R2 2 0 4K
 R3 1 0 2K
 VTEST 1 0 DC 1
 G 2 0 2 1 .2M
 .END

*******09/17/88 ********  SPICE 2G.6    3/15/83 ********13:55:00*****
SPICE 1.5 (G-NORTON)
****      SMALL SIGNAL BIAS SOLUTION        TEMPERATURE =    27.000 DEG C
*************************************************************************
  NODE    VOLTAGE     NODE    VOLTAGE
 (  1)     1.0000     (  2)     .7368
    VOLTAGE SOURCE CURRENTS
    NAME        CURRENT
    VTEST    -6.316D-04
    TOTAL POWER DISSIPATION    6.32D-04  WATTS
*******09/17/88 ********  SPICE 2G.6    3/15/83 ********13:55:00*****
SPICE 1.5 (G-NORTON)
****      OPERATING POINT INFORMATION       TEMPERATURE =    27.000 DEG C
*************************************************************************

**** VOLTAGE-CONTROLLED CURRENT SOURCES
           G
 I-SOURCE -5.26E-05

        JOB CONCLUDED
```

(b) Simulation to determine G_{No}

1.8 Maximum Power Transfer Theorem

The instantaneous power absorbed by a load network can be calculated from the voltage and current at the load terminals of a partitioned circuit [Figure 1.37(a)]:

$$p_L = v_L i_L \tag{1.9}$$

If the sign of the result is positive, then the load is receiving energy from the source network at that instant; if the sign of the result is negative, the intended source network is in fact receiving energy from the load network. At any given instant, we might expect either result, since the load may be any configuration of active and passive elements and the source network may, as we have just discussed, have a Thevenin's voltage which is zero or nonzero, and a Thevenin's resistance which is positive, negative, or zero. If we restrict our attention to a circuit in which the Thevenin's voltage is nonzero, the Thevenin's resistance is positive, and the load network is resistive, we will find that the load network will be absorbing power from the source at all times. The Thevenin's equivalent for the source network of the circuit described is a practical voltage source, and the Norton's equivalent is a practical current source. The **maximum power transfer theorem** applies to such a circuit and can be used to determine the values of Thevenin's resistance (or Norton's conductance) and load resistance (or conductance) for maximum transfer of energy between source network and load network.

Maximum Power Transfer Theorem

> A practical voltage source delivers maximum power to a resistive load when the internal resistance and the load resistance are equal, and a practical current source delivers maximum power to a resistive load when the internal conductance and the load conductance are equal.

The maximum power transfer theorem is easily proved from Equation (1.9). Referring to Figure 1.44(a), v_L can be calculated by the voltage division rule,

$$v_L = v_{\mathrm{Th}} \frac{R_L}{R_L + R_{\mathrm{Th}}}$$

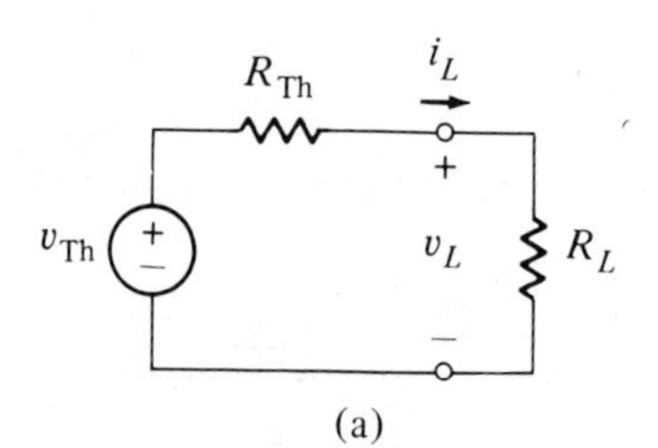

(a)

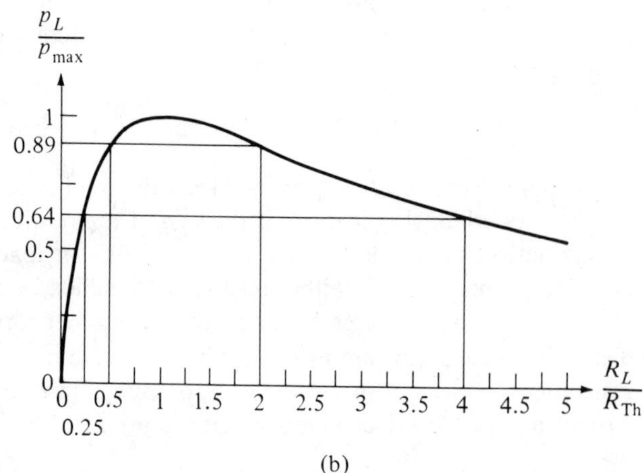

(b)

Figure 1.44 Maximum power transfer. (a) Thevenin's equivalent circuit. (b) Relationship of load power to load resistance.

and i_L can be calculated by Ohm's law,

$$i_L = \frac{v_{Th}}{R_L + R_{Th}}$$

Substitution of these results into Equation (1.9) yields

$$p_L = v_{Th}^2 \frac{R_L}{(R_L + R_{Th})^2} \tag{1.10}$$

For a given source network (i.e., v_{Th} and R_{Th} fixed), p_L is a function of R_L and varies from zero for $R_L = 0$ (short circuit) to some maximum value as R_L is increased, and back to zero as $R_L \to \infty$ (open circuit), as shown in Figure 1.44(b).

Flashback

The maximum (or minimum) value of a function, $f(x)$, occurs at a value of the independent variable, x, for which

$$\frac{d}{dx} f(x) = 0$$

The value of R_L for which p_L is maximum is found by differentiating p_L with respect to R_L, setting the result equal to zero, and solving for R_L:

$$\frac{dp_L}{dR_L} = v_{Th}^2 \left(R_L \frac{d}{dR_L} (R_L + R_{Th})^{-2} + (R_L + R_{Th})^{-2} \frac{dR_L}{dR_L} \right) = 0$$

$$v_{Th}^2 \left[-2R_L (R_L + R_{Th})^{-3} + (R_L + R_{Th})^{-2} \right] = 0$$

$$v_{Th}^2 \frac{R_{Th} - R_L}{(R_L + R_{Th})^3} = 0$$

Since v_{Th} is nonzero, and R_L and R_{Th} are both positive and nonzero, the preceding equation can be divided by v_{Th}^2 and multiplied by $(R_L + R_{Th})^3$, leaving

$$R_{Th} - R_L = 0$$

or

$$R_L = R_{Th} \tag{1.11}$$

Thus, in designing a load to utilize the maximum power available from a given source, the load resistance should be equal to the source resistance. In a circuit with the load resistance equal to the source resistance, the load is said to be **matched** to the source.

Two questions must now be asked:

1. What is the maximum power available from a given source?
2. What is the maximum power which can be dissipated by a given resistive load?

The answer is the same to both questions, and it follows directly from an analysis of Thevenin's equivalent circuit [Figure 1.44(a)] under matched conditions.

When Equation (1.11) is substituted into Equation (1.10),

$$(p_L)_{MAX} = v_{Th}^2 \frac{R_{Th}}{(R_{Th} + R_{Th})^2}$$

$$(P_L)_{MAX} = \frac{v_{Th}^2}{4R_{Th}}$$

Clearly the same result is obtained if R_L is substituted for R_{Th}, and the maximum power available is the same as the maximum power which can be dissipated. The two are often referred to simply as **available power**.

The available power from a practical current source is

$$(P_L)_{MAX} = \frac{i_{No}^2}{4G_{No}}$$

Verification of this result, which follows directly from analysis of the Norton equivalent circuit under matched conditions, is left as an exercise (Problem 1-28).

1.9 SUMMARY

We have reviewed the fundamental considerations of circuit analysis such as units, symbols, and network definitions and terminology. We have defined and described circuit elements according to terminal characteristics, that is, the relationship between voltage and current at the element terminals, and whether the element is delivering or absorbing energy.

The practical circuit analysis techniques which result from Kirchhoff's current law and Kirchhoff's voltage law—combination of elements, the voltage and current division rules, and node and mesh analysis—were discussed, and examples of the application of each technique were given. Further circuit analysis techniques based on important network theorems—the superposition principle, Thevenin's and Norton's theorems, and the maximum power transfer theorem—were discussed, and examples presented.

This chapter dealt with analysis of circuits containing only dc sources and resistors. Later we will see that the techniques presented in this chapter are applicable also to circuits containing time-dependent sources, capacitors, and inductors.

1.10 Terms

absorb
active
assumed solution
branch
circuit
circuit partitioning
closed path
common node
conductance matrix
controlled source
control voltage
control current
current-controlled current source (CCCS)
current-controlled voltage source (CCVS)
current source
dangling
dc source
dimensional analysis
dissipate
dual, duality
element
energy sink
energy source
ground node
independent node
independent source
internal conductance
internal resistance
lead
linear circuit
linear element
linear resistor
linear operating region
linearly dependent source
load
load network
load terminals
loop
matched
maximum power transfer theorem
mesh
mesh analysis
model
mutual conductance
mutual resistance
network
node
node analysis
no-load
nonlinear response
Norton's conductance
Norton's current
Norton's equivalent circuit
Norton's equivalent source
Norton's theorem
Ohm's law
open circuit
output resistance
parallel
passive
path
planar
potential difference

practical current source
practical voltage source
reference node
resistance matrix
self-conductance
self-resistance
series
short circuit
source conversion
source element
source vector
square matrix
source conductance
source conversion
source network
source resistance
superposed, superposition
superposition principle
symmetric matrix
terminal
terminal characteristic
Thevenin's equivalent source
Thevenin's equivalent circuit
Thevenin's resistance
Thevenin's theorem
Thevenin's voltage
vector of unknowns
voltage-controlled current source (VCCS)
voltage-controlled voltage source (VCVS)
voltage division rule
voltage polarity
voltage source

Problems

1-1. Identify each element as an energy source or an energy sink, and calculate the power delivered or absorbed by that element.

*a.

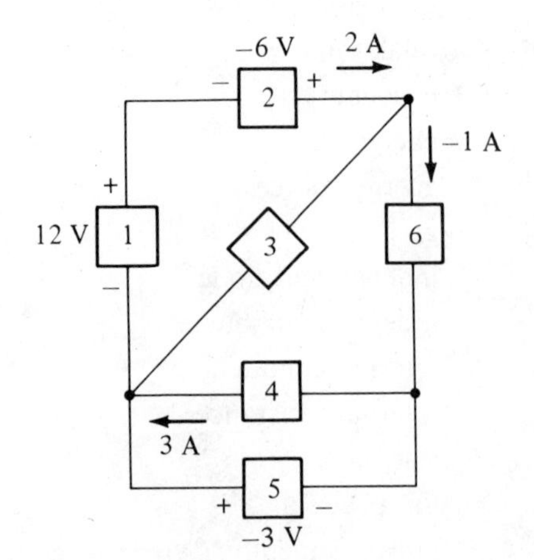

b.

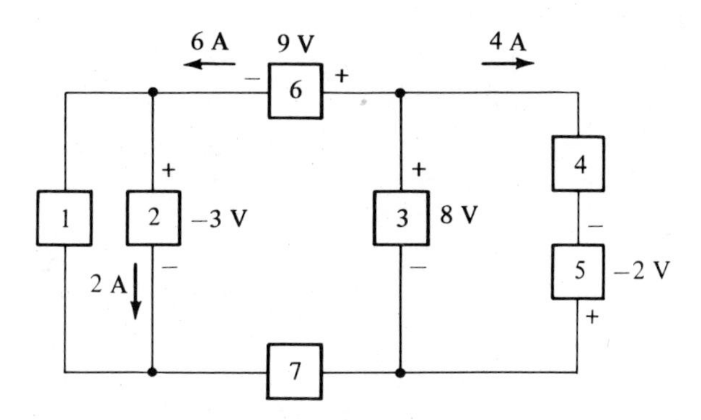

1-2. Use the Kirchhoff laws and Ohm's law to verify:

a. Equation (1.5).

b. Equation (1.6).

1-3. Simplify each circuit as much as possible by combining elements.

***a.**

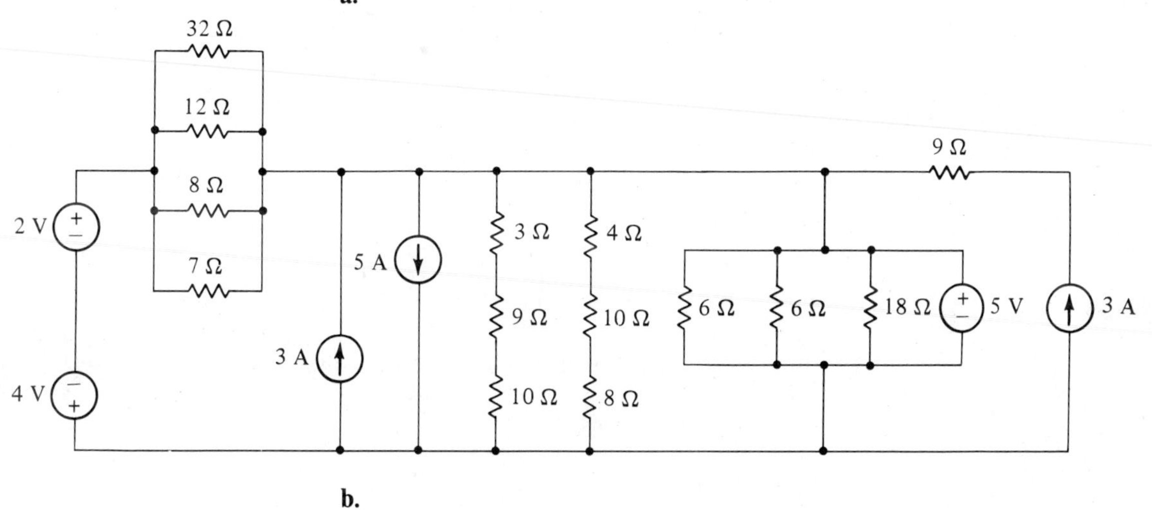

b.

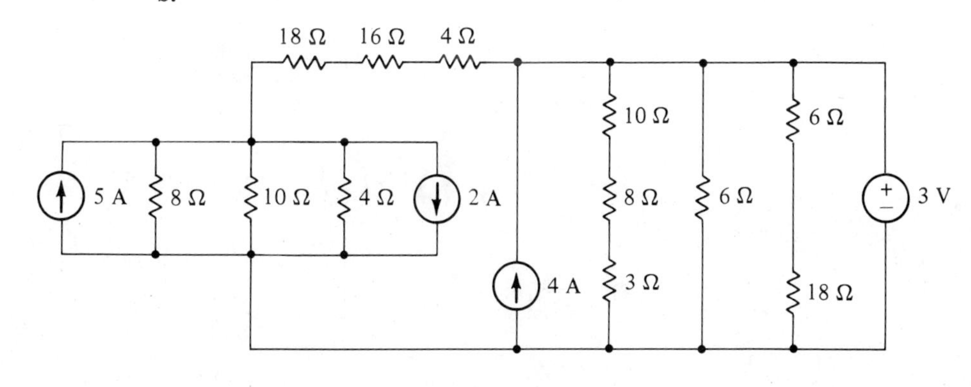

SPICE

1-4. Find the indicated voltage by use of the voltage division rule.

a.

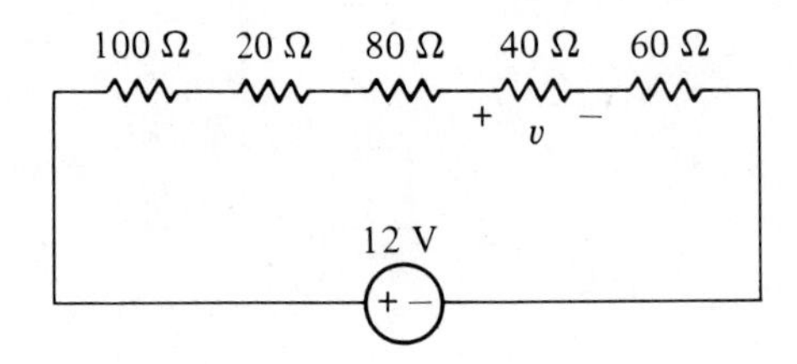

***b.**

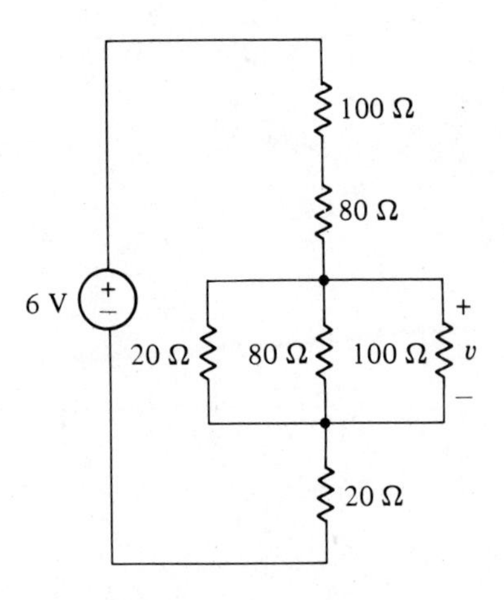

SPICE

1-5. Find the indicated voltage by use of the voltage division rule.

***a.**

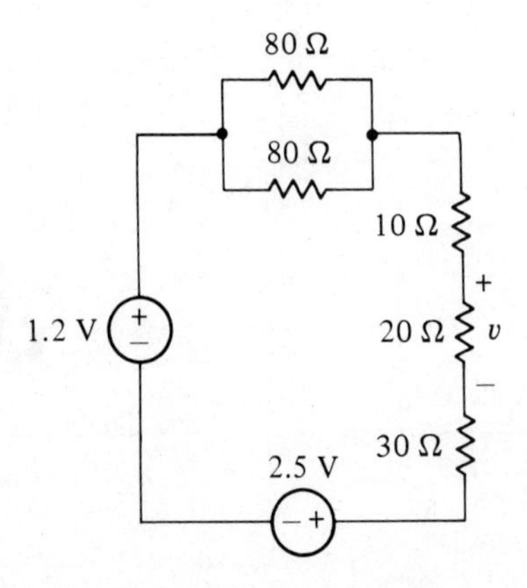

b.

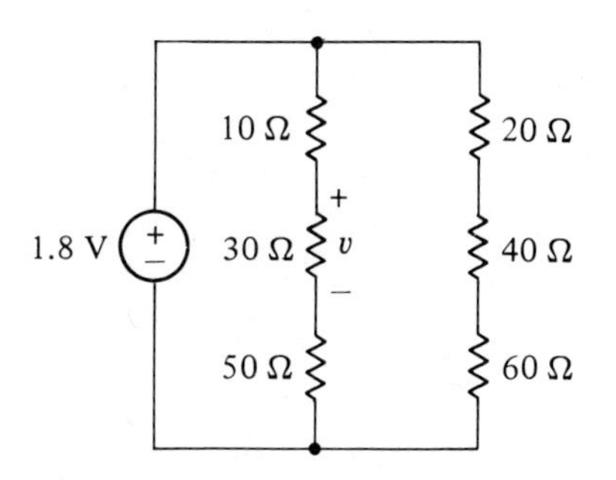

SPICE

1-6. Find the indicated current by use of the current division rule.

a.

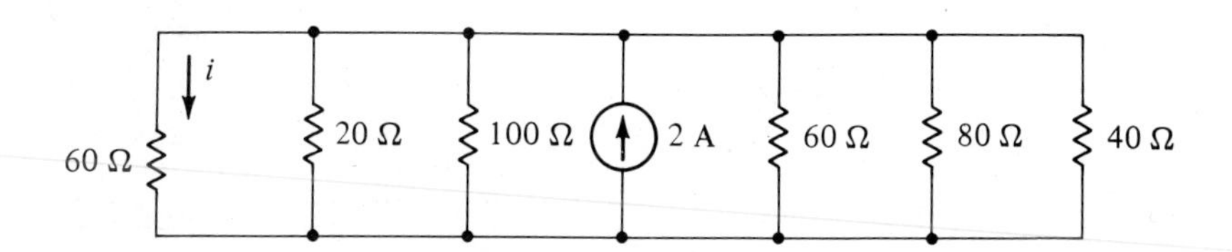

***b.**

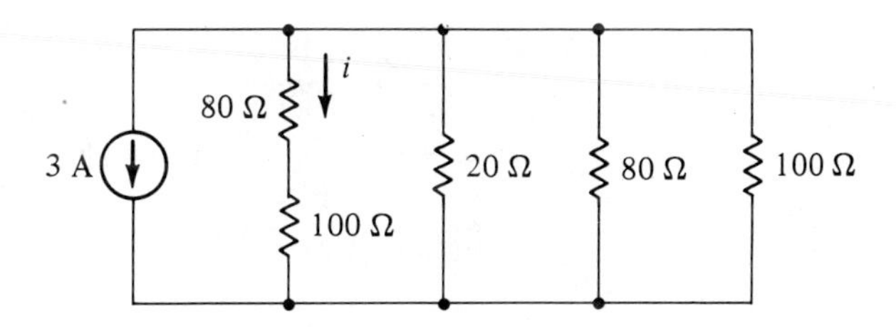

SPICE

1-7. Find the indicated current by use of the current division rule.

***a.**

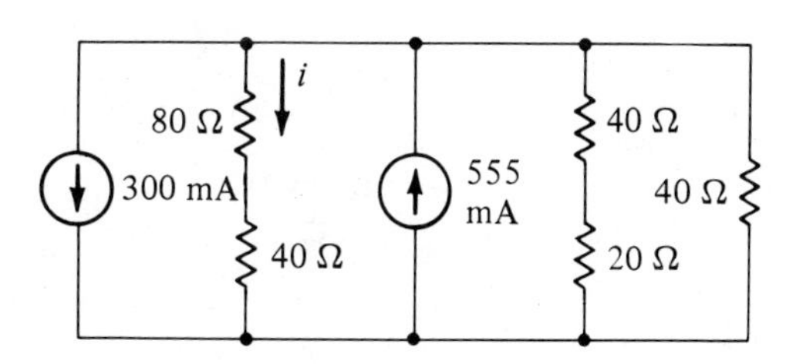

b.

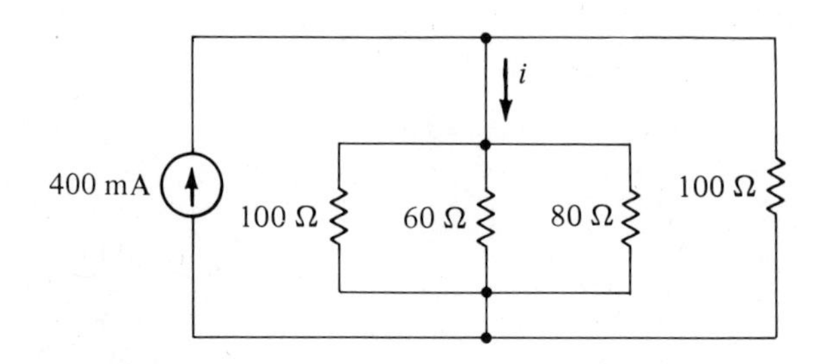

1-8. Use the Kirchhoff laws and Ohm's law to prove the:

a. Voltage division rule.

b. Current division rule.

1-9. Find v_s, i_s, R_s, and G_s.

***a.** Assume that there is no voltage dropped across the ammeter.

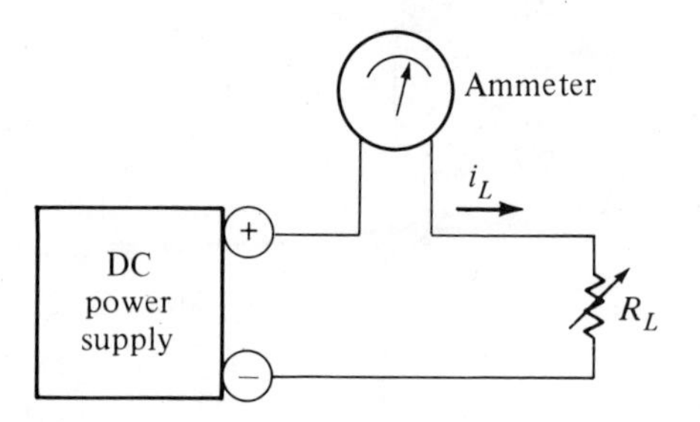

Measurement 1	Measurement 2
$R_L = 12\ \Omega$	$R_L = 6\ \Omega$
$i_L = 533.3$ mA	$i_L = 888.9$ mA

b. Assume that there is no current through the voltmeter.

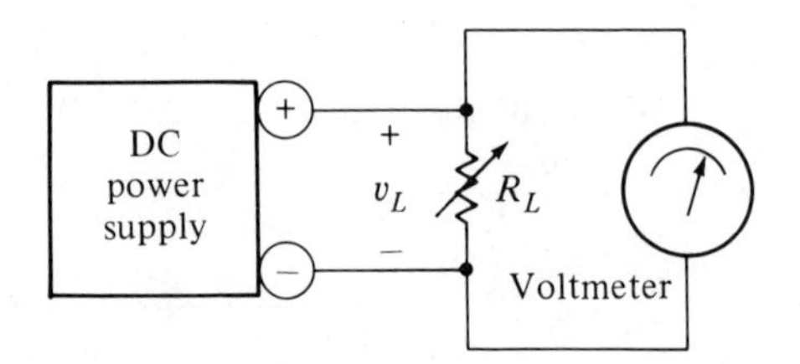

Measurement 1	Measurement 2
$R_L = 12\ \Omega$	$R_L = 6\ \Omega$
$v_L = 16.55$ V	$v_L = 14.12$ V

1-10. Verify the equivalence of the sources by finding i_L and v_L for full-load (short-circuit), no-load (open-circuit), and matched-load conditions.

a.

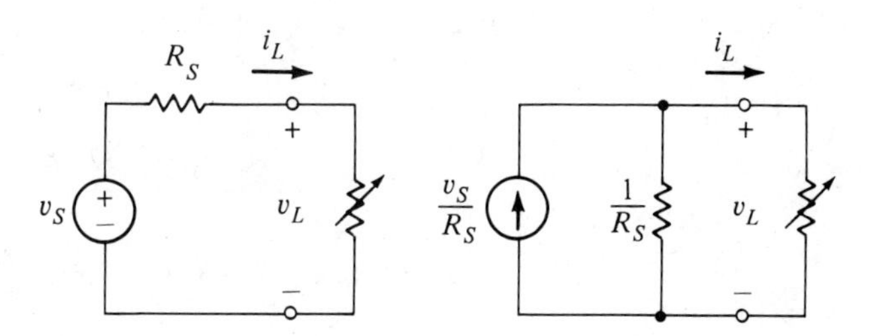

b.

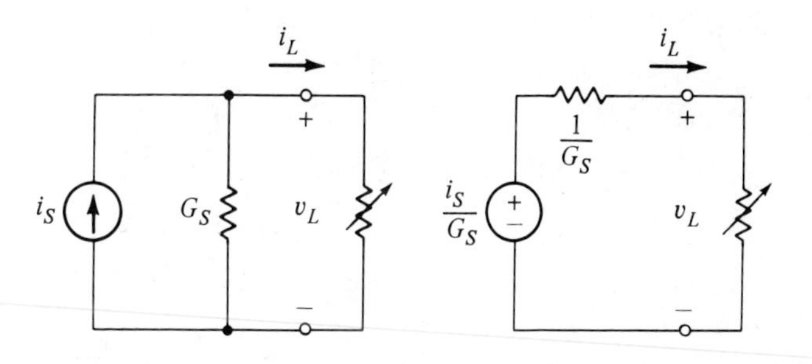

SPICE **1-11.** Solve for the voltage across and current through every element by the method of assumed solutions.

***a.** Assume $v = 1$ V.

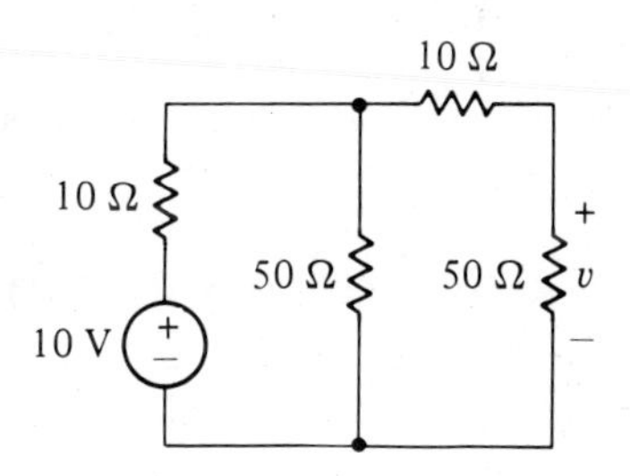

b. Assume $i = 1$ mA.

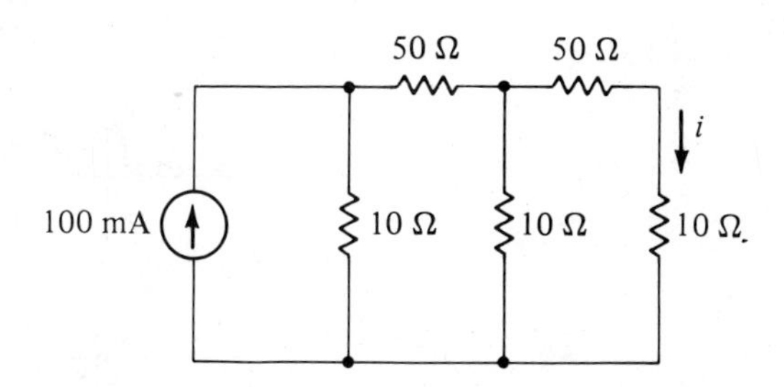

SPICE **1-12.** Use node analysis to determine v_A and v_B with respect to the reference node.

a.

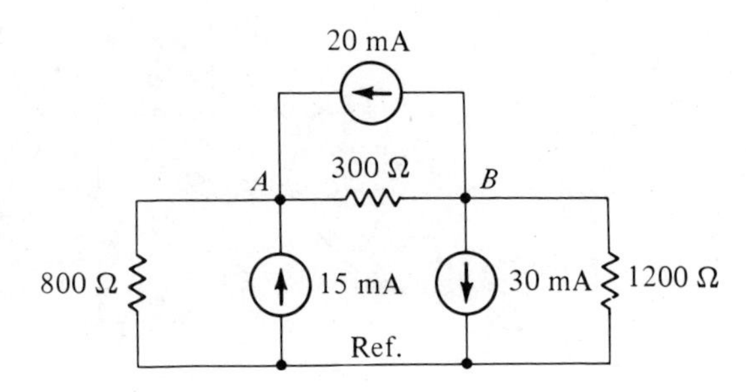

***b.**

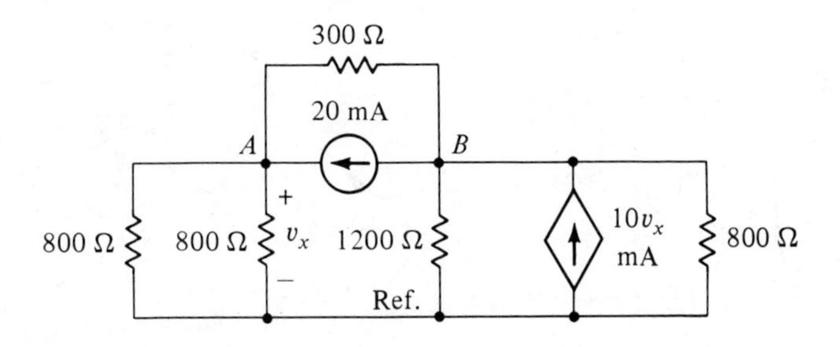

SPICE **1-13.** Use node analysis to determine v_A and v_B with respect to the reference node.

***a.**

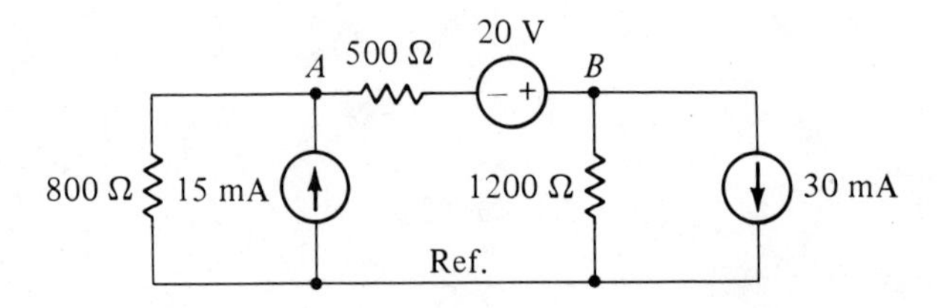

b.

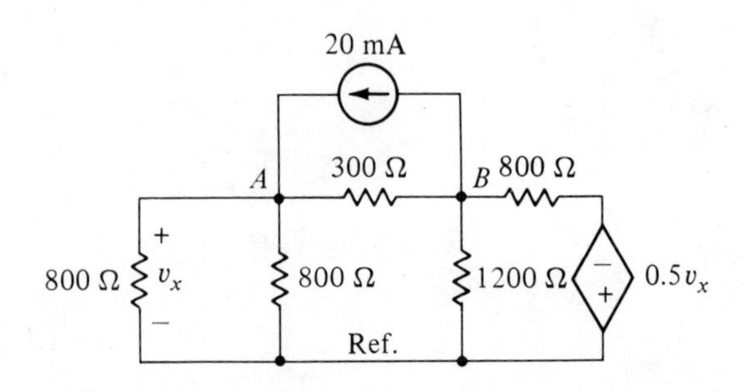

SPICE

1-14. Use node analysis to determine v_A and v_B with respect to the reference node. Find i.

a.

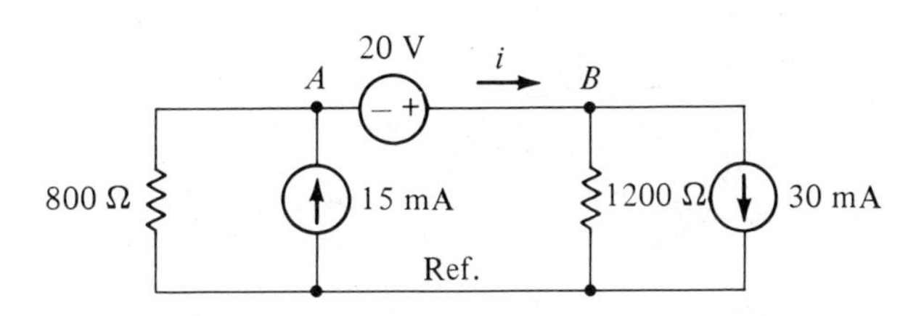

***b.**

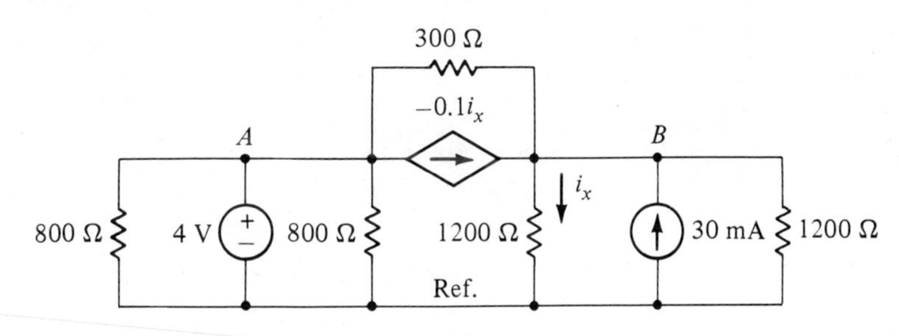

SPICE

1-15. Use the general matrix method for node analysis to determine the indicated voltages with respect to the reference node.

***a.** Find v_A, v_B, and v_C.

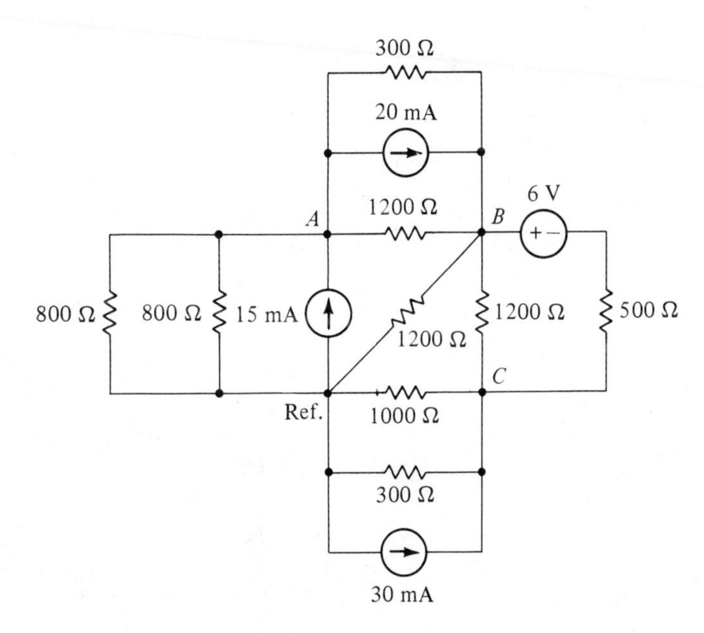

b. Find v_A, v_B, v_C, and v_D.

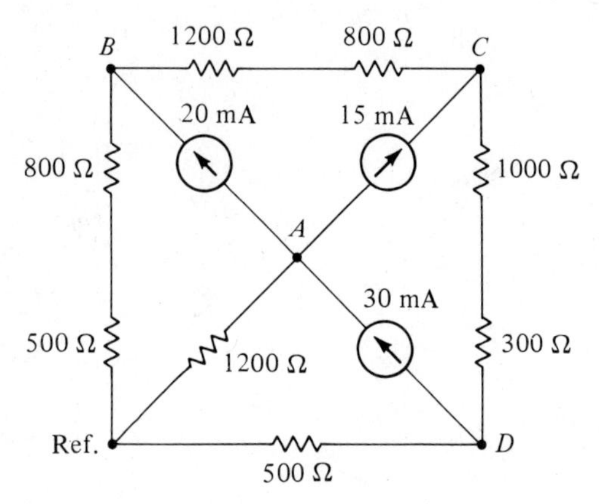

SPICE **1-16.** Use mesh analysis to determine i_1 and i_2.

a.

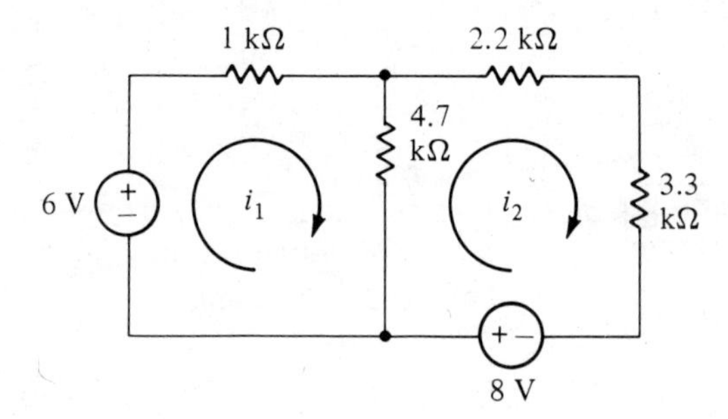

***b.**

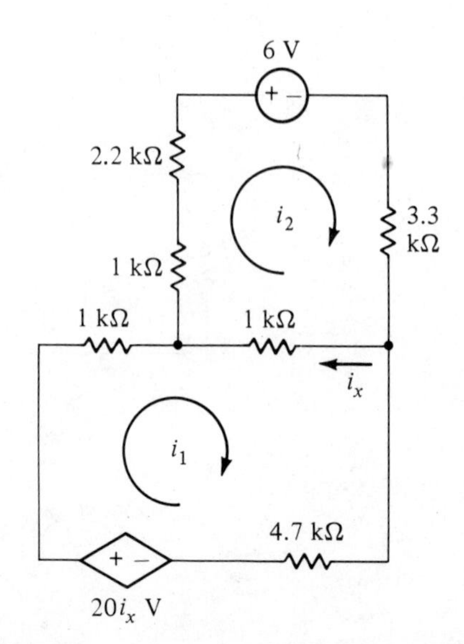

SPICE **1-17.** Use mesh analysis to determine i_1 and i_2.

***a.**

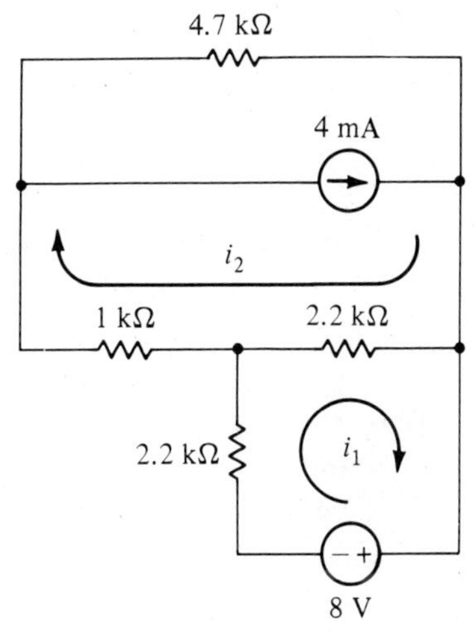

b.

v_x
12 V
3.3 kΩ
i_2
3.3 kΩ
1 kΩ
4.7 kΩ
1 kΩ
i_1
$0.6v_x$ mA
1 kΩ
3.3 kΩ

SPICE **1-18.** Use mesh analysis to determine i_1 and i_2. Find v.

a.

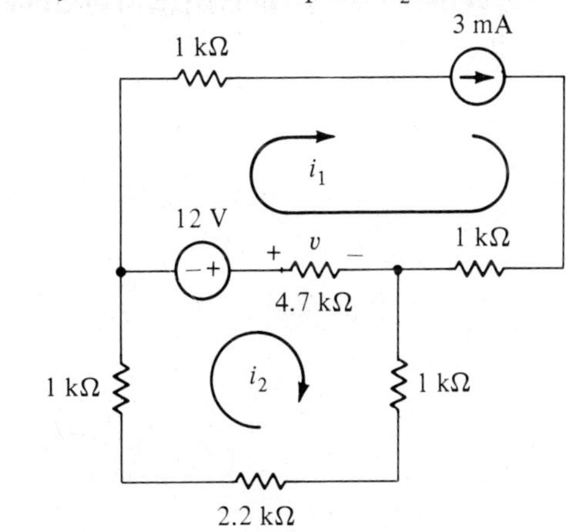

***b.**

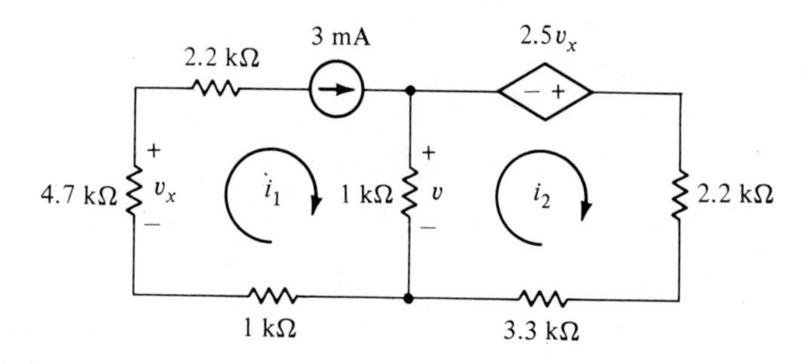

SPICE

1-19. Use the general matrix method for mesh analysis to determine the indicated currents.

***a.** Find i_1, i_2, and i_3.

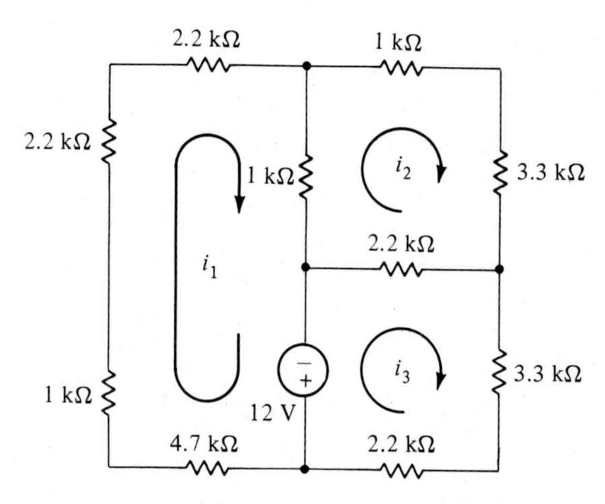

b. Find i_1, i_2, i_3, and i_4.

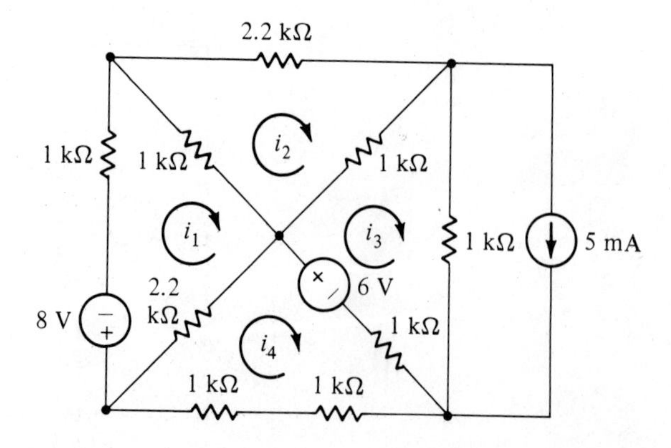

1-20. Solve the matrix equation:

a. From Example 1.14.

***b.**

$$\begin{bmatrix} 9 & -2 & 0 & -3 \\ -2 & 3 & -1 & -4 \\ 0 & -1 & 2 & -1 \\ -3 & -4 & -1 & 7 \end{bmatrix} \begin{bmatrix} v_A \\ v_B \\ v_C \\ v_D \end{bmatrix} = \begin{bmatrix} -7 \\ -15 \\ 0 \\ 14 \end{bmatrix}$$

SPICE

1-21. Solve for the indicated quantity by use of the superposition principle.

***a.** Find v.

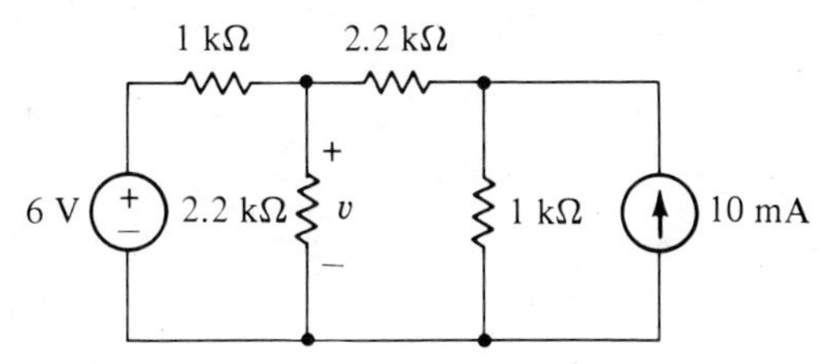

b. Find i.

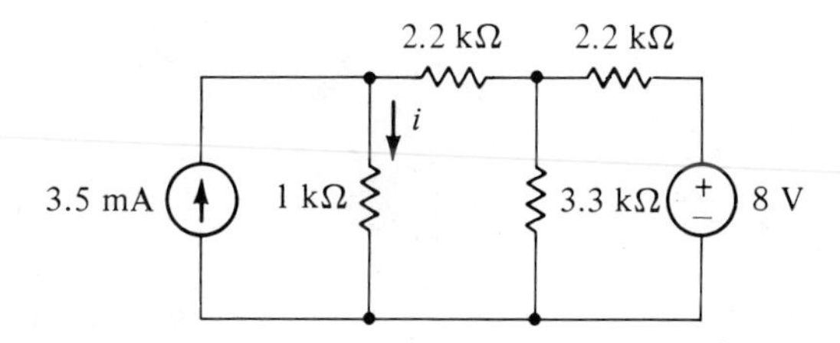

SPICE

1-22. Solve for the indicated quantity by use of the superposition principle.

a. Find i_y.

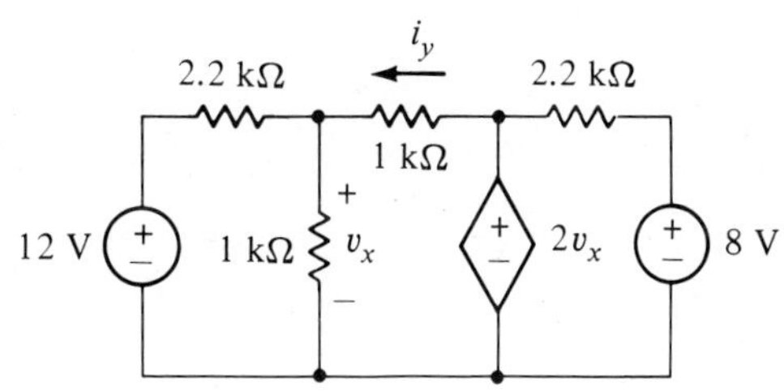

***b.** Find v_x.

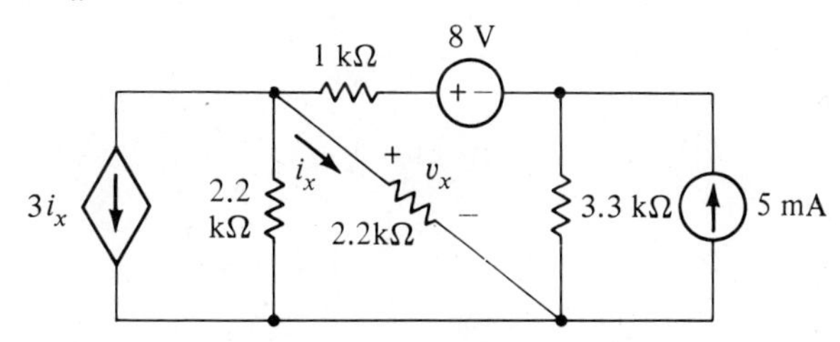

SPICE

1-23. Find Thevenin's equivalent circuit and Norton's equivalent circuit. Suggestion: Find v_{oc} *or* i_{sc}, then deactivate the source network to determine R_{Th} or G_{No}.

***a.**

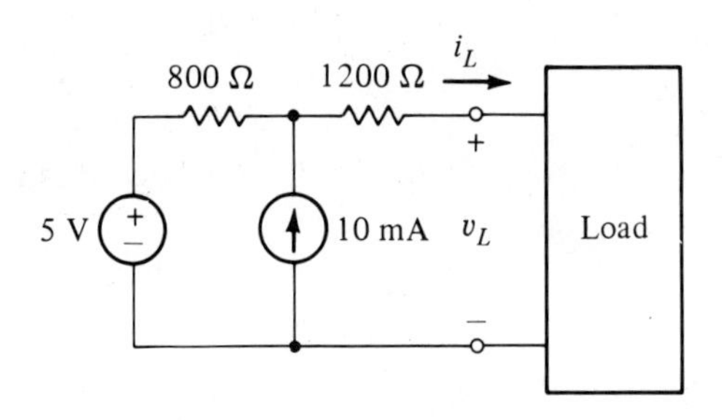

b.

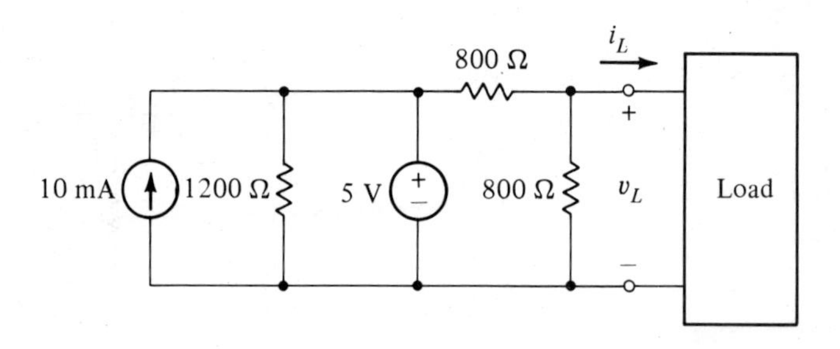

SPICE

1-24. Find Thevenin's equivalent circuit and Norton's equivalent circuit. Suggestion: Use repeated source conversions.

a.

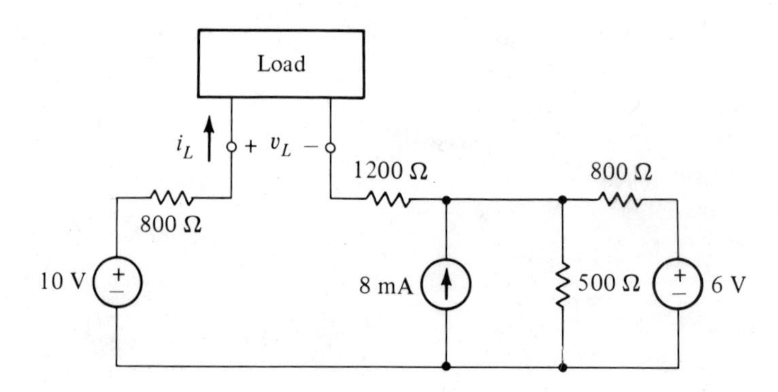

***b.**

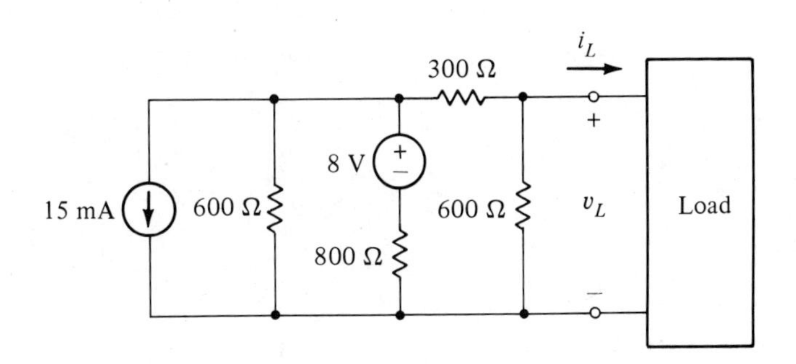

SPICE **1-25.** Find Thevenin's equivalent circuit and Norton's equivalent circuit. Suggestion: Find v_{oc} *and* i_{sc}.

***a.**

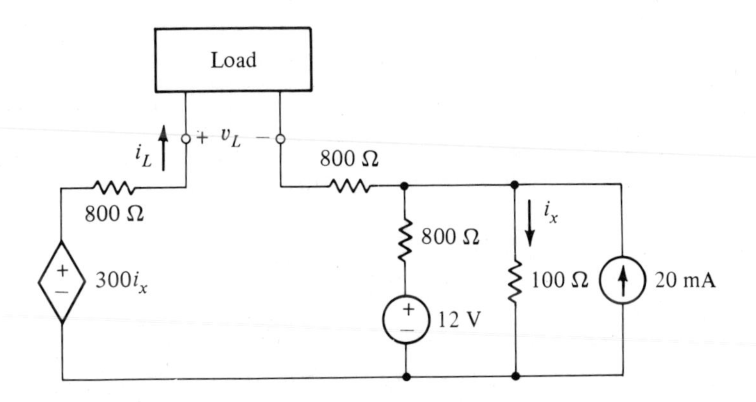

b.

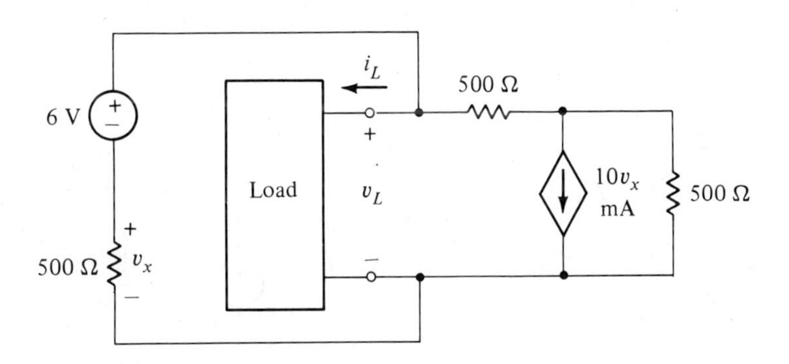

SPICE **1-26.** Find Thevenin's equivalent circuit and Norton's equivalent circuit. Suggestion: Use a test source at the load terminals.

a.

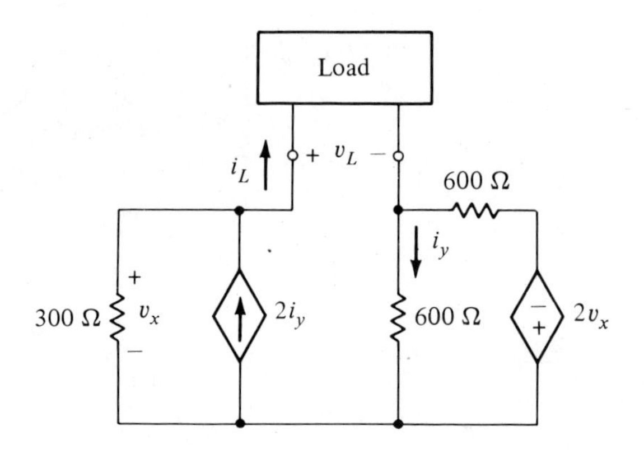

***b.**

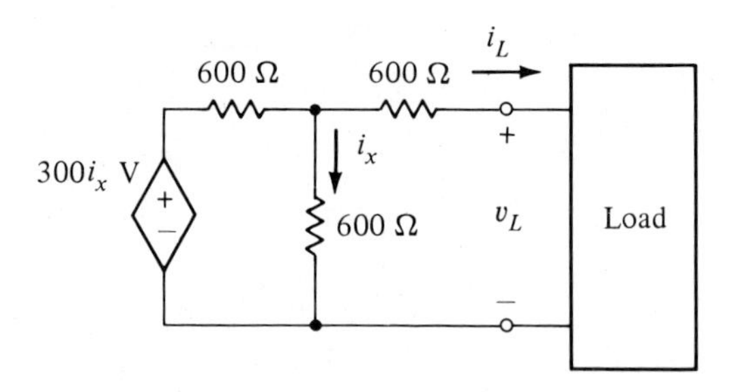

SPICE ***1-27.** Determine G_{No} for the circuit of Example 1.21 by using a 1-V test source at the load terminals.

SPICE **1-28.** Determine the value of the indicated resistor required for maximum power transfer from source to load.

a. Find R_L.

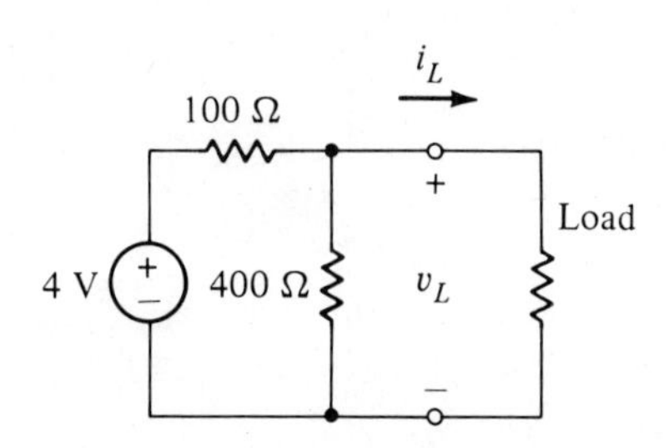

***b.** Find R_1.

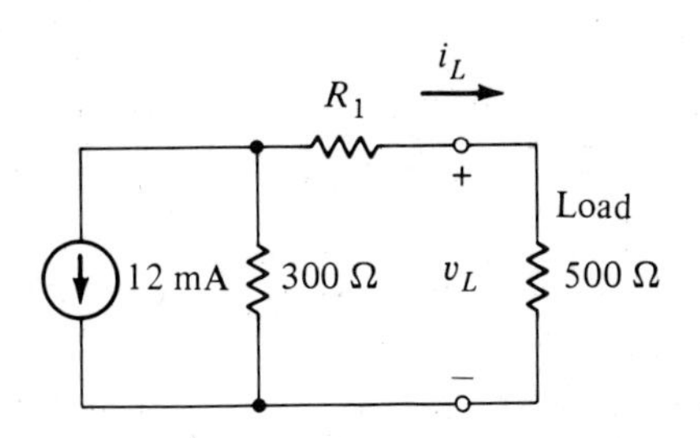

SPICE

1-29. Determine the value of the indicated resistor required for maximum power transfer from source to load.

***a.** Find R_L.

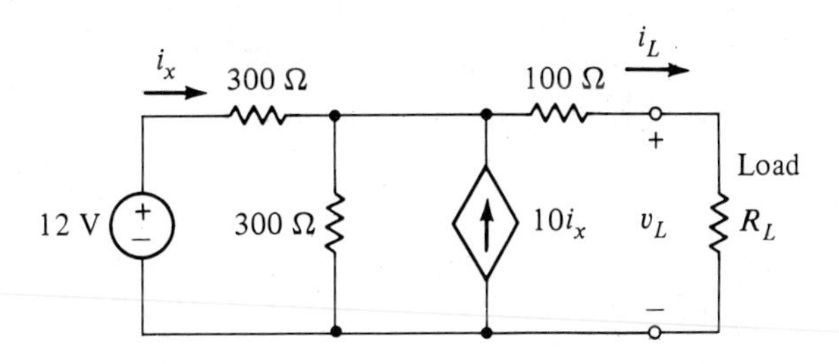

b. Find R_1.

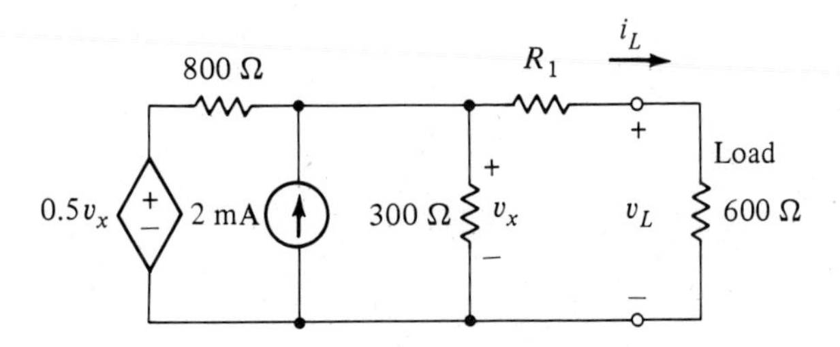

1-30. Derive the formula, in terms of i_{No} and G_{No}, for the maximum power available from a practical current source.

2 Waveform Analysis and Synthesis

2.1 INTRODUCTION

The term **waveform** will be used to refer to the depiction of time-varying voltage, current, or power. A waveform may be a sketch or graph prepared during analysis or design of a circuit, or it may be a display on an instrument, such as an oscilloscope or a plotter, observed during experimentation and testing.

Waveform analysis is the process of finding an appropriate mathematical function of time which describes the waveform; **waveform synthesis** is the process of visualizing or graphing a waveform from a given function of time. Since many waveforms display complicated variations with respect to time, waveform analysis and synthesis commonly involves **piecewise** techniques. By piecewise we mean that the waveform is separated into subdomains, or **pieces**, along the time axis [Figure 2.1(a)] so that the functional relationship of the voltage or current with respect to time can be written separately for each piece of the waveform.

In this chapter, we will discuss the techniques and mathematical tools required for piecewise analysis and synthesis of waveforms. We will consider the general concept of piecewise analysis and synthesis, but we will emphasize **piecewise-linear** [Figure 2.1(b)], **piecewise-sinusoidal** [Figure 2.1(c)], and **piecewise exponential** [Figure 2.1(d)] waveforms because of their importance as *excitation* and *response* in electric circuits.

2.2 Fundamental Considerations

Waveform functions are called **time-domain** functions, because the independent variable is time. The dependent variable is usually voltage, current, or power. Waveforms are said to be **generated** by the waveform function, and the

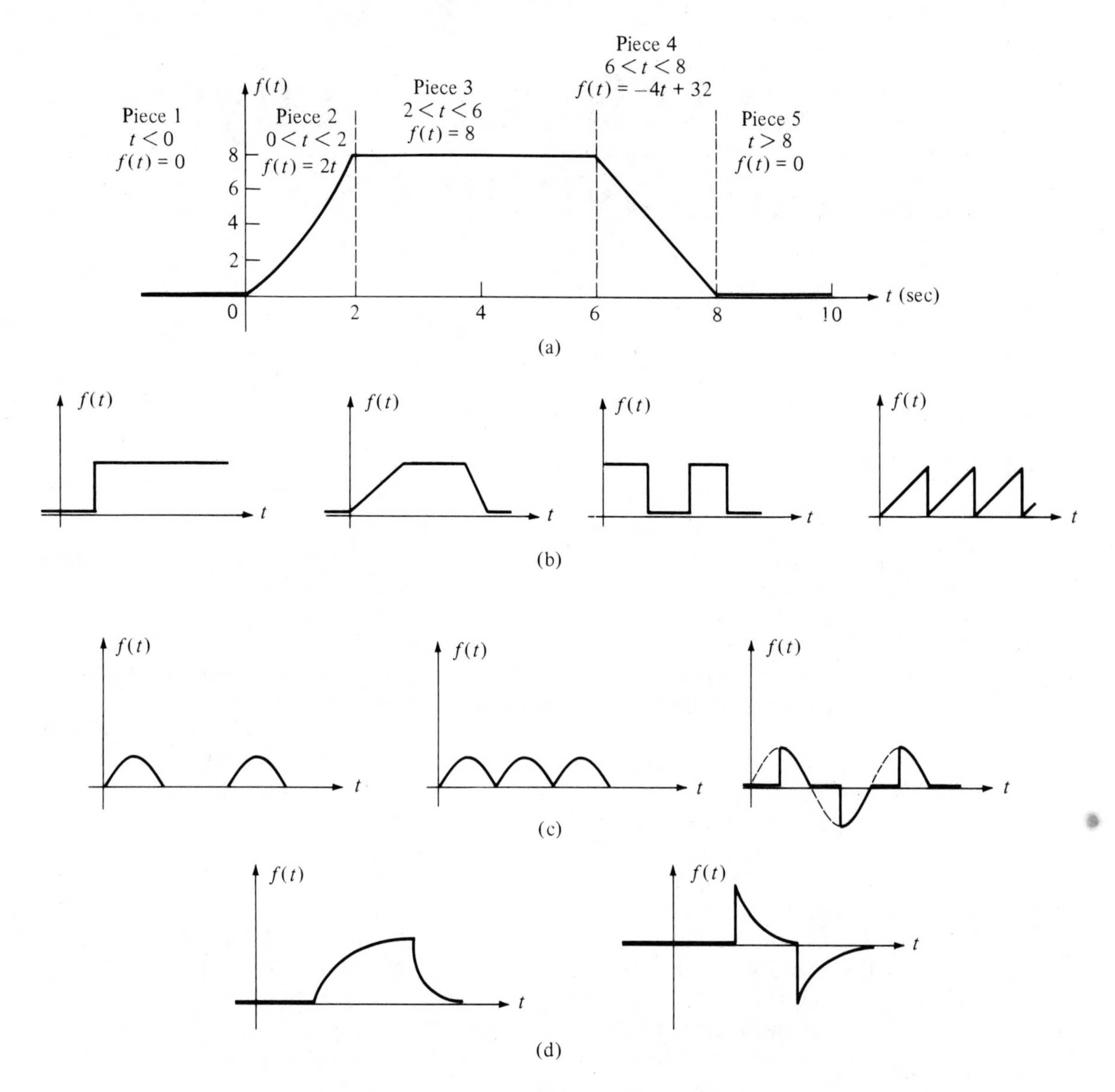

Figure 2.1 Piecewise analysis and synthesis of waveforms. (a) Waveform separated into pieces. (b) Piecewise linear waveforms. (c) Piecewise sinusoidal waveforms. (d) Piecewise exponential waveforms.

terms **waveform generation** and **generating function** are used to refer to the relationship between the waveform and its mathematical function.

The Unit Step Function

The basic mathematical tool for generation of piecewise waveforms is the **unit step function**

$$u(t - t_0) = \begin{cases} 0 & \text{for} \quad t < t_0 \\ 1 & \text{for} \quad t > t_0 \end{cases} \tag{2.1}$$

Figure 2.2
Examples of the unit step function.
(a) $f(t) = u(t + 2)$.
(b) $f(t) = u(t - 3)$.
(c) $f(t) = u(t)$.

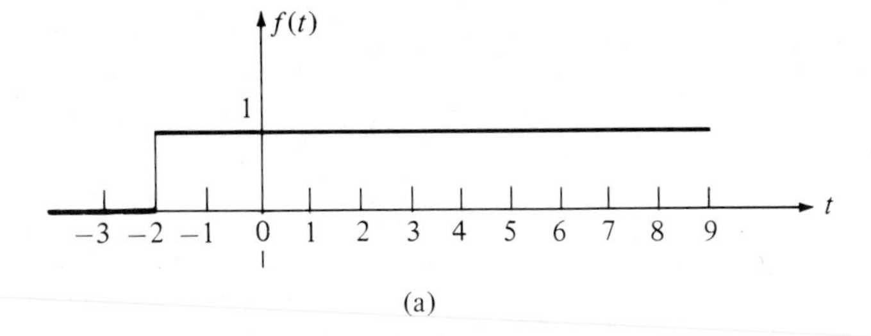

(a)

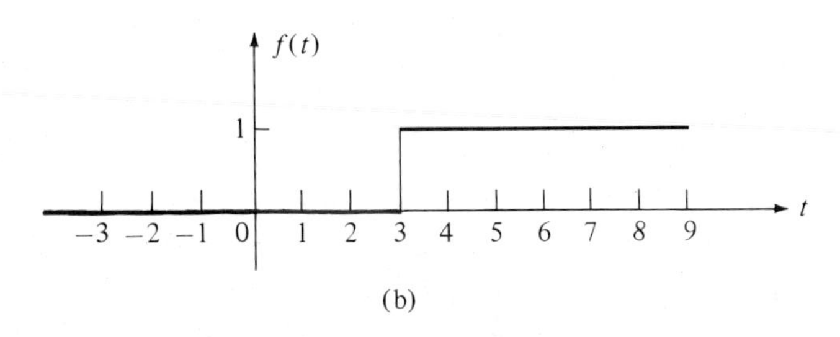

(b)

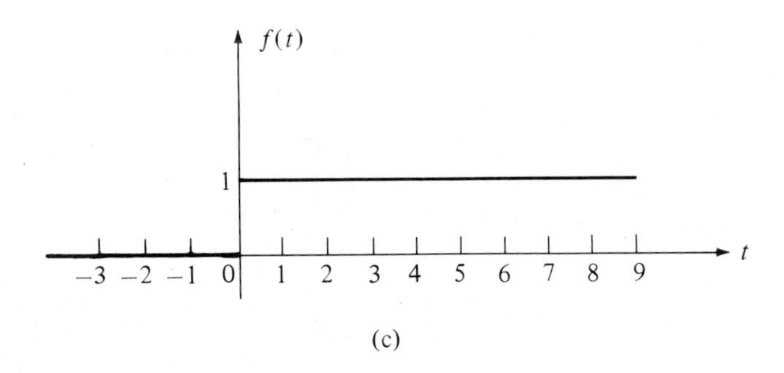

(c)

where t_0 is any specified time. Graphs of some unit step functions in Figure 2.2 make clear the origin of the name "step" function. The notation t_0^- will be used to indicate a time just prior to t_0 when the value of the unit step function is still zero, and t_0^+ will be used to indicate a time just after t_0 when the value of the unit step function is one.

The unit step function may be used alone, but more frequently it is used as a multiplier for some other time-domain function:

$$f(t) = g(t) \cdot u(t - t_0)$$

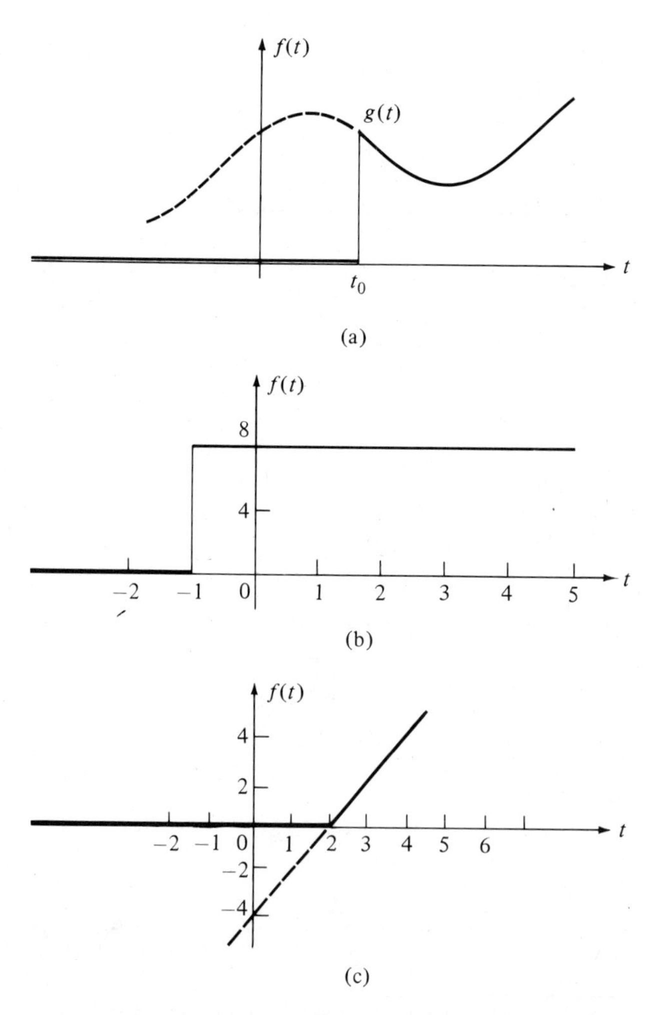

Figure 2.3 Use of the unit step function as a multiplier. (a) $f(t) = g(t) \cdot u(t - t_0)$. (b) $f(t) = 8 \cdot u(t + 1)$. (c) $f(t) = (2t - 4)u(t - 2)$.

The presence of the unit step function as a multiplier defines the function $f(t)$ in this manner:

$$f(t) = \begin{cases} 0 & \text{for} \quad t < t_0 \\ g(t) & \text{for} \quad t > t_0 \end{cases} \tag{2.2}$$

as can be seen in Figure 2.3. When used in this context, the unit step function may be thought of as a mathematical switch which turns on (multiplies by 1) a function which has been off (multiplied by 0), and t_0 as the **switching time**.

The expression $(t - t_0)$ is called the **argument** of the step function. From the definition of the unit step function [Equation (2.1)] and from the examples

Figure 2.4
Examples of the unit step function with modified argument.
(a) $f(t) = u(-2-t)$.
(b) $f(t) = u(3-t)$.
(c) $f(t) = u(-t)$.

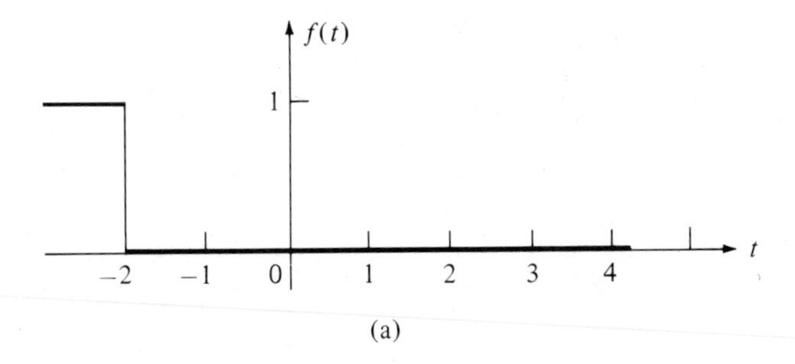

(a)

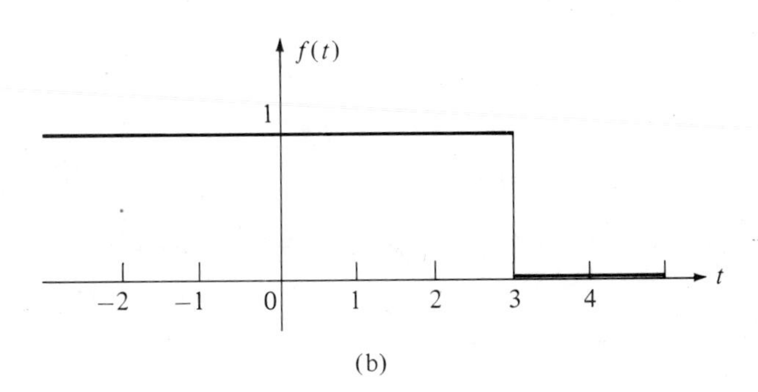

(b)

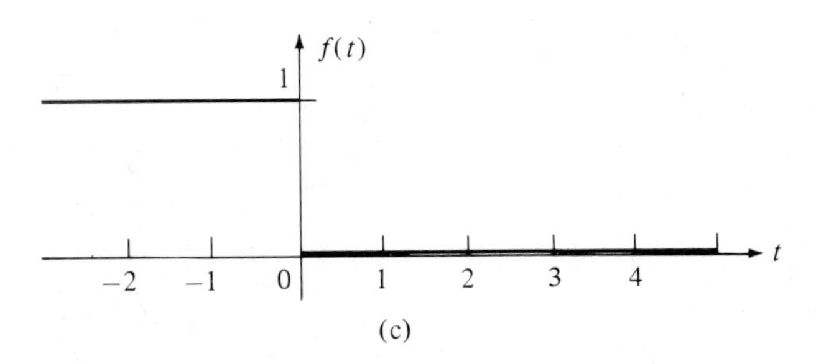

(c)

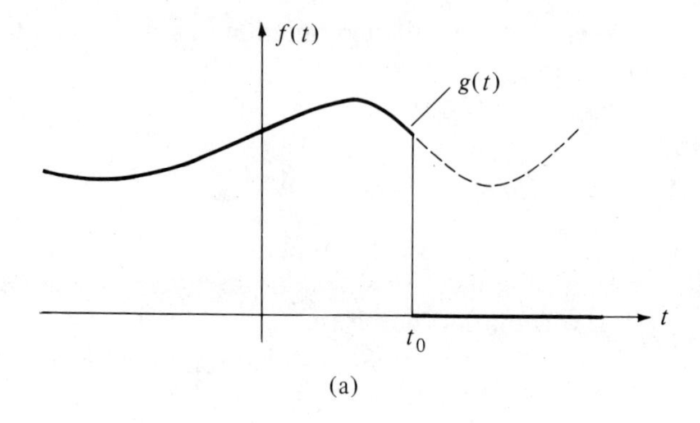

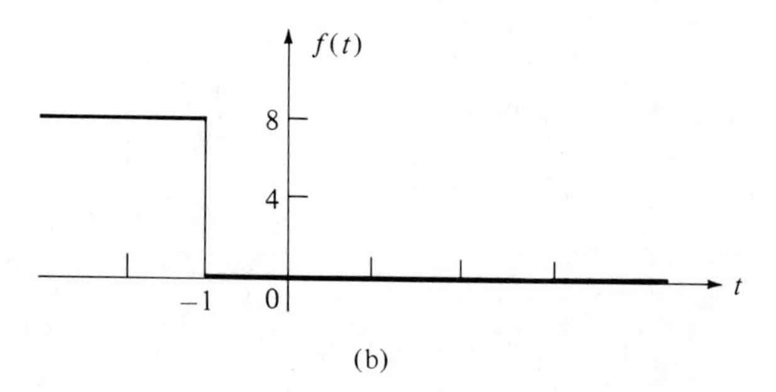

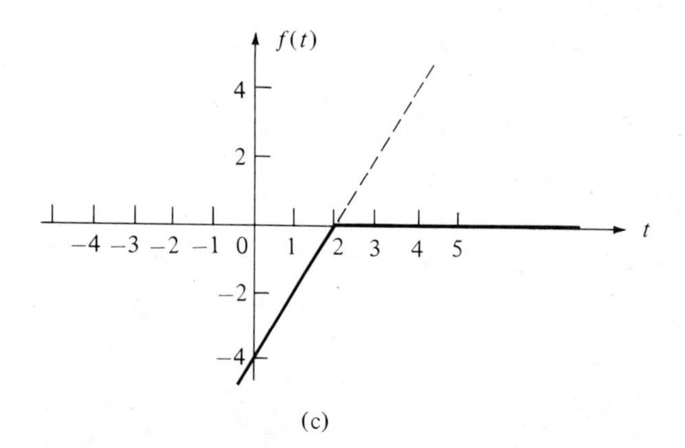

Figure 2.5 Use of the modified unit step function as a multiplier. (a) $f(t) = g(t) \cdot u(t_0 - t)$. (b) $f(t) = 8 \cdot u(-1 - t)$. (c) $f(t) = (2t - 4)u(-t + 2)$.

of Figure 2.2, it can be seen that when the argument is negative, the value of the unit step function is zero, and when the argument is positive, the value of the unit step function is one. This observation leads to a useful modification of the unit step function, obtained by reversing the order of terms in the argument:

$$u(t_0 - t) = \begin{cases} 1 & \text{for} \quad t < t_0 \\ 0 & \text{for} \quad t > t_0 \end{cases} \tag{2.3}$$

The effect of this modification, which can be seen in Figures 2.4 and 2.5, is to reverse the switching action of the step function so that a function which has been on (multiplied by 1) is turned off (multiplied by 0).

The Unit Impulse Function

The unit step function has an obvious **discontinuity** at t_0, and the **derivative** of the unit step function *does not exist* at t_0. The presence of a unit step function in a generating function makes it awkward to describe mathematically the derivative of that generating function. Rather than saying that the derivative does not exist at t_0, a special mathematical function, called the **unit impulse function**, will be used to account for the discontinuity created by the unit step function.

The graph of the function $f(t)$ in Figure 2.6(a) transitions from zero at t_0 to one in a finite amount of time Δt; the pulselike function $p(t)$ represents the derivative of $f(t)$. The width of $p(t)$ is

$$w = \Delta t$$

and the height of $p(t)$ is the slope of $f(t)$,

$$h = \frac{1 - 0}{(t_0 + \Delta t) - t_0} = \frac{1}{\Delta t}$$

The area enclosed by $p(t)$ is

$$A = w \cdot h = \Delta t \cdot \frac{1}{(\Delta t)} = 1$$

If Δt is reduced [Figure 2.6(b)], the height of $p(t)$ must increase inversely, so that the area enclosed by $p(t)$ remains constant. As Δt continues to shrink, $f(t)$ approaches a unit step function,

$$\lim_{\Delta t \to 0} f(t) = u(t - t_0)$$

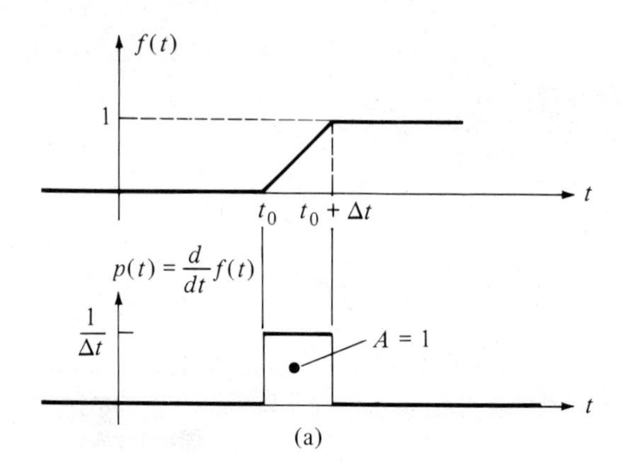

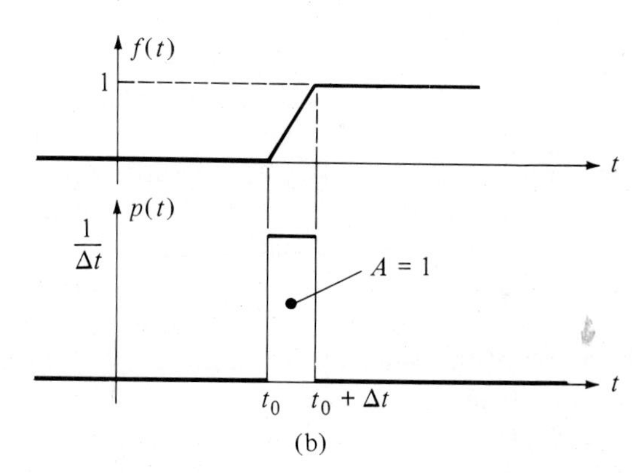

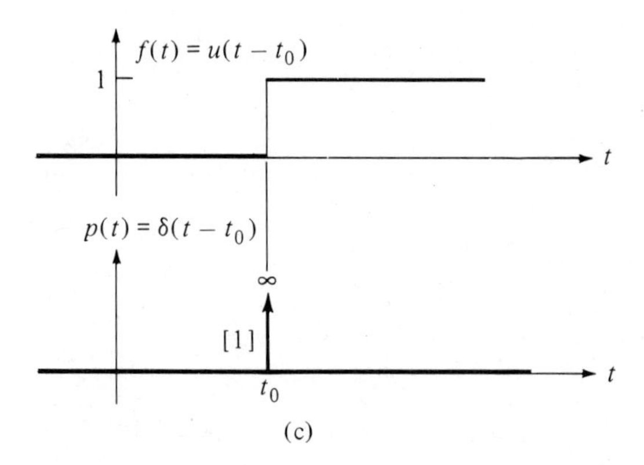

Figure 2.6
Development of the impulse function.
(a) $f(t)$ and $p(t)$.
(b) $f(t)$ and $p(t)$ with Δt reduced.
(c) $f(t)$ and $p(t)$ as $\Delta t \to 0$.

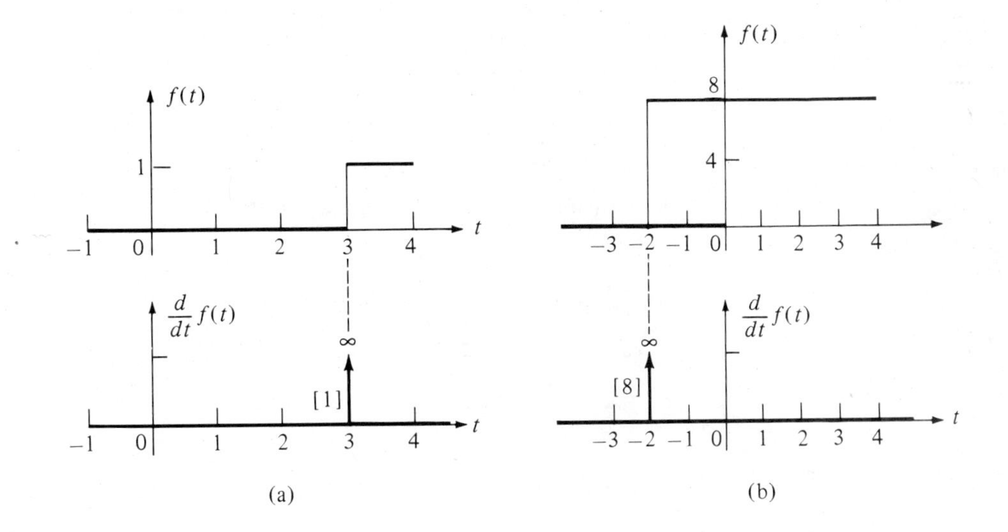

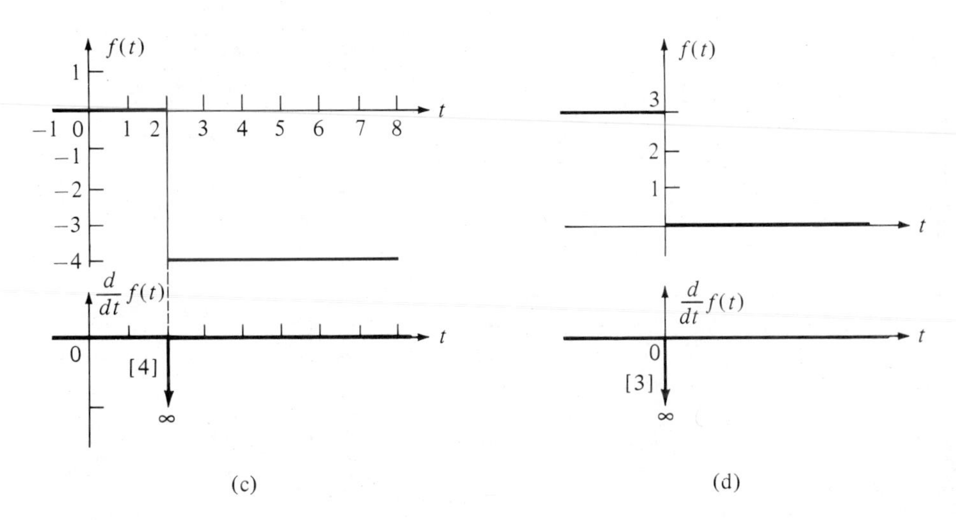

Figure 2.7 Examples of impulse functions.
(a) $f(t) = u(t - 3)$, $d\,f(t)/dt = \delta(t - 3)$.
(b) $f(t) = 8 \cdot u(t + 2)$, $d\,f(t)/dt = 8 \cdot \delta(t + 2)$.
(c) $f(t) = -4 \cdot u(t - 2)$, $d\,f(t)/dt = -4 \cdot \delta(t - 2)$.
(d) $f(t) = 3 \cdot u(-t)$, $d\,f(t)/dt = -3 \cdot \delta(t)$.

and $p(t_0)$ approaches infinity. This behavior of $p(t)$ as $\Delta t \to 0$ defines the **unit impulse**, or **delta, function**,

$$\delta(t - t_0) = \lim_{\Delta t \to 0} p(t) \tag{2.4}$$

The unit impulse function is zero everywhere except at $t = t_0$, where its value approaches infinity:

$$\delta(t - t_0) \begin{cases} = 0 & \text{for} \quad t \neq t_0 \\ \to \infty & \text{for} \quad t = t_0 \end{cases} \tag{2.5}$$

and the area enclosed by the unit impulse function is one:

$$\int_{-\infty}^{\infty} \delta(t - t_0)\, dt = 1 \tag{2.6}$$

When limits are considered as in the foregoing discussion, the unit impulse function can be related to the unit step function,

$$\delta(t - t_0) = \frac{d}{dt} u(t - t_0) \tag{2.7}$$

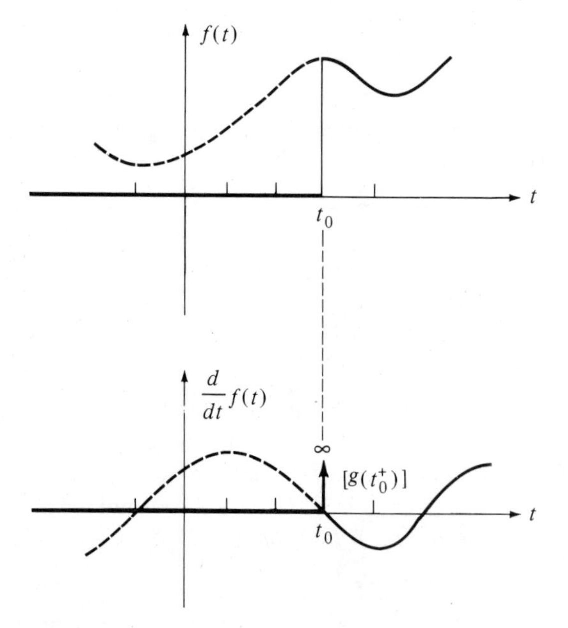

Figure 2.8 General example of the impulse function. $f(t) = g(t) \cdot u(t - t_0)$; $d\,f(t)/dt = g(t_0) \cdot \delta(t - t_0) + u(t - t_0) \cdot d\,g(t)/dt$.

and used to account for the discontinuity in the unit step function. Some examples of impulse functions and a suggested method for graphing are shown in Figures 2.7 and 2.8.

Example 2.1

SPICE

Problem:

a. Graph the function $f(t) = g(t) \cdot u(t - 4)$, where

$$g(t) = 2t - 6$$

b. Graph $\frac{d}{dt}f(t)$

Solution:

1. $g(t) = 2t - 6$ is the equation of a straight line with a slope of 2 and passing through $(0, -6)$.

Flashback

$f(x) = mx + b$ is the equation of a straight line with slope m and the vertical axis intercept b.

2. The unit-step-function multiplier causes $f(t) = 0$ for $t < 4$ and $f(t) = g(t)$ for $t > 4$, so it is only necessary to compute $f(t)$ for $t \geq 4$:

t	$f(t)$
< 4	0
4^+	2
5	4
6	6

The graph of $f(t)$ can be seen in Figure 2.9(a).

3. Compute the derivative of $f(t)$:

$$\begin{aligned}\frac{d}{dt}f(t) &= g(t) \cdot \frac{d}{dt}u(t - 4) + u(t - 4) \cdot \frac{d}{dt}g(t)\\ &= g(t) \cdot \delta(t - 4) + u(t - 4) \times 2\end{aligned}$$

Note that the value of $\delta(t - 4)$ is zero except when $t = 4$, so $g(t) \cdot \delta(t - 4)$ is also zero except when $t = 4$. Therefore,

$$g(t) \cdot \delta(t - 4) = g(4) \cdot \delta(t - 4) = 2 \cdot \delta(t - 4)$$

and

$$\frac{d}{dt}f(t) = 2 \cdot \delta(t - 4) + 2 \cdot u(t - 4)$$

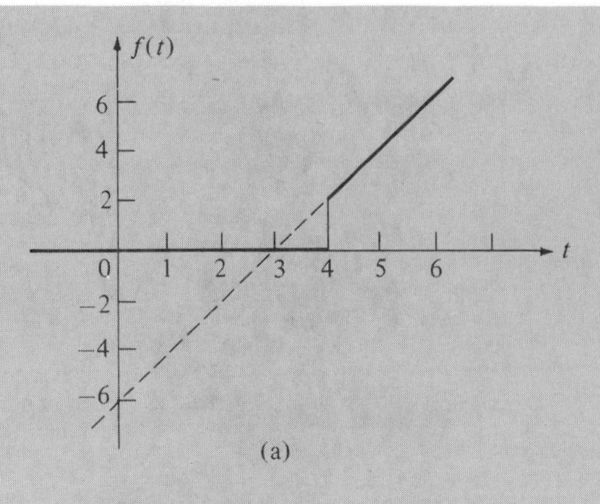

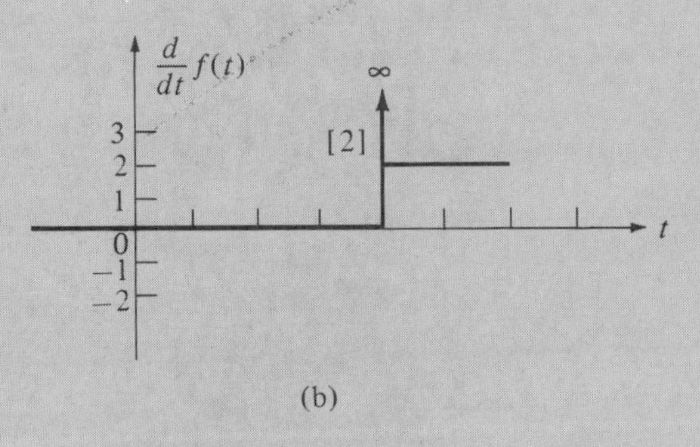

Figure 2.9 Example 2.1. (a) Graph of $f(t)$. (b) Graph of$d\ g(t)/dt$.

The value of the derivative is zero for $t < 4$, so it is only necessary to compute the derivative for $t \geq 4$:

t	$\frac{d}{dt}f(t)$
< 4	0
4	$2 \cdot \delta(t-4)$
4^+	2
5	2
6	2

The graph of $\frac{d}{dt}f(t)$ can be seen in Figure 2.9(b).

Example 2.2

Problem:

Write the function which will generate the waveform depicted in Figure 2.10.

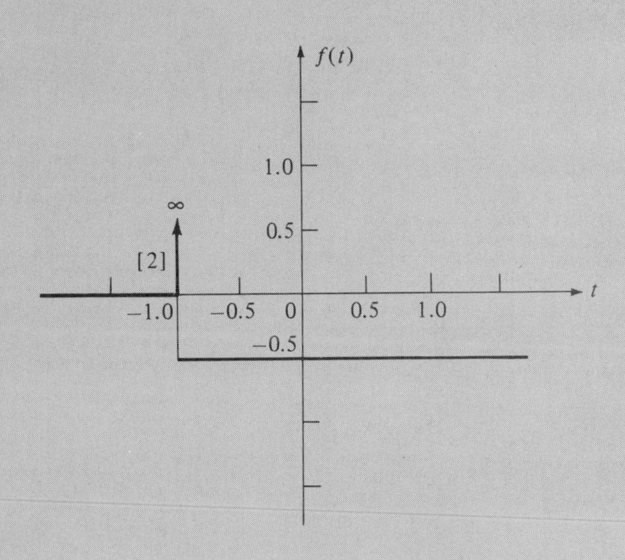

Figure 2.10
Example 2.2.

Solution:

The waveform consists of two parts, a positive impulse with an area of 2 and a negative step with a discontinuity of 0.5, both occurring at $t = -1$ sec; thus

$$f(t) = 2 \cdot \delta(t + 1) - 0.5 \cdot u(t + 1)$$

Exercise 2.1

SPICE

1. Graph the waveforms represented by:
   a. $f(t) = 9 \cdot u(t - 6)$,
   b. $f(t) = (t + 1) \cdot u(t + 1)$,
   c. $f(t) = -8 \cdot u(2 - t)$,
   d. $f(t) = 3 \cdot \delta(t) + 4t \cdot u(t)$.
2. Write the waveform functions for the waveforms shown in Figure 2.11.

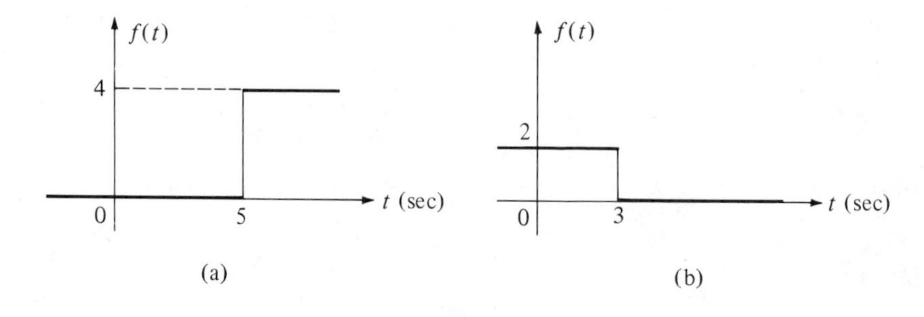

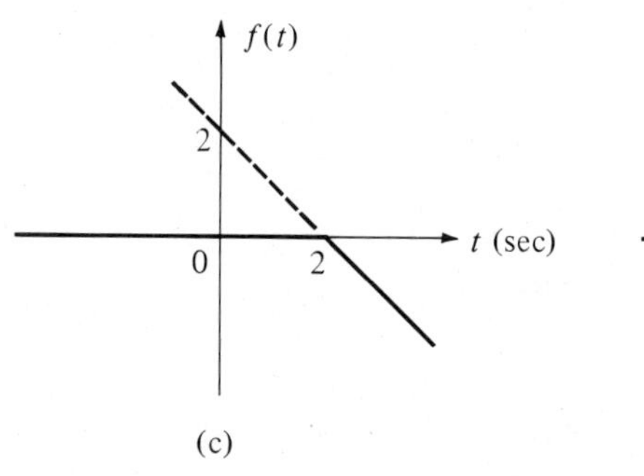

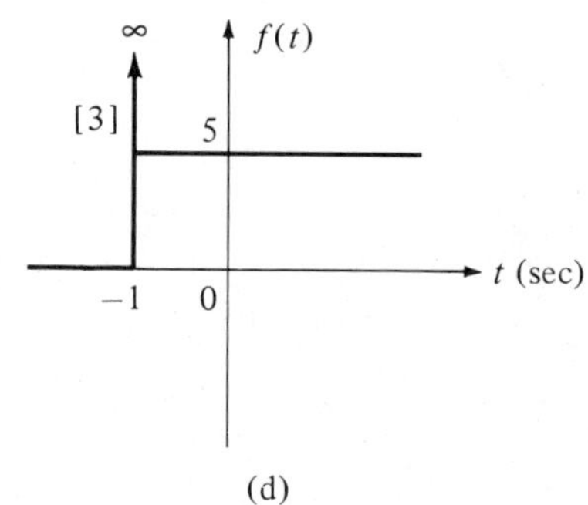

Figure 2.11
Exercise 2.1 (part 2).

Answers:

1. Refer to Figure 2.12.
2. **(a)** $4 \cdot u(t-5)$; **(b)** $2 \cdot u(3-t)$; **(c)** $-(t-2) \cdot u(t-2)$; **(d)** $3 \cdot \delta(t+1) + 5 \cdot u(t+1)$.

Computer Simulation

SPICE voltage and current sources can be specified to generate many of the waveforms commonly encountered in electrical and electronic circuits. SPICE 2.1 illustrates the method for generating a step function [from Exercise 2.1(a)] by specifying the voltage source to be a PULSE. The 1-Ω resistor is required to complete the circuit so that SPICE can be executed for this illustration. The reader should refer to Appendix D for the list of parameters associated with PULSE, and for the format of the .TRAN and .PLOT directives.

The rise time of a step function is zero; however, if the rise-time parameter (RISE) for PULSE is specified as zero, SPICE will set a default rise time equal

Figure 2.12
Answers to Exercise 2.1 (part 1).
(a) $f(t) = 9 \cdot u(t-6)$.
(b) $f(t) = (t+1) \cdot u(t+1)$.
(c) $f(t) = -8 \cdot u(2-t)$.
(d) $f(t) = 3 \cdot \delta(t) + 4t \cdot u(t)$.

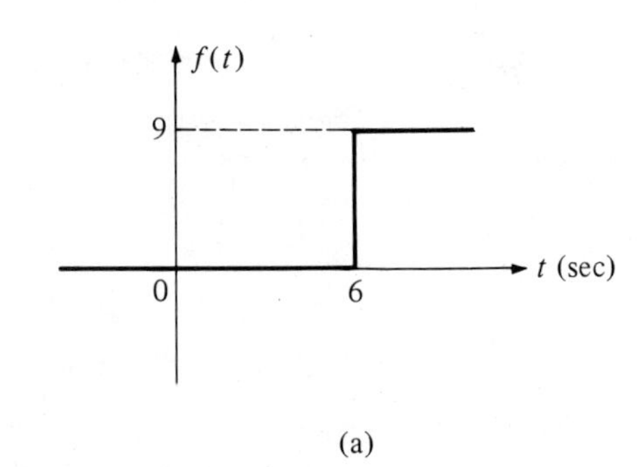

(a)

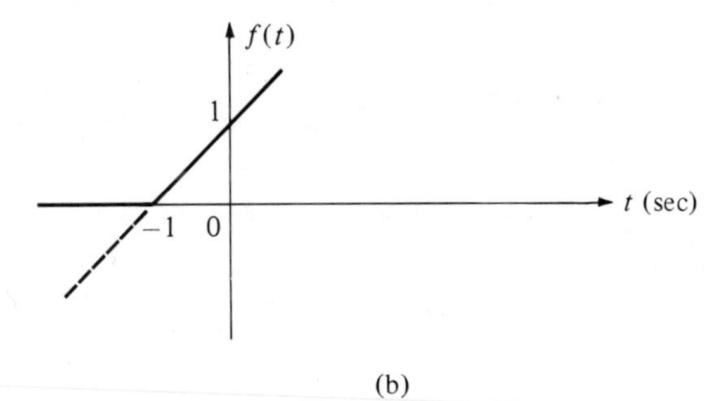

(b)

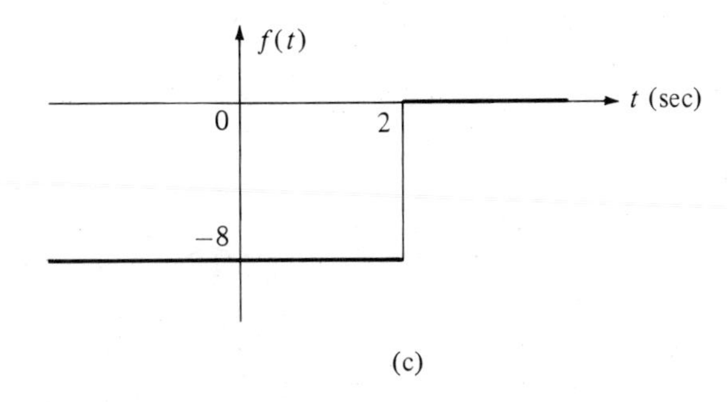

(c)

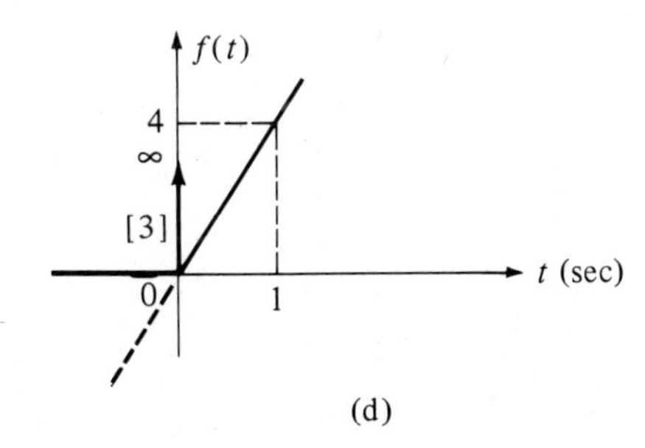

(d)

```
*******09/29/88 ********  SPICE 2G.6    3/15/83 ********09:45:43*****
SPICE 2.1
****     INPUT LISTING                       TEMPERATURE =   27.000 DEG C
******************************************************************************
 R 1 0 1
 V 1 0 PULSE(0 9 6 1F)
 .TRAN .5 10
 .PLOT TRAN V(1)
 .END
*******09/29/88 ********  SPICE 2G.6    3/15/83 ********09:45:43*****
SPICE 2.1
****     INITIAL TRANSIENT SOLUTION          TEMPERATURE =   27.000 DEG C
******************************************************************************
  NODE   VOLTAGE
 (  1)     .0000
     VOLTAGE SOURCE CURRENTS
     NAME       CURRENT
     V         0.000D+00
     TOTAL POWER DISSIPATION   0.00D+00  WATTS
*******09/29/88 ********  SPICE 2G.6    3/15/83 ********09:45:43*****
SPICE 2.1
****     TRANSIENT ANALYSIS                  TEMPERATURE =   27.000 DEG C
******************************************************************************

     TIME       V(1)
                  -5.000D+00      0.000D+00      5.000D+00      1.000D+01  1.500D+01
                            - - - - - - - - - - - - - - - - - - - - - - - - - - - -
 0.000D+00  0.000D+00 .             *             .             .             .
 5.000D-01  0.000D+00 .             *             .             .             .
 1.000D+00  0.000D+00 .             *             .             .             .
 1.500D+00  0.000D+00 .             *             .             .             .
 2.000D+00  0.000D+00 .             *             .             .             .
 2.500D+00  0.000D+00 .             *             .             .             .
 3.000D+00  0.000D+00 .             *             .             .             .
 3.500D+00  0.000D+00 .             *             .             .             .
 4.000D+00  0.000D+00 .             *             .             .             .
 4.500D+00  0.000D+00 .             *             .             .             .
 5.000D+00  0.000D+00 .             *             .             .             .
 5.500D+00  0.000D+00 .             *             .             .             .
 6.000D+00  0.000D+00 .             *             .             .             .
 6.500D+00  9.000D+00 .             .             .          *  .             .
 7.000D+00  9.000D+00 .             .             .          *  .             .
 7.500D+00  9.000D+00 .             .             .          *  .             .
 8.000D+00  9.000D+00 .             .             .          *  .             .
 8.500D+00  9.000D+00 .             .             .          *  .             .
 9.000D+00  9.000D+00 .             .             .          *  .             .
 9.500D+00  9.000D+00 .             .             .          *  .             .
 1.000D+01  9.000D+00 .             .             .          *  .             .
                            - - - - - - - - - - - - - - - - - - - - - - - - - - - -

          JOB CONCLUDED
```

to the output interval (TSTEP). In SPICE 2.1, a very small value for RISE (1 femtosecond, or 10^{-15} sec) has been specified to simulate the zero rise time of a step function. The plotted pulse appears to have a rise time equal to TSTEP because the interval between plotted points is TSTEP, but the simulation performed by SPICE is based on the specified value of RISE.

The pulse width (WIDTH) parameter defaults to the total simulation time (TSTOP) when not otherwise specified in the PULSE parameter list, so the fall time (FALL) and period (PERIOD) parameters are not a factor and are likewise omitted from the parameter list.

2.3 Piecewise-Linear Waveforms

Piecewise-linear generating functions are formed by the addition or subtraction of linear functions, each of which is multiplied by a unit step function with a different switching time (t_0).

Rectangular Waveforms

Rectangular waveforms occur in electric circuits whenever there is a rapid change, or **switching**, from one level of voltage or current to another. Rectangular waveforms can be generated by functions which consist entirely of unit step functions multiplied by constants.

The simplest rectangular waveform is a rectangular pulse [Figure 2.13(a)] of width $t_2 - t_1$ and height A. Such a waveform might represent the current from a battery to a resistive load through a series switch when the switch is initially closed at t_1 and is opened at t_2. The rising (left) edge of the pulse is generated by the step function

$$f_1(t) = A \cdot u(t - t_1)$$

shown in Figure 2.13(b), and the falling (right) edge is generated by the step function

$$f_2(t) = -A \cdot u(t - t_2)$$

shown in Figure 2.13(c). In Figure 2.13, the rectangular pulse is obtained by graphical addition of the two step functions. Therefore, the generating function for the rectangular pulse is a linear combination of these two step functions,

$$\begin{aligned} f(t) &= A \cdot u(t - t_1) + [-A \cdot u(t - t_2)] \\ &= A[u(t - t_1) - u(t - t_2)] \end{aligned}$$

To verify, consider three time subdomains,

$$t < t_1: \qquad f(t) = A(0 - 0) = 0$$

$$t_1 < t < t_2: \qquad f(t) = A(1 - 0) = A$$

$$t > t_2: \qquad f(t) = A(1 - 1) = 0$$

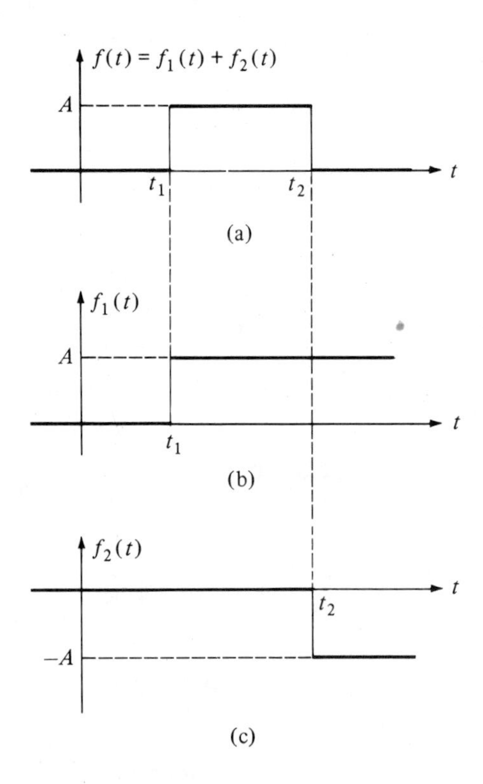

Figure 2.13 Generation of a rectangular pulse.

Example 2.3

Problem:

Write the function required to generate the voltage waveform shown in Figure 2.14(a).

Solution:

1. The first rising edge [Figure 2.14(b)] is generated by

$$v_1(t) = 4 \cdot u(t - 0.005) \text{ V}$$

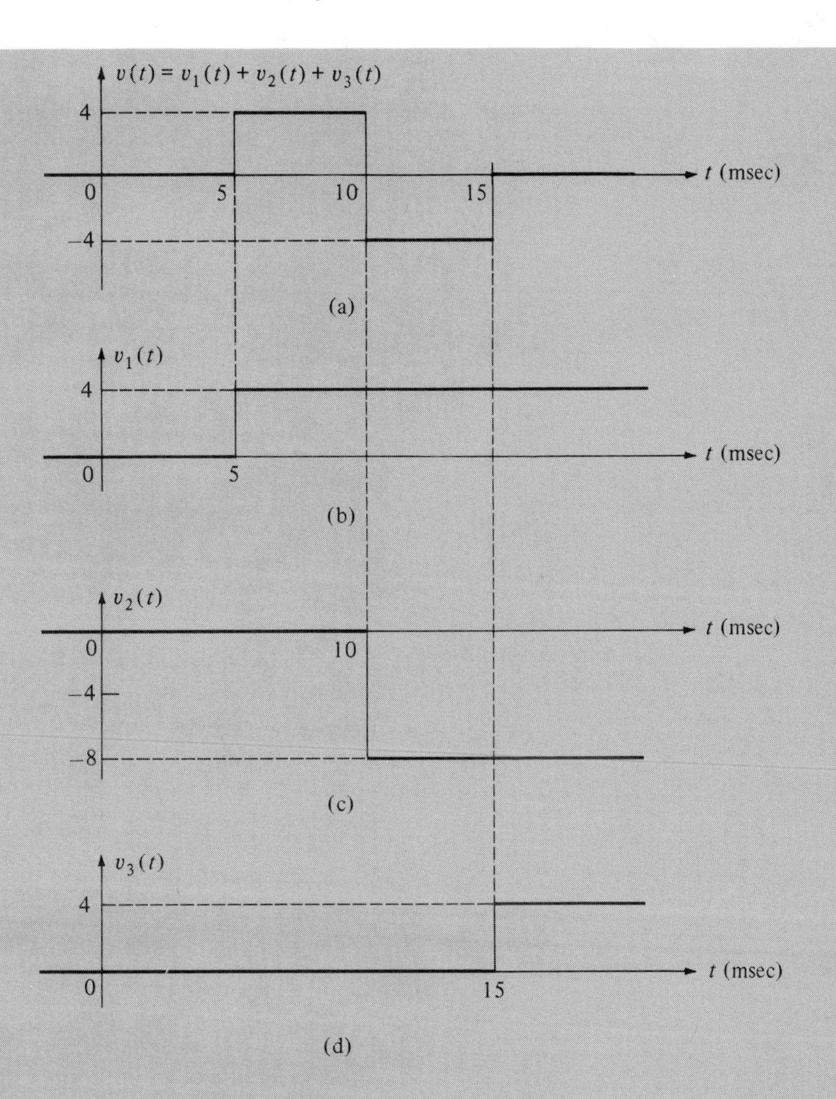

Figure 2.14
Example 2.3.

2. The falling edge [Figure 2.14(c)] is generated by

$$v_2(t) = -8 \cdot u(t - 0.01) \text{ V}$$

3. The second rising edge [Figure 2.14(d)] is generated by

$$v_3(t) = 4 \cdot u(t - 0.015) \text{ V}$$

4. The generating function is

$$v(t) = v_1(t) + v_2(t) + v_3(t)$$
$$= 4 \cdot u(t - 0.005) - 8 \cdot u(t - 0.01)$$
$$+ 4 \cdot u(t - 0.015) \text{ V}$$

5. Verify by considering the time subdomains:

$t < t_1$: $f(t) = 4 \times 0 - 8 \times 0 + 4 \times 0 = 0$

$t_1 < t < t_2$: $f(t) = 4 \times 1 - 8 \times 0 + 4 \times 0 = 4 \text{ V}$

$t_2 < t < t_3$: $f(t) = 4 \times 1 - 8 \times 1 + 4 \times 0 = -4 \text{ V}$

$t > t_3$: $f(t) = 4 \times 1 - 8 \times 1 + 4 \times 1 = 0$

Example 2.4

SPICE

Problem:

Sketch the current waveform represented by the function

$$i(t) = 3 \cdot u(1 - t) + 5 \cdot u(t - 1) - 6 \cdot u(t - 3) \text{ A}$$

Solution:

1. The generating function $i(t)$ has three components:

$$i(t) = i_1(t) + i_2(t) + i_3(t)$$

where

$$i_1(t) = 3 \cdot u(1 - t) \text{ A}$$
$$i_2(t) = 5 \cdot u(t - 1) \text{ A}$$
$$i_3(t) = -6 \cdot u(t - 3) \text{ A}$$

2. Sketch $i_1(t)$ [Figure 2.15(a)].
3. Sketch $i_2(t)$ [Figure 2.15(b)].
4. Sketch $i_3(t)$ [Figure 2.15(c)].

Figure 2.15
Example 2.4.

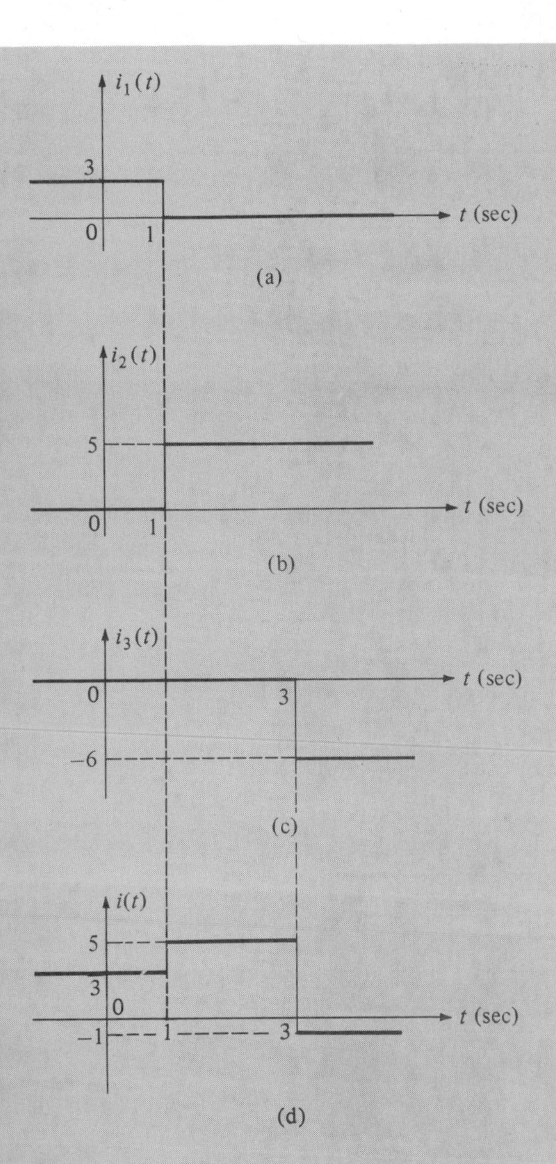

5. Add graphically the sketches for $i_1(t)$, $i_2(t)$, and $i_3(t)$, to obtain the sketch of $i(t)$ shown in Figure 2.15(d).
6. Notice that the 3-A falling edge of $i_1(t)$ partially offsets the 5-A rising edge of $i_2(t)$, and the net result is a 2-A rising edge at $t = 1$ sec.

Figure 2.16
Problems for Exercise 2.2 (part 1).

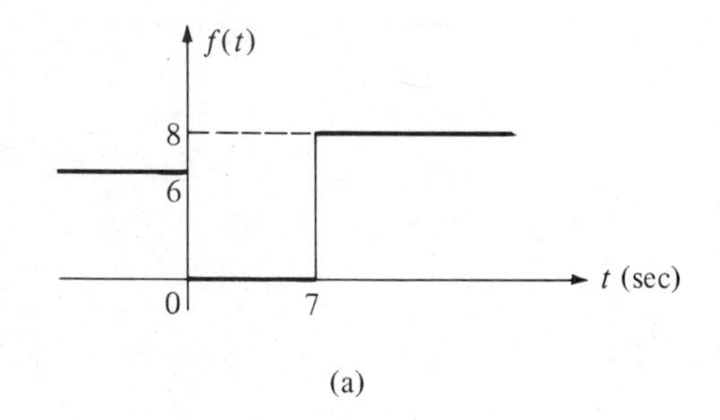

(a)

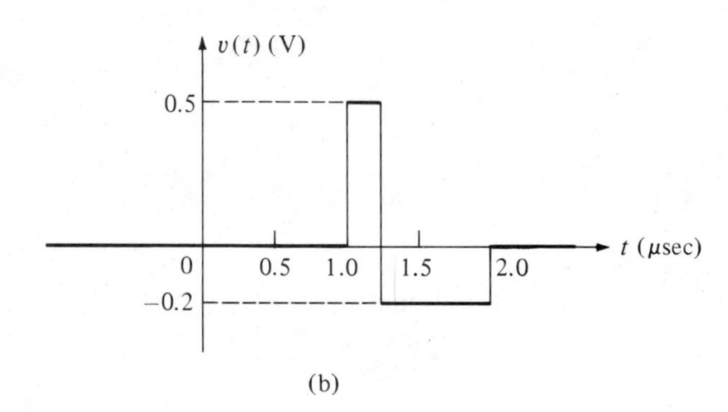

(b)

Figure 2.17
Answers to Exercise 2.2 (part 2).

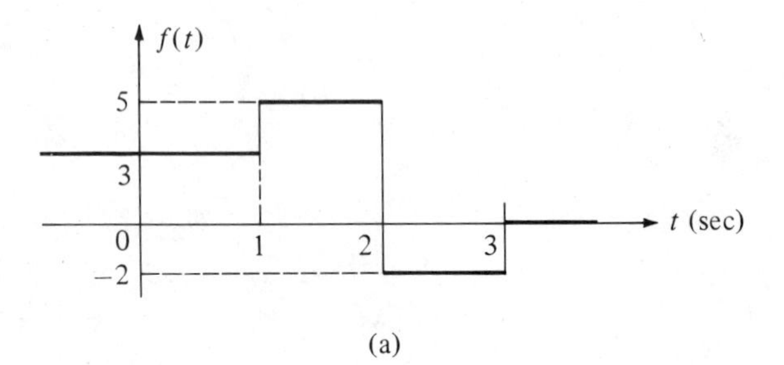

(a)

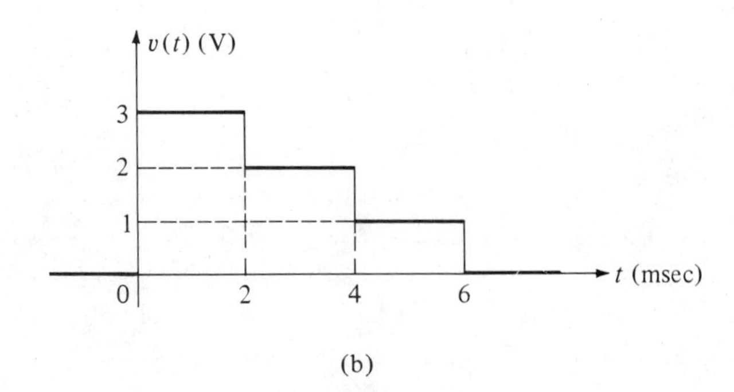

(b)

Exercise 2.2

1. Write the generating functions for the waveforms shown in Figure 2.16.

SPICE

2. Sketch the waveforms generated by the functions
   a. $f(t) = 3 \cdot u(1 - t) + 2 \cdot u(t - 1) - 7 \cdot u(t - 2) + 2 \cdot u(t - 3)$,
   b. $v(t) = 3 \cdot u(t) - u(t - 0.002) - u(t - 0.004) - u(t - 0.006)$ V.

Answers:

1. (a) $f(t) = 6 \cdot u(-t) + 8 \cdot u(t - 7)$.
   (b) $v(t) = 0.5 \cdot u(t - 1.00 \times 10^{-6}) - 0.7 \cdot u(t - 1.25 \times 10^{-6}) + 0.2 \cdot u(t - 2.00 \times 10^{-6})$ V.
2. Refer to Figure 2.17.

Ramp Functions

A **ramp function**, so named because of the shape of its graph, consists of a straight-line function multiplied by a step function. The function whose graph

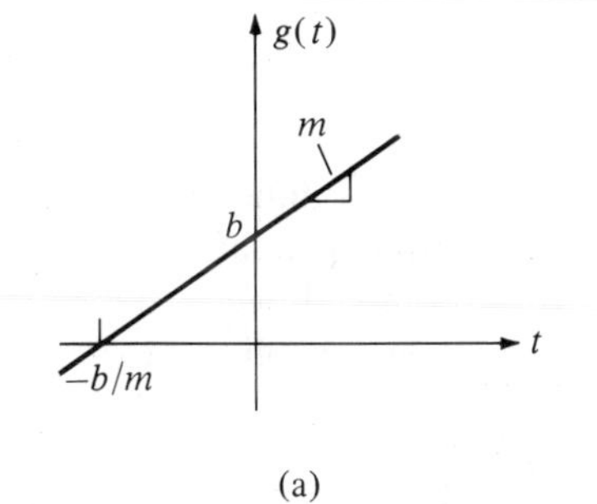

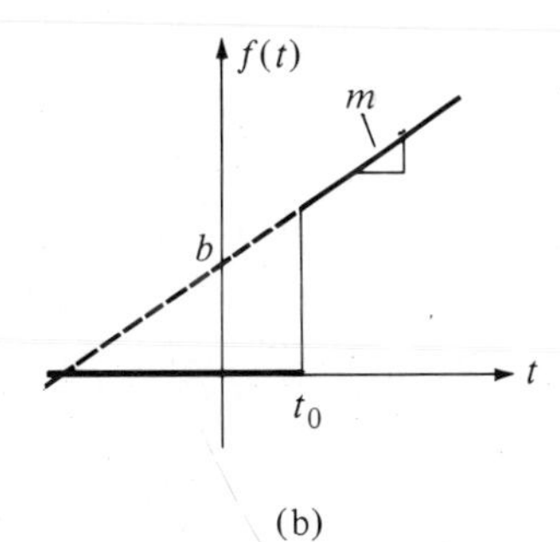

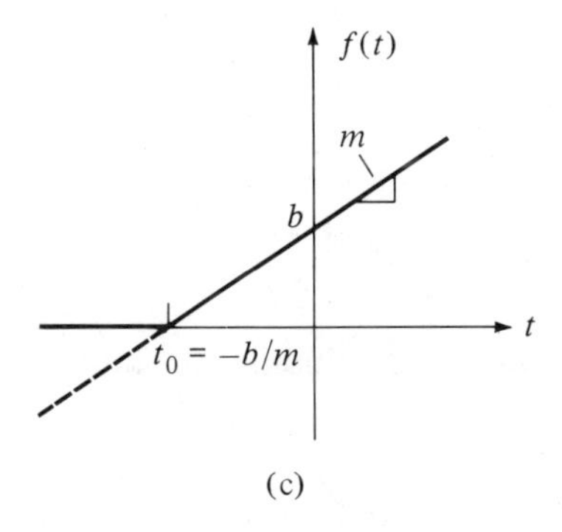

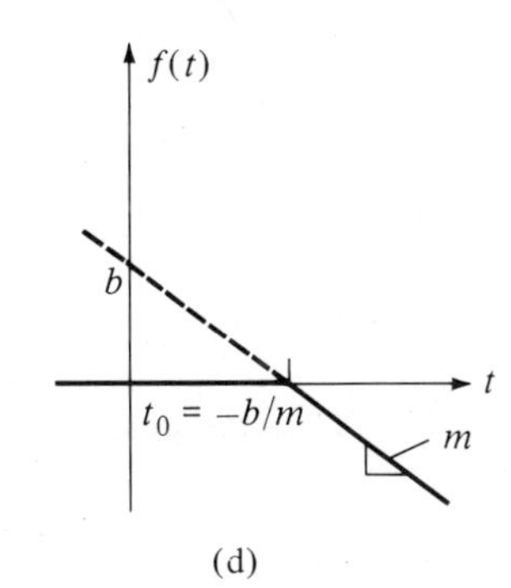

Figure 2.18 Ramp functions. (a) Straight-line function. (b) Ramp function with step. (c) Ramp function which breaks upward from t-axis. (d) Ramp function which breaks downward from t-axis.

is a straight line in the time domain [Figure 2.18(a)] is

$$g(t) = mt + b = m\left(t + \frac{b}{m}\right)$$

so the general ramp function [Figure 2.18(b)] is

$$\begin{aligned} f(t) &= g(t) \cdot u(t - t_0) \\ &= m\left(t + \frac{b}{m}\right) \cdot u(t - t_0) \end{aligned} \tag{2.8}$$

If the graph of the ramp function is to break upward (or downward) from a point on the t-axis $(t_0, 0)$, as shown in Figure 2.18(c) and (d), t_0 must equal $-b/m$, the **intercept–slope ratio**:

$$f(t) = m\left(t + \frac{b}{m}\right) \cdot u\left(t + \frac{b}{m}\right) \tag{2.9a}$$

or

$$f(t) = m(t - t_0) \cdot u(t - t_0) \tag{2.9b}$$

The expression $m(t + b/m)$ in Equation (2.9a) and the expression $m(t - t_0)$ in Equation (2.9b) are *factors* of $f(t)$; $(t + b/m)$ and $(t - t_0)$ in these factors should not be confused with the arguments of the unit step function.

In general, a ramp function [Equation (2.8)], whose graph may break upward or downward from points above or below the t-axis, is made up of a step at t_0 and a ramp with its **breakpoint** at $(t_0, 0)$:

$$f(t) = m\left(t + \frac{b}{m}\right) \cdot u(t - t_0) = m\left(\frac{b}{m} + t\right) \cdot u(t - t_0)$$

By adding and subtracting t_0 in the linear factor, $f(t)$ may be rewritten as

$$f(t) = m\left(\frac{b}{m} + t_0 + t - t_0\right) \cdot u(t - t_0)$$

After rearranging terms and distributing the unit step function,

$$f(t) = \underbrace{(mt_0 + b) \cdot u(t - t_0)}_{\text{step at } t_0} + \underbrace{m(t - t_0) \cdot u(t - t_0)}_{\text{ramp from } (t_0, 0)} \tag{2.9c}$$

The graphical composition of such a ramp function is shown in Figure 2.19. Observe in Figure 2.19(a) that the first term in $f(t)$ places the breakpoint at $(t_0, mt_0 + b)$, as expected. This form of the ramp function will be particularly useful in later chapters when *transforming* generating functions.

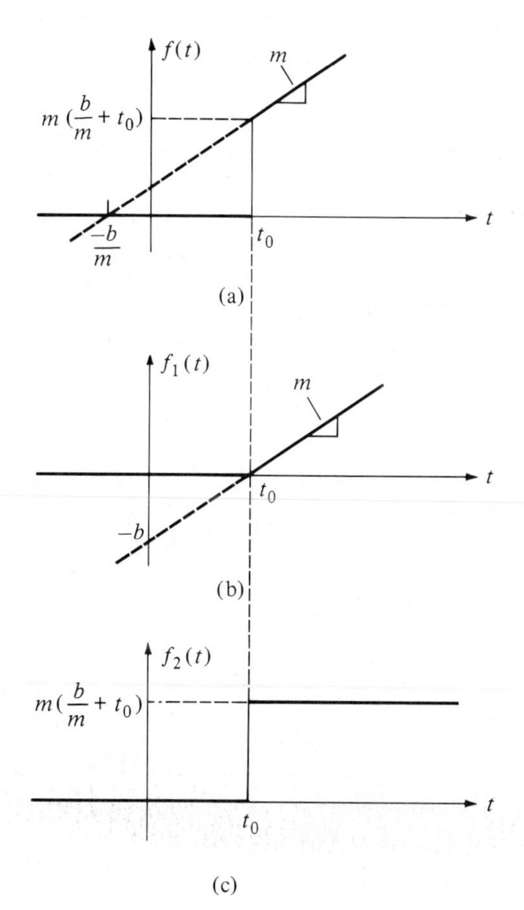

Figure 2.19
Decomposition of a general ramp function.

Trapezoidal, Triangular, and Sawtooth Waveforms

The finite rise and fall times of **trapezoidal waveforms** [Figure 2.20(a)] are sometimes better approximations of switching action in a circuit than the instantaneous rise and fall times of rectangular waveforms. **Triangular waveforms** [Figure 2.20(b)], which include **sawtooth waveforms** [Figure 2.20(c)],

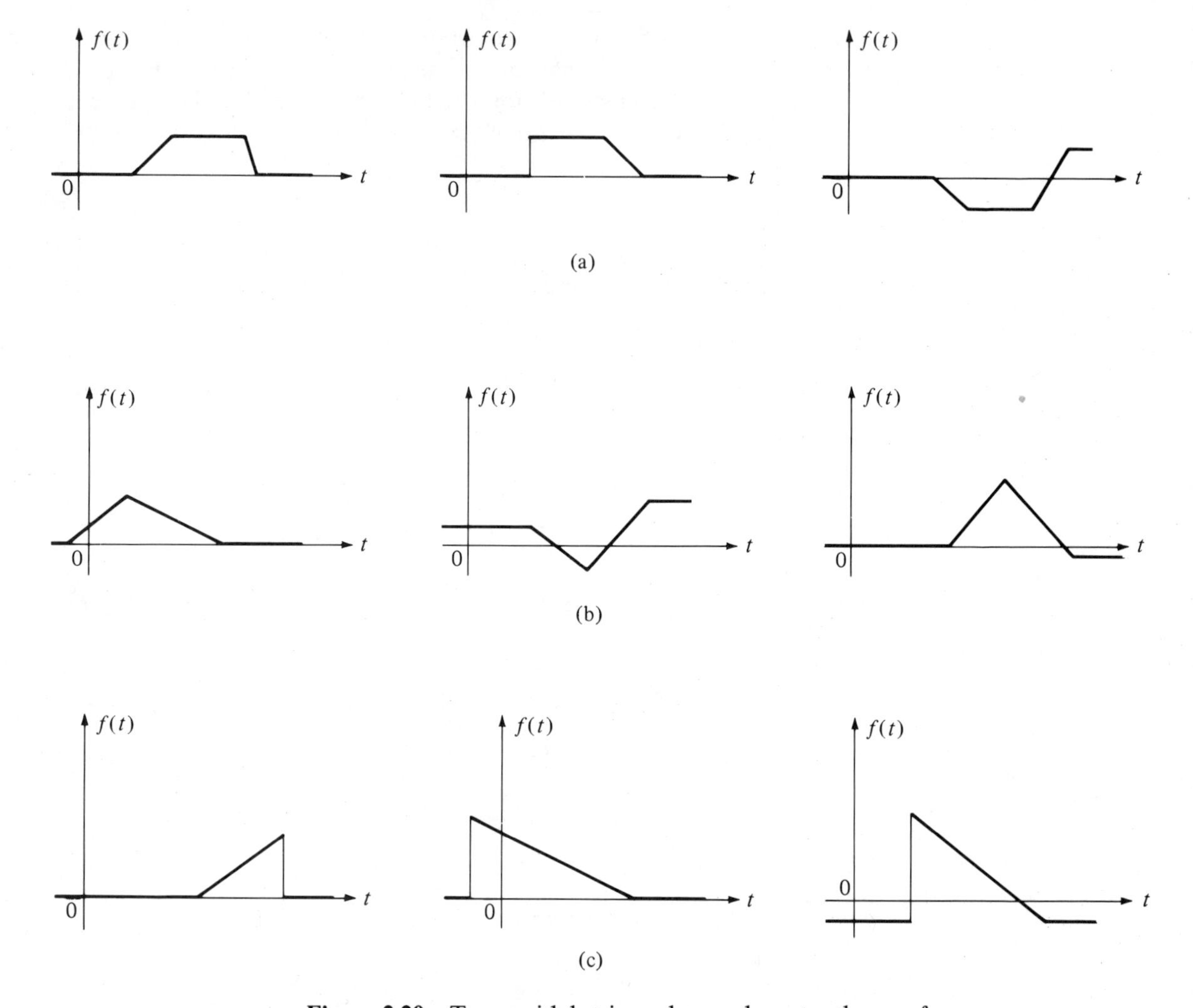

Figure 2.20 Trapezoidal, triangular, and sawtooth waveforms. (a) Trapezoidal waveforms. (b) Triangular waveforms. (c) Sawtooth waveforms.

occur frequently in electric circuits, particularly in electronic devices such as **sweep generators**, **digital-to-analog** (D/A) **converters**, etc.

A trapezoidal generating function is made up of as many as four ramp functions. The graphical composition of a trapezoidal waveform can be seen in Figure 2.21. The trapezoidal waveform [Figure 2.21(a)] begins with a ramp function,

$$f_1(t) = m_1(t - t_1) \cdot u(t - t_1)$$

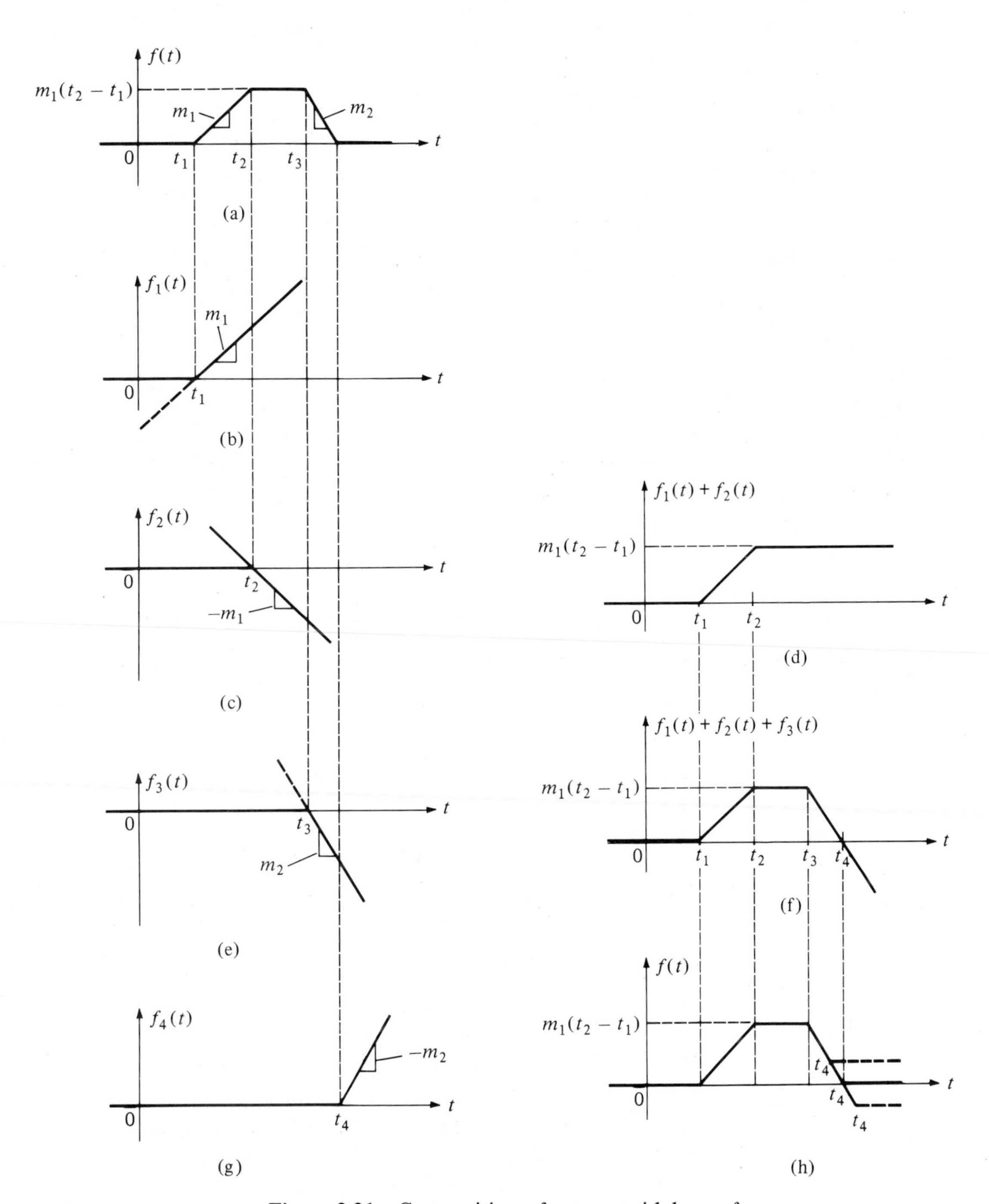

Figure 2.21 Composition of a trapezoidal waveform.

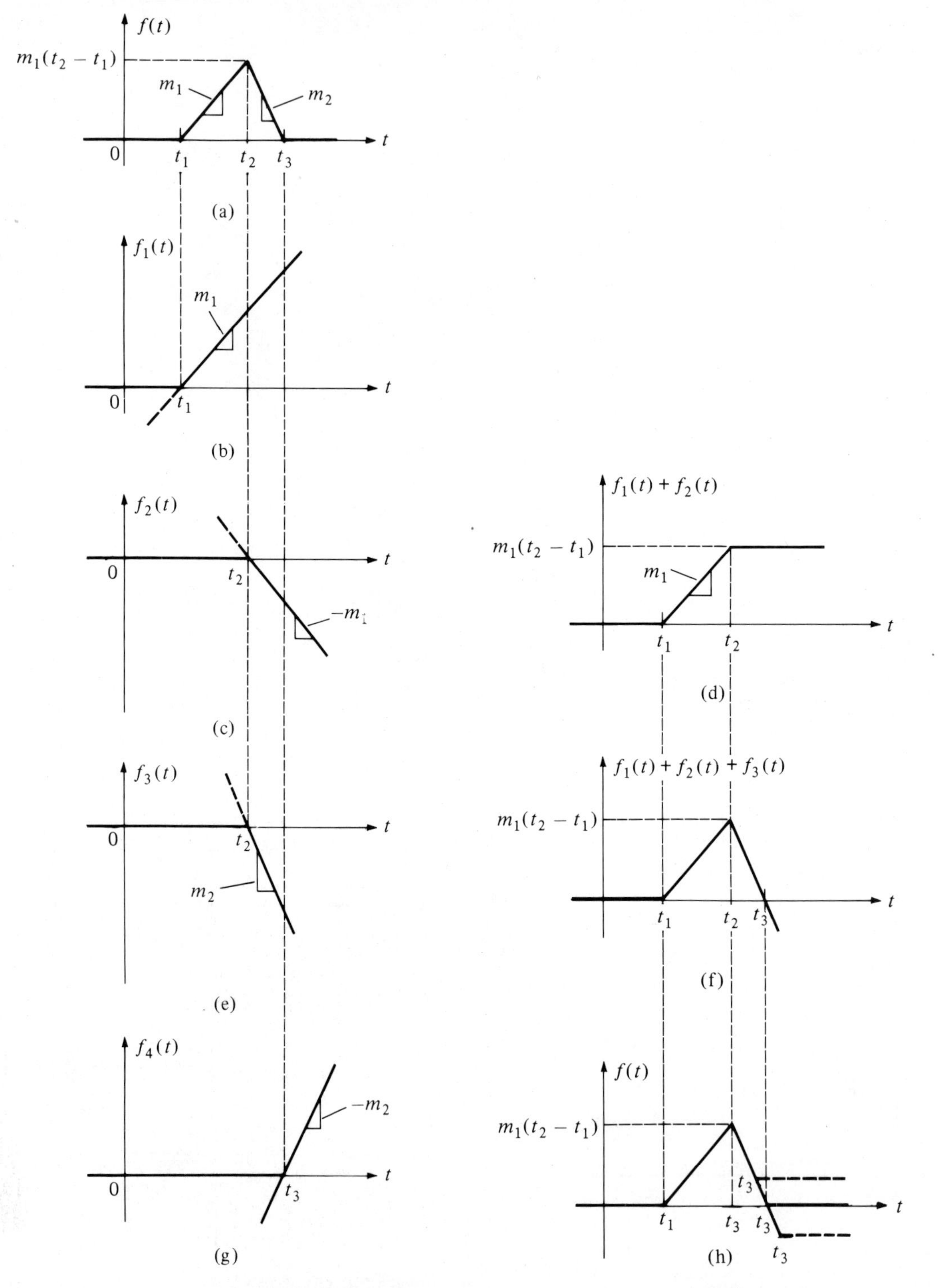

Figure 2.22 Composition of a triangular waveform.

which breaks upward from $(t_1, 0)$ with a slope m_1 [Figure 2.21(b)]. A second ramp function,

$$f_2(t) = -m_1(t - t_2) \cdot u(t - t_2)$$

which breaks downward from $(t_2, 0)$ with a slope of $-m_1$ [Figure 2.21(c)] prevents further increase in the amplitude of the waveform [Figure 2.21(d)]. A third ramp function,

$$f_3(t) = m_2(t - t_3) \cdot u(t - t_3)$$

which breaks downward from $(t_3, 0)$ with a slope of m_2 [Figure 2.21(e)], causes the amplitude to decrease for $t > t_3$ [Figure 2.21(f)]. A fourth ramp function,

$$f_4(t) = -m_2(t - t_4) \cdot u(t - t_4)$$

which breaks upward from $(t_4, 0)$ with a slope of $-m_2$ [Figure 2.21(g)], prevents any further decrease in the amplitude of the waveform.

The complete generating function for a trapezoidal function is

$$\begin{aligned} f(t) &= f_1(t) + f_2(t) + f_3(t) + f_4(t) \\ &= [m_1(t - t_1) \cdot u(t - t_1)] + [-m_1(t - t_2) \cdot u(t - t_2)] \\ &\quad + [m_2(t - t_3) \cdot u(t - t_3)] + [-m_2(t - t_4) \cdot u(t - t_4)] \end{aligned}$$

The choice of t_4 will determine whether the final value of the function is negative, zero, or positive [Figure 2.21(h)]. A triangular waveform is one in which t_2 and t_3 coincide (Figure 2.22).

Example 2.5

Problem:

Determine the generating function $v(t)$ for the trapezoidal voltage waveform shown in Figure 2.23(a).

Solution:

1. The voltage steps from zero to 1.25 V at $t = -1$ msec, then rises linearly to 5 V at 2 msec, which is a rate of (3.75 V)/(3 msec), or 1250 V/sec; therefore, the first component of $v(t)$ [Figure 2.23(b)] is a ramp function with a step,

$$\begin{aligned} v_1(t) &= 1.25 \cdot u(t + 0.001) \\ &\quad + 1250(t + 0.001) \cdot u(t + 0.001) \text{ V} \end{aligned}$$

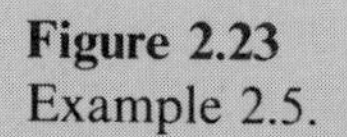
Figure 2.23
Example 2.5.

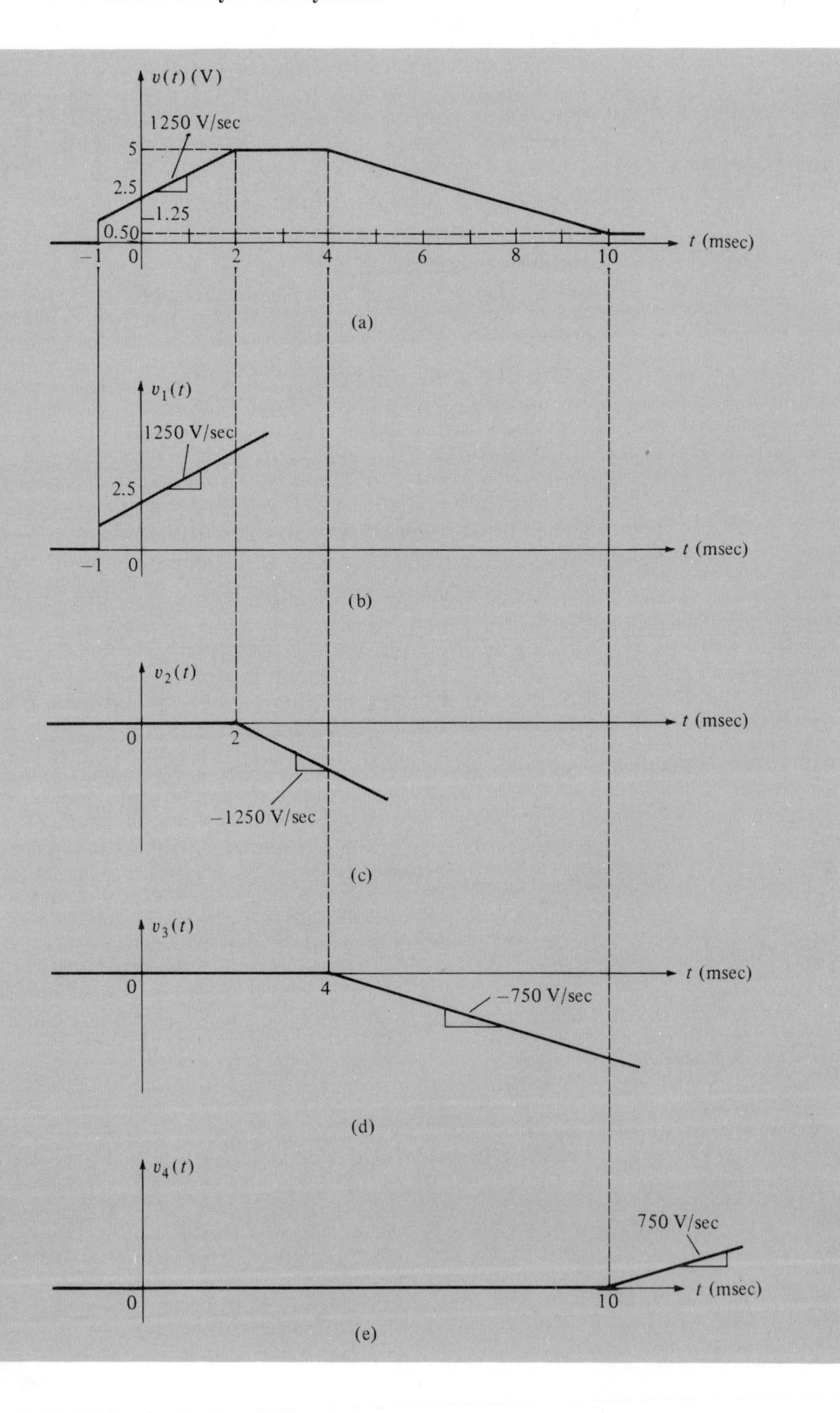

2. The second component of $v(t)$ [Figure 2.23(c)] cancels the 1250-V/sec rate of increase caused by $v_1(t)$ and prevents the voltage from rising above 5 V at $t = 2$ msec; therefore,

$$v_2(t) = -1250(t - 0.002) \cdot u(t - 0.002) \text{ V}$$

3. The third component of $v(t)$ [Figure 2.23(d)] causes the voltage to fall from 5 V at $t = 4$ msec to 0.5 V at $t = 10$ msec, which is a rate of $-(4.5 \text{ V})/(6 \text{ msec})$, or -750 V/sec; therefore,

$$v_3(t) = -750(t - 0.004) \cdot u(t - 0.004) \text{ V}$$

4. The final component of $v(t)$ [Figure 2.23(e)] cancels the rate of decrease caused by $v_3(t)$ and prevents the voltage from falling below 0.5 V at $t = 10$ msec; therefore,

$$v_4(t) = 750(t - 0.01) \cdot u(t - 0.01) \text{ V}$$

5. Add all components to obtain $v(t)$:

$$\begin{aligned} v(t) = 1.25 \cdot u(t + 0.001) \\ &+ 1250(t + 0.001) \cdot u(t + 0.001) \\ &- 1250(t - 0.002) \cdot u(t - 0.002) \\ &- 750(t - 0.004) \cdot u(t - 0.004) \\ &+ 750(t - 0.01) \cdot u(t - 0.01) \text{ V} \end{aligned}$$

This expression could be simplified by combining the first two terms, but the expanded form will prove to be more useful later in transformed circuit analysis.

Example 2.6

SPICE

Problem:

Sketch the current waveform generated by the function

$$\begin{aligned} i(t) = \big[&-240(t - 1 \times 10^{-6}) \cdot u(t - 1 \times 10^{-6}) \\ &+300(t - 1.5 \times 10^{-6}) \cdot u(t - 1.5 \times 10^{-6}) \\ &-60(t - 3.5 \times 10^{-6}) \cdot u(t - 3.5 \times 10^{-6})\big] \times 10^3 \text{ mA} \end{aligned}$$

Solution:

1. The function $i(t)$ is piecewise linear because all components of the function are ramp (linear) functions. The simplest way to sketch a piecewise-linear waveform is to evaluate the generating function at each breakpoint, plot the coordinates on the graph, and connect the points with straight lines. The breakpoints for a piecewise-linear function are easily recognized from the switching times of the unit step functions.
2. Scale the vertical axis of the plotting area in milliamperes and the horizontal axis in microseconds.
3. The first breakpoint occurs at $t = 1$ μsec, prior to which the value of the function is zero, so the coordinate of the first breakpoint is $(1, 0)$. Plot a line coincident with the t-axis to the *left* of $(1, 0)$ (Figure 2.24).

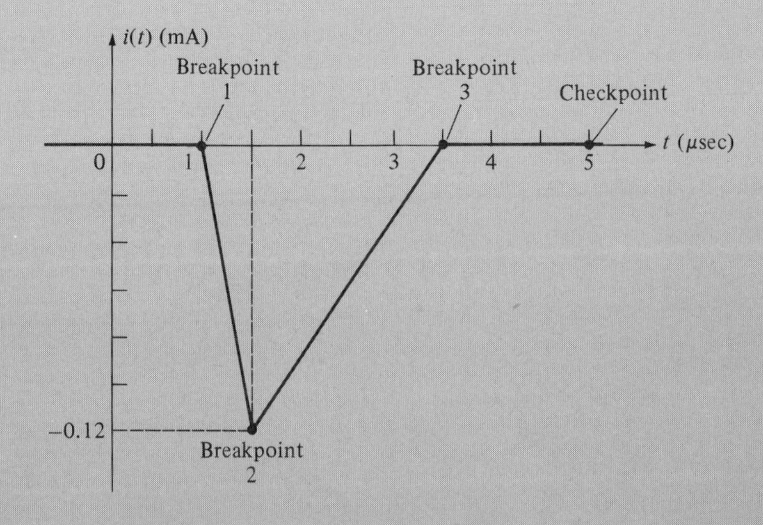

Figure 2.24
Example 2.6.

4. The second breakpoint occurs at $t = 1.5$ μsec. Evaluate $i(t)$ at $t = 1.5^-$ μsec, just prior to switching time for the second component of $i(t)$:

$$
\begin{aligned}
i(1.5 \times 10^{-6}) &= -240 \times 10^3(1.5 \times 10^{-6} - 1 \times 10^{-6}) \text{ mA} \\
&= -12 \text{ mA}
\end{aligned}
$$

Plot the second breakpoint at coordinate $(1.5, -12)$, and connect $(1, 0)$ to $(1.5, -12)$ with a straight line.

5. The final breakpoint occurs at $t = 3.5$ μsec. Evaluate $i(t)$ at $t = 3.5^-$ μsec, just prior to switching time for the third component of $i(t)$:

$$\begin{aligned} i(3.5 \times 10^{-6}) &= -240 \times 10^3(3.5 \times 10^{-6} - 1 \times 10^{-6}) \\ &\quad + 300 \times 10^3(3.5 \times 10^{-6} - 1.5 \times 10^{-6}) \text{ mA} \\ &= -60 + 60 = 0 \text{ mA} \end{aligned}$$

Plot the third breakpoint at coordinate (3.5, 0), and connect (1.5, −12) to (3.5, 0) with a straight line.

6. To determine the behavior of the waveform after the final breakpoint, evaluate the function for any $t > 3.5$ μsec, say $t = 5$ μsec:

$$\begin{aligned} i(5 \times 10^{-6}) &= \big[-240(5 \times 10^{-6} - 1 \times 10^{-6}) \\ &\quad + 300(5 \times 10^{-6} - 1.5 \times 10^{-6}) \\ &\quad - 60(5 \times 10^{-6} - 3.5 \times 10^{-6})\big] \times 10^3 \text{ mA} \\ &= [-0.096 + 0.105 - 0.009] \times 10^3 = 0 \text{ mA} \end{aligned}$$

This indicates that the waveform has zero amplitude for $t > 3.5$ μsec, as expected, because the net slope of the three components of $i(t)$ is $-240 + 300 - 60 = 0$. Once the third component becomes active, there can be no further increase or decrease in current. Plot a line coincident with the t-axis to the *right* of (3.5, 0).

Example 2.7

Problem:

Determine the generating function for the sawtooth voltage waveform shown in Figure 2.25.

Solution:

1. The voltage steps from 0 to −40 V at $t = -10$ μsec, then increases linearly from −40 V to 110 V at $t = 20$ μsec; therefore, the first component of $v(t)$ is a ramp function with a step:

$$\begin{aligned} v_1(t) &= -40 \cdot u(t + 10 \times 10^{-6}) \\ &\quad + 5 \times 10^6(t + 10 \times 10^{-6}) \cdot u(t + 10 \times 10^{-6}) \text{ V} \end{aligned}$$

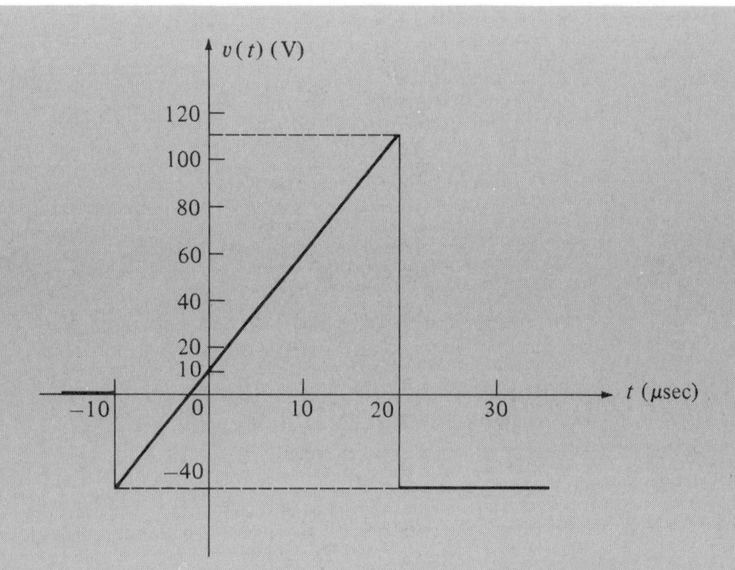

Figure 2.25
Example 2.7.

2. The next breakpoint occurs at $t = 20$ μsec, when the waveform steps from 110 V to -40 V and remains constant at -40 V thereafter. The second component, $v_2(t)$, is also a ramp function with a step, since a linear function with a negative slope is required to offset the positive slope of $v_1(t)$:

$$v_2(t) = -150 \cdot u(t - 20 \times 10^{-6})$$
$$- 5 \times 10^6(t - 20 \times 10^{-6}) \cdot u(t - 20 \times 10^{-6}) \text{ V}$$

3. Add $v_1(t)$ and $v_2(t)$ to obtain the complete generating function:

$$v(t) = -40 \cdot u(t + 10 \times 10^{-6})$$
$$+ 5 \times 10^6(t + 10 \times 10^{-6}) \cdot u(t + 10 \times 10^{-6})$$
$$- 150 \cdot u(t - 20 \times 10^{-6})$$
$$- 5 \times 10^6(t - 20 \times 10^{-6}) \cdot u(t - 20 \times 10^{-6}) \text{ V}$$

Exercise 2.3

1. Write the generating functions for the waveforms shown in Figure 2.26.

Figure 2.26
Problems for Exercise 2.3 (part 1).

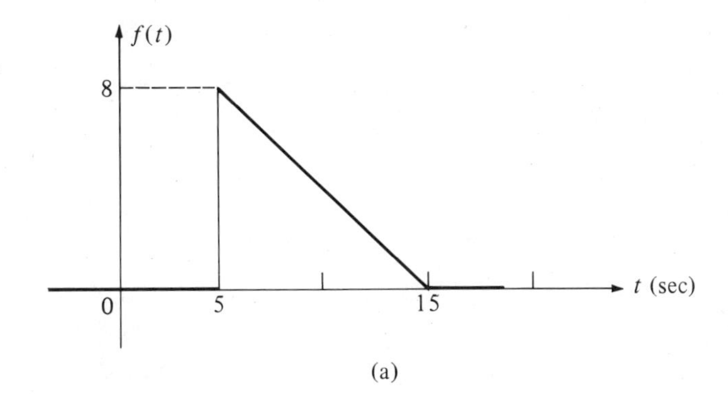

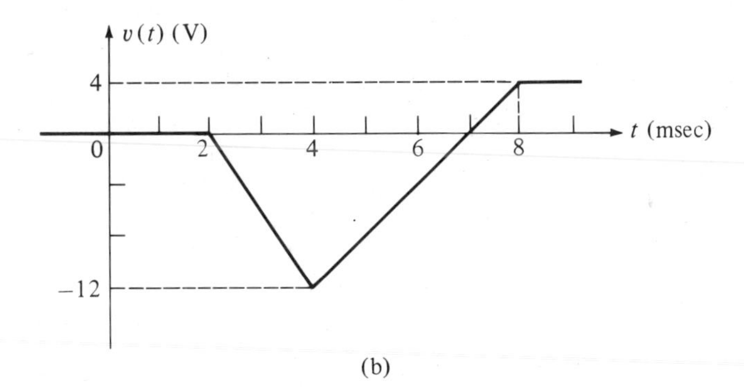

SPICE

2. Sketch the waveforms generated by the functions:
   a. $i(t) = [7.5(t + 4 \times 10^{-6})u(t + 4 \times 10^{-6}) - 15t \cdot u(t) + 7.5(t - 4 \times 10^{-6})u(t - 4 \times 10^{-6})] \times 10^{6}$ mA.
   b. $v(t) = -\ 10(t - 12 \times 10^{-6})[u(t - 12 \times 10^{-6}) - u(t - 30 \times 10^{-6})] \times 10^{6}$ V.

Answers:

1. (a) $8 \cdot u(t - 5) - 0.8(t - 5)u(t - 5) + 0.8(t - 15)u(t - 15)$.
   (b) $[-6(t - 0.002)u(t - 0.002) + 10(t - 0.004)u(t - 0.004) - 4(t - 0.008)u(t - 0.008)] \times 10^{3}$ V.

2. Refer to Figure 2.27.

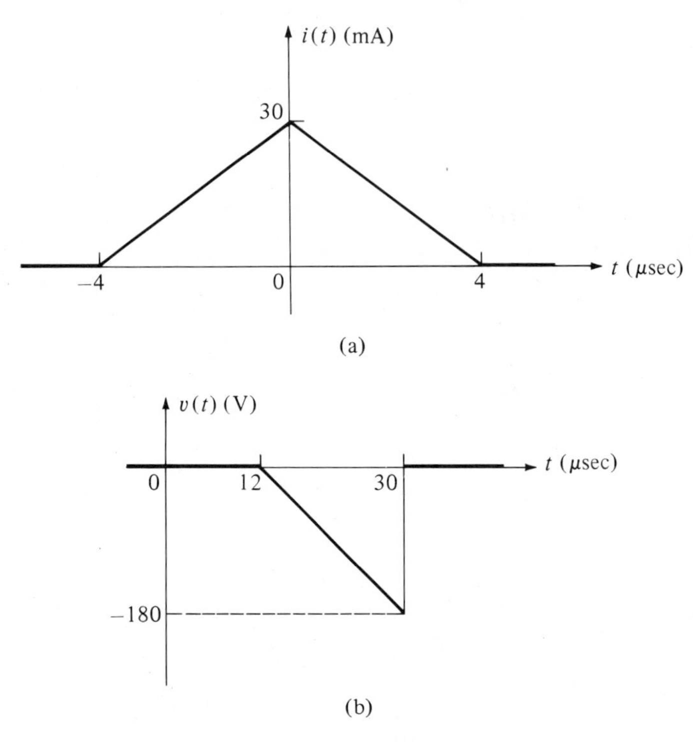

Figure 2.27 Answers to Exercise 2.3 (part 2). (a) $i(t) = [7.5(t + 4 \times 10^{-6}) \cdot u(t + 4 \times 10^{-6}) - 15t \cdot u(t) + 7.5(t - 4 \times 10^{-6}) \cdot u(t - 4 \times 10^{-6})] \times 10^6$ mA. (b) $v(t) = -10(t - 12 \times 10^{-6})\ [u(t - 12 \times 10^{-6}) - u(t - 30 \times 10^{-6})] \times 10^6$ V.

Computer Simulation

SPICE provides a piecewise-linear source (PWL) which allows the user to simulate any piecewise-linear waveform by entering the coordinates of the breakpoints as parameters of PWL. SPICE 2.2 illustrates the use of PWL for the waveform of Exercise 2.2, part 2a. Note that the breakpoints must be entered in chronological order, and there must be a small time difference between the two sets of coordinates which define a vertical edge. Trapezoidal and triangular waveforms can be defined by either PULSE or PWL. PULSE inherently provides a trapezoidal waveform; a triangular waveform can be generated by providing a very small WIDTH parameter (SPICE 2.3, for Example 2.6). WIDTH should not be set to zero, because it will then default to the output interval (TSTEP).

```
*******09/29/88 ********  SPICE 2G.6    3/15/83 ********11:15:50*****
SPICE 2.2
****     INPUT LISTING                       TEMPERATURE =   27.000 DEG C
*******************************************************************************
 R 1 0 1
 V 1 0 PWL(0 3 1 3 1.001 5 2 5 2.001 -2 3 -2 3.001 0 5 0)
 .TRAN .25 5
 .PLOT TRAN V(1)
 .END
*******09/29/88 ********  SPICE 2G.6    3/15/83 ********11:15:50*****
SPICE 2.2
****     INITIAL TRANSIENT SOLUTION          TEMPERATURE =   27.000 DEG C
*******************************************************************************
  NODE   VOLTAGE
 (  1)     3.0000
    VOLTAGE SOURCE CURRENTS
    NAME        CURRENT
    V         -3.000D+00
    TOTAL POWER DISSIPATION   9.00D+00  WATTS
*******09/29/88 ********  SPICE 2G.6    3/15/83 ********11:15:51*****
SPICE 2.2
****     TRANSIENT ANALYSIS                  TEMPERATURE =   27.000 DEG C
*******************************************************************************

    TIME       V(1)
                 -2.000D+00      0.000D+00      2.000D+00      4.000D+00  6.000D+00
                           - - - - - - - - - - - - - - - - - - - - - - - - - - - - - -
 0.000D+00  3.000D+00 .              .              .      *       .              .
 2.500D-01  3.000D+00 .              .              .      *       .              .
 5.000D-01  3.000D+00 .              .              .      *       .              .
 7.500D-01  3.000D+00 .              .              .      *       .              .
 1.000D+00  3.000D+00 .              .              .      *       .              .
 1.250D+00  5.000D+00 .              .              .              .      *       .
 1.500D+00  5.000D+00 .              .              .              .      *       .
 1.750D+00  5.000D+00 .              .              .              .      *       .
 2.000D+00  5.000D+00 .              .              .              .      *       .
 2.250D+00 -2.000D+00 *              .              .              .              .
 2.500D+00 -2.000D+00 *              .              .              .              .
 2.750D+00 -2.000D+00 *              .              .              .              .
 3.000D+00 -2.000D+00 *              .              .              .              .
 3.250D+00  0.000D+00 .              *              .              .              .
 3.500D+00  0.000D+00 .              *              .              .              .
 3.750D+00  0.000D+00 .              *              .              .              .
 4.000D+00  0.000D+00 .              *              .              .              .
 4.250D+00  0.000D+00 .              *              .              .              .
 4.500D+00  0.000D+00 .              *              .              .              .
 4.750D+00  0.000D+00 .              *              .              .              .
 5.000D+00  0.000D+00 .              *              .              .              .
                           - - - - - - - - - - - - - - - - - - - - - - - - - - - - - -

        JOB CONCLUDED
```

```
*******09/29/88 ********  SPICE 2G.6    3/15/83 ********11:30:24*****
SPICE 2.3
****     INPUT LISTING                     TEMPERATURE =   27.000 DEG C
***********************************************************************
 R 1 0 1
 V 1 0 PULSE(0 -.12 1 .5 2 1F)
 .TRAN .25 5
 .PLOT TRAN V(1)
 .END
*******09/29/88 ********  SPICE 2G.6    3/15/83 ********11:30:24*****
SPICE 2.3
****     INITIAL TRANSIENT SOLUTION        TEMPERATURE =   27.000 DEG C
***********************************************************************
  NODE   VOLTAGE
 (  1)     .0000
    VOLTAGE SOURCE CURRENTS
    NAME       CURRENT
    V        0.000D+00
    TOTAL POWER DISSIPATION   0.00D+00  WATTS
*******09/29/88 ********  SPICE 2G.6    3/15/83 ********11:30:24*****
SPICE 2.3
****     TRANSIENT ANALYSIS                TEMPERATURE =   27.000 DEG C
***********************************************************************

    TIME       V(1)
                  -1.500D-01    -1.000D-01    -5.000D-02     0.000D+00  5.000D-02
                       - - - - - - - - - - - - - - - - - - - - - - - - - - - - -
 0.000D+00  0.000D+00 .             .             .             *          .
 2.500D-01  0.000D+00 .             .             .             *          .
 5.000D-01  0.000D+00 .             .             .             *          .
 7.500D-01  0.000D+00 .             .             .             *          .
 1.000D+00  0.000D+00 .             .             .             *          .
 1.250D+00 -6.000D-02 .             .          *  .             .          .
 1.500D+00 -1.200D-01 .      *      .             .             .          .
 1.750D+00 -1.050D-01 .            *.             .             .          .
 2.000D+00 -9.000D-02 .             .  *          .             .          .
 2.250D+00 -7.500D-02 .             .      *      .             .          .
 2.500D+00 -6.000D-02 .             .          *  .             .          .
 2.750D+00 -4.500D-02 .             .             .*            .          .
 3.000D+00 -3.000D-02 .             .             .     *       .          .
 3.250D+00 -1.500D-02 .             .             .         *   .          .
 3.500D+00 -5.954D-17 .             .             .             *          .
 3.750D+00  0.000D+00 .             .             .             *          .
 4.000D+00  0.000D+00 .             .             .             *          .
 4.250D+00  0.000D+00 .             .             .             *          .
 4.500D+00  0.000D+00 .             .             .             *          .
 4.750D+00  0.000D+00 .             .             .             *          .
 5.000D+00  0.000D+00 .             .             .             *          .
                       - - - - - - - - - - - - - - - - - - - - - - - - - - - - -

        JOB CONCLUDED
```

2.4 Piecewise-Exponential Waveforms

As the reader may know, and as we will see in Chapter 3, exponential waveforms arise from the **decay** and **growth** of voltage and current in circuits containing energy storage elements, i.e., capacitors or inductors. Examples of one type of piecewise-exponential waveform, which are generated by the function

$$f(t) = A + (B - A)\left(1 - \exp\left[-\frac{t - t_0}{\tau}\right]\right) \cdot u(t - t_0) \qquad (2.10)$$

are shown in Figure 2.28(a) through (f). The value of the $f(t)$ for $t < t_0$ is A; the value of $f(t)$ changes exponentially for $t > t_0$, and approaches B as $t \to \infty$. The rate of change depends upon τ, which is known as the **time constant**.

An **exponential decay function** is one whose *absolute value* decreases as $t \to \infty$ [Figure 2.28(a), (b), (g), and (h)], and an **exponential growth function** is one whose absolute value increases as $t \to \infty$ [Figure 2.28(c), (d), (e), and (f)]. Note in Figure 2.28(e) and (f) that the absolute value of $f(t)$ first decreases, but increases as $t \to \infty$, so the functions are indeed growth functions.

Examples of a second type of piecewise-exponential waveform, generated by the function

$$g(t) = K \cdot \exp\left[-\frac{t - t_0}{\tau}\right] \cdot u(t - t_0) \qquad (2.11)$$

are shown in Figure 2.28(g) and (h). The value of the $g(t)$ for $t < t_0$ is zero; the value of $f(t)$ steps to K at t_0, then *decays* exponentially for $t > t_0$, and approaches zero as $t \to \infty$. Equation (2.11) can be shown to be the derivative of Equation (2.10):

Flashback

$$\frac{d}{dx}(u + v) = \frac{d}{dx}u + \frac{d}{dx}v$$

$$\frac{d}{dx}(uv) = v\frac{d}{dx}u + u\frac{d}{dx}v$$

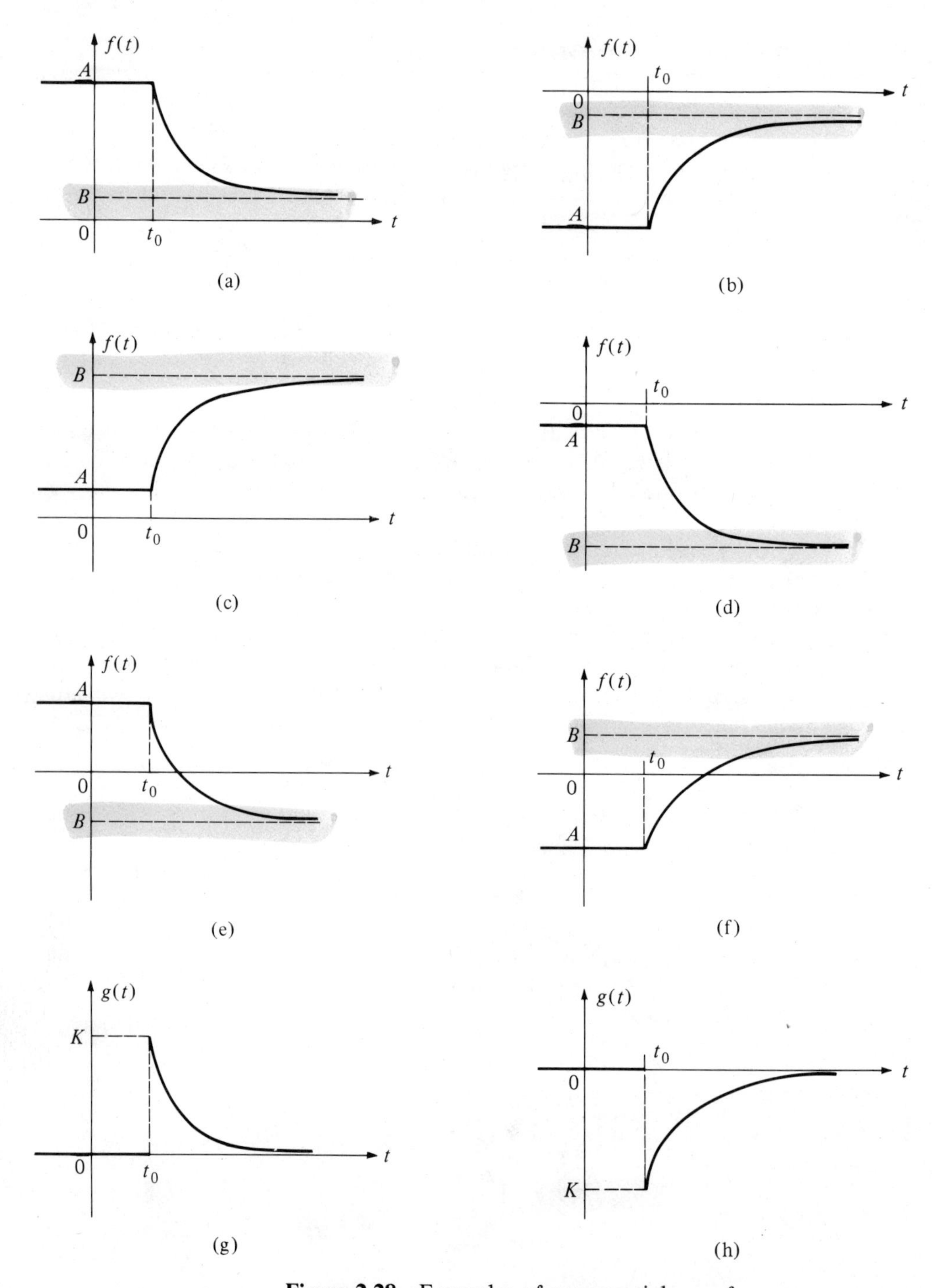

Figure 2.28 Examples of exponential waveforms.

Thus

$$\frac{d}{dt}f(t) = (B - A)\left(1 - \exp\left[-\frac{t - t_0}{\tau}\right]\right)\frac{d}{dt}u(t - t_0)$$
$$+ (B - A)u(t - t_0)\cdot\frac{d}{dt}\left(1 - \exp\left[-\frac{t - t_0}{\tau}\right]\right)$$
$$= (B - A)\left(1 - \exp\left[-\frac{t - t_0}{\tau}\right]\right)\delta(t - t_0)$$
$$+ (B - A)\left(\frac{1}{\tau}\cdot\exp\left[-\frac{(t - t_0)}{\tau}\right]\right)u(t - t_0)$$

The factor $1 - \exp[-(t - t_0)/\tau]$ is zero at t_0, and $\delta(t - t_0)$ is zero everywhere except at t_0; therefore, the first term of the expression for $df(t)/dt$ is zero everywhere, and

$$\frac{d}{dt}f(t) = \left(\frac{B - A}{\tau}\exp\left[-\frac{t - t_0}{\tau}\right]\right)u(t - t_0)$$

Since $(B - A)/\tau$ is a constant, $g(t) = df(t)/dt$.

Waveforms of the type generated by Equation (2.10) are due principally the *voltage* across a capacitor or the *current* through an inductor. Waveforms of the type generated by Equation (2.11) are due principally to the *current* through a capacitor or the *voltage* across an inductor. The basis for these observations will be made clear in Chapter 3.

Example 2.8

Problem:

Write the generating function for the waveform shown in Figure 2.29, for which $\tau = 3$ sec.

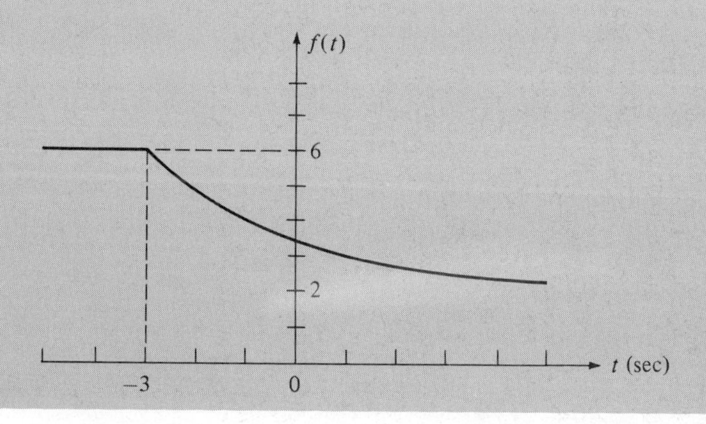

Figure 2.29
Example 2.8.

Solution:

The waveform of Figure 2.29 is of the type generated by Equation (2.10), where $t_0 = -3$ sec, $A = 6$, and $B = 2$; therefore,

$$\begin{aligned} f(t) &= 6 + (2-6)\left(1 - \exp\left[-\frac{t+3}{3}\right]\right)u(t+3) \\ &= 6 - 4\left(1 - \exp\left[-\frac{t+3}{3}\right]\right)u(t+3) \end{aligned}$$

Example 2.9

Problem:

The voltage waveform shown in Figure 2.30 was observed on an oscilloscope. Assume that the waveform is exponential, and determine the time constant τ.

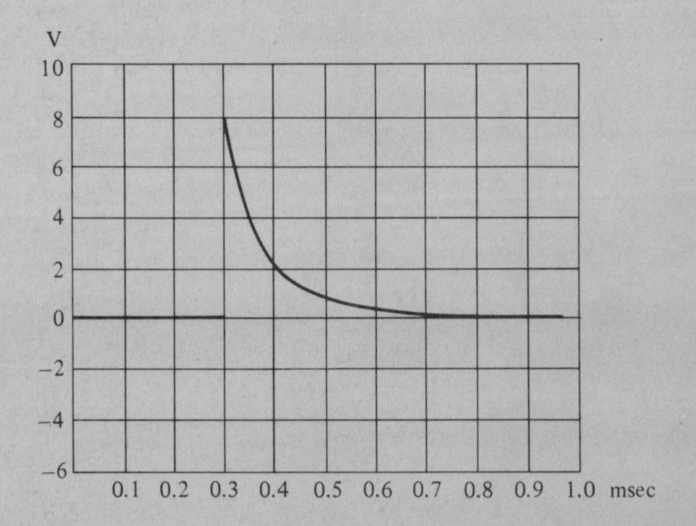

Figure 2.30 Example 2.9.

Solution:

1. The waveform in Figure 2.30 is clearly of the type generated by Equation (2.11), where $K = 8$ V and $t_0 = 0.3$ msec; therefore,

$$v(t) = 8 \cdot \exp\left[-\frac{t - 0.3 \times 10^{-3}}{\tau}\right] \cdot u(t - 0.3 \times 10^{-3}) \text{ V}$$

2. To determine τ, select a convenient point on the waveform at a time $t > t_0$, say $t = 0.35$ msec. At 0.35 msec, the voltage is 4 V; therefore,

$$v(0.35 \times 10^{-3}) = 8 \cdot \exp\left[-\frac{0.35 \times 10^{-3} - 0.3 \times 10^{-3}}{\tau}\right] = 4$$

or

$$\exp\left[-\frac{0.05 \times 10^{-3}}{\tau}\right] = 0.5$$

$$-0.05 \times 10^{-3}/\tau = \ln 0.5$$

$$\tau \simeq 0.072 \text{ msec}$$

Since an exponential function only *approaches* a final value, it is frequently necessary to terminate the exponential change in the generating

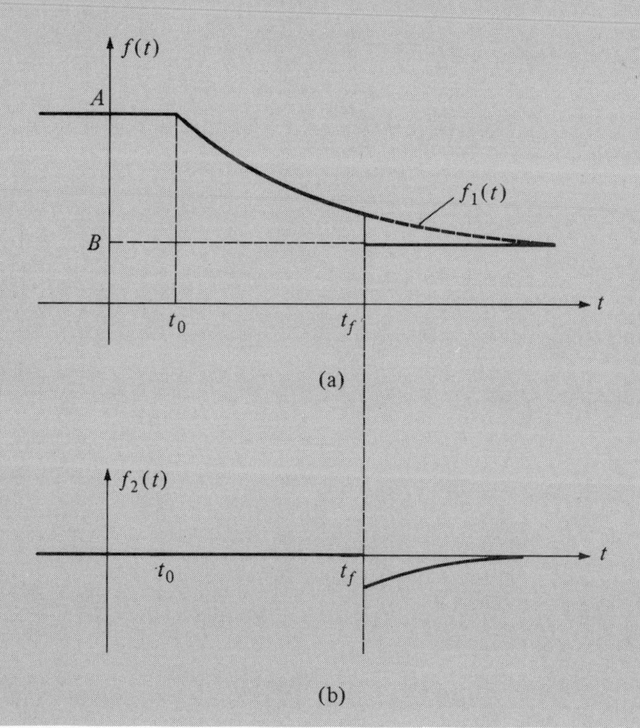

Figure 2.31 Exponential waveform with exponential terminated.

function at some finite time t_f, as shown in Figure 2.31(a). This is accomplished by adding to the generating function a second component which will step the waveform to its final value and offset further growth or decay caused by the first component. The second component must have the same time constant as the first component, but must grow or decay in the opposite direction [Figure 2.31(b)].

Example 2.10

Problem:

Write the generating function for the waveform shown in Figure 2.32(a), which is the same as in Example 2.8, except that the exponential decay terminates at $t = 1$ sec.

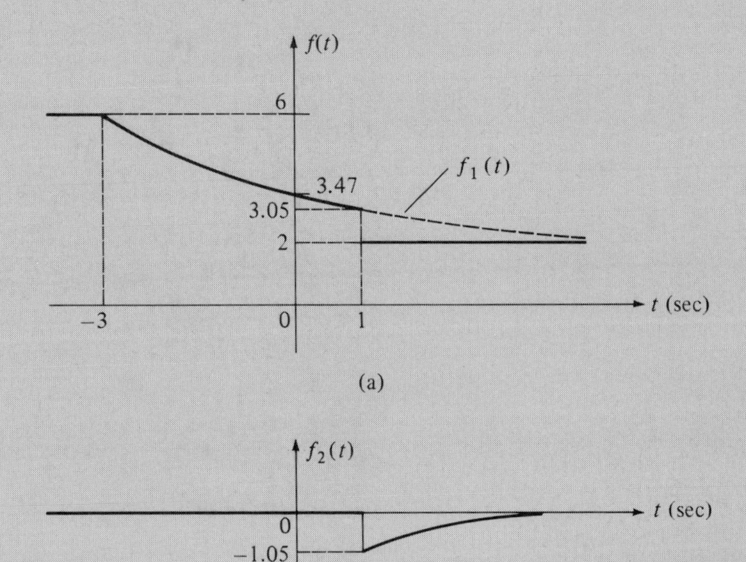

Figure 2.32 Example 2.10.

Solution:

1. From Example 2.28,

$$f_1(t) = 2 + 4 \cdot \exp\left[-\frac{t+3}{3}\right] \cdot u(t+3)$$

2. The second component, $f_2(t)$, is an exponential function switched on at $t = 1$ sec to cancel the remaining decay of $f_1(t)$. At $t = 1$ sec, $f_1(t)$

must still decay from 3.054 to 2. Therefore,

$$f_2(t) = -1.054 \cdot \exp\left[-\frac{t-1}{3}\right] \cdot u(t-1)$$

3. The generating function $f(t)$ is a linear combination of $f_1(t)$ and $f_2(t)$:

$$\begin{aligned} f(t) &= f_1(t) + f_2(t) \\ &= 2 + 4 \cdot \exp\left[-\frac{t+3}{3}\right] \cdot u(t+3) \\ &\quad - 1.054 \cdot \exp\left[-\frac{t-1}{3}\right] \cdot u(t-1) \end{aligned}$$

Example 2.11

Problem:

Write the generating function for the waveform shown in Figure 2.33(a), where $\tau = 10$ msec.

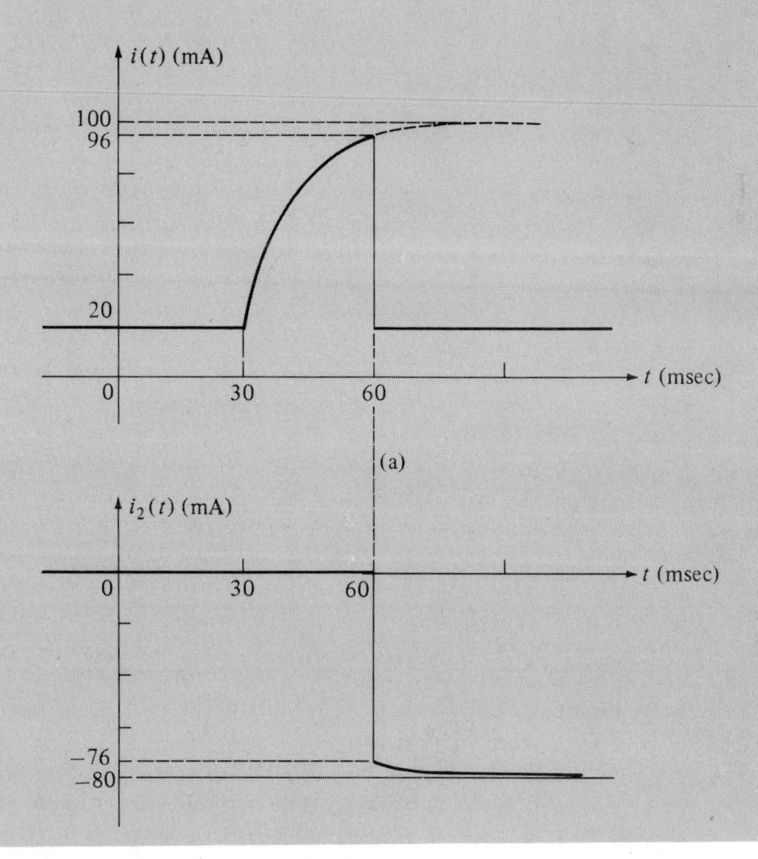

Figure 2.33
Example 2.11.

Solution:

1. The first component, $i_1(t)$, of the generating function is of the form of Equation (2.10), where $A = 20$ mA, $B = 100$ mA, and $t_0 = 30$ msec:

$$i_1(t) = 20 + (100 - 20)\left(1 - \exp\left[-\frac{t - 0.03}{\tau}\right]\right) \times u(t - 0.03)$$

$$= 20 + 80\left(1 - \exp\left[-\frac{t - 0.03}{\tau}\right]\right)u(t - 0.03) \text{ mA}$$

2. The second component, $i_2(t)$, is switched on at $t = 0.06$ sec to cancel further growth of $f_1(t)$ and reset the current to 20 mA. Therefore,

$$i_2(t) = -76.02 \cdot u(t - 0.06) - 3.983\left(1 - \exp\left[-\frac{t - 0.06}{\tau}\right]\right)u(t - 0.06)$$

$$= -\left(80 - 3.983 \cdot \exp\left[-\frac{t - 0.06}{\tau}\right]\right)u(t - 0.06) \text{ mA}$$

3. The generating function $i(t)$ is a linear combination of $i_1(t)$ and $i_2(t)$:

$$i(t) = i_1(t) + i_2(t)$$

$$= 20 + 80\left(1 - \exp\left[-\frac{t - 0.03}{\tau}\right]\right)u(t - 0.03) - 80 - 3.983 \cdot \exp\left[-\frac{t - 0.06}{\tau}\right]u(t - 0.06) \text{ mA}$$

Exercise 2.4

Problem:

1. Write the generating function for the voltage waveform shown in Figure 2.34, which is decaying exponentially toward zero.

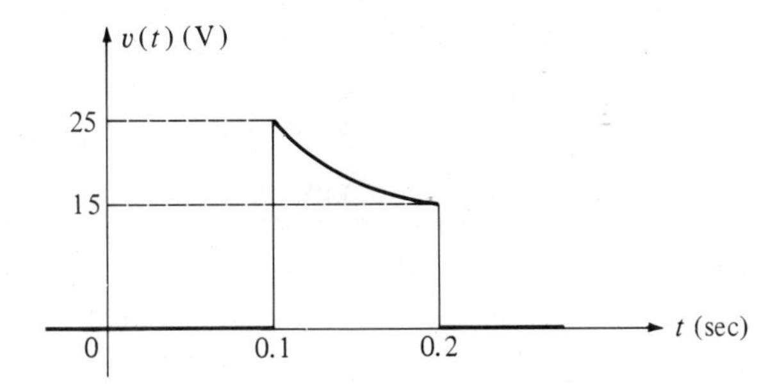

Figure 2.34
Exercise 2.4, part 1.

SPICE

2. Sketch the waveform generated by

$$f(t) = 3 + 6\left(1 - \exp\left[-\frac{t-2}{2}\right]\right) \cdot u(t-2)$$

3. Find τ for Example 2.11.

Answers:

1.

$$v(t) = 25 \cdot \exp\left[-\frac{t-0.1}{\tau}\right] \cdot u(t-0.1)$$
$$- 15 \cdot \exp\left[-\frac{t-0.2}{\tau}\right] \cdot u(t-0.2) \text{ V}$$
$$\tau \simeq 0.1958 \text{ sec}$$

2. See Figure 2.35.

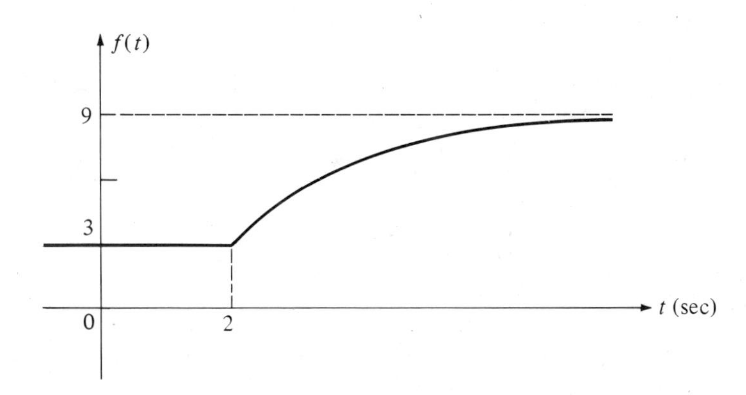

Figure 2.35
Answer to Exercise 2.4, part 2.

3. $\tau \simeq 0.01$ sec.

(See text discussion on page 141.)

```
*******09/29/88 ********  SPICE 2G.6    3/15/83 ********12:50:48*****
SPICE 2.4
****     INPUT LISTING                    TEMPERATURE =   27.000 DEG C
*******************************************************************************
 R 1 0 1
 V 1 0 EXP(3 9 2 2 10)
 .TRAN .5 10
 .PLOT TRAN V(1)
 .END
*******09/29/88 ********  SPICE 2G.6    3/15/83 ********12:50:48*****
SPICE 2.4
****     INITIAL TRANSIENT SOLUTION       TEMPERATURE =   27.000 DEG C
*******************************************************************************
  NODE   VOLTAGE
 (  1)    3.0000
     VOLTAGE SOURCE CURRENTS
     NAME       CURRENT
     V        -3.000D+00
     TOTAL POWER DISSIPATION   9.00D+00  WATTS
*******09/29/88 ********  SPICE 2G.6    3/15/83 ********12:50:48*****
SPICE 2.4
****     TRANSIENT ANALYSIS               TEMPERATURE =   27.000 DEG C
*******************************************************************************

     TIME       V(1)
                 2.000D+00     4.000D+00     6.000D+00     8.000D+00  1.000D+01
                - - - - - - - - - - - - - - - - - - - - - - - - - - - - - - -
 0.000D+00  3.000D+00 .      *      .             .             .             .
 5.000D-01  3.000D+00 .      *      .             .             .             .
 1.000D+00  3.000D+00 .      *      .             .             .             .
 1.500D+00  3.000D+00 .      *      .             .             .             .
 2.000D+00  3.000D+00 .      *      .             .             .             .
 2.500D+00  4.327D+00 .             . *           .             .             .
 3.000D+00  5.356D+00 .             .       *     .             .             .
 3.500D+00  6.166D+00 .             .             .*            .             .
 4.000D+00  6.790D+00 .             .             .      *      .             .
 4.500D+00  7.281D+00 .             .             .         *   .             .
 5.000D+00  7.660D+00 .             .             .            *.             .
 5.500D+00  7.957D+00 .             .             .             *             .
 6.000D+00  8.187D+00 .             .             .             .*            .
 6.500D+00  8.368D+00 .             .             .             .  *          .
 7.000D+00  8.507D+00 .             .             .             .   *         .
 7.500D+00  8.616D+00 .             .             .             .   *         .
 8.000D+00  8.701D+00 .             .             .             .    *        .
 8.500D+00  8.767D+00 .             .             .             .    *        .
 9.000D+00  8.819D+00 .             .             .             .     *       .
 9.500D+00  8.859D+00 .             .             .             .     *       .
 1.000D+01  8.890D+00 .             .             .             .     *       .
                - - - - - - - - - - - - - - - - - - - - - - - - - - - - - - -

          JOB CONCLUDED
```

```
*******09/29/88 ********  SPICE 2G.6    3/15/83 ********12:57:02*****
SPICE 2.5
****     INPUT LISTING                    TEMPERATURE =   27.000 DEG C
********************************************************************************
 R 1 0 1
 V 1 0 EXP(0 8 .3M 1F 1F .072M)
 .TRAN .05M 1M
 .PLOT TRAN V(1)
 .END
*******09/29/88 ********  SPICE 2G.6    3/15/83 ********12:57:02*****
SPICE 2.5
****     INITIAL TRANSIENT SOLUTION       TEMPERATURE =   27.000 DEG C
********************************************************************************
  NODE   VOLTAGE
 (  1)     .0000
    VOLTAGE SOURCE CURRENTS
    NAME       CURRENT
    V         0.000D+00
    TOTAL POWER DISSIPATION   0.00D+00  WATTS
*******09/29/88 ********  SPICE 2G.6    3/15/83 ********12:57:02*****
SPICE 2.5
****     TRANSIENT ANALYSIS               TEMPERATURE =   27.000 DEG C
********************************************************************************

    TIME      V(1)
                 0.000D+00     2.000D+00     4.000D+00     6.000D+00  8.000D+00
                    - - - - - - - - - - - - - - - - - - - - - - - - - - - - - -
 0.000D+00  0.000D+00 *             .             .             .          .
 5.000D-05  0.000D+00 *             .             .             .          .
 1.000D-04  0.000D+00 *             .             .             .          .
 1.500D-04  0.000D+00 *             .             .             .          .
 2.000D-04  0.000D+00 *             .             .             .          .
 2.500D-04  0.000D+00 *             .             .             .          .
 3.000D-04  0.000D+00 *             .             .             .          .
 3.500D-04  8.000D+00 .             .             .             .          *
 4.000D-04  3.995D+00 .             .             *             .          .
 4.500D-04  2.014D+00 .             *             .             .          .
 5.000D-04  9.961D-01 .      *      .             .             .          .
 5.500D-04  5.022D-01 .   *         .             .             .          .
 6.000D-04  2.484D-01 . *           .             .             .          .
 6.500D-04  1.252D-01 .*            .             .             .          .
 7.000D-04  6.194D-02 *             .             .             .          .
 7.500D-04  3.123D-02 *             .             .             .          .
 8.000D-04  1.544D-02 *             .             .             .          .
 8.500D-04  7.786D-03 *             .             .             .          .
 9.000D-04  3.851D-03 *             .             .             .          .
 9.500D-04  1.942D-03 *             .             .             .          .
 1.000D-03  9.602D-04 *             .             .             .          .
                    - - - - - - - - - - - - - - - - - - - - - - - - - - - - - -

        JOB CONCLUDED
```

```
*******09/29/88 ********  SPICE 2G.6    3/15/83 ********13:10:33*****
SPICE 2.6
****     INPUT LISTING                      TEMPERATURE =   27.000 DEG C
******************************************************************************
 R 1 0 1
 V1 1 2 EXP(0 25 .1 1F 1F 195.8M)
 V2 2 0 EXP(0 -15 .2 1F 1F 195.8M)
 .TRAN .01 .25
 .PLOT TRAN V(1)
 .END
*******09/29/88 ********  SPICE 2G.6    3/15/83 ********13:10:33*****
SPICE 2.6
****     INITIAL TRANSIENT SOLUTION         TEMPERATURE =   27.000 DEG C
******************************************************************************
  NODE   VOLTAGE     NODE   VOLTAGE
 (  1)     .0000    (  2)     .0000
    VOLTAGE SOURCE CURRENTS
    NAME       CURRENT
    V1        0.000D+00
    V2        0.000D+00
    TOTAL POWER DISSIPATION   0.00D+00  WATTS
*******09/29/88 ********  SPICE 2G.6    3/15/83 ********13:10:34*****
SPICE 2.6
****     TRANSIENT ANALYSIS                 TEMPERATURE =   27.000 DEG C
******************************************************************************

     TIME       V(1)
                   0.000D+00      1.000D+01      2.000D+01      3.000D+01  4.000D+01
                           - - - - - - - - - - - - - - - - - - - - - - - - - - - - - -
 0.000D+00  0.000D+00 *              .              .              .              .
 1.000D-02  0.000D+00 *              .              .              .              .
 2.000D-02  0.000D+00 *              .              .              .              .
 3.000D-02  0.000D+00 *              .              .              .              .
 4.000D-02  0.000D+00 *              .              .              .              .
 5.000D-02  0.000D+00 *              .              .              .              .
 6.000D-02  0.000D+00 *              .              .              .              .
 7.000D-02  0.000D+00 *              .              .              .              .
 8.000D-02  0.000D+00 *              .              .              .              .
 9.000D-02  0.000D+00 *              .              .              .              .
 1.000D-01  0.000D+00 *              .              .              .              .
 1.100D-01  2.500D+01 .              .              .      *       .              .
 1.200D-01  2.376D+01 .              .              .    *         .              .
 1.300D-01  2.257D+01 .              .              .   *          .              .
 1.400D-01  2.145D+01 .              .              . *            .              .
 1.500D-01  2.038D+01 .              .              .*             .              .
 1.600D-01  1.937D+01 .              .             *.              .              .
 1.700D-01  1.840D+01 .              .            * .              .              .
 1.800D-01  1.749D+01 .              .          *   .              .              .
 1.900D-01  1.662D+01 .              .         *    .              .              .
 2.000D-01  1.579D+01 .              .        *     .              .              .
 2.100D-01  1.506D-03 *              .              .              .              .
 2.200D-01  1.431D-03 *              .              .              .              .
 2.300D-01  1.360D-03 *              .              .              .              .
 2.400D-01  1.292D-03 *              .              .              .              .
 2.500D-01  1.228D-03 *              .              .              .              .
                           - - - - - - - - - - - - - - - - - - - - - - - - - - - - - -

         JOB CONCLUDED
```

Computer Simulation

Exponential waveforms can be generated by SPICE through the use of the EXP source. SPICE 2.4 (Exercise 2.4, part 2) illustrates a growth function. Note that the fall delay time (DELAY2) must be specified equal to or greater than the total simulation time (TSTOP) so that the waveform will not begin to decay before simulation is complete; in this case, the fall time constant (TAU2) need not be specified. If DELAY2 is omitted, or specified as zero, SPICE will set a default value equal to TSTEP, and decay will begin after the first output interval.

SPICE 2.5 illustrates a decay function (Example 2.9); the rise time constant (TAU1) and DELAY2 are both specified to be very short (1 fsec) to simulate the leading edge of the function followed immediately by decay. SPICE 2.6 illustrates the superposition of two EXP sources to generate the piecewise-exponential waveform of Exercise 2.4, part 1.

2.5 Periodic Waveforms

A **periodic waveform** is one which repeats itself at a regular interval called the **period** (T) of the waveform. The waveforms and their generating functions which we have discussed previously are all **aperiodic**; that is, they do not repeat themselves. In this section, we will discuss periodic piecewise-linear and piecewise-exponential waveforms and their generating functions. We will see that analysis and synthesis of periodic waveforms are simply extensions of the techniques used with aperiodic waveforms.

The generating function for a periodic waveform, called a **periodic function**, satisfies the relationship

$$f(t) = f(t + nT) \tag{2.12}$$

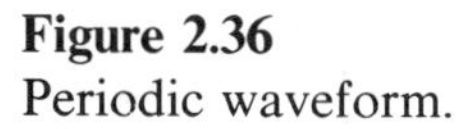
Figure 2.36
Periodic waveform.

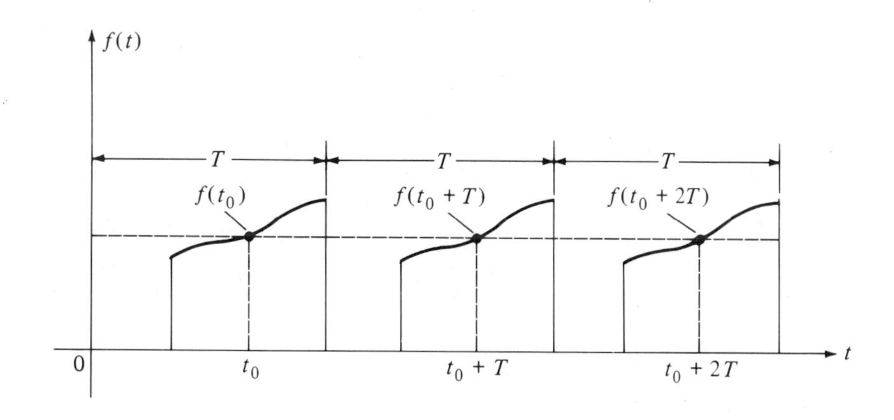

where T is the period and n is an integer $(\ldots, -2, -1, 0, 1, 2, \ldots)$. In Figure 2.36, the generating function $f(t)$ is periodic because

$$\begin{aligned} \cdots &= f(t_0 - 2T) = f(t_0 - T) = f(t_0) \\ &= f(t_0 + T) = f(t_0 + 2T) = \cdots \end{aligned}$$

Period is defined precisely as the smallest value of T for which (2.12) is true for all n. Intuitively, a periodic function is one which has a single basic behavior that is repeated indefinitely. This basic behavior is called a **cycle** of the function, and the period is the time required to complete one cycle.

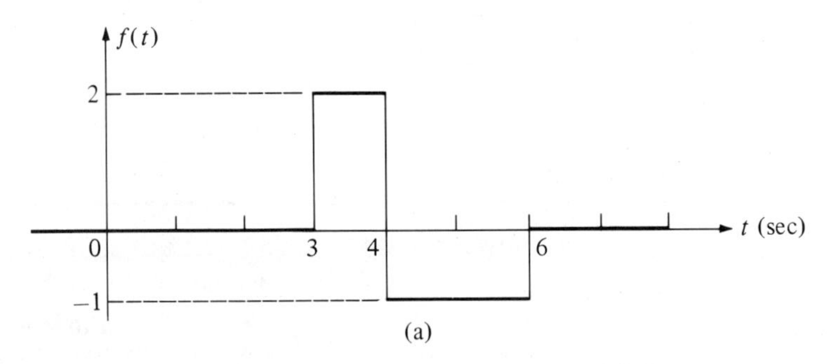

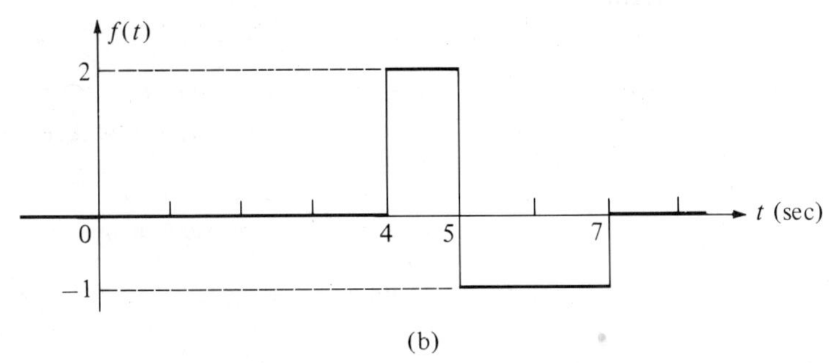

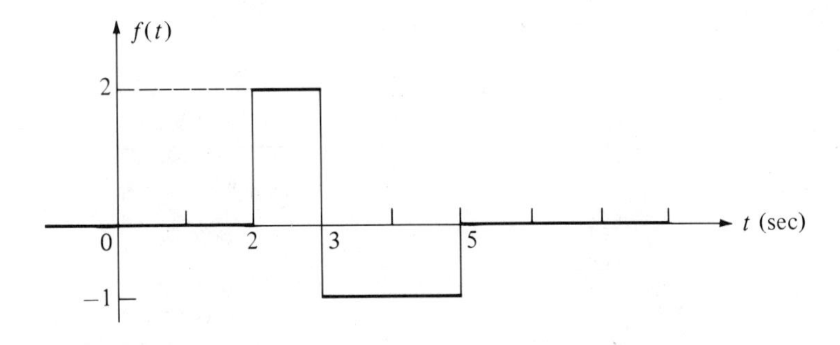

Figure 2.37
Shifted waveforms.
(a) Aperiodic waveform.
(b) Aperiodic waveform shifted right.
(c) Aperiodic waveform shifted left.

Consider the aperiodic waveform shown in Figure 2.37(a), for which the generating function is

$$f(t) = 2 \cdot u(t-3) - 3 \cdot u(t-4) + u(t-6)$$

Observe that if t is replaced by $t-1$, then

$$\begin{aligned} f(t-1) &= 2 \cdot u((t-1)-3) - 3 \cdot u((t-1)-4) \\ &\quad + u((t-1)-6) \\ &= 2 \cdot u(t-4) - 3 \cdot u(t-5) + u(t-7) \end{aligned}$$

generates the waveform shown in Figure 2.37(b). Substituting $t-1$ for t increases the switching time of each unit step function by 1 sec, which **shifts** the waveform to the *right* by 1 sec. If t is replaced by $t+1$, then

$$\begin{aligned} f(t+1) &= 2 \cdot u((t+1)-3) - 3 \cdot u((t+1)-4) + u((t+1)-6) \\ &= 2 \cdot u(t-2) - 3 \cdot u(t-3) + u(t-5) \end{aligned}$$

generates the waveform shown in Figure 2.37(c). Substituting $t+1$ for t decreases each switching time 1 sec, which shifts the waveform to the *left* by 1 sec.

Now consider the result of shifting the waveform by 3 sec, both left and right,

$$\begin{aligned} f(t+3) &= 2 \cdot u(t) - 3 \cdot u(t-1) + u(t-3) \\ f(t-3) &= 2 \cdot u(t-6) - 3 \cdot u(t-7) + u(t-9) \end{aligned}$$

and adding the three generating functions to form a new function

$$g(t) = f(t+3) + f(t) + f(t-3)$$

The function $g(t)$, which generates the waveform shown in Figure 2.38, is in fact an aperiodic function, but it suggests a method of obtaining a periodic function, $f_P(t)$:

$$\begin{aligned} f_P(t) &= \cdots + f(t+2T) + f(t+T) + f(t) + f(t-T) \\ &\quad + f(t-2T) + \cdots \\ &= \sum_{n=-\infty}^{\infty} f(t-nT) \end{aligned} \tag{2.13a}$$

where $f(t)$ is the generating function for *one* cycle of the waveform and T is the period of the waveform. Because $f_P(t)$ is an infinite sum of terms, it makes no difference which cycle of the waveform is chosen for writing $f(t)$, but usually the first complete cycle after $t = 0$ will yield the simplest expression for $f(t-nT)$. Equation (2.13a) provides an unrestricted definition for a real

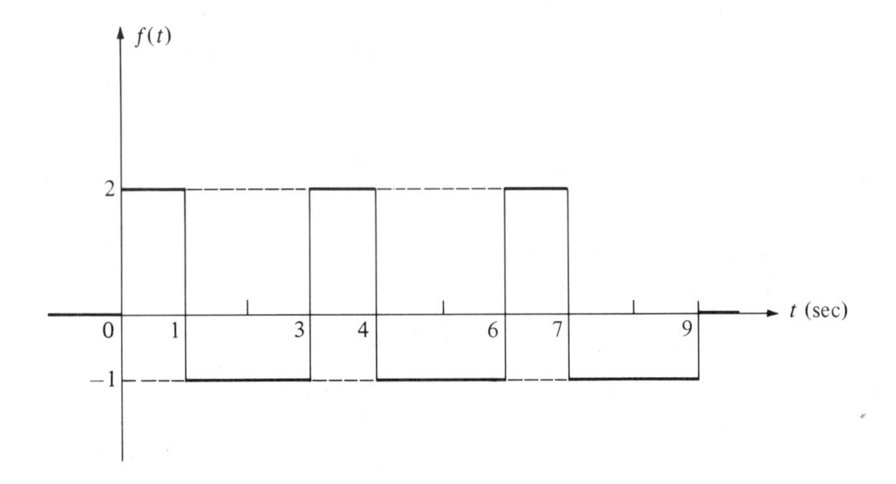

Figure 2.38 Addition of shifted waveforms.

function in the time domain. In cases where only the time $t > 0$ need be considered, a modification of Equation (2.13a) will be used:

$$f_P(t) = \sum_{n=0}^{\infty} f(t - nT) \tag{2.13b}$$

Example 2.12

Problem:

Write the generating function for the train of periodic voltage pulses shown in Figure 2.39(a). Expand the result for $n = -1, 0, 1$, and 2.

Solution:

1. Write the generating function for the first pulse [Figure 2.39(b)]:

$$\begin{aligned} v_1(t) &= 4 \cdot u(t) - 4 \cdot u(t - 0.020) \\ &= 4[u(t) - u(t - 0.020)] \text{ V} \end{aligned}$$

The period of the waveform is 50 msec (0.050 sec); therefore, from Equation (2.13a),

$$\begin{aligned} v_P(t) &= \sum_{n=-\infty}^{\infty} v(t - nT) \\ &= \sum_{n=-\infty}^{\infty} 4[u(t - 0.050n) - u(t - 0.050n - 0.020)] \text{ V} \end{aligned}$$

Figure 2.39
Example 2.12.
(a) Periodic rectangular voltage waveform.
(b) One cycle of periodic rectangular waveform.

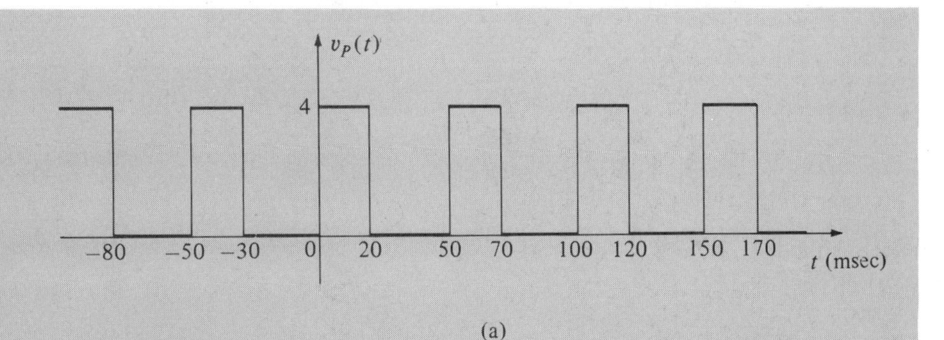

(a)

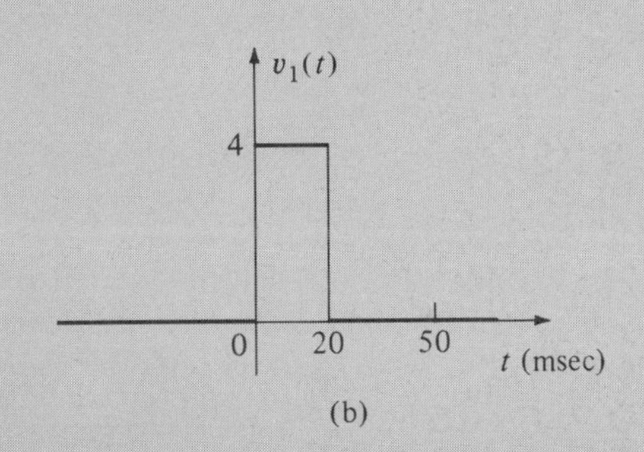

(b)

2. Expand $v_P(t)$ for $n = -1, 0, 1,$ and 2:

$$\begin{aligned} v_P(t) = \cdots &+4[u(t+0.050) - u(t+0.030)] \\ &+ 4[u(t) - u(t-0.020)] \\ &+ 4[u(t-0.050) - u(t-0.070)] \\ &+ 4[u(t-0.100) - u(t-0.120)] + \cdots \text{ V} \end{aligned}$$

Example 2.13

Problem:

Write the generating function for the periodic triangular waveform shown in Figure 2.40(a).

Solution:

1. Write the generating function for the first cycle of $f_P(t)$ [Figure 2.40(b)]:

$$\begin{aligned} f_1(t) &= -5(t-2)u(t) + 10(t-2)u(t-2) - 10 \cdot u(t-4) \\ &= -5[(t-2)u(t) + 2(t-2)u(t-2) - 2 \cdot u(t-4)] \end{aligned}$$

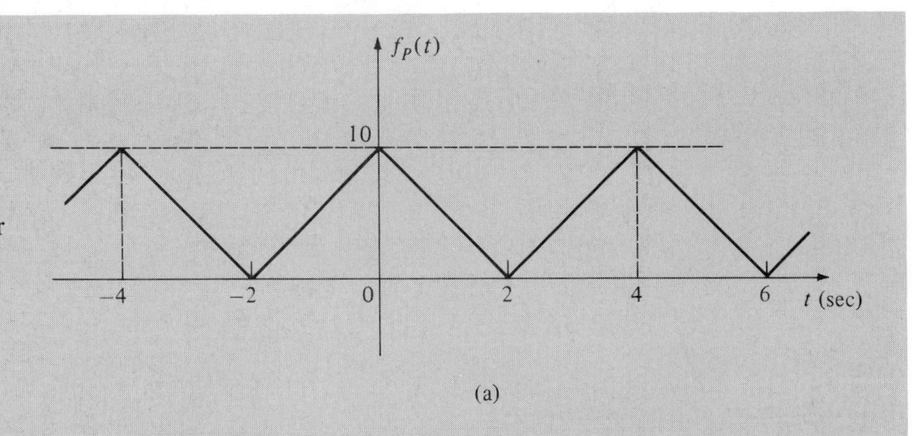

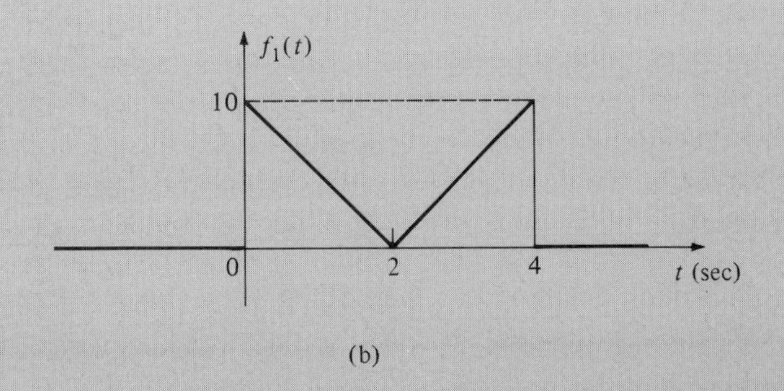

Figure 2.40 Example 2.13. (a) Periodic triangular waveform. (b) One cycle of periodic triangular waveform.

2. The period of the waveform is 4 sec. Therefore, from Equation (2.13a),

$$\begin{aligned} f_P(t) &= \sum_{n=-\infty}^{\infty} f(t - nT) \\ &= \sum_{n=-\infty}^{\infty} -5[(t - 4n - 2)u(t - 4n) \\ &\qquad + 2(t - 4n - 2)u(t - 4n - 2) \\ &\qquad - 2 \cdot u(t - 4n - 4)] \end{aligned}$$

Example 2.14

Problem:

Write the generating function for the periodic exponential current waveform shown in Figure 2.41(a).

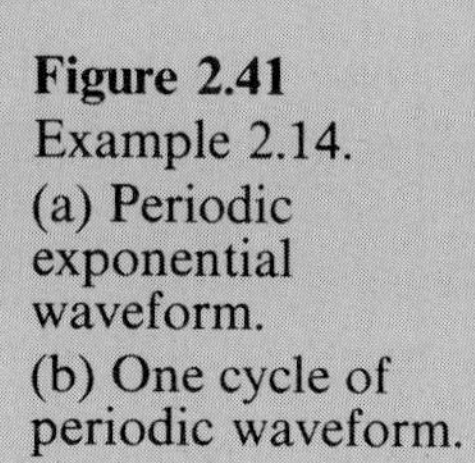
Figure 2.41
Example 2.14.
(a) Periodic exponential waveform.
(b) One cycle of periodic waveform.

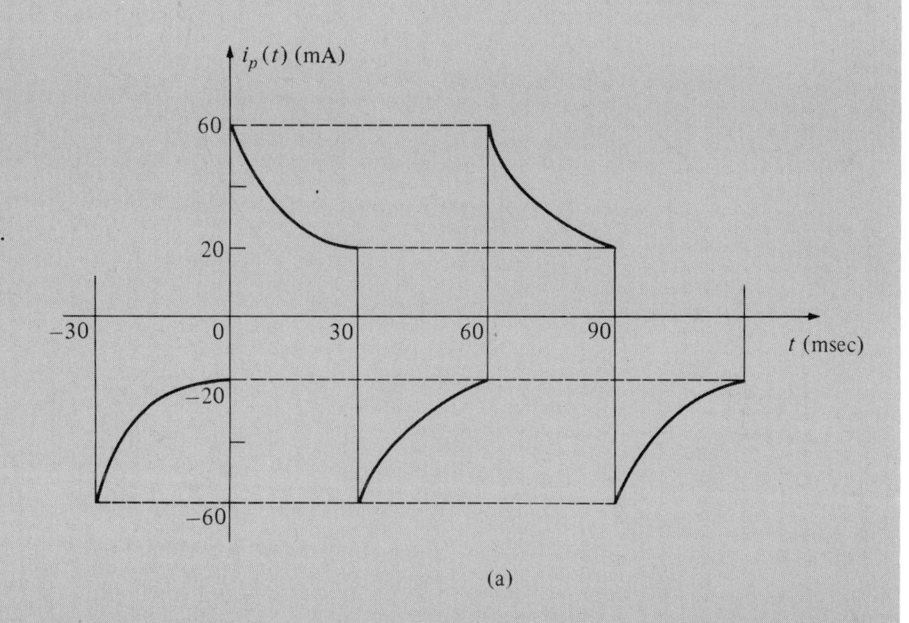

(a)

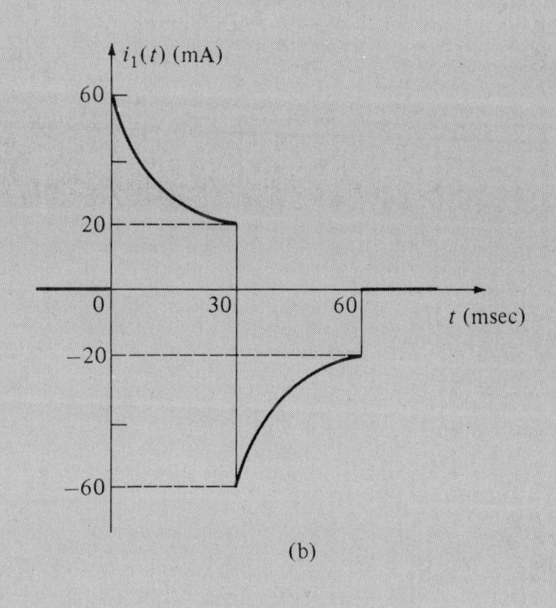

(b)

Solution:

1. Write the generating function for the first cycle of $i_P(t)$ [Figure 2.41(b)]:

$$
\begin{aligned}
i_1(t) &= 60 \cdot \exp\left[-\frac{t}{\tau}\right] \cdot u(t) - 20 \cdot \exp\left[-\frac{t-0.03}{\tau}\right] \\
&\quad \cdot u(t-0.03) \\
&\quad - 60 \cdot \exp\left[-\frac{t-0.03}{\tau}\right] \cdot u(t-0.03) \\
&\quad + 20 \cdot \exp\left[-\frac{t-0.06}{\tau}\right] \cdot u(t-0.06) \\
&= 60 \cdot \exp\left[-\frac{t}{\tau}\right] - 80 \cdot \exp\left[-\frac{t-0.03}{\tau}\right] \cdot u(t-0.03) \\
&\quad + 20 \cdot \exp\left[-\frac{t-0.06}{\tau}\right] \cdot u(t-0.06) \text{ mA} \\
\tau &\simeq 27.31 \text{ msec}
\end{aligned}
$$

2. The period of the waveform is 60 msec (0.060 sec). Therefore, from Equation (2.13a),

$$
\begin{aligned}
i_P(t) &= \sum_{n=-\infty}^{\infty} i(t-nT) \\
&= \sum_{n=-\infty}^{\infty} 60 \cdot \exp\left[-\frac{t-0.06n}{\tau}\right] \\
&\quad - 80 \cdot \exp\left[-\frac{t-0.03-0.06n}{\tau}\right] \\
&\quad \cdot u(t-0.03-0.06n) \\
&\quad + 20 \cdot \exp\left[-\frac{t-0.06-0.06n}{\tau}\right] \\
&\quad \left. \cdot u(t-0.06-0.06n) \right] \text{ mA}
\end{aligned}
$$

Exercise 2.5

1. Write the generating functions for the periodic waveforms shown in Figure 2.42(a) and (b).

Figure 2.42
Exercise 2.4, part 1.

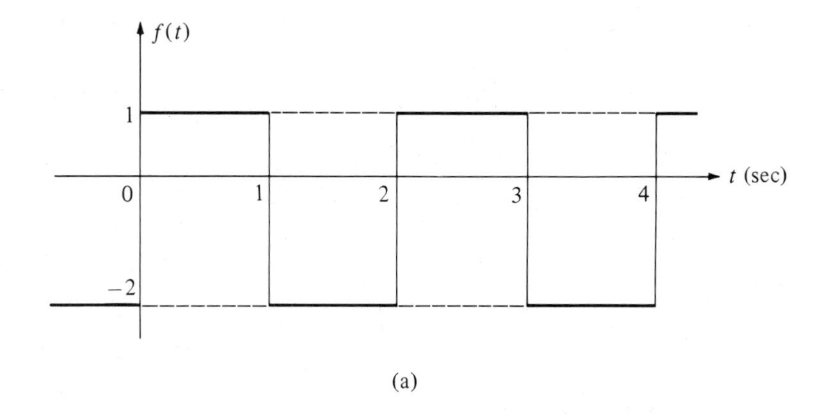

(a)

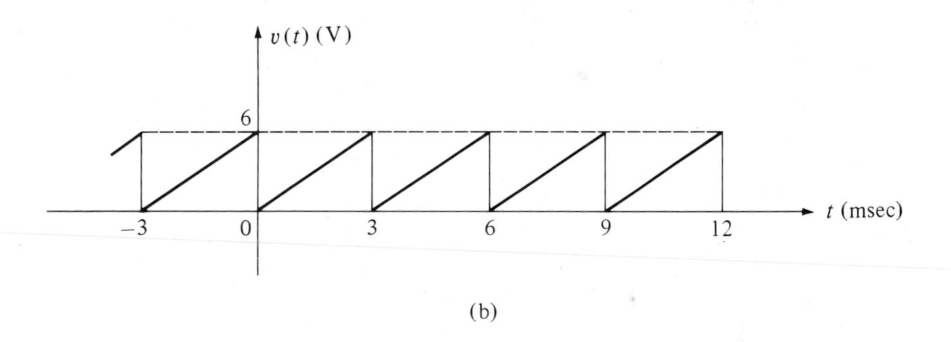

(b)

SPICE

2. Sketch the periodic waveform generated by

$$f(t) = 10 \sum_{n=-\infty}^{\infty} \left\{ \left(1 - \exp\left[-\frac{t - 10n + 0.805}{\tau}\right]\right) \times [u(t - 10n) - u(t - 10n - 5)] + \left(\exp\left[-\frac{t - 10n - 4.195}{\tau}\right]\right) \times [u(t - 10n - 5) - u(t - 10n - 10)] \right\}$$

where $\tau = 3.607$ sec. Hint: Let $n = 0$, and evaluate $f(t)$ for $t = 0^-$, 0^+, 5^-, 5^+, 10^-, and 10^+, to obtain the breakpoints for the first cycle of $f(t)$.

Answers:

1. (a)

$$f(t) = \sum_{n=-\infty}^{\infty} [u(t-2n) - 3 \cdot u(t-2n-1) + 2 \cdot u(t-2n-2)]$$

(b)

$$\begin{aligned} v(t) = \sum_{n=-\infty}^{\infty} \big[& 2000t \cdot u(t - 3 \times 10^{-3}n) \\ & -2000(t - 3 \times 10^{-3}n - 3 \times 10^{-3}) \\ & \times u(t - 3 \times 10^{-3}n - 3 \times 10^{-3}) \\ & -6 \cdot u(t - 3 \times 10^{-3}n - 3 \times 10^{-3}) \big] \text{ V} \end{aligned}$$

2. Refer to Figure 2.43.

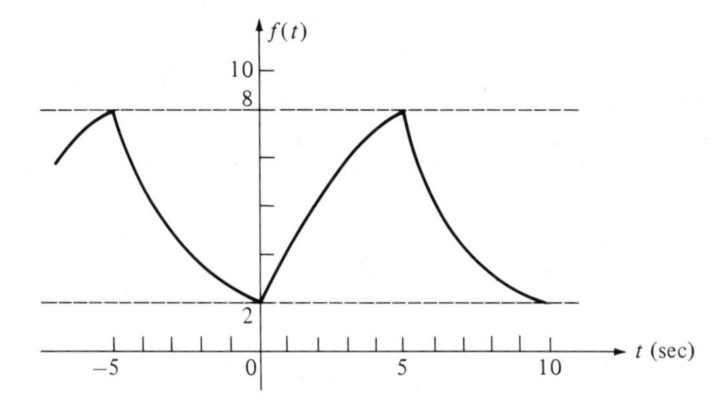

Figure 2.43
Answer to Exercise 2.4, part 2.

Computer Simulation

Simulation of periodic rectangular trapezoidal, and triangular waveforms can be achieved through use of the PULSE source. The PULSE source can be made periodic by specifying a period (PERIOD) parameter which causes the pulse to be repeated throughout the simulation (until TSTOP). Use of the PULSE source to generate a periodic waveform is illustrated in SPICE 2.7.

```
*******09/29/88 ********  SPICE 2G.6    3/15/83 ********15:14:44*****
SPICE 2.7
****     INPUT LISTING                        TEMPERATURE =   27.000 DEG C
******************************************************************************
 R 1 0 1
 V 1 0 PULSE(-1 2 0 1 2 2 6)
 .TRAN .5 12
 .PLOT TRAN V(1)
 .END
*******09/29/88 ********  SPICE 2G.6    3/15/83 ********15:14:44*****
SPICE 2.7
****     INITIAL TRANSIENT SOLUTION           TEMPERATURE =   27.000 DEG C
******************************************************************************
  NODE   VOLTAGE
 (  1)   -1.0000
     VOLTAGE SOURCE CURRENTS
     NAME       CURRENT
     V         1.000D+00
     TOTAL POWER DISSIPATION   1.00D+00  WATTS
*******09/29/88 ********  SPICE 2G.6    3/15/83 ********15:14:44*****
SPICE 2.7
****     TRANSIENT ANALYSIS                   TEMPERATURE =   27.000 DEG C
******************************************************************************

     TIME       V(1)
                  -1.000D+00       0.000D+00       1.000D+00       2.000D+00  3.000D+00
                            - - - - - - - - - - - - - - - - - - - - - - - - - - - - - -
 0.000D+00 -1.000D+00 *              .              .              .              .
 5.000D-01  5.000D-01 .              .      *       .              .              .
 1.000D+00  2.000D+00 .              .              .              *              .
 1.500D+00  2.000D+00 .              .              .              *              .
 2.000D+00  2.000D+00 .              .              .              *              .
 2.500D+00  2.000D+00 .              .              .              *              .
 3.000D+00  2.000D+00 .              .              .              *              .
 3.500D+00  1.250D+00 .              .              .   *          .              .
 4.000D+00  5.000D-01 .              .      *       .              .              .
 4.500D+00 -2.500D-01 .          *   .              .              .              .
 5.000D+00 -1.000D+00 *              .              .              .              .
 5.500D+00 -1.000D+00 *              .              .              .              .
 6.000D+00 -1.000D+00 *              .              .              .              .
 6.500D+00  5.000D-01 .              .      *       .              .              .
 7.000D+00  2.000D+00 .              .              .              *              .
 7.500D+00  2.000D+00 .              .              .              *              .
 8.000D+00  2.000D+00 .              .              .              *              .
 8.500D+00  2.000D+00 .              .              .              *              .
 9.000D+00  2.000D+00 .              .              .              *              .
 9.500D+00  1.250D+00 .              .              .   *          .              .
 1.000D+01  5.000D-01 .              .      *       .              .              .
 1.050D+01 -2.500D-01 .          *   .              .              .              .
 1.100D+01 -1.000D+00 *              .              .              .              .
 1.150D+01 -1.000D+00 *              .              .              .              .
 1.200D+01 -1.000D+00 *              .              .              .              .
                            - - - - - - - - - - - - - - - - - - - - - - - - - - - - - -

          JOB CONCLUDED
```

The PWL source can also be used to generate periodic waveforms, but the user must enter breakpoints through several cycles of the waveform sufficient to fill the total simulation time.

2.6 Sinusoidal Waveforms

Readers of this text will recognize immediately the importance of sinusoidal waveforms in electric circuits. A large body of the work involved with analysis and design of electric circuits is devoted to sinusoidal waveforms.

Sinusoidal-Waveform Fundamentals

Sinusoidal waveforms (Figure 2.44) are generated by the function

$$f(t) = A \cdot \sin[\omega(t - t_0)] \tag{2.14}$$

Aperiodic sinusoidal waveforms may be generated by appropriate combinations of the sine function and unit step functions; however, the sine function alone is periodic by definition.

The parameters associated with sinusoidal functions are:

A = amplitude (or peak value),
T = period in seconds,
ω = angular frequency in radians per second (rad/sec),
$= 2\pi/T$,
t_0 = displacement time in seconds.

With a displacement time t_0 of zero, the waveform passes through the origin with a positive slope [Figure 2.44(a)]. A negative displacement time shifts the sinusoid to the left [Figure 2.44(b)], and a positive displacement time shifts it to the right [Figure 2.44(c)]. The terms **lead** and **lag** are commonly used to relate sinusoids, based on the time the waveform passes through zero with a positive slope. The sinusoid in Figure 2.44(a) leads the sinusoid in Figure 2.44(c) and lags the sinusoid in Figure 2.44(b). Correspondingly, (b) leads (a) and (c), and (c) lags (a) and (b).

Displacement times of $\pm T/4$ [Figure 2.44(d) and (e)] and $\pm T/2$ [Figure 2.44(f)] have special significance:

Flashback

$$\sin\left[x \pm \frac{\pi}{2}\right] = \pm\cos x$$

$$\sin[x \pm \pi] = -\sin x$$

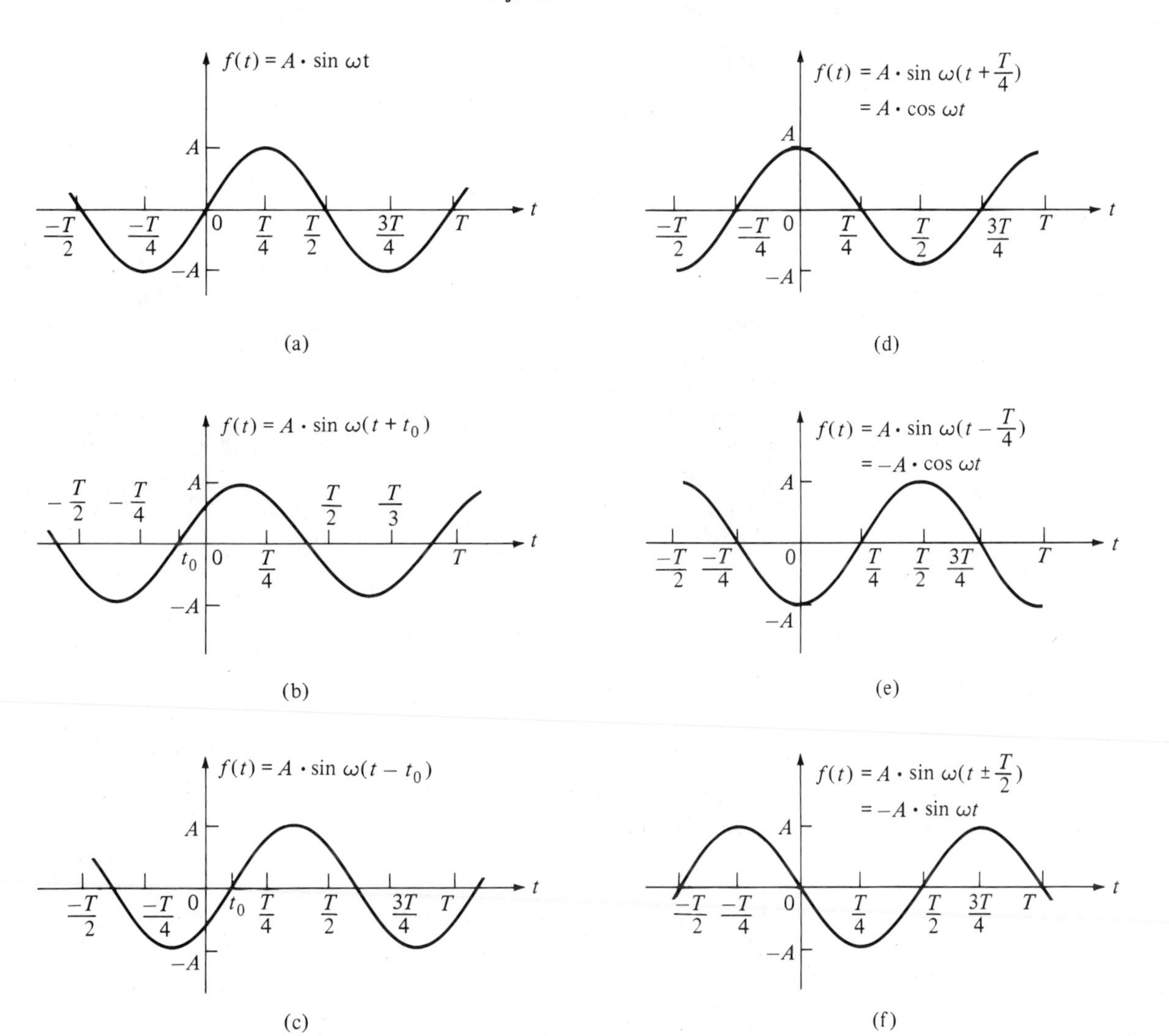

Figure 2.44 Sinusoidal waveforms.

Thus

$$f(t) = A \cdot \sin\left[\omega\left(t \pm \frac{T}{4}\right)\right] = A \sin\left[\omega t \pm \frac{\omega T}{4}\right]$$

Substitution of $2\pi/T$ for ω in the term $\omega T/4$ yields

$$f(t) = A \cdot \sin\left[\omega t \pm \frac{\pi}{2}\right] = \pm A \cdot \cos \omega t$$

Similarly, for a displacement time $t_0 = \pm T/2$,

$$f(t) = A \cdot \sin[\omega t \pm \pi] = -A \cdot \sin \omega t$$

The sine function is also written

$$f(t) = A \cdot \sin[\omega t + \phi] \tag{2.15}$$

where $\phi = -\omega t_0$ is called the **phase angle**. Phase angle is usually specified in degrees, which leads to confusion because of the mixed units in the argument of the sine function. The reader is cautioned to convert both parts of the argument (ωt and ϕ) to common units before evaluating a sine function.

Frequency is often specified in terms of **cycle frequency** f, rather than angular frequency ω. The symbol f in this context should not be confused with the function $f(t)$. The standard unit of cyclic frequency is the **hertz** (Hz); one hertz equals one cycle per second. Cyclic frequency is related to period and angular frequency:

$$f = \frac{1}{T} = \frac{\omega}{2\pi}$$

so

$$f(t) = A \cdot \sin[2\pi f t + \phi] \tag{2.16}$$

Observe in Figure 2.45(a) that the greatest rate of change (slope) in a sinusoidal waveform occurs when the waveform is crossing the t-axis, that is, when the generating function has a value of zero. The magnitude of the slope is the same at the crossing point whether the waveform is crossing from negative to positive or from positive to negative. The slope is zero when the waveform is at its peak, either positive or negative. Analytically, the slope of a function at a point is equal to its derivative at that point, and since

Flashback

$$\frac{d}{dx}\sin au = a \cdot \cos au \cdot \frac{d}{dx}u$$

$$\frac{d}{dx}\cos au = -a \cdot \sin au \cdot \frac{d}{dx}u$$

then

$$\frac{d}{dt}A \cdot \sin[\omega(t - t_0)] = A\omega \cdot \cos[\omega(t - t_0)] \tag{2.17}$$

Figure 2.45
Derivative of a sinusoidal waveform.

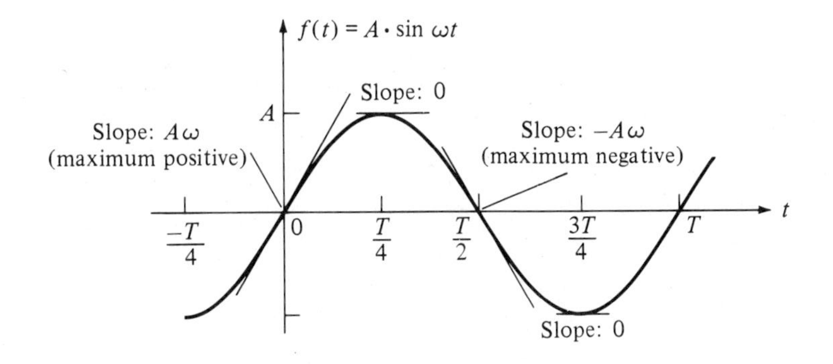

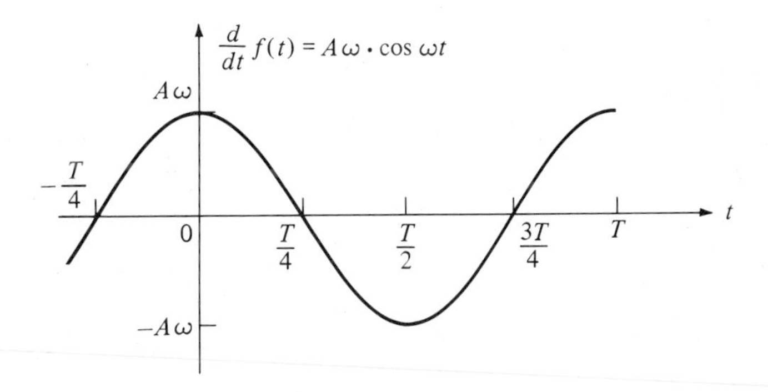

From Equation (2.16), the slope of a sinusoid at its t-axis crossing is $A\omega$, the product of its amplitude and its frequency [Figure 2.45(b)].

Example 2.15

Problem:

1. Write the generating function for the waveform shown in Figure 2.46 in the form of Equation (2.14) and Equation (2.15).
2. Calculate frequency in hertz.
3. Write the expression for the derivative of the generating function with respect to time.

Solution:

1. From the waveform graph,

$$A = 140 \text{ V}$$
$$t_0 = 10 \text{ msec}$$
$$T = 70 - 10 \text{ msec} = 60 \text{ msec}$$

Figure 2.46
Example 2.15.

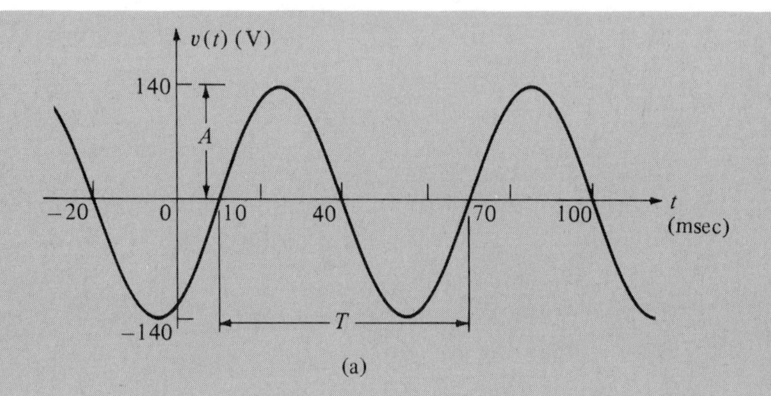

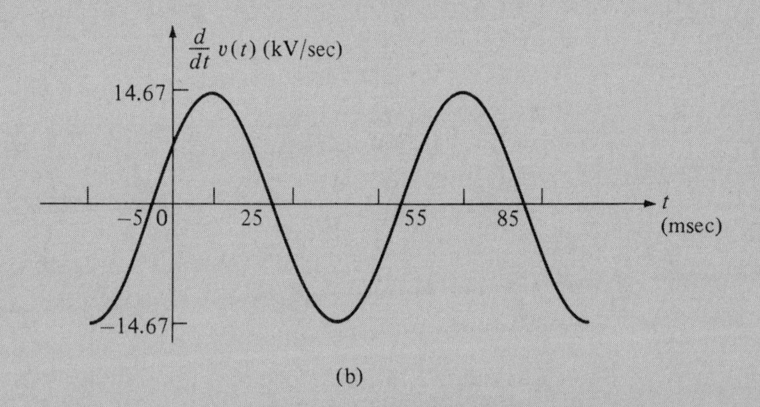

a. Calculate the angular frequency:

$$\omega = 2\pi/T = 104.7 \text{ rad/sec}$$

b. Write $v(t)$ in the form of Equation (2.14):

$$v(t) = 140 \cdot \sin[104.7(t - 0.010)] \text{ V}$$

c. Calculate the phase angle:

$$\phi = -\omega t_0 = -\pi/3 \text{ rad} = -60°$$

d. Write $v(t)$ in the form of Equation (2.15):

$$v(t) = 140 \cdot \sin[104.7t - 630°] \text{ V}$$

2. Calculate frequency in hertz:

$$f = 1/T = 16.67 \text{ Hz}$$

3. Calculate the derivative of $f(t)$:

$$\begin{aligned}\frac{d}{dt}f(t) &= 140 \cdot \cos[104.7(t - 0.010)] \cdot \frac{d}{dt}[104.7(t - 0.010)] \\ &= 140(104.7) \cdot \cos[104.7(t - 0.010)] \\ &= 14.67 \cdot \cos[104.7(t - 0.010)] \text{ kV/sec}\end{aligned}$$

Figure 2.47 Switched sinusoidal waveforms.

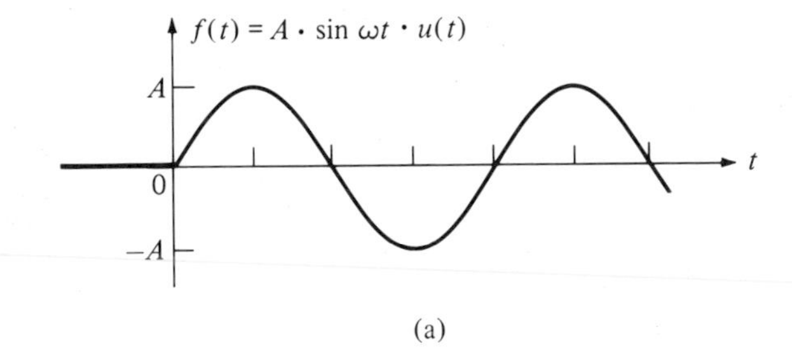

(a)

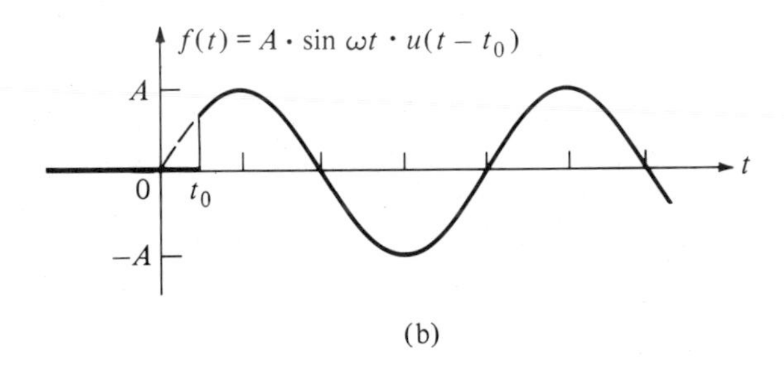

(b)

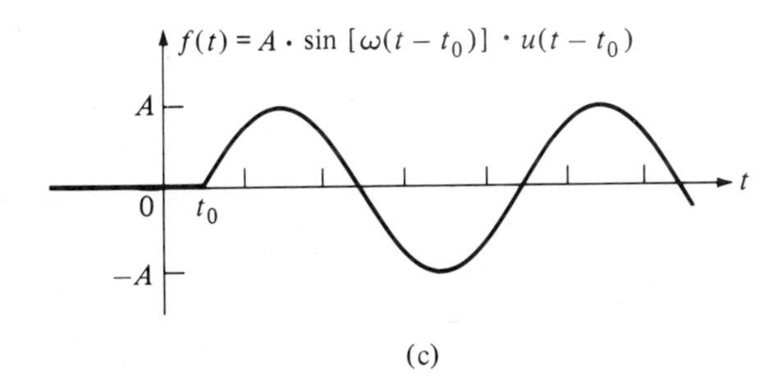

(c)

Piecewise-Sinusoidal Functions

A sinusoidal function can be multiplied by a unit step function with the result that the value of the function is zero prior to the switching time t_0 (Figure 2.47). Such a function is technically aperiodic, but it possesses periodic properties for $t > t_0$. The value of a sinusoidal function can be made zero beyond some switching time by adding a second sinusoidal component which cancels the first (Figure 2.48). The second component leads or lags the first component by $T/2$, or π radians, and the two components are said to be **180° out of phase**.

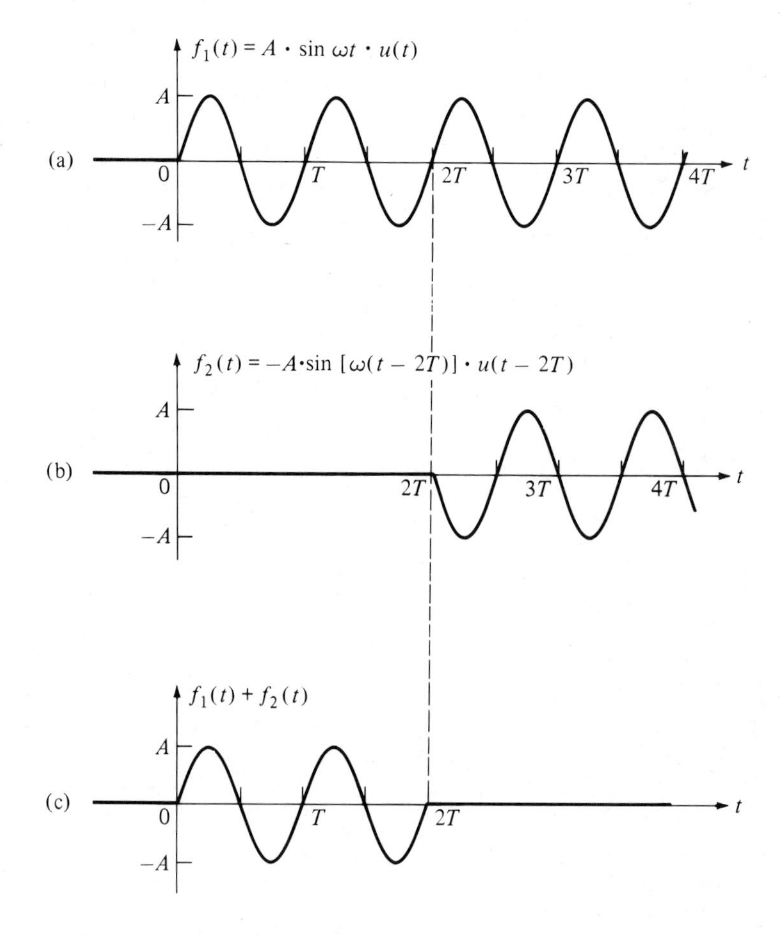

Figure 2.48
Combining switched sinusoidal generating functions.

Example 2.16

Problem:

Write the generating function for a 50-Hz, 30-V peak, sinusoidal voltage waveform which begins at $t = 15$ msec and continues for 2.5 cycles.

Solution:

1. Calculate the parameters of the sinusoid:

$$T = 1/f = 0.02 \text{ sec} = 20 \text{ msec}$$

$$\omega = 2\pi f = 314 \text{ rad/sec}$$

2. Write the function for the first component [Figure 2.49(a)]:

$$v_1(t) = 30 \cdot \sin[314(t - 0.015)] \cdot u(t - 0.015) \text{ V}$$

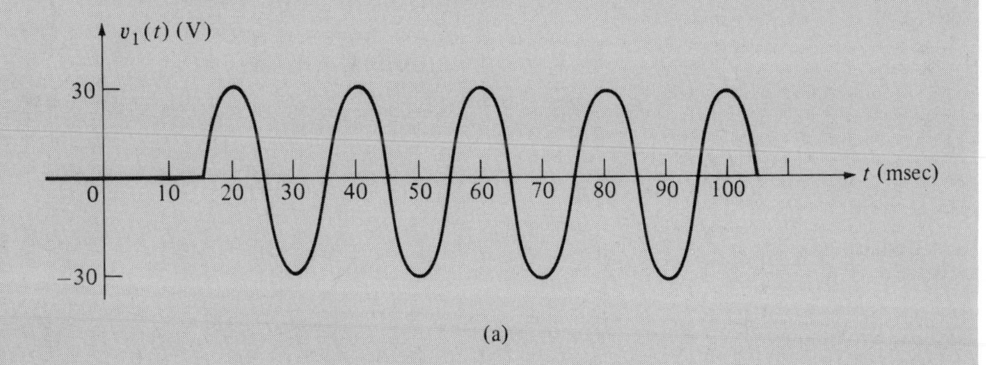

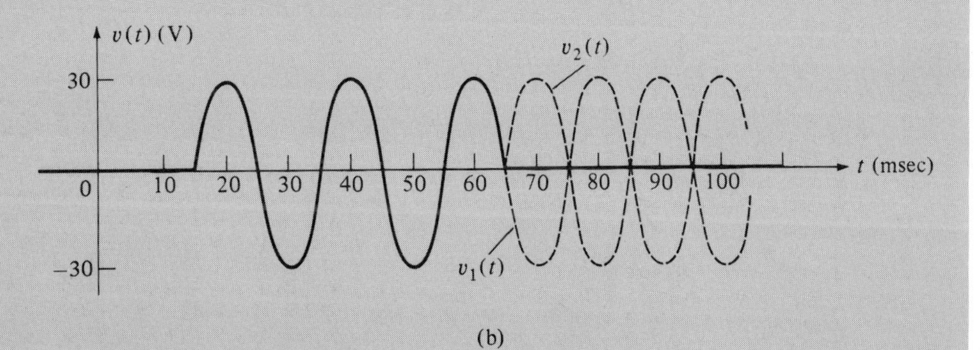

Figure 2.49 Example 2.16.

3. The second component [Figure 2.49(b)] must be 180° out of phase with and have a switching time 2.5 cycles, or 50 msec, later than the first component:

$$v_2(t) = 30 \cdot \sin[314(t - 0.065)] \cdot u(t - 0.065) \text{ V}$$

4. Add the two components to obtain $v(t)$:

$$v(t) = 30\{\sin[314(t - 0.015)] \cdot u(t - 0.015) + \sin[314(t - 0.065)] \cdot u(t - 0.065)\} \text{ V}$$

Another type of piecewise-sinusoidal waveform is the **rectified** sinusoid (Figure 2.50) associated with **diodes** and other electronic **rectifiers**. The composition of one cycle of a **half-wave** rectified sinusoid is shown in Figure 2.51. The first peak of $f_{HW}(t)$ is generated by $f_1(t)$ [Figure 2.51(a)],

$$f_1(t) = A \cdot \sin \omega t \cdot u(t)$$

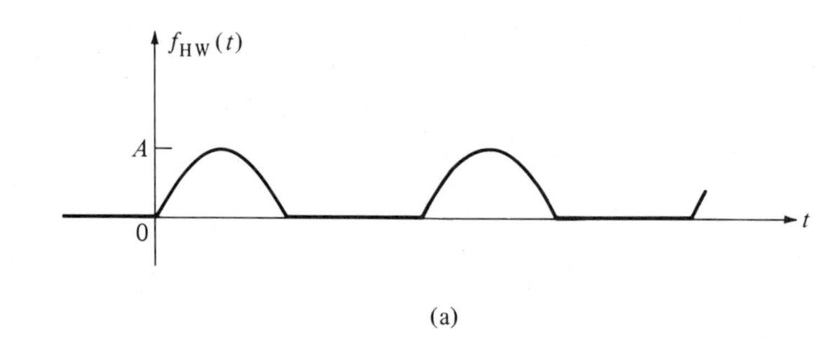

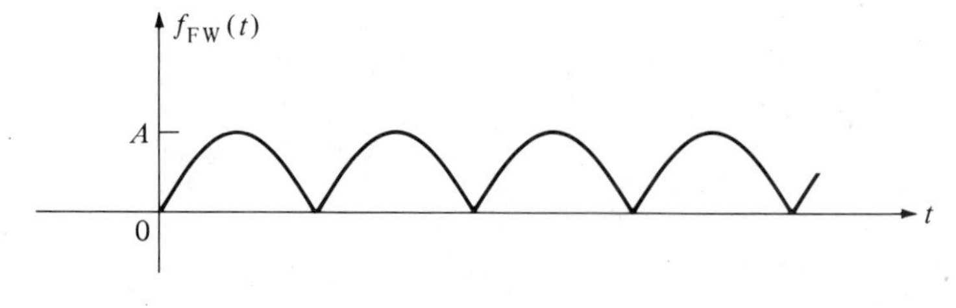

Figure 2.50
Rectified sinusoidal waveforms.
(a) Half-wave.
(b) Full-wave.

Figure 2.51
Composition of a half-wave rectified sinusoid.

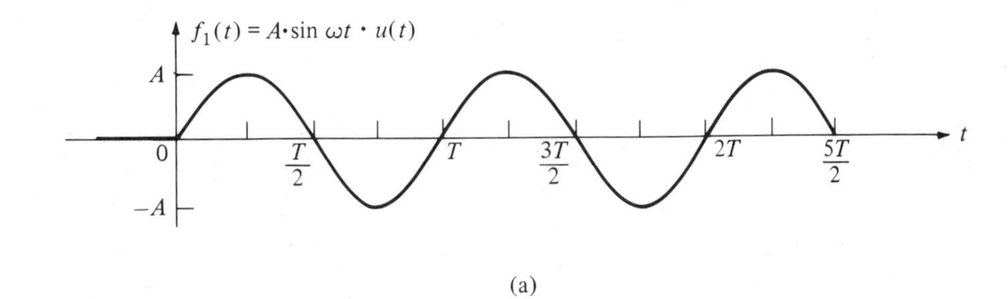

(a)

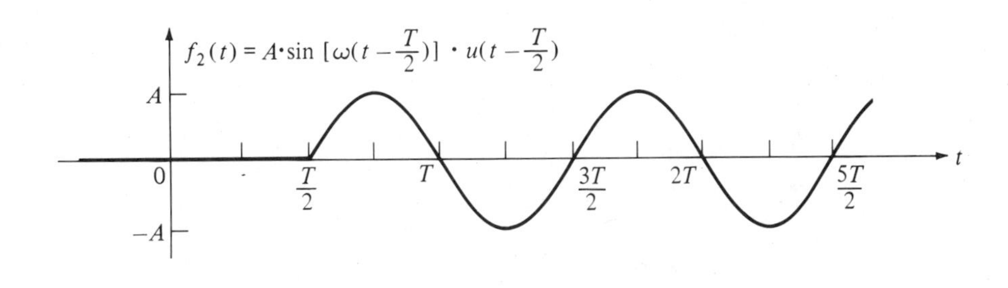

(b)

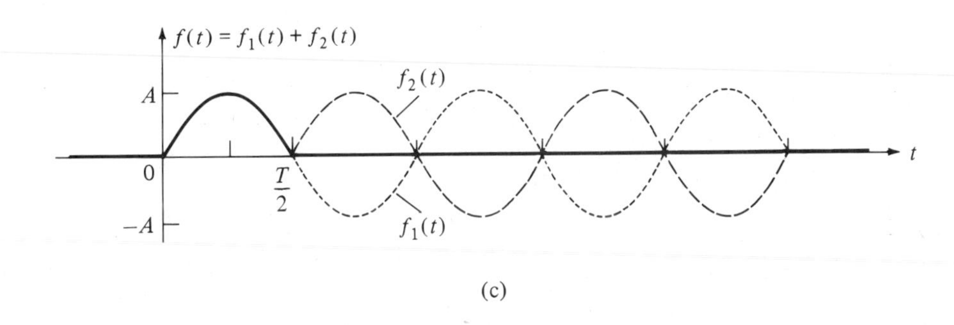

(c)

For $t > T/2$, cancellation of $f_1(t)$ is provided by $f_2(t)$ [Figure 2.51(b)]:

$$f_2(t) = A \cdot \sin\left[\omega\left(t - \frac{T}{2}\right)\right] \cdot u\left(t - \frac{T}{2}\right)$$

Therefore, for the first cycle of a half-wave rectified sinusoid,

$$f(t) = A \cdot \sin \omega t \cdot u(t) + A \cdot \sin\left[\omega\left(t - \frac{T}{2}\right)\right] \cdot u\left(t - \frac{T}{2}\right) \qquad (2.18)$$

Figure 2.52
Composition of a full-wave rectified sinusoid.

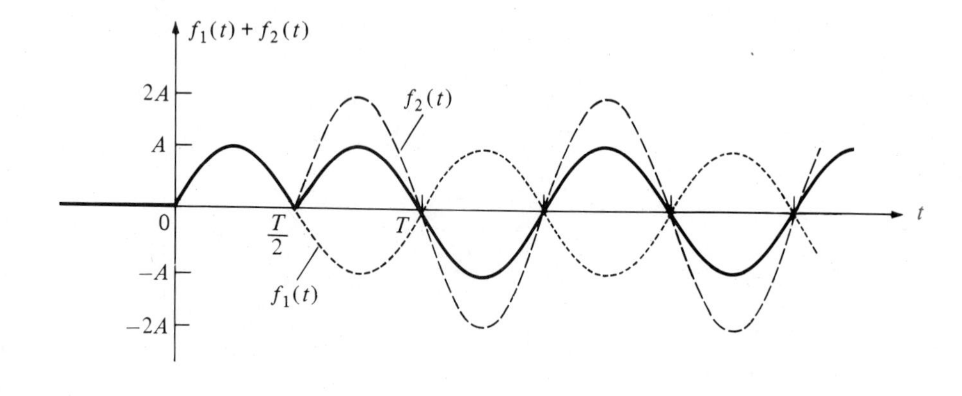

(a)

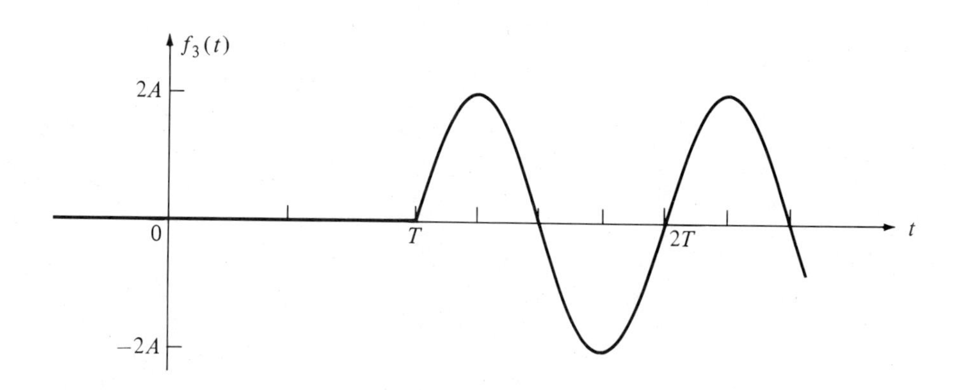

(b)

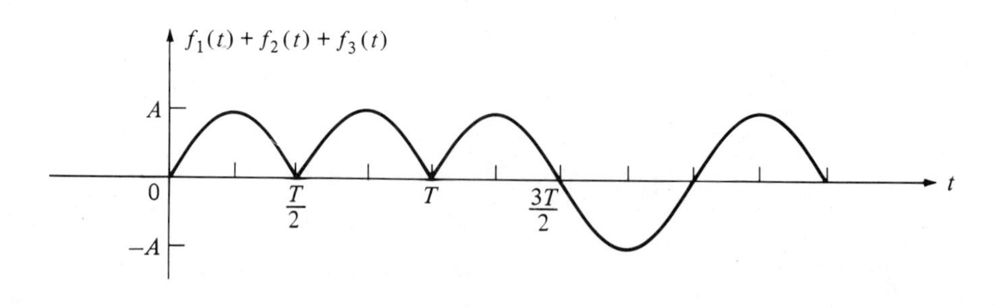

(c)

From Equations (2.13a) and (2.18), the function which generates a periodic half-wave rectified sinusoid in the domain $t > 0$ is

$$f_{\text{HW}}(t) = A \sum_{n=0}^{\infty} \sin\left[\omega\left(t - \frac{nT}{2}\right)\right] \cdot u\left(t - \frac{nT}{2}\right) \tag{2.19}$$

The composition of the **full-wave** rectified sinusoid is shown in Figure 2.52. The first peak of $f_{\text{FW}}(t)$ is generated by $f_1(t)$:

$$f_1(t) = A \cdot \sin \omega t \cdot u(t)$$

The second component, $f_2(t)$, must not only cancel $f_1(t)$ for $t > T/2$, it must also generate the second peak [Figure 2.52(a)], so

$$f_2(t) = 2A \cdot \sin\left[\omega\left(t - \frac{T}{2}\right)\right] \cdot u\left(t - \frac{T}{2}\right)$$

However, as can be seen in Figure 2.52(a), addition of $f_2(t)$ with its double amplitude introduces negative peaks for $t > T$, and it is necessary to add $f_3(t)$, and so on. The function which generates a periodic full-wave rectified sinusoid in the domain $t > 0$ is

$$f_{\text{FW}}(t) = A \cdot \sin \omega t \cdot u(t) + 2A \sum_{n=1}^{\infty} \sin\left[\omega\left(t - \frac{nT}{2}\right)\right] \cdot u\left(t - \frac{nT}{2}\right) \tag{2.20}$$

Note that the index n for the summation begins with 0 for the half-wave generating function and with 1 for the full-wave generating function.

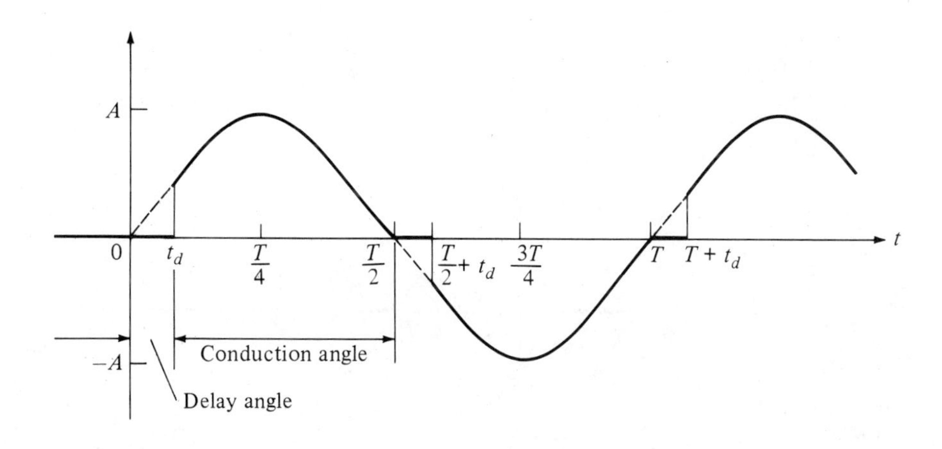

Figure 2.53 Sinusoidal waveform with variable conduction angle.

Figure 2.54
Composition of a sinusoid with variable conduction angle

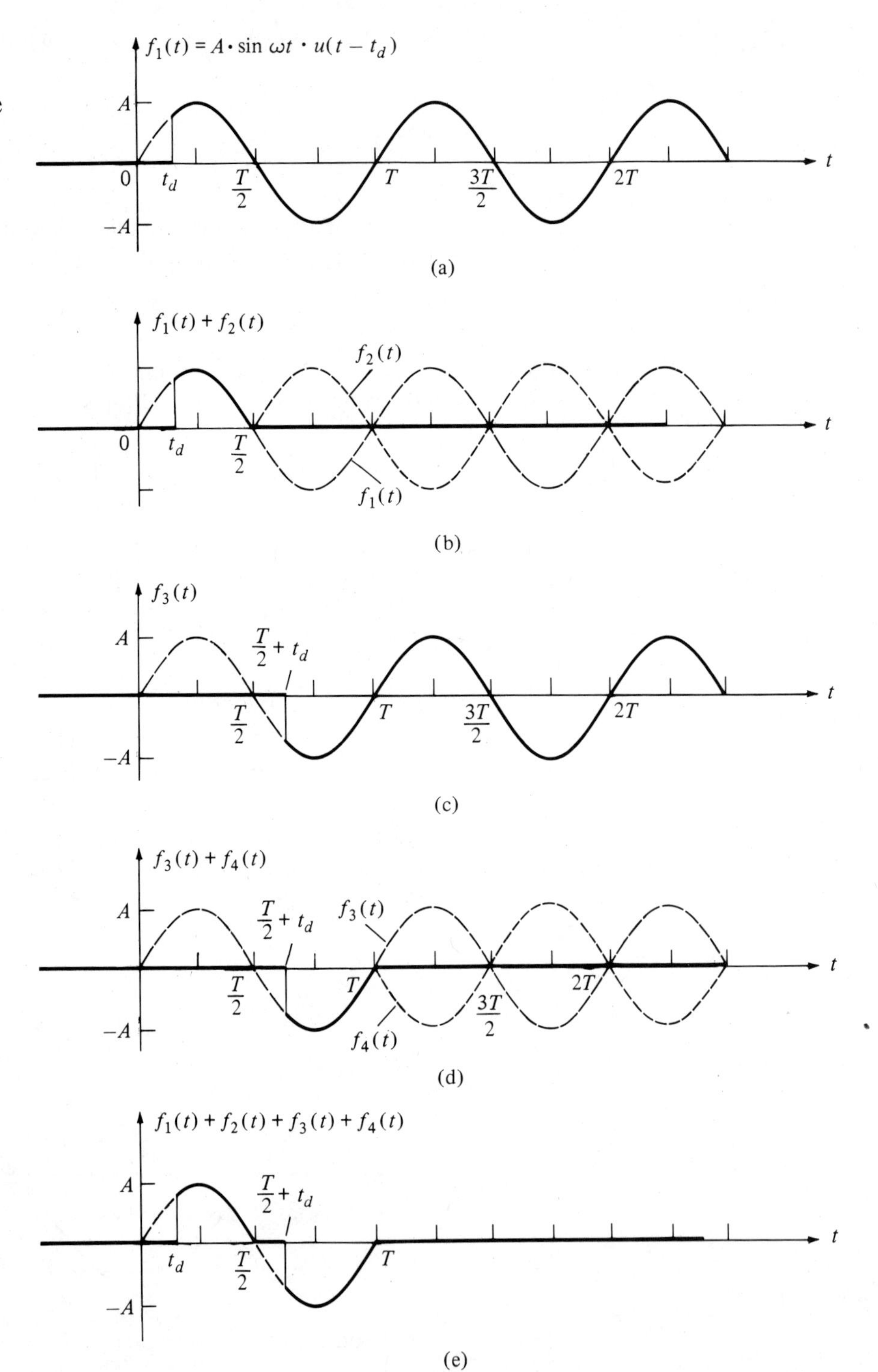

The last type of piecewise-sinusoidal waveform to be considered is that associated with electronic devices, such as the **triac**, which have **variable conduction angles**. This type of waveform has a value of zero for part of each half cycle of the waveform (Figure 2.53). The angular equivalent of the time between zero and t_d is called the **delay angle**, and the angular equivalent of the time between t_d and $T/2$ is called the **conduction angle**. The composition of a sinusoid with variable conduction angle is shown in Figure 2.54. The first component, $f_1(t)$, in Figure 2.54(a) is an unshifted sinusoid with a switching time of t_d,

$$f_1(t) = A \cdot \sin[\omega t] \cdot u(t - t_d)$$

The second component [Figure 2.54(b)] cancels the first component for $t > T/2$:

$$f_2(t) = A \cdot \sin\left[\omega\left(t - \frac{T}{2}\right)\right] \cdot u\left(t - \frac{T}{2}\right)$$

By trignometric identity,

$$\sin\left[\omega\left(t - \frac{T}{2}\right)\right] = -\sin \omega t$$

so

$$f_2(t) = A \cdot \sin\left[\omega - \frac{T}{2}\right]$$

The third component [Figure 2.54(c)] has a switching time of $T/2 + t_d$:

$$f_3(t) = A \cdot \sin[\omega t] \cdot u\left(t - \left(\frac{T}{2} + t_d\right)\right)$$

The fourth component [Figure 2.54(d)] cancels $f_3(t)$ for $t > T$:

$$f_4(t) = A \cdot \sin[\omega(t - T)] \cdot u(t - T)$$

By trignometric identity, $\sin[\omega(t - T)] = \sin \omega t$, so

$$f_4(t) = A \cdot \sin \omega t \cdot u(t - T)$$

The generating function for one cycle of a sinusoid with variable conduction angle is

$$f(t) = f_1(t) + f_2(t) + f_3(t) + f_4(t)$$
$$= A \cdot \sin \omega t \cdot \left[u(t) + u(t - t_d) + u\left(t - \left(\frac{T}{2} + t_d\right)\right) - u(t - T) \right]$$

and the generating function for the periodic waveform in the domain $t > 0$ is

$$f(t) = A \sum_{n=0}^{\infty} \sin[\omega(t - nT)] \cdot \left[u(t - nT) + u(t - (nT + t_d)) + u\left(t - \left(nT + \frac{T}{2} + t_d\right)\right) - u(t - (nT + T)) \right]$$
$$= A \cdot \sin \omega t \cdot \sum_{n=0}^{\infty} \left[u(t - nT) + u(t - (nT + t_d)) + u\left(t - \left(nT + \frac{T}{2} + t_d\right)\right) - u(t - (nT + T)) \right] \quad (2.21)$$

Example 2.17

Problem:

Write the generating function for

a. a half-wave rectified sinusoidal waveform,
b. a full-wave rectified sinusoidal waveform,
c. a sinusoidal waveform with a 135° conduction angle, based on a 60-Hz sinusoid with a 160-V amplitude.

Solution:

1. Calculate parameters of the sinusoid:

$$T = 1/f \simeq 16.67 \text{ msec}$$
$$\omega = 2\pi f \simeq 377 \text{ rad/sec}$$
$$t_d = \frac{180° - 135°}{180°} T \simeq 2.083 \text{ msec}$$

2. Use Equation (2.19) to write the generating function for the half-wave rectified sinusoidal waveform:

$$v_{HW}(t) = 160 \sum_{n=0}^{\infty} \sin[377(t - 8.333 \times 10^{-3}n)] \cdot u(t - 8.333 \times 10^{-3}n) \text{ V}$$

3. Use Equation (2.20) to write the generating function for the full-wave rectified sinusoidal waveform:

$$v_{FW}(t) = 160 \cdot \sin 377t \cdot u(t) + 320 \sum_{n=1}^{\infty} \sin[377(t - 8.333 \times 10^{-3}n)] \cdot u(t - 8.333 \times 10^{-3}n) \text{ V}$$

4. Use Equation (2.21) to write the generating function for the sinusoidal waveform with a conduction angle of 135°:

$$v(t) = 160 \cdot \sin 377t \cdot \sum_{n=0}^{\infty} \big[u(t - 16.67 \times 10^{-3}n) + u(t - (16.67n + 2.083) \times 10^{-3}) + u(t - (16.67n + 10.42) \times 10^{-3}) - u(t - (16.67n + 16.67) \times 10^{-3})\big] \text{ V}$$

Computer Simulation

The sinusoidal (SIN) source is inherently periodic and will be repeated throughout the simulation. SPICE 2.8 illustrates use of the SIN source to generate a delayed, offset sinusoid. SPICE 2.9 illustrates the superposition of three SIN sources to generate three cycles of a full-wave rectified sinusoid. In both examples the damping factor (DAMPING) has been omitted, so that it defaults to zero, and the waveforms do not decay. The beginning SPICE user should be careful not to confuse the SIN source, used for transient analysis, with the AC source used for steady-state sinusoidal analysis in a later chapter.

```
*******09/29/88 ********  SPICE 2G.6    3/15/83 ********15:22:35*****
SPICE 2.8
****     INPUT LISTING                      TEMPERATURE =   27.000 DEG C
*******************************************************************************
 R 1 0 1
 V 1 0 SIN(2 1 1K .5M)
 .TRAN .1M 2M
 .PLOT TRAN V(1)
 .END
*******09/29/88 ********  SPICE 2G.6    3/15/83 ********15:22:36*****
SPICE 2.8
****     INITIAL TRANSIENT SOLUTION         TEMPERATURE =   27.000 DEG C
*******************************************************************************
  NODE   VOLTAGE
 (  1)    2.0000
     VOLTAGE SOURCE CURRENTS
     NAME       CURRENT
     V        -2.000D+00
     TOTAL POWER DISSIPATION   4.00D+00  WATTS
*******09/29/88 ********  SPICE 2G.6    3/15/83 ********15:22:36*****
SPICE 2.8
****     TRANSIENT ANALYSIS                 TEMPERATURE =   27.000 DEG C
*******************************************************************************

     TIME       V(1)
                    1.000D+00     1.500D+00     2.000D+00     2.500D+00  3.000D+00
                          - - - - - - - - - - - - - - - - - - - - - - - - - - - - -
 0.000D+00  2.000D+00 .             .             *             .           .
 1.000D-04  2.000D+00 .             .             *             .           .
 2.000D-04  2.000D+00 .             .             *             .           .
 3.000D-04  2.000D+00 .             .             *             .           .
 4.000D-04  2.000D+00 .             .             *             .           .
 5.000D-04  2.000D+00 .             .             *             .           .
 6.000D-04  2.588D+00 .             .             .             . *         .
 7.000D-04  2.944D+00 .             .             .             .         * .
 8.000D-04  2.951D+00 .             .             .             .          *.
 9.000D-04  2.583D+00 .             .             .             . *         .
 1.000D-03  2.000D+00 .             .             *             .           .
 1.100D-03  1.417D+00 .           * .             .             .           .
 1.200D-03  1.049D+00 .*            .             .             .           .
 1.300D-03  1.056D+00 . *           .             .             .           .
 1.400D-03  1.412D+00 .           * .             .             .           .
 1.500D-03  2.000D+00 .             .             *             .           .
 1.600D-03  2.588D+00 .             .             .             . *         .
 1.700D-03  2.944D+00 .             .             .             .         * .
 1.800D-03  2.951D+00 .             .             .             .          *.
 1.900D-03  2.583D+00 .             .             .             . *         .
 2.000D-03  2.000D+00 .             .             *             .           .
                          - - - - - - - - - - - - - - - - - - - - - - - - - - - - -

          JOB CONCLUDED
```

```
*******09/29/88 ********  SPICE 2G.6    3/15/83 ********15:54:31*****
SPICE 2.9
****     INPUT LISTING                    TEMPERATURE =   27.000 DEG C
********************************************************************************
 R 1 0 1
 V1 1 2 SIN(0 1 1 0)
 V2 2 3 SIN(0 2 1 .5)
 V3 3 0 SIN(0 2 1 1)
 .TRAN .05 1.5
 .PLOT TRAN V(1)
 .END
*******09/29/88 ********  SPICE 2G.6    3/15/83 ********15:54:31*****
SPICE 2.9
****     INITIAL TRANSIENT SOLUTION       TEMPERATURE =   27.000 DEG C
********************************************************************************
  NODE   VOLTAGE     NODE   VOLTAGE     NODE   VOLTAGE
 (  1)     .0000    (  2)     .0000    (  3)     .0000
    VOLTAGE SOURCE CURRENTS
    NAME       CURRENT
    V1        0.000D+00
    V2        0.000D+00
    V3        0.000D+00
    TOTAL POWER DISSIPATION   0.00D+00  WATTS
*******09/29/88 ********  SPICE 2G.6    3/15/83 ********15:54:31*****
SPICE 2.9
****     TRANSIENT ANALYSIS               TEMPERATURE =   27.000 DEG C
********************************************************************************

    TIME       V(1)
               -5.000D-01     0.000D+00     5.000D-01     1.000D+00  1.500D+00
                          - - - - - - - - - - - - - - - - - - - - - - - - - - - -
 0.000D+00  0.000D+00 .            *            .             .             .
 5.000D-02  3.077D-01 .            .        *   .             .             .
 1.000D-01  5.872D-01 .            .            . *           .             .
 1.500D-01  8.062D-01 .            .            .        *    .             .
 2.000D-01  9.470D-01 .            .            .            *.             .
 2.500D-01  9.991D-01 .            .            .             *             .
 3.000D-01  9.476D-01 .            .            .            *.             .
 3.500D-01  8.056D-01 .            .            .        *    .             .
 4.000D-01  5.873D-01 .            .            . *           .             .
 4.500D-01  3.078D-01 .            .        *   .             .             .
 5.000D-01 -4.554D-18 .            *            .             .             .
 5.500D-01  3.082D-01 .            .        *   .             .             .
 6.000D-01  5.864D-01 .            .            . *           .             .
 6.500D-01  8.054D-01 .            .            .        *    .             .
 7.000D-01  9.487D-01 .            .            .            *.             .
 7.500D-01  9.975D-01 .            .            .             *             .
 8.000D-01  9.468D-01 .            .            .            *.             .
 8.500D-01  8.071D-01 .            .            .        *    .             .
 9.000D-01  5.863D-01 .            .            . *           .             .
 9.500D-01  3.076D-01 .            .        *   .             .             .
 1.000D+00 -1.349D-18 .            *            .             .             .
 1.050D+00  3.082D-01 .            .        *   .             .             .
 1.100D+00  5.864D-01 .            .            . *           .             .
 1.150D+00  8.054D-01 .            .            .        *    .             .
 1.200D+00  9.487D-01 .            .            .            *.             .
 1.250D+00  9.975D-01 .            .            .             *             .
 1.300D+00  9.468D-01 .            .            .            *.             .
 1.350D+00  8.071D-01 .            .            .        *    .             .
 1.400D+00  5.863D-01 .            .            . *           .             .
 1.450D+00  3.076D-01 .            .        *   .             .             .
 1.500D+00 -9.304D-18 .            *            .             .             .
                          - - - - - - - - - - - - - - - - - - - - - - - - - - - -
```

2.7 SUMMARY

Depictions of the time variations of electrical quantities such as voltage, current, and power are commonly called waveforms. The analysis of waveforms, or the development of mathematical functions to generate the waveforms, is frequently carried out in a piecewise fashion by dealing with the waveform during discrete intervals of the time domain.

The principal mathematical tool for piecewise analysis and synthesis is the unit step function $u(t)$ and its derivative, the unit impulse function $\delta(t)$. The unit step function serves as the mathematical equivalent of a switch, selecting or rejecting certain terms of the function at indicated times.

Piecewise-linear waveforms are those generated by functions composed of a linear combination of terms containing only linear or unit step function factors. Piecewise linear waveforms include rectangular, trapezoidal, and triangular waveforms.

Piecewise-exponential waveforms are those generated by functions composed of a linear combination of terms containing only exponential or unit-step-function factors. Exponential waveforms usually represent the growth and decay of voltage or current within a circuit containing capacitors and inductors.

Sinusoidal waveforms appear in many situations in electrical circuit analysis and design. The basis for generation of sinusoidal waveforms is the sine function, which may be multiplied by the unit step function to provide switching of the sinusoidal waveform. Use of the unit step function, and the principle of shifting, permits special variations of the sinusoid such as the waveform obtained from rectifiers or devices with variable conduction angle.

Waveforms may be aperiodic, occurring only once in the time domain, or they may be periodic, repeating themselves at regular intervals, called the period, throughout the time domain. Periodic waveforms are generated by functions which consist of an infinite sum of terms, each of which represents the aperiodic waveform shifted one period from the previous term.

2.8 Terms

amplitude
aperiodic
argument
breakpoint
conduction angle
cycle
cyclic frequency
decay
delay angle
delta function
discontinuity
exponential decay function

exponential growth function
full-wave rectified sinusoid
generating function
growth
half-wave rectified sinusoid
intercept–slope ratio
lag
lead
period
periodic function
periodic waveform
phase angle
piecewise
piecewise-exponential
piecewise-linear
piecewise-sinusoidal
radian frequency
ramp function
rectangular waveform
rectified sinusoid
sawtooth waveform
shift
sinusoidal waveform
switching time
time constant
time domain
trapezoidal waveform
triangular waveform
unit impulse function
unit step function
variable conduction angles
waveform
waveform analysis
waveform generation
waveform synthesis

Problems

SPICE

2-1. Graph each waveform generated by $f(t) = g(t) \cdot u(t - t_0)$.

- ***a.** $g(t) = 2,\ t_0 = 0.$
- **b.** $g(t) = 2,\ t_0 = 2.$
- ***c.** $g(t) = 2t,\ t_0 = 0.$
- **d.** $g(t) = 2t,\ t_0 = 2.$
- ***e.** $g(t) = 2t - 5,\ t_0 = 0.$
- **f.** $g(t) = 2t - 5,\ t_0 = 2.$

2-2. Write the generating function $f(t) = g(t) \cdot u(t - t_0)$ for each waveform:

a.

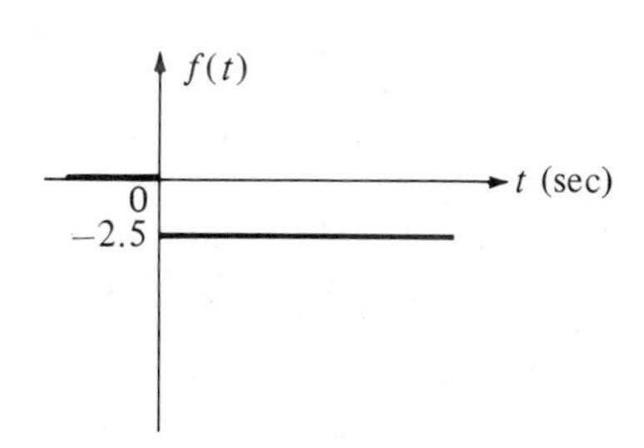

*b.

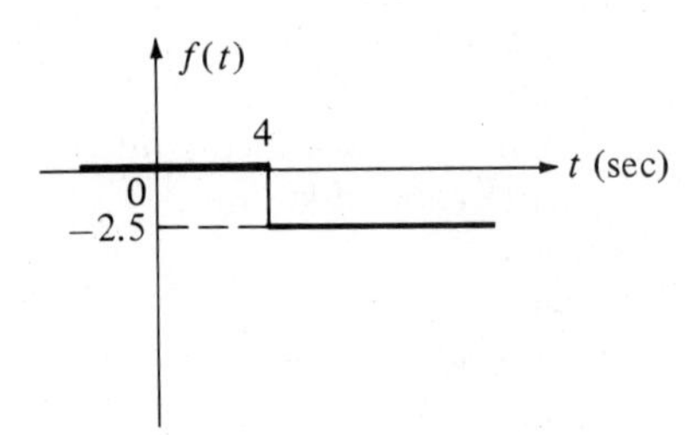

c.

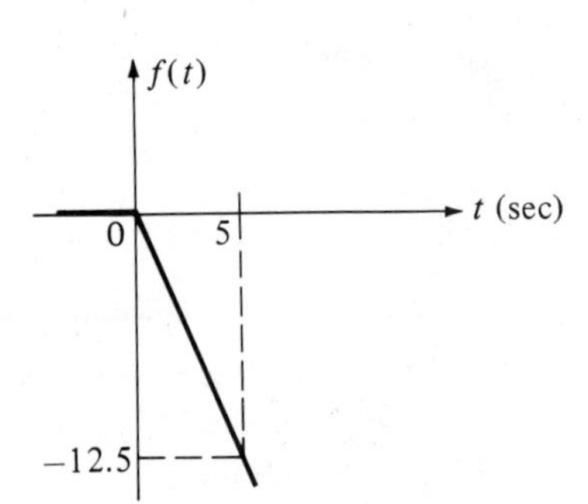

*d.

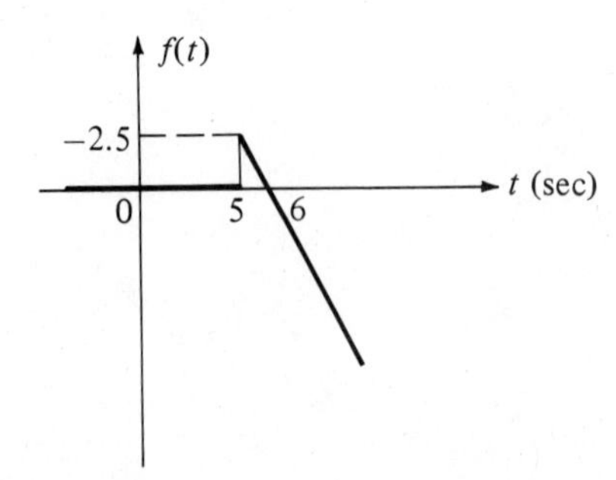

***e.**

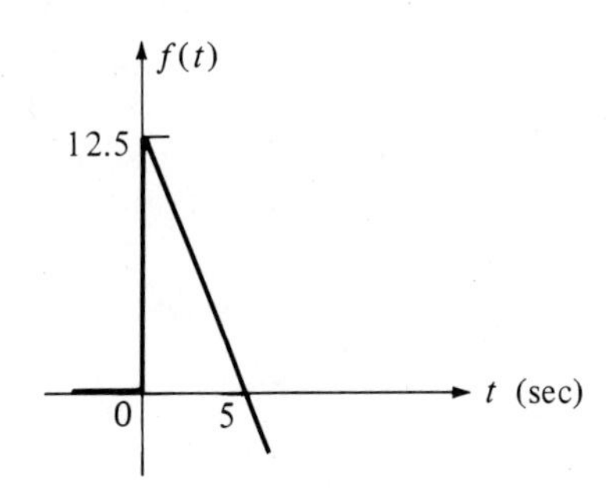

f.

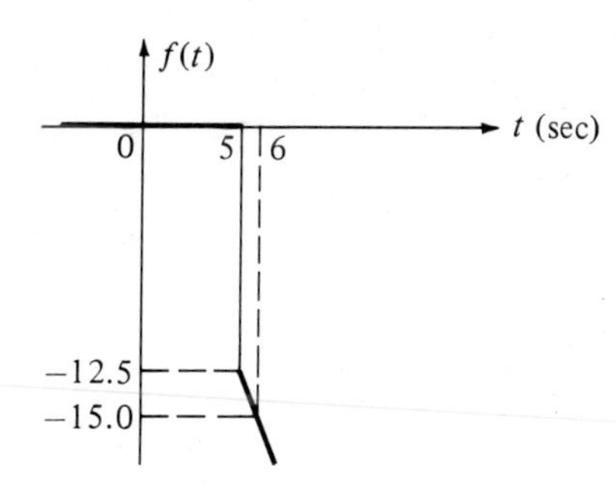

***2-3.** Graph the derivative, $df(t)/dt$, of each generating function in Problem 2-2.

2-4. Graph the waveform generated by each impulse function.

a. $f(t) = 2 \cdot \delta(t)$.
***b.** $f(t) = \delta(t - 2)$.
c. $f(t) = 4 \cdot \delta(t + 2)$.
***d.** $f(t) = 2[\delta(t + 2) - 0.5 \cdot \delta(t) + \delta(t - 2)]$.

2-5. Write the generating function for each waveform:

***a.**

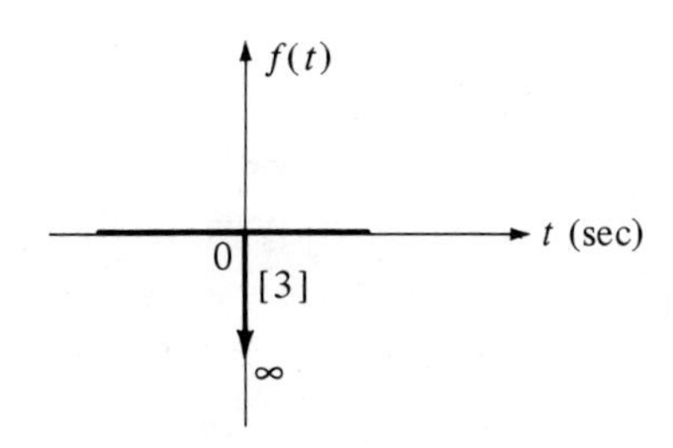

b.

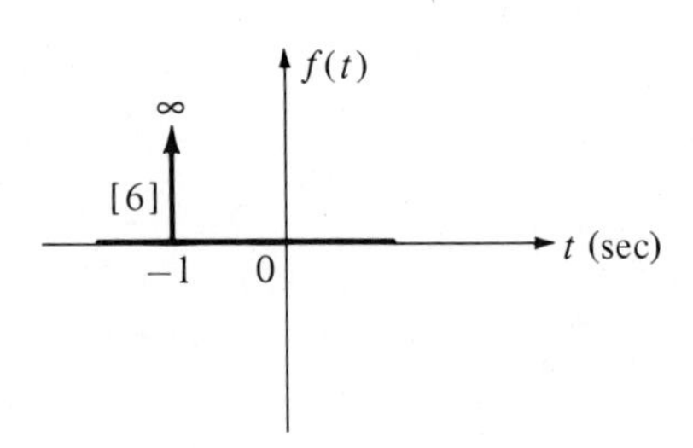

***c.**

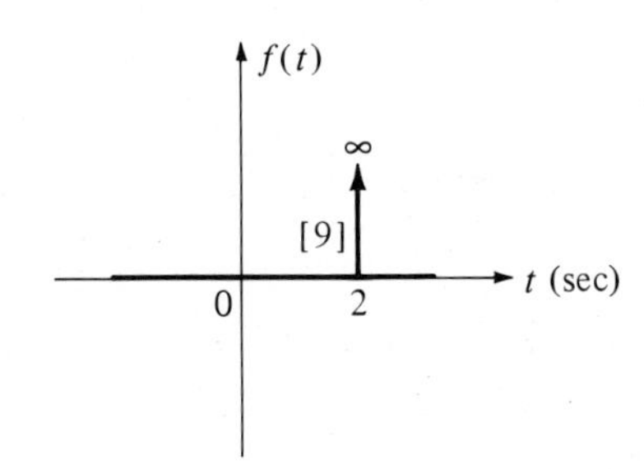

d.

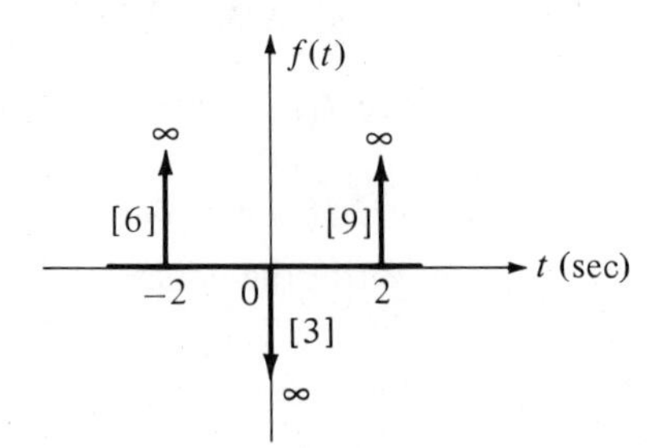

SPICE

2-6. Graph the waveform generated by each step function:

a. $f(t) = -4 \cdot u(-3 - t)$.

***b.** $f(t) = -4 \cdot u(-t)$.

c. $f(t) = -4 \cdot u(3 - t)$.

2-7. Write the generating function for each waveform.

***a.**

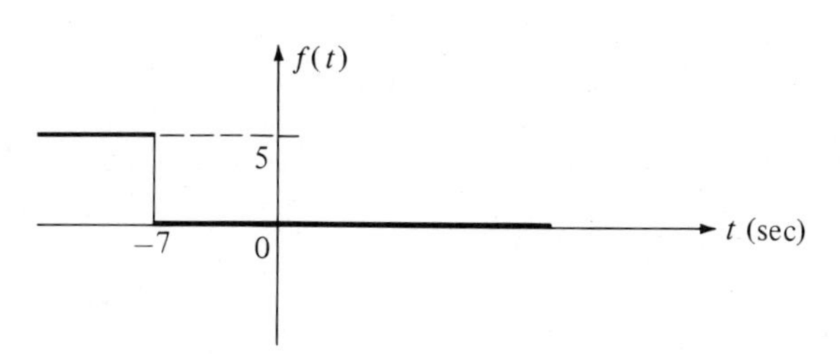

b.

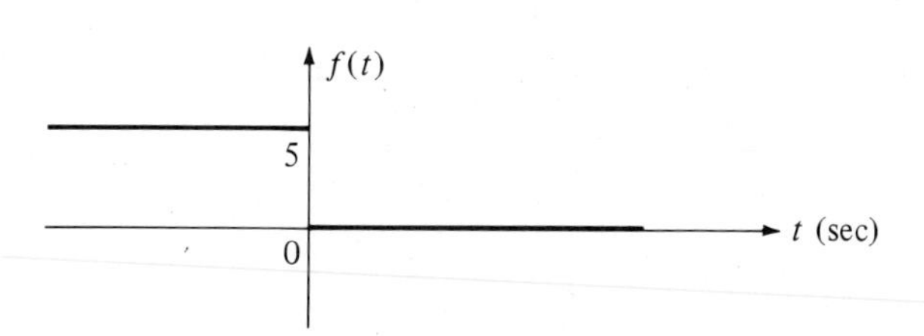

***c.**

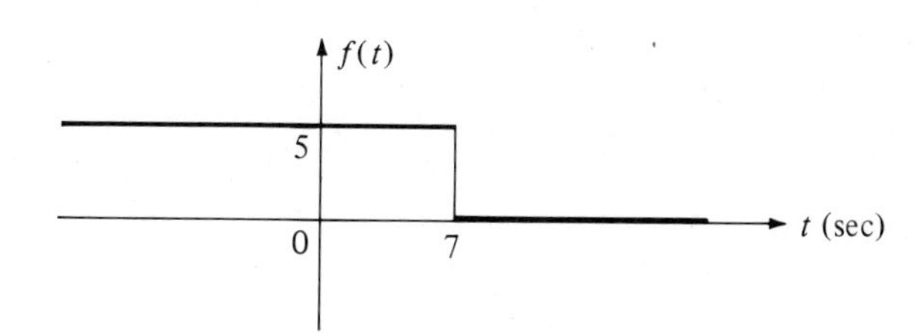

SPICE

2-8. Graph the rectangular waveform generated by each function:

a. $f(t) = 30[u(t - 2) - 2 \cdot u(t - 5) + u(t - 7)]$.

***b.** $v(t) = 5[u(t) - u(t - 6 \times 10^{-6})]$ V.

c. $i(t) = -150[u(t - 0.003) - 2 \cdot u(t - 0.005)]$ mA.

2-9. Write the generating function for each waveform.

***a.**

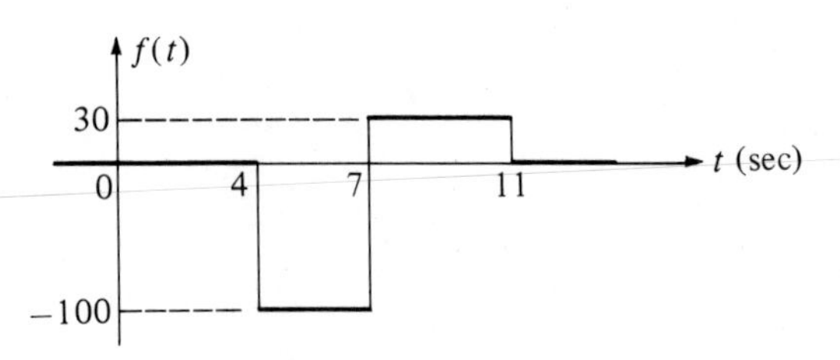

b.

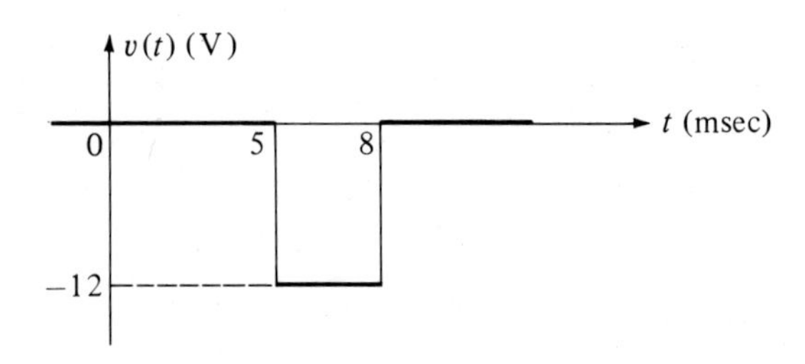

***c.**

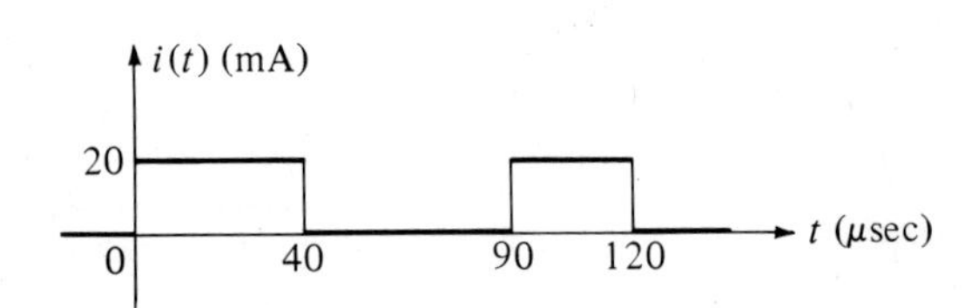

SPICE

2-10. Graph the piecewise-linear waveform generated by each function.

a. $f(t) = 4t \cdot u(t) - 4(t - 1.5)u(t - 1.5) - 6(t - 5)u(t - 5) + 6(t - 6)u(t - 6)$.

***b.** $v(t) = 2.2 \times 10^6[(t - 2 \times 10^{-6})u(t - 2 \times 10^{-6}) - 2(t - 3.5 \times 10^{-6})u(t - 3.5 \times 10^{-6}) + (t - 5 \times 10^{-6})u(t - 5 \times 10^{-6})]$ V.

c. $i(t) = -60 \cdot u(t - 0.02) + 1000[(t - 0.02)u(t - 0.02) - (t - 0.08)u(t - 0.08)]$ mA.

2-11. Write the generating function for each waveform:

***a.**

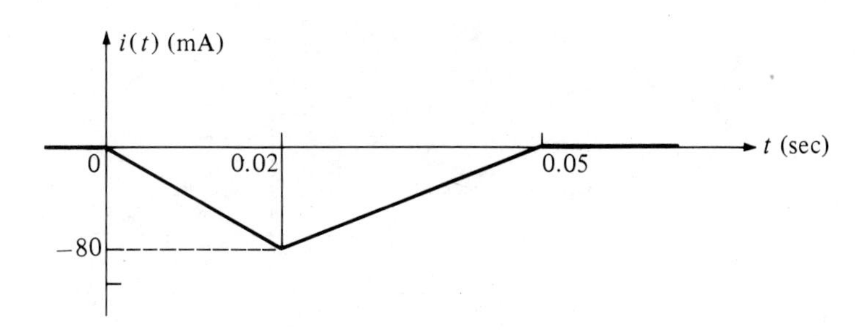

b.

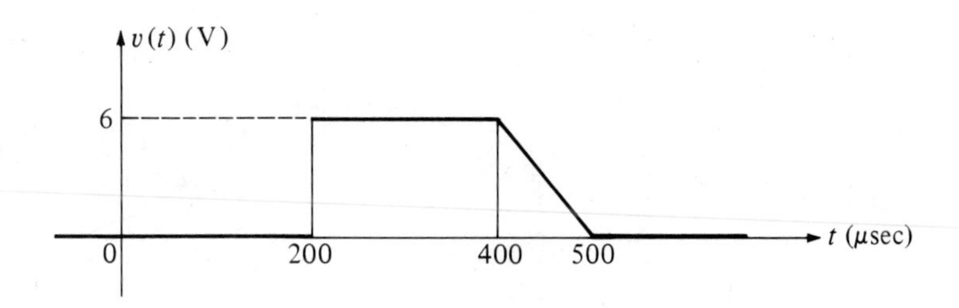

***c.**

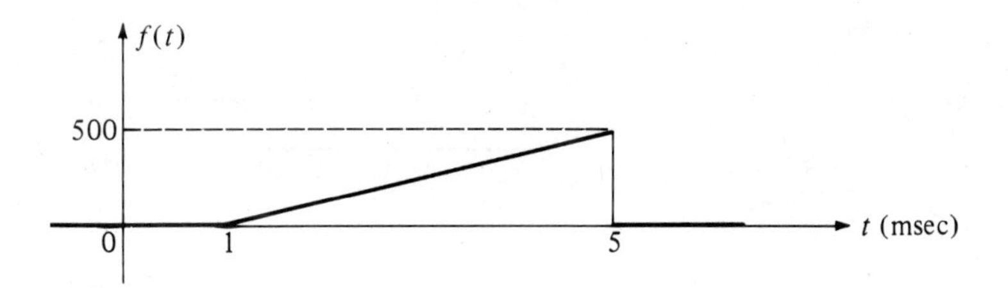

SPICE

2-12. Graph the exponential waveform generated by each function:

a. $i(t) = -88(1 - \exp[-200t])u(t - 0.05)$ mA.

***b.** $v(t) = (1 + 5 \cdot \exp[-300 \times 10^3 t])u(t - 10 \times 10^{-6})$ V.

c. $f(t) = -200 \cdot \exp[-t] \cdot u(t - 1)$.

2-13. Write the generating function for each waveform:

***a.**

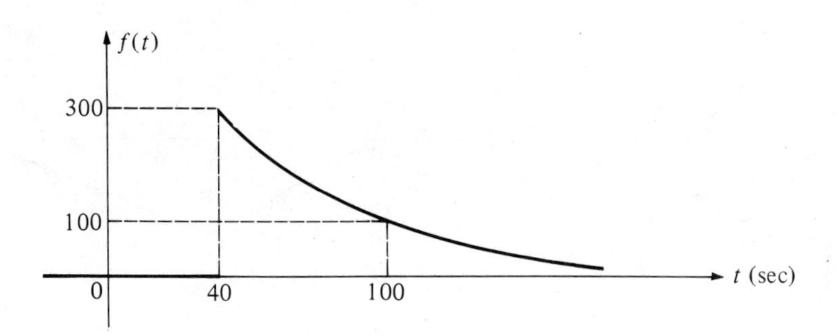

b.

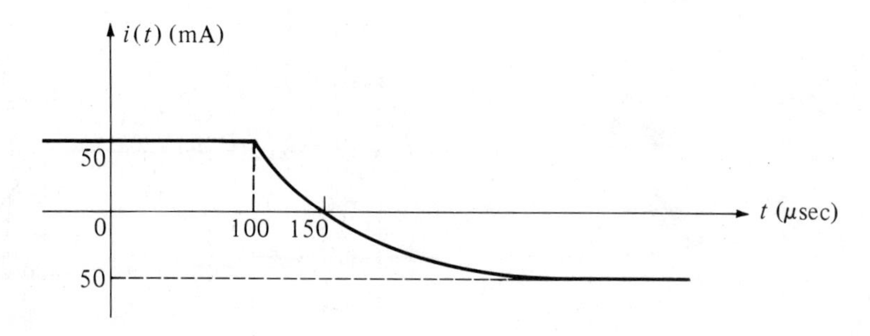

***c.**

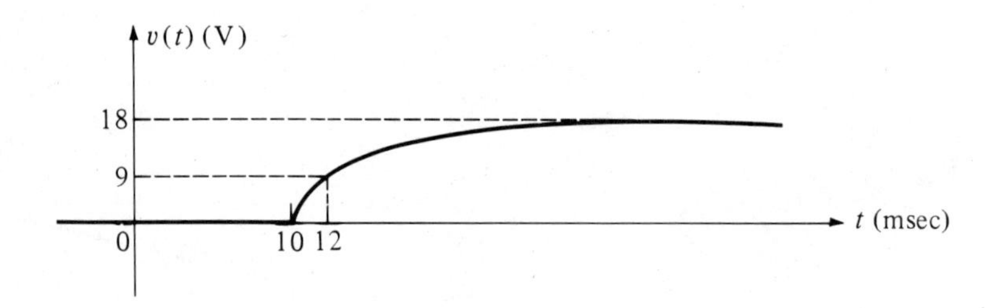

SPICE

2-14. Graph the exponential waveform generated by each function:

a. $f(t) = 3 + 7\{1 - \exp[-(t - 5)/3]\}u(t - 5)$
$-14\{1 - \exp[-(t - 10)/3]\}u(t - 10)$.

***b.** $v(t) = -5\{1 - \exp[-(t - 0.02)/0.01]\}u(t - 0.02)$
$+5\{1 - \exp[-2]\}u(t - 0.04) + 5 \cdot \exp[-2]$
$\cdot\{1 - \exp[-(t - 0.04)/0.01]\}u(t - 0.04)$ V.

2-15. Write the generating function for each waveform:

***a.**

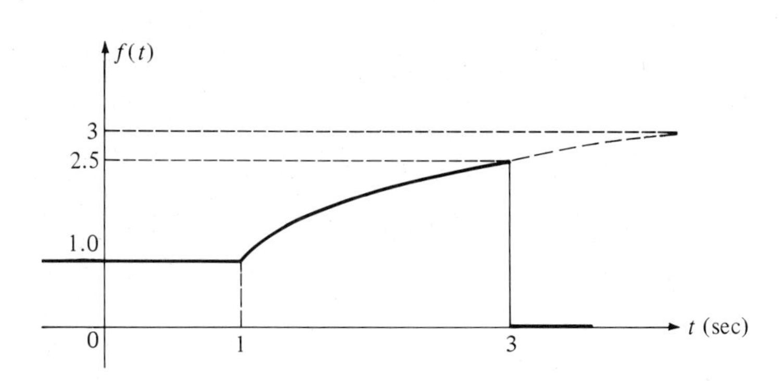

b.

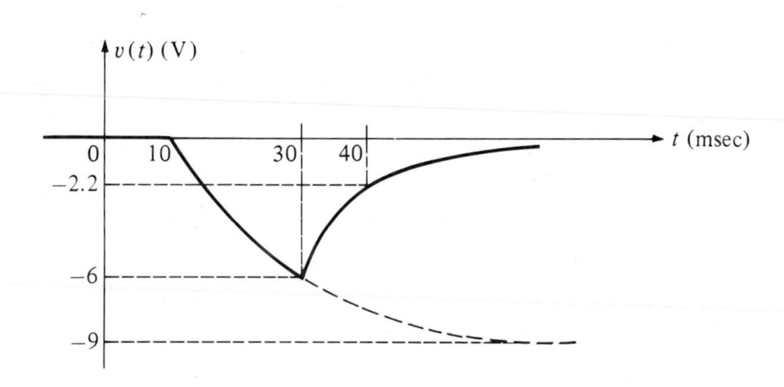

SPICE

2-16. Graph the periodic waveform generated by each function:

a. $i(t) = 2.5\Sigma_{n=1}^{\infty}[u(t + 0.1n) - 2 \cdot u(t - 0.05 + 0.1n) + u(t - 0.1 + 0.1n)]$ mA.

***b.** $f(t) = \Sigma_{n=1}^{\infty}[(t + 6n)u(t + 6n) - (t - 1 + 6n)u(t - 1 + 6n) - u(t - 3 + 6n)]$.

c. $v(t) \simeq 10\Sigma_{n=1}^{\infty}\{(1 - \exp[-(t + 40n)])u(t + 40n) - [1 - \exp[-(t - 20 + 40n)])u(t - 20 + 40n)\}$ V.

This approximation assumes that the voltage grows to maximum value and decays to zero during each cycle.

2-17. Write the generating function for each waveform:

*__a.__

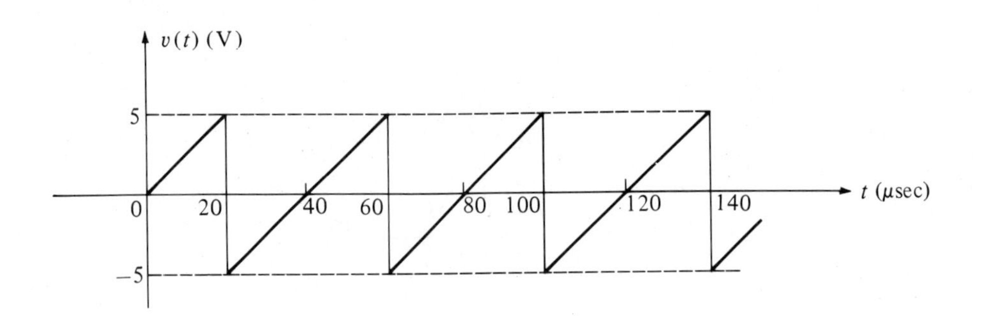

b.

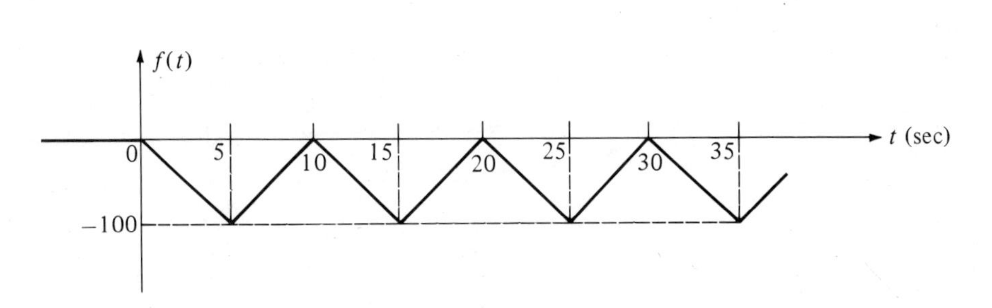

*__c.__ Assume that the current decays to zero during each half cycle.

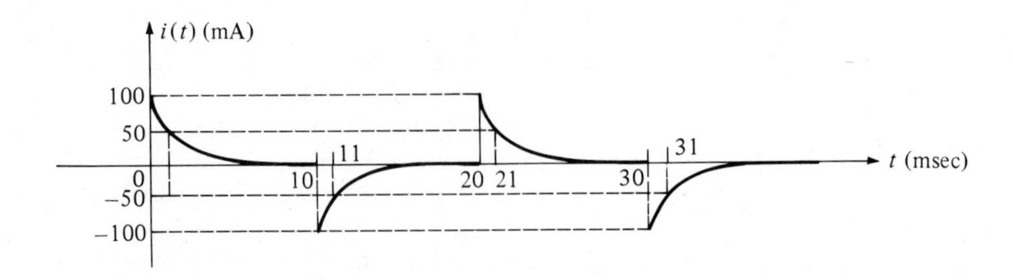

SPICE

2-18. Graph the sinusoidal function generated by each function:
a. $f(t) = 8 \cdot \sin[0.6(t - 7)] \cdot u(t - 7)$.
***b.** $v(t) = 200 \cdot \sin[3000t + 30°]$ V.
c. $i(t) = -60 + 30 \cdot \cos 500t \cdot u(t - 0.005)$ A.

2-19. Write the generating function for each waveform:
***a.**

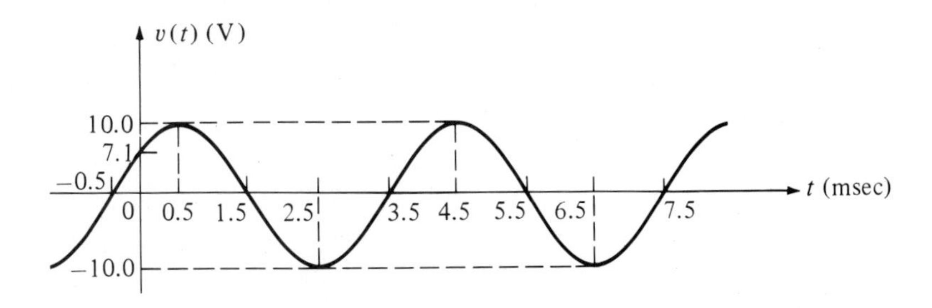

b.

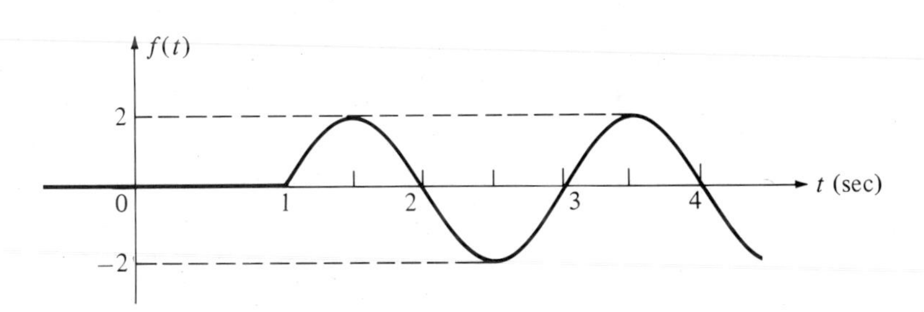

***c.**

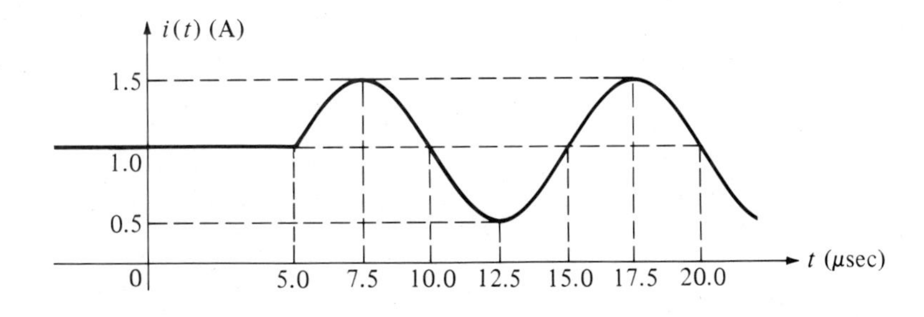

***2-20.** Graph and write the generating function for $v(t)$ in the circuit shown. Assume $t_d = T/3$, and no forward resistance in the variable-conduction-angle device.

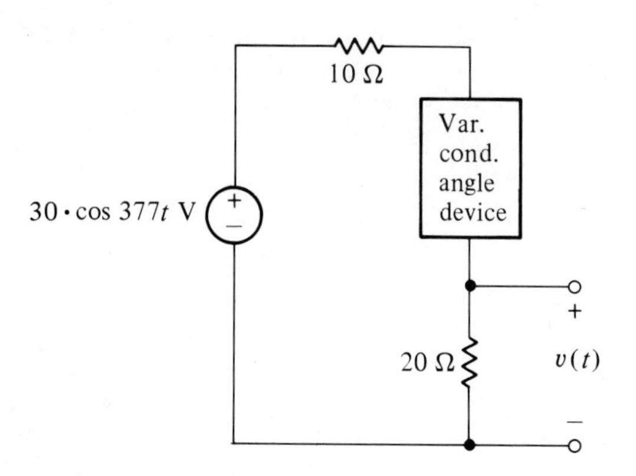

3 Time-Domain Analysis

3.1 INTRODUCTION

Although the topic of this book is circuit analysis by transform methods, some attention will be given in this chapter to **time-domain analysis** of simple resistor–capacitor (R–C) and resistor–inductor (R–L) circuits. Time-domain analysis refers to determining the circuit response (either voltage or current) by direct solution of the time-dependent differential equations which result when the Kirchhoff laws are applied to circuits containing resistors, energy storage elements (i.e., capacitors and inductors), switches (either mechanical or electronic), and time-dependent sources. This brief investigation of time-domain analysis is motivated by two considerations:

1. Time-domain analysis provides a straightforward procedure for determining the response of a large number of R–C and R–L circuits.
2. Application of the Kirchhoff laws in the time domain is a first step in developing the transform methods which will be presented in the following chapters.

3.2 Terminal Characteristics of Energy Storage Elements

We saw in Chapter 1 that the terminal characteristic of an element is the relationship between voltage and current at the terminals of the element. The reader will recall that Ohm's law defines the terminal characteristic of a linear resistor. In this section, we develop and examine in some detail the terminal characteristic of the energy storage elements—namely the linear capacitor and the linear inductor.

Terminal Characteristics of a Capacitor

The **capacitance** (C) of a capacitor is the ratio between the charge stored in the capacitor and the voltage at the terminals of the capacitor at any instant, and is written

$$C = \frac{q}{v}, \qquad \text{or} \quad q = Cv$$

If C, the ratio of charge to voltage, remains constant when the capacitor is operated over a range of charge and voltage values, then the capacitor is said to be **linear** for that operating range [Figure 3.1(b)], so

$$q(t) = C \cdot v(t) \tag{3.1}$$

as charge and voltage vary with time within the linear operating range of the capacitor.

Differentiation of Equation (3.1) with respect to time yields

$$\frac{d}{dt}q(t) = C\frac{d}{dt}v(t)$$

The derivative (or rate of change) of charge with respect to time is *current*;

Figure 3.1 Terminal characteristics of a capacitor.

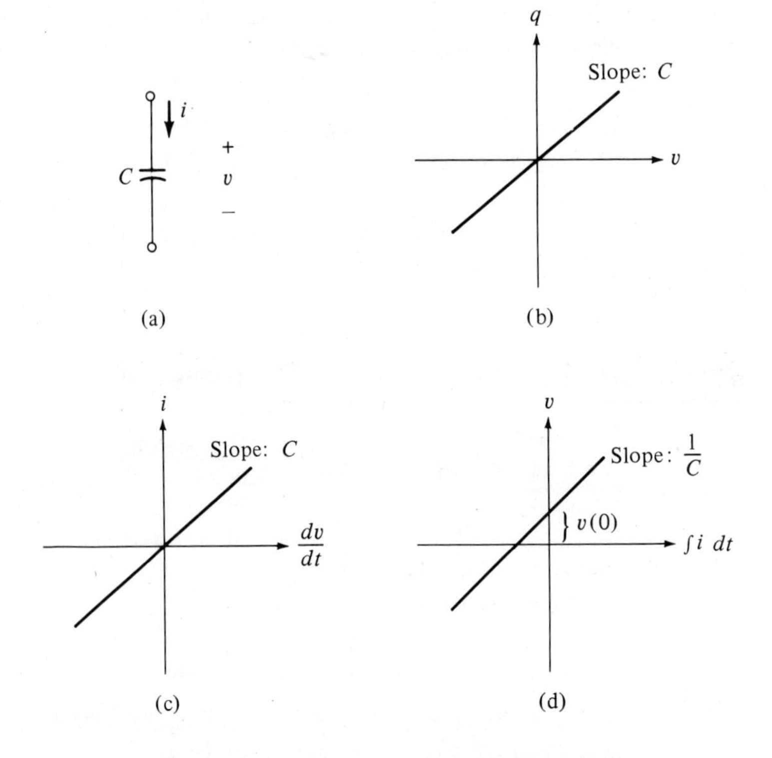

hence

$$i(t) = C\frac{d}{dt}v(t) \tag{3.2}$$

The usual unit of capacitance is the **farad**; Equation (3.2) indicates that a farad is also an *ampere-second per volt*.

Example 3.1

Problem:

Find the expression for capacitor current when a voltage ramp, $v(t) = 3t \cdot u(t)$ V (perhaps the rise of a sawtooth voltage), is present at the terminals of a 3-μF capacitor.

Solution:

1. Use Equation (3.2) to obtain $i(t)$:

$$i(t) = 3 \times 10^{-6}\frac{d}{dt}[3t \cdot u(t)] \text{ A}$$

Flashback

$$\frac{d}{dt}[f(t) \cdot u(t)] = u(t) \cdot \frac{d}{dt}f(t) + f(t) \cdot \delta(t)$$

$$\frac{d}{dt}(at) = a$$

2. Calculate $i(t)$:

$$i(t) = (3 \times 10^{-6})[3 \cdot u(t) + 0 \cdot \delta(0)] = 9 \cdot u(t)\ \mu\text{A}$$

This result indicates that a linear increase in voltage produces a constant current through a capacitor.

Equation (3.2) reveals some important operating characteristics of the capacitor:

1. If the terminal voltage is not changing over a particular time interval, then during that time $dv(t)/dt = 0$ V/sec, which causes $i(t) = 0$. The conclusion is that there is current flow through a capacitor only when the terminal voltage is changing.
2. If there is a finite change in terminal voltage during a time interval, then the current is finite and nonzero for that interval. When voltage is

increasing according to the passive sign convention, the capacitor is charging, and current flows into the positive terminal. When voltage is decreasing, the capacitor is discharging, and current flows from the positive terminal.

3. The more rapid the change in terminal voltage, the larger the current. Very rapid changes in voltage, such as at the edge of a pulse, produce large, fast-rising excursions in current which are commonly called spikes. The reader will recall from Chapter 2 that $du(t)/dt = \delta(t)$; thus, as the change in voltage approaches a step, the current must approach an impulse. In practice, there cannot be a discontinuity, or "jump," in capacitor terminal voltage unless an impulse of current can be supplied by an external source. This observation has important consequences in circuits with switches or step sources.

4. If there is an instantaneous change (discontinuity) in the *slope* (derivative) of the terminal voltage at a particular time (such as at the beginning of a ramp), then at that time the current is also discontinuous.

These characteristics of the capacitor are illustrated in Figure 3.2.

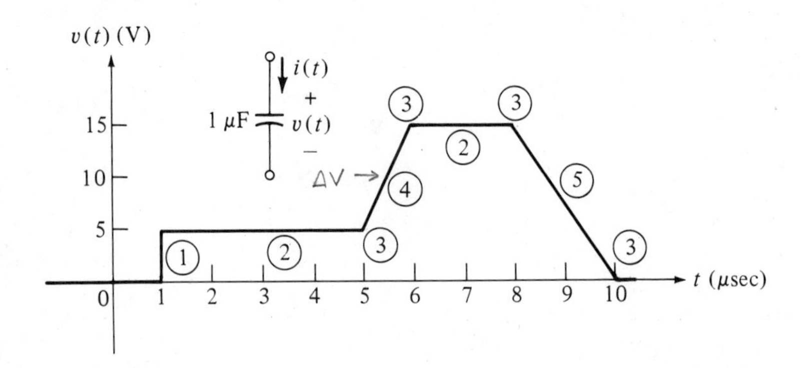

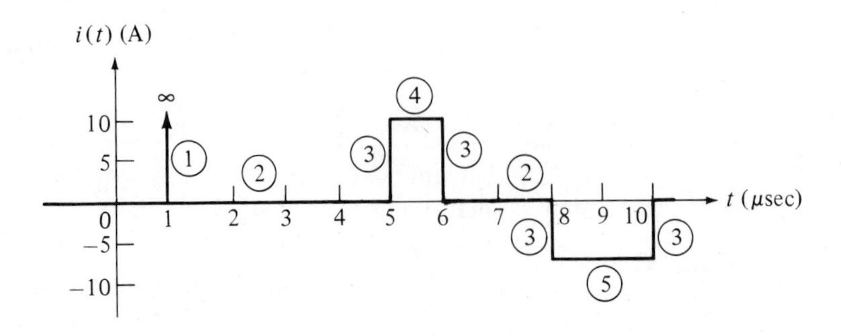

Figure 3.2
Operating characteristics of a capacitor.
(1) Voltage discontinuity requires current impulse.
(2) Voltage constant; current zero.
(3) Voltage *slope* discontinuity; current discontinuity.
(4) Voltage slope $+10^7$ V/sec (capacitor charging linearly); current constant 10 A.
(5) Voltage slope -7.5×10^6 V/sec (capacitor discharging linearly); current constant -7.5 A.

Equation (3.2) may be rearranged and integrated to express the terminal voltage of a capacitor in terms of the current through the capacitor:

$$dv(t) = \frac{1}{C} \cdot i(t)\, dt$$

$$\int_{v(t_0)}^{v(t_1)} dv(t) = \frac{1}{C} \int_{t_0}^{t_1} i(t)\, dt$$

$$v(t_1) - v(t_0) = \frac{1}{C} \int_{t_0}^{t_1} i(t)\, dt$$

$$v(t_1) = \frac{1}{C} \int_{t_0}^{t_1} i(t)\, dt + v(t_0) \qquad (3.3)$$

Equation (3.3) indicates that if the voltage across the capacitor terminals is known for some initial time t_0, then the voltage at any time t_1 may be calculated. Integration of current over the interval t_0 to t_1, and division of the result by C, yields the change in capacitor voltage over the interval. Adding this change to the initial voltage yields the total capacitor voltage at time t_1.

Equations (3.2) and (3.3), as depicted in Figure 3.1(c) and (d), express the terminal characteristics of a capacitor, and are to a capacitor what Ohm's law is to a resistor.

Example 3.2

Problem:

Calculate the voltage which develops across a 200-pF capacitor when an exponentially decaying current,

$$i(t) = 2 \cdot \exp[-10^6 t] \cdot u(t) \text{ mA}$$

flows into the positive terminal of the capacitor during the interval $t = 0$ and $t = 2\ \mu\text{sec}$. The capacitor is charged to 5.0 V at $t = 0$.

Solution:

Use Equation (3.3) to calculate $v(2 \times 10^{-6})$:

$$v(2 \times 10^{-6}) = 5 \times 10^9 \int_0^{2\ \mu\text{sec}} 2 \times 10^{-3} \cdot \exp[-10^6 t]\, dt + 5.0$$

initial conditions 5V at t=ø

The unit step function need not be written into the calculation, since it has a value of 1 for $t > 0$.

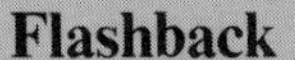

Flashback

$$\int \exp[at]\,dt = \frac{1}{a} \cdot \exp[at] + K$$

Thus

$$v(2 \times 10^{-2}) = (5 \times 10^{9})(2 \times 10^{-3})(-10^{-6}) \times \exp[-10^{6}t]\Big|_{0}^{2\ \mu\text{sec}}$$

$$= -10(\exp[-2] - \exp[0]) + 5.0 \simeq 13.7 \text{ V}$$

Terminal Characteristics of an Inductor

The **inductance** L of an inductor is the ratio between the magnetic flux produced by the inductor and the current through the inductor at any instant, and is written

$$L = \frac{\phi}{i}, \qquad \text{or} \quad \phi = Li$$

If L, the ratio of flux to current, remains constant when the inductor is operated over a range of flux and current values, then the inductor is said to be **linear** for that operating range (Figure 3.3), so

$$\phi(t) = L \cdot i(t) \tag{3.4}$$

as flux and current vary with time within the linear operating range of the inductor.

Differentiation of Equation (3.4) with respect to time yields

$$\frac{d}{dt}\phi(t) = L\frac{d}{dt}i(t)$$

The derivative (or rate of change) of flux with respect to time is *voltage* (Faraday's law); hence

$$v(t) = L\frac{d}{dt}i(t) \tag{3.5}$$

The usual unit of inductance is the **henry**; Equation (3.5) indicates that a henry is an *volt-second per ampere*.

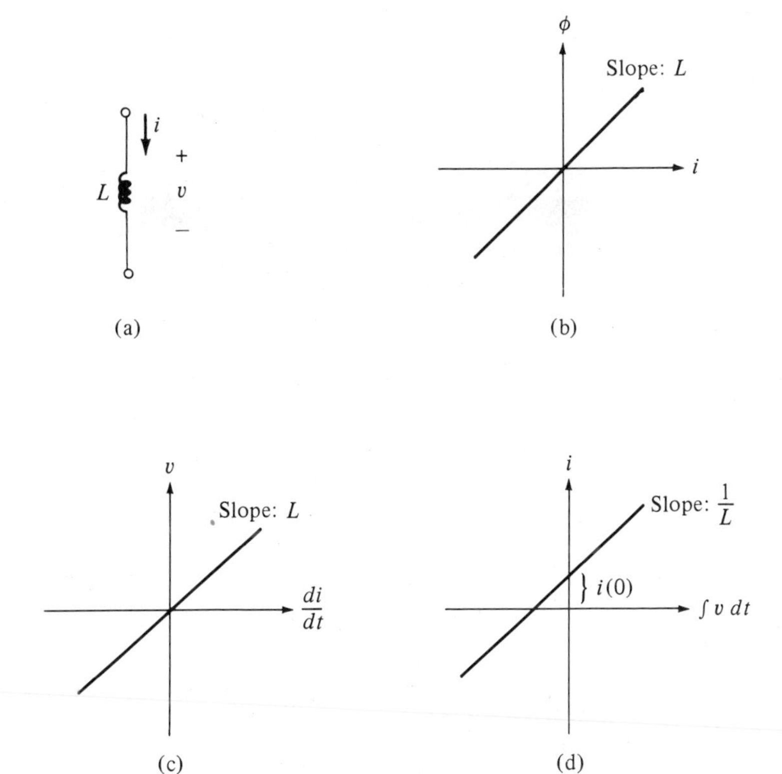

Figure 3.3 Terminal characteristics of an inductor.

Example 3.3

Problem:

Calculate the voltage across the terminals of a 40-mH inductor when the current through the inductor is decreasing linearly at a rate of 50 A/sec.

Solution:

Use Equation (3.5) to calculate $v(t)$:

$$v(t) = (4 \times 10^{-2})(-50) = -2 \text{ V}$$

Note that the current is, by convention, flowing *into* the positive terminal and is decreasing. The calculated voltage drop is negative; hence, the current is actually flowing from the positive terminal, indicating that the inductor is discharging when current is decreasing.

Equation (3.2) reveals some important operating characteristics of the inductor:

1. If the inductor current is not changing over a particular time interval, then during that time $di(t)/dt = 0$ A/sec, which causes $v(t) = 0$ V. The conclusion is that there is voltage across an inductor only when the current through the inductor is changing.
2. If there is a finite change in inductor current during a time interval, then the voltage is finite and nonzero for that interval. When current is increasing, the inductor is charging, and the voltage drop across the inductor terminals is in the direction of current flow. When current is decreasing, the inductor is discharging, and voltage rises in the direction of current flow.
3. The more rapid the change in inductor current, the larger the voltage. Very rapid changes in current produce voltage spikes. As the change in current approaches a step, the voltage must approach an impulse. In practice, there cannot be a discontinuity, or "jump," in inductor current unless an impulse of voltage can be supplied by an external source. This observation has important consequences in circuits with switches or step sources.
4. If there is an instantaneous change (discontinuity) in the *slope* (derivative) of the inductor current at a particular time (such as at the corner of a pulse), then at that time the voltage is also discontinuous.

These characteristics of the inductor are illustrated in Figure 3.4.

Equation (3.5) may be rearranged and integrated to express the inductor current in terms of the terminal voltage:

$$di(t) = \frac{1}{L} \cdot v(t)\,dt$$

$$\int_{i(t_0)}^{i(t_1)} di(t) = \frac{1}{L}\int_{t_0}^{t_1} v(t)\,dt$$

$$i(t_1) - i(t_0) = \frac{1}{L}\int_{t_0}^{t_1} v(t)\,dt$$

$$i(t_1) = \frac{1}{L}\int_{t_0}^{t_1} v(t)\,dt + i(0) \tag{3.6}$$

Equation (3.6) indicates that if the inductor current is known for some initial time t_0, then the voltage at any time t_1 may be calculated. Integration of voltage over the interval t_0 to t_1, and division of the result by L, yields the change in inductor current over the interval. Adding this change to the initial current yields the total inductor current at time t_1.

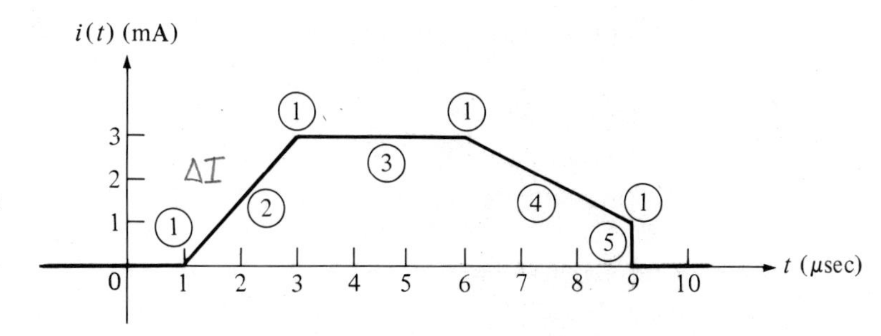

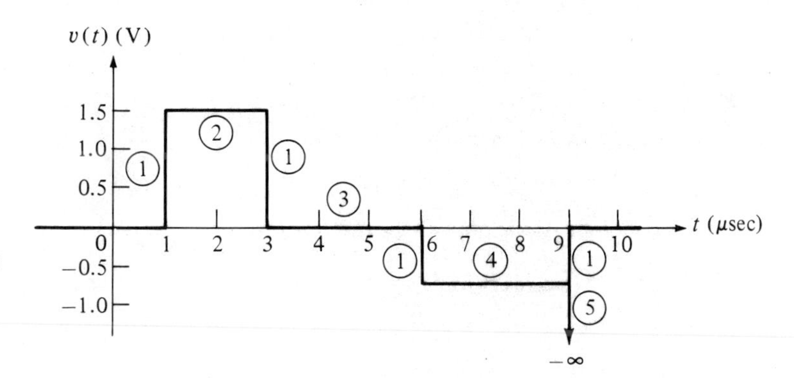

Figure 3.4
Operating characteristics of an inductor.
(1) Current *slope* discontinuity; voltage discontinuity.
(2) Current slope +1.5 mA/sec (inductor charging); voltage constant at 1.5 V.
(3) Current constant; voltage zero.
(4) Current slope −0.667 mA/sec (inductor discharging); voltage constant at −0.667 V.
(5) Current discontinuity requires voltage impulse.

Equations (3.5) and (3.6), as depicted in Figure 3.3(c) and (d), express the terminal characteristics of an inductor, and are to an inductor what Ohm's law is to a resistor.

We have been careful to develop the terminal characteristics for the inductor and the capacitor separately. At this point, however, we should point out that the inductor is the electrical dual of the capacitor. Equations (3.4), (3.5), and (3.6) are duals of Equations (3.1), (3.2), and (3.3), respectively, in that they have the same form, but incorporate dual electrical quantities, q and ϕ, i and v, and L and C.

Example 3.4

Problem:

Calculate the current into the positive terminal of a 2-H inductor at $t = 2$ sec when a voltage

$$v(t) = 2t^2 \cdot u(t) \text{ V}$$

is applied across the inductor terminals between $t = 1$ sec and $t = 2$ sec. The inductor current at $t = 1$ sec is 1 A.

Solution:

Use Equation (3.6) to obtain $i(2)$:

$$i(2) = 0.5\int_1^2 2t^2\,dt + 1.0$$

Flashback

$$\int t^n\,dt = \frac{t^{n+1}}{n+1} + K$$

Thus

$$i(2) = (0.5)(2)\frac{t^3}{3}\bigg|_1^2 + 1.0$$
$$= \tfrac{1}{3}(8 - 1) + 1.0 \simeq 3.33 \text{ A}$$

3.3 Initial Conditions

We can see from Equations (3.3) and (3.6) that the initial values of the capacitor voltage and inductor current must be known before the voltage or current, respectively, can be determined for some later time. These initial values, or **initial conditions**, as they are sometimes called, may be obvious or easily calculated in many circuits. Obtaining initial conditions may be somewhat more of a challenge in circuits which include several energy storage elements, and perhaps several switches or a sequence of switching actions. For that reason, we will take time for a brief discussion of techniques for determining initial conditions.

Consider first a circuit which contains only one switch, or a group of switches whose operation is synchronized, and for which the charging source is constant (a dc source). As a further condition, assume that no switching action has occurred in a long time. A "long time" in this context is sufficient time for the capacitors and inductors to become fully charged. For a fully charged capacitor, there is no change in terminal voltage $[dv(t)/dt = 0]$, and since the terminal characteristic of the capacitor requires that $i(t) = C\,dv(t)/dt$, the capacitor current is zero. Similarly, for a fully charged inductor, $di(t)/dt = 0$, and since $v(t) = L\,di(t)/dt$, then the inductor voltage is also zero. The fully charged capacitor can therefore be represented in the time domain by an open circuit, and the fully charged inductor can be represented by a short circuit. Replacement of capacitors by open circuits and inductors by short circuits leads to a simplified circuit from which initial conditions may be more easily determined.

Example 3.5

Problem:

Determine the initial conditions for the capacitor and the inductor; i.e., determine $v_C(0^-)$ and $i_L(0^-)$ [Figure 3.5(a)].

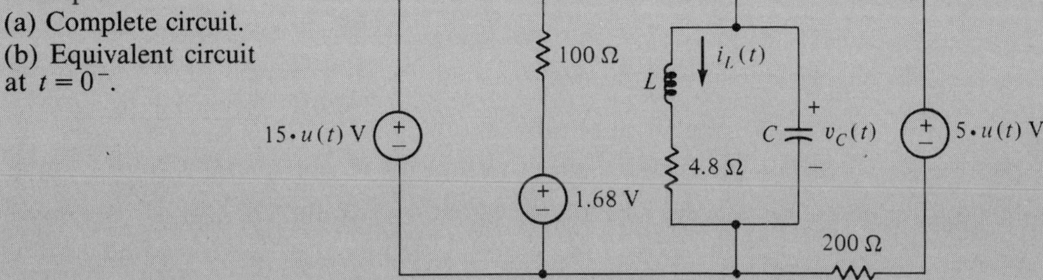

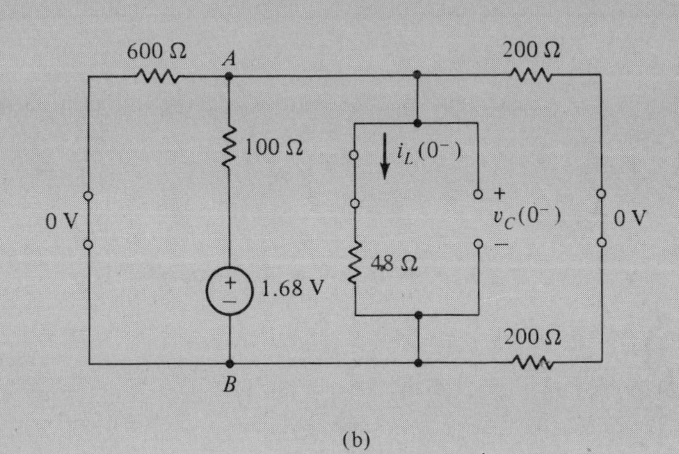

Figure 3.5
Example 3.5.
(a) Complete circuit.
(b) Equivalent circuit at $t = 0^-$.

Solution:

1. Note that the step functions associated with the 5- and 15-V sources serve as synchronized switches causing the source voltages to be zero until $t > 0$. Thus, the 5- and 15-V sources may be replaced by short circuits in the time-domain equivalent circuit for $t = 0^-$ [Figure 3.5(b)].
2. To determine $v_C(0^-)$, first determine the equivalent resistance in parallel with terminal pair A–B:

$$R_{A-B} = \frac{1}{\dfrac{1}{(200+200)} + \dfrac{1}{48} + \dfrac{1}{600}} = 40\ \Omega$$

3. Use voltage division to calculate $v_{A-B}(0^-)$:

$$v_{A-B}(0^-) = v_C(0^-) = \frac{1.68(40)}{100+40} = 0.48\ \text{V}$$

4. Use Ohm's law to calculate $i_L(0^-)$:

$$i_L(0^-) = 0.48/48 = 10\ \text{mA}$$

Exercise 3.1

Problem:

Find the capacitor voltage and the inductor current at $t = 0^-$ (Figure 3.6).

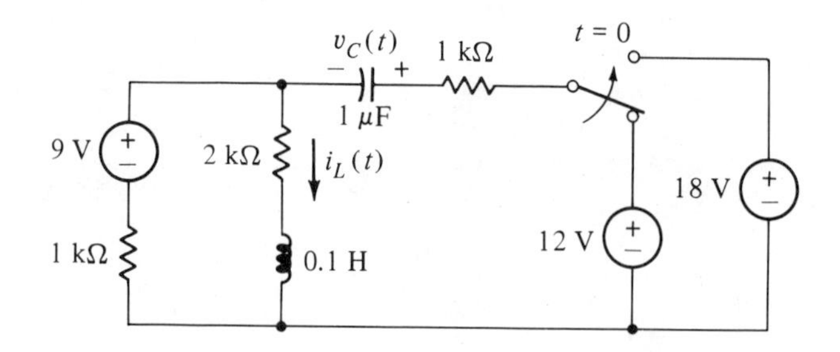

Figure 3.6
Exercise 3.1.

Answer:

$i_L(0^-) = 3$ mA; $v_C(0^-) = 6$ V.

It must be emphasized that the technique presented in this section is limited to circuits in which

1. the charging source is constant (dc) or has a rectangular waveform, and
2. the time since the last switching action is sufficient for all inductors and capacitors to become fully charged.

Figure 3.7 provides a flow chart to assist in determining which circuits are compatible with the technique presented in this section. Other circuits must be analyzed for $t < t_0$, and the values of capacitor voltage and inductor current calculated for $t = t_0^-$. This, of course, requires knowledge of the analysis procedure itself, which is the topic of later sections of this chapter and of later chapters.

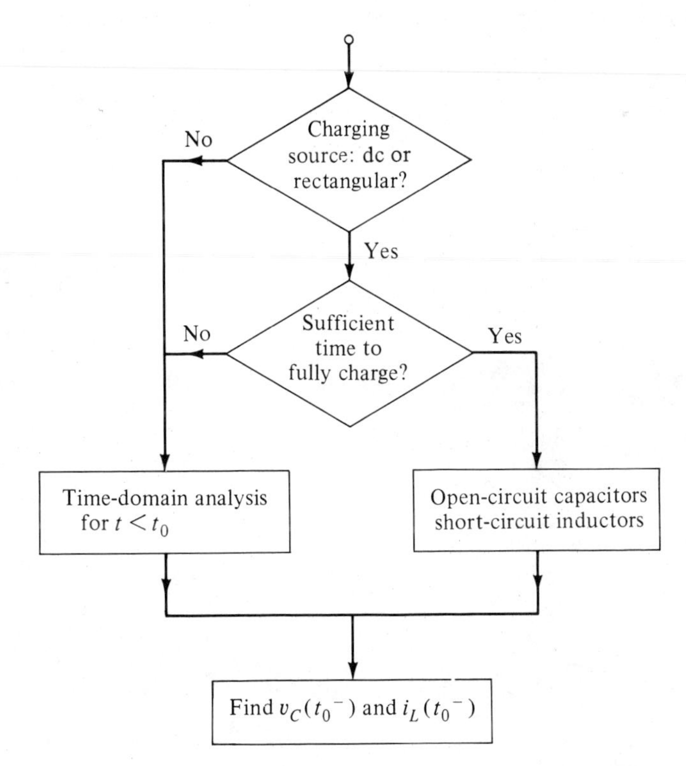

Figure 3.7
Flow chart for determining initial conditions. The time-domain analysis block represents the requirement to derive an expression for $v_C(t)$ or $i_L(t)$ by other means which will be discussed later in this chapter and in other chapters. This expression must then be evaluated at $t = t_0$ (next block).

Further examples for determining initial conditions are provided in Example 3.6, which involves multiple unsynchronized switches; Example 3.7, which involves sequential complementary action (OPEN–CLOSE–OPEN, etc); and Example 3.8, which involves a charging source with rectangular waveform.

Example 3.6

Problem:

Refer to Figure 3.8(a). Switch S1 moves from position 1 to position 2 at $t = 0$; $\frac{1}{2}$ sec later, with S1 still in position 2, S2 closes. Assuming that $\frac{1}{2}$ sec is sufficient time for the capacitor and inductor to become fully charged, determine the initial conditions for the time interval $0 \le t < \frac{1}{2}$ sec, and for the interval $t \ge \frac{1}{2}$ sec.

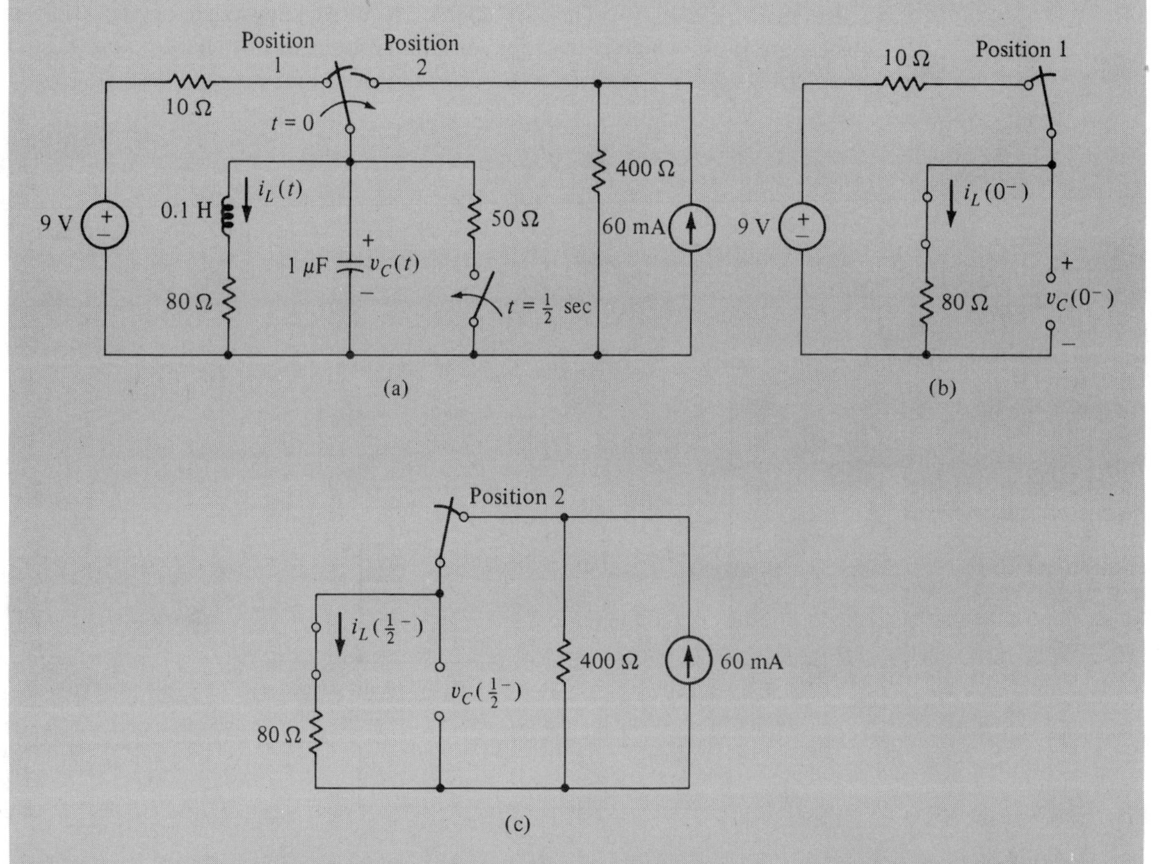

Figure 3.8 Example 3.6. (a) Complete circuit. (b) Equivalent circuit at $t = 0^-$. (c) Equivalent circuit at $t = \frac{1}{2}^-$ sec.

Solution:

1. Determine capacitor voltage and inductor current for $t = 0$ from the time-domain equivalent circuit for $t = 0^-$ [Figure 3.8(b)]:

$$i_L(0^-) = \frac{9}{10 + 80} = 0.1 \text{ A}$$
$$v_C(0^-) = (0.1)(80) = 8 \text{ V}$$

2. Determine initial conditions for $t = \frac{1}{2}$ sec from the time-domain equivalent circuit for $t = \frac{1}{2}^-$ sec [Figure 3.8(c)]:

$$i_L\left(\tfrac{1}{2}^-\right) = (60 \times 10^{-3})\frac{1/80}{(1/400) + (1/80)} = 50 \text{ mA}$$
$$v_C\left(\tfrac{1}{2}^-\right) = (50 \times 10^{-3})(80) = 4 \text{ V}$$

Example 3.7

Problem:

Refer to Figure 3.9(a). The switch cycles between position 1 and position 2 at a 1-Hz rate, beginning at $t = 0$ sec. Determine the initial conditions for the time interval $0 \le t < 1$ sec and for $t \ge 1$ sec. Assume that the period of the cycle (1 sec) is sufficient for the capacitor and inductor to become fully charged.

Solution:

1. Determine the inductor current and capacitor voltage at $t = 0$ from the time-domain equivalent circuit at $t = 0^-$ [Figure 3.9(b)]:

$$i_L(0^-) = 150 \times 10^{-3} \frac{1/40}{(1/80) + (1/40)} = 100 \text{ mA}$$
$$v_C(0^-) = 100 \times 10^{-3}(40) = 4 \text{ V}$$

2. Determine the inductor current and capacitor voltage at $t = 1$ sec from the time-domain equivalent circuit at $t = 1^-$ sec [Figure 3.9(c)]:

$$i_L(1^-) = \frac{10}{160 + 40} = 50 \text{ mA}$$
$$v_C(1^-) = 50 \times 10^{-3}(40) = 2 \text{ V}$$

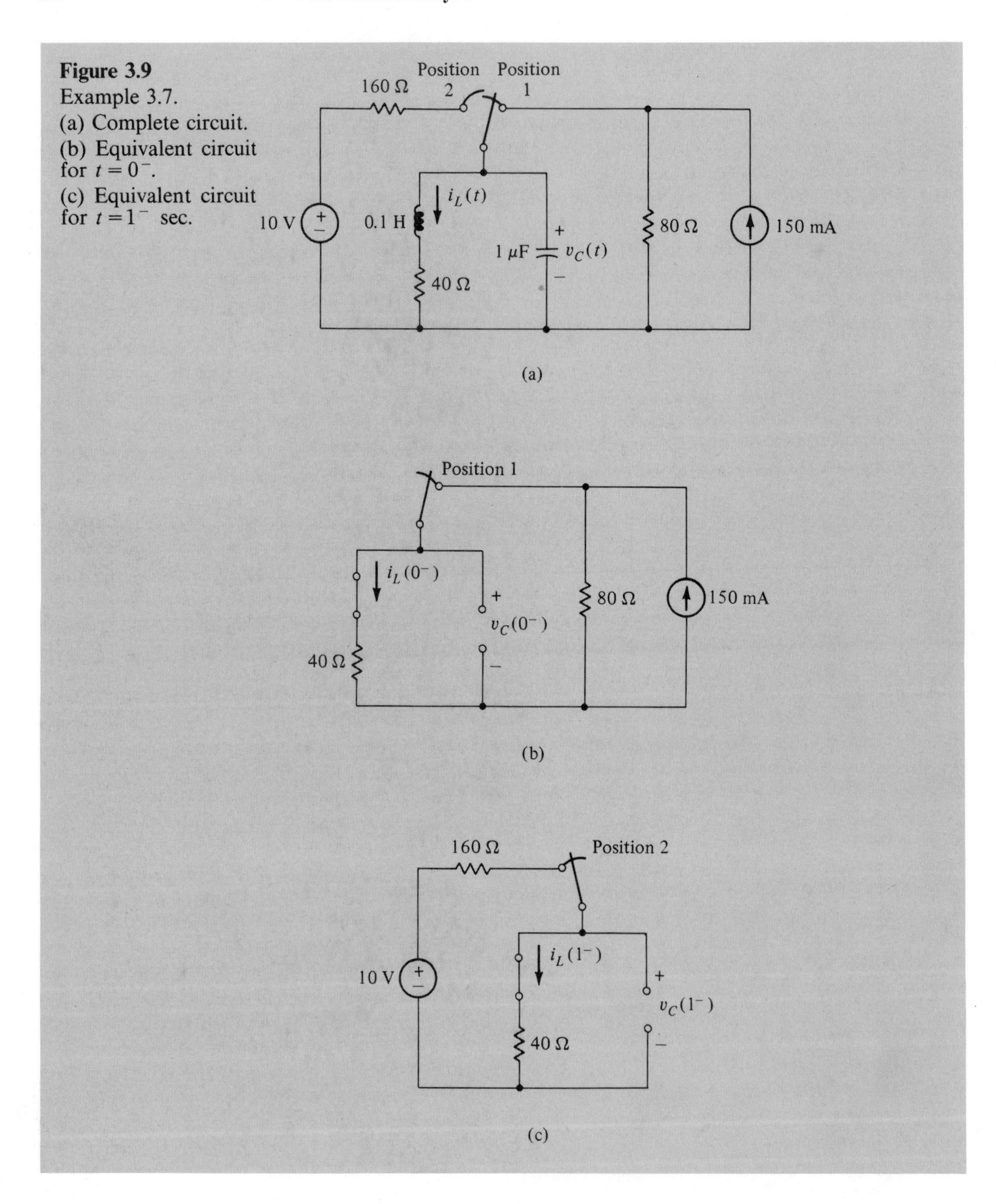

Figure 3.9
Example 3.7.
(a) Complete circuit.
(b) Equivalent circuit for $t = 0^-$.
(c) Equivalent circuit for $t = 1^-$ sec.

Example 3.8

Problem:

Determine the inductor current and capacitor voltage at $t = 0$ and at $t = 1$ sec [Figure 3.10(a)]. Assume that 1 sec is sufficient time for the inductor and capacitor to charge fully.

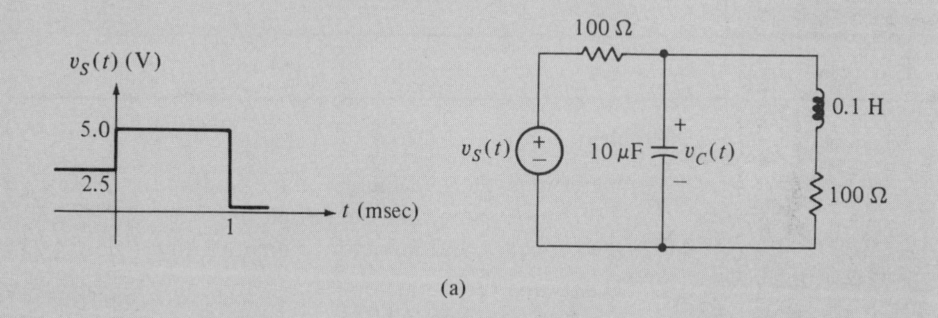

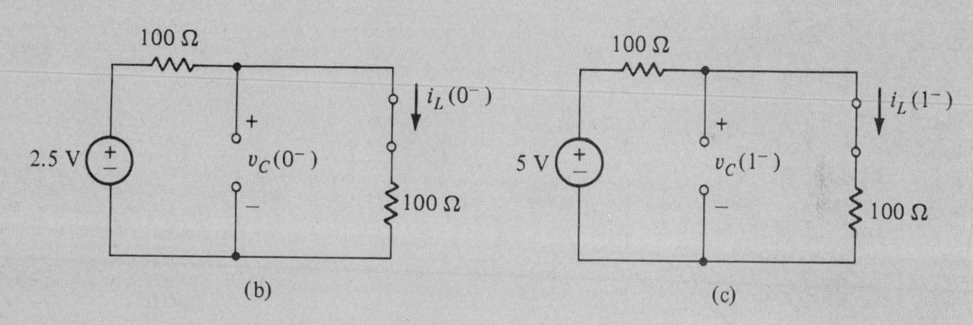

Figure 3.10 Example 3.8. (a) Complete circuit. (b) Equivalent circuit for $t = 0^-$. (c) Equivalent circuit for $t = 1^-$ msec.

Solution:

1. Prior to $t = 0$, the source produces a constant 2.5 V. Determine $v_C(0^-)$ and $i_L(0^-)$ from the equivalent circuit at $t = 0^-$ [Figure 3.10(b)]:

$$v_C(0^-) = 1.25 \text{ V}$$

$$i_L(0^-) = 12.5 \text{ mA}$$

2. Between 0 and 1 sec, the source produces a constant 5 V. Determine $v_C(1^-)$ and $i_L(1^-)$ from the equivalent circuit at $t = 1^-$ sec [Figure 3.10(c)]:

$$v_C(1^-) = 2.5 \text{ V}$$

$$i_L(1^-) = 25 \text{ mA}$$

Exercise 3.2

Problem:

1. Determine the capacitor voltage and inductor current at $t = 0$ [Figure 3.11(a)]. Assume that the switch has been closed for a time sufficient for the capacitor and inductor to charge fully.
2. Determine the capacitor voltage and inductor current at $t = 1$ sec and at $t = 5$ sec [Figure 3.11(b)]. Assume that 2 sec is sufficient time for the capacitor and inductor to charge fully.

Figure 3.11
Exercise 3.2.

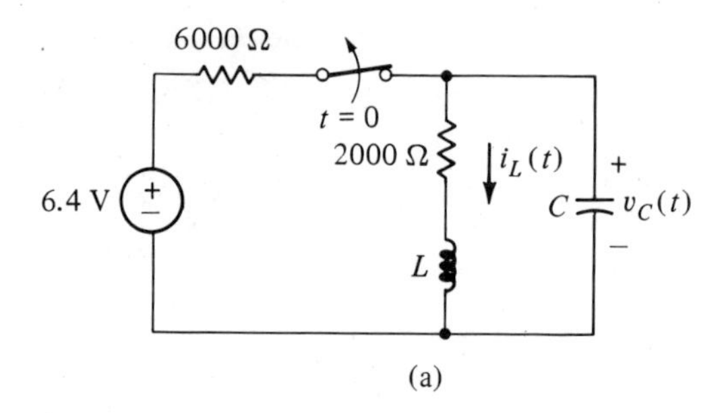

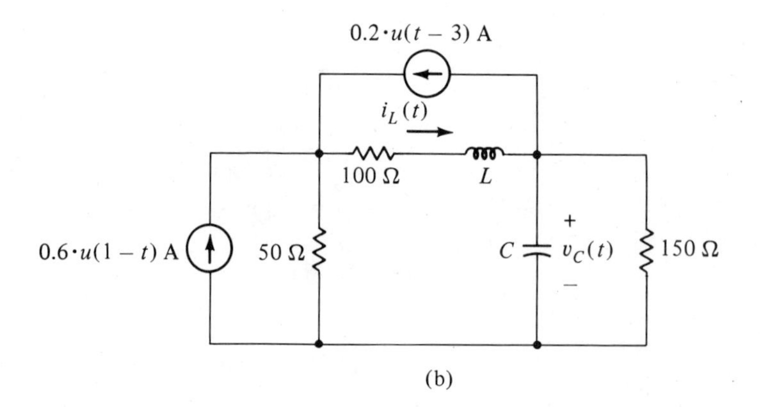

Answer:

1. $v_C(0^-) = 1.6$ V; $i_L(0^-) = 0.8$ mA
2. $v_C(1^-) = 15$ V; $i_L(1^-) = 0.1$ A; $v_C(5^-) = 10$ V; $i_L(5^-) = 0.1333$ A.

3.4 Differential Equations

Equations of the form

$$c_0 \frac{d^n y}{dx^n} + c_1 \frac{d^{n-1} y}{dx^{n-1}} + \cdots + c_{n-1} \frac{dy}{dx} + c_n y = F(x) \tag{3.7}$$

arise during analysis of linear electrical circuits when Kirchhoff's current law is applied at a node or Kirchhoff's voltage law is applied around a mesh. Equation (3.7) is an ordinary linear **differential equation**, with constant coefficients $(c_0, c_1, \ldots, c_n)$. The derivatives are **ordinary**, rather than partial, derivatives. The dependent variable y and its derivatives enter into the equation in a linear manner; that is, there are no terms which contain products of y and derivatives of y, or in which y or its derivatives are raised to a power other than 1. Equation (3.7) is said to be of **order** n, which is the order of the highest derivative. Note that Equation (3.7) has been arranged so that the right member of the equation, called the **forcing function**, depends only upon the independent variable x, and not upon y. If $F(x) = 0$, the differential equation is said to be **homogeneous**.

Some examples of differential equations which arise during analysis of linear circuits are given below. Note that the independent variable is t and the dependent variable is $v(t)$ or $i(t)$, and that the equations may or may not be homogeneous:

$$\frac{d}{dt} i(t) + \tfrac{1}{5} \cdot i(t) = 8 \cdot u(t)$$

$$\frac{d^2}{dt^2} v(t) + 8000 \frac{d}{dt} v(t) + 6 \times 10 \cdot 6v(t) = 0$$

The differential equation and initial conditions describing a given system constitute an **initial value problem**. The solution of initial value problems involving first-order differential equations will be presented in the next section. The Laplace transform method for solution of initial value problems involving differential equations of any order will be presented in the next chapter.

3.5 Analysis of Source-Free Circuits

The term **source-free circuit** refers to a circuit in which a charged capacitor or inductor is removed from its charging source and allowed to discharge through other passive elements. The simplest source-free circuit, the R–C or R–L circuit, which occurs frequently in electrical and electronic networks, will be analyzed in this section.

Source-Free R–C Circuit

The capacitor shown in Figure 3.12(a) is charged to V volts at time $t = t_0^-$. After the switch is closed at $t = t_0$, the capacitor can discharge through the resistor R [Figure 3.12(c)]. By Kirchhoff's current law,

$$i_C(t) = i_R(t) \tag{3.8}$$

Clearly, the voltage is the same across both elements, so by substitution of Equation (3.2) and Ohm's law, Equation (3.8) can be written

$$-C\frac{d}{dt}v_C(t) = \frac{1}{R}v_C(t)$$

The minus sign in the left member of the equation indicates that the capacitor

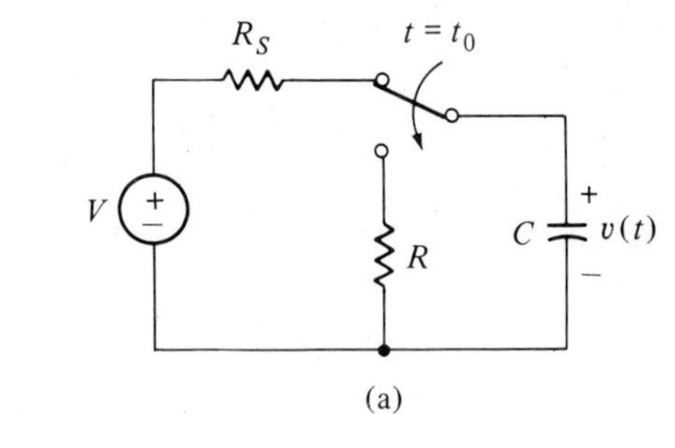

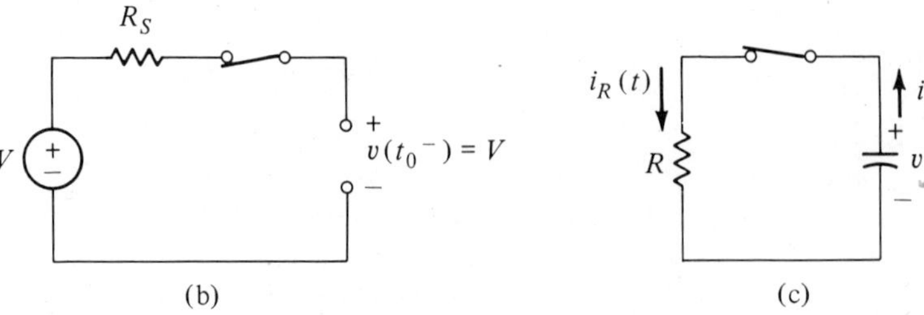

Figure 3.12
Source-free R–C circuit.
(a) Capacitor charging circuit.
(b) Equivalent charging circuit at $t = t_0^-$.
(c) Capacitor discharge circuit.

is discharging. After rearranging terms,

$$C\frac{d}{dt}v_C(t) + \frac{1}{R}v_C(t) = 0 \tag{3.9}$$

which is an ordinary, first-order, linear, homogenous differential equation with constant coefficients. This particular type of initial value problem can be solved directly by **separation of variables.**

Separation of variables refers to the rearrangement of the differential equation so that the dependent variable and its derivative appear only in the left member of the equation, and the independent variable appears only in the right member. Usually, the coefficients are combined and placed in the right member of the equation. Thus,

$$\frac{d}{dt}v_C(t) = -\frac{1}{RC}v_C(t) \tag{3.10}$$

or

$$\frac{dv_C(t)}{v_C(t)} = -\frac{1}{RC}\,dt$$

Integration of both members of the equation will yield the solution $v_C(t)$. Note in the steps that follow that the limits of integration for the right member are t_0 and t_1 because the variable of integration is t, whereas the limits of integration for the left member are $v_C(t_0)$ and $v_C(t_1)$ because the variable of integration is $v_C(t)$:

$$\int_{v_C(t_0)}^{v_C(t_1)} \frac{dv_C(t)}{v_C(t)} = \int_{t_0}^{t_1} -\frac{1}{RC}\,dt$$

Flashback

$$\int \frac{dx}{x} = \ln x + K$$

Thus

$$\ln[v_C(t)]\Big|_{v_C(t_0)}^{v_C(t_1)} = -\frac{1}{RC}t\Big|_{t_0}^{t_1}$$

$$\ln[v_C(t_1)] - \ln[v_C(t_0)] = -\frac{1}{RC}(t_1 - t_0)$$

Flashback

$$\log a - \log b = \log \frac{a}{b}$$

Thus

$$\ln\left[\frac{v_C(t_1)}{v_C(t_0)}\right] = -\frac{1}{RC}(t_1 - t_0)$$

In the next step, each member of the equation will be exponentiated:

Flashback

$$\exp[\ln a] = a$$

Thus

$$\frac{v_C(t_1)}{v_C(t_0)} = \exp\left[-\frac{t_1 - t_0}{RC}\right]$$

$$v_C(t_1) = v_C(t_0) \cdot \exp\left[-\frac{t_1 - t_0}{RC}\right]$$

The product RC, called the **time constant** τ_c of the circuit, has the dimensions of time (see Problem 3-14), which means that the argument of the exponential is dimensionless, as required. Since t_1 is any time $t > t_0$ and $v_C(t_0) = V$, then

$$v_C(t > t_0) = V \cdot \exp\left[-\frac{t - t_0}{\tau_C}\right] \tag{3.11}$$

The factor $\exp[-(t - t_0)/\tau_C]$ is the **decay function** associated with many other physical phenomena such as heat transfer, biological extinction, radioactive decay, etc., and expresses the fraction of a quantity present at t_0 which remains at a later time t. The decay function, often called the **discharge function** in circuit analysis, has a value of 1 when $t = t_0$ and a value that approaches zero as time increases:

$$\lim_{t \to \infty} \exp\left[-\frac{t - t_0}{\tau_C}\right] \to 0$$

It is important to note that Equation (3.11) represents the capacitor voltage only for the domain $t > t_0$. The capacitor voltage at $t = t_0$ is specified as V, but capacitor voltage at any prior time cannot be determined from Equation (3.11).

Table 3.1
Values of the Discharge (Decay) Function

Elapsed Time	Fraction (Percent) of Charge Remaining
$0.1\tau_C$	0.90484 (90.484)
$0.3\tau_C$	0.74082 (74.082)
$0.7\tau_C$	0.49659 (49.659)
$1.0\tau_C$	0.36788 (36.788)
$1.4\tau_C$	0.24660 (24.660)
$2.3\tau_C$	0.10026 (10.026)
$5.0\tau_C$	0.00673 (0.673)
$10.0\tau_C$	0.00004 (0.004)

Table 3.1 lists the fraction (or percentage) of charge remaining with elapsed time $(t - t_0)$ expressed in terms of the time constant. Full discharge is not possible theoretically, since the value of the decay function only approaches zero in the limit; however, an interval equal to five times the time constant $(5\tau_C)$ is considered sufficient from a practical viewpoint for full discharge.

Example 3.9

SPICE

Problem:

Find and sketch the voltage $v_C(t)$ as the capacitor discharges [Figure 3.13(a)].

Solution:

1. Use the equivalent circuit for $t = 0^-$ [Figure 3.13(b)] and voltage division to calculate $v_C(0^-)$:

$$v_C(0^-) = \frac{12(1200)}{600 + 1200} = 8 \text{ V}$$

2. The voltage across a capacitor must be continuous, so

$$v_C(0) = v_C(0^-) = 8 \text{ V}$$

3. The equivalent resistance of the discharge circuit [Figure 3.13(c)] is 3000 Ω; therefore,

$$\tau_C = (3000)(3 \times 10^{-6}) = 9 \text{ msec}$$

4. Use Equation (3.11) to determine $v_C(t)$ for $t > 0$:

$$v_C(t > 0) = 8 \cdot \exp\left[-\frac{t}{9 \times 10^{-3}}\right] \text{ V}$$

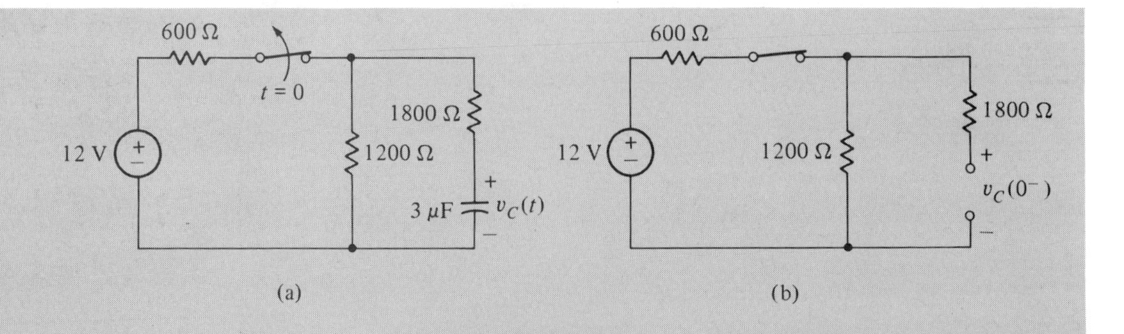

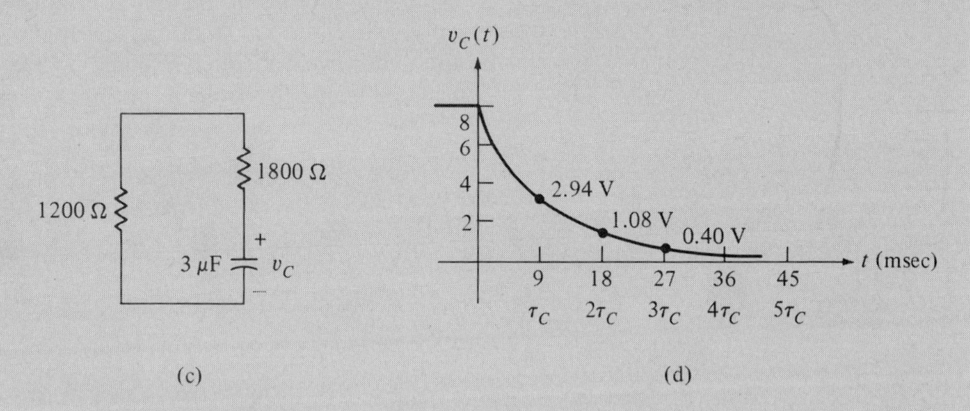

Figure 3.13 Example 3.9. (a) Charging circuit. (b) Equivalent circuit at $t = 0^-$. (c) Discharge circuit. (d) $v_C(t)$ vs. t.

5. Combine $v_C(0^-)$ and $v_C(t > 0)$ to obtain $v_C(t)$:

$$v_C(t) = 8 - 8 \cdot u(t) + 8 \cdot \exp\left[-\frac{t}{9 \times 10^{-3}}\right] \cdot u(t)$$

$$= 8 - 8\left(1 - \exp\left[-\frac{t}{9 \times 10^{-3}}\right]\right)u(t) \text{ V}$$

Figure 3.13(d) is a sketch of $v_C(t)$. For practical purposes, the capacitor will be fully discharged in approximately 45 msec ($5\tau_C$).

Exercise 3.3

SPICE

Problem:

a. How long will it take the voltage across a 50-μF capacitor, initially charged to 10 V, to reach 4 V as the capacitor discharges through a 400-Ω resistor?

b. What is the value of capacitor voltage after 70 msec of discharge?

Answer:

a. 18.3 msec.

b. 0.302 V.

Source-Free *R–L* Circuit

The source-free R–L circuit is the electrical dual of the source-free R–C circuit. Consider the inductor shown in Figure 3.14(a) which is charged to I amperes at $t = t_0^-$. After the switch is moved to the right at $t = t_0$, the inductor can discharge through the resistor R, which has conductance G [Figure 3.14(b)]. By Kirchhoff's voltage law,

$$v_L(t) = v_R(t) \tag{3.12}$$

The same current flows through both elements, so by substitution of Equation

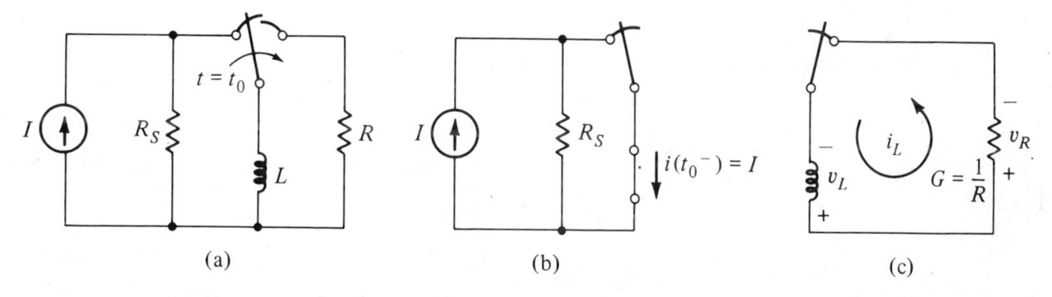

Figure 3.14 Source-free R–L circuit. (a) Inductor charging circuit. (b) Equivalent charging circuit at $t = t_0^-$. (c) Inductor discharge circuit.

(3.5) and Ohm's law, Equation (3.12) can be written

$$L\frac{d}{dt}i_L(t) + \frac{1}{G}i_L(t) = 0 \tag{3.13}$$

Clearly, this is a differential equation of the same type as Equation (3.9), so the solution must take the same form as Equation (3.11):

$$i_L(t > t_0) = I \cdot \exp\left[-\frac{t - t_0}{\tau_C}\right] \tag{3.14}$$

where $\tau_C = GL$. Since $G = 1/R$, τ_C is frequently expressed as L/R. Equation (3.14) represents inductor current only for the domain $t > t_0$.

Example 3.10

SPICE

Problem:

Find and sketch the inductor discharge current (Figure 3.15). At what time will the inductor current be 12 mA?

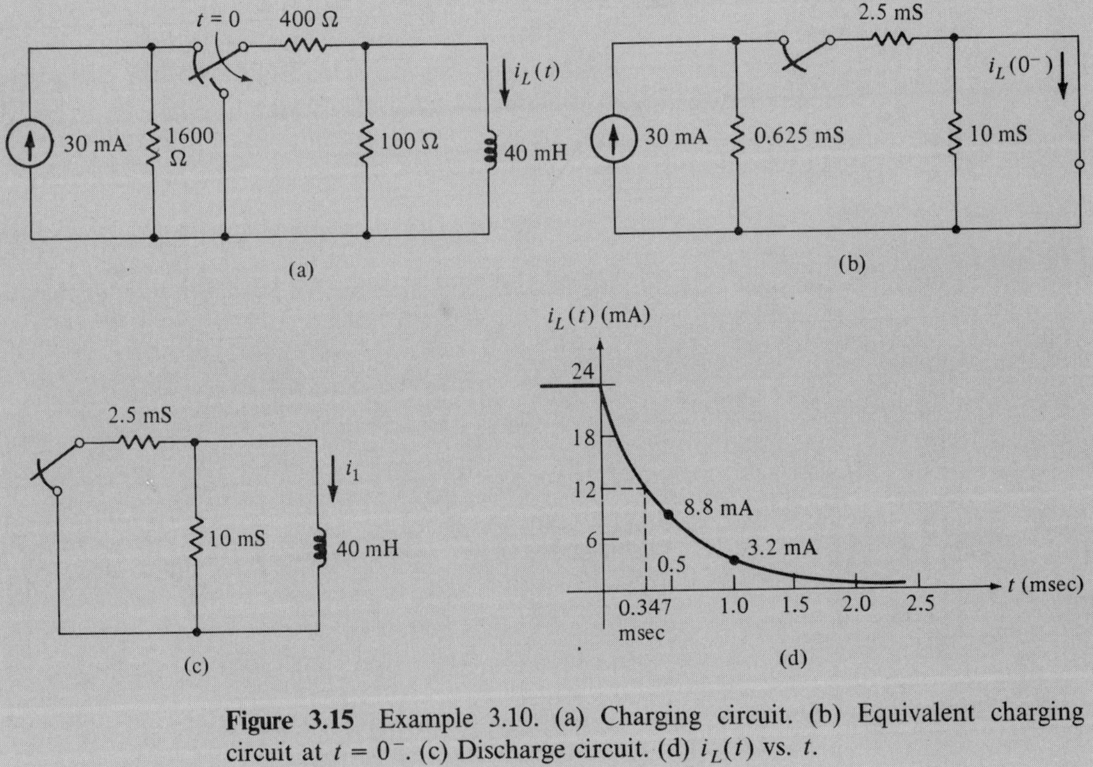

Figure 3.15 Example 3.10. (a) Charging circuit. (b) Equivalent charging circuit at $t = 0^-$. (c) Discharge circuit. (d) $i_L(t)$ vs. t.

Solution:

1. Use the equivalent circuit for $t = 0^-$ [Figure 3.15(b)] and current division to calculate $i_L(0^-)$:

$$i_L(0^-) = \frac{30(2.5)}{2.5 + 0.625} = 24 \text{ mA}$$

2. The conductance of the discharge circuit is 12.5 mS [Figure 3.15(c)], so

$$\tau_C = (12.5 \times 10^{-3})(40 \times 10^{-3}) = 0.5 \text{ msec}$$

3. Use Equation (3.14) to determine $i_L(t)$ for $t > 0$:

$$\begin{aligned} i_L(t > 0) &= 24 \cdot \exp[-t/0.5 \times 10^{-3}] \\ &= 24 \cdot \exp[-2000t] \text{ mA} \end{aligned}$$

4. Combine $i_L(0^-)$ and $i_L(t > 0)$ to obtain $i_L(t)$:

$$\begin{aligned} i_L(t) &= 24 - 24 \cdot u(t) + 24 \cdot \exp[-2000t] \cdot u(t) \\ &= 24 - 24(1 - \exp[-2000t])u(t) \text{ mA} \end{aligned}$$

5. Figure 3.15(d) is a sketch of the inductor discharge current. To determine the time at which $i_L(t) = 12$ mA, substitute 12 mA for $i_L(t)$ in the foregoing expression:

$$\begin{aligned} 12 &= 24 - 24(1 - \exp[-2000t]) \\ \exp[-2000t] &= 12/24 = 0.5 \end{aligned}$$

and take the natural logarithm of both members of the equation:

$$\begin{aligned} -2000t &= \ln 0.5 = -0.6931 \\ t &= 0.3466 \text{ msec} \end{aligned}$$

Exercise 3.4

SPICE

Problem:

a. How long will it take the current through a 30-mH inductor, initially charged to 4 A, to decay to 1A as the inductor discharges through a 500-Ω resistor?

b. What is the value of inductor current after 100 μsec of discharge?

Answer:

a. 83.18 μsec.
b. 0.7555 A.

3.6 Analysis of Driven $R-C$ and $R-L$ Circuits

The term **driven circuit** refers to a circuit which includes a source. The analysis presented here involves only the simple circuits shown in Figure 3.16, but the source can have any time dependence (e.g., step, ramp, sinusoid, etc.). We shall show that the circuit response (voltage or current) is determined by an ordinary differential equation which is first-order, linear, and with constant coefficients, but not homogeneous because of the presence of the source. Preliminary solutions of the differential equations are presented in the first part of this section, and completed solutions for step-function sources are obtained in the second part of this section.

These circuit configurations, though simple, are quite important because of Thevenin's and Norton's theorems, which ensure that any network of sources and resistors can be reduced to a single equivalent source and a single

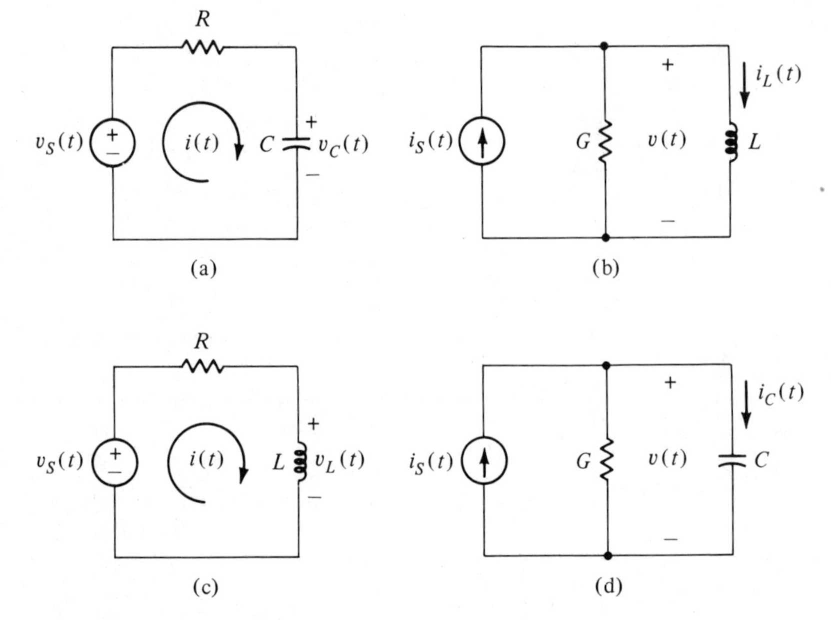

Figure 3.16
Simple $R-C$ and $R-L$ circuits with time-dependent sources.
(a) Series $R-C$ circuit.
(b) Parallel $R-L$ circuit.
(c) Series $R-L$ circuit.
(d) Parallel $R-C$ circuit.

equivalent resistance. As a consequence, circuits in which the source is resistive and the load contains only one energy storage element can always be reduced to one of the simple configurations of Figure 3.16.

Preliminary Solutions

Consider first the series R–C circuit shown in Figure 3.16(a). An obvious consequence of Kirchhoff's current law is that

$$i_C(t) = i_R(t)$$

From Equation (3.2) and Ohm's law, the capacitor voltage is determined by the differential equation

$$C\frac{d}{dt}v_C(t) = \frac{1}{R}[v_S(t) - v_C(t)]$$

or, when terms are rearranged,

$$\frac{d}{dt}v_C(t) + \frac{1}{\tau_C}v_C(t) = \frac{1}{\tau_C}v_S(t) \tag{3.15}$$

where $\tau_C = RC$. Note that Equation (3.15) is similar to Equation (3.9), except that the forcing function in Equation (3.15) is not zero.

Multiplication of all terms of Equation (3.15) by the **integrating factor** $\exp[t/\tau_C]$ will make the right member an **exact derivative**:

$$\exp\left[\frac{t}{\tau_C}\right]\cdot\frac{d}{dt}v_C(t) + \frac{1}{\tau_C}\exp\left[\frac{t}{\tau_C}\right]\cdot v_C(t) = \frac{1}{\tau_C}\exp\left[\frac{t}{\tau_C}\right]\cdot v_S(t) \tag{3.16}$$

Flashback

$$\frac{d}{dx}(y\cdot\exp[ax]) = \exp[ax]\cdot\frac{dy}{dx} + ay\cdot\exp[ax]$$

Therefore,

$$\frac{d}{dt}\left\{\exp\left[\frac{t}{\tau_C}\right]\cdot v_C(t)\right\} = \frac{1}{\tau_C}\exp\left[\frac{t}{\tau_C}\right]\cdot v_S(t)$$

and

$$d\left\{\exp\left[\frac{t}{\tau_C}\right]\cdot v_C(t)\right\} = \frac{1}{\tau_C}\exp\left[\frac{t}{\tau_C}\right]\cdot v_S(t)\,dt \tag{3.17}$$

Integration of both members of Equation (3.17) yields the solution $v(t_1)$, where t_1 is any time greater than t_0:

$$\int_{\exp[t_0/\tau_C]\cdot v_C(t_0)}^{\exp[t_1/\tau_C]\cdot v_C(t_1)} d(\exp[t/\tau_C]\cdot v_C(t)) = \frac{1}{\tau_C}\int_{t_0}^{t_1}\exp\left[\frac{t}{\tau_C}\right]\cdot v_S(t)\,dt$$

$$\exp\left[\frac{t_1}{\tau_C}\right]\cdot v_C(t_1) - \exp\left[\frac{t_0}{\tau_C}\right]\cdot v_C(t_0) = \frac{1}{\tau_C}\int_{t_0}^{t_1}\exp\left[\frac{t}{\tau_C}\right]\cdot v_S(t)\,dt$$

$$v_C(t_1) = \frac{1}{\tau_C}\exp\left[-\frac{t_1}{\tau_C}\right]\cdot\int_{t_0}^{t_1}\exp[t/\tau_C]\cdot v_S(t)\,dt + \exp\left[-\frac{t_1-t_0}{\tau_C}\right]\cdot v_C(t_0) \tag{3.18}$$

The parallel R–L circuit [Figure 3.16(b)] is the electrical dual of the series R–C circuit [Figure 3.16(a)]; thus

$$L\frac{d}{dt}i_L(t) = \frac{1}{G}[i_S(t) - i_L(t)] \tag{3.19}$$

which is the dual of Equation (3.15), determines the inductor current. The solution for inductor current is the dual of the solution for capacitor voltage given by Equation (3.18):

$$i_L(t_1) = \frac{1}{\tau_C}\cdot\exp\left[-\frac{t_1}{\tau_C}\right]\cdot\int_{t_0}^{t_1}\exp\left[\frac{t}{\tau_C}\right]\cdot i_S(t)\,dt + \exp\left[-\frac{t_1-t_0}{\tau_C}\right]\cdot i_L(t_0) \tag{3.20}$$

where $\tau_C = GL$ (or L/R).

For the series R–L shown in Figure 3.16(c), Kirchhoff's voltage law requires that

$$v_L(t) + v_R(t) = v_S(t)$$

Substitution of Equation (3.5) and Ohm's law results in a differential equation which determines the circuit current, $i_L(t)$:

$$L\frac{d}{dt}i_L(t) + R\cdot i_L(t) = v_S(t)$$

or

$$\frac{d}{dt}i_L(t) + \frac{1}{\tau_C}i_L(t) = \frac{1}{L}\cdot v_S(t) \tag{3.21}$$

where $\tau_C = GL$ (or L/R). Use of the integrating factor will lead to the solution:

$$i_L(t_1) = \frac{1}{L}\cdot \exp\left[-\frac{t_1}{\tau_C}\right]\cdot \int_{t_0}^{t_1} \exp\left[\frac{t}{\tau_C}\right]\cdot v_S(t)\,dt + \exp\left[-\frac{t_1 - t_0}{\tau_C}\right]\cdot i_L(t_0) \tag{3.22}$$

Since the parallel R–C circuit [Figure 3.16(d)] is the electrical dual of the series R–L circuit, then

$$\frac{d}{dt}v_C(t) + \frac{1}{\tau_C}v_C(t) = \frac{1}{L}i_S(t) \tag{3.23}$$

where $\tau_C = RC$. The solution is the dual of the solution given by Equation (3.22):

$$v_C(t_1) = \frac{1}{C}\cdot \exp\left[-\frac{t_1}{\tau_C}\right]\cdot \int_{t_0}^{t_1} \exp\left[\frac{t}{\tau_C}\right]\cdot i_S(t)\,dt + \exp\left[-\frac{t_1 - t_0}{\tau_C}\right]\cdot v_C(t_0) \tag{3.24}$$

We have in Equations (3.18), (3.20), (3.22), and (3.24) a set of preliminary solutions for simple series and parallel R–C and R–L circuits which contain a general time-dependent source. We will now turn our attention to completing the solutions for specified source functions.

Response to a Step-Function Source

The step-function source, though the simplest type of time-dependent source to consider, is very important and occurs frequently in electronics circuits. The reader will recall from Chapter 2 that the step-function source, in which $v_S(t) = V_S \cdot u(t - t_0)$ or $i_S(t) = I_S \cdot u(t - t_0)$, is used to represent a dc source switched into a circuit at t_0, or to represent sources with rectangular waveforms (pulses, square waves, etc.) with leading or trailing edges occurring at t_0.

Consider a series R–C circuit [Figure 3.16(a)] with a step-function source $v_S(t) = V_S \cdot u(t - t_0)$. The capacitor voltage at any time $t_1 > t_0$ is determined by Equation (3.18). Substitution for $v_S(t)$ in the integrand leads to

$$v_C(t_1) = \frac{1}{\tau_C}\exp\left[-\frac{t_1}{\tau_C}\right] \cdot \int_{t_0}^{t_1} \exp\left[\frac{t}{\tau_C}\right] \cdot V_S\, dt$$
$$+ \exp\left[-\frac{t_1 - t_0}{\tau_C}\right] \cdot v_C(t_0)$$

The integral is for the interval $t_0 < t < t_1$, and $u(t - t_0)$ is 1 in that interval, so only the constant V_S enters into the integrand.

Flashback

$$\int \exp[ax]\, dx = \frac{1}{a} \cdot \exp[ax] + K$$

Thus,

$$v_C(t_1) = \frac{1}{\tau_C}\exp\left[-\frac{t_1}{\tau_C}\right] \cdot V_S\tau_C \cdot \exp\left[\frac{t}{\tau_C}\right]\Bigg|_{t_0}^{t_1}$$
$$+ \exp\left[-\frac{t_1 - t_0}{\tau_C}\right] \cdot v_C(t_0)$$
$$= V_S \cdot \exp\left[-\frac{t_1}{\tau_C}\right] \cdot \left(\exp\left[\frac{t_1}{\tau_C}\right] - \exp\left[\frac{t_0}{\tau_C}\right]\right)$$
$$+ \exp\left[-\frac{t_1 - t_0}{\tau_C}\right] \cdot v_C(t_0)$$
$$= V_S - [V_S - v_C(t_0)] \cdot \exp\left[-\frac{t_1 - t_0}{\tau_C}\right]$$

Since t_1 is any time $t > t_0$, then

vC

$$v_C(t > t_0) = V_S - [V_S - v_C(t_0)] \cdot \exp\left[-\frac{t - t_0}{\tau_C}\right]$$

By recognizing that the same solution procedures apply to Equations (3.20), (3.22), and (3.24), and by exploiting electrical duality, the remaining solutions may be written without actually performing the solution procedures:

Series R–C circuit [Figure 3.16(a)]:

$$v_C(t > t_0) = V_S - [V_S - v_C(t_0)]\exp\left[-\frac{t - t_0}{\tau_C}\right] \tag{3.25}$$

Parallel R–L circuit [Figure 3.16(b)]:

$$i_L(t > t_0) = I_S - [I_S - i_L(t_0)]\exp\left[-\frac{t - t_0}{\tau_C}\right] \tag{3.26}$$

Series R–L circuit [Figure 3.16(c)]:

$$i(t > t_0) = \frac{V_S}{R} - \left(\frac{V_S}{R} - i(t_0)\right)\exp\left[-\frac{t - t_0}{\tau_C}\right] \tag{3.27}$$

Parallel R–C circuit [Figure 3.16(c)]:

$$v(t > t_0) = \frac{I_S}{G} - \left(\frac{I_S}{G} - v(t_0)\right)\exp\left[-\frac{t - t_0}{\tau_C}\right] \tag{3.28}$$

If the initial conditions are zero, then Equations (3.25) through (3.28) become

Series R–C circuit:

$$v_C(t > t_0) = V_S\left(1 - \exp\left[-\frac{t - t_0}{\tau_C}\right]\right) \tag{3.29}$$

Parallel R–L circuit:

$$i_L(t > t_0) = I_S\left(1 - \exp\left[-\frac{t - t_0}{\tau_C}\right]\right) \tag{3.30}$$

Series R–L circuit:

$$i(t > t_0) = \frac{V_S}{R}\left(1 - \exp\left[-\frac{t - t_0}{\tau_C}\right]\right) \tag{3.31}$$

Parallel R–C circuit:

$$v(t > t_0) = \frac{I_S}{G}\left(1 - \exp\left[-\frac{t - t_0}{\tau_C}\right]\right) \tag{3.32}$$

The factor $1 - \exp[-(t - t_0)/\tau_C]$ in each of Equations (3.29) through (3.32) is the **growth function**, and, like the decay function, is associated with many physical phenomena other than electrical circuits. The growth function, often called the **charge function** in circuit analysis, expresses the fraction of full charge present at any time. The charge function has a value of 0 when

Table 3.2
Values of the Charge (Growth) Function

Elapsed Time	Fraction (Percent) of Charge
$0.1\tau_C$	0.09516 (9.516)
$0.3\tau_C$	0.25918 (25.918)
$0.7\tau_C$	0.50341 (50.341)
$1.0\tau_C$	0.63212 (63.212)
$1.4\tau_C$	0.75340 (75.340)
$3.0\tau_C$	0.95021 (95.021)
$5.0\tau_C$	0.99326 (99.326)
$10.0\tau_C$	0.99995 (99.995)

$t = t_0$ and a value that approaches 1 as time increases:

$$\lim_{t \to \infty} \left(1 - \exp\left[-\frac{t - t_0}{\tau_C}\right]\right) = 1$$

Table 3.2 lists the fraction (or percentage) of full charge with elapsed time $(t - t_0)$ expressed in terms of the time constant. Although full charge is not possible theoretically, since the growth function only *approaches* unity, an interval equal to five times the time constant ($5\tau_C$) is conventionally considered sufficient for full charge.

As with Equations (3.11) and (3.14) in Section 3.5, Equations (3.25) through (3.32) are valid only in the domain $t > t_0$, and cannot be used to determine voltage or current, as the case may be, for any time prior to t_0.

Example 3.11

SPICE

Problem:

Find and sketch $v_C(t)$ for the circuit of Figure 3.17(a).

Solution:

1. From the equivalent circuit for $t = 0^-$ [Figure 3.17(b)],

$$v_C(0^-) = 4 \text{ V}$$

2. Calculate the time constant for charging from the equivalent circuit for $t > 0$ [Figure 3.17(c)]:

$$\tau_C = (200)(1.5 \times 10^{-6}) = 3 \text{ msec}$$

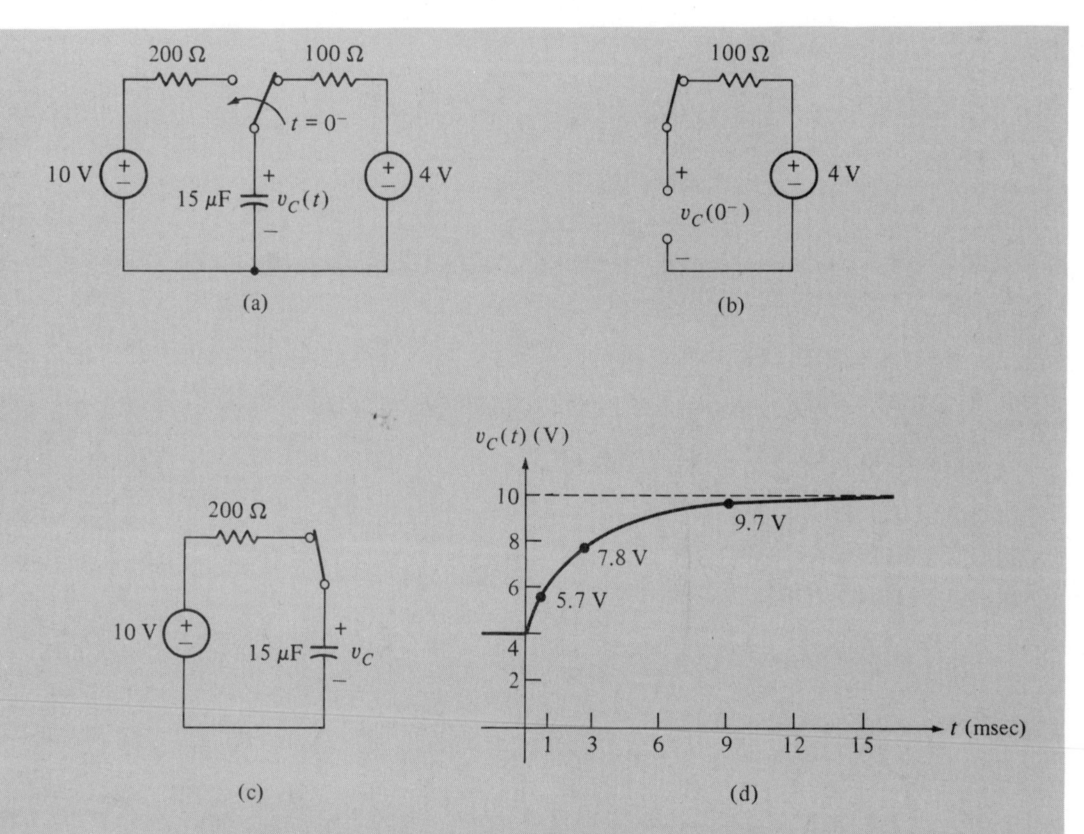

Figure 3.17 Example 3.11. (a) Complete circuit. (b) Equivalent circuit at $t = 0^-$. (c) Equivalent circuit for $t > 0$. (d) $v_C(t)$ vs. t.

3. Use Equation 3.25 to determine $v_C(t)$ for $t > 0$:

$$v_C(t > 0) = 10 - (10 - 4)\exp\left[-\frac{t}{3 \times 10^{-3}}\right]$$

$$= 10 - 6 \cdot \exp[-333.3t] \text{ V}$$

4. Combine $v_C(0^-)$ and $v_C(t > 0)$ to obtain $v_C(t)$:

$$v_C(t) = 4 - 4 \cdot u(t) + (10 - 6 \cdot \exp[-333.3t])u(t)$$

$$= 4 + 6(1 - \exp[-333.3t])u(t) \text{ V}$$

Figure 3.17(d) is a sketch of $v_C(t)$ vs. t.

 Source conversion

Example 3.12

SPICE

Problem:

An inductor with inductance of 0.3 mH and winding resistance r_W of 10 Ω is connected to a step-function current source, $60 \cdot u(t)$ mA, with internal resistance of 50 Ω [Figure 3.18(a)]. Find the voltage $v_L(t)$ across the inductor.

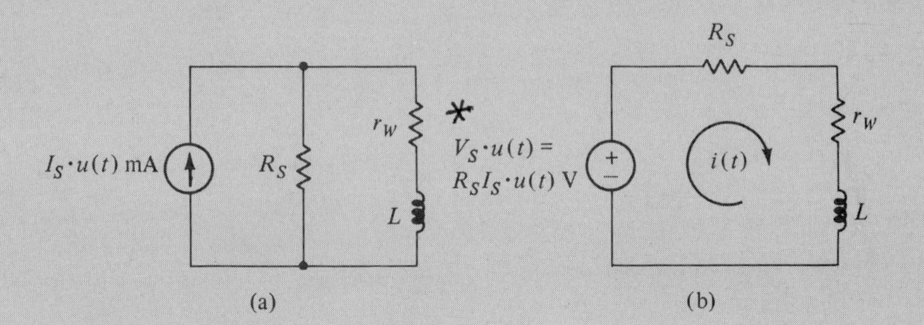

Figure 3.18 Example 3.12. (a) Inductor charging from current source. (b) Equivalent circuit after source conversion.

Solution:

1. The circuit as shown is not a simple parallel circuit, so none of Equations (3.25) through (3.28) can be used to obtain the solution directly.
2. *Use *source conversion*, as indicated in Figure 3.18(b), to obtain a simple series R–L circuit for which the current can be determined from Equation (3.27).
3. Because the only source is a step-function source, there is no energy in the circuit prior to $t = 0$, and the initial inductor current $i_L(0)$ is zero, as is the initial inductor voltage. With zero initial conditions, Equation (3.27) reduces to Equation (3.31), and

$$i(t > 0) = \frac{3}{50 + 10}\left(1 - \exp\left[-\frac{t}{\tau_C}\right]\right)$$

where $\tau_C = 0.3 \times 10^{-3}/(50 + 10) = 5\ \mu\text{sec}$. Therefore,

$$\begin{aligned} i(t) &= i(t > t_0) \cdot u(t) \\ &= 50(1 - \exp[-200 \times 10^3 t]) \cdot u(t) \text{ mA} \end{aligned}$$

4. Calculate inductor voltage from KVL:

$$v_L(t) = r_W \cdot i(t) + L\frac{d}{dt}i(t)$$

$$v_L(t > 0) = 10 \cdot i(t > 0) + 0.3 \times 10^{-3}\frac{d}{dt}i(t > 0)$$

$$= 0.5(1 - \exp[-200 \times 10^3 t])$$

$$+ 15 \times 10^{-6}\{-(-200 \times 10^3)\exp[-200 \cdot 10^3 t]\}$$

$$= 0.5 + 2.5 \cdot \exp[-200 \times 10^3 t] \text{ V}$$

$$v_L(t) = v_L(t > 0) \cdot u(t)$$

$$= (0.5 + 2.5 \cdot \exp[-200 \times 10^3 t])u(t) \text{ V}$$

Example 3.13

SPICE

Problem:

Find and sketch the voltage $v_C(t)$ across the capacitor in Figure 3.19(a). Determine the time at which $v_C(t) = 0$ V.

Solution:

1. Determine the initial capacitor voltage from the equivalent circuit at $t = 0^-$ [Figure 3.19(b)]:

$$v_C(0) = v_C(0^-) = \frac{15(100)}{400 + 100} = 3 \text{ V}$$

2. The equivalent circuit for $t > 0$ is shown in Figure 3.19(c).
3. Obtain a series R–C circuit [Figure 3.19(d)] by applying Thevenin's theorem to the network to the left of the capacitor terminals. Use Equation (3.25) to obtain the solution $v_C(t)$ for $t > 0$:

$$v_C(t > 0) = -1 - (-1 - 3)\exp[-t/\tau_C] \text{ V}$$

where

$$\tau_C = (180) \cdot 50 \times 10^{-6} = 9 \text{ msec}$$

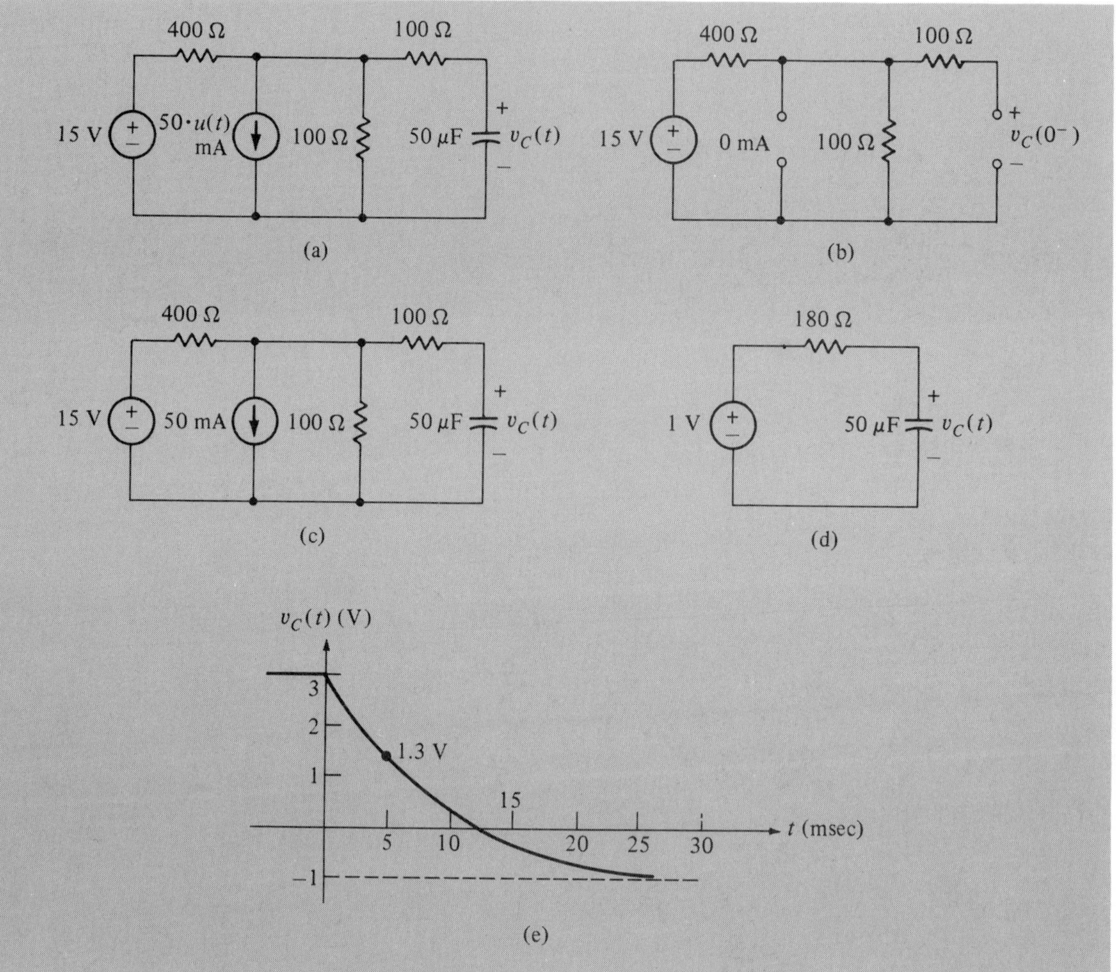

Figure 3.19 Example 3.13. (a) Complete circuit. (b) Equivalent circuit at $t = 0^-$. (c) Equivalent circuit for $t > 0$. (d) Thevenin's equivalent circuit. (e) $v_C(t)$ vs. t.

so

$$v_C(t > 0) = -1 + 4 \cdot \exp[-111.1t] \text{ V}$$

4. Combine $v_C(0^-)$ and $v_C(t > 0)$ to obtain $v_C(t)$:

$$v_C(t) = 3 - 3 \cdot u(t) + (-1 + 4 \cdot \exp[-111.1t])u(t)$$
$$= 3 - 4(1 - \exp[-111.1t]) \cdot u(t) \text{ V}$$

5. Figure 3.19(e) is a sketch of $v_C(t)$ vs. t. Solve the equation $v_C(t) = 0$ V for t:

$$\begin{aligned} 3 - 4(1 - \exp[-111t]) &= 0 \\ 4 \cdot \exp(-111t) &= 1 \\ -111t &= \ln\tfrac{1}{4} \\ t &= 12.5 \text{ msec} \end{aligned}$$

Example 3.14 SPICE

Problem:

Sketch the capacitor voltage $v_C(t)$ [Figure 3.20(a)] when

$$v_S(t) = 5 \cdot u(t) - 5 \cdot u(t-5) + 5 \cdot u(t-8) \text{ V}$$

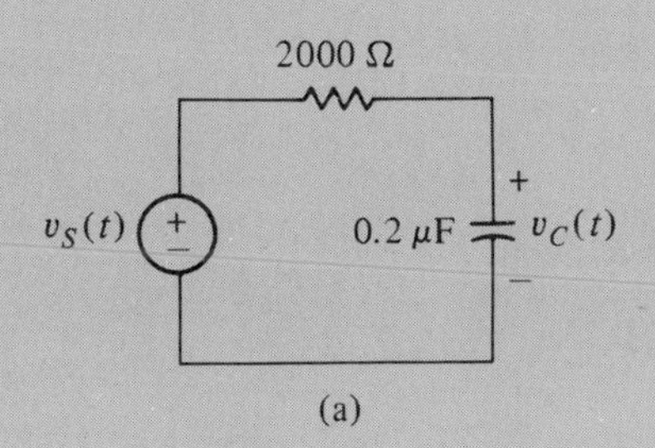

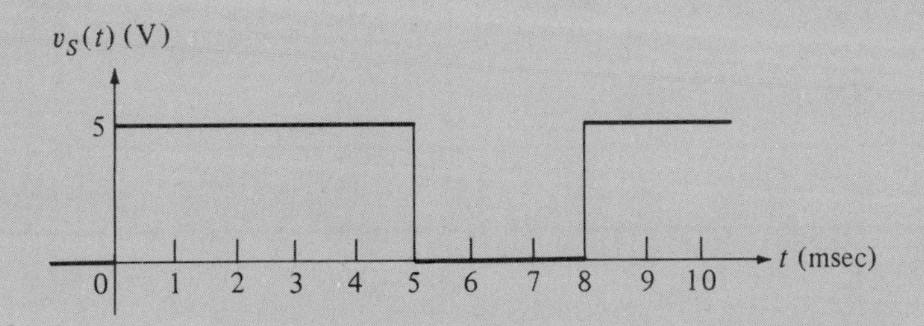

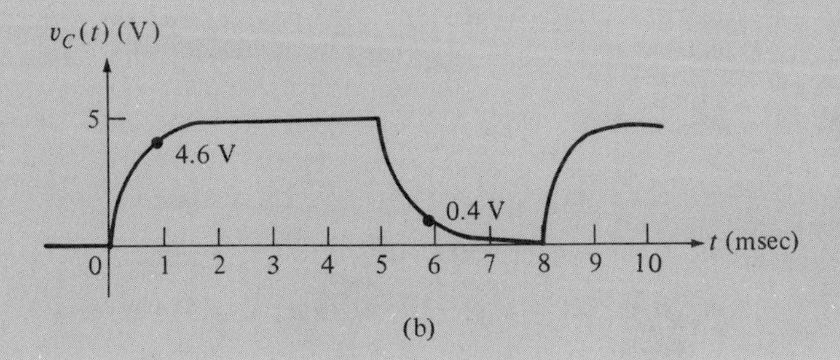

Figure 3.20
Example 3.14.
(a) R–C circuit with pulse source.
(b) Sketch of $v_S(t)$ and $v_C(t)$ vs. t.

Solution:

1. Since the source voltage is zero prior to $t = 0$, then $v_C(0^-)$ is zero.
2. Use Equation (3.29) to determine $v_C(t)$ for the interval $0 < t < 5$ msec:

$$v_C(0 < t < 0.005) = 5(1 - \exp[-t/\tau_C])\ \text{V}$$

3. Calculate the time constant for the circuit:

$$\tau_C = (2000)(0.2 \times 10^{-6}) = 400\ \mu\text{sec}$$

Therefore,

$$v_C(0 < t < 0.005) = 5(1 - \exp[-2500t])\ \text{V}$$

4. Calculate $v_C(t)$ at 5 msec:

$$v_C(0.005) = 5\{1 - \exp[-(2500)(5 \times 10^{-3})]\} \simeq 5\ \text{V}$$

which is the initial capacitor charge for the discharge period $5 < t < 8$ msec.

5. Use Equation (3.11) to determine $v_C(t)$ during discharge:

$$v_C(0.005 < t < 0.008) = 5 \cdot \exp[-2500(t - 0.005)]\ \text{V}$$

Figure 3.20(b) is a sketch of the capacitor voltage.

Exercise 3.5

SPICE

Problem:

For each of the circuits in Figure 3.21, write the expression for $v_C(t)$ or $i_L(t)$, as appropriate. Determine the time required for capacitor voltage or inductor current to reach a value halfway between initial value and final value.

Answers:

a. $6(1 - \exp[-45.45t])u(t)$ V; 15.25 msec.

b. $0.1(1 - \exp[-6000t])u(t)$ A; 115.5 μsec.

c. $2 + 10(1 - \exp[-106.4t])u(t)$ V; 6.516 msec.

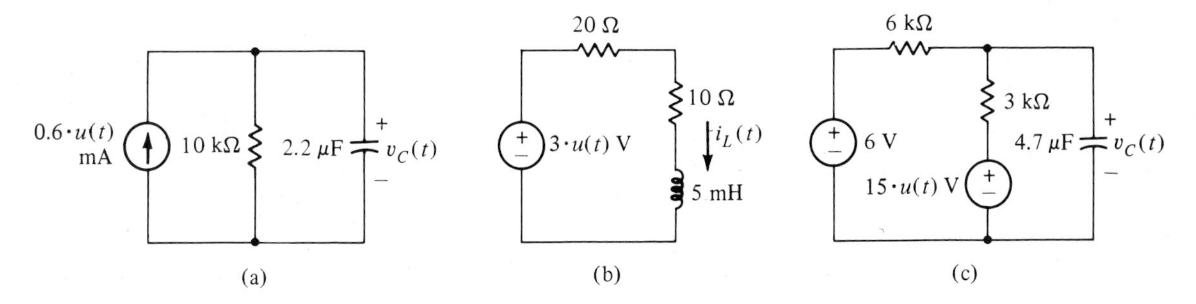

Figure 3.21 Exercise 3.5.

Computer Simulation

SPICE 3.1 illustrates time-domain simulation of a driven R–C circuit (Example 3.13). Circuits containing inductors, or both inductors and capacitors, are simulated in a similar manner. The .TRAN directive specifies that a time-domain analysis is to be performed, and controls timing and scaling of the analysis. The reader should refer to Appendix D for the format of the .TRAN directive. The output interval (TSTEP) and the final time (TSTOP) must be specified. If the initial time (TSTART) is not specified, then TSTART is set to zero by default. If no maximum computing interval (TMAX) is specified, then TMAX defaults to the smaller of TSTEP and (TSTOP − TSTART)/50. In SPICE 3.1, TSTEP is 2.5 msec, but TMAX is 0.6 msec [or (30 − 0)/50]. If a TMAX of 0.5 msec is desired, then the .TRAN directive must be modified in one of the following ways:

```
.TRAN .5M 30M
```

or

```
.TRAN 2.5M 30M 0 .5M
```

The first modification causes both TSTEP and TMAX to be 0.5 msec, and thus provides five times as many output points. The second modification causes TMAX to be 0.5 msec, but does not increase the number of output points. Note, in the second modification, that when TMAX is specified, then TSTART must also be specified.

Appending the keyword UIC (for user initial conditions) to the .TRAN directive allows the user to force the initial conditions, or to analyze a source-free circuit. Omission of UIC from the .TRAN statement (as in SPICE 3.1) causes SPICE to compute the initial conditions by default before beginning the

```
*******10/14/88 ********  SPICE 2G.6    3/15/83 ********08:22:11*****
SPICE 3.1
****     INPUT LISTING                       TEMPERATURE =   27.000 DEG C
*******************************************************************************

 R1 1 2 400
 R2 2 0 100
 R3 2 3 100
 C 3 0 50U
 V 1 0 DC 15
 I 2 0 PULSE(0 50M 0 1F)
 .TRAN 2.5M 30M
 .PLOT TRAN V(3)
 .END
*******10/14/88 ********  SPICE 2G.6    3/15/83 ********08:22:12*****
SPICE 3.1
****     INITIAL TRANSIENT SOLUTION          TEMPERATURE =   27.000 DEG C
*******************************************************************************

  NODE   VOLTAGE     NODE   VOLTAGE     NODE   VOLTAGE
 (  1)   15.0000    (  2)    3.0000    (  3)    3.0000
     VOLTAGE SOURCE CURRENTS
     NAME       CURRENT
     V        -3.000D-02
     TOTAL POWER DISSIPATION   4.50D-01  WATTS
*******10/14/88 ********  SPICE 2G.6    3/15/83 ********08:22:12*****
SPICE 3.1
****     TRANSIENT ANALYSIS                  TEMPERATURE =   27.000 DEG C
*******************************************************************************


     TIME       V(3)
                  -1.000D+00      0.000D+00      1.000D+00      2.000D+00  3.000D+00
                - - - - - - - - - - - - - - - - - - - - - - - - - - - - - - - - - -
 0.000D+00  3.000D+00 .              .              .              .              *
 2.500D-03  2.030D+00 .              .              .              *              .
 5.000D-03  1.295D+00 .              .              .   *          .              .
 7.500D-03  7.386D-01 .              .        *     .              .              .
 1.000D-02  3.170D-01 .              .  *           .              .              .
 1.250D-02 -2.531D-03 .              *              .              .              .
 1.500D-02 -2.446D-01 .           *  .              .              .              .
 1.750D-02 -4.280D-01 .        *     .              .              .              .
 2.000D-02 -5.668D-01 .      *       .              .              .              .
 2.250D-02 -6.718D-01 .     *        .              .              .              .
 2.500D-02 -7.514D-01 .   *          .              .              .              .
 2.750D-02 -8.117D-01 .   *          .              .              .              .
 3.000D-02 -8.575D-01 .  *           .              .              .              .
                - - - - - - - - - - - - - - - - - - - - - - - - - - - - - - - - - -

        JOB CONCLUDED
```

time-domain analysis. When UIC is present, SPICE looks to the capacitor and inductor records for the entry IC = ... for initial conditions. The IC = ... entry specifies the initial voltage for capacitors and initial current for inductors. A secondary method for entering user initial conditions is to include an .IC statement, which specifies initial conditions in terms of node voltages. SPICE 3.2 illustrates the application of user initial conditions in time-domain analysis of a source-free circuit (Example 3.10).

```
*******10/14/88 ********  SPICE 2G.6    3/15/83 ********08:27:23*****
SPICE 3.2
****     INPUT LISTING                      TEMPERATURE =   27.000 DEG C
*****************************************************************************
 R1 1 0 400
 R2 1 0 100
 L 2 0 40M IC=24M
 VAM 1 2
 .TRAN .125M 2.5M UIC
 .PLOT TRAN I(VAM)
 .END
*******10/14/88 ********  SPICE 2G.6    3/15/83 ********08:27:24*****
SPICE 3.2
****     TRANSIENT ANALYSIS                 TEMPERATURE =   27.000 DEG C
*****************************************************************************

    TIME      I(VAM)
                 0.000D+00     1.000D-02     2.000D-02     3.000D-02  4.000D-02
                 - - - - - - - - - - - - - - - - - - - - - - - - - - - - - - - -
 0.000D+00   2.400D-02 .            .             .     *       .            .
 1.250D-04   1.870D-02 .            .            * .            .            .
 2.500D-04   1.457D-02 .            .      *       .            .            .
 3.750D-04   1.134D-02 .            . *            .            .            .
 5.000D-04   8.832D-03 .          * .              .            .            .
 6.250D-04   6.875D-03 .        *   .              .            .            .
 7.500D-04   5.354D-03 .      *     .              .            .            .
 8.750D-04   4.168D-03 .     *      .              .            .            .
 1.000D-03   3.246D-03 .    *       .              .            .            .
 1.125D-03   2.527D-03 .   *        .              .            .            .
 1.250D-03   1.968D-03 .  *         .              .            .            .
 1.375D-03   1.532D-03 . *          .              .            .            .
 1.500D-03   1.193D-03 . *          .              .            .            .
 1.625D-03   9.289D-04 .*           .              .            .            .
 1.750D-03   7.234D-04 .*           .              .            .            .
 1.875D-03   5.632D-04 .*           .              .            .            .
 2.000D-03   4.386D-04 .*           .              .            .            .
 2.125D-03   3.414D-04 *            .              .            .            .
 2.250D-03   2.659D-04 *            .              .            .            .
 2.375D-03   2.070D-04 *            .              .            .            .
 2.500D-03   1.611D-04 *            .              .            .            .
                 - - - - - - - - - - - - - - - - - - - - - - - - - - - - - - - -

        JOB CONCLUDED
```

```
*******10/14/88 ********  SPICE 2G.6    3/15/83 ********08:49:59*****
SPICE 3.3
****     INPUT LISTING                      TEMPERATURE =   27.000 DEG C
********************************************************************************
 R 1 2 2K
 C 2 0 .2U
 V 1 0 PULSE(0 5 0 1F 1F 5M 8M)
 .TRAN .5M 10M
 .PLOT TRAN V(2) V(1)
 .END
*******10/14/88 ********  SPICE 2G.6    3/15/83 ********08:49:59*****
SPICE 3.3
****     INITIAL TRANSIENT SOLUTION         TEMPERATURE =   27.000 DEG C
********************************************************************************
  NODE   VOLTAGE     NODE   VOLTAGE
 (  1)     .0000    (  2)     .0000
     VOLTAGE SOURCE CURRENTS
     NAME       CURRENT
     V         0.000D+00
     TOTAL POWER DISSIPATION   0.00D+00  WATTS
*******10/14/88 ********  SPICE 2G.6    3/15/83 ********08:49:59*****
SPICE 3.3
****     TRANSIENT ANALYSIS                 TEMPERATURE =   27.000 DEG C
********************************************************************************
LEGEND:
 *: V(2)
 +: V(1)

     TIME       V(2)
(*+)----------- 0.000D+00     2.000D+00     4.000D+00      6.000D+00  8.000D+00
                   - - - - - - - - - - - - - - - - - - - - - - - - - - - - - - -
 0.000D+00  0.000D+00 X             .             .             .             .
 5.000D-04  3.562D+00 .             .          *  .      +      .             .
 1.000D-03  4.596D+00 .             .             .   *  +      .             .
 1.500D-03  4.888D+00 .             .             .     *+      .             .
 2.000D-03  4.969D+00 .             .             .      X      .             .
 2.500D-03  4.991D+00 .             .             .      X      .             .
 3.000D-03  4.998D+00 .             .             .      X      .             .
 3.500D-03  4.999D+00 .             .             .      X      .             .
 4.000D-03  5.000D+00 .             .             .      X      .             .
 4.500D-03  5.000D+00 .             .             .      X      .             .
 5.000D-03  5.000D+00 .             .             .      X      .             .
 5.500D-03  1.410D+00 +         *   .             .             .             .
 6.000D-03  4.061D-01 +   *         .             .             .             .
 6.500D-03  1.096D-01 +*            .             .             .             .
 7.000D-03  3.158D-02 X             .             .             .             .
 7.500D-03  8.526D-03 X             .             .             .             .
 8.000D-03  2.387D-03 X             .             .             .             .
 8.500D-03  3.549D+00 .             .          *  .      +      .             .
 9.000D-03  4.603D+00 .             .             .   *  +      .             .
 9.500D-03  4.887D+00 .             .             .     *+      .             .
 1.000D-02  4.969D+00 .             .             .      X      .             .
                   - - - - - - - - - - - - - - - - - - - - - - - - - - - - - - -

        JOB CONCLUDED
```

SPICE 3.3 illustrates time-domain analysis of a circuit with a periodic source.

3.7 Final Conditions and a More General Technique

Equations (3.25) through (3.28) represent solutions for the differential equations which arise from R–C or R–L circuits with arbitrary initial conditions and driven by a switched dc source. A closer examination of these equations suggests a more general result.

If each of Equations (3.25) through (3.28) is evaluated for a large value of t ($t \to \infty$), then

Series R–C circuit [Figure 3.16(a)]:

$$\lim_{t \to \infty} v_C(t) = V_S \tag{3.33}$$

Parallel R–L circuit [Figure 3.16(b)]:

$$\lim_{t \to \infty} i_L(t) = I_S \tag{3.34}$$

Series R–L circuit [Figure 3.16(c)]:

$$\lim_{t \to \infty} i(t) = \frac{V_S}{R} \tag{3.35}$$

Parallel R–C circuit [Figure 3.16(d)]:

$$\lim_{t \to \infty} v(t) = \frac{I_S}{G} \tag{3.36}$$

These are **final conditions** of capacitor voltage or inductor current. If initial conditions [$v(t_0)$ or $i(t_0)$] are designated F_i, and final conditions are designated F_f, then the single equation

$$f(t) = F_i + (F_f - F_i)\left(1 - \exp\left[-\frac{t - t_0}{\tau_C}\right]\right)u(t - t_0) \tag{3.37}$$

represents Equations (3.25) through (3.28). The symbol $f(t)$ represents either capacitor voltage or inductor current.

Equation (3.37) is not restricted to the simple series or parallel circuits of Figure 3.16, but applies to a single capacitor or a single inductor imbedded anywhere in a circuit in which all other elements are resistors or switched dc sources. In short, if the initial conditions, final conditions, and equivalent resistance of the charge or discharge path can be determined, then the capacitor voltage or inductor current at any time greater than initial time can be calculated. Equivalent resistance is required to determine the time constant τ.

In source-free circuits, $F_f = 0$, so Equation (3.37) becomes

$$f(t) = F_i - F_i\left(1 - \exp\left[-\frac{t - t_0}{\tau_C}\right]\right)u(t - t_0)$$

$$= F_i \cdot \exp\left[\frac{t - t_0}{\tau_C}\right], \quad \text{for } t > t_0 \tag{3.38}$$

which confirms Equation (3.11) for the source-free R–C circuit and Equation (3.14) for the source-free R–L circuit. For circuits in which initial conditions are zero, $F_i = 0$, so Equation (3.38) becomes

$$f(t) = F_f\left(1 - \exp\left[-\frac{t - t_0}{\tau}\right]\right)u(t - t_0) \tag{3.39}$$

which confirms Equations (3.29) through (3.32).

Obtaining Final Conditions

The final conditions for one region of the time domain are the initial conditions for the region which immediately follows. Thus, the initial conditions for the region which begins with t_0 are actually the final conditions for the region preceding t_0. Thus, final conditions for the region which begins with t_0 are obtained in exactly the same way as initial conditions (Section 3.3)—that is, by replacing capacitors with open circuits to obtain the final capacitor voltage, and replacing inductors with short circuits to obtain the final inductor current.

Example 3.15

Problem:

Use Equation (3.37) to confirm the results obtained in Examples 3.11 through 3.14.

Solution:

1. Example 3.11: Obtain V_i from Figure 3.17(b):

$$V_i = 4 \text{ V}$$

Replace the capacitor in Figure 3.17(c) with an open circuit to obtain V_f:

$$V_f = 10 \text{ V}$$

From Equation (3.37),

$$\begin{aligned} v_C(t) &= 4 + (10 - 4)(1 - \exp[-t/\tau_C])u(t) \\ &= 4 + 6(1 - \exp[-t/\tau_C])u(t) \text{ V} \end{aligned}$$

and

$$\begin{aligned} v_C(t > 0) &= 4 + 6(1 - \exp[-t/\tau_C]) \\ &= 10 - 6 \cdot \exp[-t/\tau_C] \text{ V} \end{aligned}$$

which agrees with the result obtained in Example 3.11.

2. Example 3.12: $I_i = 0$ because the only source is an unshifted step function. Replace the inductor in Figure 3.18(a) with a short circuit, and use the current division rule to obtain I_f:

$$I_f = \frac{60(0.1)}{0.1 + 0.02} = 50 \text{ mA}$$

From Equation (3.37),

$$\begin{aligned} i_L(t) &= 0 + (50 - 0)(1 - \exp[-t/\tau_C])u(t) \\ &= 50(1 - \exp[-t/\tau_C])u(t) \text{ mA} \end{aligned}$$

which agrees with step 3 of Example 3.12.

3. Example 3.13: Obtain V_i from Figure 3.19(b), as in Example 3.13:

$$V_i = 3 \text{ V}$$

Replace the capacitor in Figure 3.17(d) with an open circuit to obtain V_f:

$$V_f = -1 \text{ V}$$

From Equation (3.37),

$$\begin{aligned} v_C(t) &= 3 + (-1 - 3)(1 - \exp[-t/\tau_C])u(t) \\ &= 3 - 4(1 - \exp[-t/\tau_C])u(t) \text{ V} \end{aligned}$$

which agrees with the result obtained in Example 3.13.

4. Example 3.14: For the interval $0 < t < 5$ msec, $V_i = 0$ because the only source is an unshifted step function. Replace the capacitor in Figure 3.20(a) with an open circuit to obtain V_f at $t \to 5$ msec:

$$V_f = 5 \text{ V}$$

From Equation (3.37),

$$\begin{aligned} v_C(t) &= 0 + (5 - 0)(1 - \exp[-t/\tau_C])u(t) \\ &= 5(1 - \exp[-t/\tau_C])u(t) \text{ V} \end{aligned}$$

which agrees with step 2 of Example 3.14. For the interval $5 < t < 8$ msec, the initial condition is the same as the final condition from the previous interval,

$$V_i = 5 \text{ V}$$

Furthermore, the circuit is source-free during this interval, so

$$V_f = 0$$

From Equation (3.37),

$$\begin{aligned} v_C(t) &= 5 + (0 - 5)\left(1 - \exp\left[-\frac{t - 0.005}{\tau_C}\right]\right)u(t - 0.005) \\ &= 5 \cdot \exp\left[-\frac{t - 0.005}{\tau_C}\right] \cdot u(t - 0.005) \text{ V} \end{aligned}$$

which agrees with step 5 of Example 3.14.

Equation (3.37) is very useful in the analysis and design of switched dc circuits which contain only resistors and a single capacitor or inductor. Proper application of Equation (3.37) avoids having to solve a first-order differential equation or to recall a formula [such as Equation (3.11), (3.14), or (3.25) through (3.32)] for each circuit model.

3.8 Higher-Order Circuits

The objective of this section is to investigate the existence and the origin of higher (than first) order differential equations in circuit analysis. The solution of initial value problems involving these differential equations will be deferred to the chapters involved with Laplace transform techniques.

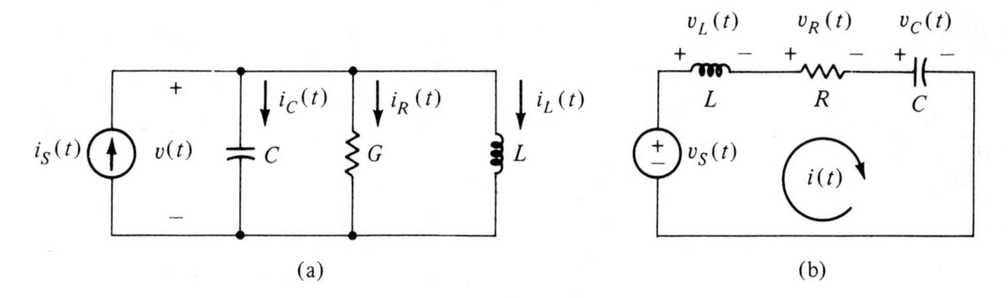

Figure 3.22 *R–L–C* circuits. (a) Parallel. (b) Series.

Consider first the *R–L–C* parallel circuit shown in Figure 3.22(a). By Kirchhoff's current law,

$$i_C(t) + i_R(t) + i_L(t) = i_S(t) \tag{3.40}$$

When Equation (3.2), Ohm's law and Equation (3.6) are substituted for $i_C(t)$, $i_R(t)$, and $i_L(t)$, respectively, the Equation (3.40) becomes

$$C\frac{d}{dt}v(t) + G \cdot v(t) + \frac{1}{L}\int_{t_0}^{t_1} v(t)\, dt + i_L(0) = i_S(t)$$

Division by C and differentiation of both members yields

$$\frac{d^2}{dt^2}v(t) + \frac{G}{C} \cdot \frac{d}{dt}v(t) + \frac{1}{LC} \cdot v(t) = \frac{1}{C} \cdot \frac{d}{dt}i_S(t) \tag{3.41}$$

Application of Kirchhoff's voltage law and electrical duals to the series *R–L–C* circuit [Figure 3.22(b)] leads to the differential equation for the circuit current:

$$\frac{d^2}{dt^2}i(t) + \frac{R}{L} \cdot \frac{d}{dt}i(t) + \frac{1}{CL} \cdot i(t) = \frac{1}{L} \cdot \frac{d}{dt}v_S(t) \tag{3.42}$$

Equations (3.41) and (3.42) demonstrate that even the simplest circuit configurations which involve both a capacitor and an inductor will give rise to a second-order response. Circuit configurations which involve more than one capacitor or more than one inductor, not in parallel or series, will also give rise to a second-order response. The circuit shown in Figure 3.23(a), for example,

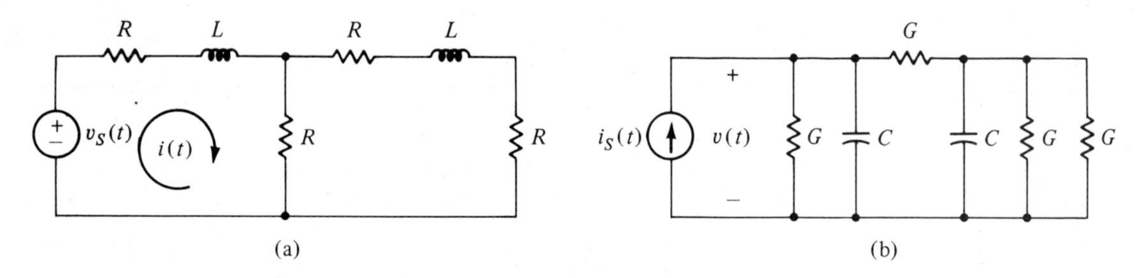

Figure 3.23 Circuits incorporating more than one energy storage element. (a) Inductor tee circuit. (b) Capacitor pi circuit.

has a current response determined by the equation

$$\frac{d^2}{dt^2}i(t) + \frac{5R}{L} \cdot \frac{d}{dt}i(t) + \frac{5R^2}{L^2} \cdot i(t)$$
$$= \frac{1}{L} \cdot \frac{d}{dt}v_S(t) + \frac{3R}{L^2} \cdot v_S(t)$$

and the dual circuit in Figure 3.23(b) has a voltage response determined by

$$\frac{d^2}{dt^2}v(t) + \frac{5G}{C} \cdot \frac{d}{dt}v(t) + \frac{5G^2}{C^2} \cdot v(t)$$
$$= \frac{1}{C} \cdot \frac{d}{dt}i_S(t) + \frac{3G}{C^2} \cdot i_S(t)$$

In general, the order of the response is equal to the number of energy storage elements which are not connected in parallel or series with another energy storage element of the same type. The output voltage of the L–C pi circuit shown in Figure 3.24, for example, is a third-order response determined by

$$\frac{d^3}{dt^3}v_0(t) + \frac{1}{RC} \cdot \frac{d^2}{dt^2}v_0(t) + \frac{2}{LC} \cdot \frac{d}{dt}v_0(t) + \frac{1}{RLC^2} \cdot v_0(t)$$
$$= \frac{1}{RLC^2} \cdot v_S(t)$$

Obtaining solutions to these higher-order differential equations can be quite difficult working in the time domain. Fortunately, the Laplace transform, to be introduced in the next chapter, will reduce the solution procedure to relatively routine algebraic operations.

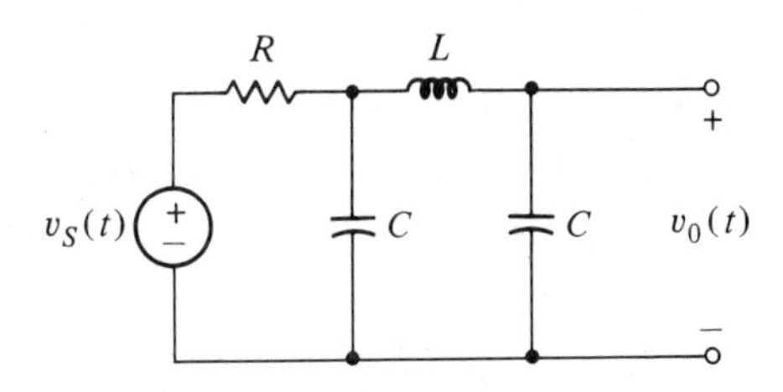

Figure 3.24 Circuit with third-order response.

3.9 SUMMARY

We have examined the time-domain characteristics of the resistor, inductor, and capacitor, and have demonstrated that their incorporation into electrical circuits introduces differential equations during analysis of the response of those circuits. Analysis of those circuits containing only one energy storage element (i.e., first-order circuits) can be accomplished by direct solution of the differential equation using methods of integral calculus. Further, the relatively few possible circuit configurations make it possible to develop a manageable set of formulas for use in analysis of first-order circuits excited by dc or rectangular waveform sources.

Incorporation of additional energy storage elements into a circuit results in higher-order differential equations. The difficulty of solving these equations by methods of integral calculus leads us to the next chapter, where the use of Laplace transform methods for solving differential equations will be introduced.

3.10 Terms

capacitance
charge function
decay function
differential equation
discharge function
driven circuit
exact derivative
forcing function
growth function
homogeneous
inductance
initial conditions
initial value problem
integrating factor
linear capacitor
linear differential equation
linear inductor
ordinary differential equation
separation of variables
source-free circuit
time constant
time-domain analysis

Problems

3-1. Find and sketch the terminal current $i(t)$ through a 0.1-μF capacitor for the given terminal voltage. Evaluate $i(100\ \mu\text{sec})$ in each case.

*__a.__ $3 \cdot u(t)$ V.

b. $6000t \cdot u(t)$ V.

*__c.__ $200 \cdot \exp[-3000t] \cdot u(t)$ V.

d. $5\sum_{n=0}^{\infty}(-1)^n \cdot u(t - n \times 10^{-3})$ V.

3-2. Find and sketch the terminal voltage $v(t)$ for a 2200-pF capacitor for the given capacitor current. Assume that $v(0^-) = 5$ V, and evaluate $v(1\ \mu\text{sec})$ in each case.

a. $400 \cdot u(t)\ \mu$A.

*__b.__ $-10 \times 10^3 t \cdot u(t)$ A.

c. $2 \cdot \exp[-10^6 t] \cdot u(t)$ mA.

*__d.__ $5\sum_{n=0}^{\infty}(-1)^n \cdot \delta(t - n \cdot 10 \times 10^{-6})$ nA.

3-3. Find C for the given terminal voltage and capacitor current.

*__a.__ $300t$ V and 30 mA.

b. $-2(1 - \exp[-5000t])$ V and $400 \cdot \exp[-5000 \cdot t]\ \mu$A.

*__c.__ $20t^2$ V and $i(20\ \text{msec}) = 17.6\ \mu$A.

3-4. Find and sketch the terminal voltage $v(t)$ across a 10-mH inductor for the given inductor current. Evaluate $v(t)$ for $t = 3\ \mu$sec.

a. $5 \cdot u(t)$ mA.

*__b.__ $-10^4 t \cdot u(t)$ A.

c. $-3 \cdot \exp[-10^6 t] \cdot u(t)$ mA.

d. $300[t \cdot u(t) - 10^{-6}\sum_{n=0}^{\infty}u(t - (n+1) \times 10^{-6})]$
$\qquad -10^{-6}\sum_{n=0}^{\infty}u(t - (n+1) \times 10^{-6})$ A.

3-5. Find and sketch the current $i(t)$ through a 0.2-H inductor for the given terminal voltage. Assume that $i(0^-) = 1$ mA, and evaluate $i(30\ \mu\text{sec})$.

*__a.__ $10 \cdot u(t)$ V

b. $-3 \times 10^6 t \cdot u(t)$ V

*__c.__ $4 \cdot \exp[-2000t] \cdot u(t)$ V

d. $2000\sum_{n=1}^{\infty}\delta(t - n \cdot 400 \times 10^{-6})$ V

3-6. Find L for the given terminal voltage and inductor current.

a. 8 V and $400t$ mA.

*__b.__ $18 \cdot \exp[-10 \times 10^3 t]$ V and $-60 \cdot \exp[-10 \times 10^3 t]$ mA.

c. $v(5\ \text{msec}) = 12$ V and $96 \times 10^6 t^3$ A.

3-7. Show from Equations (3.2) and (3.5), respectively, that

a. 1 farad = 1 ampere · second/volt, and

b. 1 henry = 1 volt · second/ampere.

***3-8.** The voltage across a 47-μF capacitor is given by

$$v(t) = 2500\big[t \cdot u(t) - (t - 2 \times 10^{-3})u(t - 2 \times 10^{-3}) - (t - 6 \times 10^{-3})u(t - 6 \times 10^{-3}) + (t - 8 \times 10^{-3})u(t - 8 \times 10^{-3})\big] \text{ V}$$

Sketch the capacitor voltage and current.

***3-9.** The voltage across a 6-mH inductor is given by

$$v(t) = 10 \cdot u(t - 2 \times 10^{-6}) - 10^{6}\big[(t - 2 \times 10^{-6})u(t - 2 \times 10^{-6}) + (t - 12 \times 10^{-6})u(t - 12 \times 10^{-6})\big] \text{ V}$$

Sketch the inductor voltage and current.

3-10. Find the initial conditions for each L and C.

a.

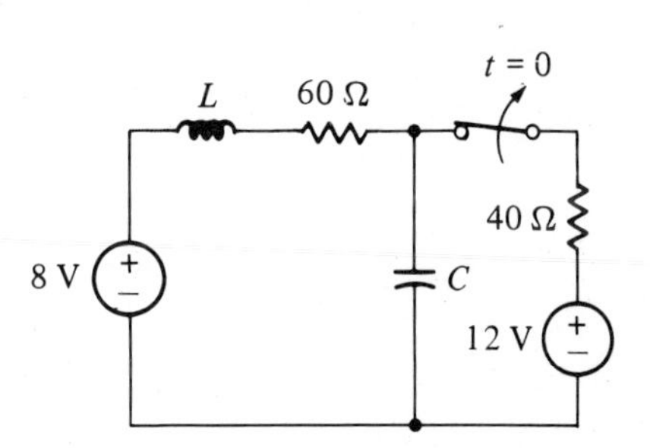

***b.**

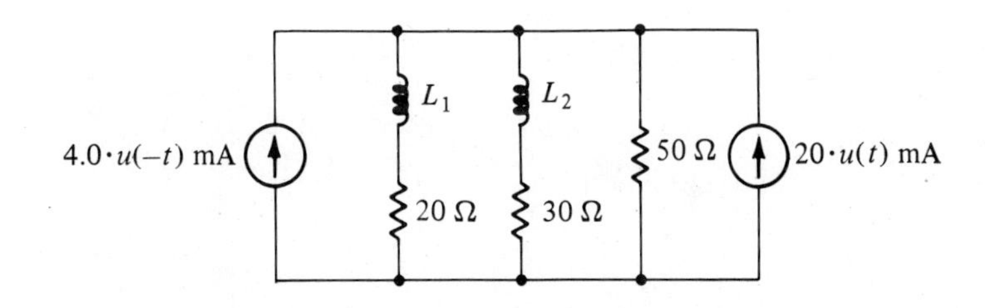

c.

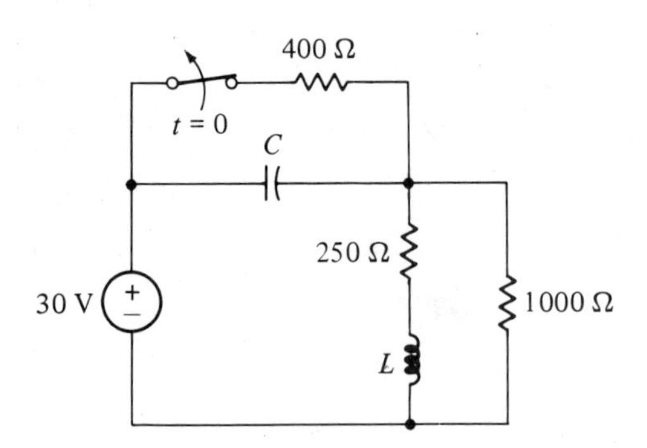
400 Ω
t = 0
C
250 Ω
30 V
1000 Ω
L

***d.**

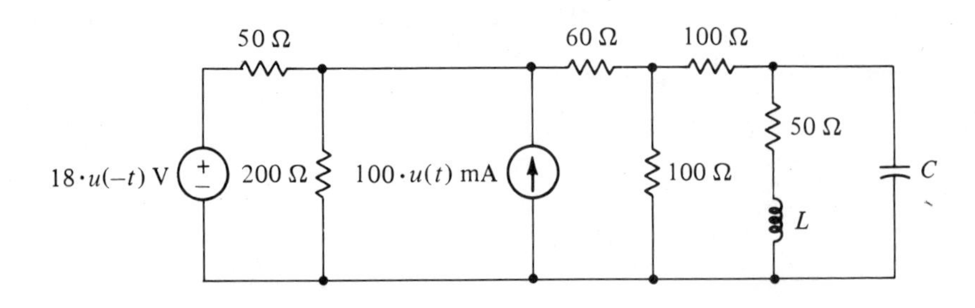
50 Ω
60 Ω
100 Ω
50 Ω
18·u(−t) V
200 Ω
100·u(t) mA
100 Ω
C
L

e.

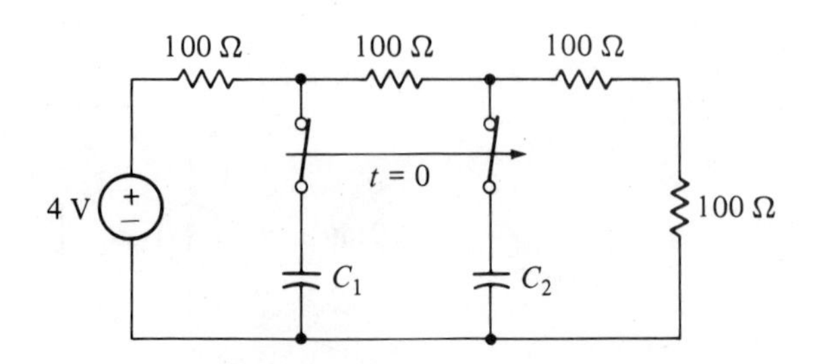
100 Ω
100 Ω
100 Ω
t = 0
4 V
100 Ω
C1
C2

SPICE

3-11. Find and sketch the unknown currents and/or voltages:

***a.**

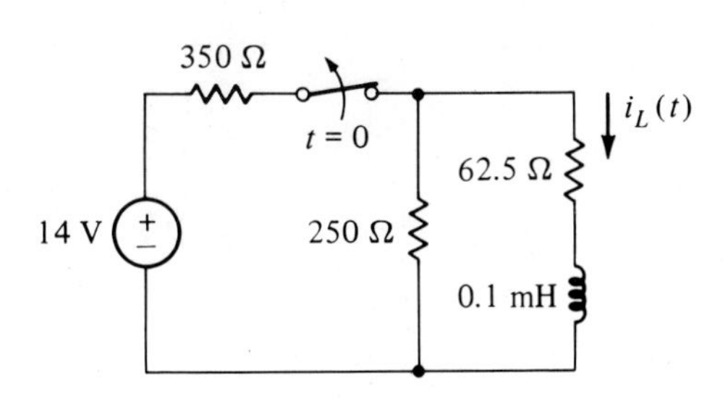

b.

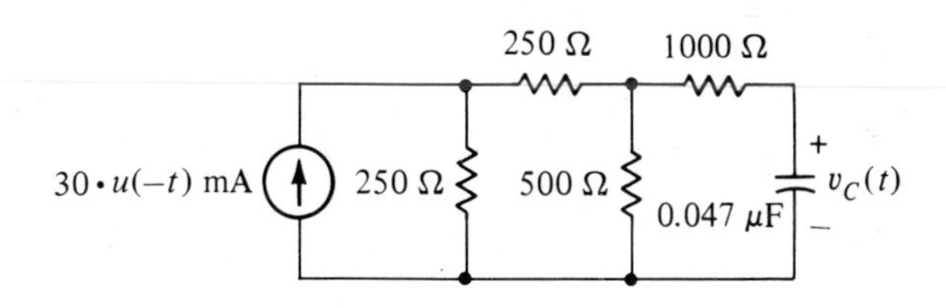

***c.** For how long after $t = 0$ will $i_x(t)$ be greater than 20 mA?

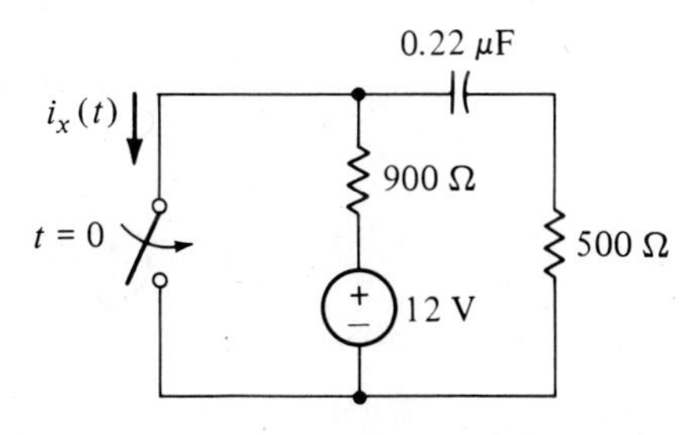

d.

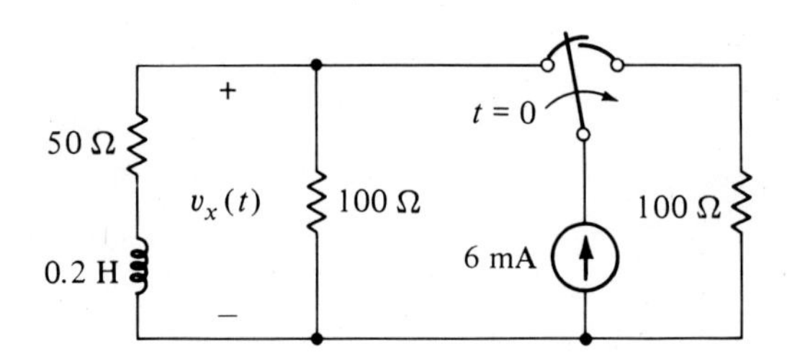

*e. At what time will $v_x(t) = -1$ V?

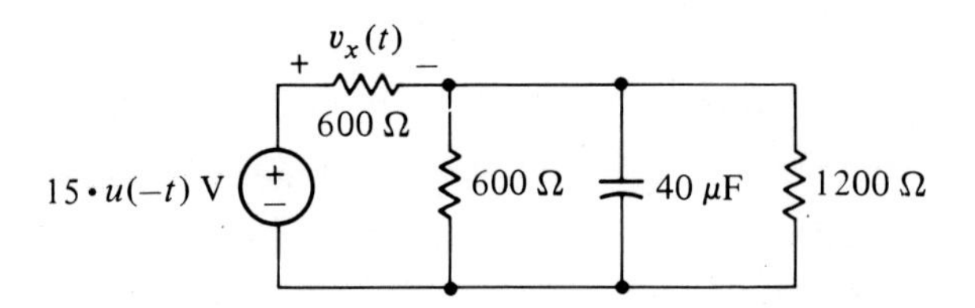

SPICE

3-12. Find and sketch the unknown currents and/or voltages:

a.

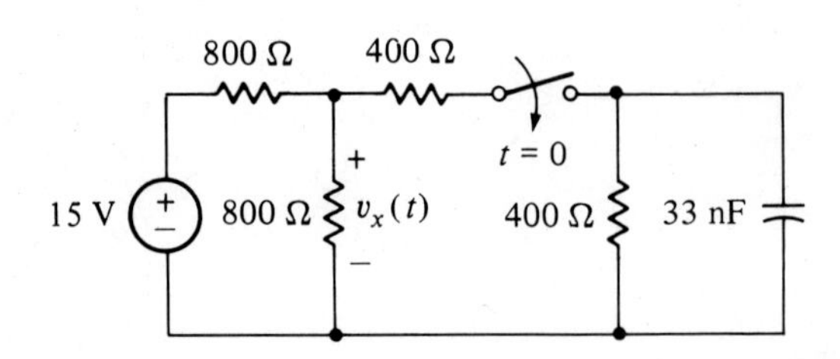

***b.** Calculate i (500 μsec).

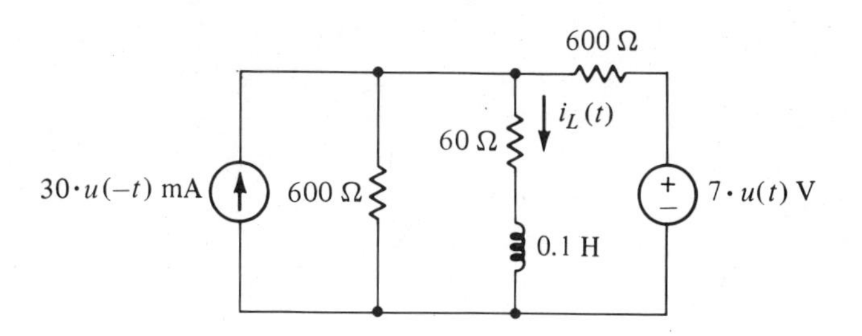

c.

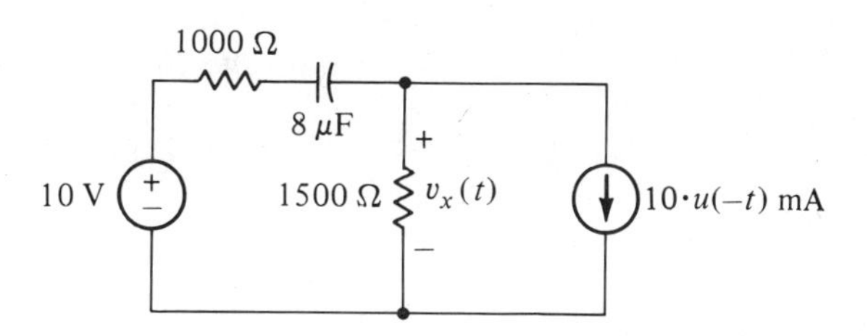

***d.** At what time after $t = 0$ will $v_0(t) = -1.5$ V?

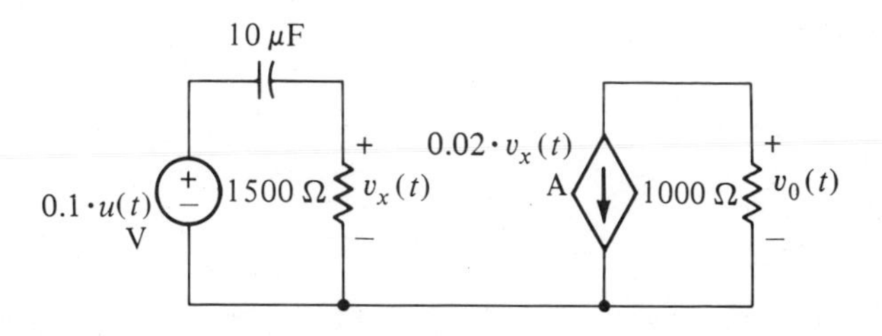

e. At what time after $t = 0$ will $v_C(t) = 0$ V?

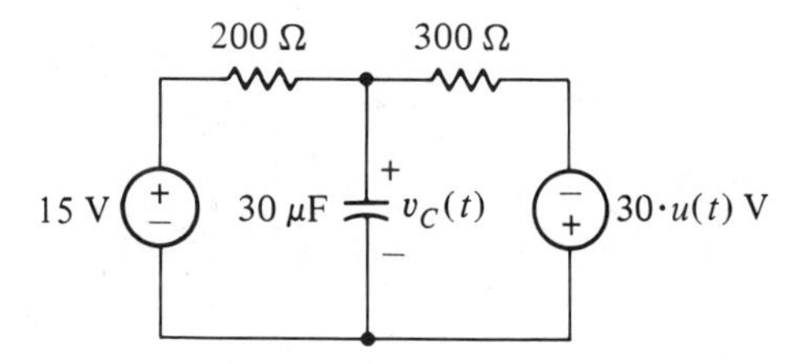

SPICE ***3-13.** Find and sketch the capacitor voltage when

$$v_S(t) = 4.5 \sum_{n=0}^{\infty} (-1)^{n-1} u(t - (n-1) \cdot 400 \times 10^{-6}) \text{ V}$$

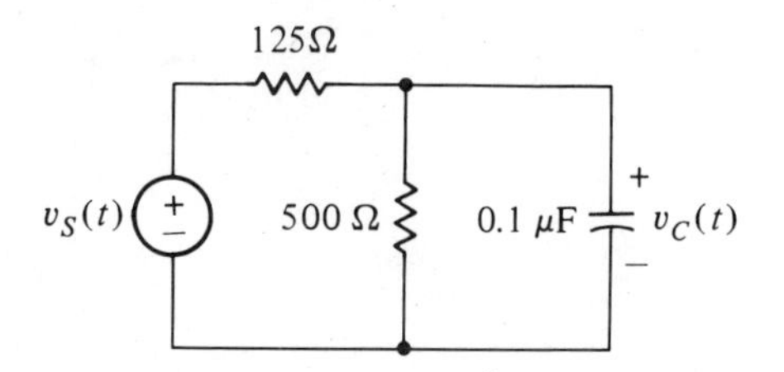

3-14. Use the fundamental SI units to show that τ_C has dimensions of seconds for both the R–C and the R–L circuit.

4 Laplace Transform Solutions of Initial Value Problems

4.1 INTRODUCTION

In the most general sense, a mathematical **transformation** is simply a change of variables. The usual reason for a transformation is to make a difficult problem easier to solve [Figure 4.1(a)].

The logarithm is frequently used to illustrate the concept of mathematical transformation. Assume that we are in the precalculator era, when multiplication of numbers with a large number of significant figures was a difficult operation, and consider the problem

$$x = yz$$

where y and z are known values. If the equation is **transformed** by means of logarithms, we have

$$\log x = \log yz$$

and from the properties of the logarithm,

$$\log x = \log y + \log z$$

We can look up the logarithm of y and the logarithm of z in a table of logarithms, and *add* the result to determine the **logarithm** of x. We can then obtain the value of x from a table of **antilogarithms**, or **inverse logarithms**. Thus, the difficult problem (multiplication) was transformed into an easy problem (addition).

One type of transformation, the **integral transform**, is aimed primarily at obtaining solutions to differential equations. As the name implies, integral

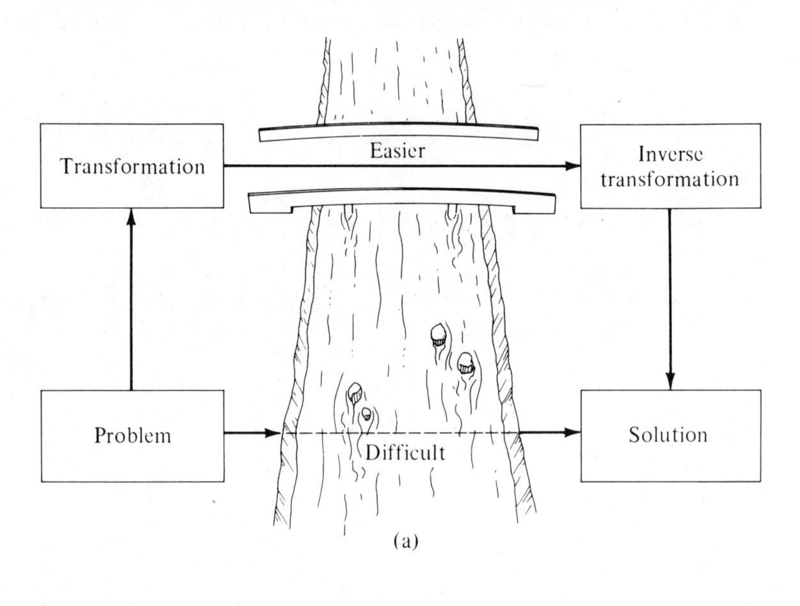

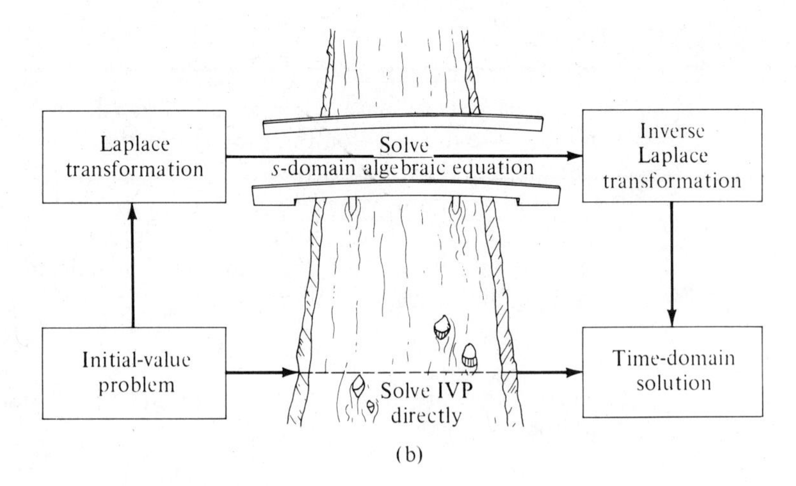

Figure 4.1 Mathematical transformation. (a) Concept of mathematical transformation. (b) Laplace transformation.

transforms accomplish the change of variables by integration. Probably the most widely used integral transform is the **Laplace transform**, which is the subject of this chapter. Figure 4.1(b) illustrates the way in which the Laplace transform is used in the solution of a time-domain initial value problem (i.e., a time-dependent differential equation and the associated initial conditions). Several subsequent chapters deal with application of the Laplace transform to circuit analysis and design.

4.2 Laplace Transform Fundamentals

We will now define the Laplace transform and introduce the notation and symbols associated with Laplace transformation. **Laplace transform pairs** will be derived in a logical order in this and subsequent sections of this chapter. A table of all the results may be found in Appendix C and inside the back cover.

Definition, Notation, and Symbols

The function $F(s)$ denotes the **Laplace transform** of a time-dependent function $f(t)$ and is defined

$$F(s) = \int_0^\infty \exp[-st] \cdot f(t)\, dt$$

where s is a **complex variable** ($s = \sigma + j\omega$). The fact that s is complex will not require any special attention at this time. The domain of the Laplace transform is called the s-**domain** or the **Laplace domain**. The argument of the exponential, $-st$, must be dimensionless, so the dimensions of s must be $1/\text{sec}$, or the dimensions of frequency. For this reason, the term **complex frequency domain** is also used for the domain of the Laplace transform. The exponential factor $\exp[-st]$ is called the **kernel** of the integral; other integral transforms have different kernels.

A special symbol $\mathscr{L}$, called the **Laplace operator**, is commonly used to avoid writing out the integral:

$$F(s) = \mathscr{L}\{f(t)\}$$

The equation is read as "$F(s)$ is the Laplace transform of $f(t)$," and $\mathscr{L}\{f(t)\}$ is read as "take the Laplace transform of $f(t)$."

As a result of integrating the time-domain function with respect to time, Laplace transforms have the dimensions of the time-domain function multiplied by seconds (e.g., A · sec, V · sec, etc.).

Impulse Function. For the purposes of circuit analysis, only a few Laplace transform pairs need be derived from the definition of the Laplace transform, i.e., obtained by integration. The simplest pair is the impulse (delta) function occurring at $t = 0$ and its Laplace transform

$$\mathscr{L}\{\delta(t)\} = \int_0^\infty \exp[-st] \cdot \delta(t)\, dt$$

By definition (Equation 2.5), $\delta(t)$ is zero everywhere except at $t = 0$, which means that $\exp[-st] \cdot \delta(t)$ is zero everywhere except at $t = 0$. However, at

$t = 0$, $\exp[-st] = \exp[0] = 1$, so

$$\mathscr{L}\{\delta(t)\} = \int_0^\infty \delta(t)\, dt$$

From Equation (2.6),

$$\int_0^\infty \delta(t)\, dt = 1$$

so

$$\mathscr{L}\{\delta(t)\} = 1$$

Pair I

$f(t) = \delta(t) \qquad F(s) = 1$

Unit Step Function. The Laplace transform of an unshifted unit step function is also easy to obtain directly:

$$\mathscr{L}\{u(t)\} = \int_0^\infty \exp[-st] \cdot u(t)\, dt$$

By definition (Equation 2.1), $u(t) = 1$ for $t > 0$, so

$$\begin{aligned}\mathscr{L}\{u(t)\} &= \int_0^\infty \exp[-st]\, dt = -\frac{1}{s} \cdot \exp[-st]\Big|_0^\infty \\ &= -\frac{1}{s}(\exp[-\infty] - \exp[0]) \\ &= 1/s\end{aligned}$$

Since $u(t) = 1$ for the domain of integration ($0 < t < \infty$), then clearly the Laplace transform of 1 (a constant) is the same as the Laplace transform of $u(t)$:

$$\mathscr{L}\{1\} = \mathscr{L}\{u(t)\} = 1/s$$

Pair II

$f(t) = u(t)$ or 1 $\qquad F(s) = 1/s$

Ramp Function. For an unshifted ramp function,

$$\mathscr{L}\{t \cdot u(t)\} = \int_0^{\infty} \exp[-st] \cdot t \cdot u(t)\, dt$$

For $t > 0$, $t \cdot u(t) = t$, so

$$\mathscr{L}\{t \cdot u(t)\} = \int_0^{\infty} \exp[-st] \cdot t\, dt$$

The following result is found in many tables of definite integrals:

$$\int_0^{\infty} x^n \cdot \exp[-ax]\, dx = \frac{n!}{a^{n+1}} \tag{4.1}$$

for $a > 0$ and n a positive integer. By substitution

$$\mathscr{L}\{t \cdot u(t)\} = 1!/s^{1+1} = 1/s^2$$

Pair IIIa

$$f(t) = t \cdot u(t) \qquad F(s) = 1/s^2$$

Equation (4.1) can be used to obtain a Laplace transform pair which is more general than Pair IIIa:

$$\begin{aligned} \mathscr{L}\{t^n \cdot u(t)\} &= \int_0^{\infty} \exp[-st] \cdot t^n \cdot u(t)\, dt \\ &= \int_0^{\infty} \exp[-st] \cdot t^n\, dt \\ &= n!/s^{n+1} \end{aligned}$$

Pair IIIb

$$f(t) = t^n \cdot u(t) \qquad F(s) = n!/s^{n+1}$$

Pair IIIc

$$f(t) = \frac{t^{n-1}}{(n-1)!} \cdot u(t) \qquad F(s) = \frac{1}{s^n}$$

4.3 Properties of the Laplace Transform

In this section we will state several properties of the Laplace transform which are particularly useful in circuit analysis, and we will use these properties to continue development of the table of Laplace transform pairs. We will not take time to prove the properties or theorems which we will use here; the interested reader may find these and related proofs, along with conditions for the existence of the Laplace transform of a function, in Appendix B.

Linearity Properties

Laplace transformation is a *linear* operation, so

$$\mathscr{L}\{k \cdot f(t)\} = k \cdot \mathscr{L}\{f(t)\} = k \cdot F(s), \quad \text{where } k \text{ is a constant}$$

and

$$\begin{aligned}\mathscr{L}\{f_1(t) \pm f_2(t)\} &= \mathscr{L}\{f_1(t)\} \pm \mathscr{L}\{f_2(t)\} \\ &= F_1(s) \pm F_2(s)\end{aligned}$$

where $f_1(t)$ and $f_2(t)$ are functions for which Laplace transforms exist. The Linearity Properties are often expressed jointly:

Linearity Property

$$\mathscr{L}\{k_1 \cdot f_1(t) \pm k_2 \cdot f_2(t)\} = k_1 \cdot F_1(s) \pm k_2 \cdot F_2(s)$$

Example 4.1

Problem:

Determine the Laplace transform $V(s)$ of the function which generates the voltage waveform in Figure 4.2.

Solution:

1. From Equation (2.9c), the generating function for the waveform is

$$v(t) = 10 \cdot u(t) + 50t \cdot u(t) \text{ V}$$

2. The Laplace transform of $v(t)$ is

$$V(s) = \mathscr{L}\{v(t)\} = \mathscr{L}\{10 \cdot u(t) + 50t \cdot u(t)\}$$

Figure 4.2
Example 4.1.

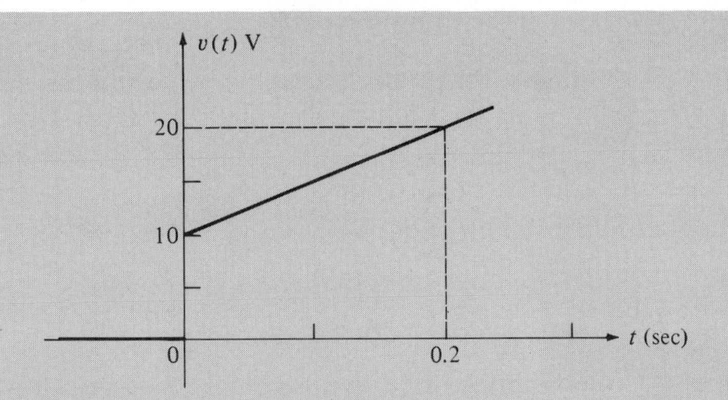

3. Use the linearity properties to obtain

$$V(s) = 10 \cdot \mathscr{L}\{u(t)\} + 50 \cdot \mathscr{L}\{t \cdot u(t)\}$$

4. From Pairs II and IIIa,

$$V(s) = 10\frac{1}{s} + 50\frac{1}{s^2} = 10\left(\frac{1}{s} + \frac{5}{s^2}\right) \text{ V} \cdot \text{sec}$$

The s-domain function should not be expressed with a common denominator. The reasons for this will become apparent when performing *inverse* Laplace transformations.

Exercise 4.1

Problem:

Determine the Laplace transform of

$$f(t) = 8 \cdot \delta(t) - 5 \cdot u(t) + 35 \cdot t \cdot u(t)$$

Answer:

$$F(s) = 8 - 5\left(\frac{1}{s} - \frac{7}{s^2}\right)$$

s-Domain Shifting Property

The **s-domain shifting property**, also called the **complex translation property**, is used to find the Laplace transform of functions which have an exponential factor.

s-Domain Shifting Property

$$\mathscr{L}\{\exp[-at] \cdot f(t)\} = F(s + a)$$

where $F(s)$ is the Laplace transform of $f(t)$.

To apply the s-domain shifting property, first find the Laplace transform of the function without the exponential factor, then substitute $(s + a)$ for s everywhere in $F(s)$.

Decay Function. Laplace transformation of the decay function associated with R–C and R–L source-free circuits will serve to illustrate application of the s-domain shifting property:

$$\mathscr{L}\{\exp[-at] \cdot u(t)\} = F(s + a)$$

where

$$F(s) = \mathscr{L}\{u(t)\}$$

From Pair II,

$$\mathscr{L}\{u(t)\} = 1/s$$

therefore $\mathscr{L}\{\exp[-at] \cdot u(t)\} = 1/(s + a)$.

Pair IV

$$f(t) = \exp[-at] \cdot u(t) \qquad F(s) = \frac{1}{s + a}$$

Example 4.2

Problem:

Determine the Laplace transform $I(s)$ of the function which generates the current waveform in Figure 4.3. The decay is exponential with a time constant τ_C of 250 μsec.

Figure 4.3
Example 4.2.

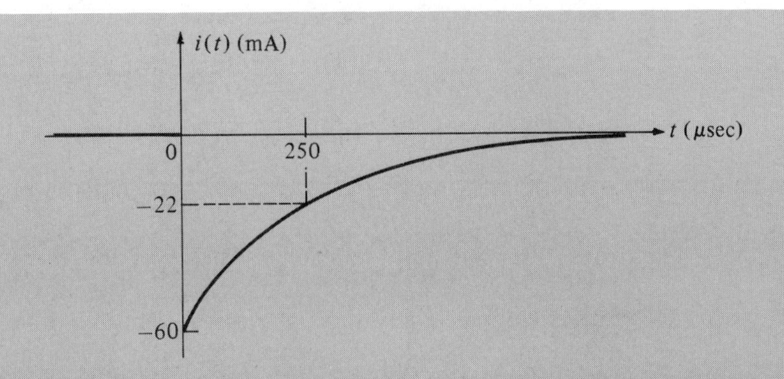

Solution:

1. From Equation (2.11), the generating function for the waveform is

$$i(t) = -60 \cdot \exp\left[-\frac{t}{250 \times 10^{-6}}\right] \cdot u(t) \text{ mA}$$

2. The Laplace transform of $i(t)$ is

$$\begin{aligned} I(s) &= \mathscr{L}\left\{-60 \cdot \exp\left[-\frac{t}{250 \times 10^{-6}}\right] \cdot u(t)\right\} \\ &= \mathscr{L}\{-60 \cdot \exp[-4000t] \cdot u(t)\} \end{aligned}$$

3. Use the linearity properties and Pair IV to obtain

$$\begin{aligned} I(s) &= -60 \cdot \mathscr{L}\{\exp[-4000t] \cdot u(t)\} \\ &= \frac{-60}{s + 4000} \text{ mA} \cdot \text{sec} \end{aligned}$$

Exercise 4.2

Problem:

Determine the Laplace transform of the function which generates a waveform which rises instantaneously from 0 to 5 V at $t = 0$, then decays exponentially toward zero with $\tau_C = 40$ msec.

Answer:

$$v(t) = 5 \cdot \exp[-25t] \cdot u(t) \text{ V}$$

$$V(s) = \frac{5}{s + 25} \text{ V} \cdot \text{sec}$$

Growth Function. The Laplace transform of the growth function associated with R–C and R–L circuits with a dc or rectangular source is given as Pairs Va and Vb. The derivation of Pairs Va and Vb follows directly from the linearity properties, Pair I, and Pair IV, and is left as an exercise for the reader (Problem 4-5).

Pair Va

$$f(t) = (1 - \exp[-at])u(t) \qquad F(s) = \frac{a}{s(s + a)}$$

Pair Vb

$$f(t) = \frac{1}{a}(1 - \exp[-at])u(t) \qquad F(s) = \frac{1}{s(s + a)}$$

Example 4.3

Problem:

Determine the Laplace transform of the function which generates the voltage waveform shown in Figure 4.4. The growth is exponential with $\tau_C = 50\ \mu\text{sec}$.

Solution:

1. From Equation (2.10), the generating function for the waveform is

$$v(t) = 2.5 + (5 - 2.5)\left(1 - \exp\left[-t/50 \times 10^{-6}\right]\right)u(t)$$
$$= 2.5 + 2.5\left(1 - \exp\left[-20 \times 10^{3}t\right]\right)u(t) \text{ V}$$

2. The Laplace transform of $v(t)$ is

$$V(s) = \mathscr{L}\left\{2.5 + 2.5\left(1 - \exp\left[-20 \times 10^{3}t\right]\right)u(t)\right\} \text{ V} \cdot \text{sec}$$

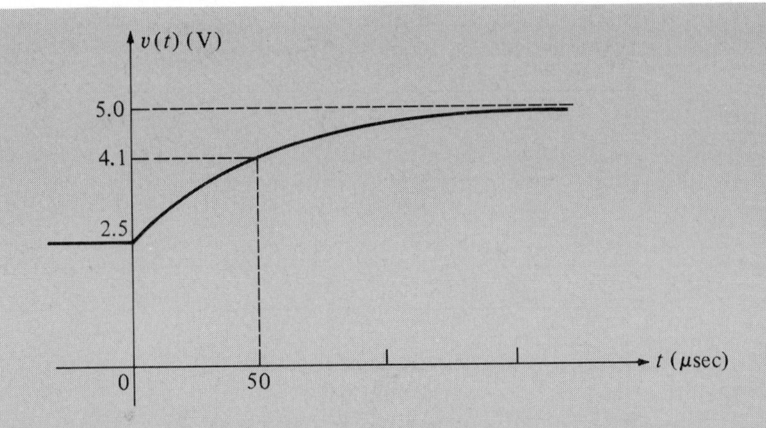

Figure 4.4
Example 4.3.

3. Use the linearity properties to obtain

$$V(s) = 2.5\left(\mathscr{L}\{1\} + \mathscr{L}\{(1 - \exp[-20 \times 10^3 t])u(t)\}\right)$$

4. From Pairs II and Va,

$$V(s) = 2.5\left(\frac{1}{s} + \frac{20 \times 10^3}{s(s + 20 \times 10^3)}\right) \text{ V} \cdot \text{sec}$$

Exercise 4.3

Problem:

Determine the Laplace transform for a waveform which is 0 at $t = 0$, then grows exponentially toward -12 mA with $\tau_C = 30$ msec.

Answer:

$$i(t) = -12(1 - \exp[-33.33t])u(t) \text{ mA}$$

$$I(s) = \frac{-400}{s(s + 33.33)} \text{ mA} \cdot \text{sec}$$

Application of the s-domain shifting property to Pairs IIIa and IIIc leads to Pairs VIa and VIb, two additional Laplace transform pairs involving exponentials in the time-domain function. Derivation of Pairs VIa and VIb is another exercise which the reader should undertake (Problem 4-6).

Pair VIa

$$f(t) = t \cdot \exp[-at] \cdot u(t) \qquad F(s) = \frac{1}{(s+a)^2}$$

Pair VIb

$$f(t) = \frac{t^{n-1}}{(n-1)!} \cdot \exp[-at] \cdot u(t) \qquad F(s) = \frac{1}{(s+a)^n}$$

Sinusoidal Functions. The Laplace transforms of the sinusoidal functions, sine and cosine, follow from **Euler's identity**, the linearity properties, and Pair IV.

Flashback

Euler's identity:

$$\exp[jx] = \cos x + j \cdot \sin x$$

where $j^2 = -1$.

From Euler's identity,

$$\exp[jx] = \cos x + j \cdot \sin x \tag{4.2a}$$

and

$$\exp[-jx] = \cos[-x] + j \cdot \sin[-x]$$

$$= \cos x - j \cdot \sin x \tag{4.2b}$$

If Equations (4.2) and (4.2a) are added,

$$\exp[jx] + \exp[-jx] = 2 \cdot \cos[x]$$

or

$$\cos x = \frac{\exp[jx] + \exp[-jx]}{2} \tag{4.3a}$$

and if Equation (4.2b) is subtracted from (4.2a),

$$\exp[jx] - \exp[-jx] = j2 \cdot \sin x$$

or

$$\sin x = \frac{\exp[jx] - \exp[-jx]}{j2} \tag{4.3b}$$

The Laplace transform of the time-domain sine function, $\sin \omega t$, is

$$\mathscr{L}\{\sin[\omega t] \cdot u(t)\} = \mathscr{L}\left\{\frac{\exp[j\omega t] - \exp[-j\omega t]}{j2} \cdot u(t)\right\}$$

and from the linearity properties,

$$\mathscr{L}\{\sin[\omega t] \cdot u(t)\}$$
$$= \frac{1}{j2}(\mathscr{L}\{\exp[j\omega t] \cdot u(t)\} - \mathscr{L}\{\exp[-j\omega t] \cdot u(t)\})$$

The imaginary exponent requires no special treatment, so from Pair IV,

$$\begin{aligned}\mathscr{L}\{\sin[\omega t] \cdot u(t)\} &= \frac{1}{j2} \cdot \frac{1}{s - j\omega} - \frac{1}{s + j\omega} \\ &= \frac{1}{j2} \cdot \frac{(s + j\omega) - (s - j\omega)}{(s - j\omega)(s + j\omega)} \\ &= \frac{1}{j2} \cdot \frac{j2\omega}{s^2 - j\omega s + j\omega s - j^2\omega^2} \\ &= \frac{\omega}{s^2 + \omega^2}\end{aligned}$$

Pair VII

$$f(t) = \sin[\omega t] \cdot u(t) \qquad F(s) = \frac{\omega}{s^2 + \omega^2}$$

Example 4.4

Problem:

Determine the Laplace transform of the generating function for the 5-Hz sinusoidal waveform depicted in Figure 4.5.

Figure 4.5
Example 4.4.

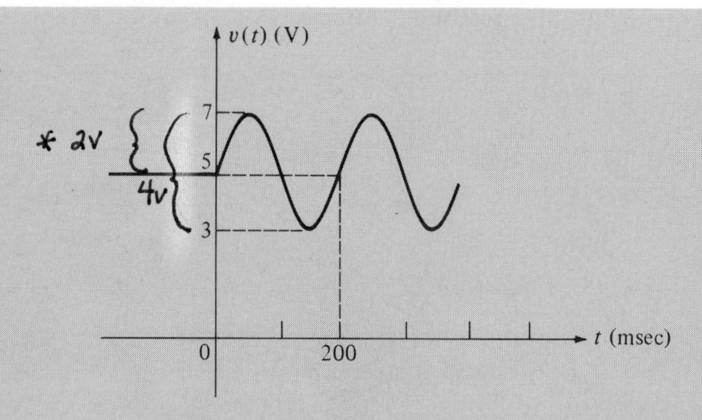

Solution:

1. The generating function is

$2\pi f$
$2\pi \cdot T_c = \omega$

$$v(t) = 5 + 2 \cdot \sin[31.42t] \cdot u(t)\ \text{V}$$

2. Use the linearity properties to obtain

$$V(s) = 5 \cdot \mathscr{L}\{1\} + 2 \cdot \mathscr{L}\{\sin[31.42t] \cdot u(t)\}\ \text{V} \cdot \text{sec}$$

3. From Pairs II and VII,

$$V(s) = \frac{5}{s} + \frac{62.83}{s^2 + 987.0}\ \text{V} \cdot \text{sec}$$

Exercise 4.4

Problem:

Determine the Laplace transform of a 60-Hz unshifted sine wave which begins abruptly at $t = 0$ and which has a peak current of 300 mA.

Answer:

$$I(s) = \frac{113.1}{s^2 + 142.1 \times 10^3}\ \text{A} \cdot \text{sec}$$

The Laplace transform of the time-domain cosine function, $\cos[\omega t] \cdot u(t)$, is given by Pair VIII. The derivation follows directly from Equation (4.3a) and Pair IV (Problem 4-8).

Pair VIII

$$f(t) = \cos[\omega t] \cdot u(t) \qquad F(s) = \frac{s}{s^2 + \omega^2}$$

Example 4.5

Problem:

Determine the Laplace transform of

$$f(t) = \sin[300t - 30°] \cdot u(t)$$

Solution:

1. Rewrite $f(t)$ using the trignometric identity

$$\sin[x - a] = \sin x \cos a - \cos x \sin a$$

$$\begin{aligned} f(t) &= (\sin 300t \cdot \cos 30° - \cos 300t \cdot \sin 30°)u(t) \\ &= (0.8660 \cdot \sin 300t - 0.5 \cdot \cos 300t)u(t) \end{aligned}$$

2. Use the linearity properties to obtain

$$\begin{aligned} F(s) = 0.8660 \cdot \mathscr{L}\{\sin[300t] \cdot u(t)\} \\ - 0.5 \cdot \mathscr{L}\{\cos[300t] \cdot u(t)\} \end{aligned}$$

3. From Pairs VII and VIII,

$$F(s) = \frac{259.8}{s^2 + 90 \times 10^3} - \frac{0.5s}{s^2 + 90 \times 10^3}$$

Exercise 4.5

Problem:

Determine the Laplace transform of the generating function for the 25-kHz sinusoidal waveform depicted in Figure 4.6.

Figure 4.6
Example 4.5.

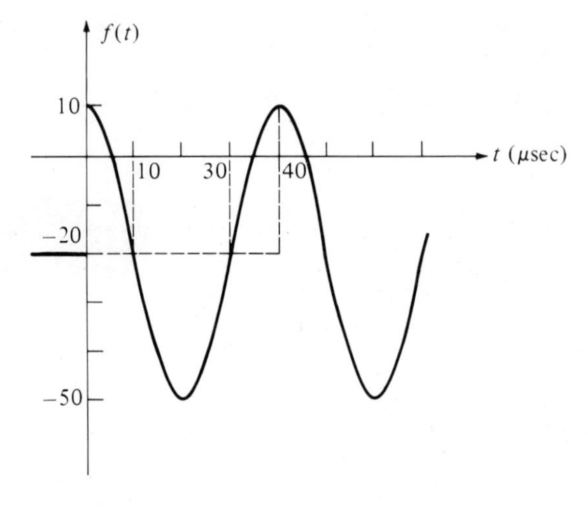

Answer:

$$f(t) = -20 + 30 \cdot \cos[157.1 \times 10^3 t] \cdot u(t)$$

$$F(s) = -10\left(\frac{2}{s} - \frac{3s}{s^2 + 24.67 \times 10^9}\right)$$

The Laplace transforms of decaying sinusoidal functions are obtained by applying the s-domain shifting property to Pairs VII and VIII.

Pair IX

$$f(t) = \exp[-at] \cdot \sin[\omega t] \cdot u(t) \qquad F(s) = \frac{\omega}{(s+a)^2 + \omega^2}$$

Pair X

$$f(t) = \exp[-at] \cdot \cos[\omega t] \cdot u(t) \qquad F(s) = \frac{s+a}{(s+a)^2 + \omega^2}$$

Example 4.6

Problem:

Determine the Laplace transform for

$$f(t) = 120 \cdot \exp[-2 \times 10^3 t] \cdot \cos[10 \times 10^3] \cdot u(t)$$

Solution:

Use the linearity properties and Pair X to obtain

$$F(s) = \frac{120(s + 2 \times 10^3)}{(s + 2 \times 10^3)^2 + 0.1 \times 10^9}$$

Exercise 4.6

Problem:

Determine the Laplace transform for

$$i(t) = 80 \cdot \exp[-50t] \cdot \sin[2 \times 10^3 t] \cdot u(t) \text{ mA}$$

Answer:

$$I(s) = \frac{160 \times 10^3}{(s + 50)^2 + 4 \times 10^6} \text{ mA} \cdot \text{sec}$$

Time-Domain Shifting Property

The **time-domain shifting property**, sometimes called the **real-axis translation property**, facilitates the Laplace transformation of functions which are shifted along the t-axis.

Time-Domain Shifting Property

$$\mathcal{L}\{f(t - t_0)\} = \exp[-t_0 s] \cdot F(s)$$

where $F(s)$ is the Laplace transform of $f(t)$.

The time-domain shifting property applies to every Laplace transform pair and effectively creates an auxiliary table of Laplace transforms:

Pair	$f(t)$	$F(s)$
I	$\delta(t-t_0)$	$\exp[-st_0]$
II	$u(t-t_0)$	$\exp[-st_0]\cdot 1/s$
IIIa	$(t-t_0)u(t-t_0)$	$\exp[-st_0]\cdot 1/s^2$
IIIb	$(t-t_0)^n u(t-t_0)$	$\exp[-st_0]\cdot n!/s^n$
IV	$\exp[-a(t-t_0)]\cdot u(t-t_0)$	$\exp[-st_0]\cdot 1/(s+a)$

and so forth.

Note that in order to apply the t-shifting property, the arguments of all the factors which make up the function must agree.

Example 4.7

Problem:

Determine the Laplace transform of the function which generates the waveform in Figure 4.7.

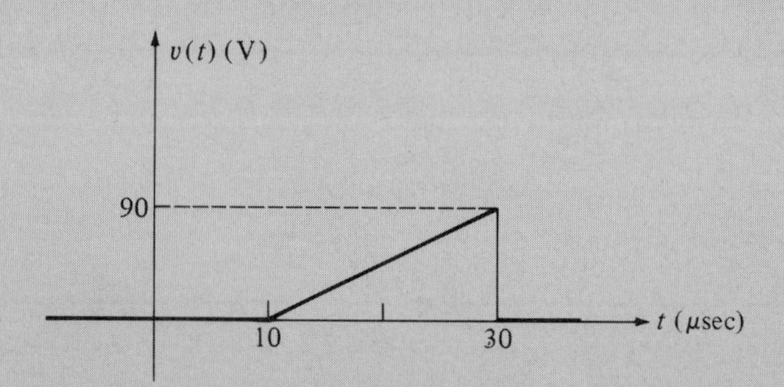

Figure 4.7
Example 4.7.

Solution:

1. The generating function is

$$f(t) = 4.5(t-10)u(t-10) - 4.5(t-30)u(t-30) - 90\cdot u(t-30)$$

2. Use the linearity properties to obtain

$$F(s) = 4.5[\mathscr{L}\{(t-10)u(t-10)\} \\ -\mathscr{L}\{(t-30)u(t-30)\} \\ -20 \cdot \mathscr{L}\{u(t-30)\}]$$

3. Use the time-domain shifting property to obtain

$$F(s) = 4.5(\exp[-10s] \cdot \mathscr{L}\{t \cdot u(t)\} \\ -\exp[-30s] \cdot \mathscr{L}\{t \cdot u(t)\} \\ -20 \cdot \exp[-30s] \cdot \mathscr{L}\{u(t)\})$$

4. From Pairs IIIa and II,

$$F(s) = 4.5\left(\frac{\exp[-10s]}{s^2} - \frac{\exp[-30s]}{s^2} - \frac{20 \cdot \exp[-30s]}{s}\right)$$

This expression can be simplified, but it is not usually advantageous to do so.

Exercise 4.7

Problem:

Determine the Laplace transform of the function which generates a single 0-to-5-V rectangular pulse with leading edge at 30 msec and a pulse width of 15 msec.

Answer:

$$v(t) = 5 \cdot u(t-0.030) - 5 \cdot u(t-0.045) \text{ V}$$

$$V(s) = 5\left(\frac{\exp[-0.030s]}{s} - \frac{\exp[-0.045s]}{s}\right) \text{ V} \cdot \text{sec}$$

Periodicity Property

All of the Laplace transform pairs developed thus far apply to aperiodic functions. Through use of the **periodicity property**, the Laplace transform of a periodic function can be obtained if the function which generates the first period is known.

Periodicity Property

$$\mathscr{L}\{f_P(t)\} = \frac{1}{1 - \exp[-Ts]} \cdot F_1(s)$$

where $F_1(s)$ is the Laplace transform of $f_1(t)$, the generating function for the first period of periodic function $f_P(t)$.

Example 4.8

Problem:

If the voltage pulse described in Exercise 4.7 is the first in a train of periodic pulses, determine the Laplace transform for the function $v_P(t)$ which generates the pulse train.

Solution:

1. The periodic function $v_P(t)$ could be written using the form suggested by Equation (2.13),

$$v_P(t) = \sum_{n=0}^{\infty} v(t - nT)$$

but this is not necessary, since the periodicity property requires only the generating function $v_1(t)$ for the first period.

2. From Exercise 4.7,

$$v_1(t) = 5 \cdot u(t - 0.030) - 5 \cdot u(t - 0.045)\ \text{V}$$

$$T = 45\ \text{msec}$$

$$V_1(s) = 5\left(\frac{\exp[-0.030s]}{s} - \frac{\exp[-0.045s]}{s}\right)\ \text{V} \cdot \text{sec}$$

3. Use the periodicity property to obtain

$$V_P(s) = \frac{5}{1 - \exp[-0.045s]}\left(\frac{\exp[-0.030s]}{s} - \frac{\exp[-0.045s]}{s}\right)\ \text{V} \cdot \text{sec}$$

Exercise 4.8

Problem:

Determine the Laplace transform of the function which generates the periodic triangular waveform of Figure 4.8.

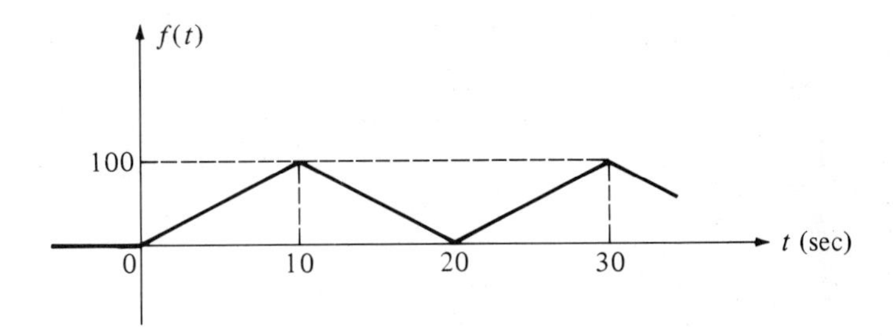

Figure 4.8
Exercise 4.8.

Answer:

$$f_1(t) = 10t \cdot u(t) - 20(t-10)\,u(t-10) + 10(t-20)\,u(t-20)$$

$$F_1(s) = 10\left(\frac{1}{s^2} - \frac{\exp[-10s]}{s^2} - \frac{\exp[-20s]}{s^2}\right)$$

$$F_P(s) = \frac{10}{1-\exp[-20s]}\left(\frac{1}{s^2} - \frac{\exp[-10s]}{s^2} - \frac{\exp[-20s]}{s^2}\right)$$

4.4 Laplace Transformation of Derivatives and Integrals

We now turn our attention to Laplace transformation of the initial value problem itself. Recall that the initial value problem consists of a differential equation and the associated initial conditions. In circuit analysis, the differential equation results from application of one of the Kirchhoff laws, and the initial conditions are the initial voltages across the capacitors and/or the initial currents through the inductors.

Laplace Transformation of Derivatives

Laplace transformation of the derivative of a function is obtained from the defining integral through integration by parts (Appendix B).

Differentiation Property

$$\mathscr{L}\left\{\frac{d}{dt}f(t)\right\} = s \cdot F(s) - f(0)$$

where $F(s) = \mathscr{L}\{f(t)\}$ and $f(0)$ is the initial value of $f(t)$.

The differentiation property can be extended to second and higher derivatives. For the second derivative,

$$\mathscr{L}\left\{\frac{d^2}{dt^2}f(t)\right\} = \mathscr{L}\left\{\frac{d}{dt}\left(\frac{d}{dt}f(t)\right)\right\} = s[s\cdot F(s) - f(0)] - \left.\frac{d}{dt}f(t)\right|_{t=0}$$

$$= s^2\cdot F(s) - s\cdot f(0) - \left.\frac{d}{dt}f(t)\right|_{t=0}$$

In general, for any order derivative,

$$\mathscr{L}\left\{\frac{d^n}{dt^n}f(t)\right\} = s^n\cdot F(s) - s^{n-1}\cdot f(0) - s^{n-2}\left.\frac{d}{dt}f(t)\right|_{t=0}$$

$$- \cdots - s\left.\frac{d^{n-2}}{dt^{n-2}}f(t)\right|_{t=0} - \left.\frac{d^{n-1}}{dt^{n-1}}f(t)\right|_{t=0}$$

Laplace Transformation of Integrals

Laplace transformation of the integral of a function is also obtained from the defining integral through integration by parts (Appendix B).

Integration Property

$$\mathscr{L}\left\{\int_0^{t_1} f(t)\,dt\right\} = \frac{1}{s}\cdot F(s)$$

where $F(s) = \mathscr{L}\{f(t)\}$.

Laplace Transformation of Initial Value Problems

Laplace transformation of initial value problems is essential to much of the material regarding transform circuit analysis presented in later chapters. The reader should devote particular attention to the examples which follow.

Example 4.9 **Problem:**

Solve for $F(s)$ in the initial value problem

$$8\frac{d}{dt}f(t) + 3\cdot f(t) = 2t\cdot u(t) \qquad f(0) = -1$$

Solution:

1. Perform a Laplace transformation on both members of the equation to obtain

$$\mathscr{L}\left\{8\frac{d}{dt}f(t) + 3 \cdot f(t)\right\} = \mathscr{L}\{2t \cdot u(t)\}$$

2. Use the linearity properties to obtain

$$8 \cdot \mathscr{L}\left\{\frac{d}{dt}f(t)\right\} + 3 \cdot \mathscr{L}\{f(t)\} = 2 \cdot \mathscr{L}\{t \cdot u(t)\}$$

3. Use the differentiation property, the definition of $F(s)$, and Pair III to obtain

$$8[s \cdot F(s) - f(0)] + 3 \cdot F(s) = 2\frac{1}{s^2}$$

4. Substitute for $f(0)$ and solve for $F(s)$:

$$8s \cdot F(s) + 8 + 3 \cdot F(s) = \frac{2}{s^2}$$

$$(8s + 3)F(s) = \frac{2}{s^2} - 8$$

$$F(s) = \frac{2}{s^2(8s + 3)} - \frac{8}{8s + 3}$$

Several important points should be emphasized regarding Example 4.9:

1. Both left and right members of the equation were transformed; it is meaningless, and entirely incorrect, to transform one member only.
2. In transforming the term $3 \cdot \mathscr{L}\{f(t)\}$ in the left member of the equation, $\mathscr{L}\{f(t)\}$ was simply replaced with its definition, $F(s)$.
3. The initial condition, $f(0) = -1$, written as a separate equation in the initial value problem, was incorporated into the *transformed* differential equation. As a result, the initial value problem, which consists of a differential equation and the initial condition, was transformed into a single algebraic equation.

Example 4.10

Problem:

Solve for $I(s)$ in the initial value problem

$$0.04\frac{d^2}{dt^2}i(t) + 200\frac{d}{dt}i(t) + 25 \times 10^3 \cdot i(t) = 12 \cdot u(t)\text{ V}$$

$$\left.\frac{d}{dt}i(t)\right|_{t=0} = 100\text{ A/sec}, \qquad i(0) = 0.02\text{ A}$$

Solution:

1. Perform a Laplace transformation on both members of the equation and use the linearity properties to obtain

$$0.04 \cdot \mathscr{L}\left\{\frac{d^2}{dt^2}i(t)\right\} + 200 \cdot \mathscr{L}\left\{\frac{d}{dt}i(t)\right\}$$
$$+25 \times 10^3 \cdot \mathscr{L}\{i(t)\} = 12 \cdot \mathscr{L}\{u(t)\}$$

2. Use the differentiation property and Pair II to obtain

$$0.04\left(s^2 \cdot I(s) - s \cdot i(0) - \left.\frac{d}{dt}i(t)\right|_{t=0}\right)$$
$$+200[s \cdot I(s) - i(0)] + 25 \times 10^3 \cdot I(s) = \frac{12}{s}$$

3. Substitute the initial conditions and solve for $I(s)$:

$$0.04[s^2 \cdot I(s) - 0.020s - 100]$$
$$+ 200[s \cdot I(s) - 0.020] + 25 \times 10^3 \cdot I(s) = \frac{12}{s}$$

$$(0.04s^2 + 200s + 25 \times 10^3)I(s) = 0.8 \cdot 10^{-3}s + 8 + \frac{12}{s}$$

$$I(s) = \frac{0.8 \times 10^{-3}s + 8 + 12/s}{0.04(s^2 + 5 \times 10^3 s + 625 \times 10^3)}$$

$$= \frac{0.02(s^2 + 10 \times 10^3 s + 15 \times 10^3)}{s(s^2 + 5 \times 10^3 s + 625 \times 10^3)}\text{ A} \cdot \text{sec}$$

Example 4.11 **Problem:**

Solve for $V(s)$ in the initial value problem

$$0.5\frac{d}{dt}v(t) + 0.2 \cdot v(t) + 2\int_0^{t_1} v(t)\,dt + 10 = 0.5 \cdot \sin[10t] \cdot u(t)\ A$$

$$v(0) = 20\ \mathrm{V}$$

Solution:

1. Perform a Laplace transformation on both members of the equation and use the linearity properties to obtain

$$0.5 \cdot \mathscr{L}\left\{\frac{d}{dt}v(t)\right\} + 0.2 \cdot \mathscr{L}\{v(t)\} + 2 \cdot \mathscr{L}\left\{\int_0^{t_1} v(t)\,dt\right\} + 10 \cdot \mathscr{L}\{1\} = 0.5 \cdot \mathscr{L}\{\sin[10t] \cdot u(t)\}$$

2. Use the differentiation property, the integration property, Pair VII, and Pair II to obtain

$$0.5[s \cdot V(s) - v(0)] + 0.2 \cdot V(s) + 2\frac{1}{s} \cdot V(s) + 10\frac{1}{s} = 0.5\frac{10}{s^2 + 100}$$

3. Substitute the initial condition and solve for $V(s)$:

$$0.5[s \cdot V(s) - 20] + 0.2 \cdot V(s) + 2\frac{1}{s} \cdot V(s) + 10\frac{1}{s} = 0.5\frac{10}{s^2 + 100}$$

$$\left(0.5s + 0.2 + \frac{2}{s}\right)V(s) = 10 - \frac{10}{s} + \frac{5}{s^2 + 100}$$

$$\frac{0.5(s^2 + 0.4s + 4)}{s} \cdot V(s) = \frac{10(s^3 - s^2 + 100.5s - 100)}{s(s^2 + 100)}$$

$$V(s) = \frac{20(s^3 - s^2 + 100.5s - 100)}{(s^2 + 100)(s^2 + 0.4s + 4)}\ \mathrm{V \cdot sec}$$

Exercise 4.9

Problem:

Solve for the Laplace transform of the time-domain function in each of the following initial value problems:

1. $0.1\frac{d}{dt}v(t) + 60 \cdot v(t) = 0,\ v(0) = 1$ V.
2. $2\frac{d}{dt}f(t) + 2 \cdot f(t) + \int_0^{t_1} f(t)\,dt + 5 = u(t),\ f(0) = 10.$
3. $\frac{d^2}{dt^2}i(t) + \frac{d}{dt}i(t) + i(t) = \frac{t^3}{6} \cdot u(t),\ \frac{d}{dt}i(t)\Big|_{t=0} = -1$ A/sec, $i(0) = 2$ A.

Answers:

1. $V(s) = \dfrac{1}{s + 600}$ V · sec.
2. $F(s) = \dfrac{10s - 2}{s^2 + s + 0.5}.$
3. $I(s) = \dfrac{2(s^5 + 0.5s^4 + 0.5)}{s^4(s^2 + s + 1)}$ A · sec.

4.5 Inverse Laplace Transformation

The **inverse Laplace transform** is denoted

$$f(t) = \mathscr{L}^{-1}\{F(s)\}$$

where

$$F(s) = \mathscr{L}\{f(t)\}$$

The inverse Laplace transform is also an integral transform, defined by

$$f(t) = \int_{c-j\infty}^{c+j\infty} \exp[st] \cdot F(s)\,ds$$

Computation of inverse Laplace transforms from the definition, which requires use of complex-variable theory, is seldom necessary in circuit analysis. Instead, we utilize tables of Laplace transform pairs, such as we are in the process of compiling, and simply look up the inverse Laplace transform.

Example 4.12

Problem:

Determine $f(t)$ for the s-domain function

$$F(s) = \frac{12}{s + 100}$$

Solution:

1. $f(t)$ is the inverse Laplace transform of $F(s)$:

$$f(t) = \mathscr{L}^{-1}\left\{\frac{12}{s + 100}\right\}$$

2. Use the linearity properties to obtain

$$f(t) = 12 \cdot \mathscr{L}^{-1}\left\{\frac{1}{s + 100}\right\}$$

3. From Pair V,

$$f(t) = 12 \cdot \exp[-100t] \cdot u(t)$$

Example 4.13

Problem:

Determine $v(t)$ for the s-domain function

$$V(s) = \frac{8s^2 - 3}{s^3} \cdot \exp[-2s] \text{ V} \cdot \text{sec}$$

Solution:

1. As written, $V(s)$ does not appear to match any $F(s)$ in the table of Laplace transform pairs until the numerator of $V(s)$ is decomposed:

$$V(s) = \left(\frac{8s^2}{s^3} - \frac{3}{s^3}\right)\exp[-2s] = \frac{8}{s} \cdot \exp[-2s] - \frac{3}{s^3} \cdot \exp[-2s]$$

2. $v(t)$ is the inverse Laplace transform of $V(s)$:

$$v(t) = \mathscr{L}^{-1}\left\{\frac{8}{s} \cdot \exp[-2s] - \frac{3}{s^3} \cdot \exp[-2s]\right\}$$

3. Use the linearity properties to obtain

$$v(t) = 8 \cdot \mathscr{L}^{-1}\left\{\frac{1}{s} \cdot \exp[-2s]\right\} - 3 \cdot \mathscr{L}^{-1}\left\{\frac{1}{s^3} \cdot \exp[-2s]\right\}$$

4. Use the time-domain shifting property and Pairs II and IV to obtain

$$\begin{aligned} v(t) &= 8 \cdot u(t-2) - 3\frac{(t-2)^{3-1}}{(3-1)!} \cdot u(t-2) \\ &= -(1.5t^2 - 6t - 2)u(t-2) \text{ V} \end{aligned}$$

Example 4.14

Problem:

Determine $f(t)$ for the s-domain function

$$F(s) = \frac{s+35}{(s+25)^2}$$

Solution:

1. Decompose the numerator of $F(s)$ so that the variable s can be canceled from the numerator:

$$\begin{aligned} F(s) &= \frac{s+25}{(s+25)^2} + \frac{10}{(s+25)^2} \\ &= \frac{1}{s+25} + \frac{10}{(s+25)^2} \end{aligned}$$

2. $f(t)$ is the inverse Laplace transform of $F(s)$:

$$f(t) = \mathscr{L}^{-1}\left\{\frac{1}{s+25} + \frac{10}{(s+25)^2}\right\}$$

3. Use the linearity properties to obtain

$$f(t) = \mathcal{L}^{-1}\left\{\frac{1}{s+25}\right\} + 10 \cdot \mathcal{L}^{-1}\left\{\frac{1}{(s+25)^2}\right\}$$

4. From Pairs IV and VIa,

$$\begin{aligned} f(t) &= \exp[-25t] \cdot u(t) + 10t \cdot \exp[-25t] \cdot u(t) \\ &= (10t+1)\exp[-25t] \cdot u(t) \end{aligned}$$

Exercise 4.10

Problem:

Determine the time-domain function for each s-domain function:

1. $I(s) = \dfrac{120}{s^2 + 225}$ mA · sec.
2. $F(s) = \dfrac{6s + 200}{s(s+80)}\exp[-20s]$.
3. $F(s) = \dfrac{10s^2 - 10s + 1000}{s^2 + 100}$.

Answers:

1. $i(t) = 8 \cdot \sin[15t] \cdot u(t)$ mA.
2. $f(t) = \{2.5 + 3.5 \cdot \exp[-80(t-20)]\}u(t-20)$.
3. $f(t) = 10[\delta(t) - \cos[10t] \cdot u(t)]$.

4.6 Partial Fraction Decomposition of s-Domain Functions

In the examples of inverse Laplace transformation in the preceding section, the s-domain functions were immediately associated with a Laplace transform pair from our table (Example 4.12), or were easily decomposed into an algebraic sum of terms, each of which could be associated with a Laplace transform pair (Examples 4.13 and 4.14). Clearly this requires intuition, experience, and familiarity with the Laplace transform table. In this section, we will introduce a method for decomposing s-domain functions known as

partial fraction decomposition. The method is systematic, thus alleviating the need for intuition, and reduces our reliance on an extensive table of Laplace transforms. Our modest table of Laplace transforms, with the addition of one essential Laplace transform pair, will be quite serviceable for circuit analysis. Actually, we will add several transform pairs to the table for convenience rather than necessity.

Polynomial Theory

The review of some essentials from the theory of polynomials is pertinent to the technique of partial fraction decomposition as well as to several topics which arise in later chapters.

Consider the polynomial $P(x)$ of **degree** n:

$$P(x) = C_n x^n + C_{n-1} x^{n-1} + \cdots + C_1 x + C_0 \tag{4.4}$$

where the coefficients $(C_n, C_{n-1}, \ldots, C_1, C_0)$ may be any real constants (positive, negative, or zero). $P(x)$ may be written with the coefficient of the highest-degree term **normalized**:

$$P(x) = C_n\left(x^n + \frac{C_{n-1}}{C_n} x^{n-1} + \cdots + \frac{C_1}{C_n} x + \frac{C_0}{C_n}\right)$$

$$P(x) = C_n\left(x_n + c_{n-1} x^{n-1} + \cdots c_1 x + c_0\right) \tag{4.5}$$

Polynomial theory ensures that $P(x)$ can also be written as a product of n factors:

$$P(x) = C_n(x - r_1)(x - r_2)\cdots(x - r_{n-1})(x - r_n) \tag{4.6}$$

where $r_1, r_2, \ldots, r_{n-1}, r_n$ are the **roots** of $P(x) = 0$. Recall that a **root** is a value of x which satisfies the equation $P(x) = 0$. Clearly, if $x = r_1$, then the factor $x - r_1 = 0$, and $P(x) = 0$; likewise, if $x = r_2$, then $x - r_2 = 0$, and so on.

The *nature* of the roots, that is, real or complex, distinct or repeated, is an extremely important characteristic of the polynomial, as will be seen in the next section and in later chapters. For the purpose of partial fraction decomposition, the nature of the roots determines which decomposition procedure to use; more will said about this very soon. An examination of the possibilities is appropriate.

1. Real (positive, negative, or zero), distinct roots. For example, the roots of

$$P(x) = 9x^3 + 9x^2 - 18x = 9x(x - 1)(x + 2)$$

are $r_1 = 0$, $r_2 = 1$, $r_3 = -2$, and are all real and distinct.

2. Real, repeated roots. For example, the roots of

$$P(x) = x^6 - 2x^5 - 12x^4 - 14x^3 - 5x^2$$
$$= x^2(x+1)^3(x-5)$$

are $r_1 = r_2 = 0$, $r_3 = r_4 = r_5 = -1$, $r_6 = 5$. The root $r = 0$ is repeated, and $r = -1$ is repeated twice; the root r_6 is distinct.

3. Complex **conjugate** roots. If $P(x)$ has a complex root, then the conjugate of the complex root must also be a root of $P(x)$. This observation is correct if and only if all coefficients are real, which was a condition imposed on Equation (4.4). For example, the roots of

$$P(x) = 2x^3 - 4x^2 + 10x = 2x(x - 1 - j2)(x - 1 + j2)$$

are $r_1 = 0$, $\mathbf{r}_2 = 1 + j2$, and $\mathbf{r}_3 = \mathbf{r}_2^* = 1 - j2$.

4. Complex roots, repeated in conjugate pairs. For example, the roots of

$$P(x) = x^4 + 4x^3 + 8x^2 + 8x + 4$$
$$= (x + 1 - j1)^2(x + 1 + j1)^2$$

are $\mathbf{r}_1 = \mathbf{r}_2 = -1 + j1$ and $\mathbf{r}_3 = \mathbf{r}_4 = \mathbf{r}_1^* = \mathbf{r}_2^* = -1 - j1$. A given polynomial can have any combination of roots with the characteristics just outlined.

Why this interest in polynomials? The reader may have already noticed that all of the s-domain functions which represent Laplace transforms of unshifted, aperiodic time-domain functions are **polynomial fractions**, that is,

$$F(s) = \frac{N(s)}{D(s)}$$

where $N(s)$ is the *numerator* polynomial and $D(s)$ is the *denominator* polynomial. $N(s)$ may be a constant, but a constant is merely a polynomial of degree zero. If the degree of $N(s)$ is less than the degree of $D(s)$, $F(s)$ is said to be a **proper polynomial fraction.**

Partial Fraction Decomposition

For the purpose of partial fraction decomposition, $F(s)$ must be a proper polynomial fraction with $D(s)$ written in factored form:

$$F(s) = \frac{K \cdot N(s)}{(s - p_1)(s - p_2) \cdots (s - p_{n-1})(s - p_n)} \tag{4.7}$$

The constant K results from normalizing the coefficient of the highest degree term in $D(s)$, so $K = 1/C_n$, where C_n is the leading coefficient in $D(s)$. The roots of $D(s)$ are called **poles** of $F(s)$ and are indicated as p_k. The function $F(s)$ becomes undefined for any $s = p_k$, that is, when $D(s) = 0$.

Partial fraction decomposition means that $F(s)$ can be written as a sum of fractions, each having a constant for a numerator and one factor of $D(s)$ (possibly repeated) for a denominator. The procedure will be explained in detail in the following sections.

If $F(s)$ is not a proper polynomial fraction, **polynomial long division** (not as ominous as it sounds) must be performed to obtain

$$F(s) = Q(s) + \frac{R(s)}{D(s)} \tag{4.8}$$

where $Q(s)$ is the quotient, $R(s)$ is the remainder, and the degree of $R(s)$ is less than the degree of $D(s)$. The partial fraction decomposition then applies to $R(s)/D(s)$. The degree of $N(s)$ in the s-domain functions representing circuit responses is usually no more than one degree greater than the degree of $D(s)$, so at most $Q(s)$ is a first-degree polynomial $[Q(s) = as + b]$, and $R(s)/D(s)$ is a proper polynomial fraction after two steps of long division.

4.7 Distinct Real Poles (Case 1)

In the case where $F(s)$ has only distinct real poles, the partial fraction decomposition of Equation (4.7) is

$$F(s) = \frac{K_1}{s - p_1} + \frac{K_2}{s - p_2} + \cdots + \frac{K_n}{s - p_n} \tag{4.9}$$

where each K_k is a constant. Now $\mathscr{L}^{-1}\{F(s)\}$ can be readily evaluated by invoking the linearity properties and Pair IV:

$$\begin{aligned} f(t) &= K_1 \cdot \mathscr{L}^{-1}\left\{\frac{1}{s - p_1}\right\} + K_2 \cdot \mathscr{L}^{-1}\left\{\frac{1}{s - p_2}\right\} \\ &\quad + \cdots + K_n \cdot \mathscr{L}^{-1}\left\{\frac{1}{s - p_n}\right\} \\ &= K_1 \cdot \exp[p_1 t] \cdot u(t) \\ &\quad + K_2 \cdot \exp[p_2 t] \cdot u(t) \\ &\quad + \cdots + K_n \cdot \exp[p_n t] \cdot u(t) \\ &= (K_1 \cdot \exp[p_1 t] + K_2 \cdot \exp[p_2 t] \\ &\quad + \cdots + K_n \cdot \exp[p_n t])u(t) \end{aligned} \tag{4.10}$$

All that remains at this point is to evaluate $K_1, K_2, \ldots, K_n$, which can be done systematically as follows:

$$K_k = (s - p_k) \cdot F(s)|_{s=p_k} \tag{4.11}$$

For any pole p_k equal to zero, the associated term in Equation (4.10) is

$$K_k \cdot \exp[p_k t] \cdot u(t) = K_k \cdot \exp[0] \cdot u(t) = K_k \cdot u(t)$$

Equivalently, for $p_k = 0$, the associated term in $F(s)$ is

$$K_k \frac{1}{s}$$

and Pair II is invoked with the same result.

If all poles are zero or negative, then the time-domain function decays to a constant value, possibly zero, as $t \to \infty$. This case occurs frequently in circuit analysis, and the result is called an **overdamped response**. Note that it is the *roots* of $D(s)$ [the *poles* of $F(s)$] which determine the characteristics of the time-domain response. This observation is useful in predicting how a physical system, such as an electric circuit, will respond to a given excitation without actually obtaining the time-domain function which describes the response.

Example 4.15

Problem:

Determine $f(t)$ for the s-domain function

$$F(s) = \frac{3s + 4}{5s^3 + 15s^2 + 10s}$$

Solution:

1. $F(s)$ is a proper polynomial fraction, i.e., the degree of $N(S)$ is less than the degree of $D(s)$, so $F(s)$ can be decomposed as in Equation (4.7).
2. Rewrite $F(s)$ with $D(s)$ in factored form:

$$F(s) = \frac{3s + 4}{5s(s^2 + 3s + 2)} = \frac{0.6s + 0.8}{s(s+1)(s+2)}$$

3. Decompose $F(s)$ as in Equation (4.9):

$$F(s) = \frac{K_1}{s} + \frac{K_2}{s+1} + \frac{K_3}{s+2}$$

4. Solve for K_1, K_2, and K_3:

$$K_1 = s\frac{0.6s + 0.8}{s(s+1)(s+2)}\bigg|_{s=0} = \frac{0.8}{(1)(2)} = 0.4$$

$$K_2 = (s+1)\frac{0.6s + 0.8}{s(s+1)(s+2)}\bigg|_{s=-1}$$

$$= \frac{-0.6 + 0.8}{(-1)(-1+2)} = -0.2$$

$$K_3 = (s+2)\frac{0.6s + 0.8}{s(s+1)(s+2)}\bigg|_{s=-2}$$

$$= \frac{-1.2 + 0.8}{(-2)(-2+1)} = -0.2$$

5. Substitute the results in $F(s)$ to obtain

$$F(s) = \frac{0.4}{s} - \frac{0.2}{s+1} - \frac{0.2}{s+2}$$

6. Determine $f(t)$ by inverse Laplace transformation:

$$f(t) = \mathcal{L}^{-1}\{F(s)\}$$

$$= \mathcal{L}^{-1}\left\{\frac{0.4}{s}\right\} - \mathcal{L}^{-1}\left\{\frac{0.2}{s+1}\right\} - \mathcal{L}^{-1}\left\{\frac{0.2}{s+2}\right\}$$

7. Use the linearity properties and Pairs II and IV to obtain

$$f(t) = \{0.4 \cdot u(t) - 0.2 \cdot \exp[-t] \cdot u(t) - 0.2 \cdot \exp[-2t]\}u(t)$$

$$f(t) = (0.4 - 0.2 \cdot \exp[-t] - 0.2 \cdot \exp[-2t])u(t)$$

Example 4.16

Problem:

Determine $f(t)$ for the s-domain function

$$F(s) = \frac{1}{s^2 + (a+b)s + ab}$$

Solution:

1. $F(s)$ is a proper polynomial fraction, so $F(s)$ can be decomposed as in Equation (4.9):

$$F(s) = \frac{1}{(s+a)(s+b)} = \frac{K_1}{s+a} + \frac{K_2}{s+b}$$

2. Solve for K_1 and K_2:

$$K_1 = \cancel{(s+a)}\frac{1}{\cancel{(s+a)}(s+b)}\bigg|_{s=-a} = \frac{1}{-a+b} = \frac{1}{b-a}$$

$$K_2 = \cancel{(s+b)}\frac{1}{(s+a)\cancel{(s+b)}}\bigg|_{s=-b} = \frac{1}{-b+a} = \frac{1}{a-b}$$

3. Substitute the results into $F(s)$ to obtain

$$F(s) = \frac{1/(b-a)}{s+a} + \frac{1/(a-b)}{s+b}$$

4. Determine $f(t)$ by inverse Laplace transformation; use the linearity properties and Pair IV to obtain

$$f(t) = \frac{1}{b-a} \cdot \exp[-at] \cdot u(t) + \frac{1}{a-b}\exp[-bt] \cdot u(t)$$

$$f(t) = \frac{\exp[-at] - \exp[-bt]}{b-a} u(t)$$

Example 4.16 represents a *second-order*, *overdamped response*, which occurs frequently in circuit analysis. For that reason, the result of Example 4.16 makes a useful addition to the table of Laplace transforms.

Pair XI

$$f(t) = \frac{\exp[-at] - \exp[-bt]}{b-a} \cdot u(t) \qquad F(s) = \frac{1}{(s+a)(s+b)}$$

Example 4.17

Problem:

Determine $i(t)$ for the s-domain function

$$I(s) = \frac{3s^2 + 2}{s^2 + 5s + 6} \text{ mA} \cdot \text{sec}$$

Solution:

1. $I(s)$ is not a proper polynomial fraction, so polynomial long division must be performed:

$$\begin{array}{r|l} & \overbrace{3}^{Q(s)} \\ \underbrace{s^2 + 5s + 6}_{D(s)} & 3s^2 \qquad\qquad + 2 \\ & -(3s^2 \quad + 15s + 18) \\ \hline & \underbrace{-15s - 16}_{R(s)} \end{array}$$

2. Rewrite $I(s)$ in the form of Equation (4.8):

$$I(s) = 3 + \frac{-15s - 16}{s^2 + 5s + 6}$$

$$= 3 + \frac{-15s - 16}{(s + 2)(s + 3)} \text{ mA} \cdot \text{sec}$$

3. Decompose the fractional part of $I(s)$ as in Equation (4.9):

$$I(s) = 3 + \frac{K_1}{s + 2} + \frac{K_2}{s + 3} \text{ mA} \cdot \text{sec}$$

4. Evaluate K_1 and K_2:

$$K_1 = \cancel{(s + 2)} \frac{-15s - 16}{\cancel{(s + 2)}(s + 3)}\bigg|_{s=-2} = \frac{30 - 16}{-2 + 3} = 14$$

$$K_2 = \cancel{(s + 3)} \frac{-15s - 16}{(s + 2)\cancel{(s + 3)}}\bigg|_{s=-3} = \frac{45 - 16}{-3 + 2} = -29$$

5. Substitute K_1 and K_2 in $I(s)$:

$$I(s) = 3 + \frac{14}{s+2} - \frac{29}{s+3} \text{ mA} \cdot \text{sec}$$

6. Use the linearity properties and Pairs I and IV to obtain

$$i(t) = 3 \cdot \delta(t) + (14 \cdot \exp[-2t] - 29 \cdot \exp[-3t])u(t) \text{ mA}$$

Exercise 4.11

Problem:

Determine the time-domain function for each s-domain function:

1. $V(s) = \dfrac{144s}{(s+3)(12s^2 + 144s + 384)}$ V · sec.
2. $F(s) = \dfrac{9s + 12}{3s^3 + 30s^2 + 72s}$.
3. $I(s) = \dfrac{8s^2 + 3100s + 305 \times 10^3}{(s+100)(s+200)(s+250)}$ A · sec.

Answers:

1. $v(t) = (-7.2 \cdot \exp[-3t] + 12 \cdot \exp[-4t] - 4.8 \cdot \exp[-8t])u(t)$ V.
2. $f(t) = \frac{1}{6}(1 + 6 \cdot \exp[-4t] - 7 \cdot \exp[-6t])u(t)$.
3. $i(t) = (5 \cdot \exp[-100t] - \exp[-200t] + 4 \cdot \exp[-250t])u(t)$ A.

4.8 Repeated Real Poles (Case 2)

In the case where $F(s)$ has only one repeated real pole, the partial fraction decomposition of Equation (4.7) is

$$\begin{aligned} F(s) &= \frac{K \cdot N(s)}{(s-p)^n} \\ &= \frac{K_1}{(s-p)^n} + \frac{K_2}{(s-p)^{n-1}} + \cdots + \frac{K_{n-1}}{(s-p)^2} + \frac{K_n}{s-p} \end{aligned} \quad (4.12)$$

where each K_k is a constant. Notice that the partial fractions are written with the exponent of the denominator in descending order, beginning with n and progressing downward to 1, while the subscript for K_k in the numerator progresses upward from 1 to n. Now $\mathscr{L}^{-1}\{F(s)\}$ can be evaluated by invoking the linearity properties and Pairs VIb, VIa, and IV:

$$\begin{aligned}
f(t) &= K_1 \cdot \mathscr{L}^{-1}\left\{\frac{1}{(s-p)^n}\right\} + K_2 \cdot \mathscr{L}^{-1}\left\{\frac{1}{(s-p)^{n-1}}\right\} \\
&\quad + \cdots + K_{n-1} \cdot \mathscr{L}^{-1}\left\{\frac{1}{(s-p)^2}\right\} + K_n \cdot \mathscr{L}^{-1}\left\{\frac{1}{s-p}\right\} \\
&= K_1 \frac{t^{n-1}}{(n-1)!} \cdot \exp[pt] \cdot u(t) \\
&\quad + K_2 \frac{t^{n-2}}{(n-2)!} \cdot \exp[pt] \cdot u(t) \\
&\quad + \cdots + K_{n-1} t \exp[pt] \cdot u(t) + K_n \exp[pt] \cdot u(t) \\
&= \left(K_1 \frac{t^{n-1}}{(n-1)!} + K_2 \frac{t^{n-2}}{(n-2)!} + \cdots + K_{n-1} t + K_n \right) \exp[pt] \cdot u(t)
\end{aligned} \tag{4.13}$$

$K_1, K_2, \ldots, K_n$ can be evaluated systematically as follows:

$$K_k = \frac{1}{(k-1)!} \frac{d^{k-1}}{ds^{k-1}} \left[(s-p)^n \cdot F(s)\right]\Bigg|_{s=p} \tag{4.14}$$

Notice that the order of the derivative is always one less than the subscript of K_k. Also notice that $F(s)$ is always multiplied by $(s-p)^n$ before differentiation, regardless of which K_k is being evaluated. Thus,

$$\begin{aligned}
K_1 &= \left[(s-p)^n \cdot F(s)\right]\Big|_{s=p} \\
K_2 &= \frac{d}{ds}\left[(s-p)^n \cdot F(s)\right]\Bigg|_{s=p} \\
K_3 &= \frac{1}{2} \frac{d^2}{ds^2}\left[(s-p)^n \cdot F(s)\right]\Bigg|_{s=p}
\end{aligned}$$

and so forth.

If the pole is negative, the time-domain function will decay to zero as $t \to \infty$; if the pole is positive (including zero) the time-domain function will grow without bounds as $t \to \infty$. Again, it is the *poles* of $F(s)$ which determine the characteristics of the time-domain response.

Example 4.18

Problem:

Determine $f(t)$ for the s-domain function

$$F(s) = \frac{2s^3 + 3}{(s+3)^4}$$

Solution:

1. $F(s)$ is a proper polynomial fraction, i.e., the degree of $N(S)$ is less than the degree of $D(s)$, so $F(s)$ can be decomposed as in Equation (4.12):

$$F(s) = \frac{K_1}{(s+3)^4} + \frac{K_2}{(s+3)^3} + \frac{K_3}{(s+3)^2} + \frac{K_4}{s+3}$$

2. Solve for K_1, K_2, K_3, and K_4:

$$K_1 = \cancel{(s+3)^4}\frac{2s^3+3}{\cancel{(s+3)^4}}\Bigg|_{s=-3} = -54 + 3 = -51$$

Flashback

$$\frac{d}{dx}x^n = nx^{n-1}$$

$$K_2 = \frac{d}{ds}\left(\cancel{(s+3)^4}\frac{2s^3+3}{\cancel{(s+3)^4}}\right)\Bigg|_{s=-3} = 6s^2\Big|_{s=-3} = 54$$

$$K_3 = \frac{1}{2}\frac{d^2}{ds^2}\left(\cancel{(s+3)^4}\frac{2s^3+3}{\cancel{(s+3)^4}}\right)\Bigg|_{s=-3} = \frac{12s}{2}\Bigg|_{s=-3} = -18$$

$$K_4 = \frac{1}{6}\frac{d^3}{ds^3}\left(\cancel{(s+3)^4}\frac{2s^3+3}{\cancel{(s+3)^4}}\right)\Bigg|_{s=-3} = \frac{12}{6} = 2$$

3. Substitute the results in $F(s)$ to obtain

$$F(s) = -\frac{51}{(s+3)^4} + \frac{54}{(s+3)^3} - \frac{18}{(s+3)^2} + \frac{2}{s+3}$$

4. Determine $f(t)$ by inverse Laplace transformation:

$$f(t) = \mathscr{L}^{-1}\left\{-\frac{51}{(s+3)^4}\right\} + \mathscr{L}^{-1}\left\{\frac{54}{(s+3)^3}\right\} - \mathscr{L}^{-1}\left\{\frac{18}{(s+3)^2}\right\} + \mathscr{L}^{-1}\left\{\frac{2}{s+3}\right\}$$

5. Use the linearity properties, Pairs VI and II, and Equation (4.13) to obtain

$$\begin{aligned} f(t) = &-51\frac{t^{4-1}}{(4-1)!}\cdot \exp[-3t]\cdot u(t) \\ &+54\frac{t^{3-1}}{(3-1)!}\cdot \exp[-3t]\cdot u(t) \\ &-18\frac{t^{2-1}}{(2-1)!}\cdot \exp[-3t]\cdot u(t) + 2\cdot \exp[-3t]\cdot u(t) \end{aligned}$$

$$f(t) = -(8.5t^3 - 27t^2 + 18t - 2)\exp[-3t]\cdot u(t)$$

Frequently s-domain functions have a combination of distinct and repeated real poles. In this case the decomposition is expressed as a combination of Equations (4.9) and (4.12). For example,

$$\begin{aligned} F(s) &= \frac{K\cdot N(s)}{(s-p_1)^3(s-p_4)(s-p_5)} \\ &= \frac{K_1}{(s-p_1)^3} + \frac{K_2}{(s-p_1)^2} + \frac{K_3}{s-p_1} + \frac{K_4}{s-p_4} + \frac{K_5}{s-p_5} \end{aligned}$$

K_1, K_2, and K_3 are evaluated according to Equation (4.14), and K_4 and K_5 are evaluated according to Equation (4.11). It is not necessary to write the terms arising from the repeated pole first, but it is helpful, since the subscript of K_k determines the value of the coefficient $1/(k-1)!$ and the order of the derivative d^k/ds^k in evaluating K_k.

Example 4.19

Problem:

Determine $i(t)$ for the s-domain function

$$I(s) = \frac{200s}{(s+10^3)(s+400)^2}$$

Solution:

1. $I(s)$ is a proper polynomial fraction, so $I(s)$ can be decomposed as in Equations (4.9) and (4.12):

$$I(s) = \frac{K_1}{(s+400)^2} + \frac{K_2}{s+400} + \frac{K_3}{s+10^3} \text{ A} \cdot \text{sec}$$

2. Solve for K_1 and K_2 according to Equation (4.14):

$$K_1 = \cancel{(s+400)^2}\frac{200s}{(s+10^3)\cancel{(s+400)^2}}\bigg|_{s=-400}$$

$$= \frac{-80 \times 10^3}{-400+10^3} \simeq -133.3$$

Flashback

$$\frac{d}{dx}\frac{f(x)}{g(x)} = \frac{1}{g(x)}\frac{d}{dx}f(x) - \frac{f(x)}{[g(x)]^2}\frac{d}{dx}g(x)$$

$$K_2 = \frac{d}{ds}\left(\cancel{(s+400)^2} \cdot \frac{200s}{(s+10^3)\cancel{(s+400)^2}}\right)\bigg|_{s=-400}$$

$$= \frac{200}{s+10^3} - \frac{200s}{(s+10^3)^2}\bigg|_{s=-400}$$

$$= \frac{200}{-400+10^3} - \frac{-80 \times 10^3}{(-400+10^3)^2} \simeq 0.5556$$

3. Solve for K_3 according to Equation (4.11):

$$K_3 = \cancel{(s+10^3)}\frac{200s}{\cancel{(s+10^3)}(s+400)^2}\bigg|_{s=-1000}$$

$$= \frac{-200 \times 10^3}{(-10^3+400)^2} \simeq -0.5556$$

4. Substitute the results into $I(s)$ to obtain

$$I(s) = -\frac{133.3}{(s+400)^2} + \frac{0.5556}{s+400} - \frac{0.5556}{s+10^3} \text{ A} \cdot \text{sec}$$

5. Determine $f(t)$ by inverse Laplace transformation. Use the linearity properties and Pair IV to obtain

$$i(t) = -\{(133.3t - 0.5556)\exp[-400t] + 0.5556 \cdot \exp[-10^3 t]\} \cdot u(t) \text{ A} \cdot \text{sec}$$

Exercise 4.12

Problem:

Determine the time-domain function for each s-domain function.

1. $V(s) = \dfrac{6s+9}{(3s+9)(s^2+6s+9)}$ V · sec.
2. $F(s) = \dfrac{3s}{(s+1)(s+2)^3}$.

Answers:

1. $v(t) = -0.5t(3t+4)\exp[-3t] \cdot u(t)$ V.
2. $f(t) = 3\{[t^2+t+1]\exp[-2t] - \exp[-t]\} \cdot u(t)$.

4.9 Complex Conjugate Poles (Case 3)

Before discussing the decomposition of s-domain functions with complex conjugate poles, we will examine the **quadratic**, or second order, polynomial to determine the origin of complex conjugate roots.

Flashback

$$P(x) = ax^2 + bx + c = (x - r_1)(x - r_2)$$

$$r_1 = \frac{-b + \sqrt{b^2 - 4ac}}{2a}$$

$$r_2 = \frac{-b - \sqrt{b^2 - 4ac}}{2a}$$

For the quadratic

$$P(x) = x^2 + bx + c$$

the roots are

$$r_1 = -\frac{b}{2} + \sqrt{\left(\frac{b}{2}\right)^2 - c}$$

$$r_2 = -\frac{b}{2} - \sqrt{\left(\frac{b}{2}\right)^2 - c}$$

The expression $(b/2)^2 - c$, which is called the **discriminant** of the quadratic, determines the nature of the roots. If the discriminant is greater than zero $[c < (b/2)^2]$, then r_1 and r_2 are real and distinct because $\sqrt{(b/2)^2 - c}$ is real and is added to (in r_1) and subtracted from (in r_2) the real number $-b/2$. If $c = (b/2)^2$, then $\sqrt{(b/2)^2 - c}$ is zero, and $r_1 = r_2 = -b/2$.

The third possibility is that the discriminant is negative $[c > (b/2)^2]$, in which case $\sqrt{(b/2)^2 - c}$ is imaginary, and $\mathbf{r}_1$ and $\mathbf{r}_2$ are complex conjugates. When -1 is factored from the discriminant,

$$\mathbf{r}_1 = -\frac{b}{2} + \sqrt{(-1)\left[c - \left(\frac{b}{2}\right)^2\right]}$$

$$= -\frac{b}{2} + j\sqrt{c - \left(\frac{b}{2}\right)^2}$$

$$\mathbf{r}_2 = -\frac{b}{2} - \sqrt{(-1)\left(c - \left(\frac{b}{2}\right)^2\right)}$$

$$= -\frac{b}{2} - j\sqrt{c - \left(\frac{b}{2}\right)^2}$$

Now it is clear that $\mathbf{r}_1$ is complex with a real part $-b/2$ (which may be zero if $b = 0$), and an imaginary part $\sqrt{c - (b/2)^2}$; $\mathbf{r}_2$ is the conjugate of $\mathbf{r}_1$.

The standard way of writing an s-domain function with complex conjugate poles is

$$F(s) = \frac{K \cdot N(s)}{s^2 + 2\alpha s + \omega_0^2} = \frac{K \cdot N(s)}{(s - \mathbf{p})(s - \mathbf{p}^*)} \tag{4.15}$$

where

$$\mathbf{p} = -\alpha + j\sqrt{\omega_0^2 - \alpha^2} = -\alpha + j\omega_d$$

$$\mathbf{p}^* = -\alpha - j\sqrt{\omega_0^2 - \alpha^2} = -\alpha - j\omega_d$$

$$\omega_d = \sqrt{\omega_0^2 - \alpha^2}$$

In Equation (4.15), the discriminant of $D(s)$ is $\alpha^2 - \omega_0^2$; for p to be complex, ω_0^2 must be greater than α^2, or $\omega_0 > \alpha$. Physical significance will be given to this choice of symbols later in this section.

The partial fraction decomposition of Equation (4.15) is

$$\begin{aligned} F(s) &= \frac{\mathbf{K}_1}{s - p} + \frac{\mathbf{K}_1^*}{s - p^*} \\ &= \frac{\mathbf{K}_1}{s - (-\alpha + j\omega_d)} + \frac{\mathbf{K}_1^*}{s - (-\alpha - j\omega_d)} \\ &= \frac{\mathbf{K}_1}{s + \alpha - j\omega_d} + \frac{\mathbf{K}_1^*}{s + \alpha + j\omega_d} \end{aligned} \tag{4.16}$$

where $\mathbf{K}_1$ is a *complex* constant. In order to properly apply the decomposition method presented here, the term with the pole containing the *positive imaginary part* must be written first. Since the denominator of the first term is $s - \mathbf{p}$, then $-j\omega_d$ must appear in the first term in Equation (4.16).

$\mathscr{L}^{-1}\{F(s)\}$ can be evaluated by invoking the linearity properties and Pair IV:

$$\begin{aligned} f(t) &= \mathbf{K}_1 \cdot \mathscr{L}^{-1}\left\{\frac{1}{s - \mathbf{p}}\right\} + \mathbf{K}_1^* \cdot \mathscr{L}^{-1}\left\{\frac{1}{s - \mathbf{p}^*}\right\} \\ &= \mathbf{K}_1 \cdot \exp[\mathbf{p}t] \cdot u(t) + \mathbf{K}_1^* \cdot \exp[\mathbf{p}^*t] \cdot u(t) \\ &= (\mathbf{K}_1 \cdot \exp[\mathbf{p}t] + \mathbf{K}_1^* \cdot \exp[\mathbf{p}^*t])u(t) \end{aligned}$$

Recall that $\mathbf{p} = -\alpha + j\omega_d$, so

$$f(t) = \{\mathbf{K}_1 \cdot \exp[(-\alpha + j\omega_d)t] + \mathbf{K}_1^* \cdot \exp[(-\alpha - j\omega_d)t]\} \cdot u(t)$$

Flashback

$$x^{a+b} = x^a x^b$$

Thus

$$f(t) = \{\exp[-\alpha t] \cdot (\mathbf{K}_1 \cdot \exp[j\omega_d t] + \mathbf{K}_1^* \cdot \exp[-j\omega_d t])\} u(t)$$

$\mathbf{K}_1$ and $\mathbf{K}_1^*$ are also complex numbers, and when they are written in **polar form**,

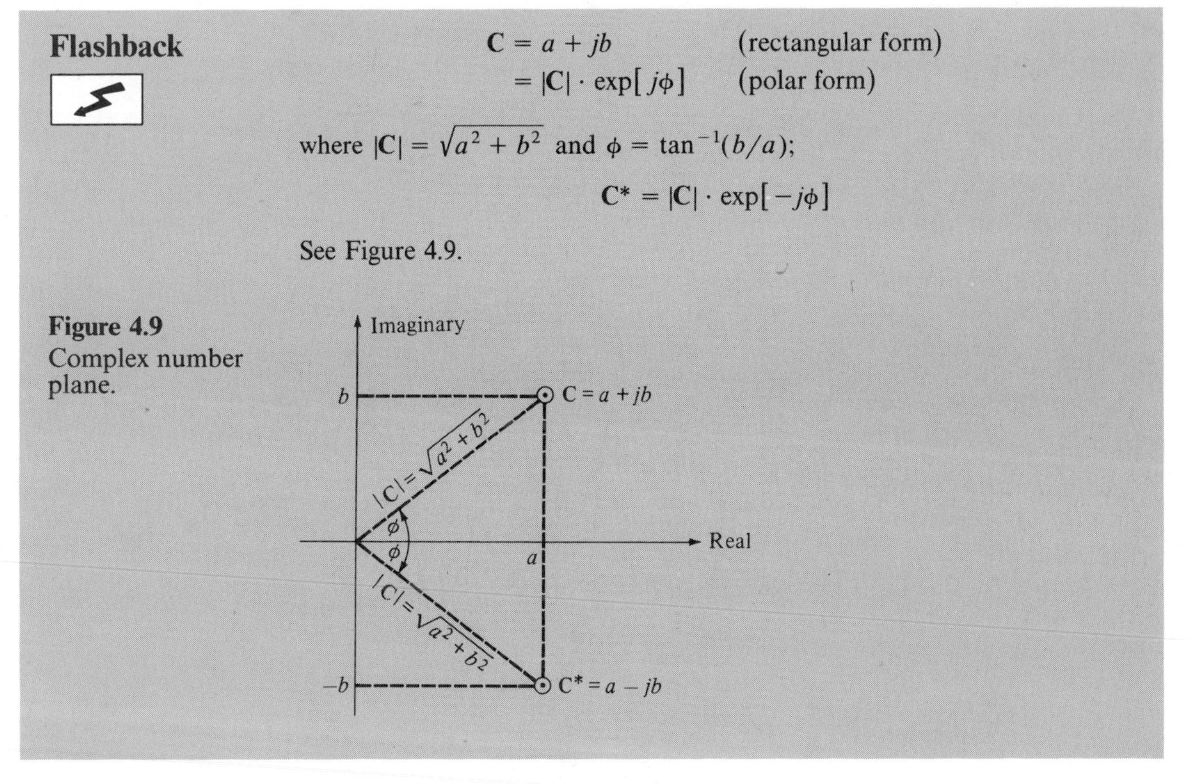

Flashback

$$\mathbf{C} = a + jb \qquad \text{(rectangular form)}$$
$$= |\mathbf{C}| \cdot \exp[j\phi] \qquad \text{(polar form)}$$

where $|\mathbf{C}| = \sqrt{a^2 + b^2}$ and $\phi = \tan^{-1}(b/a)$;

$$\mathbf{C}^* = |\mathbf{C}| \cdot \exp[-j\phi]$$

See Figure 4.9.

Figure 4.9 Complex number plane.

Thus

$$f(t) = \{\exp[-\alpha t] \cdot (|\mathbf{K}_1| \cdot \exp[j\phi] \cdot \exp[j\omega_d t] + |\mathbf{K}_1| \cdot \exp[-j\phi] \cdot \exp[-j\omega_d t])\}\, u(t)$$

where $|\mathbf{K}_1|$ is the **magnitude** and ϕ is the **polar angle** of $\mathbf{K}_1$.

When $|\mathbf{K}_1|$ is factored and the imaginary exponents combined,

$$f(t) = [|\mathbf{K}_1| \cdot \exp[-\alpha t] \cdot \{\exp[j(\omega_d t + \phi)] + \exp[-j(\omega_d t + \phi)]\}] \cdot u(t)$$

From Equation (4.3a),

$$\begin{aligned} f(t) &= |K_1| \cdot \exp[-\alpha t] \cdot (2 \cdot \cos[\omega_d t + \phi])u(t) \\ &= 2|K_1| \cdot \exp[-\alpha t] \cdot \cos[\omega_d t + \phi] \cdot u(t) \end{aligned} \qquad (4.17)$$

This type of time-domain function is said to be **oscillatory** or **underdamped**. The graphs of $f(t)$ in Figure 4.10 indicate that $f(t)$ is oscillating at a

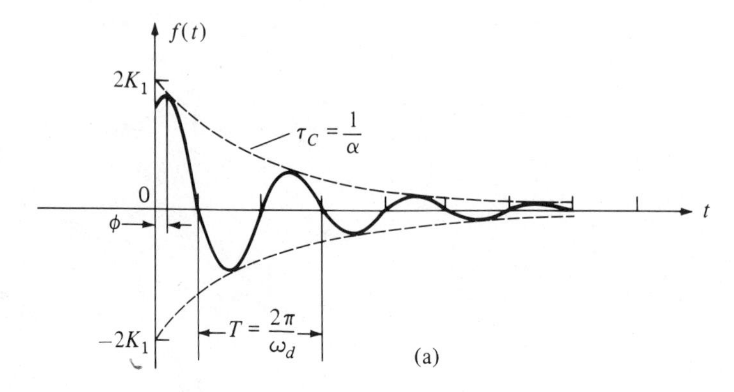

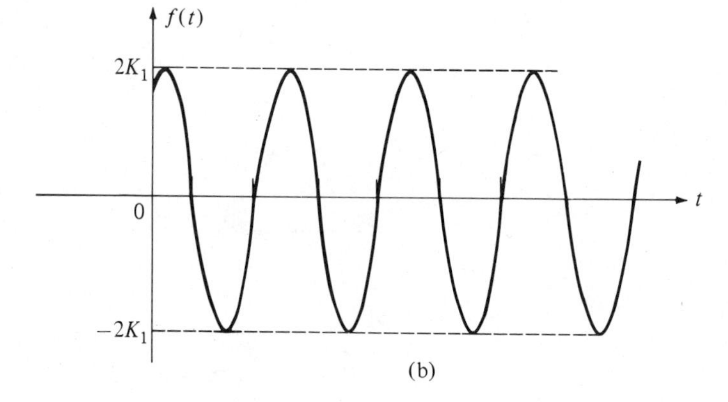

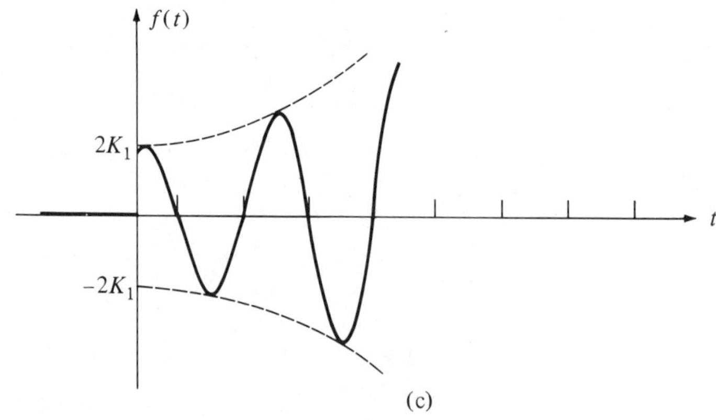

Figure 4.10
Oscillatory time-domain function.
(a) Decaying oscillation ($\alpha > 0$ and $\omega_d < \omega_0$).
(b) Sustained oscillation ($\alpha = 0$ and $\omega_d = \omega_0$).
(c) Growing oscillation ($\alpha < 0$ and $\omega_d < \omega_0$).

radian frequency ω_d, called the **damped frequency** (ω_0 is called the **natural** or **resonant frequency**). The magnitude of $\mathbf{K}_1$ is the **amplitude**, and the polar angle of $\mathbf{K}_1$ is the **phase angle** of $f(t)$. Whether the amplitude of the function decays [Figure 4.10(a)], is sustained [Figure 4.10(b)], or grows [Figure 4.10(c)] depends upon α, which is called the **damping coefficient**.

$\mathbf{K}_1$ must now be evaluated using Equation (4.11) and the result expressed in polar form to obtain a final result for $f(t)$.

Example 4.20

Problem:

Determine $f(t)$ for the s-domain function

$$F(s) = \frac{2s + 30}{s^2 + 4s + 162}$$

Solution:

1. Compare $F(s)$ with Equation (4.15):

$$\omega_0 = \sqrt{162} \simeq 12.73 \quad \text{and} \quad \alpha = 2$$

2. Since $\omega_0 > \alpha$, $F(s)$ will have complex conjugate poles

$$\mathbf{p} = -\alpha + j\omega_d = -2 + j12.57$$

$$\mathbf{p}^* = -\alpha - j\omega_d = -2 - j12.57$$

and may be written in the form of Equation (4.16):

$$F(s) = \frac{\mathbf{K}_1}{s + 2 - j12.57} + \frac{\mathbf{K}_1^*}{s + 2 + j12.57}$$

3. From Equation (4.17),

$$f(t) = 2|\mathbf{K}_1| \cdot \exp[-2t] \cdot \cos[12.57t + \phi] \cdot u(t)$$

and from Equation (4.11),

$$\mathbf{K}_1 = \cancel{(s + 2 - j12.57)} \frac{2s + 30}{\cancel{(s + 2 - j12.57)}(s + 2 + j12.57)}\Bigg|_{s=-2+j12.57}$$

$$= \frac{2(-2 + j12.57) + 30}{-2 + j12.57 + 2 + j12.57}$$

$$= \frac{26 + j25.14}{j25.14} = \frac{36.17 \cdot \exp[j44.04°]}{25.14 \cdot \exp[j90°]}$$

$$= 1.439 \cdot \exp[-j45.96°]$$

4. Substitute $|\mathbf{K}_1|$ and ϕ in $f(t)$ to obtain

$$f(t) = 2.877 \cdot \exp[-2t] \cdot \cos[12.57t - 45.96°] \cdot u(t)$$

Two useful Laplace transform pairs can be obtained by applying Equations (4.16) and (4.17). Let

$$F(s) = \frac{1}{s^2 + 2\alpha s + \omega_0^2}$$

where $\omega_0 > \alpha$. From Equation (4.16),

$$F(s) = \frac{\mathbf{K}_1}{s + \alpha + j\omega_d} + \frac{\mathbf{K}_1^*}{s + \alpha + j\omega_d}$$

and from Equation (4.11),

$$\mathbf{K}_1 = \cancel{(s + \alpha - j\omega_d)} \frac{1}{\cancel{(s + \alpha - j\omega_d)}(s + \alpha + j\omega_d)} \bigg|_{s = -\alpha + j\omega d}$$

$$= \frac{1}{(-\alpha + j\omega_d + \alpha + j\omega_d)} = \frac{1}{j2\omega_d} = \frac{0.5 \cdot \exp[-j90°]}{\omega_d}$$

From Equation (4.17),

$$f(t) = 2|\mathbf{K}_1| \cdot \exp[-\alpha t] \cdot \cos[\omega_d t + \phi] \cdot u(t)$$

$$= \frac{2(0.5)}{\omega_d} \cdot \exp[-\alpha t] \cdot \cos[\omega_d t - 90°] \cdot u(t)$$

Flashback

$$\cos[x - 90°] = \sin x$$

Thus

$$f(t) = \frac{1}{\omega_d} \cdot \exp[-\alpha t] \cdot \sin \omega_d t \cdot u(t)$$

This result is entered into the table of Laplace transform pairs as Pair XII. The derivation of Pair XIII is very similar and is given as an exercise (Problem 4-17).

Pair XII

$$f(t) = \frac{1}{\omega_d} \cdot \exp[-\alpha t] \cdot \sin \omega_d t \cdot u(t) \qquad F(s) = \frac{1}{s^2 + 2\alpha s + \omega_0^2}$$

where $\omega_d = \sqrt{\omega_0^2 - \alpha^2}$

Pair XIII

$$f(t) = \frac{\omega_0}{\omega_d} \cdot \exp[-\alpha t] \cdot \cos[\omega_d t + \phi] \cdot u(t) \qquad F(s) = \frac{s}{s^2 + 2\alpha s + \omega_0^2}$$

where $\omega_d = \sqrt{\omega_0^2 - \alpha^2}$ and $\phi = \tan^{-1}[\alpha/\omega_d]$

Notice that if $\alpha = 0$, Pairs XII and XIII reduce to Pairs VII and VIII, respectively. Also, notice the similarity between Pairs XII and XIII and Pairs IX and X, respectively (Problems 4-18, 4-19, and 4-20).

An s-domain function which has a combination of real poles (distinct and/or repeated) and complex poles can be expressed as a combination of Equations (4.9), (4.12), and (4.16). For example,

$$\begin{aligned} F(s) &= \frac{K \cdot N(s)}{(s - p_1)^2(s - p_3)(s + 2\alpha s + \omega_0^2)} \\ &= \frac{K_1}{(s - p_1)^2} + \frac{K_2}{s - p_1} + \frac{K_3}{s - p_3} \\ &\quad + \frac{\mathbf{K}_4}{s + \alpha - j\omega_d} + \frac{\mathbf{K}_4^*}{s + \alpha + j\omega_d} \end{aligned}$$

K_1 and K_2 are evaluated according to Equation (4.14), and K_3 and $\mathbf{K}_4$ are evaluated according to Equation (4.11).

Example 4.21

Problem:

Determine $v(t)$ for the s-domain function

$$V(s) = \frac{0.8s^2}{(1.25s + 2.5)(s^2 + 6s + 25)} \text{ V} \cdot \text{sec}$$

Solution:

1. $V(s)$ can be rewritten as

$$V(s) = \frac{0.8s^2}{1.25(s+2)(s+3-j4)(s+3+j4)}$$

$$= \frac{0.64s^2}{(s+2)(s+3-j4)(s+3+j4)} \text{ V} \cdot \text{sec}$$

2. Decompose $V(s)$ according to Equations (4.9) and (4.16):

$$V(s) = \frac{K_1}{s+2} + \frac{\mathbf{K}_2}{s+3-j4} + \frac{\mathbf{K}_2^*}{s+3+j4} \text{ V} \cdot \text{sec}$$

3. Use Pair IV and Equation (4.17) to obtain

$$v(t) = (K_1 \cdot \exp[-2t] + 2|\mathbf{K}_2| \cdot \exp[-3t]$$

$$\cdot \cos[4t + \phi])u(t) \text{ V}$$

4. Solve for K_1 and $\mathbf{K}_2$ according to Equation (4.11):

$$K_1 = \cancel{(s+2)} \frac{0.64s^2}{\cancel{(s+2)}(s+3-j4)(s+3+j4)}\Bigg|_{s=-2}$$

$$= \frac{0.64(4)}{(-2+3-j4)(-2+3+j4)} \simeq 0.1506$$

$$\mathbf{K}_2 = \cancel{(s+3-j4)} \frac{0.64s^2}{(s+2)\cancel{(s+3-j4)}(s+3+j4)}\Bigg|_{s=-3+j4}$$

$$= \frac{0.64(-3+j4)^2}{(-3+j4+2)(-3+j4+3+j4)}$$

$$= 0.48512 \cdot \exp[j59.70°]$$

5. Substitute the results into $v(t)$ to obtain

$$v(t) = (0.1506 \cdot \exp[-2t]$$

$$+0.9701 \cdot \exp[-3t] \cdot \cos[4t + 59.70°])u(t) \text{ V}$$

Exercise 4.13

Problem:

Determine the time-domain function for each s-domain function.

1. $F(s) = \dfrac{2.6s - 200}{s^2 + 1000s + 1.25 \cdot 10^6}$.
2. $I(s) = \dfrac{12}{s^2 + 5s + 100}$ A · sec.
3. $V(s) = \dfrac{5s^2 + 12s + 15}{(s+1)(s^2 + 2s + 2)}$ V · sec.

Answers:

1. $f(t) = 3.002 \cdot \exp[-500t] \cdot \cos[1000t + 29.98°] \cdot u(t)$.
2. $i(t) = 1.239 \cdot \exp[-2.5t] \cdot \sin[9.682t] \cdot u(t)$.
3. $v(t) = (8 - 3.606 \cdot \cos[t + 33.69°]) \cdot \exp[-t] \cdot u(t)$ V · sec.

4.10 Repeated Complex Conjugate Poles (Case 4)

In the case where $F(s)$ has only repeated complex conjugate poles, the partial fraction decomposition of Equation (4.7) is

$$\begin{aligned} F(s) &= \frac{K \cdot N(s)}{\left(s^2 + 2\alpha s + \omega_0^2\right)^n} = \frac{K \cdot N(s)}{(s + \alpha - j\omega_d)^n (s + \alpha + j\omega_d)^n} \\ &= \frac{\mathbf{K}_1}{(s + \alpha - j\omega_d)^n} + \frac{\mathbf{K}_2}{(s + \alpha - j\omega_d)^{n-1}} \\ &\quad + \cdots + \frac{\mathbf{K}_n}{s + \alpha - j\omega_d} \\ &\quad + \frac{\mathbf{K}_1^*}{(s + \alpha + j\omega_d)^n} + \frac{\mathbf{K}_2^*}{(s + \alpha + j\omega_d)^{n-1}} \\ &\quad + \cdots + \frac{\mathbf{K}_n^*}{s + \alpha + j\omega_d} \end{aligned} \tag{4.18}$$

where $\mathbf{K}_k$ is a constant which may be complex. $\mathscr{L}^{-1}\{F(s)\}$ can be evaluated by invoking the linearity properties and Pairs VIb, VIa:

$$f(t) = \mathbf{K}_1 \cdot \mathcal{L}^{-1}\left\{\frac{1}{(s+\alpha-j\omega_d)^n}\right\}$$

$$+ \mathbf{K}_1^* \cdot \mathcal{L}^{-1}\left\{\frac{1}{(s+\alpha-j\omega_d)^n}\right\}$$

$$+ \mathbf{K}_2 \cdot \mathcal{L}^{-1}\left\{\frac{1}{(s+\alpha+j\omega_d)^{n-1}}\right\}$$

$$+ \mathbf{K}_2^* \cdot \mathcal{L}^{-1}\left\{\frac{1}{(s+\alpha+j\omega_d)^{n-1}}\right\}$$

$$+ \cdots + \mathbf{K}_n \cdot \mathcal{L}^{-1}\left\{\frac{1}{s+\alpha+j\omega_d}\right\}$$

$$+ \mathbf{K}_n^* \cdot \mathcal{L}^{-1}\left\{\frac{1}{s+\alpha-j\omega_d}\right\}$$

The same manipulation of the exponential terms used to obtain Equation (4.17) can be used to obtain

$$\begin{aligned} f(t) = 2\Bigg(&|\mathbf{K}_1|\frac{t^{n-1}}{(n-1)!} \cdot \cos[\omega_d t + \phi_1] \\ &+|\mathbf{K}_2|\frac{t^{n-2}}{(n-2)!} \cdot \cos[\omega_d t + \phi_2] \\ &+ \cdots + |\mathbf{K}_{n-1}|t \cdot \cos[\omega_d t + \phi_{n-1}] \\ &+|\mathbf{K}_n| \cdot \cos[\omega_d t + \phi_n]\Bigg)\exp[-\alpha t] \cdot u(t) \end{aligned} \tag{4.19}$$

where $|\mathbf{K}_k|$ is the magnitude and ϕ_k is the polar angle of $\mathbf{K}_k$. The constants $\mathbf{K}_1, \mathbf{K}_2, \ldots, \mathbf{K}_n$ can be evaluated according to Equation (4.14).

Example 4.22 **Problem:**

Determine $f(t)$ for the s-domain function

$$F(s) = \frac{s}{(s^2+2s+2)^2}$$

Solution:

1. Rewrite $F(S)$ with $D(s)$ factored:

$$F(s) = \frac{s}{(s+1-j1)^2(s+1+j1)^2}$$

2. From Equation (4.19),

$$f(t) = 2(|\mathbf{K}_1|t \cdot \cos[t+\phi_1] + |\mathbf{K}_2| \cdot \cos[t+\phi_2])\exp[-t] \cdot u(t)$$

3. Solve for $\mathbf{K}_1$ and $\mathbf{K}_2$:

$$\mathbf{K}_1 = \cancel{(s+1-j1)^2}\frac{s}{\cancel{(s+1-j1)^2}(s+1+j1)^2}\Bigg|_{s=-1+j1}$$

$$= \frac{-1+j1}{(-1+j1+1+j1)^2} = \frac{1-j1}{4} = \frac{\sqrt{2}}{4}\exp[-j45°]$$

$$\mathbf{K}_2 = \frac{d}{ds}\left(\cancel{(s+1-j1)^2}\frac{s}{\cancel{(s+1-j1)^2}(s+1+j1)^2}\right)\Bigg|_{s=-1+j1}$$

$$= \frac{1}{(s+1+j1)^2} - \frac{2s}{(s+1+j1)^3}\Bigg|_{s=-1+j}$$

$$= \frac{1}{(-1+j1+1+j1)^2} - \frac{2(-1+j1)}{(-1+j1+1+j1)^3} = \frac{j}{4}$$

$$= \tfrac{1}{4} \cdot \exp[j90°]$$

4. Substitute in $f(t)$ to obtain

$$f(t) = 2\left(\tfrac{1}{4}\sqrt{2}\,t \cdot \cos[t-45°] + \tfrac{1}{4} \cdot \cos[t+90°]\right)\exp[-t] \cdot u(t)$$
$$= 0.5\left(\sqrt{2}\,t \cdot \cos[t-45°] - \sin t\right)\exp[-t] \cdot u(t)$$

The derivation of Pairs XIV and XV, given below, is left as an exercise (Problems 4-21 and 4-22).

Pair XIV

$$f(t) = \frac{1}{2\omega_d^2}\left(\frac{1}{\omega_d} \cdot \sin \omega_d t - t \cdot \cos \omega_d t\right)\exp[-\alpha t] \cdot u(t)$$

where $\omega_d = \sqrt{\omega_0^2 - \alpha^2}$

$$F(s) = \frac{1}{\left(s^2 + 2\alpha s + \omega_0^2\right)^2}$$

Pair XV

$$f(t) = -\frac{1}{2\omega_d^2}\left(\frac{\alpha}{\omega_d} \cdot \sin \omega_d t - \omega_0 t \cdot \cos(\omega_d t + \phi)\right)\exp[-\alpha t] \cdot u(t)$$

where $\omega_d = \sqrt{\omega_0^2 - \alpha^2}$ and $\phi = \tan^{-1}(-\omega_d/\alpha)$

$$F(s) = \frac{s}{\left(s^2 + 2\alpha s + \omega_0^2\right)^2}$$

4.11 Additional Inverse Laplace Transform Techniques

The amount of text devoted to partial fraction decomposition might lead the reader to believe that partial fraction decomposition is the first and last resort for determining an inverse Laplace transform. The first resort is always, of course, to look in the most available table of Laplace transform pairs for the s-domain function, as is, among the pairs. Failing this, some additional inspection of $F(s)$ is appropriate before turning to partial fraction decomposition.

Inverse Laplace Transformation When $N(s)$ Is Multiplied by $\exp[-t_0 s]$

Recall from the time-domain shifting property that the Laplace transform of a time-domain function shifted by $t = t_0$ is the Laplace transform of the unshifted function multiplied by $\exp[-t_0 s]$. To determine the inverse Laplace transform of $F(s)$ when the numerator $N(s)$ is multiplied by $\exp[-t_0 s]$:

1. Remove $\exp[-t_0 s]$ from $N(s)$, and determine the inverse Laplace transform of what remains of $F(s)$.
2. Substitute $t - t_0$ for t everywhere in the resulting time-domain function.

Example 4.23 **Problem:**

Determine $f(t)$ for the s-domain function

$$F(s) = \frac{3\exp[-5s]}{(s+2)^2}$$

Solution:

1. Remove $\exp[-5s]$ from $F(s)$, and find the inverse Laplace transformation of what remains of $F(s)$:

$$\mathscr{L}^{-1}\left\{\frac{3}{(s+2)^2}\right\} = 3t \cdot \exp[-2t] \cdot u(t)$$

2. Substitute $t-5$ for t:

$$\mathscr{L}^{-1}\left\{\frac{3 \cdot \exp[-5s]}{(s+2)^2}\right\} = 3(t-5)\exp[-2(t-5)] \cdot u(t-5)$$

Algebraic Decomposition

All of the s-domain functions $F(s)$ considered thus far have had polynomial numerators $N(s)$ and denominators $D(s)$, which means that inverse Laplace transformation yielded aperiodic, unshifted time-domain functions. In these cases, partial fraction decomposition is usually the technique of choice.

The Laplace transform of a time-domain function composed of several components, each with a different amount of shifting, will have one or more terms of $N(s)$ multiplied by factors of the form $\exp[-t_0 s]$. In these cases, the first step in inverse Laplace transformation is to perform an **algebraic decomposition**, which was used without formal comment in Examples 4.13 and 4.14, to isolate each term of $N(s)$ which contains an exponential factor.

Consider the s-domain function $F(s)$, in which $N(s)$ consists of more than one term:

$$\begin{aligned} F(s) &= \frac{K \cdot N(s)}{D(s)} = K \cdot \left(\frac{N_1(s)}{D(s)} + \frac{N_2(s)}{D(s)} + \cdots + \frac{N_n(s)}{D(s)}\right) \\ &= K\left[F_1(s) + F_2(s) + \cdots + F_n(s)\right] \end{aligned} \tag{4.20}$$

where $N_k(s)$ is a term of $N(s)$. Each term $F_k(s)$ of $F(s)$ is an s-domain function which, because of the linearity properties, can be considered separately in determining $\mathscr{L}^{-1}\{F(s)\}$.

Example 4.24

Problem:

Determine $i(t)$ for the s-domain function

$$I(s) = \frac{20s + 2000 + 8000 \cdot \exp[-0.03s] + (1520s - 8000) \cdot \exp[-0.06s]}{s(s+100)} \text{ A} \cdot \text{sec}$$

Solution:

1. Use algebraic decomposition to rewrite $I(s)$:

$$I(s) = \frac{20s + 2000}{s(s+100)} + \frac{8000}{s(s+100)} \cdot \exp[-0.03s] + \frac{1520s - 8000}{s(s+100)} \cdot \exp[-0.06s] \text{ A} \cdot \text{sec}$$

2. Use the linearity properties to obtain

$$i(t) = 20 \cdot \mathscr{L}^{-1}\left\{\frac{s+100}{s(s+100)}\right\} + 8000 \cdot \mathscr{L}^{-1}\left\{\frac{1}{s(s+100)} \cdot \exp[-0.03s]\right\} + 80 \cdot \mathscr{L}^{-1}\left\{\frac{19s - 100}{s(s+100)} \cdot \exp[-0.06s]\right\} \text{ A}$$

The last two terms are shifted time-domain functions, as evidenced by the exponential factors of the form $\exp[-t_0 s]$ in the s-domain terms. The inverse Laplace transforms for the first two terms are available in the table of Laplace transform pairs (Pairs II and Vb, respectively). The inverse Laplace transform of the last term can be obtained by further algebraic decomposition of that term only.

3. Rewrite $i(t)$ with the last term decomposed:

$$i(t) = 20 \cdot \mathscr{L}^{-1}\left\{\frac{1}{s}\right\} + 8000 \cdot \mathscr{L}^{-1}\left\{\frac{1}{s(s+100)} \cdot \exp[-0.03s]\right\} + 80(19) \cdot \mathscr{L}^{-1}\left\{\frac{1}{s+100} \exp[-0.06s]\right\} + 80(100) \cdot \mathscr{L}^{-1}\left\{\left(\frac{1}{s(s+100)}\right) \cdot \exp[-0.06s]\right\} \text{ A}$$

4. Use Pair II (first term), Pair Vb (second and fourth terms), Pair IV (third term), and the time-domain shifting property (second, third, and

fourth terms) to obtain

$$\begin{aligned}
i(t) &= 20 \cdot u(t) \\
&\quad + \frac{8000}{100}\{1 - \exp[-100(t - 0.03)]\}u(t - 0.03) \\
&\quad - 1520 \cdot \exp[-100(t - 0.06)] \cdot u(t - 0.06) \\
&\quad + \frac{8000}{100}\{1 - \exp[-100(t - 0.06)]\}u(t - 0.06) \\
&= 20 \cdot u(t) + 80\{1 - \exp[-100(t - 0.03)]\}u(t - 0.03) \\
&\quad - 80\{1 - 20 \cdot \exp[-100(t - 0.06)]\}u(t - 0.06) \text{ A}
\end{aligned}$$

Inverse Laplace Transformation when $D(s)$ Contains $1 - \exp[-Ts]$

The Laplace transform $F_P(s)$ of a periodic time-domain function $f_P(t)$ is characterized by the factor $1 - \exp[-Ts]$ in $D_P(s)$. If $1 - \exp[-Ts]$ is extracted from $D_P(s)$, what remains of $F_P(s)$ is $F(s)$, the Laplace transform of the time-domain function for the first period of $f_P(t)$:

$$\begin{aligned}
F_P(s) &= \frac{K \cdot N_P(s)}{D_P(s)} = \frac{K \cdot N_P(s)}{(1 - \exp[-Ts])D(s)} \\
&= \frac{1}{1 - \exp[-Ts]} \cdot F(s)
\end{aligned}$$

The time-domain function for the first period of $f_P(t)$ is

$$f(t) = \mathscr{L}^{-1}\{F(s)\}$$

and

$$f_P(t) = \sum_{n=0}^{\infty} f(t - nT)$$

Any appropriate technique may be used to obtain the inverse transformation of $F(s)$; however, most periodic time-domain functions have shifted components, so $N_P(s)$ is likely to have terms containing $\exp[-t_0 s]$.

Example 4.25

Problem:

Determine $f_P(t)$ for the s-domain function

$$F_P(s) = \frac{10s - 5 + 10 \cdot \exp[-2s] - 10s \cdot \exp[-4s]}{(1 - \exp[-4s])s^2}$$

Solution:

1. Extract $1 - \exp[-4s]$ from the denominator:

$$F(s) = \frac{10s - 5 + 10 \cdot \exp[-2s] - 10s \cdot \exp[-4s]}{s^2}$$

2. Use algebraic decomposition to rewrite $F(s)$:

$$\begin{aligned} F(s) &= \frac{10s}{s^2} - \frac{5}{s^2} + \frac{10}{s^2} \cdot \exp[-2s] - \frac{10s}{s^2} \cdot \exp[-4s] \\ &= \frac{10}{s} - \frac{5}{s^2} + \frac{10}{s^2} \cdot \exp[-2s] - \frac{10}{s} \cdot \exp[-4s] \end{aligned}$$

3. Use the linearity properties, Pairs II and IIIa, and the time-domain shifting property to obtain

$$\begin{aligned} f(t) &= 10 \cdot u(t) - 5t \cdot u(t) + 10(t-2)u(t-2) - 10 \cdot u(t-4) \\ &= -5(t-2)u(t) + 10(t-2)u(t-2) - 10 \cdot u(t-4) \end{aligned}$$

4. The periodic time-domain function is

$$\begin{aligned} f_P(t) = -5 \sum_{n=0}^{\infty} & (t - 2 - 4n)u(t - 4n) \\ & - 2(t - 2 - 4n)u(t - 2 - 4n) + 2 \cdot u(t - 4 - 4n) \end{aligned}$$

Additional Laplace Transform Pairs

Derivation of the following three Laplace transform pairs, which completes the table of Laplace transform pairs, is given as an exercise (Problem 4-23 and 4-24).

Pair XVI

$$f(t) = \delta(t) - a \cdot \exp[-at] \cdot u(t) \qquad F(s) = \frac{s}{s+a}$$

Pair XVIIa

$$f(t) = (1 - at)\exp[-at] \cdot u(t) \qquad F(s) = \frac{s}{(s+a)^2}$$

Pair XVIIb

$$f(t) = \frac{t^{n-2}}{(n-2)!}\left(1 - \frac{a}{n-1}t\right)\exp[-at] \cdot u(t)$$

$$F(s) = \frac{s}{(s+a)^n} \qquad n = 2, 3, \ldots$$

ServoSoftware
Laplace Transforms
denominator

ROOT NUMBER	REAL PART	IMAGINARY PART	ROOT ORDER
1	2	0	1
2	3	4	1
3	3	-4	1

(a)

ServoSoftware
Laplace Transforms
numerator

POWER OF s	COEFFICIENT
0	0
1	0
2	.64

(b)

Figure 4.11 Computer program for inverse Laplace transformation program. (By permission of ServoSoftware.) (a) Data entry screen (denominator). (b) Data entry screen (numerator). (c) Tabular output. (d) Graphical output.

```
ROOT  REAL                      IMAGINARY             NO . S   COEFF
---  ----------------------  -------------------- -- . -  --------------------
                                                     .
 0    400                      0                    2 . 0   0
 1    1000                     0                    1 . 1   200
                                                      . 2   0
```

T	F(T)
0	-1.387778780781446E-17
.0005	6.330689055332819E-02
.001	7.864655211980209E-02
.0015	7.117182608430407E-02
.002	5.461987672463486E-02
.0025	3.614820991371017E-02
.003	1.919295053752931E-02
.0035	5.143406004632552E-03
.004	-5.688765649415534E-03
.0045	-1.351828199776071E-02
.005	-1.878055758133783E-02
.0055	-2.196876784133946E-02
.006	-.02355258423556
.0065	-2.394257944055817E-02
.007	-2.347929141705717E-02
.0075	-2.243485503579954E-02
.008	-2.102038349333312E-02
.0085	-1.939537173763886E-02
.009	-1.767718213497012E-02
.0095	-1.594968878098879E-02
.01	-1.427071909666138E-02
.0105	-1.267822956470889E-02
.011	-1.119529952323479E-02
.0115	-9.834089446752141E-03
.012	-8.598927035839495E-03
.0125	-7.488678139635062E-03
.013	-6.49854272301106 E-03
.0135	-5.621395705854911E-03
.014	-4.84877216645494 E-03
.0145	-4.171577842330362E-03
.015	-3.580589756473939E-03
.0155	-3.066798261033727E-03
.016	-2.62163066166041 E-03
.0165	-2.237087581764204E-03
.017	-1.905816030410675E-03
.0175	-1.621137444314466E-03
.018	-1.377044507661304E-03
.0185	-1.168177075408564E-03
.019	-9.897848365805731E-04
.0195	-8.376822893963621E-04
.02	-7.082000262128791E-04
.0205	-5.981351358020429E-04
.021	-5.047026377436672E-04
.0215	-4.254892008681537E-04
.022	-3.584099113085875E-04
.0225	-3.016685039839941E-04
.023	-2.537212216235914E-04
.0235	-2.132442924579029E-04
.024	-1.791049019154525E-04
.0245	-1.503354600702089E-04
.025	-1.261109237224286E-04

(c)

Figure 4.11 Continued

```
ROOT  REAL                         IMAGINARY                 NO . S   COEFF
---  ----------------------  ----------------------  --  .  -  ----------------------
                                                         .
                                                         .
 0    400                      0                           2 . 0   0
 1    1000                     0                           1 . 1   200
                                                             . 2   0

T DIVISION   .0025                       F(T) DIVISION   .013125
                   INITIAL T VALUE   0

==========:==========*==========:==========:==========:==========:==========:==========:
          :         !         :         :         :         :       * :         :
          :         !         :         :         :         :         :         *
          :         !         :         :         :         :         :   *     :
          :         !         :         :         :         : *       :         :
------------------------------------------------*--------------------------------
          :         !         :    *    :         :         :         :         :
          :         !   *     :         :         :         :         :         :
          :     *   !         :         :         :         :         :         :
          *         !         :         :         :         :         :         :
------*--------------------------------------------------------------------------
   *      :         !         :         :         :         :         :         :
  *       :         !         :         :         :         :         :         :
  *       :         !         :         :         :         :         :         :
  *       :         !         :         :         :         :         :         :
---*-----------------------------------------------------------------------------
    *     :         !         :         :         :         :         :         :
     *    :         !         :         :         :         :         :         :
       *  :         !         :         :         :         :         :         :
        * :         !         :         :         :         :         :         :
---------*-----------------------------------------------------------------------
          *         !         :         :         :         :         :         :
          :*        !         :         :         :         :         :         :
          :  *      !         :         :         :         :         :         :
          :  *      !         :         :         :         :         :         :
--------------*------------------------------------------------------------------
          :   *     !         :         :         :         :         :         :
          :     *   !         :         :         :         :         :         :
          :     *   !         :         :         :         :         :         :
          :      *  !         :         :         :         :         :         :
-----------------*---------------------------------------------------------------
          :       * !         :         :         :         :         :         :
          :       * !         :         :         :         :         :         :
          :       * !         :         :         :         :         :         :
          :        *!         :         :         :         :         :         :
-------------------*-------------------------------------------------------------
          :        *!         :         :         :         :         :         :
          :        *!         :         :         :         :         :         :
          :        *!         :         :         :         :         :         :
          :        *!         :         :         :         :         :         :
-------------------*-------------------------------------------------------------
          :         *         :         :         :         :         :         :
          :         *         :         :         :         :         :         :
          :         *         :         :         :         :         :         :
          :         *         :         :         :         :         :         :
--------------------*------------------------------------------------------------
          :         *         :         :         :         :         :         :
          :         *         :         :         :         :         :         :
          :         *         :         :         :         :         :         :
          :         *         :         :         :         :         :         :
--------------------*------------------------------------------------------------
```

(d)

Figure 4.11 Continued

4.12 Computer Applications

A number of computer programs are capable of performing the inverse Laplace transformation. Normally these programs do not provide an analytic expression for the resulting time-domain function, but instead provide a tabular or graphic display of $f(t)$ vs. t. These programs provide a means of obtaining the time-domain response of a circuit for which the s-domain response is known, without the use of a sophisticated circuit simulation program such as SPICE. Inverse Laplace transformation programs are found in the sophisticated mathematical libraries of mainframes and minicomputers, but easy-to-use and relatively inexpensive programs, such as the one described in this section,[1] are available for use on personal microcomputers.

For the program described here, the data for the s-domain function (Example 4.21) are entered in two parts. First, the denominator must be factored so that the roots of the denominator (poles of the s-domain function) may be entered. The program prompts for entry of the roots, and displays them as entered [Figure 4.11(a)], but the piece of data that the program requires is in fact the *negative* of the root. Next, the program prompts for entry of the coefficients of the numerator polynomial, in ascending degree of s [Figure 4.11(b)], rather than descending degree, as the polynomial is usually written. The program then computes the time-domain response and prints either a table or a graph of $f(t)$ vs. t [Figure 4.11(c) and (d)]. The data entry format provides for entering repeated roots by specifying a root order greater than 1, and for entering complex conjugate roots by specifying a real and an imaginary part. The number of output data points, the interval between points, and the scaling of the graph may all be controlled by the user.

4.13 SUMMARY

The Laplace transform of a time-domain function can be computed from the defining integral, but in practice is usually obtained from a table of Laplace-transform pairs. The utility of such a table is increased greatly by the properties of linearity, s-domain and time-domain shifting, and periodicity.

The primary importance of the Laplace transform is that it provides a method of solving initial value problems. Initial value problems consist of a differential equation and a number of initial conditions equal to the order of the differential equation. Transformation of the differential equation results in a single algebraic equation in the s-domain which incorporates the initial

[1] Laplace Transform Inversions, ServoSoftware, Box 72, West Covina, CA 91793.

conditions. The solution to the algebraic equation is inverse-transformed to obtain the solution to the initial value problem.

Inverse transformation is also accomplished through the use of a table of Laplace transform pairs. The greatest amount of effort involved in the solution of initial value problems by the Laplace transform method is usually associated with inverse transformation. The s-domain solution must be manipulated into a form which can be associated with one or more Laplace transform pairs in the table. A combination of several techniques may be used to this end, among which are algebraic decomposition and partial fraction decomposition.

4.14 Terms

algebraic decomposition
complex conjugate pole
complex frequency domain
complex variable
complex translation property
conjugate root
damped frequency
damping coefficient
degree of a polynomial
differentiation property
discriminant
distinct real pole
Euler's identity
initial value problem
integral transform
integration property
inverse Laplace transform
inverse Laplace transform operator
kernel
Laplace domain
Laplace operator
Laplace transform
Laplace transform pair
linearity properties
magnitude of a complex number
natural frequency
normalized
oscillatory response
overdamped response
partial fraction decomposition
periodicity property
polar angle of a complex number
pole
polynomial fraction
proper polynomial fraction
quadratic
real-axis translation property
repeated complex conjugate pole
repeated real pole
resonant frequency
root
s-domain
s-domain shifting property
time-domain shifting property
transformation
underdamped response

Problems

4-1. Find the Laplace transform of each time-domain function.

***a.** $f(t) = 7t \cdot u(t)$.

b. $v(t) = 37[1 + (1 + t)u(t)]$ V.

***c.** $f(t) = 41[\delta(t) + 1 + t \cdot u(t)]$.

d. $v(t) = 59[53 \cdot \delta(t) + 49 \cdot u(t)]$ V.
***e.** $i(t) = 67[29 \cdot \delta(t) + 23t \cdot u(t)]$ μA.
f. $i(t) = 7(13 + 11)u(t)$ mA.

4-2. Find the Laplace transform of each time-domain function.
a. $v(t) = 2 \cdot \exp[-97t] \cdot u(t)$ V.
***b.** $f(t) = (5 \cdot \exp[-83t] + 7 \cdot \exp[-83t])u(t)$.
c. $i(t) = (13 \cdot \exp[-73t] + 71 \cdot \exp[-17]t)u(t)$ mA.

4-3. Show that

$$\mathscr{L}\left\{\left(\exp\left[-\frac{t}{\tau_1}\right] - \exp\left[-\frac{t}{\tau_2}\right]\right)u(t)\right\} = \frac{\tau_1 - \tau_2}{\tau_1\tau_2}\left(\frac{1}{s + 1/\tau_1} \cdot \frac{1}{s + 1/\tau_2}\right)$$

4-4. Show that

$$\mathscr{L}\{(b \cdot \exp[-at] - a \cdot \exp[-bt])u(t)\} = (b - a)\frac{s + a + b}{(s + a)(s + b)}$$

4-5. Derive Pairs Va and Vb.

4-6. Derive Pairs VIa and VIb.

4-7. Find the Laplace transform of each time-domain function.
***a.** $i(t) = 37(1 - \exp[-97t])u(t)$ mA.
b. $f(t) = 89(2 - \exp[-83t])u(t)$.
***c.** $v(t) = 7(-6 + 5 \cdot \exp[-5t])u(t)$ V.
d. $v(t) = 31t \cdot \exp[-73t] \cdot u(t)$ V.
***e.** $f(t) = 29t^4 \cdot \exp[-49t] \cdot u(t)$.
f. $i(t) = 37(1 - t \cdot \exp[-97t])u(t)$ mA.

4-8. Derive Pair VIII.

4-9. Find the Laplace transform of each time-domain function.
***a.** $v(t) = 19 \cdot \sin[10 \times 10^3 t] \cdot u(t)$ V.
b. $i(t) = 31 \cdot \cos 300t \cdot u(t)$ mA.
***c.** $f(t) = 5 \cdot \sin[377t + 60°] \cdot u(t)$.
Suggestion: Use a trignometric identity to expand $f(t)$.

4-10. Find the Laplace transform of each time-domain function.
a. $f(t) = 19 \cdot \exp[-97t] \cdot \sin[10 \times 10^3 t] \cdot u(t)$.
***b.** $v(t) = 31 \cdot \exp[-7t] \cdot \cos 300t \cdot u(t)$ V.
c. $i(t) = 5 \cdot \exp[-5t] \cdot \sin[377t + 60°] \cdot u(t)$ mA.

4-11. Find the Laplace transform of each time-domain function.

***a.** $f(t) = 2 \cdot \delta(t - 7)$.

b. $v(t) = 3 \cdot u(t - 5)$ V.

***c.** $v(t) = 31[\delta(t - 19) + 1 + (t - 31)u(t - 31)]$ V.

d. $f(t) = 3\{\exp[-89(t - 2)] - 83 \cdot \exp[-2(t - 1)]\}u(t - 2)$.

***e.** $i(t) = 61\{1 - \exp[-97(t - 83)]\}u(t - 83)$ mA.

f. $i(t) = 5(t - 3)^3 \exp[-49(t - 3)] \cdot u(t - 3)$ A.

***g.** $f(t) = 43 \cdot \cos[10^3(t - 0.001)] \cdot u(t - 0.001)$.

4-12. Find the Laplace transform of each time-domain function.

a. $v_P(t) = 5 \sum_{n=0}^{\infty} [u(t - 18n) - u(t - 7 - 18n)]$ V.

***b.** $f_P(t) = 73t \cdot u(t) - 219 \sum_{n=0}^{\infty} u(t - 3(n + 1))$.

***c.** $i_P(t) = 11 \sum_{n=0}^{\infty} (u(t - 3n) + (t - 2 - 3n)u(t - 2 - 3n)$
$+(t - 3 - 3n)u(t - 3 - 3n)$ mA.

4-13. Write the s-domain equation for each initial value problem.

***a.** $\frac{d}{dt}f(t) + 2 \cdot f(t) = 0;\ f(0) = 3$.

b. $\frac{d}{dt}v(t) + 7 \cdot v(t) = 11 \cdot u(t);\ v(0) = 13$ V.

***c.** $\frac{d}{dt}f(t) + 19 \cdot f(t) + 23\int_0^{t_1} f(t)\,dt = 37t^3 \cdot u(t);\ f(0) = 31$.

d. $\frac{d^2}{dt^2}i(t) + 0.041\frac{d}{dt}i(t) + 0.43 \cdot i(t) = 0.5 \cdot \sin(10 \times 10^3 t);$
$\left.\frac{d}{dt}i(t)\right|_{t=0} = 0;\ i(0) = 0.47$ A.

***e.** $97\frac{d^2}{dt^2}v(t) + 91\frac{d}{dt}v(t) + 93 \cdot v(t) + 79\int_0^{t_1} v(t)\,dt + 71 = 7 \cdot \delta(t);$
$\left.\frac{d}{dt}v(t)\right|_{t=0} = 7;\ v(0) = 11$ V.

4-14. Find the inverse Laplace transform for each s-domain function. These results should be obtained from the table of Laplace transform pairs using only the linearity properties.

a. $F(s) = 31/s^6$.

***b.** $V(s) = \frac{7}{(s + 11)^4}$ V · sec.

c. $V(s) = \dfrac{300}{(s+2)^2 + 225}$ V · sec.

*d. $F(s) = \dfrac{s+7}{(s+7)^2 + 6400}$.

e. $F(s) = \dfrac{12}{(s+36)(s+3)}$.

*f. $V(s) = \dfrac{85}{s^2 + 17s + 30}$ V · sec.

g. $I(s) = \dfrac{300s}{s^2 + 650s + 10^9}$ mA · sec.

*h. $F(s) = \dfrac{7s}{s+11}$.

i. $F(s) = \dfrac{31}{(s+13)^2}$.

4-15. Find the inverse Laplace transform for each s-domain function. These results should be obtained from the table of Laplace transform pairs using only the linearity properties.

*a. $V(s) = 2 + 3/s$.

b. $I(s) = \dfrac{5}{s^2} + \dfrac{23}{s+93}$ mA · sec.

*c. $F(s) = \dfrac{48}{s^5} + \dfrac{158}{s(s+79)}$.

d. $V(s) = \dfrac{91}{s(s+97)} - \dfrac{3}{(s+2)^2}$ V · sec.

*e. $I(s) = \dfrac{60}{s^2 + 900} + \dfrac{10s}{s^2 + 10^4}$ mA · sec.

4-16. Find the inverse Laplace transform for each s-domain function. These results should be obtained by partial fraction expansion.

a. $F(s) = \dfrac{7}{s(s+3)(s+5)}$.

*b. $V(s) = \dfrac{17}{s^2(s+3)(s+5)}$ V · sec.

c. $V(s) = \dfrac{23s^2 + 1}{(s+1)(s+2)(s+3)^3}$ V · sec.

*d. $I(s) = \dfrac{31}{s^2 + s + 11}$ A · sec.

e. $F(s) = \dfrac{1}{(s+1)(s^2+3s+17)}$.

***f.** $V(s) = \dfrac{7s^2+11}{s(s^2+3s+2)(s^2+6s+8)}$ V · sec.

g. $F(s) = \dfrac{5s}{(s+1)(s^2+6s+9)^2}$.

***h.** $I(s) = \dfrac{s}{(s^2+7)(s^2+5s+91)}$ mA · sec.

i. $V(s) = \dfrac{97s+91}{(s+3)^2(s+5)^2}$ V · sec.

***j.** $F(s) = \dfrac{19s^3}{s(s+3)^2(s^2+5s+31)}$.

4-17. Derive Pair XIII.

4-18. Show that for $\alpha = 0$,

a. Pair XII is equivalent to Pair VII, and

b. Pair XIII is equivalent to Pair VIII.

4-19. Show that Pair XII is equivalent to Pair IX.

4-20. Under what condition(s) is Pair XIII equivalent to Pair X?

4-21. Derive Pair XIV.

4-22. Derive Pair XV.

4-23. Derive Pair XVI.

4-24. Derive Pairs XVIIa and XVIIb.

4-25. Find the inverse Laplace transform for each s-domain function.

***a.** $F(s) = \dfrac{7s^3+23s+1}{s^3+s^2+5s}$.

b. $F(s) = \dfrac{19s^2 \cdot \exp[-2s] + 17s \cdot \exp[-s] + 11s^3 \cdot \exp[-3s]}{s^3+s^2+5s}$.

***c.** $F(s) = \dfrac{3s \cdot \exp[-2s] + 2 \cdot \exp[-17s]}{s^2(s+1)}$.

d. $F(s) = \dfrac{17s^2-11s-3}{3s^2+24s+36}$.

***e.** $F(s) = \dfrac{7 \cdot \exp[-97s]}{(2s^2+8s+8)}$.

***f.** $F(s) = \dfrac{7s^2 \cdot \exp[-s]}{(1-\exp[-10s])(s+5)^3}$.

5 Circuit Transformation

5.1 INTRODUCTION

We saw in Chapter 3 that application of the Kirchhoff laws to R–L, R–C, or R–L–C circuits results in an initial value problem in which the differential equation is ordinary and linear and has constant coefficients. In Chapter 4, we saw that Laplace transform methods can be used to solve this type of problem. We have therefore already suggested a technique for transform circuit analysis:

1. Apply the appropriate Kirchhoff law to the time-domain circuit to obtain an initial value problem in which dependent variable of the differential equation is the desired response (current or voltage).
2. Perform a Laplace transformation of the differential equation to obtain an algebraic equation which incorporates the initial conditions.
3. Solve the algebraic equation for the s-domain response.
4. Obtain the time-domain response by inverse transformation of the s-domain response.

A modification of this technique which is favored in circuit analysis is **circuit transformation**, which involves element-by-element transformation of the time-domain circuit to obtain an s-domain circuit. Circuit transformation changes the first two steps in the procedure described above as follows:

1. Perform the circuit transformation and sketch the transformed circuit.
2. Apply the appropriate Kirchhoff law to the s-domain circuit to obtain an algebraic equation for the desired s-domain response.

Then proceed with steps 3 and 4, as above.

The principal advantage of circuit transformation, as we will show in this and later chapters, is that the rich set of analysis techniques and circuit theorems from dc analysis (Chapter 1) can be applied to the transformed circuit. Further, in Chapter 9 we will show that the phasor technique commonly used with sinusoidal steady-state (ac) analysis is only a specialization of the more general circuit transformation technique.

5.2 Transformation of Ideal Sources

Equivalent ideal sources may be obtained by Laplace transformation of the time-domain function which describes the ideal source:

$$V(s) = \mathscr{L}\{v(t)\}$$

and

$$I(s) = \mathscr{L}\{i(t)\}$$

Figure 5.1 shows the symbolic transformation of ideal sources, and Table 5.1 provides a list of the transformed equivalents of some common time-domain sources.

As was noted in Chapter 4, Laplace transformation has the effect of multiplying all dimensions by time. Thus, the unit for transformed voltage is the volt-second (V · s), and the unit for transformed current is the ampere-second (A · s).

Figure 5.1
Transformed equivalent sources.

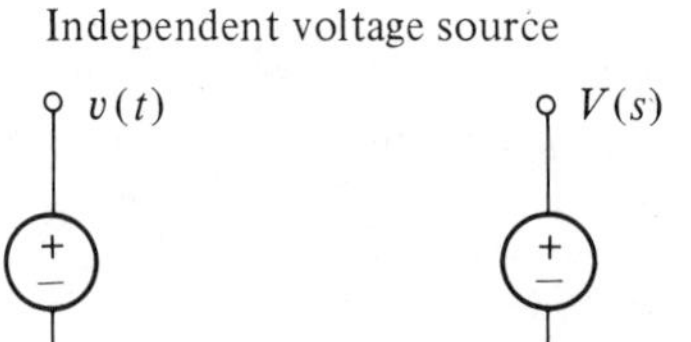

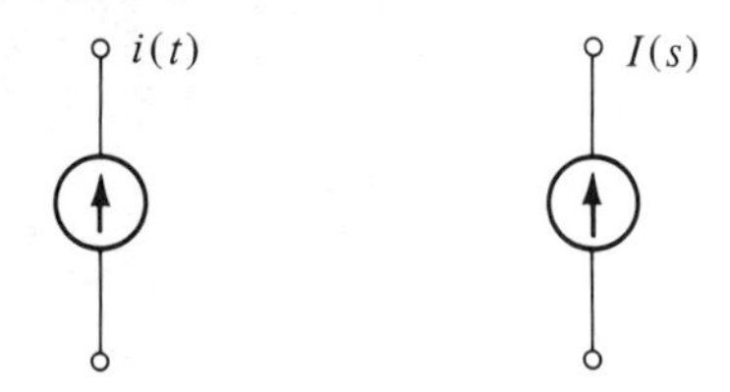

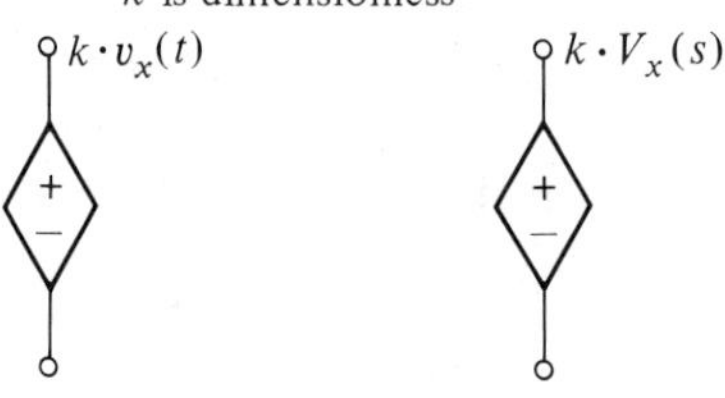

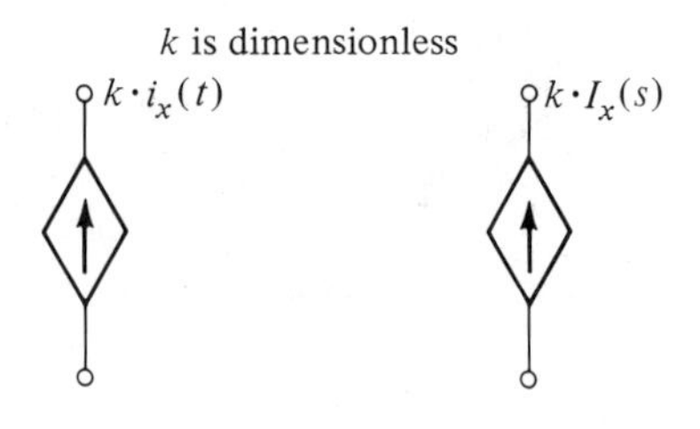

Current-controlled
voltage source (CCVS)

k is in ohms

$k \cdot i_x(t)$ $k \cdot I_x(s)$

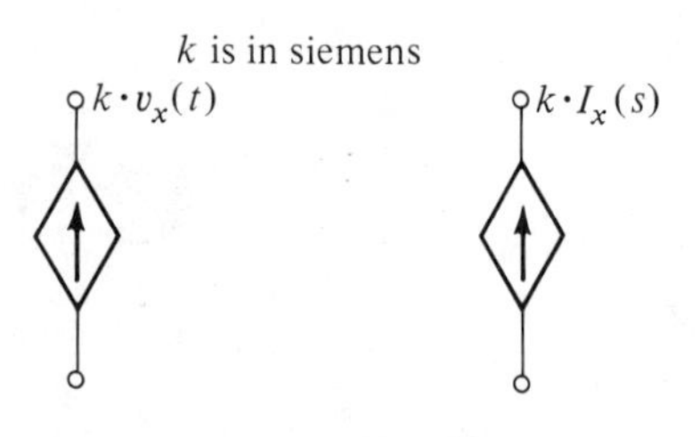

Table 5.1 Transformed Equivalent Sources

	Time Domain	s-Domain
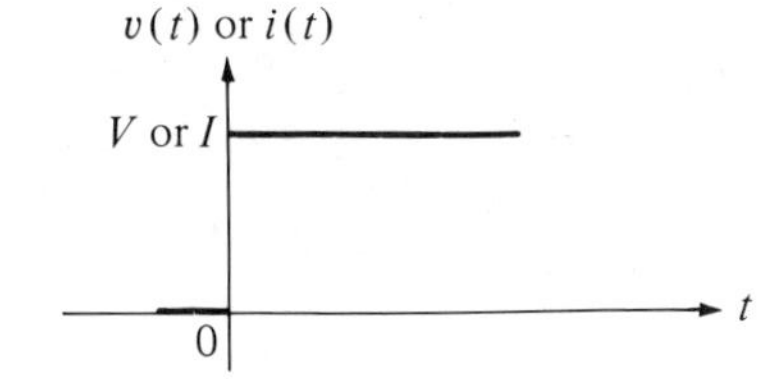	***Switched dc or step*** $V \cdot u(t)$ or $I \cdot u(t)$	V/s or I/s
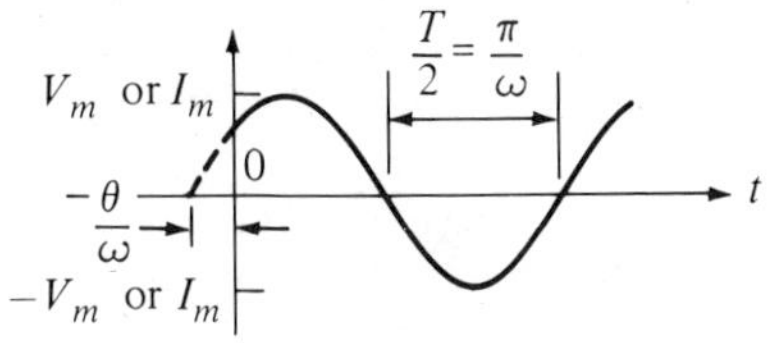	***Sinusoid*** $(V_m \text{ or } I_m) \cdot \sin[\omega t + \theta] \cdot u(t)$	$(V_m \text{ or } I_m)\dfrac{\omega}{\sqrt{a^2+\omega^2}} \cdot \dfrac{s+a}{s^2+\omega^2}$, where $a = \omega/\tan\theta$
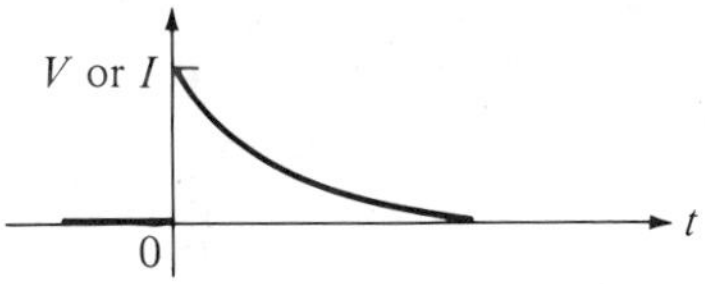	***Exponential decay with time constant*** τ $(V \text{ or } I) \cdot \exp[-t/\tau] \cdot u(t)$	$\dfrac{(V \text{ or } I)}{s+a}$, where $a = 1/\tau$
	***Exponential growth with time constant*** τ $(V \text{ or } I)(1 - \exp[t/\tau])u(t)$	$\dfrac{a \cdot (V \text{ or } I)}{s(s+a)}$
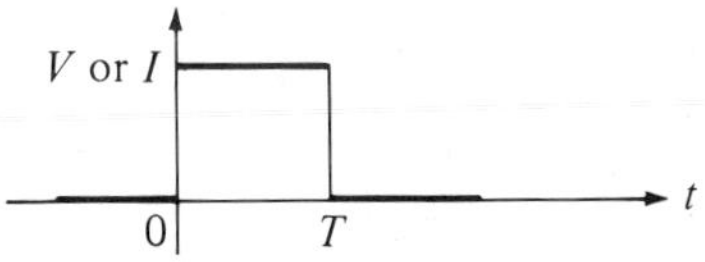	***Single pulse of width T*** $(V \text{ or } I)[u(t) - u(t-T)]$	$(V \text{ or } I)\dfrac{1 - \exp[-Ts]}{s}$
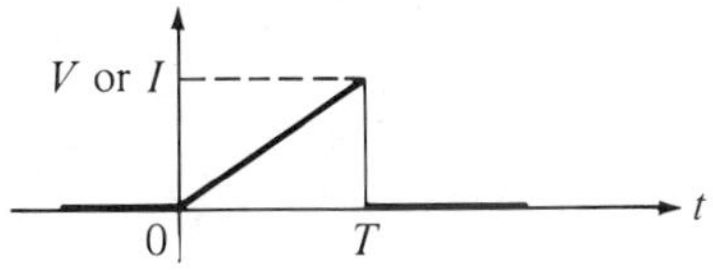	***Single sawtooth of base width T*** $(V \text{ or } I)\dfrac{t}{T}[u(t) - u(t-T)]$	$\dfrac{(V \text{ or } I)}{T} \cdot \dfrac{1 - (Ts+1)\exp[-Ts]}{s^2}$

5.3 Transformation of Linear Passive Elements

The terminal characteristic of any element is the fundamental relationship between voltage and current at the terminals of that element. The Laplace transform of the equation which expresses the terminal characteristic in the time domain is an equation which expresses the terminal characteristic in the

s-domain. Examination of the transformed equation will suggest a circuit model, called a **transformed equivalent element**, which will satisfy the terminal characteristic in the s-domain.

For each passive element, the ratio of transformed voltage across the element terminals to the transformed current entering the positive terminal $[V(s)/I(s)]$ is called the Laplace or s-domain **impedance** $Z(s)$ for that element. The reciprocal ratio is called the Laplace or s-domain **admittance** of the element:

$$Z(s) = \frac{V(s)}{I(s)} \tag{5.1a}$$

$$Y(s) = \frac{I(s)}{V(s)} \tag{5.1b}$$

Recall that Laplace transformation has the dimensional effect of multiplication by time which means that the unit of voltage in the s-domain is the volt-second (V · sec), and the unit of current is the ampere-second (A · sec). Therefore, impedance in the s-domain has the dimension volt/ampere, or ohm (Ω), and admittance has the dimension ampere/volt, or siemens (S).

Resistor Transformation

The terminal characteristic of a resistor in the time domain is given by Ohm's law:

$$R = \frac{v(t)}{i(t)}$$

from which

$$v(t) = R \cdot i(t) \tag{5.2a}$$

and

$$i(t) = \frac{1}{R} \cdot v(t) = G \cdot v(t) \tag{5.2b}$$

After Laplace transformation, Equations (5.2a) and (5.2b) become

$$V(s) = R \cdot I(s) \tag{5.3a}$$

and

$$I(s) = G \cdot V(s) \tag{5.3b}$$

Thus, from Equations (5.1a), (5.1b), (5.3a), and (5.3b),

$$Z_R(s) = R \tag{5.4a}$$

and

$$Y_R(s) = G \tag{5.4b}$$

The transformed equivalent resistor is shown in Figure 5.2.

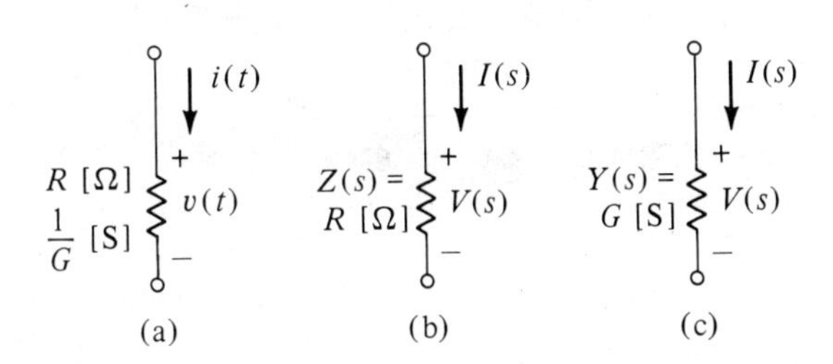

Figure 5.2 Transformed equivalent resistor. (a) Time-domain resistor. (b) Transformed resistor: impedance model. (c) Transformed resistor: admittance model.

Capacitor Transformation

The equations which give the terminal characteristic of a capacitor in the time domain were derived in Chapter 3 and are repeated here:

$$v(t_1) = \frac{1}{C}\int_{t_0}^{t_1} i(t)\,dt + v(t_0) \tag{5.5a}$$

$$i(t) = C\frac{d}{dt}v(t) \tag{5.5b}$$

After Laplace transformation, Equations (5.5a) and (5.5b) become

$$V(s) = \frac{I(s)}{Cs} + \frac{v(t_0)}{s} \tag{5.6a}$$

$$I(s) = C\left[s \cdot V(s) - v(t_0)\right] = Cs \cdot V(s) - C \cdot v(t_0) \tag{5.6b}$$

The s-domain impedance and admittance can be determined by letting the initial capacitor voltage be zero in Equations (5.6a) and (5.6b):

$$V(s) = \frac{I(s)}{Cs} \tag{5.7a}$$

$$I(s) = Cs \cdot V(s) \tag{5.7b}$$

Then from the definitions of impedance and admittance, and Equations (5.7a) and (5.7b),

$$Z_C(s) = \frac{1}{Cs}\ \Omega \tag{5.8a}$$

$$Y_C(s) = Cs\ \mathrm{S} \tag{5.8b}$$

The expression given in Equation (5.6a) for the s-domain voltage $V(s)$ across a capacitor consists of two components—one due to the s-domain current $I(s)$ through the impedance of the capacitor $(1/Cs)$, and a second due to the initial voltage $v(t_0)$. A linear combination of two components in an expression for voltage suggests the series model for the transformed equivalent capacitor shown in Figure 5.3(b).

The expression for the s-domain current $I(s)$ through a capacitor given in Equation (5.6b) also consists of two components—one due to the s-domain voltage $V(s)$ across the admittance of the capacitor (Cs), and a second due to the capacitance and the initial charge $[C \cdot v(t_0)]$. A linear combination of two components in an expression for current suggests the parallel model for the transformed equivalent capacitor shown in Figure 5.3(c).

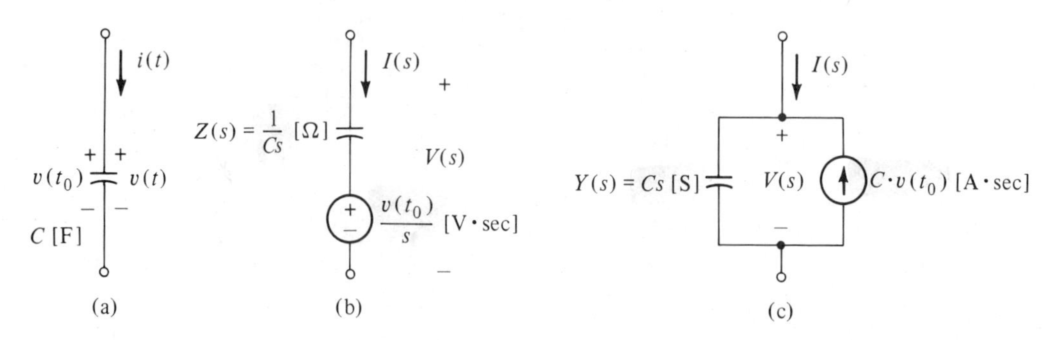

Figure 5.3 Transformed equivalent capacitor. (a) Time-domain capacitor. (b) Transformed capacitor: series model. (c) Transformed capacitor: parallel model.

Example 5.1

Problem:

Determine the series and parallel transformed equivalent models for a 2.5-μF capacitor charged to 5 V at $t = 0^-$ (Figure 5.4).

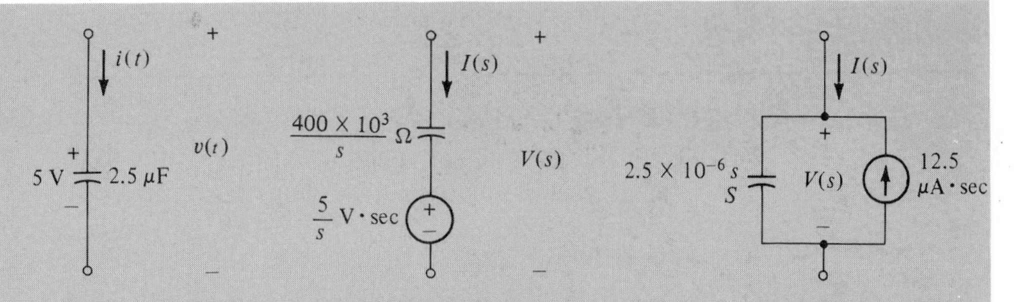

Figure 5.4 Example 5.1.

Solution:

1. From Equation (5.8a),

$$Z(s) = \frac{1}{2.5 \times 10^{-6}s} = \frac{400 \times 10^3}{s}\ \Omega$$

2. The impedance, in series with a voltage source

$$\frac{v(0)}{s} = \frac{5}{s}\ \text{V} \cdot \text{sec}$$

makes up the series model.

3. From Equation (5.8b),

$$Y(s) = 2.5 \times 10^{-6}s\ \text{S}$$

4. The admittance, in parallel with a current source

$$C \cdot v(0) = (2.5 \times 10^{-6}\ \text{F})(5\ \text{V}) = 12.5\ \mu\text{A} \cdot \text{sec}$$

makes up the parallel model. Recall from Chapter 3 that 1 farad is equivalent to 1 A · sec/V, so the result is dimensionally correct.

See Figure 5.4 for the models of the transformed capacitor.

Exercise 5.1

Problem:

Sketch and label the series and parallel transformed equivalent models for a 0.001-μF capacitor charged to 30 V at $t = 0^-$.

Answer:

Series: $Z(s) = 10^9/s\ \Omega$; $v(0)/s = 30/s$ V · sec.

Parallel: $Y(s) = 10^{-9}s$ S; $C \cdot v(0) = 30$ nA · sec.

Inductor Transformation

The transformed equivalent inductor is the electrical dual of the transformed equivalent capacitor. The equations which give the terminal characteristics of an inductor in the time domain are

$$i(t_1) = \frac{1}{L}\int_{t_0}^{t_1} v(t)\,dt + i(t_0) \tag{5.9a}$$

$$v(t) = L\frac{d}{dt}i(t) \tag{5.9b}$$

After Laplace transformation, Equations (5.9a) and (5.9b) become

$$I(s) = \frac{V(s)}{Ls} + \frac{i(t_0)}{s} \tag{5.10a}$$

$$V(s) = L[s \cdot I(s) - i(t_0)] = Ls \cdot I(s) - L \cdot i(t_0) \tag{5.10b}$$

The s-domain impedance and admittance are

$$Y_L(s) = \frac{1}{Ls}\ \text{S} \tag{5.11a}$$

$$Z_L(s) = Ls\ \Omega \tag{5.11b}$$

Equation (5.10a) leads to the parallel model for the transformed equivalent inductor [Figure 5.5(b)], and Equation (5.10b) to the series model [Figure 5.5(c)].

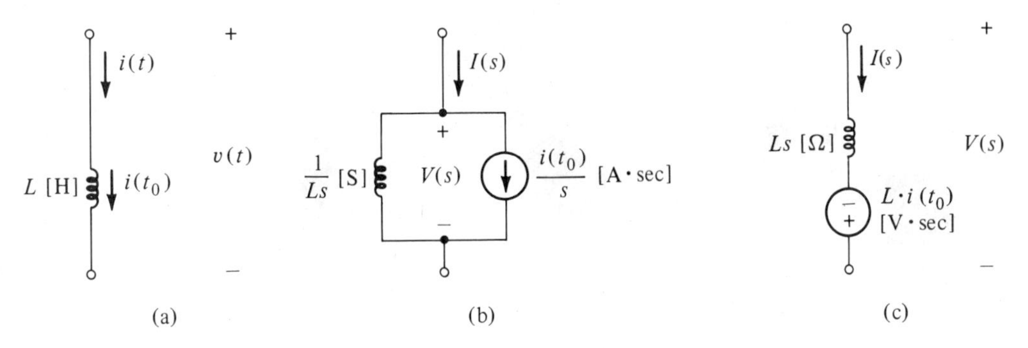

Figure 5.5 Transformed equivalent inductor. (a) Time-domain inductor. (b) Transformed inductor: parallel model. (c) Transformed inductor: series model.

Example 5.2

Problem:

Determine the series and parallel transformed equivalent models for a 20-mH inductor charged to 0.3 A at $t = 0^-$ (Figure 5.6).

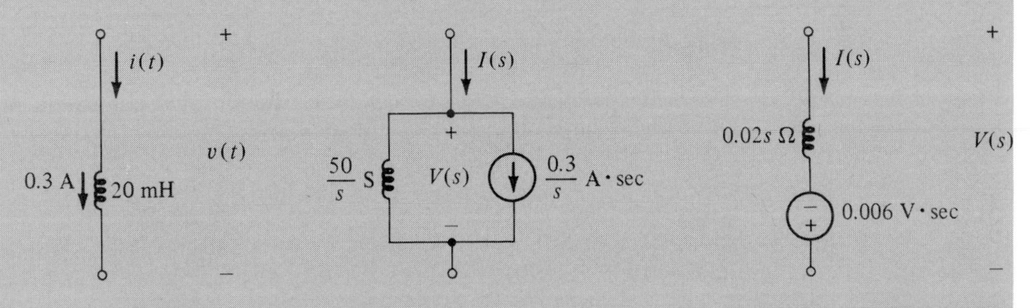

Figure 5.6 Example 5.2.

Solution:

1. From Equation (5.11b),

$$Z(s) = 20 \times 10^{-3} s \ \Omega$$

2. The impedance, in series with a voltage source,

$$L \cdot i(0) = (20 \times 10^{-3}\ \text{H})(0.3\ \text{A}) = 6\ \text{mV} \cdot \text{sec}$$

makes up the series model. Recall from Chapter 3 that 1 henry is equivalent to 1 V · sec/A, so the result is dimensionally correct.

3. From Equation (5.11a),

$$Y(s) = \frac{1}{20 \times 10^{-3} s} = \frac{50}{s}\ \text{S}$$

4. The admittance, in parallel with a current source,

$$\frac{i(0)}{s} = \frac{0.3}{s}\ \text{A} \cdot \text{sec}$$

makes up the parallel model.

See Figure 5.6 for the transformed equivalent inductor.

Exercise 5.2

Problem:

Sketch and label the series and parallel transformed equivalent models for a 0.47-mH inductor charged to 2 mA at $t = 0^-$.

Answer:

Series: $Z(s) = 0.47 \times 10^{-3} s\ \Omega$; $L \cdot i(0) = 940$ nA · sec. Parallel: $Y(s) = 2.128 \times 10^3/s$ S; $i(0)/s = 2/s$ mV · sec.

5.4 The Circuit Transformation Technique

In this section we will use several examples to illustrate the technique for obtaining the transformed equivalent of a time-domain circuit. In each example, we will determine initial conditions for the capacitors and inductors from the time-domain circuit just before the switching occurs. The reader should review Section 3.3 for the methods used to determine initial conditions.

Example 5.3

Problem:

In the circuit shown in Figure 5.7(a), the switch has been open long enough for the inductor to charge fully. Find the transformed equivalent circuit after the switch is closed at $t = 0$.

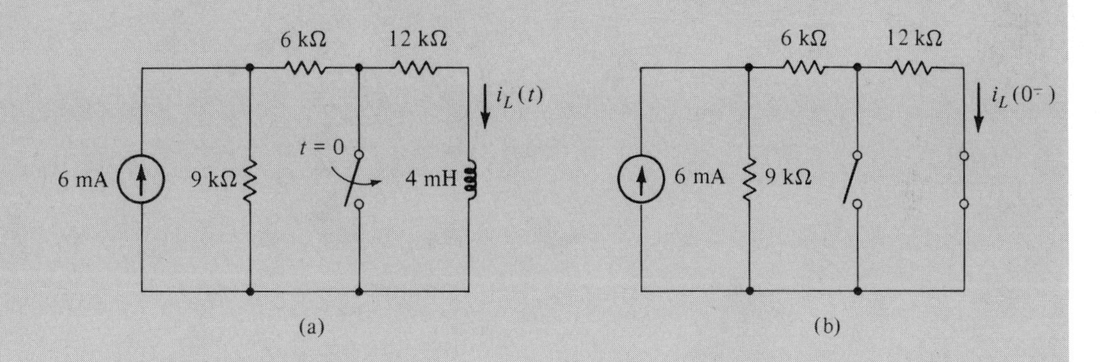

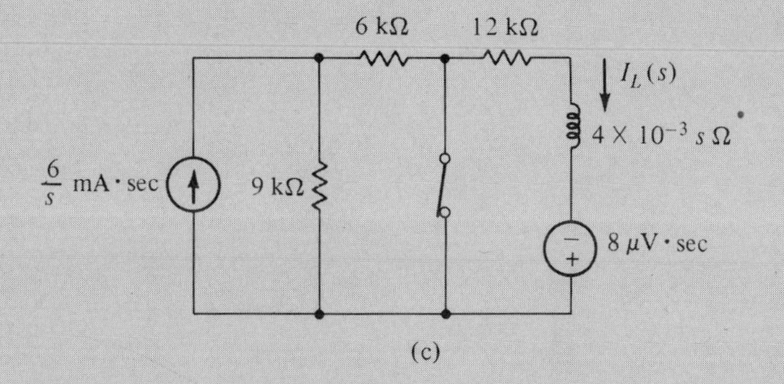

Figure 5.7 Example 5.3. (a) Time-domain circuit. (b) Equivalent time-domain circuit for $t = 0^-$. (c) Transformed circuit for $t > 0$.

Solution:

1. Determine the initial current through the inductor from the time-domain equivalent circuit for $t = 0^-$ [Figure 5.7(b)]. By current division,

$$i_L(0) = 6 \times 10^{-3} \frac{1/(6 + 12)}{1/(6 + 12) + 1/9} = 2 \text{ mA}$$

2. The transformed equivalent circuit is shown in Figure 5.7(c). The series model was chosen for the transformed inductor. More will be said later about the choice between series and parallel models for transformed passive elements.

Example 5.4

Problem:

In the circuit shown in Figure 5.8(a), the capacitor and inductor charge from the 10-V source through the switch. At $t = 0$, the switch is opened, removing the source and allowing the capacitor and inductor to discharge. Find the transformed equivalent circuit after the switch is opened.

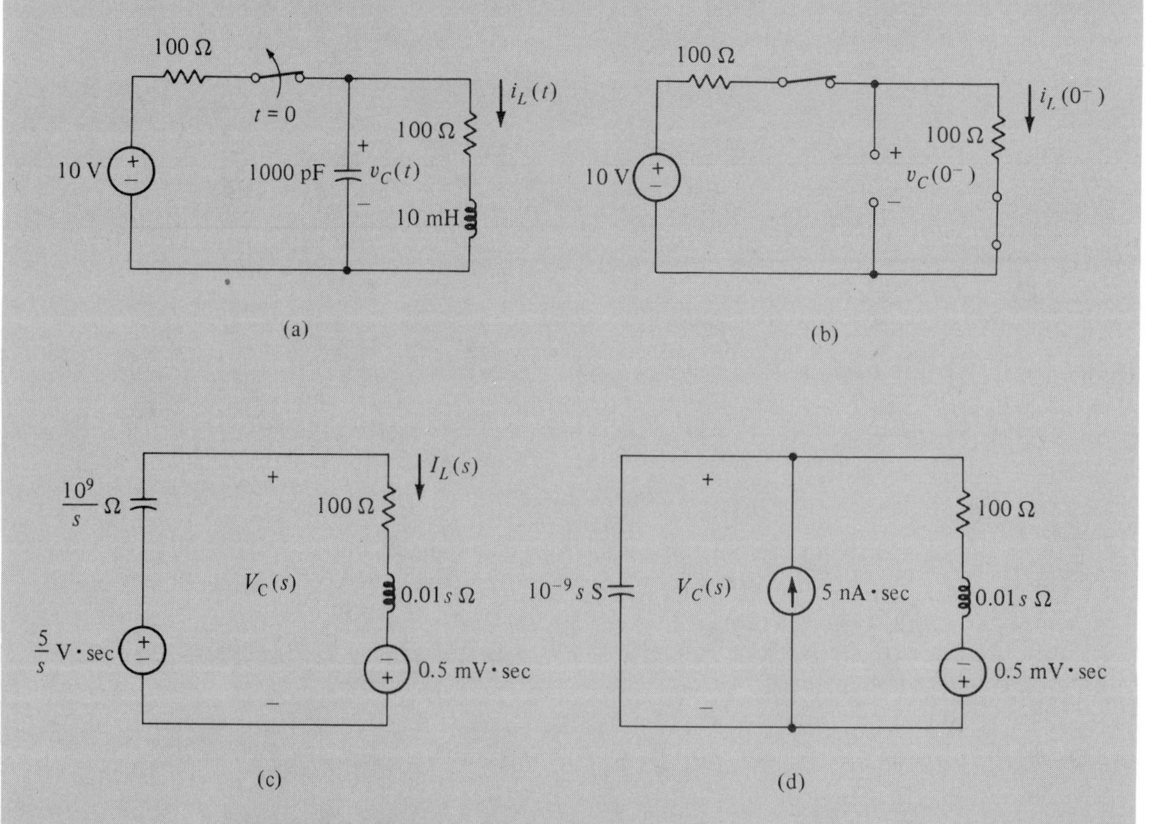

Figure 5.8 Example 5.4. (a) Time-domain circuit. (b) Time-domain circuit at $t = 0^-$. (c) Transformed circuit. (d) Transformed circuit.

Solution:

1. Determine initial conditions for the capacitor and inductor from the time-domain equivalent circuit for $t = 0^-$ shown in Figure 5.8(b). By voltage division,

$$v_C(0^-) = 10\frac{100}{100 + 100} = 5 \text{ V}$$

and by Ohm's law,

$$i_L(0^-) = 5/100 = 50 \text{ mA}$$

2. The transformed equivalent circuit shown in Figure 5.8(c), which uses the series models for both capacitor and inductor, would be the most advantageous transformed circuit for determining the inductor current. The circuit which uses the parallel model for the capacitor and the series model for the inductor [Figure 5.8(d)] would be most advantageous for determining the capacitor voltage. The reason that the parallel model was not chosen for the inductor will become clear in the next chapter.

Exercise 5.3

Problem:

Obtain a transformed equivalent circuit suitable for determining $v_C(t)$ after the switch is operated at $t = 0$ in the circuit shown in Figure 5.9(a).

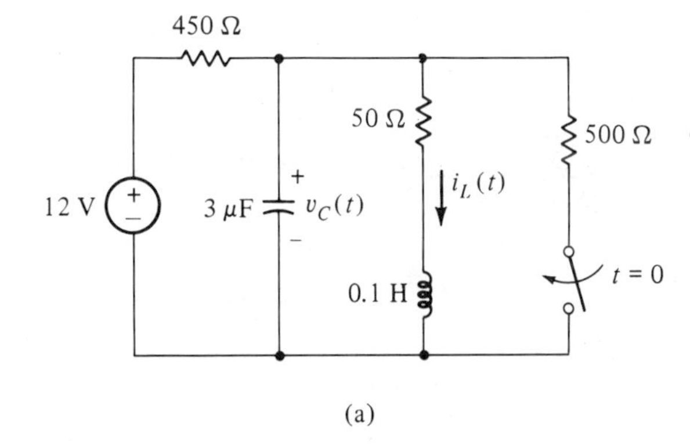

Figure 5.9
Circuits for Exercise 5.3.
(a) Time-domain circuit.

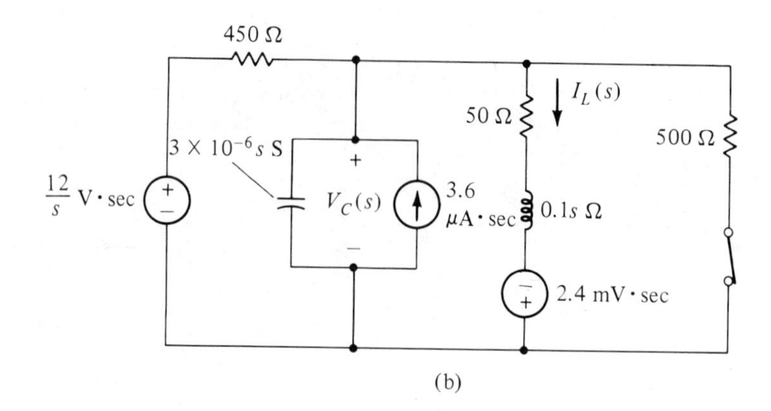

Figure 5.9 Continued (b) Transformed circuit.

Answer:

See Figure 5.9.

5.5 Analysis of Transformed Circuits

We have seen that the concepts of impedance and admittance, which relate transformed voltage to transformed current at the terminals of an element, permit extension of Ohm's law and the terminal characteristics of the capacitor and the inductor into the s-domain. In this section, we will extend the Kirchhoff laws to the s-domain and apply some direct consequences of the Kirchhoff laws to transformed circuits. We will then be in a position to proceed with analysis of the transformed circuit.

The Kirchhoff Laws

Kirchhoff's voltage law states that the algebraic sum of voltage rises and drops around any loop is zero, or

$$\sum_{k=1}^{N} v_k(t) = 0$$

where $v_k(t)$ is the time-dependent voltage (with the correct sign) across branch k, and N is the number of branches in the loop. Linearity of the Laplace transform operation ensures that

$$\mathscr{L}\left\{\sum_{k=1}^{N} v_k(t)\right\} = \sum_{k=1}^{N} V_k(s) = 0$$

which is Kirchhoff's voltage law in the s-domain.

Kirchhoff's current law states that the algebraic sum of currents entering or leaving a node is zero, or

$$\sum_{k=1}^{N} i_k(t) = 0$$

where $i_k(t)$ is the time-dependent current (with the correct sign) in branch k, and N is the number of branches attached to the node. Thus,

$$\sum_{k=1}^{N} I_k(s) = 0$$

which is Kirchhoff's current law in the s-domain.

Combining Ideal Sources

Combination of transformed ideal voltage and current sources follows directly from the s-domain Kirchhoff laws. Voltage sources in series can be combined by algebraic addition and replaced by one equivalent voltage source. Likewise, current sources in parallel can be combined by algebraic addition and replaced by one equivalent current source.

Exercise 5.4

Problem:

What is the total transformed source current provided by parallel current sources

$$I_1(s) = 5 \text{ nA} \cdot \text{sec} \quad \text{and} \quad I_2(s) = \frac{5}{s + 10^4} \text{ mA} \cdot \text{sec?}$$

Answer:

$$5 + \frac{5 \times 10^6}{s + 10^4} \text{ nA} \cdot \text{sec}$$

Example 5.5

Problem:

Combine the transformed voltage sources in the transformed circuit from Example 5.4 [Figure 5.8(c)].

Solution:

$$V_T(s) = \left(0.5 + \frac{5000}{s}\right) \text{ mV} \cdot \text{sec}$$

In most cases the expression for the combined voltage sources, $V_T(s)$, will be involved in an inverse Laplace transformation while solving for the circuit response. Consequently, there is usually no advantage to simplifying the expression through the use of a common denominator.

Combining Impedance and Admittance

The series circuit shown in Figure 5.10(a) consists of any number N of branches, each with a known impedance $Z_k(s)$, and a voltage source $V(s)$. The response current $I(s)$ causes a voltage drop $V_k(s)$ across each element. By Kirchhoff's voltage law,

$$V(s) = \sum_{k=1}^{N} V_k(s) \tag{5.12}$$

The total impedance of the circuit is

$$Z(s) = \frac{V(s)}{I(s)} \tag{5.13}$$

By combining Equations 5.12 and Equation 5.13,

$$Z(s) = \frac{\sum_{k=1}^{N} V_k(s)}{I(s)} = \sum_{k=1}^{N} \frac{V_k(s)}{I(s)}$$

The term $V_k(s)/I(s)$ in the summation is simply the impedance $Z_k(s)$, so

$$Z(s) = \sum_{k=1}^{N} Z_k(s) \tag{5.14}$$

which means that the impedance of a series circuit is the algebraic sum of the impedances of its individual branches.

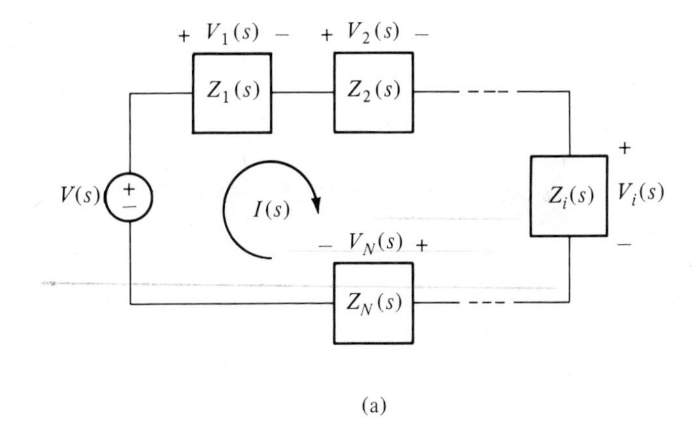

(a)

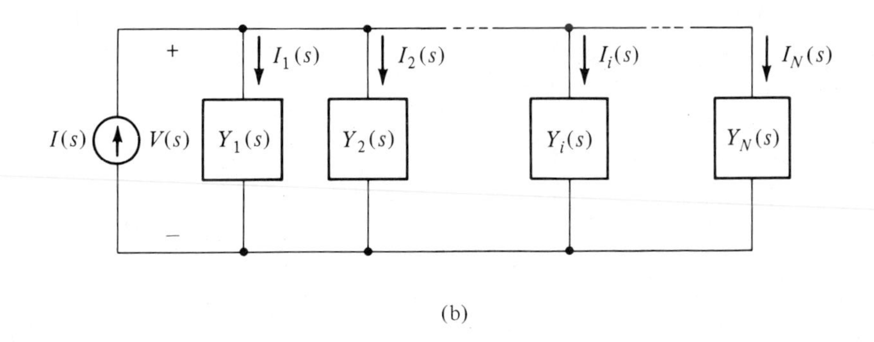

(b)

Figure 5.10 General s-domain circuits.
(a) Series s-domain circuit.
(b) Parallel s-domain circuit.

Similarly, for the parallel circuit of Figure 5.10(b),

$$I(s) = \sum_{k=1}^{N} I_k(s) \tag{5.15}$$

The admittance of the circuit is

$$Y(s) = \frac{I(s)}{V(s)} \tag{5.16}$$

By combining Equations (5.15) and (5.16),

$$Y(s) = \frac{\sum_{k=1}^{N} I_k(s)}{V(s)} = \sum_{k=1}^{N} \frac{I_k(s)}{V(s)}$$

Since $I_k(s)/V(s)$ is the admittance $Y_k(s)$, then

$$Y(s) = \sum_{k=1}^{N} Y_k(s) \tag{5.17}$$

which means that the admittance of a parallel circuit is the algebraic sum of the admittances of its individual branches.

With the ability to apply Ohm's law and the Kirchhoff laws, and to combine sources, impedances, and admittances, we now have sufficient tools to determine the s-domain response of simple circuits. Let us begin with some examples.

Example 5.6 **Problem:**

SPICE

Find and sketch $i_L(t)$ for the circuit shown in Figure 5.11(a).

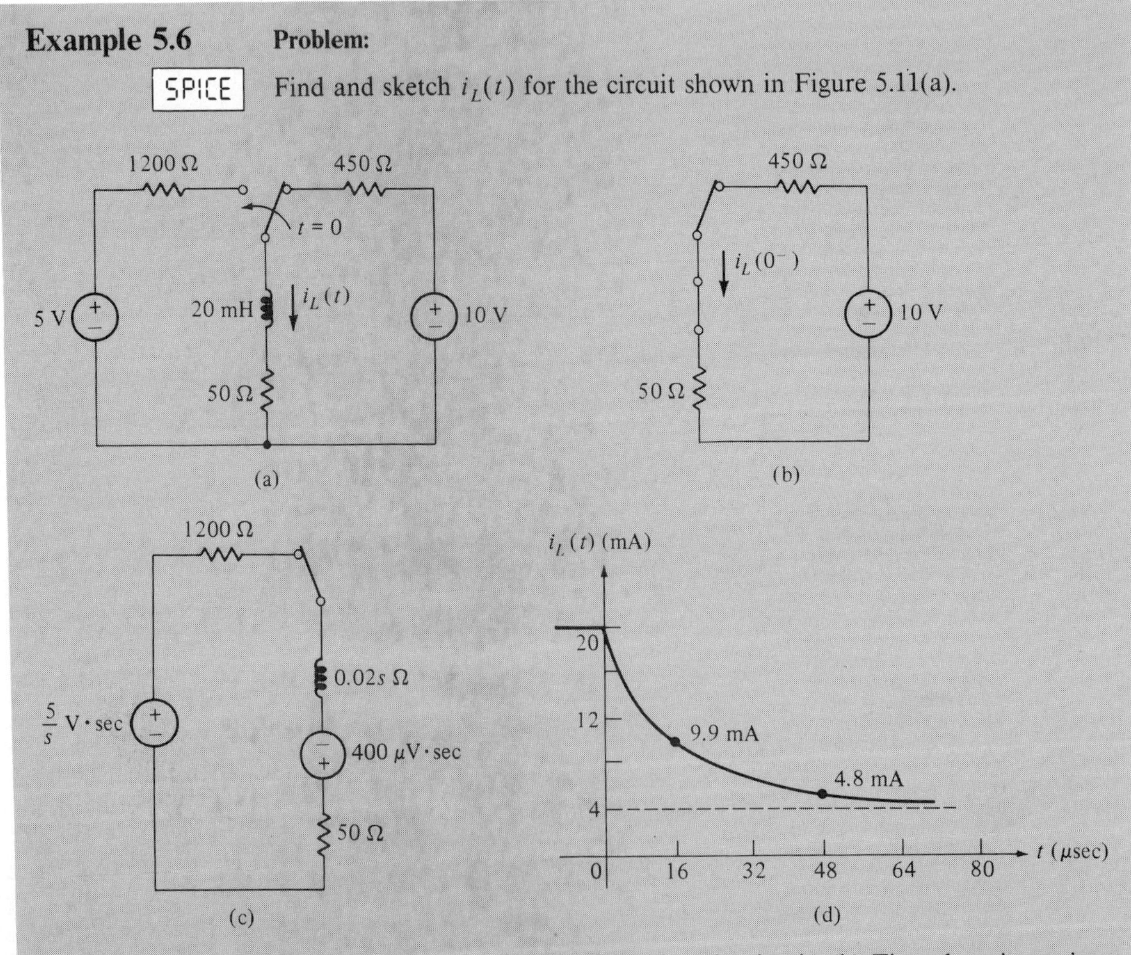

Figure 5.11 Example 5.6. (a) Time-domain circuit. (b) Time-domain equivalent circuit at $t = 0^-$. (c) Transformed equivalent circuit for $t > 0$. (d) $i_L(t)$ vs. t.

Solution:

1. Determine the initial inductor current from the equivalent circuit shown in Figure 5.11(b):

$$i_L(0) = \frac{10}{450 + 50} = 20 \text{ mA}$$

2. Use the series model for the transformed inductor to obtain the transformed circuit of Figure 5.11(c).
3. Combine the two voltage sources in Figure 5.11(c):

$$V_T(s) = \frac{5}{s} + 400 \times 10^{-6} \text{ V} \cdot \text{sec}$$

4. Combine the impedance of the three elements:

$$Z_T(s) = 1200 + 0.02s + 50 = 0.02(s + 62.5 \times 10^3)\ \Omega$$

5. Compute $I_L(s) = V_T(s)/Z_T(s)$:

$$I_L(s) = \frac{5}{0.02s(s + 62.5 \times 10^3)} + \frac{400 \times 10^{-6}}{0.02(s + 62.5 \times 10^3)}$$

$$= \frac{250}{s(s + 62.5 \times 10^3)} + \frac{0.02}{s + 62.5 \times 10^3} \text{ A} \cdot \text{sec}$$

6. Determine the time-domain response $i_L(t)$ by inverse Laplace transformation of $I_L(s)$:

$$i_L(t) = \mathscr{L}^{-1}\left\{\frac{250}{s(s + 62.5 \times 10^3)}\right\} + \mathscr{L}^{-1}\left\{\frac{0.02}{s + 62.5 \times 10^3}\right\} \text{ A}$$

7. From the linearity properties, and Pairs Vb and IV,

$$i_L(t) = \frac{250}{62.5 \times 10^3}(1 - \exp[-62.5 \times 10^3 t])u(t)$$

$$+ 0.02 \cdot \exp[-62.5 \times 10^3 t] \cdot u(t)$$

$$= 4(1 + 4 \cdot \exp[-62.5 \times 10^3 t])u(t) \text{ mA}$$

The current exhibits an exponential response, decaying from an initial 20 mA to a final 4 mA [Figure 5.11(d)].

Example 5.7

SPICE

Problem:

Find and sketch $v_C(t)$ for the circuit of Figure 5.12(a).

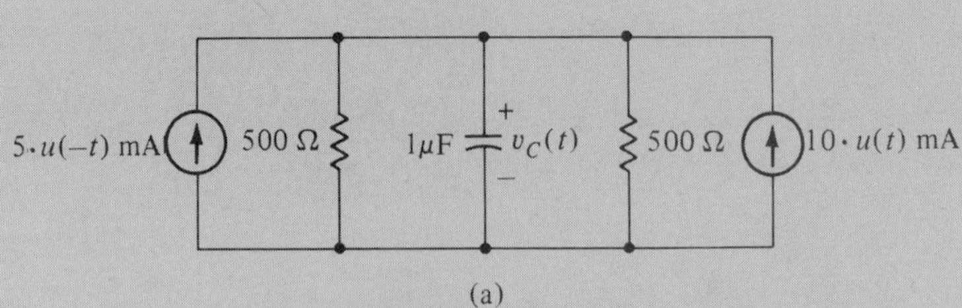

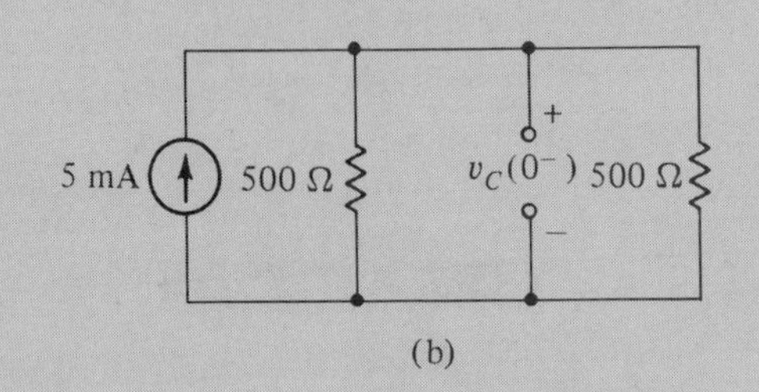

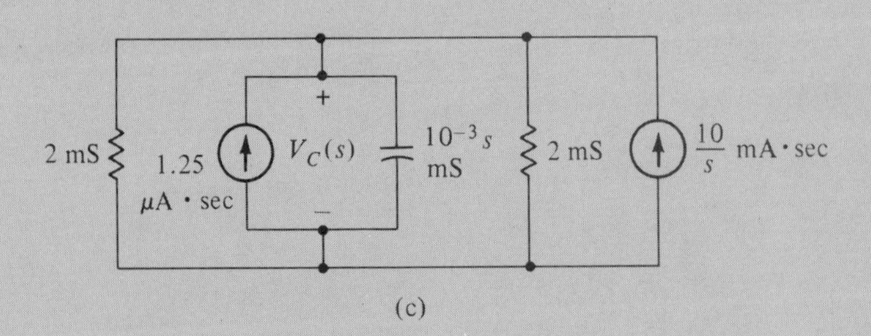

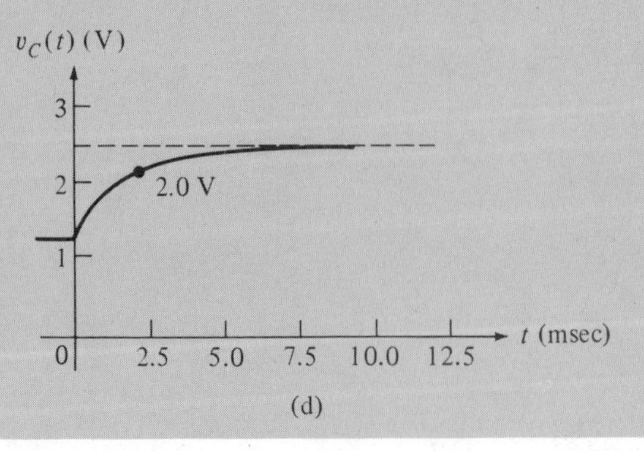

Figure 5.12
Example 5.7.
(a) Time-domain circuit.
(b) Time-domain equivalent circuit at $t = 0^-$.
(c) Transformed equivalent circuit for $t > 0$.
(d) $i_L(t)$ vs. t.

Solution:

1. Determine the initial capacitor voltage from the equivalent circuit shown in Figure 5.12(b):

$$v_C(0) = \frac{5 \times 10^{-3}}{1/500 + 1/500} = 1.25 \text{ V}$$

2. Use the parallel model for the transformed capacitor to obtain the transformed circuit of Figure 5.12(c).
3. Combine the two current sources in Figure 5.12(c):

$$I_T(s) = \left(1.25 \times 10^{-3} + \frac{10}{s}\right) \text{mA} \cdot \text{sec}$$

4. Combine the admittance of the three elements:

$$Y_T(s) = 2 + 10^{-3}s + 2 = 10^{-3}(s + 4000) \text{ mS}$$

5. Compute $V_C(s) = I_T(s)/Y_T(s)$:

$$V_C(s) = \frac{1.25 \times 10^{-3}}{10^{-3}(s + 4000)} + \frac{10}{10^{-3}s(s + 4000)}$$
$$= \frac{1.25}{s + 4000} + \frac{10 \times 10^3}{s(s + 4000)} \text{ V} \cdot \text{sec}$$

6. Determine the time-domain response $v_C(t)$ by inverse Laplace transformation of $V_C(s)$:

$$v_C(t) = \mathscr{L}^{-1}\left\{\frac{1.25}{s + 4000}\right\} + \mathscr{L}^{-1}\left\{\frac{10 \times 10^3}{s(s + 4000)}\right\} \text{ V}$$

7. From the linearity properties, and Pairs IV and Vb,

$$v_C(t) = 1.25 \cdot \exp[-4000t] \cdot u(t)$$
$$+ 2.5(1 - \exp[-4000t]) \cdot u(t)$$
$$= 2.5(1 - 0.5 \cdot \exp[-4000t])u(t) \text{ V}$$

The capacitor voltage exhibits an exponential response, growing from an initial 1.25 V to a final 2.5 V [Figure 5.12(d)].

Exercise 5.5

Problem:

1. What is the total impedance of a branch containing a 50-μF capacitor connected in series with a 50-mH inductor which has a 50-Ω winding resistance?
2. What is the total admittance of the same two elements connected in parallel?

Answers:

1. $0.05\dfrac{s^2 + 1000s + 400 \times 10^3}{s}$ Ω.

2. $50 \times 10^{-6}\dfrac{s^2 + 1000s + 400 \times 10^3}{s + 1000}$ S.

Voltage Division and Current Division Rules

The **voltage division rule** for transformed circuits permits calculation of the voltage across any branch of a series circuit if the source voltage, total impedance, and branch impedance are known. Consider the voltage across the kth branch of the series circuit shown in Figure 5.10(a):

$$V_k(s) = I(s) \cdot Z_k(s)$$

The circuit current is

$$I(s) = \frac{V(s)}{Z_T(s)}$$

so

$$V_k(s) = V(s) \cdot \frac{Z_k(s)}{Z_T(s)} \tag{5.18}$$

Similarly, the **current division rule** for transformed circuits permits calculation of the current through any branch of a parallel circuit if the source current, total admittance, and branch admittance are known. Consider the current through the kth branch of the parallel circuit shown in Figure 5.10(b):

$$I_k(s) = V(s) \cdot Y_k(s)$$

The circuit voltage is

$$V(s) = \frac{I(s)}{Y_T(s)}$$

so

$$I_k(s) = I(s) \cdot \frac{Y_k(s)}{Y_T(s)} \tag{5.19}$$

Example 5.8

SPICE

Problem:

Use the current division rule to find the capacitor current $i_C(t)$ in the circuit shown in Figure 5.13(a).

Solution:

1. Determine the initial capacitor voltage from the time-domain equivalent circuit of Figure 5.13(b). By the voltage division rule,

$$v_C(0) = 10\frac{250}{500 + 250} = 3.333 \text{ V}$$

2. In the transformed circuit [Figure 5.13(c)], the parallel model was chosen for the transformed capacitor in order to maintain the simplicity of the parallel circuit. Combine the current sources to obtain

$$I(s) = 0.01667 + \frac{0.2}{s} \text{ mA} \cdot \text{sec}$$

3. Compute the total admittance:

$$\begin{aligned} Y_T(s) &= 2 + 5 \times 10^{-3}s + 4 \\ &= 5 \times 10^{-3}(s + 1200) \text{ mS} \end{aligned}$$

4. Use the current division rule to determine $I_x(s)$.

$$\begin{aligned} I_x(s) &= \left(0.01667 + \frac{0.2}{s}\right)\frac{5 \times 10^{-3}s}{5 \times 10^{-3}(s + 1200)} \\ &= \frac{0.01667(s + 12)}{s + 1200} = \frac{0.01667(s + 1200 - 1188)}{s + 1200} \\ &= 0.01667 - \frac{19.8}{s + 1200} \text{ mA} \cdot \text{sec} \end{aligned}$$

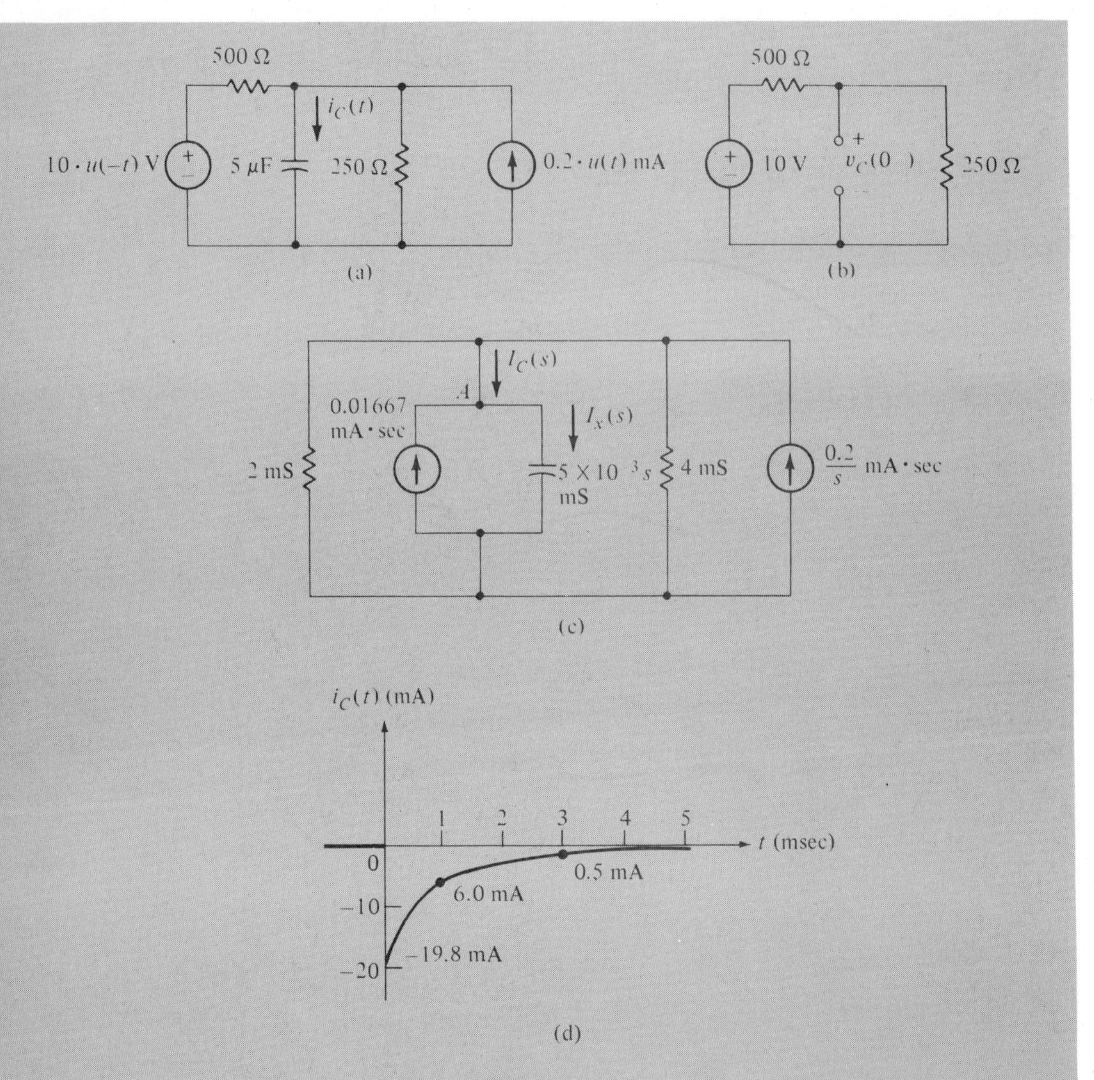

Figure 5.13 Example 5.8. (a) Time-domain circuit. (b) Time-domain equivalent circuit at $t = 0^-$. (c) Transformed circuit for $t > 0$. (d) $i_C(t)$ vs. t.

5. Use Kirchhoff's current law at node A to determine $I_C(s)$:

$$I_C(s) = I_X(s) - 0.01667 \text{ mA} \cdot \text{sec}$$

$$= -\frac{19.8}{s + 1200} \text{ mA} \cdot \text{sec}$$

6. From the linearity properties and Pair IV,

$$i_C(t) = -19.8 \cdot \exp[-1200t] \cdot u(t) \text{ mA}$$

Figure 5.13(d) is a sketch of the capacitor current versus time.

Example 5.9

SPICE

Problem:

Use voltage division to find the voltage $v(t)$ across the 25-Ω resistor in the circuit shown in Figure 5.14(a). Sketch the response, and calculate the time at which $v(t) = 0$.

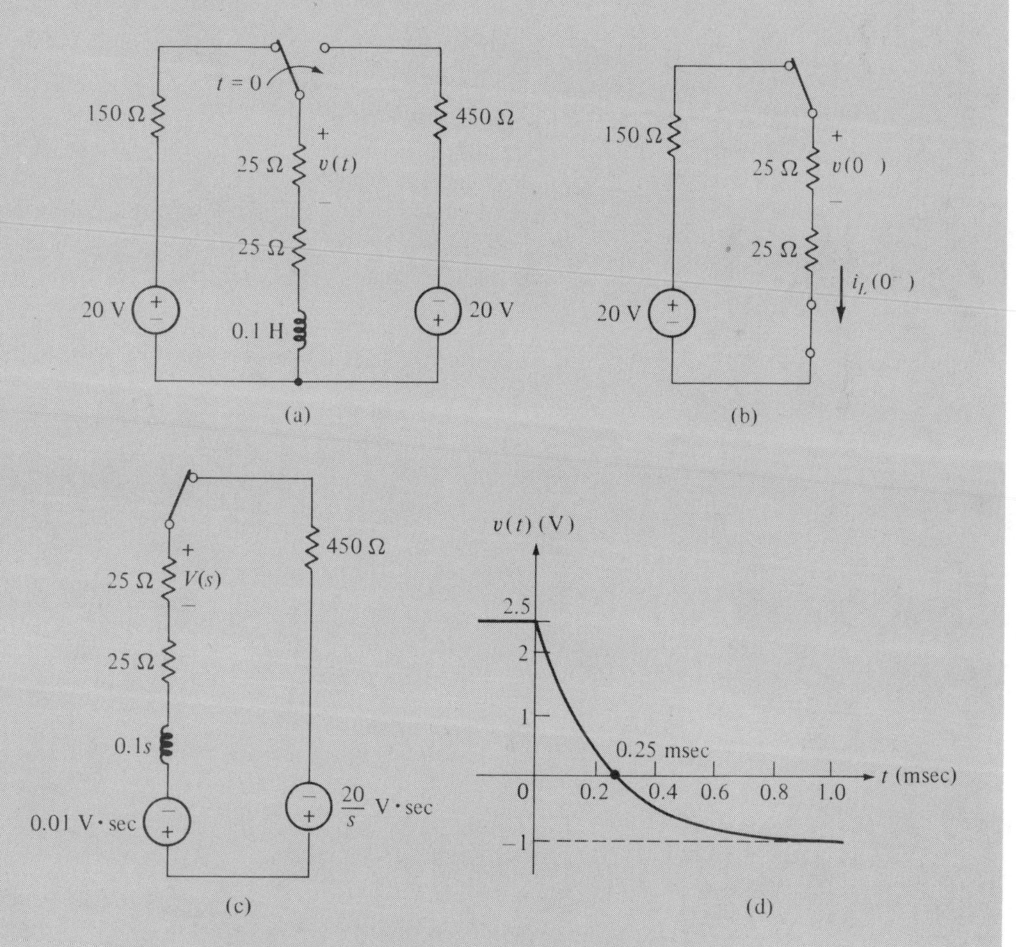

Figure 5.14 Example 5.9. (a) Complete circuit. (b) Equivalent circuit at $t = 0^-$. (c) Transformed equivalent circuit for $t > 0$. (d) $v(t)$ vs. t.

Solution:

1. Determine the initial inductor current from Figure 5.14(b):

$$i_L(0) = \frac{20}{150 + 25 + 25} = 0.1 \text{ A}$$

Therefore, by Ohm's law,

$$v(0) = 0.1(25) = 2.5 \text{ V}$$

2. In the transformed circuit [Figure 5.14(c)], the series model was chosen for the transformed inductor in order to maintain the simplicity of the series circuit. Combine the voltage sources to obtain

$$V_T(s) = 0.01 - \frac{20}{s} \text{ V}$$

3. Calculate the total impedance:

$$Z_T(s) = 450 + 25 + 25 + 0.1s = 0.1(s + 5000)\ \Omega$$

4. Use the voltage division rule to determine $V(s)$:

$$V(s) = \left(0.01 - \frac{20}{s}\right)\frac{25}{0.1(s + 5000)}$$
$$= \frac{2.5}{s + 5000} - \frac{5000}{s(s + 5000)} \text{ V}$$

5. Use Pairs IV and Vb to determine $v(t)$:

$$v(t) = 2.5 \cdot \exp[-5000t] \cdot u(t) - \frac{5000}{5000}(1 - \exp[-5000t])u(t)$$
$$= (-1 + 3.5 \cdot \exp[-5000t])u(t) \text{ V}$$

6. Solve for the time at which $v(t) = 0$:

$$-1 + 3.5 \cdot \exp[-5000t] = 0$$
$$3.5 \cdot \exp[-5000t] = 1$$
$$-5000t = \ln(1/3.5)$$
$$t = 0.251 \text{ msec}$$

Figure 5.14(d) is a sketch of $v(t)$ vs. t.

Exercise 5.6

Problem:

1. In Figure 5.15(a), find the admittance of the branch containing the inductor and resistor, then use current division to calculate the current through that branch.

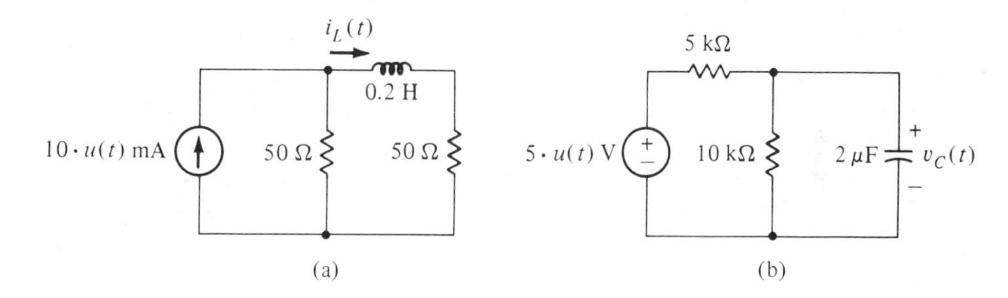

Figure 5.15 Exercise 5.6.

2. In Figure 5.15(b), find the impedance of the parallel combination of capacitor and resistor, then use voltage division to calculate the voltage across that combination.

Answers:

1. $i(t) = 5(1 - \exp[-500t])u(t)$ mA.
2. $v(t) = 3.333(1 - \exp[-150t])u(t)$ V.

5.6 Computer Simulations

SPICE may be used in a transient analysis mode to simulate any of the circuit examples, exercises, or problems in this chapter in the same way it was used for the circuits in Chapter 3. SPICE 5.1 and 5.2 are both used to simulate the circuit in Example 5.8 in order to illustrate some further features of SPICE.

In SPICE 5.1, the initial condition of the capacitor due to the step-function voltage source $[10 \cdot u(-t)$ V], which is switched off at $t = 0$, was calculated analytically (Example 5.8). The initial condition is entered into SPICE through use of the keyword UIC in the .TRAN directive and the specification IC = 3.333

```
*******10/14/88 ********  SPICE 2G.6    3/15/83 ********08:39:46*****
SPICE 5.1
****     INPUT LISTING                      TEMPERATURE =   27.000 DEG C
*****************************************************************************
 R1 1 0 500
 R2 1 0 250
 C 2 0 5U IC=3.333
 I 0 2 PULSE(0 .2M 0 1F)
 VAM 1 2
 .TRAN .25M 5M UIC
 .PLOT TRAN I(VAM)
 .END
*******10/14/88 ********  SPICE 2G.6    3/15/83 ********08:39:46*****
SPICE 5.1
****     TRANSIENT ANALYSIS                 TEMPERATURE =   27.000 DEG C
*****************************************************************************

     TIME       I(VAM)
                  -2.000D-02     -1.500D-02     -1.000D-02     -5.000D-03  0.000D+00
                          - - - - - - - - - - - - - - - - - - - - - - - - - - - - - -
 0.000D+00 -2.000D-02 *             .              .              .            .
 2.500D-04 -1.488D-02 .             *              .              .            .
 5.000D-04 -1.108D-02 .             .          *   .              .            .
 7.500D-04 -8.251D-03 .             .              .      *       .            .
 1.000D-03 -6.164D-03 .             .              .           *  .            .
 1.250D-03 -4.616D-03 .             .              .              .*           .
 1.500D-03 -3.471D-03 .             .              .              .   *        .
 1.750D-03 -2.622D-03 .             .              .              .      *     .
 2.000D-03 -1.994D-03 .             .              .              .       *    .
 2.250D-03 -1.528D-03 .             .              .              .         *  .
 2.500D-03 -1.184D-03 .             .              .              .          * .
 2.750D-03 -9.283D-04 .             .              .              .          * .
 3.000D-03 -7.395D-04 .             .              .              .           *.
 3.250D-03 -5.994D-04 .             .              .              .           *.
 3.500D-03 -4.959D-04 .             .              .              .            *.
 3.750D-03 -4.190D-04 .             .              .              .            *.
 4.000D-03 -3.623D-04 .             .              .              .            *.
 4.250D-03 -3.201D-04 .             .              .              .            *.
 4.500D-03 -2.890D-04 .             .              .              .            *.
 4.750D-03 -2.659D-04 .             .              .              .            *.
 5.000D-03 -2.487D-04 .             .              .              .            *.
                          - - - - - - - - - - - - - - - - - - - - - - - - - - - - - -

         JOB CONCLUDED


*******10/14/88 ********  SPICE 2G.6    3/15/83 ********08:44:54*****
SPICE 5.2
****     INPUT LISTING                      TEMPERATURE =   27.000 DEG C
*****************************************************************************
 R1 1 3 500
 R2 1 0 250
 C 2 0 5U
 V 3 0 PULSE(10 0 0 1F)
 I 0 2 PULSE(0 .2M 0 1F)
 VAM 1 2
```

```
 .TRAN .25M 5M
 .PLOT TRAN I(VAM)
 .END
*******10/14/88 ********  SPICE 2G.6    3/15/83 ********08:44:54*****
SPICE 5.2
****      INITIAL TRANSIENT SOLUTION        TEMPERATURE =   27.000 DEG C
****************************************************************************
  NODE    VOLTAGE      NODE    VOLTAGE      NODE    VOLTAGE
 (  1)     3.3333     (  2)     3.3333     (  3)    10.0000
     VOLTAGE SOURCE CURRENTS
     NAME        CURRENT
     V         -1.333D-02
     VAM        0.000D+00
     TOTAL POWER DISSIPATION    1.33D-01  WATTS
*******10/14/88 ********  SPICE 2G.6    3/15/83 ********08:44:54*****
SPICE 5.2
****      TRANSIENT ANALYSIS                TEMPERATURE =   27.000 DEG C
****************************************************************************

     TIME        I(VAM)
                   -1.500D-02      -1.000D-02      -5.000D-03      0.000D+00   5.000D-03
                  - - - - - - - - - - - - - - - - - - - - - - - - - - - - - - - - - -
 0.000D+00  0.000D+00 .               .               .               *               .
 2.500D-04 -1.488D-02 *               .               .               .               .
 5.000D-04 -1.108D-02 .           *   .               .               .               .
 7.500D-04 -8.252D-03 .               .     *         .               .               .
 1.000D-03 -6.165D-03 .               .          *    .               .               .
 1.250D-03 -4.616D-03 .               .               .*              .               .
 1.500D-03 -3.471D-03 .               .               .   *           .               .
 1.750D-03 -2.622D-03 .               .               .       *       .               .
 2.000D-03 -1.994D-03 .               .               .        *      .               .
 2.250D-03 -1.528D-03 .               .               .          *    .               .
 2.500D-03 -1.184D-03 .               .               .           *   .               .
 2.750D-03 -9.284D-04 .               .               .           *   .               .
 3.000D-03 -7.395D-04 .               .               .            *  .               .
 3.250D-03 -5.995D-04 .               .               .            *  .               .
 3.500D-03 -4.959D-04 .               .               .             * .               .
 3.750D-03 -4.191D-04 .               .               .             *.                .
 4.000D-03 -3.623D-04 .               .               .             *.                .
 4.250D-03 -3.201D-04 .               .               .             *.                .
 4.500D-03 -2.890D-04 .               .               .             *.                .
 4.750D-03 -2.659D-04 .               .               .             *.                .
 5.000D-03 -2.487D-04 .               .               .             *.                .
                  - - - - - - - - - - - - - - - - - - - - - - - - - - - - - - - - - -

          JOB CONCLUDED
```

in the capacitor record. Note that the voltage source is replaced by a short circuit for the simulation, because the voltage source is providing no energy to the circuit after $t = 0$. A PULSE source is used to simulate the step-function current source.

Nearly identical results are obtained in SPICE 5.2 by using a two PULSE sources and allowing SPICE to determine the initial condition of the capacitor at $t = 0$. Note that the PULSE voltage source has an initial level of 10 V, which falls to zero, simulating the function $10 \cdot u(-t)$; whereas, the PULSE current source has an initial level of zero, which rises to 0.2 mA, simulating $0.2 \cdot u(t)$.

5.7 SUMMARY

In Chapter 3 we demonstrated that application of the Kirchhoff laws to a circuit containing energy storage elements results in an initial value problem with an ordinary linear differential equation having constant coefficients. In Chapter 4 we used the Laplace transform to solve this type of initial value problem. In this chapter we have presented a technique for transforming a circuit to the s-domain and have demonstrated that the Kirchhoff laws are valid in the s-domain. Now, by applying the Kirchhoff laws directly to the transformed circuit, we can obtain an algebraic equation which is the Laplace transform of the initial value problem which would have obtained had we applied the Kirchhoff laws to the original time-domain circuit.

In dc analysis, we characterize dc sources, resistors, and circuit response as constants. In analysis of transformed circuits, we characterize time-dependent sources, energy storage elements, and circuit response as functions of the Laplace variable s. Our preliminary work in this chapter with transformed circuits shows that transformed-circuit analysis follows directly from dc analysis. We must concede that the algebra is more difficult, but the analysis procedures are the same.

The reader has probably noted that we could reduce all of the circuits analyzed in this chapter to simple parallel or simple series R–C or R–L circuits and analyze them using the formulas developed in Chapter 3. In other words, all the circuits in this chapter exhibit first-order response. The author assures the reader that this is intentional. The curious reader may cross-check the solutions for circuit response and thereby build confidence in transformed-circuit analysis methods.

5.8 Terms

circuit transformation
current division rule
s-domain impedance
s-domain admittance
transformed equivalent element
voltage division rule

Problems

5-1. Sketch and label both the series model and the parallel model of the transformed equivalent element:

***a.** 50-μF capacitor; $v_C(0^-) = 50$ V.

b. 220-pF capacitor; $v_C(0^-) = 9$ V.
***c.** 40-mH inductor; $i_L(0^-) = 300$ mA.
d. 0.8-H inductor; $i_L(0^-) = 1.2$ A.

***5-2.** Sketch and label the transformed equivalent circuit for $t > 0$ for each time-domain circuit in Problem 3.10 (Chapter 3).

5-3. Calculate the impedance $Z(s)$ and the admittance $Y(s)$ for each network. The result should be expressed as a polynomial fraction with a leading coefficient of 1 in both numerator and denominator.

***a.**

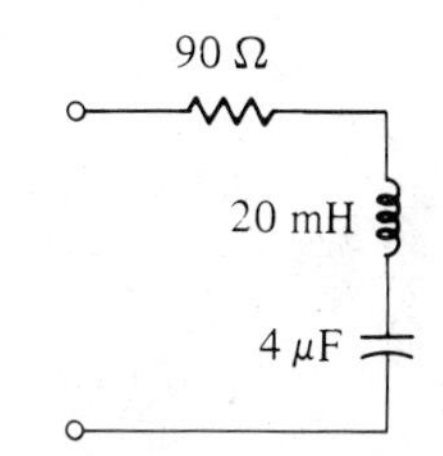

b.

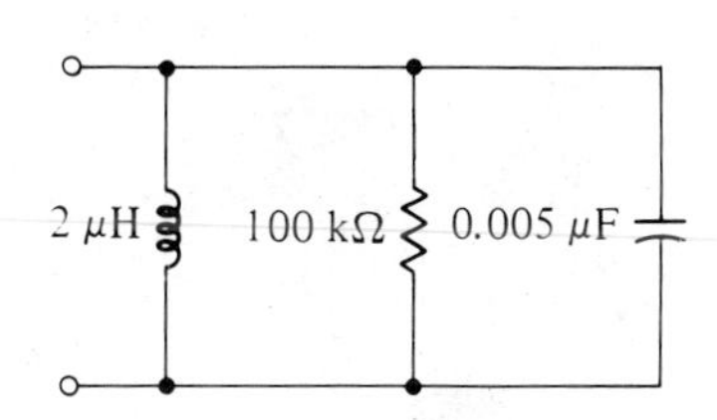

***c.**

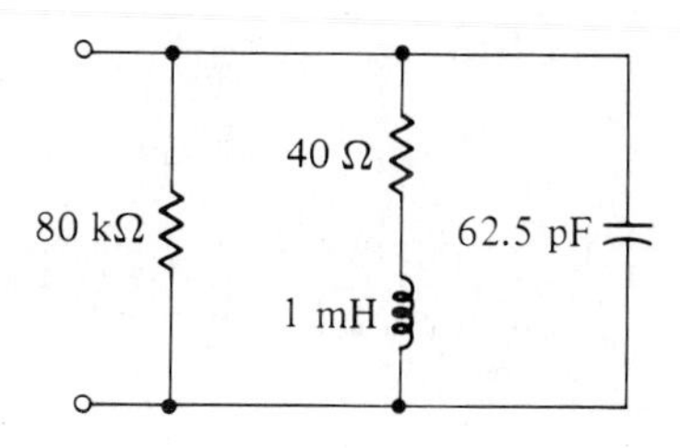

d.

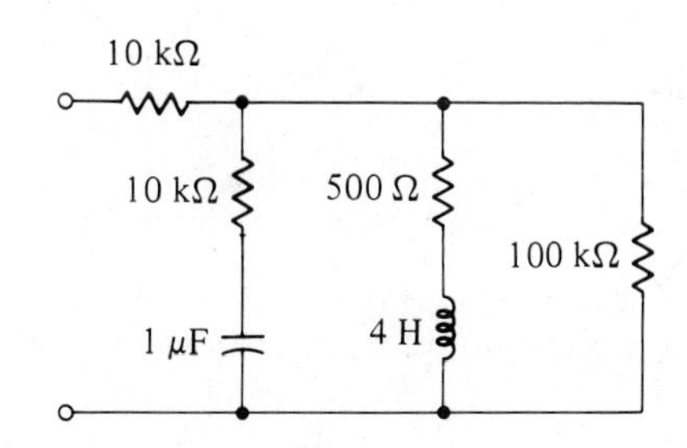

5-4. Use voltage division to calculate the unknown voltage in each circuit. Perform the inverse Laplace transformation to obtain the time-domain response.

a.

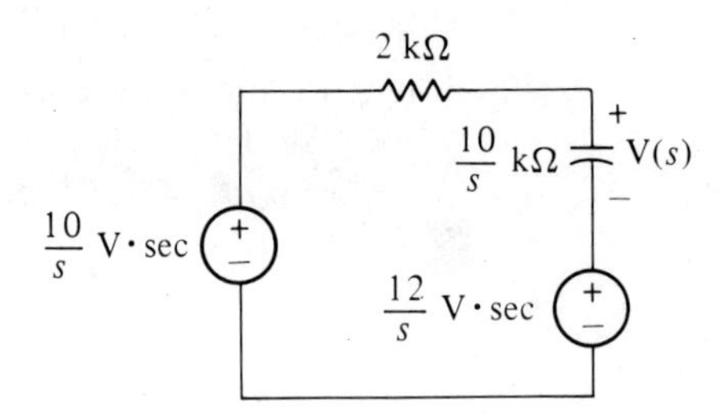

***b.**

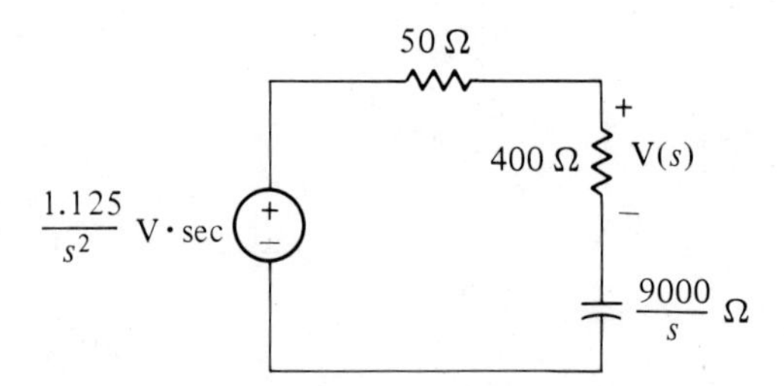

c.

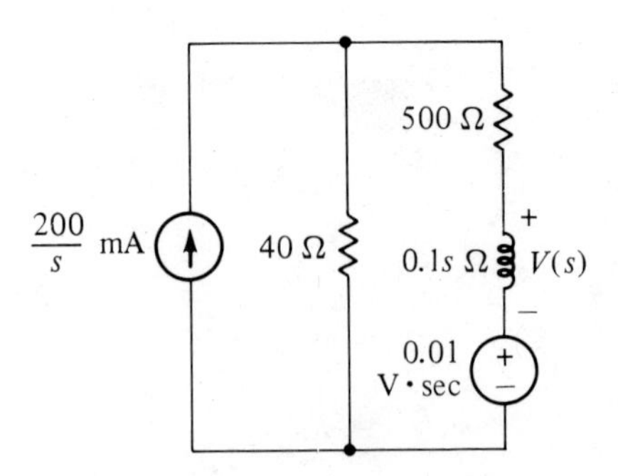

***d.**

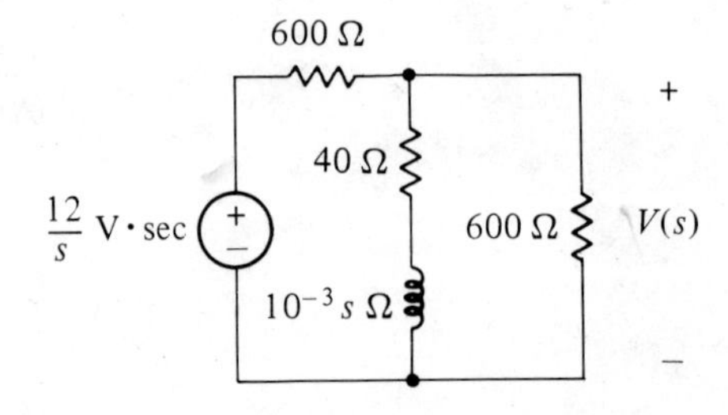

SPICE **5-5.** Use current division to calculate the unknown current in each circuit. Perform the inverse Laplace transformation to obtain the time-domain response.

***a.**

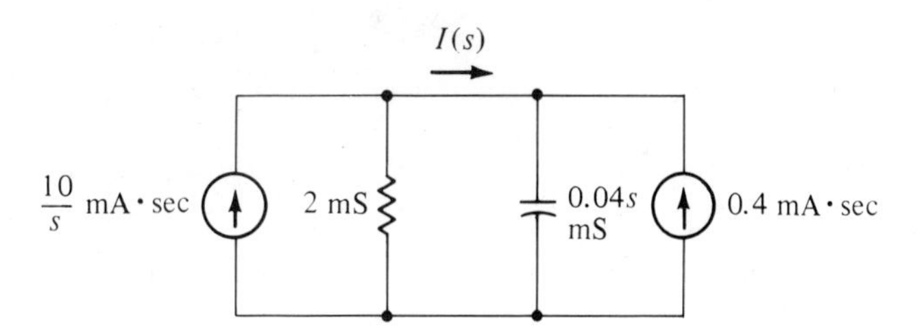

b.

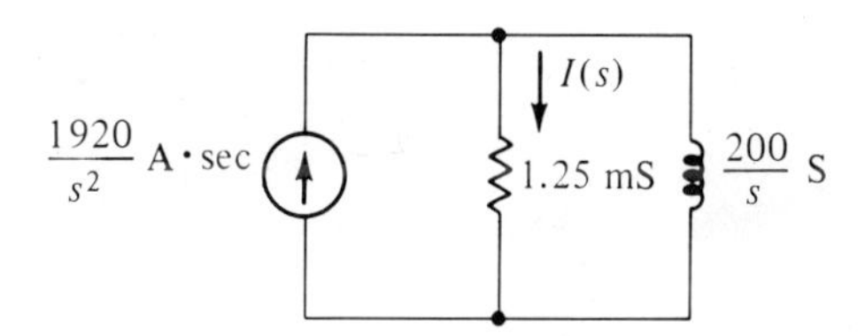

***c.**

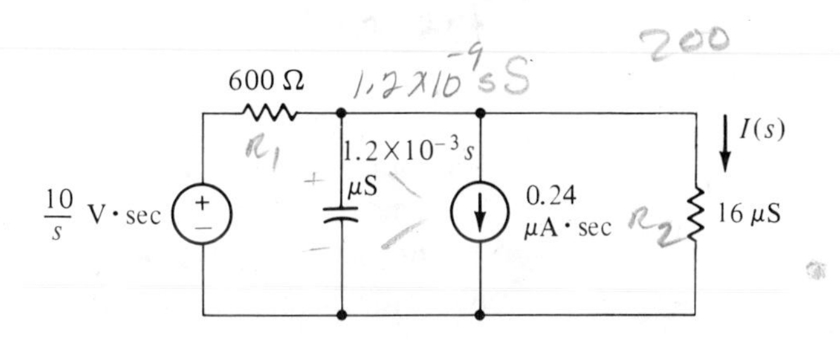

d.

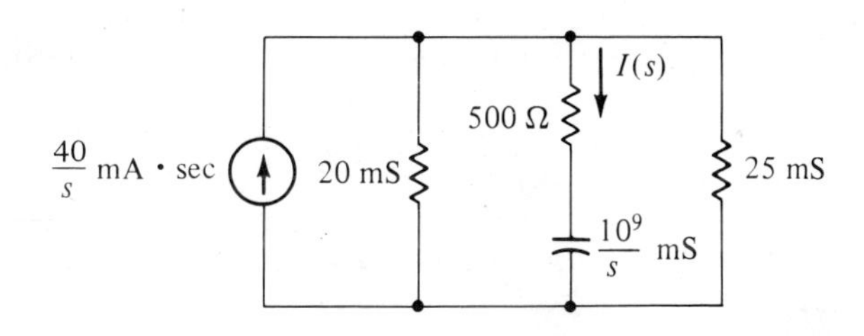

SPICE **5-6.** Repeat Problem 3-11 (Chapter 3) using transformed-circuit analysis.

SPICE **5-7.** Repeat Problem 3-12 (Chapter 3) using transformed-circuit analysis.

***5-8.** The BJT in the driver circuit has a collector–emitter saturation voltage V_{CES} of 0.2 V and a collector cutoff current I_{CO} approximately equal to

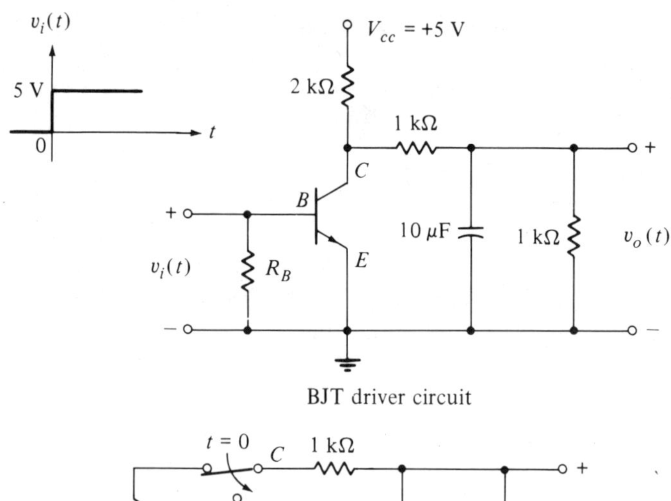

BJT driver circuit

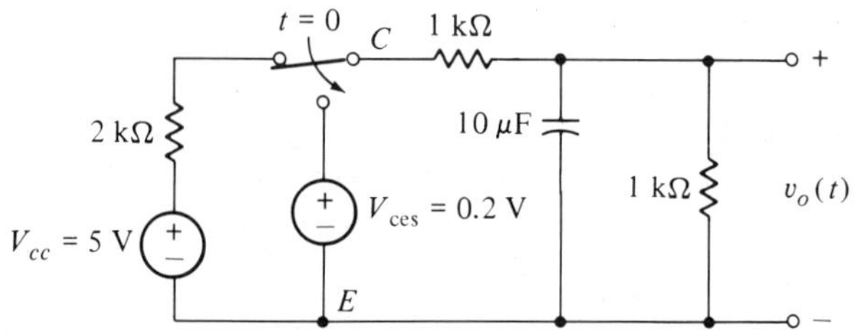

Equivalent circuit for BJT driver (Switch models transition from cutoff to saturation)

zero. With 0 V at the base, the BJT is cut off; the 5-V step at the base will saturate the collector–emitter junction. Transform the circuit model, and analyze the output voltage $v_o(t)$.

***5-9.** The n-channel MOSFET has a very large drain–source resistance r_{DS} when the gate voltage is zero; the step input to the gate will make r_{DS} negligibly small. The relay winding K has an inductance of 0.4 H and a resistance of 75 Ω; the relay pull-in current is 120 mA. Transform the circuit model, and determine the time of relay pull-in. What is the effect on pull-in time if a 220-Ω resistor is connected drain to source?

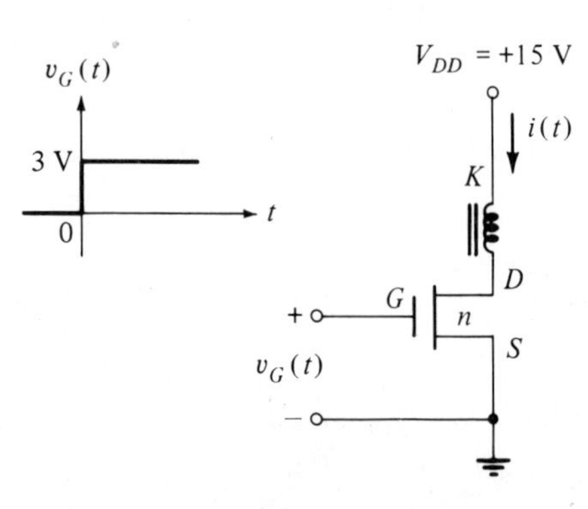

MOSFET relay control

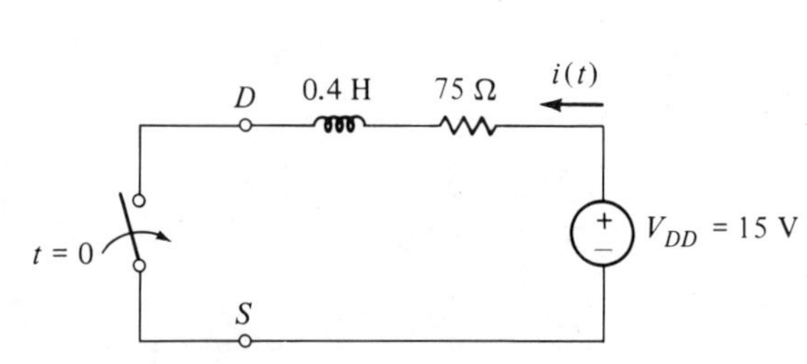

Equivalent circuit for MOSFET relay control (Switch models transition from very large r_{DS} to negligible r_{DS}.)

6 Transform Circuit Analysis

6.1 INTRODUCTION

In this chapter we will examine some important applications of transform circuit analysis, relying only on the Kirchhoff laws and the concepts of the Laplace impedance $Z(s)$ and admittance $Y(s)$, and those techniques derived directly from them in Chapter 5. The extension of transform analysis to more general network analysis methods must wait until Chapter 7.

The reader may have noticed that the brief introduction to analysis of transformed circuits in Chapter 5 (Section 5.5) dealt only with circuits containing a source, resistor(s), and no more than one energy storage element (capacitor or inductor). As we have previously noted, such circuits are called **first-order** circuits because the initial value problem whose solution yields the circuit response consists of a first-order differential equation and one initial condition. In addition, the circuits considered in Chapter 5 contained only switched dc or step-function sources. We will now expand our interest to circuits which contain other than dc or step-function sources and to **second-order** circuits, that is, circuits which contain two energy storage elements.[1] A circuit with two energy storage elements is said to be of second order because the response is the solution to an initial value problem consisting of a second-order differential equation and two initial conditions.

Since it is not possible to consider every circuit configuration and every type of source, we will emphasize the transform circuit analysis technique rather than the development and application of formulas. We will establish only a few formulas—those which relate to the simple series and simple parallel combinations of source, resistor, capacitor, and inductor. In addition,

[1] If the elements are both capacitors or both inductors, and they are connected in parallel or in series, the circuit is not second order, because the two elements can be combined into one equivalent element.

we will also limit our illustrations to circuits containing only step-function, piecewise-linear, or sinusoidal sources.

A major topic in this chapter is the *nature* (overdamped, critically damped, or underdamped) of a circuit response and its decomposition into a **transient** (or short-lived) component and a **steady-state** (or continuing) component.

6.2 Response of *R–L–C* Circuits

Virtually every electrical or electronic network incorporates switching and time-dependent sources. Even in networks where capacitors and inductors are not incorporated, we find inductance and capacitance associated with the metallic and dielectric structure and with the wires or printed circuits required to assemble the network. Analysis of *R–L–C* circuits is essential in designing a network to produce a desired response, in confirming the existence of a designed response, or in determining the source of an unwanted response.

A circuit response, referred to as a **complete response**, can be separated into a **transient response** (sometimes called the **natural response**) and a **steady-state response** (sometimes called the **forced response**). The transient response is present immediately after the switching time t_0, but vanishes as $t \to \infty$. The steady-state response is present for all $t > t_0$, and is all that remains as $t \to \infty$.

In this section we will categorize *R–L–C* circuit responses as **overdamped**, **critically damped**, or **underdamped** according to the nature of the transient response. We will see that the type of transient response is associated with the factors of the denominator of the s-domain function which represents the circuit response (Chapter 4).

Series *R–L–C* Circuit

The circuit shown in Figure 6.1(a) is the transformed equivalent of a series *R–L–C* circuit excited by a voltage source with an unspecified time dependence. The time-domain response current $i(t)$ is obtained by inverse Laplace transformation of the s-domain circuit current $I(s)$. When Kirchhoff's voltage law is applied to the transformed circuit,

$$V_L(s) + V_R(s) + V_C(s) = V_S(s)$$

When the element voltages are expressed in terms of circuit current,

$$Ls \cdot I(s) - L \cdot i(t_0) + R \cdot I(s) + \frac{1}{Cs} \cdot I(s) + \frac{v_C(t_0)}{s} = V_S(s)$$

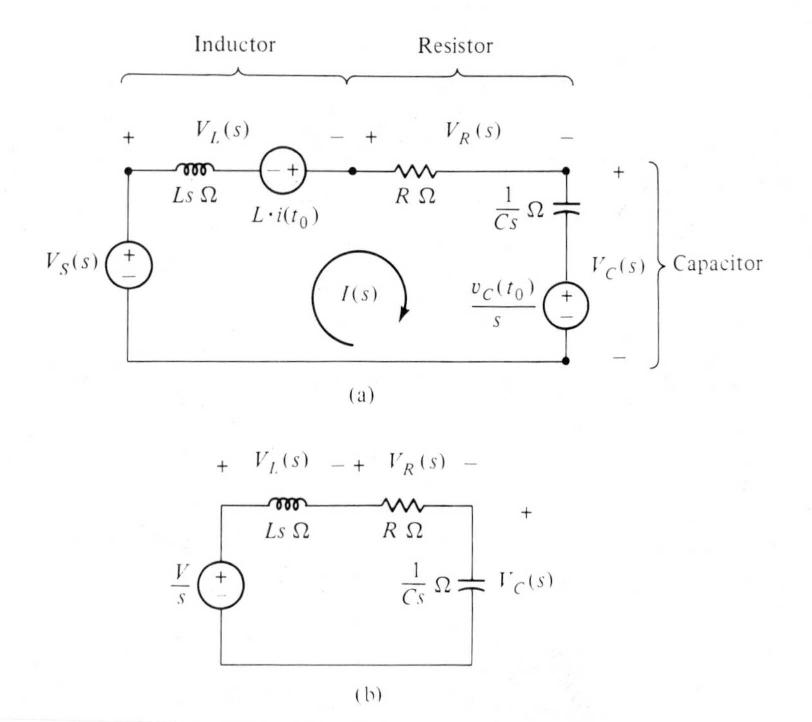

Figure 6.1
Series $R-L-C$ circuit.
(a) Arbitrary excitation and initial conditions.
(b) Unshifted step excitation and zero initial conditions.

or

$$Ls \cdot I(s) + R \cdot I(s) + \frac{1}{Cs} \cdot I(s) = V_S(s) + L \cdot i(t_0) - \frac{v_C(t_0)}{s}$$

Multiplying by s/L and factoring $I(s)$ from the left member yields

$$\left(s^2 + \frac{R}{L}s + \frac{1}{LC}\right)I(s) = \frac{s \cdot V_S(s)}{L} + i(t_0) - \frac{v_C(t_0)}{L}$$

or

$$I(s) = \frac{s\left(\frac{V_S(s)}{L} + i(t_0)\right) - \frac{v_C(t_0)}{L}}{s^2 + \frac{R}{L}s + \frac{1}{LC}} \tag{6.1}$$

Response to a Step Excitation. If the excitation is an unshifted step voltage source of arbitrary amplitude,

$$v(t) = V \cdot u(t)$$

and the circuit is previously inactive [$i(0^-)$ and $v_C(0^-)$ both zero], then Equation (6.1) becomes

$$I(s) = \frac{V/L}{s^2 + (R/L)s + 1/(LC)}$$

$$= \frac{V/L}{(s - p_1)(s - p_2)} \text{ A} \cdot \text{sec} \tag{6.2}$$

where

$$p_1 = -\frac{R}{2L} + \sqrt{\left(\frac{R}{2L}\right)^2 - \frac{1}{LC}}$$

$$p_2 = -\frac{R}{2L} - \sqrt{\left(\frac{R}{2L}\right)^2 - \frac{1}{LC}}$$

The transformed equivalent circuit is shown in Figure 6.1(b).

Case 1

If the discriminant (the expression under the radical in p_1 and p_2) is greater than zero, the poles of $I(s)$ are real and distinct, and $i(t)$ is the **overdamped** response associated with Case 1 in Chapter 4. Pair XI is used to obtain

$$i(t) = \frac{V/L}{p_1 - p_2}(\exp[p_1 t] - \exp[p_2 t])u(t)$$

$$= \frac{V/L}{p_2 - p_1}(\exp[p_2 t] - \exp[p_1 t])u(t) \tag{6.3}$$

Case 2

If the discriminant is zero, the poles are real and repeated, and Equation (6.2) becomes

$$I(s) = \frac{V/L}{(s - p)^2} \tag{6.4}$$

where $p = p_1 = p_2 = -R/(2L)$. Pair VIa is used to obtain

$$i(t) = \frac{V}{L}t \cdot \exp\left[-\frac{R}{2L}t\right] \cdot u(t) \tag{6.5}$$

which is the **critically damped** response of Case 2 in Chapter 4.

Case 3
Finally, if the discriminant is negative, p_1 and p_2 are complex conjugates, and Equation (6.2) becomes

$$I(s) = \frac{V/L}{s^2 + 2\alpha s + \omega_0^2} = \frac{V/L}{(s + \alpha - j\omega_d)(s + \alpha + j\omega_d)} \tag{6.6}$$

where $\alpha = R/(2L)$, $\omega_0^2 = 1/(LC)$, $\omega_d = \sqrt{\omega_0^2 - \alpha^2}$. Note that $\mathbf{p}_1 = -\alpha + j\omega_d$ and $\mathbf{p}_2 = \mathbf{p}_1^* = -\alpha - j\omega_d$. Pair XII is used to obtain

$$i(t) = \frac{V/L}{\omega_d} \cdot \exp[-\alpha t] \cdot \sin[\omega_d t] \cdot u(t) \text{ A} \tag{6.7}$$

which is the **underdamped**, or **oscillatory**, response described for Case 3 in Chapter 4.

Critical damping, which occurs when the discriminant equals zero, is the demarcation between underdamping and overdamping, so the solution for R in the equation

$$\left(\frac{R}{2L}\right)^2 - \frac{1}{LC} = 0$$

yields R_C, the **critical value** of R:

$$\left(\frac{R_C}{2L}\right)^2 = \frac{1}{LC}$$

$$R_C^2 = \frac{4L^2}{LC}$$

$$R_C = 2\sqrt{L/C} \tag{6.8}$$

The circuit designer can use this result, once L and C have been determined for a particular series $R–L–C$ circuit design, to produce a desired response (or eliminate an undesired response). The circuit analyst can also use this result to predict the nature of the response of a given $R–L–C$ circuit from a knowledge of element values alone.

Example 6.1

SPICE

Problem:

For the series $R–L–C$ circuit in Figure 6.1, assume that $L = 0.1$ mH, $C = 0.01\ \mu$F, and R is adjustable.

a. Find the critical value of R.

b. Find and graph the response current to a 4-V step excitation for $R = 1.25R_C$.

Solution:

1. Use Equation (6.8) to calculate R_C:

$$R_C = 2\sqrt{0.1 \times 10^{-3}/0.01 \times 10^{-6}}$$
$$= 200\ \Omega$$

2. For $R = 1.25R_C = 250\ \Omega$,

$$\frac{R}{2L} = \frac{250}{2(0.1 \times 10^{-3})} = 1.25 \times 10^6$$

and

$$\frac{1}{LC} = \frac{1}{(0.1 \times 10^{-3})(0.01 \times 10^{-6})} = 10^{12}$$

Therefore,

$$p_1 = -1.25 \times 10^6 + \sqrt{(-1.25 \times 10^6)^2 - 10^{12}} = -0.5 \times 10^6$$

$$p_2 = -1.25 \times 10^6 - \sqrt{(-1.25 \times 10^6)^2 - 10^{12}} = -2 \times 10^6$$

Equation (6.2) becomes

$$I(s) = \frac{40 \times 10^3}{(s + 0.5 \times 10^6)(s + 2 \times 10^6)}\ \text{A} \cdot \text{sec}$$

which yields an overdamped response obtained from Pair XI or Equation (6.3):

$$i(t) = \frac{40 \times 10^3}{-0.5 \times 10^6 + 2 \times 10^6}$$
$$\times \left(\exp[-0.5 \times 10^6 t] - \exp[-2 \times 10^6 t]\right) u(t)$$
$$= 26.67\left(\exp[-0.5 \times 10^6 t] - \exp[-2 \times 10^6 t]\right) u(t)\ \text{mA}$$

3. Determine the maximum value of $i(t)$ and the time at which the maximum value occurs. The derivative of a function is zero at the maximum (or minimum) value of the function, so set the derivative of

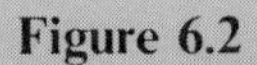

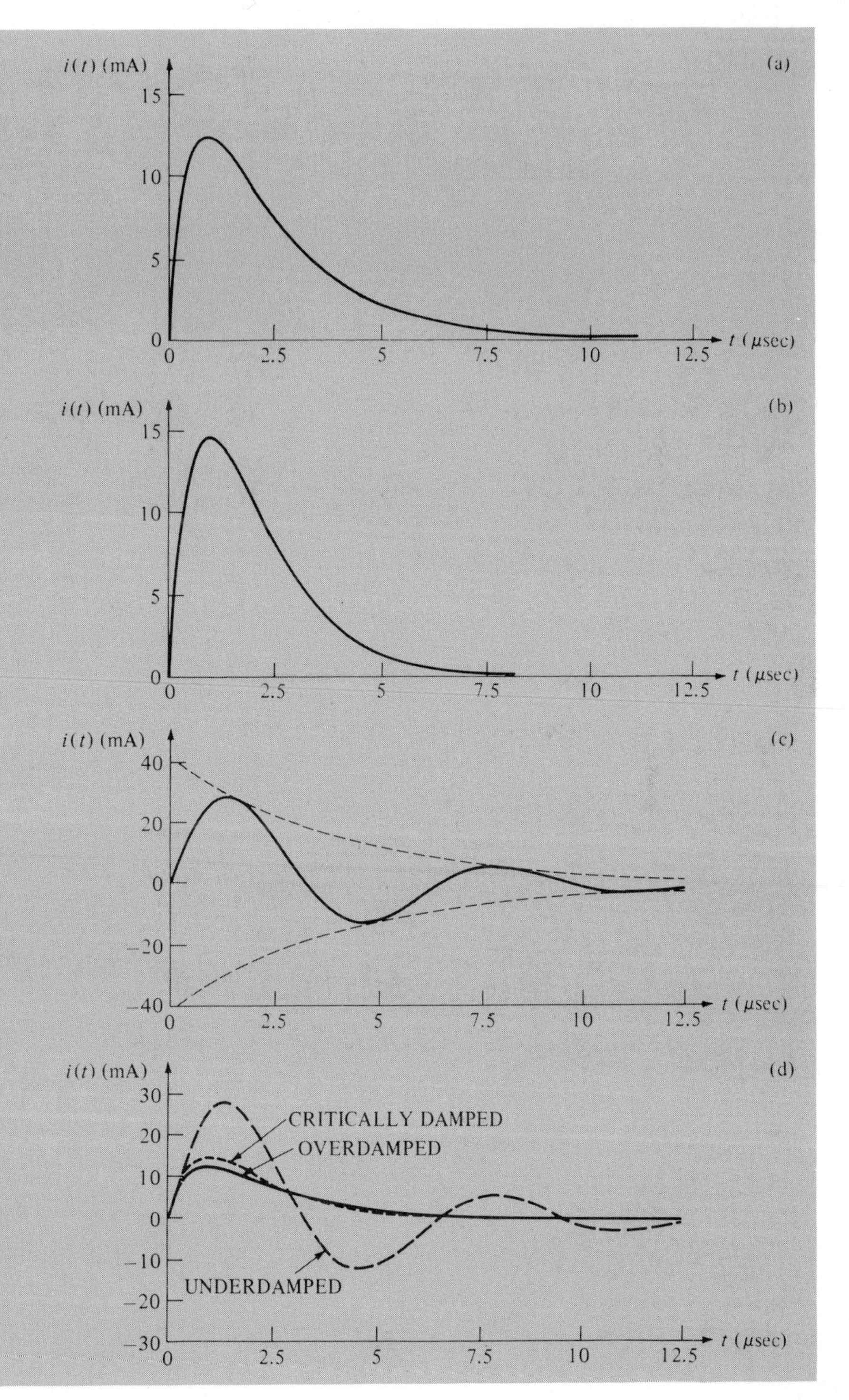

Figure 6.2
Response of a series $R-L-C$ circuit to step excitation.
(a) Overdamped response.
(b) Critically damped response.
(c) Underdamped response.
(d) Comparison of responses.

$i(t)$ equal to zero and solve for t:

$$\frac{di(t)}{dt} = 26.67(-0.5 \times 10^6 \cdot \exp[-0.5 \times 10^6 t] + 2 \times 10^6 \cdot \exp[-2 \times 10^6 t])u(t) = 0$$

$$0.5 \times 10^6 \cdot \exp[-0.5 \times 10^6 t] = 2 \times 10^6 \cdot \exp[-2 \times 10^6 t]$$

$$\frac{\exp[-0.5 \times 10^6 t]}{\exp[-2 \times 10^6 t]} = \frac{2 \times 10^6}{0.5 \times 10^6}$$

$$\exp[1.5 \times 10^6 t] = 4$$

$$1.5 \times 10^6 t = \ln 4 \simeq 1.386$$

$$t_{\mathrm{MAX}} \simeq 0.9242\ \mu\mathrm{sec}$$

Substitute $t = 0.9242\ \mu$sec into the expression for $i(t)$ to determine the maximum value of $i(t)$:

$$i(0.9242 \times 10^{-6}) = 26.67\{\exp[-0.5 \times 10^6(0.9242 \times 10^{-6})] - \exp[-2 \times 10^6(0.9242 \times 10^{-6})]\}.$$

$$i_{\mathrm{MAX}} \simeq 12.60\ \mathrm{mA}$$

A graph of $i(t)$ is given in Figure 6.2(a).

Example 6.2

SPICE

Problem:

Repeat Example 6.1 for $R = R_C$.

Solution:

1. For $R = R_C = 200\ \Omega$,

$$p = -\frac{R}{2L} = -\frac{200}{2(0.1 \times 10^{-3})} = -10^6$$

Equation (6.4) becomes

$$I(s) = \frac{40 \times 10^3}{(s + 10^6)^2}\ \mathrm{A \cdot sec}$$

which yields a critically damped response obtained from Pair VIa or

Equation (6.5):

$$i(t) = 40 \times 10^6 t \cdot \exp[-10^6 t] \cdot u(t) \text{ mA} \cdot \text{sec}$$

2. Set the derivative of $i(t)$ equal to zero and solve for t to determine the time at which the maximum value of $i(t)$ occurs:

$$\frac{di(t)}{dt} = 40 \times 10^6(\exp[-10^6 t] - 10^6 t \cdot \exp[-10^6 t])u(t) = 0$$

$$\exp[-10^6 t] = -10^6 t \cdot \exp[-10^6 t]$$

$$t_{\text{MAX}} = 1 \ \mu\text{sec}$$

Substitute $t = 1 \ \mu$sec into the expression for $i(t)$ to determine the maximum value of $i(t)$:

$$i(10^{-6}) = 40 \times 10^6(10^{-6}) \cdot \exp\left[-10^6(10^{-6})\right]$$

$$i_{\text{MAX}} \simeq 14.72 \text{ mA}$$

A graph of $i(t)$ is given in Figure 6.2(b).

Example 6.3

SPICE

Problem:

Repeat Example 6.1 for $R = 0.25R_C$.

Solution:

1. For $R = 0.25R_C = 50 \ \Omega$,

$$\alpha = \frac{R}{2L} = \frac{50}{2(0.1 \times 10^{-3})} = 250 \times 10^3$$

$$\omega_0^2 = \frac{1}{LC} = \frac{1}{(0.1 \times 10^{-3} 0.01 \times 10^{-6})} = 10^{12}$$

$$\omega_d = \sqrt{10^{12} - (250 \times 10^3)^2} \simeq 968.2 \times 10^3$$

Equation (6.6) becomes

$$I(s) = \frac{40 \times 10^3}{s^2 + 500 \times 10^3 s + 10^{12}} \text{ A} \cdot \text{sec}$$

which yields an underdamped response obtained from Pair XII or

Equation (6.7):

$$i(t) = \frac{40 \times 10^3}{968.2 \times 10^3} \cdot \exp[-250 \times 10^3 t] \cdot \sin[968.2 \times 10^3 t] \cdot u(t)$$
$$= 41.31 \cdot \exp[-250 \times 10^3 t] \cdot \sin[968.2 \times 10^3 t] \cdot u(t) \text{ mA}$$

2. Determine the maximum value of $i(t)$ and the time at which the maximum value occurs. As a good estimate, the maximum value of $i(t)$ occurs after one-quarter cycle, or $\pi/2$ radians, of oscillation. Based on this estimate,

$$\sin[968.2 \times 10^3 t] = \sin[\pi/2]$$
$$t_{MAX} \simeq 1.622\ \mu\text{sec}$$
$$i(1.622 \times 10^{-6}) = 41.31 \exp\left[-250 \times 10^3(1.622 \times 10^{-6})\right] \cdot \sin[\pi/2]$$
$$i_{MAX} \simeq 27.54 \text{ mA}$$

In fact, the damping of the sinusoid causes the maximum value to occur in slightly less than one-quarter cycle of oscillation. To demonstrate, set the derivative of $i(t)$ equal to zero and solve for t:

$$\frac{di(t)}{dt} = 41.31(-250 \times 10^3 \cdot \exp[-250 \times 10^3 t] \cdot \sin[968.2 \times 10^3 t]$$
$$+968.2 \times 10^3 \cdot \exp[-250 \times 10^3 t] \cdot \cos[968.2 \times 10^3 t])$$
$$= 0$$
$$250 \times 10^3 \cdot \exp[-250 \times 10^3 t] \cdot \sin[968.2 \times 10^3 t]$$
$$= 968.2 \times 10^3 \cdot \exp[-250 \times 10^3 t] \cdot \cos[968.2 \times 10^3 t]$$
$$\frac{\sin[968.2 \times 10^3 t]}{\cos[968.2 \times 10^3 t]} = \frac{968.2 \times 10^3}{250 \times 10^3}$$
$$\tan[968.2 \times 10^3 t] = 3.873$$
$$t_{MAX} \simeq 1.361\ \mu\text{sec}$$
$$i(1.361 \times 10^{-6}) = 41.31 \exp\left[-250 \times 10^3(1.361 \times 10^{-6})\right]$$
$$\times \sin\left[968.2 \times 10^3(1.361 \times 10^{-6})\right]$$
$$i_{MAX} \simeq 28.46 \text{ mA}$$

In graphing the underdamped response [Figure 6.2(c)], it is helpful to note that the positive peaks (maxima) of the oscillation lie on a curve defined by $41.31 \cdot \exp[-250 \times 10^3 t] \cdot u(t)$, and the negative peaks (minima) lie on the mirror-image curve, $-41.31 \cdot \exp[-250 \times 10^3 t] \cdot u(t)$. These two curves create an **envelope** which contains the oscillatory response.

Figure 6.2(d), a composite graph of all three waveforms, indicates that current rises toward a peak value most slowly when $R > R_C$ and most rapidly when $R < R_C$. To achieve fastest *initial response* in a series R–L–C circuit, R should be as small as possible. After reaching its peak value, the current falls toward zero most slowly when $R > R_C$ and most rapidly when $R < R_C$. However, when $R < R_C$, the current **overshoots** zero during its fall and becomes negative and then positive again as it oscillates about zero. Oscillation increases the **settling time**, which is the time required for the response to remain within some specified range of its steady-state value, so it may be necessary to increase the resistance to improve the settling time. The fastest initial response, without the oscillation which increases the settling time, occurs when $R = R_C$.

In all three cases, the **steady-state** value of $i(t)$ is zero, so the current response of a series R–L–C circuit to a step-function source is entirely transient.

Response of a Parallel R–L–C Circuit to Step Excitation

The circuit shown in Figure 6.3(a) is the dual of the series R–L–C circuit in Figure 6.1(a). The time-domain response voltage $v(t)$ is obtained by inverse Laplace transformation of the s-domain voltage $V(s)$. When Kirchhoff's

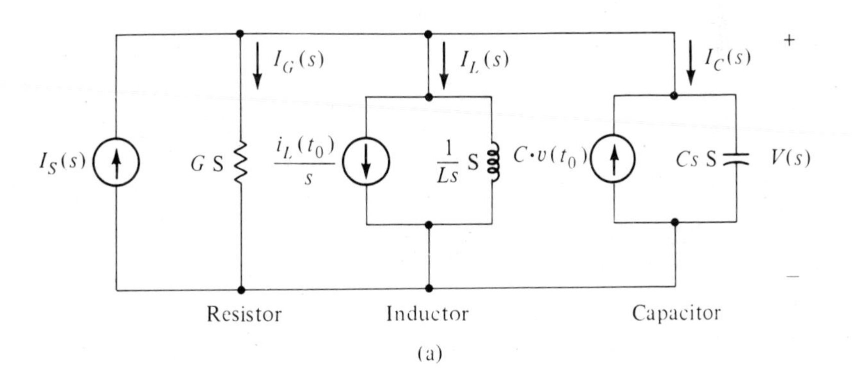

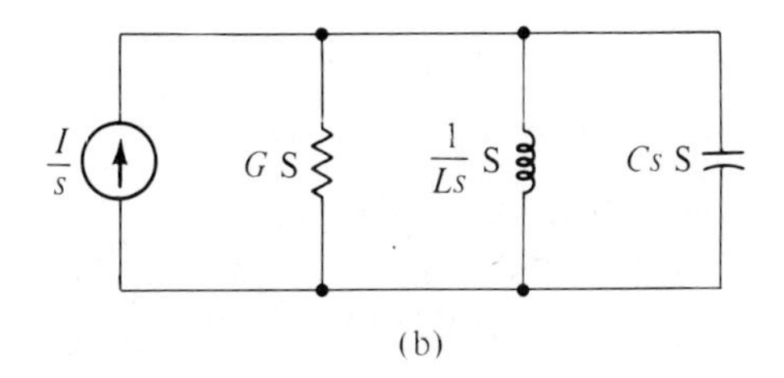

Figure 6.3
Parallel R–L–C circuit.
(a) Arbitrary excitation and initial conditions.
(b) Unshifted step excitation and zero initial conditions.

current law is applied to the transformed circuit,

$$I_L(s) + I_G(s) + I_C(s) = I_S(s)$$

When the element currents are expressed in terms of voltage,

$$Cs \cdot V(s) - C \cdot v(t_0) + G \cdot V(s) + \frac{1}{Ls} \cdot V(s) + \frac{i_L(t_0)}{s} = I_S(s)$$

The solution for $V(s)$ is the dual of Equation (6.1):

$$V(s) = \frac{s\left(\dfrac{I_S(s)}{C} + v(t_0)\right) - \dfrac{i_L(t_0)}{C}}{s^2 + \dfrac{G}{C}s + \dfrac{1}{LC}} \tag{6.9}$$

Duality can be used further to obtain the response to excitation by an unshifted step current source of arbitrary amplitude,

$$i(t) = I \cdot u(t)$$

then Equation (6.9) becomes

$$\begin{aligned} V(s) &= \frac{I/C}{s^2 + (G/C)s + 1/(LC)} \\ &= \frac{I/C}{(s - p_1)(s - p_2)} \end{aligned} \tag{6.10}$$

where

$$p_1 = -\frac{G}{2C} + \sqrt{\left(\frac{G}{2C}\right)^2 - \frac{1}{LC}}$$

$$p_2 = -\frac{G}{2C} - \sqrt{\left(\frac{G}{2C}\right)^2 - \frac{1}{LC}}$$

The transformed equivalent circuit is shown in Figure 6.3(b).

The three types of response—overdamped, critical, and underdamped—are based on the relationship of G, L, and C, and a critical value of G can be calculated for given values of L and C:

$$\left(\frac{G_C}{2C}\right)^2 = \frac{1}{LC}$$

$$G_C = 2\sqrt{C/L} \tag{6.11a}$$

$$R_C = \tfrac{1}{2}\sqrt{L/C} \tag{6.11b}$$

The reader should compare carefully the effect of varying parallel resistance with the effect of varying series resistance. In a series circuit, if $R < R_C$, the response is underdamped; increasing R beyond R_C will cause the response to become overdamped. The reverse is true for a parallel circuit. If $R > R_C$ ($G < G_C$), the response will be underdamped, and R must be decreased (G increased) to less than R_C ($G > G_C$) to obtain an overdamped response. Clearly, this reflects the duality of the series and parallel circuits.

The overdamped response (Case 1) of a parallel R–L–C circuit is given by the dual of Equation (6.3):

$$v(t) = \frac{I/C}{p_2 - p_1}(\exp[p_2 t] - \exp[p_1 t])u(t) \tag{6.12}$$

The critical response is given by the dual of Equation (6.5):

$$v(t) = \frac{I}{C}t \cdot \exp\left[-\frac{G}{2C}t\right] \cdot u(t) \tag{6.13}$$

The underdamped response is given by the dual of Equation (6.7):

$$v(t) = \frac{I/C}{\omega_d} \cdot \exp[-\alpha t] \cdot \sin[\omega_d t] \cdot u(t) \text{ V} \tag{6.14}$$

where $\alpha = G/(2C)$, $\omega_0^2 = 1/(LC)$, $\omega_d = \sqrt{\omega_0^2 - \alpha^2}$.

Practical *R–L–C* Circuits

In practice, the series R–L–C circuit model is more useful than the parallel circuit model. A discussion of the circuit models in Figure 6.4 will provide more insight into the reason why this is true.

A practical inductor exhibits capacitance (c_L) and resistance (r_L). The capacitance is mainly associated with the metallic structure (coils and leads) and the dielectric materials used for coil insulation and for the core; the resistance is associated with the wire used for the coils and leads. Similarly, a practical capacitor exhibits inductance (l_C) associated with its metallic struc-

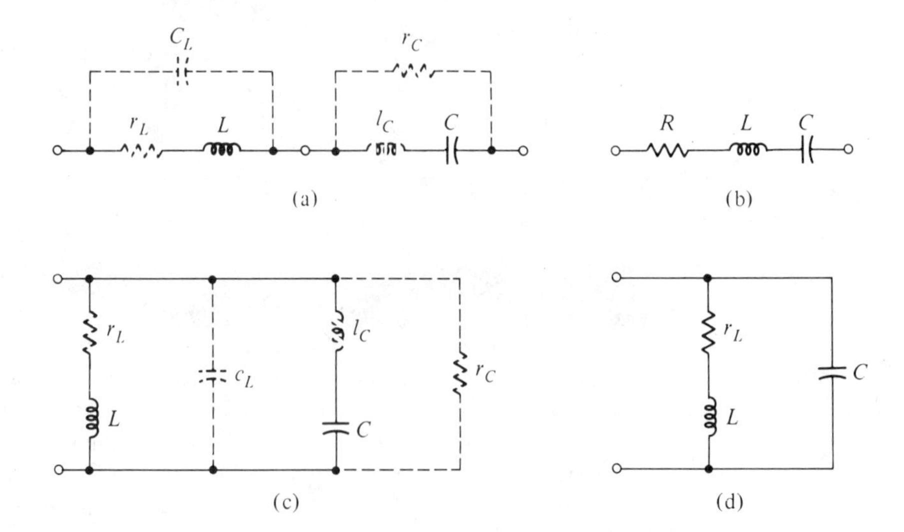

Figure 6.4
Practical capacitors and inductors.

ture (leads and plates) and resistance (r_C) associated with the dielectric material. Except for some high-frequency applications, such as communications circuits, c_L for the inductor, and l_C and r_C for the capacitor, are usually neglected, leaving only r_L for the inductor to be considered.

The series connection of a practical inductor and a practical capacitor in Figure 6.4(a) is usually represented by the model shown in Figure 6.4(b). The resistance of the inductor, r_L, can then be combined with any other resistance in the series R–L–C circuit, such as that associated with the excitation source, without complicating the analysis which generates Equations (6.1) through (6.8). It is important to note that r_L must be considered when calculating R_C and always increases the amount of damping present in a series R–L–C circuit. In some cases, r_L will be large enough so that a critical (or underdamped) response cannot be obtained even if all other series resistance is reduced to zero.

The model shown in Figure 6.4(d), which represents the parallel connection of a practical inductor and a practical capacitor [Figure 6.4(c)], is not a parallel circuit, and r_L cannot be combined with any other resistance. The presence of r_L complicates the analysis, and Equations (6.1) through (6.8) cannot be used except for an approximation which neglects r_L. It is appropriate at this point to present an example in which r_L is not neglected.

Example 6.4

SPICE

Problem:

Find and graph the voltage $v(t)$ for the circuit shown in Figure 6.5(a).

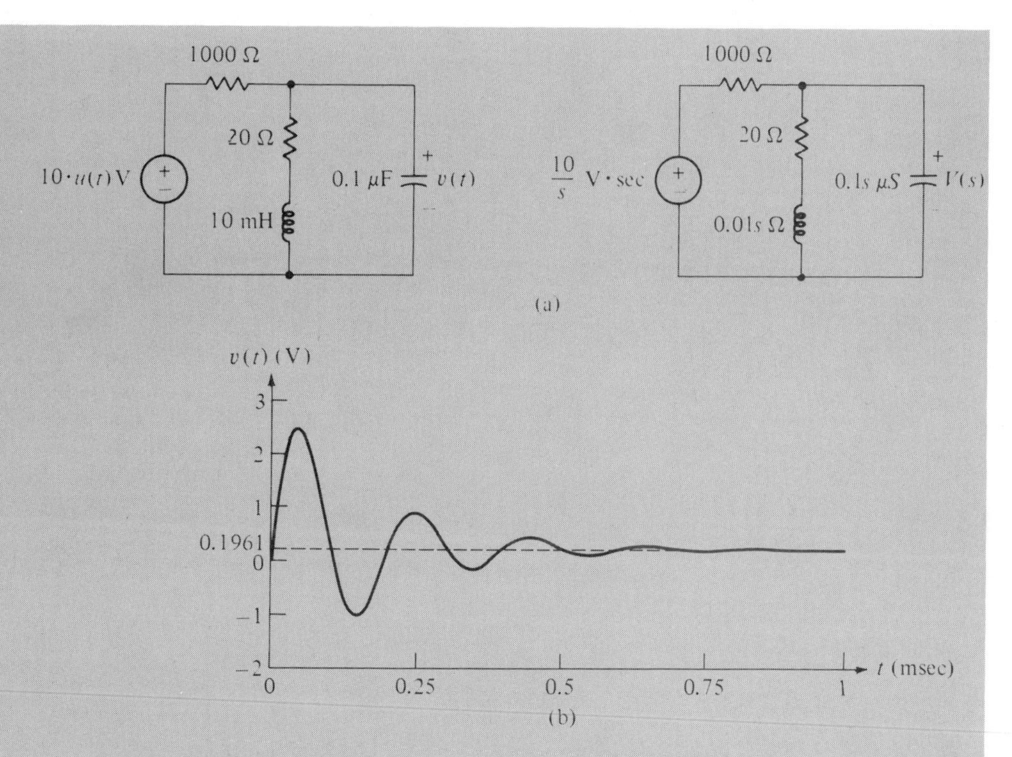

Figure 6.5 Example 6.4.

Solution:

1. Calculate the admittance of the inductor:

$$Z_L = 10 \times 10^{-3}s + 20 = 0.01(s + 2000)\ \Omega$$

$$Y_L = \frac{1}{Z_L} = \frac{100}{s + 2000}\ \text{S}$$

2. Calculate the impedance of the inductor–capacitor network:

$$Y_{L\text{-}C} = Y_L + Y_C = \frac{100}{s + 2000} + 0.1 \times 10^{-6}s$$

$$= 0.1 \times 10^{-6}\frac{s^2 + 2000s + 10^9}{s + 2000}\ \text{S}$$

$$Z_{L\text{-}C} = \frac{1}{Y_{L\text{-}C}} = 10 \times 10^{6}\frac{s + 2000}{s^2 + 2000s + 10^9}\ \Omega$$

3. Use the voltage division rule to calculate $V(s)$:

$$V(s) = V_S(s) \cdot \frac{Z_{L-C}}{R_S + Z_{L-C}}$$

$$= \frac{10}{s} \cdot \frac{10 \times 10^6(s + 2000)/(s^2 + 2000s + 10^9)}{1000 + 10 \times 10^6(s + 2000)/(s^2 + 2000s + 10^9)}$$

$$= 0.1 \times 10^6 \frac{s + 2000}{s(s^2 + 12\,000s + 1.02 \times 10^9)} \text{ V} \cdot \text{sec}$$

4. Solve for $v(t)$ by partial fraction expansion:

$$V(s) = \frac{K_1}{s} + \frac{\mathbf{K}_2}{s + 6000 - j31\,370} + \frac{\mathbf{K}_2^*}{s + 6000 + j31\,370} \text{ V} \cdot \text{sec}$$

where $K_1 = s \cdot V(s)|_{s=0} \simeq 0.1961$, and

$$\mathbf{K}_2 = (s + 6000 - j31\,370) \cdot V(s)|_{s=-6000+j31\,370}$$

$$\simeq 1.578 \cdot \exp[-j93.53°]$$

5. Use Pair II and Equation (4.17) to obtain $v(t)$:

$$v(t) = (0.1961 + 3.156 \cdot \exp[-6000t] \cdot \cos[31\,370t - 93.53°])u(t)$$

$$= (0.1961 + 3.156 \cdot \exp[-6000t] \cdot \sin[31\,370t - 3.53°])u(t) \text{ V}$$

Refer to Figure 6.5(b) for a graph of $v(t)$. The reader should confirm that if r_L were not present, the 0.1961-V dc level would not be present (Problem 6-5).

The problem of Example 6.4 can be formulated for an arbitrary excitation $V_S(s)$ with arbitrary internal resistance R_S:

$$V(s) = V_S(s) \cdot \frac{\dfrac{s + (r_L/L)}{R_S C}}{s^2 + \left(\dfrac{r_L}{L} + \dfrac{1}{R_S C}\right)s + \dfrac{1 + (r_L/R_S)}{LC}} \tag{6.15}$$

Derivation of Equation (6.15) is left as an exercise for the reader (Problem 6-14).

Exercise 6.1

SPICE

Problem:

For the circuit shown in Figure 6.6(a),

a. find $V(s)$ and $v(t)$;

b. find the time at which the peak (maximum or minimum) voltage occurs and the value of $v(t)$ at that time;

c. graph $v(t)$.

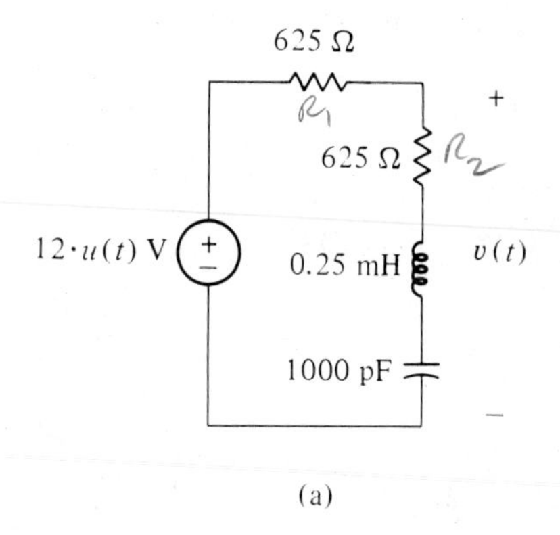

(a)

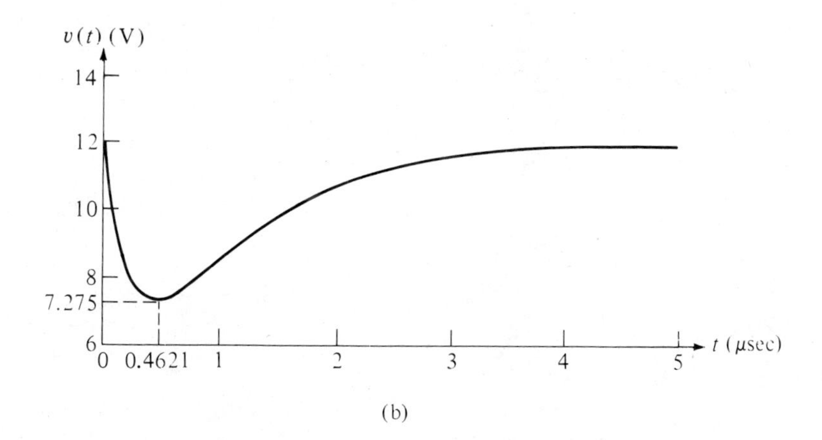

(b)

Figure 6.6
Exercise 6.1.

Answers:

$$V(s) = \frac{12(s^2 + 2.5 \times 10^6 s + 4 \times 10^{12})}{s(s^2 + 5 \times 10^6 s + 4 \times 10^{12})} \text{ V} \cdot \text{sec}$$

$$v(t) = \{12 - 10(\exp[-10^6 t] - \exp[-4 \times 10^6 t])\} u(t) \text{ V}$$

$$t_{\text{MIN}} \simeq 0.4621\ \mu\text{sec}$$

$$v_{\text{MIN}} \simeq 7.275 \text{ V}$$

a graph of $v(t)$ is given in Figure 6.6(b).

6.3 Characterization of the Response of an R–L–C Circuit

Now that we have examined some circuit responses in Chapter 5 and in the previous section, we can make some general observations concerning these responses. These observations will promote a better understanding of circuit response and facilitate our later discussions of circuit response. In addition, we will introduce two useful Laplace transform theorems—the **final value theorem** and the **initial value theorem.**

Example 6.5

Problem:

Identify the transient part and the steady-state part of the response current of the R–L circuit in Example 5.6.

Solution:

$$\begin{aligned} i_L(t) &= 4(1 + 4 \cdot \exp[-62.5 \times 10^3 t])u(t) \\ &= 4 \cdot u(t) + 16 \cdot \exp[-62.5 \times 10^3 t] \cdot u(t) \text{ mA} \end{aligned}$$

1. The transient part of $i_L(t)$ is

$$16 \cdot \exp[-62.5 \times 10^3 t] \text{ mA}$$

2. The steady-state part is 4 mA.
3. The initial response at $t = 0^+$ is

$$4 + 16 \cdot \exp[-0] = 20 \text{ mA}$$

which satisfies the initial condition and the requirement for continuity of inductor current.

In Example 6.5, for any $t > 0$, both parts contribute to the complete response, but as t gets larger, the transient response contributes less and less. In theory, the transient response is never zero, but as a practical matter, it may be neglected after a time equal to several (usually 4 or 5) time constants.

The reader should verify that if the initial inductor current is caused to be 25 mA in Example 5.6 (perhaps by reducing the resistance of the charging source to 350 Ω), then the response current is

$$i_L(t) = 4 \cdot u(t) + 21 \cdot \exp[-62.5 \times 10^3 t] \cdot u(t) \text{ mA}$$

This leads to an important conclusion—the transient response depends upon both the initial conditions and the forcing function (source), whereas the steady-state response is independent of initial conditions.

Example 6.6

Problem:

Identify the transient part and the steady-state part of the response current of the R–L–C circuit in Example 6.3.

Solution:

$$i(t) = 41.31 \cdot \exp[-250t] \cdot \sin[968.2 \times 10^3 t] \cdot u(t) \text{ mA}$$

The complete response in this case is a transient, and the steady-state response is zero.

The results from Examples 6.5 and 6.6 are not conclusive regarding the relationship between the steady-state response and the excitation. In fact, it is difficult to generalize this relationship except in the following cases:

1. If the excitation is a dc or a step-function source, the steady-state response will be a constant (possibly zero).
2. If the excitation is a sinusoidal source, the steady-state response will also be sinusoidal, differing from the excitation only in amplitude and phase angle.

The first case leads to that part of circuit analysis which we reviewed in Chapter 1, commonly called **dc analysis**, and the second case leads to that part of circuit analysis commonly called **ac** (or **phasor**) **analysis**, which we will review and examine thoroughly in Chapter 9.

The reader has certainly already made note that the transient response is characterized by the presence of an exponential factor $\exp[-t/\tau]$, and the steady-state response is characterized by its absence.

Final Value Theorem

The final value theorem provides a means of determining the steady-state time-domain response of a circuit without performing the inverse transformation of the s-domain function. The final value theorem is limited in its application, as will be seen shortly, but it is particularly useful when the circuit excitation is a dc or a step-function source.

Final Value Theorem

If $f(t)$ exists for $t \to \infty$, then

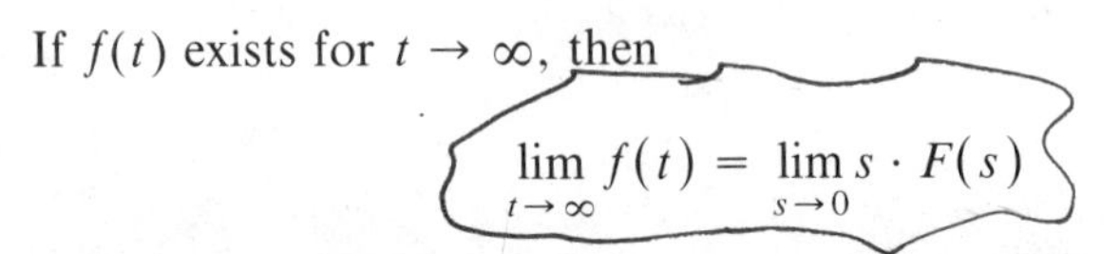

$$\lim_{t \to \infty} f(t) = \lim_{s \to 0} s \cdot F(s)$$

The value of $f(t)$ as $t \to \infty$ is, of course, the steady-state value of $f(t)$.

The question which now arises is how can the existence of $f(t)$ as $t \to \infty$, required for application of the final value theorem, be verified if $f(t)$ is not known? The answer, stated here without proof, is that all poles of $F(s)$ must lie in the left half of the complex s-plane (Figure 6.7). Note that the left half plane *does not* include the imaginary axis. One exception is that $F(s)$ may contain a simple (unrepeated) pole at zero, since that pole will be canceled when $F(s)$ is multiplied by s in applying the final value theorem.

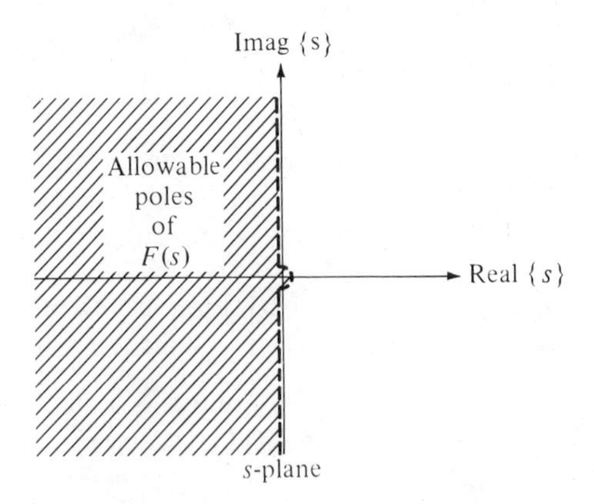

Figure 6.7
Criterion for validity of final value theorem.

Example 6.7

Problem:

Apply the final value theorem to determine the steady-state response of the R–L circuit in Examples 5.6 and 6.5.

Solution:

$$I(s) = \frac{250}{s(s + 62.5 \times 10^3)} + \frac{0.02}{s + 62.5 \times 10^3} \text{ A} \cdot \text{sec}$$

The denominator of $I(s)$ has a simple pole at zero and a pole at -62.5×10^3, so the final value theorem may be applied to determine the steady-state value of $i(t)$:

$$\lim_{t \to \infty} i(t) = \lim_{s \to 0} s \cdot I(s)$$

$$s \cdot I(s) = \frac{250}{s + 62.5 \times 10^3} + \frac{0.02s}{s + 62.5 \times 10^3} \text{ A}$$

$$\lim_{s \to 0} s \cdot I(s) = \frac{250}{0 + 62.5 \times 10^3} + \frac{0.02(0)}{0 + 62.5 \times 10^3} = 4 \text{ mA}$$

Therefore, $\lim_{t \to \infty} i(t) = 4$ mA, which confirms the result obtained in Example 6.5.

Example 6.8

Problem:

Apply the final value theorem to determine the steady-state response of the R–L–C circuit in Example 6.3.

Solution:

$$I(s) = \frac{40 \times 10^3}{s^2 + 500 \times 10^3 s + 10^{12}} \text{ A} \cdot \text{sec}$$

$I(s)$ has poles only at $-250 \times 10^3 \pm j968.2 \times 10^3$, so the final value theorem is applicable:

$$\lim_{t \to \infty} i(t) = \lim_{s \to 0} sI(s)$$

$$s \cdot I(s) = \frac{40 \times 10^3 s}{s^2 + 500 \times 10^3 s + 10^{12}} \text{ A}$$

$$\lim_{s \to 0} s \cdot I(s) = \frac{40 \times 10^3(0)}{(0)^2 + 500 \times 10^3(0) + 10^{12}} = 0$$

Therefore,

$$\lim_{t \to \infty} i(t) = 0$$

which confirms the result obtained in Example 6.6.

Example 6.9

Problem:

Find the steady-state value of $v(t)$ if

$$V(s) = \frac{15s^2 + 90 \times 10^6}{s(s^2 + 9 \times 10^6)} \text{ V} \cdot \text{sec}$$

Solution:

1. If the final value theorem is applied without regard to the existence of $v(t)$ as $t \to \infty$,

$$\lim_{t \to \infty} v(t) = \lim_{s \to 0} s \cdot V(s) = \lim_{s \to 0} \frac{15s^2 + 90 \times 10^6}{s^2 + 9 \times 10^6} = 10 \text{ V}$$

2. If $v(t)$ is obtained by inverse Laplace transformation,

$$v(t) = (10 + 5 \cdot \cos 3000t)u(t) \text{ V}$$

   The steady-state response is clearly $5 \cdot \cos 3000t$ V, which continues to oscillate, rather than approach a constant, at $t \to \infty$.

3. Examination of $V(s)$ reveals that the poles of $V(s)$ are 0 and $\pm j3000$. The pole at zero is canceled, but the poles on the imaginary axis remain, so step 1 was not valid. This example serves to point out the principal limitation of the final value theorem.

Initial Value Theorem

The initial value theorem provides a means of determining the initial value of a response without performing the inverse transformation of the s-domain function. This theorem is particularly useful in checking the correctness of the s-domain function obtained when analyzing circuit response, since application of the initial value theorem will recover the initial value of the response used to obtain the s-domain function. While this is not absolute confirmation of the correctness of the s-domain function, it provides a reassuring checkpoint in a complicated analysis problem.

Initial Value Theorem

$$\lim_{t \to 0} f(t) = \lim_{s \to \infty} s \cdot F(s)$$

The expression $t \to 0$ means that t is decreasing toward zero, or approaching zero from the right. The value of $f(t)$ as $t \to 0$ is the initial value of $f(t)$, or $f(0^+)$.

Example 6.9

Problem:

Use the initial value theorem to determine the initial current in the R–L circuit of Example 5.6.

Solution:

$$I(s) = \frac{250}{s(s + 62.5 \times 10^3)} + \frac{0.02}{s + 62.5 \times 10^3} \text{ A} \cdot \text{sec}$$

1. According to the initial value theorem,

$$\lim_{t \to 0} i(t) = \lim_{s \to \infty} s \cdot I(s)$$

$$s \cdot I(s) = \frac{250}{s + 62.5 \times 10^3} + \frac{0.02s}{s + 62.5 \times 10^3} \text{ A}$$

$$\lim_{s \to \infty} s \cdot I(s) = \lim_{s \to \infty} \left(\frac{250/s}{1 + 62.5 \times 10^3/s} + \frac{0.02}{1 + 62.5 \times 10^3/s} \right)$$

$$= \frac{0}{1 + 0} + \frac{0.02}{1 + 0} = 20 \text{ mA}$$

Therefore,

$$i(0^+) = \lim_{t \to 0} i(t) = 20 \text{ mA}$$

Exercise 6.2

Problems:

1. Identify the transient part and the steady-state part:
   a. $v(t) = 4(2 + \exp[-250t] \cdot \cos[5000t - 30°])u(t)$ V.
   b. $f(t) = 12(1 + 2 \cdot \exp[-t] + 0.03 \cdot \sin[3t + 45°])u(t)$.

2. Use the final value theorem to determine the steady-state response:

a. $I(s) = \dfrac{0.03s^2 + 7s + 300}{s(s^2 + 800s + 150 \times 10^3)}$ A · sec.

b. $V(s) = \dfrac{8}{s^2} + \dfrac{4s}{s^2 + 800s + 150 \times 10^3}$ V · sec.

3. Use the initial value theorem to determine the initial value of the response for the problems of part 2 above.

Answers:

1. a. Transient: $4 \cdot \exp[-250t] \cdot \cos[5000t - 30°]$ V,
   Steady state: 8 V.
   b. Transient: $24 \cdot \exp[-t]$,
   Steady state: $12 + 0.36 \cdot \sin[3t + 45°]$.
2. a. $i(t \to \infty) = 2$ mA.
   b. Cannot apply final value theorem, due to repeated pole at zero.
3. a. 30 mA.
   b. 4 V.

6.4 Response to Sinusoidal Excitation

Most introductory circuit analysis textbooks and courses present what is commonly called **ac analysis**, which treats only the steady-state response of a circuit to sinusoidal excitation.[2] The large body of important circuit analysis techniques which come under the umbrella of ac analysis makes it easy to overlook the presence and effects of transients in sinusoidally excited circuits. The transient response is insignificant in an electric heater which is activated by a mechanical switch and allowed to operate for many minutes, but the transient response of a R–L–C circuit may be quite significant when a high-frequency sinusoidal source (such as a communications carrier wave) is switched on and off electronically many times per second.

Response of *R*–*C* and *R*–*L* Circuits to Sinusoidal Excitation

The following example and exercise show that the response of a R–C or R–L circuit is easily determined by straightforward application of the transform circuit analysis methods previously discussed. Preparing a graph of the re-

[2] We will see in Chapter 9 that ac analysis is a specialization of the more general Laplace transform analysis which we are now pursuing.

sponse is a bit more difficult, and the reader is encouraged to use a computer to graph the response.[3] Manual graphs, unless carefully prepared, may not reveal important aspects of the response.

The graphs indicate that the peak-to-peak value of the response in these first-order circuits is the same throughout the transient phase as it is at steady state. The average value and the peak values (both negative and positive peaks) follow the same exponential decay or growth during the transient phase. The time constant of the transient is determined by RC or L/R.

Example 6.10

SPICE

Problem:

Determine the source current and the capacitor voltage for $t > 0$ in the circuit shown in Figure 6.8(a).

Solution:

1. Calculate $I(s)$:

$$I(s) = \frac{V(s)}{Z(s)} = \frac{30 \times 10^6/(s^2 + 10^{12})}{50 + 20 \times 10^6/s}$$

$$= \frac{600 \times 10^3 s}{(s + 400 \times 10^3)(s^2 + 10^{12})} \text{ A} \cdot \text{sec}$$

2. Determine $i(t)$ from a partial fraction expansion of $I(s)$:

$$I(s) = \frac{K_1}{s + 400 \times 10^3} + \frac{\mathbf{K}_2}{s - j10^6} + \frac{\mathbf{K}_2^*}{s + j10^6}$$

From Pair IV and Equation (4.17),

$$i(t) = \big(K_1 \cdot \exp[-400 \times 10^3 t]$$
$$+ 2|\mathbf{K}_2| \cdot \cos[10^6 t + \phi]\big)u(t) \text{ A}$$

[3] We must be clear here that the author is only advocating the use of the computer as a graphics aid to plot the response once the equation of the response has been obtained analytically. The use of computer simulations (such as SPICE), which conduct the analysis as well as present a graphical display of the response, should be used *only after the analytical techniques are fully understood and have been practiced*.

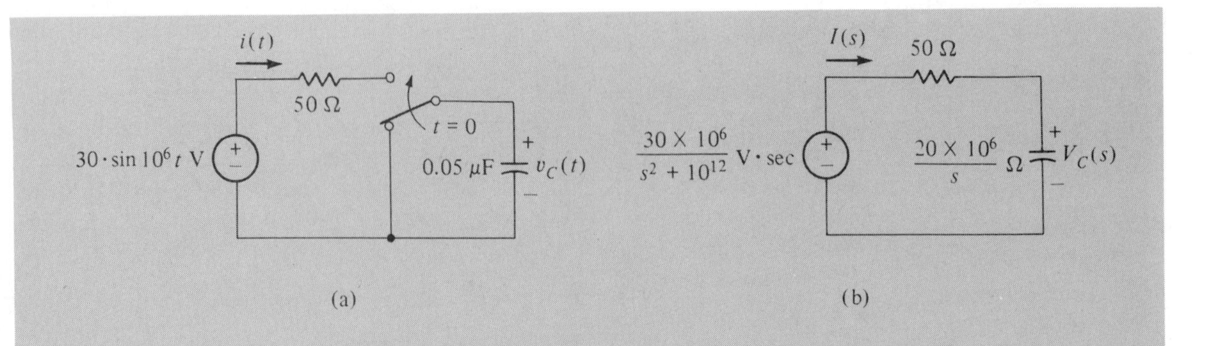

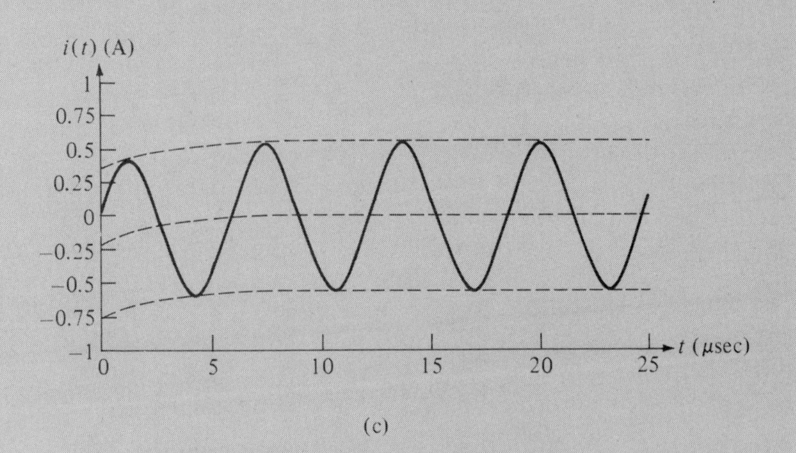

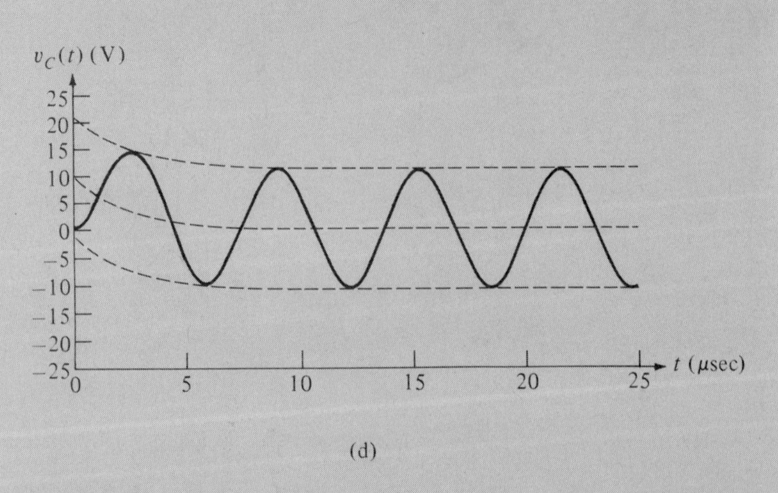

Figure 6.8 Example 6.10.

where

$$K_1 = (s + 400 \times 10^3) I(s)\Big|_{s=-400\times10^3}$$

$$= \left.\frac{600 \times 10^3 s}{s^2 + 10^{12}}\right|_{s=-400\times10^3} \simeq -0.2069$$

and

$$\mathbf{K}_2 = (s - j10^6) I(s)\Big|_{s=j10^6}$$

$$= \left.\frac{600 \times 10^3 s}{(s + 400 \times 10^3)(s + j10^6)}\right|_{s=j10^6}$$

$$\simeq 0.2785 \cdot \exp[-j68.20°]$$

Therefore,

$$i(t) = (-0.2069 \cdot \exp[-400 \times 10^3 t] + 0.5571 \cdot \cos[10^6 t - 68.20°])u(t) \text{ A}$$

See Figure 6.8(c) for a graph of $i(t)$.

3. Calculate $V_C(s)$:

$$V_C(s) = \frac{1}{Cs} \cdot I(s) = \frac{(20 \times 10^6/s)(600 \times 10^3 s)}{(s + 400 \times 10^3)(s^2 + 10^{12})}$$

$$= \frac{12 \times 10^{12}}{(s + 400 \times 10^3)(s^2 + 10^{12})} \text{ V} \cdot \text{sec}$$

4. Determine $v_C(t)$ from a partial fraction decomposition of $V_C(s)$:

$$V_C(s) = \frac{K_1}{s + 400 \times 10^3} + \frac{\mathbf{K}_2}{s - j10^6} + \frac{\mathbf{K}_2^*}{s + j10^6}$$

From Pair IV and Equation (4.17),

$$v_C(t) = (K_1 \cdot \exp[-400 \times 10^3 t] + 2|\mathbf{K}_2| \cdot \cos[10^6 t + \phi])u(t) \text{ V}$$

where

$$K_1 = (s + 400 \times 10^3) V_C(s)\Big|_{s=-400\times10^3} \simeq 10.34$$

and

$$\mathbf{K}_2 = (s - j10^6)V_C(s)\big|_{s=j10^6} \simeq 5.571 \cdot \exp[-j158.2°]$$

Therefore,

$$\begin{aligned} v_C(t) &= (10.34 \cdot \exp[-400 \times 10^3 t] \\ &\quad + 11.14 \cdot \cos[10^6 t - 158.2°])u(t) \\ &= (10.34 \cdot \exp[-400 \times 10^3 t] \\ &\quad + 11.14 \cdot \sin[10^6 t - 68.2°])u(t) \text{ V} \end{aligned}$$

See Figure 6.8(d) for a graph of $v_C(t)$.

Exercise 6.3

Problem:

Find and graph $v_R(t)$ for the circuit shown in Figure 6.9(a).

Figure 6.9
Exercise 6.3.

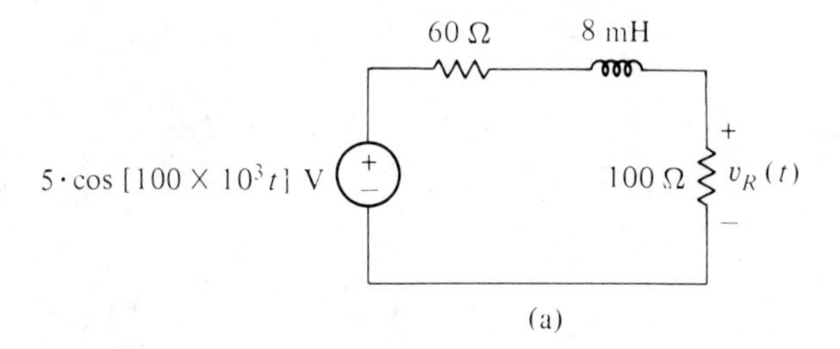

(a)

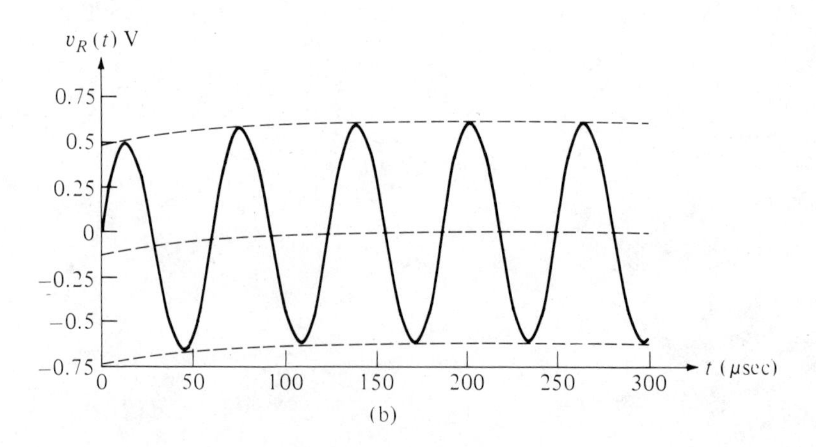

(b)

Answer:

$$v_R(t) = -(0.1202 \cdot \exp[-20 \times 10^3 t] - 0.6129 \cdot \cos[100 \times 10^3 t - 78.69°])u(t) \text{ V}$$

See Figure 6.9(b) for a graph of $v_R(t)$.

Response of *R–L–C* Circuits to Sinusoidal Excitation

Examples 6.11, 6.12, and 6.13 show that the transient response of a R–L–C circuit excited by a sinusoidal source depends on the relationship between resistance, capacitance, and inductance in precisely the same way as in an R–L–C circuit excited by a step-function source. In Example 6.11, the circuit is heavily damped by the large resistance, so the response is slow to reach steady state. The resistance is equal to its critical value in Example 6.12, and the response reaches steady state more quickly. In Example 6.13, the resistance is below its critical value, so the circuit oscillates at its damped frequency. Two distinct oscillations can be seen in the response—one due to the source and another due to the damped oscillation created by the circuit elements. In Exercise 6.4, the reader is asked to analyze the response when the source frequency is the same as the damped frequency of the circuit.

Example 6.11

SPICE

Problem:

Find and graph the response current for the circuit shown in Figure 6.10 when $R = 300\ \Omega$.

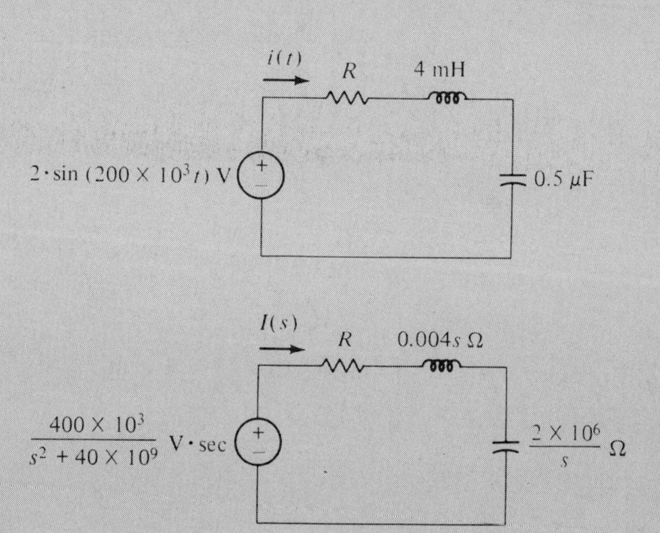

Figure 6.10 Circuit for Examples 6.11, 6.12, 6.13 and Exercise 6.4.

Solution:

1. Calculate $I(S)$:

$$I(s) = \frac{400 \times 10^3/(s^2 + 40 \times 10^9)}{0.004s + 300 + 2 \times 10^6/s}$$
$$= \frac{100 \times 10^6 s}{(s^2 + 75 \times 10^3 s + 500 \times 10^6)(s^2 + 40 \times 10^9)}$$
$$= \frac{100 \times 10^6 s}{(s + 67\,600)(s + 7396)(s - j200 \times 10^3)(s + j200 \times 10^3)} \text{ A} \cdot \text{sec}$$

2. Determine $i(t)$ from a partial fraction decomposition of $I(s)$:

$$I(s) = \frac{K_1}{s + 67\,600} + \frac{K_2}{s + 7396} + \frac{\mathbf{K}_3}{s - j200 \times 10^3} + \frac{\mathbf{K}_3^*}{s + j200 \times 10^3}$$

From Pair IV and Equation (4.17),

$$i(t) = \left(K_1 \cdot \exp[-67\,600t] + K_2 \cdot \exp[-7396t] + 2|\mathbf{K}_3| \cdot \cos[200 \times 10^3 t + \phi]\right) u(t) \text{ A}$$

where

$$K_1 = (s + 67\,600) I(s)|_{s=-67\,600} \simeq 2.519 \times 10^{-3}$$
$$K_2 = (s + 7396) I(s)|_{s=-7396} \simeq -0.3067 \times 10^{-3}$$

and

$$\mathbf{K}_3 = (s - j200 \times 10^3) I(s)|_{s=j200 \times 10^3} \simeq 1.183 \times 10^{-3} \cdot \exp[-j159.2°]$$

Therefore,

$$i(t) = \left(2.519 \cdot \exp[-67\,600t] - 0.3067 \cdot \exp[-7396t] + 2.367 \cdot \cos[200 \times 10^3 t - 159.2°]\right) u(t)$$
$$= \left(2.519 \cdot \exp[-67\,600t] - 0.3067 \cdot \exp[-7396t] + 2.367 \cdot \sin[200 \times 10^3 t - 69.20°]\right) u(t) \text{ mA}$$

See Figure 6.11 for a graph of $i(t)$.

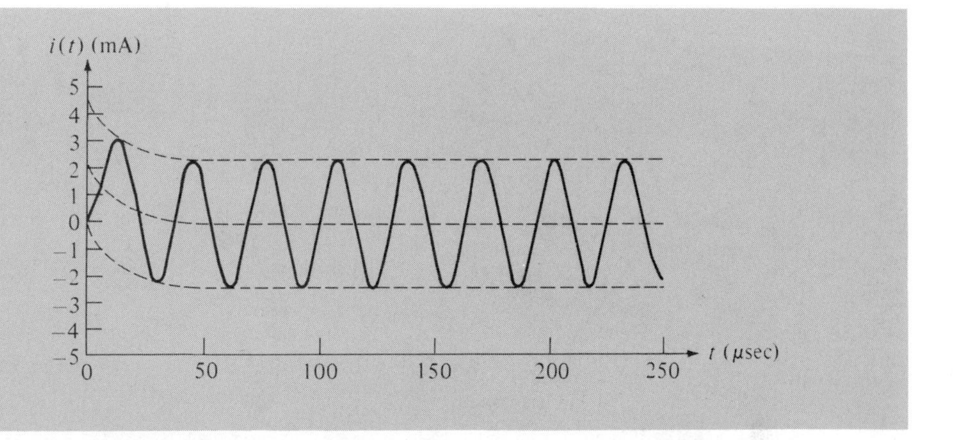

Figure 6.11
Response for Example 6.11.

Example 6.12

SPICE

Problem:

Find and graph the response current for the circuit shown in Figure 6.10 when $R = R_C$.

Solution:

1. Calculate R_C from Equation (6.8):

$$R_C = 2\sqrt{L/C} \simeq 178.9\ \Omega$$

2. Calculate $I(s)$:

$$I(s) = \frac{400 \times 10^3/(s^2 + 40 \times 10^9)}{0.004s + 178.9 + 2 \times 10^6/s}$$

$$= \frac{100 \times 10^6 s}{(s^2 + 44\,720s + 500 \times 10^6)(s^2 + 40 \times 10^9)}$$

$$= \frac{100 \times 10^6 s}{(s + 22\,360)^2(s - j200 \times 10^3)(s + j200 \times 10^3)}\ \text{A} \cdot \text{sec}$$

3. Determine $i(t)$ from a partial fraction decomposition of $I(s)$:

$$I(s) = \frac{K_1}{(s + 22\,360)^2} + \frac{K_2}{s + 22\,360}$$

$$+ \frac{\mathbf{K}_3}{s - j200 \times 10^3} + \frac{\mathbf{K}_3^*}{s + j200 \times 10^3}$$

From Pair VIa, Pair IV, and Equation (4.17),

$$i(t) = \{(K_1 t + K_2)\exp[-22\,360t] + 2|\mathbf{K}_3| \cdot \cos[200 \times 10^3 t + \phi]\} \cdot u(t) \text{ A}$$

where

$$K_1 = (s + 22\,360)^2 I(s)\big|_{s=-22\,360} \simeq -55.21$$

$$K_2 = \frac{d}{ds}(s + 22\,360)^2 I(s)\bigg|_{s=-22\,360} \simeq 2.408 \times 10^{-3}$$

and

$$\mathbf{K}_3 = (s - j200 \times 10^3) I(s)\big|_{s=j200\times 10^3} \simeq 1.235 \times 10^{-3} \cdot \exp[-j167.2°]$$

Therefore,

$$i(t) = \{-(55\,210t - 2.408)\exp[-22\,360t] + 2.469 \cdot \sin[200 \times 10^3 t - 77.20°]\} u(t) \text{ mA}$$

See Figure 6.12 for a graph of $i(t)$.

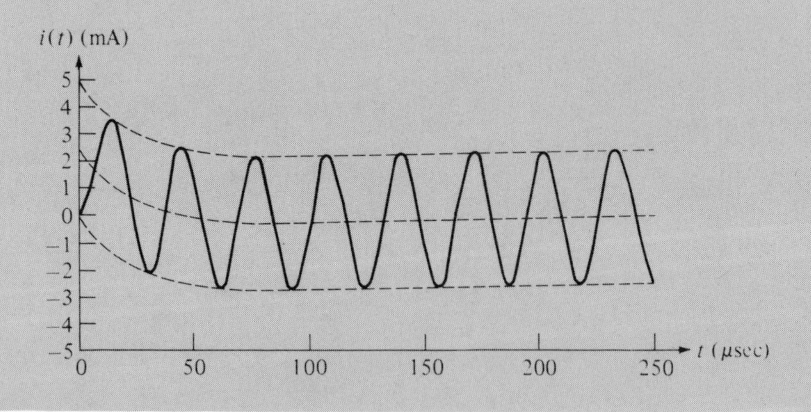

Figure 6.12 Response for Example 6.12.

Example 6.13

SPICE

Problem:

Find and graph the response current for the circuit shown in Figure 6.10 when $R = 50\ \Omega$.

Solution:

1. Calculate $I(S)$:

$$I(s) = \frac{400 \times 10^3/(s^2 + 40 \times 10^9)}{0.004s + 50 + 2 \times 10^6/s}$$

$$= \frac{100 \times 10^6 s}{(s^2 + 12\,500s + 500 \times 10^6)(s^2 + 40 \times 10^9)}$$

$$= \frac{100 \times 10^6 s}{(s + 6250 - j21\,470)(s + 6250 + j21\,470)(s - j200 \times 10^3)(s + j200 \times 10^3)} \text{ A} \cdot \text{sec}$$

2. Determine $i(t)$ from a partial fraction decomposition of $I(s)$:

$$I(s) = \frac{\mathbf{K}_1}{s + 6250 - j21\,470} + \frac{\mathbf{K}_1^*}{s + 6250 + j21\,470} + \frac{\mathbf{K}_2}{s - j200 \times 10^3} + \frac{\mathbf{K}_2^*}{s + j200 \times 10^3}$$

From Equation (4.17),

$$i(t) = \big(2|\mathbf{K}_1| \cdot \exp[-6250t] \cdot \cos[21\,470t + \phi_1] + 2|\mathbf{K}_2| \cdot \cos[200 \times 10^3 t + \phi_2]\big)u(t) \text{ A}$$

where

$$\mathbf{K}_1 = (s + 6250 - j21\,470)I(s)\big|_{s=-6250+j21\,470} \simeq 1.316 \times 10^{-3} \cdot \exp[j16.59°]$$

and

$$\mathbf{K}_2 = (s - j200 \times 10^3)I(s)\big|_{s=j200\times 10^3} \simeq 1.263 \times 10^{-3} \cdot \exp[-j176.4°]$$

Therefore,

$$i(t) = (2.632 \cdot \exp[-6250t] \cdot \cos[21\,470t + 16.59°] + 2.526 \cdot \sin[200 \times 10^3 t - 86.40°])u(t) \text{ mA}$$

See Figure 6.13 for a graph of $i(t)$.

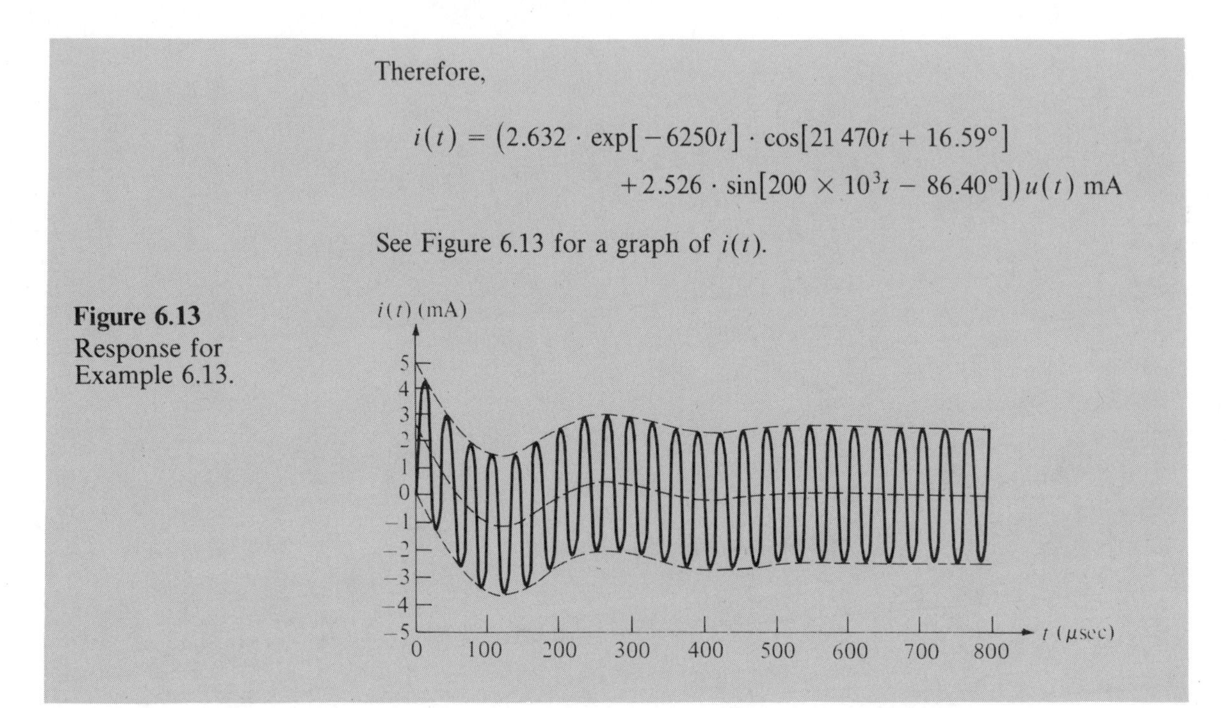

Figure 6.13 Response for Example 6.13.

Exercise 6.4

SPICE

Problem:

Find and graph the response current for the circuit shown in Figure 6.10 when $R = 50\ \Omega$ and the frequency of the excitation voltage is the same as the damped frequency of the circuit ($v_S(t) = 2 \cdot \sin[21\,470t]$ V).

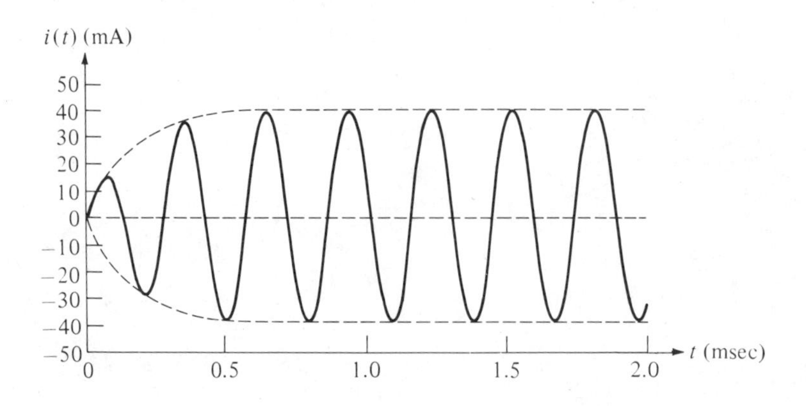

Figure 6.14 Response for Exercise 6.4.

Answer:

$$i(t) = 40(\exp[-6250t] \cdot \cos[21\,470t + 98.10°] + \cos[21\,470t - 81.90°])u(t) \text{ mA}$$

See Figure 6.14 for a graph of $i(t)$.

6.5 Response to Piecewise-Linear Periodic Excitation

The response of a R–L, R–C, or R–L–C circuit to piecewise-linear periodic excitation, such as a train of voltage pulses or a sawtooth current, is not truly periodic, although the response function *approaches* a periodic function after several time constants. We will refer to the time required for the response to become sufficiently periodic for practical purposes as the transient component of the response, even though there are transients occurring within each period of the response. Rather than attempt to categorize responses to piecewise-linear periodic excitation, we will instead offer examples which illustrate several possible types of response.

The circuit shown in Figure 6.15(a) with the excitation shown in Figure 6.15(b) will be used in Examples 6.14, 6.15, and 6.16; τ will be adjusted to provide a different response in each example. The waveform function which generates the first cycle of excitation is given by

$$v_1(t) = 5[u(t) - u(t - 0.5)]$$

so the periodic waveform function is

$$v(t) = \sum_{n=0}^{\infty} v_1(t - 1.5n)$$

$$= 5 \sum_{n=0}^{\infty} [u(t - 1.5n) - u(t - 0.5 - 1.5n)] \text{ V}$$

From Pair II and the time-domain shifting property,

$$V_1(s) = 5\left(\frac{1}{s} - \frac{\exp[-0.5s]}{s}\right) = \frac{5}{s}(1 - \exp[-0.5s]) \text{ V} \cdot \text{sec}$$

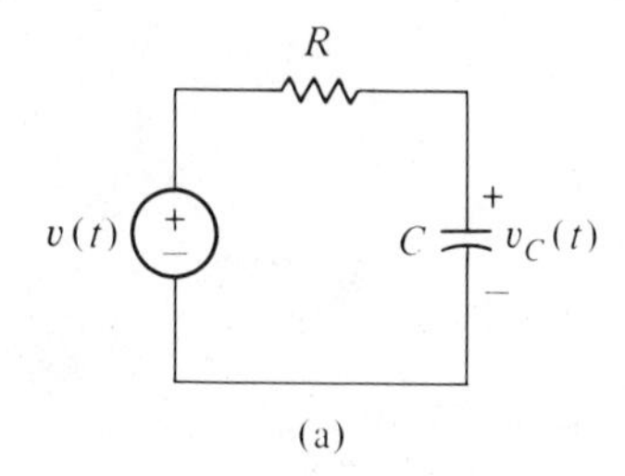

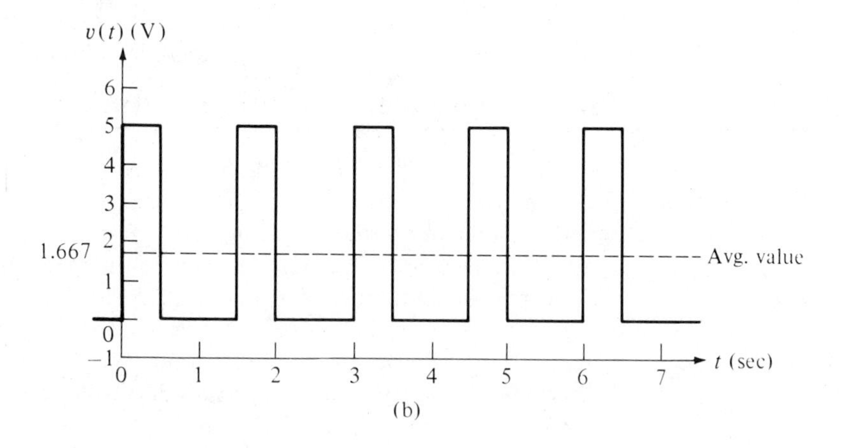

Figure 6.15
(a) Circuit and
(b) excitation for
Examples 6.14, 6.15,
and 6.16.

From the periodicity property,

$$V(s) = \frac{5}{s} \frac{1 - \exp[-0.5s]}{1 - \exp[-1.5s]} \text{ V} \cdot \text{sec}$$

By the voltage division rule,

$$V_C(s) = V(s) \cdot \frac{\dfrac{1}{Cs}}{R + \dfrac{1}{Cs}}$$

$$= \frac{\dfrac{5}{RC}}{s\left(s + \dfrac{1}{RC}\right)} \cdot \frac{1 - \exp[-0.5s]}{1 - \exp[-1.5s]} \text{ V} \cdot \text{sec}$$

By inverse Laplace transformation (Pair Vb, periodicity property, and time-domain shifting property),

$$v_C(t) = \sum_{n=0}^{\infty} v_{C1}(t - 1.5n)$$

$$= 5 \sum_{n=0}^{\infty} \left(\left\{ 1 - \exp\left[-\frac{t - 1.5n}{\tau} \right] \right\} u(t - 1.5n) - \left\{ 1 - \exp\left[-\frac{t - 0.5 - 1.5n}{\tau} \right] \right\} u(t - 0.5 - 1.5n) \right) \quad (6.16)$$

where $\tau = RC$.

Example 6.14

SPICE

Problem:

Find and graph the capacitor voltage $v_C(t)$ for the circuit shown in Figure 6.15 when $R = 10\ \text{k}\Omega$ and $C = 5\ \mu\text{F}$.

Solution:

1. Calculate τ:

$$\tau = RC = (10 \times 10^3)(5 \times 10^{-6}) = 0.05 \text{ sec}$$

Note that $1/\tau = 20/\text{sec}$.

2. From Equation (6.16),

$$v_C(t) = 5 \sum_{n=0}^{\infty} (\{1 - \exp[-20(t - 1.5n)]\}\, u(t - 1.5n) - \{1 - \exp[-20(t - 0.5 - 1.5n)]\}\, u(t - 0.5 - 1.5n)) \text{ V}$$

3. A graph of $v_C(t)$ can be seen in Figure 6.16. At each breakpoint, $v_C(t)$ is very close to steady state (5 V for breakpoints at $t = 0.5, 2.0, 3.5, \ldots$ sec, and zero for breakpoints at $t = 1.5, 3.0, 4.5, \ldots$ sec). Further, the steady-state, or final, value of the response voltage (zero in this case) for each period corresponds to the initial value of the excitation voltage for the next period. Whenever this last condition is present, the re-

sponse has no transient component, since the response appears periodic beginning with the first cycle.

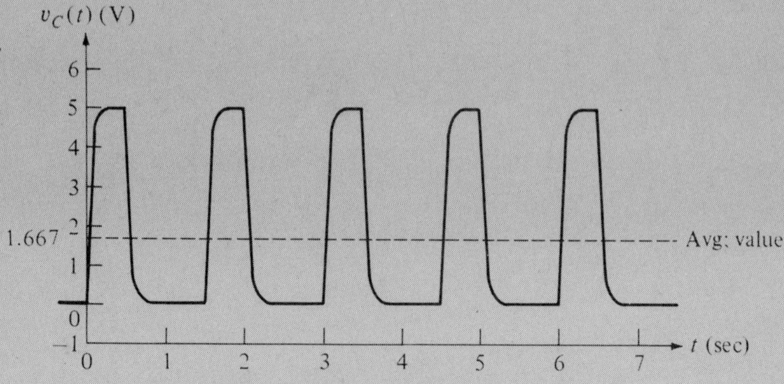

Figure 6.16 Response voltage for Example 6.14.

Example 6.15

SPICE

Problem:

Find and graph the capacitor voltage $v_C(t)$ for the circuit shown in Figure 6.15 when $R = 20\ \text{k}\Omega$ and $C = 10\ \mu\text{F}$.

Solution:

1. Calculate τ:

$$\tau = RC = (20 \times 10^3)(10 \times 10^{-6}) = 0.2 \text{ sec}$$

2. From Equation (6.16),

$$v_C(t) = 5 \cdot \sum_{n=0}^{\infty} (\{1 - \exp[-5(t - 1.5n)]\} u(t - 1.5n)$$

$$- \{1 - \exp[-5(t - 0.5 - 1.5n)]\} u(t - 0.5 - 1.5n)) \text{ V}$$

3. A graph of $v_C(t)$ can be seen in Figure 6.17. At the breakpoints at $t = 0.5, 2.0, 3.5, \ldots$ sec, $v_C(t)$ has not reached its steady-state value of 5 V; however, for breakpoints at $t = 1.5, 3.0, 4.5, \ldots$ sec, $v_C(t)$ is very close to its steady-state value of zero. As in Example 6.15, there is no

transient component, since the response appears to be periodic beginning with the first cycle.

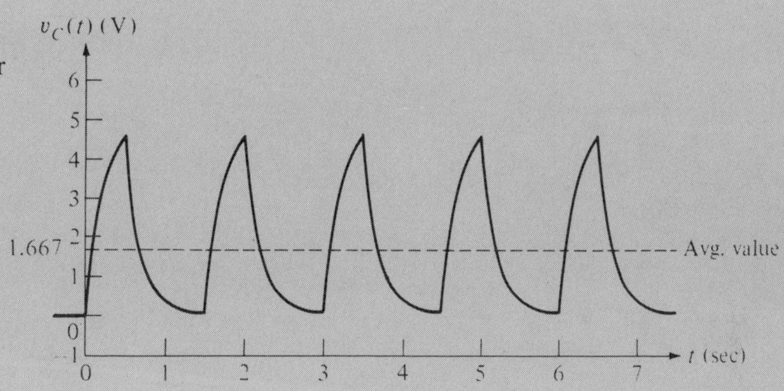

Figure 6.17 Response voltage for Example 6.15.

Example 6.16

SPICE

Problem:

Find and graph the capacitor voltage $v_C(t)$ for the circuit shown in Figure 6.15 when $R = 50$ kΩ and $C = 20$ μF.

Solution:

1. Calculate τ:

$$\tau = RC = (50 \times 10^3)(20 \times 10^{-6}) = 1 \text{ sec}$$

2. From Equation (6.16),

$$v_C(t) = 5 \sum_{n=0}^{\infty} (\{1 - \exp[-(t - 1.5n)]\} u(t - 1.5n) - \{1 - \exp[-(t - 0.5 - 1.5n)]\} u(t - 0.5 - 1.5n)) \text{ V}$$

3. A graph of $v_C(t)$ can be seen in Figure 6.18. The response is still changing at the breakpoints, and there is an obvious transient component in the response, during which the response is clearly not periodic. The transient component is only present for a few time constants, after which the response appears to become periodic. The response is then

said to be at steady state, even though the response within each period of excitation never reaches a steady-state dc level.

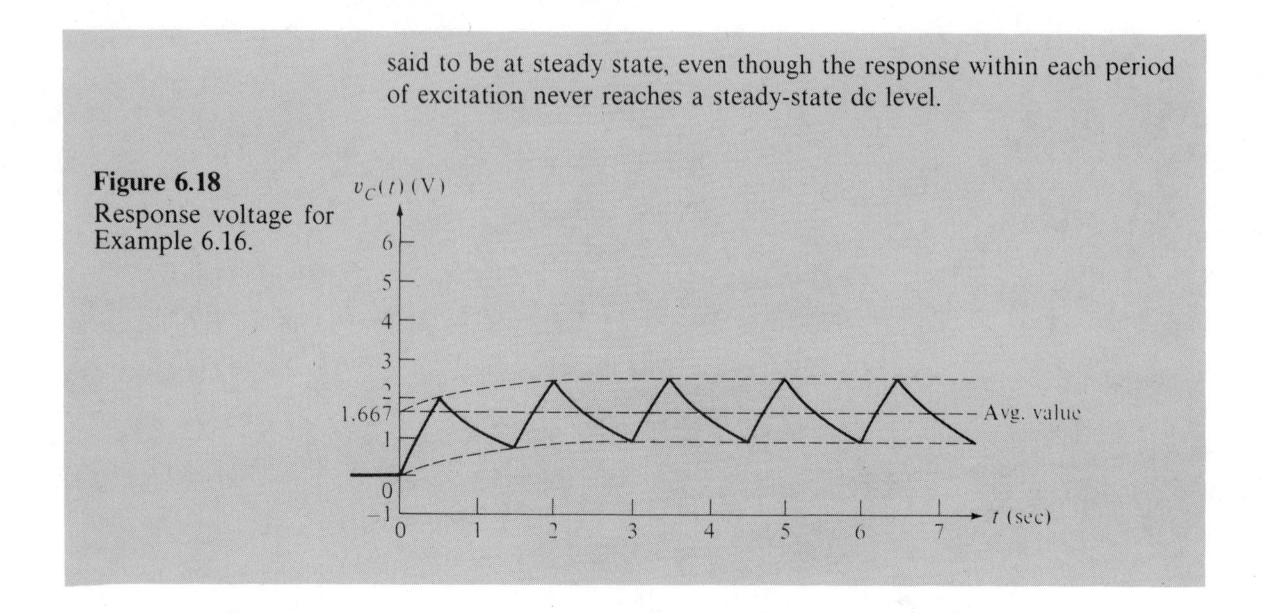

Figure 6.18
Response voltage for Example 6.16.

An analytical approach to the steady-state response of a simple R–C or R–L circuit to periodic rectangular excitation, as is the case in Example 6.14, 6.15, and 6.16, yields relatively simple formulas for predicting the maximum and minimum values (peaks) of the response. Figure 6.19 shows the steady-state response waveform superposed on the periodic rectangular excitation waveform. Equation (3.37) can be used to describe the excursion from F_{MIN} to

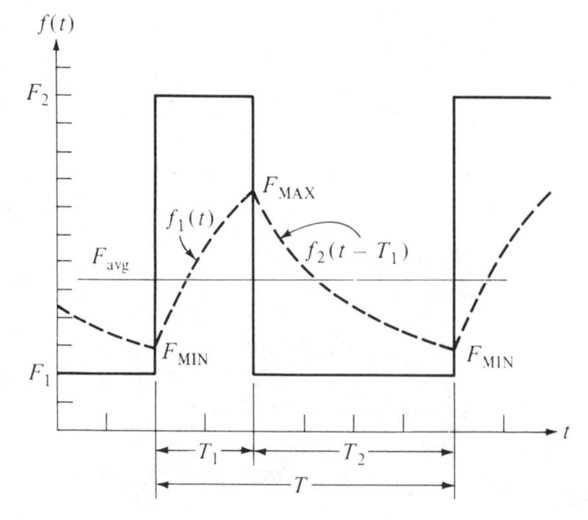

Figure 6.19
Peak-to-peak response to periodic rectangular excitation.

F_{MAX} and from F_{MAX} to F_{MIN},

$$f_1(t) = F_{MIN} + (F_2 - F_{MIN})\left(1 - \exp\left[-\frac{t}{\tau_C}\right]\right)$$

$$f_2(t - T_1) = F_{MAX} + (F_1 - F_{MAX})\left(1 - \exp\left[-\frac{t - T_1}{\tau_C}\right]\right)$$

Clearly, $f_1(T_1) = F_{MAX}$ and $f_2(T - T_1) = f_2(T_2) = F_{MIN}$, so the equations for $f_1(t)$ and $f_2(t - T_1)$ can be solved simultaneously (Problem 6-15) for F_{MIN} and F_{MAX},

$$F_{MIN} = \frac{F_1 + (F_2 - F_1)\exp[-T_2/\tau_C] - F_2 \cdot \exp[-T/\tau_C]}{1 - \exp[-T/\tau_C]} \tag{6.17a}$$

$$F_{MAX} = \frac{F_2 + (F_1 - F_2)\exp[-T_1/\tau_C] - F_1 \cdot \exp[-T/\tau_C]}{1 - \exp[-T/\tau_C]} \tag{6.17b}$$

where $T = T_1 + T_2$. These analytical results apply only to simple R–C or R–L circuits with rectangular periodic sources; there is no convenient analytical method for obtaining the peak values of the steady-state response in general R–L–C circuits with arbitrary periodic excitation.

Equations (6.17a) and (6.17b) may be used to evaluate the steady-state peaks in Examples 6.14, 6.15, and 6.16, with the following results.

Example 6.14:

$$V_{MIN} = \frac{0 + (5 - 0)\exp[-1/0.05] - 5 \cdot \exp[-1.5/0.05]}{1 - \exp[-1.5/0.05]} \simeq 0$$

$$V_{MAX} = \frac{5 + (0 - 5)\exp[-0.5/0.05] - 0 \cdot \exp[-1.5/0.05]}{1 - \exp[-1.5/0.05]} \simeq 5.000 \text{ V}$$

Example 6.15:

$$V_{MIN} = \frac{0 + (5 - 0)\exp[-1/0.2] - 5 \cdot \exp[-1.5/0.2]}{1 - \exp[-1.5/0.2]} \simeq 0.03092 \text{ V}$$

$$V_{MAX} = \frac{5 + (0 - 5)\exp[-0.5/0.2] - 0 \cdot \exp[-1.5/0.2]}{1 - \exp[-1.5/0.2]} \simeq 4.587 \text{ V}$$

Example 6.16:

$$V_{\text{MIN}} = \frac{0 + (5 - 0)\exp[-1] - 5 \cdot \exp[-1.5]}{1 - \exp[-1.5]} \simeq 0.9316 \text{ V}$$

$$V_{\text{MAX}} = \frac{5 + (0 - 5)\exp[-0.5] - 0 \cdot \exp[-1.5]}{1 - \exp[-1.5]} \simeq 2.532 \text{ V}$$

A comparison of the responses for Example 6.14, 6.15 and 6.16 reveals that the peak-to-peak steady-state response, that is, the difference between its minimum and maximum values, depends upon the time constant. In Example 6.14, the response has the full range of the source (0 to 5 V) because there is time to charge and discharge the capacitor almost completely between breakpoints in the excitation. In Example 6.15, there is insufficient time to charge to 5 V, but sufficient time to discharge to near zero, so the steady-state response ranges between approximately zero and 4.587 V. In Example 6.16, there is neither sufficient time to charge fully nor to discharge fully, and the response ranges from 0.9316 to 2.532 V. The range of the response has no bearing on the *average* value of the steady-state response, which is the same (1.667 V) in each of the examples cited.

As a final example of the response to piecewise-linear excitation, consider the current through an inductor excited by a sawtooth voltage source (Example 6.17).

Example 6.17

SPICE

Problem:

Determine the response current for the circuit shown in Figure 6.20(a).

Solution:

1. Write the generating function for the voltage waveform shown in Figure 6.20(b):

$$v(t) = 20\,000t - \sum_{n=1}^{\infty} 50 \cdot u(t - 0.025n) \text{ V}$$

2. Determine the Laplace transform of the waveform generating function:

$$V(s) = \frac{20\,000}{s^2} - \sum_{n=1}^{\infty} \frac{50 \cdot \exp[-(0.0025n)s]}{s}$$

The periodicity property will not be invoked here, because the lower index of the summation is one rather than zero. Adjusting the index would add another term to the Laplace transform and introduce an unnecessary complication.

Figure 6.20
Example 6.17.

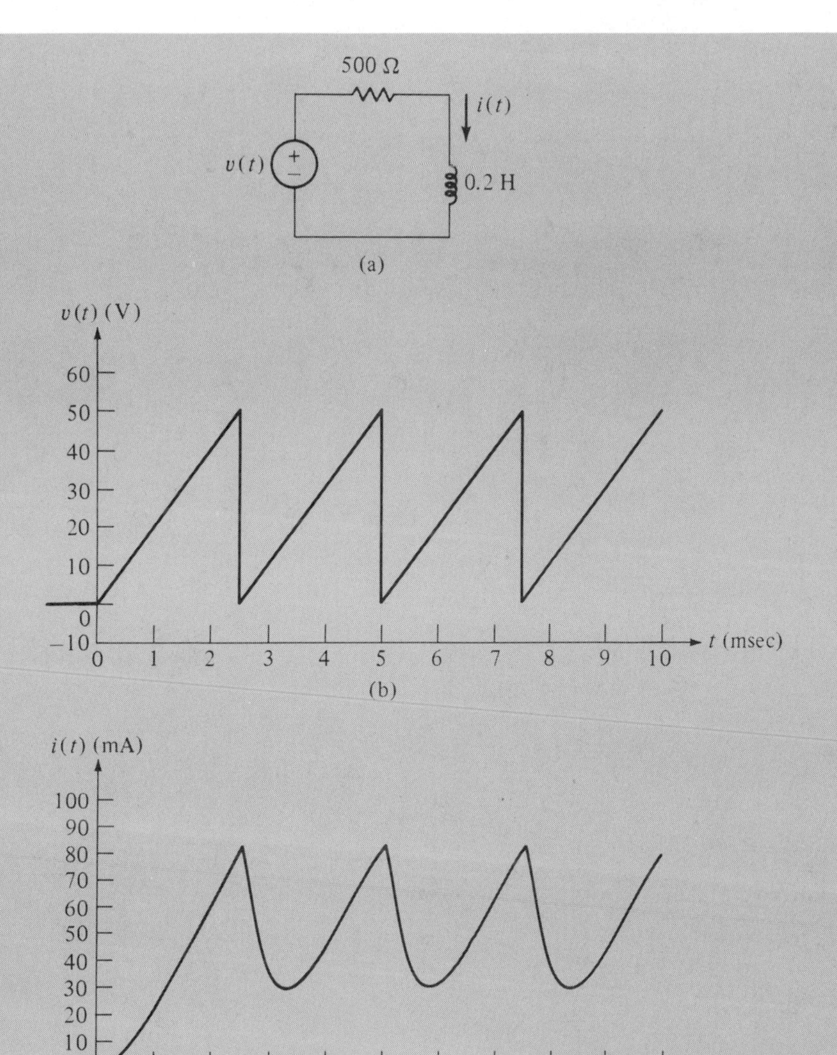

3. Compute $I(s)$:

$$I(s) = \frac{V(s)}{R + Ls} = \frac{5}{s + 2500} \cdot V(s)$$

$$= \frac{100 \times 10^3}{s^2(s + 2500)} - \sum_{n=1}^{\infty} \frac{250 \cdot \exp[-(0.0025)s]}{s(s + 2500)} \text{ A} \cdot \text{sec}$$

4. Perform a partial fraction expansion of the first term of $I(s)$:

$$\frac{100 \times 10^3}{s^2(s + 2500)} = \frac{K_1}{s^2} + \frac{K_2}{s} + \frac{K_3}{s + 2500}$$

$$K_1 = \left.\frac{100 \times 10^3}{s + 2500}\right|_{s=0} = 40$$

$$K_2 = \left.\frac{d}{ds}\left(\frac{100 \times 10^3}{s + 2500}\right)\right|_{s=0} = -\left.\frac{100 \times 10^3}{(s + 2500)^2}\right|_{s=0} = -0.016$$

$$K_3 = \left.\frac{100 \times 10^3}{s^2}\right|_{s=-2500} = 0.016$$

Therefore,

$$I(s) = \frac{40}{s^2} - \frac{0.016}{s} + \frac{0.016}{s + 2500} + \sum_{n=1}^{\infty} \frac{250 \cdot \exp[-(0.0025n)s]}{s(s + 2500)} \text{ A} \cdot \text{sec}$$

5. Determine the inverse Laplace transform of $I(s)$. By Pairs IIIa, II, IV, Vb and the linearity properties,

$$i(t) = \{40 \times 10^3 t - 16(1 - \exp[-2500t])\}\, u(t) - \sum_{n=1}^{\infty} 100\{1 - \exp[-2500(t - 0.0025n)]\}\, u(t - 0.0025n) \text{ mA}$$

A graph of $i(t)$ may be seen in Figure 6.20(c).

Exercise 6.5

SPICE

Problem:

Graph the response voltage of the circuit shown in Figure 6.21(a) for the excitation current shown in Figure 6.21(b).

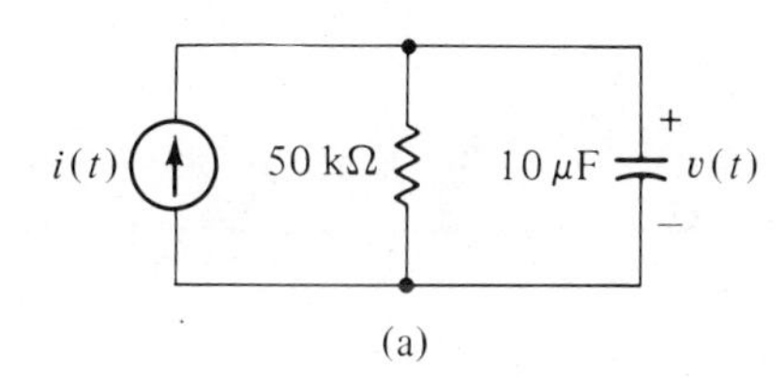

(a)

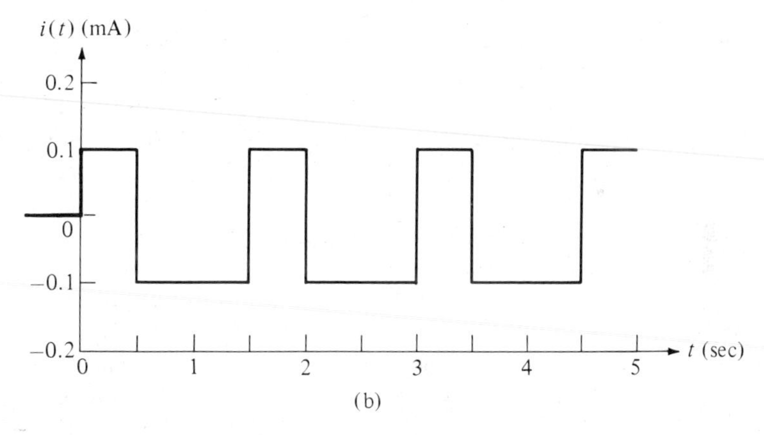

(b)

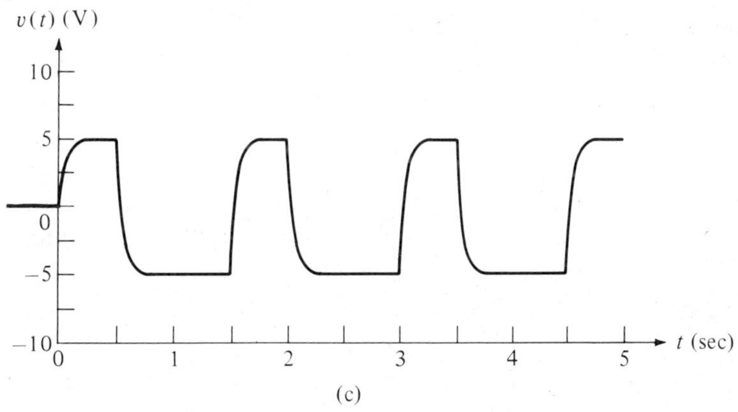

(c)

Figure 6.21
Exercise 6.5.

Answer:

$$i(t) = \sum_{n=0}^{\infty} u(t - 1.5n) - 2 \cdot u(t - 0.5 - 1.5n) + u(t - 1.5 - 1.5n) \text{ mA}$$

$$\begin{aligned} v(t) = 5 \cdot \sum_{n=0}^{\infty} (&\{1 - \exp[-20(t - 1.5n)]\} u(t - 1.5n) \\ &-2 \cdot \{1 - \exp[-20(t - 0.5 - 1.5n)]\} u(t - 0.5 - 1.5n) \\ &+ \{1 - \exp[-20(t - 1.5 - 1.5n)]\} u(t - 1.5 - 1.5n)) \text{ V} \end{aligned}$$

The graph of $v(t)$ is shown in Figure 6.21(c).

6.6 Computer Simulations

As was the case in Chapter 5, most of the problems in this chapter can be simulated with SPICE. SPICE is particularly useful in simulating circuits with periodic excitation, because time-domain or Laplace transform analysis is so tedious.

SPICE 6.1 illustrates transient analysis of a circuit with sinusoidal excitation (Example 6.8). The reader should not confuse this type of analysis with ac analysis—or *steady-state* sinusoidal analysis—which will be illustrated in a later chapter. The time-delay (DELAY) parameter permits shifting of the sinusoid so that sources with phase displacement can be simulated.[4] For example, the source in Exercise 6.3 can be simulated by

SIN(0 5 15.92K 15.71U)

Note that 1000 rad/sec must be converted to 15.92 kHz, and the displacement angle, which is 90°, or one-fourth of a period, in this case must be converted to a time delay of 15.71 μsec.

SPICE 6.2 illustrates the use of the piecewise-linear (PWL) source to analyze the circuit of Example 6.17.

[4]Some microcomputer versions of SPICE do not have this feature.

```
*******10/27/88 ********  SPICE 2G.6    3/15/83 ********16:31:18*****
SPICE 6.1
****     INPUT LISTING                    TEMPERATURE =   27.000 DEG C
*****************************************************************************
 R 1 2 50
 C 2 0 .05U
 V 1 0 SIN(0 30 159.2K)
 .TRAN 1U 20U
 .PLOT TRAN V(2)
 .END
*******10/27/88 ********  SPICE 2G.6    3/15/83 ********16:31:18*****
SPICE 6.1
****     INITIAL TRANSIENT SOLUTION       TEMPERATURE =   27.000 DEG C
*****************************************************************************
  NODE   VOLTAGE     NODE   VOLTAGE
 (  1)     .0000    (  2)     .0000
     VOLTAGE SOURCE CURRENTS
     NAME        CURRENT
     V         0.000D+00
     TOTAL POWER DISSIPATION   0.00D+00  WATTS
*******10/27/88 ********  SPICE 2G.6    3/15/83 ********16:31:18*****
SPICE 6.1
****     TRANSIENT ANALYSIS               TEMPERATURE =   27.000 DEG C
*****************************************************************************

TIME        V(2)
                   -2.000D+01    -1.000D+01      0.000D+00      1.000D+01  2.000D+01
                   - - - - - - - - - - - - - - - - - - - - - - - - - - - - - - - -
 0.000D+00  0.000D+00 .              .              *              .              .
 1.000D-06  4.815D+00 .              .              .      *       .              .
 2.000D-06  1.243D+01 .              .              .              .   *          .
 3.000D-06  1.365D+01 .              .              .              .     *        .
 4.000D-06  5.652D+00 .              .              .       *      .              .
 5.000D-06 -5.316D+00 .              .      *       .              .              .
 6.000D-06 -9.860D+00 .              *              .              .              .
 7.000D-06 -4.343D+00 .              .       *      .              .              .
 8.000D-06  5.832D+00 .              .              .       *      .              .
 9.000D-06  1.112D+01 .              .              .              . *            .
 1.000D-05  6.466D+00 .              .              .        *     .              .
 1.100D-05 -3.945D+00 .              .       *      .              .              .
 1.200D-05 -1.056D+01 .             *.              .              .              .
 1.300D-05 -7.395D+00 .              .   *          .              .              .
 1.400D-05  2.631D+00 .              .              .   *          .              .
 1.500D-05  1.030D+01 .              .              .              *              .
 1.600D-05  8.503D+00 .              .              .           *  .              .
 1.700D-05 -1.103D+00 .              .            * .              .              .
 1.800D-05 -9.657D+00 .              *              .              .              .
 1.900D-05 -9.346D+00 .              .*             .              .              .
 2.000D-05 -4.103D-01 .              .             *.              .              .
                   - - - - - - - - - - - - - - - - - - - - - - - - - - - - - - - -

                JOB CONCLUDED
```

```
*******10/27/88 ********  SPICE 2G.6    3/15/83 ********16:45:40*****
SPICE 6.2
****     INPUT LISTING                        TEMPERATURE =   27.000 DEG C
******************************************************************************
 R 1 2 500
 L 3 0 .2
 V 1 0 PWL(0 0 2.5M 50 2.501M 0 5M 50 5.001M 0 7.5M 50 7.501M 0 10M 50)
 VAM 2 3
 .TRAN .5M 10M
 .PLOT TRAN I(VAM)
 .END
*******10/27/88 ********  SPICE 2G.6    3/15/83 ********16:45:41*****
SPICE 6.2
****     INITIAL TRANSIENT SOLUTION           TEMPERATURE =   27.000 DEG C
******************************************************************************
  NODE   VOLTAGE     NODE   VOLTAGE     NODE   VOLTAGE
 (  1)     .0000    (  2)     .0000    (  3)     .0000
     VOLTAGE SOURCE CURRENTS
     NAME       CURRENT
     V         0.000D+00
     VAM       0.000D+00
     TOTAL POWER DISSIPATION   0.00D+00  WATTS
*******10/27/88 ********  SPICE 2G.6    3/15/83 ********16:45:41*****
SPICE 6.2
****     TRANSIENT ANALYSIS                   TEMPERATURE =   27.000 DEG C
******************************************************************************

     TIME       I(VAM)
                  -5.000D-02      0.000D+00      5.000D-02      1.000D-01  1.500D-01
                         - - - - - - - - - - - - - - - - - - - - - - - - - - - - - -
 0.000D+00  0.000D+00 .             *             .             .             .
 5.000D-04  8.627D-03 .             . *           .             .             .
 1.000D-03  2.529D-02 .             .     *       .             .             .
 1.500D-03  4.436D-02 .             .          *  .             .             .
 2.000D-03  6.410D-02 .             .             .   *         .             .
 2.500D-03  8.403D-02 .             .             .         *   .             .
 3.000D-03  3.218D-02 .             .       *     .             .             .
 3.500D-03  3.210D-02 .             .       *     .             .             .
 4.000D-03  4.617D-02 .             .           * .             .             .
 4.500D-03  6.462D-02 .             .             .   *         .             .
 5.000D-03  8.416D-02 .             .             .         *   .             .
 5.500D-03  3.222D-02 .             .       *     .             .             .
 6.000D-03  3.211D-02 .             .       *     .             .             .
 6.500D-03  4.618D-02 .             .           * .             .             .
 7.000D-03  6.462D-02 .             .             .   *         .             .
 7.500D-03  8.416D-02 .             .             .         *   .             .
 8.000D-03  3.222D-02 .             .       *     .             .             .
 8.500D-03  3.211D-02 .             .       *     .             .             .
 9.000D-03  4.618D-02 .             .           * .             .             .
 9.500D-03  6.462D-02 .             .             .   *         .             .
 1.000D-02  8.416D-02 .             .             .         *   .             .
                         - - - - - - - - - - - - - - - - - - - - - - - - - - - - - -

                  JOB CONCLUDED
```

6.7 SUMMARY

In this chapter we have utilized transform circuit analysis techniques to take a more detailed look at the transient response of simple circuits to commonly encountered types of excitation. In all cases the analysis was based solely on the Kirchhoff laws and the concepts of Laplace impedance and admittance.

The analysis presented in Section 6.2 applies directly to a wide range of circuit analysis and design problems. Understanding the response of a second-order circuit to step-function excitation will serve as a basis for understanding the response of more complex circuits to other excitation.

The initial and final value theorems introduced in Section 6.3 permit evaluation of initial and steady-state response, respectively, in many circuits without carrying out the entire transform circuit analysis process.

Section 6.4 served to introduce concepts involved with analysis of circuit response to a periodic excitation by pointing out the presence of transients in circuits which include switched sinusoidal (ac) sources. In Section 6.4 and 6.5, examples were used to illustrate and discuss the response of circuits to periodic sources, including sinusoidal and piecewise-linear sources.

6.8 Terms

ac (or phasor) analysis
complete response
critical value
critically damped
dc analysis
envelope
final value theorem
first-order circuit
forced response
initial response
initial value theorem
natural response
overdamped
overshoots
second-order circuit
settling time
steady-state response
transient response
underdamped (or oscillatory)

Problems

SPICE

6-1. **a.** Determine $I(s)$ in terms of R.
b. Determine $i(t)$ for $t > 0$ when $R = 0$, $R = 300\ \Omega$, and $R = 3000\ \Omega$. Graph $i(t)$ in each case.

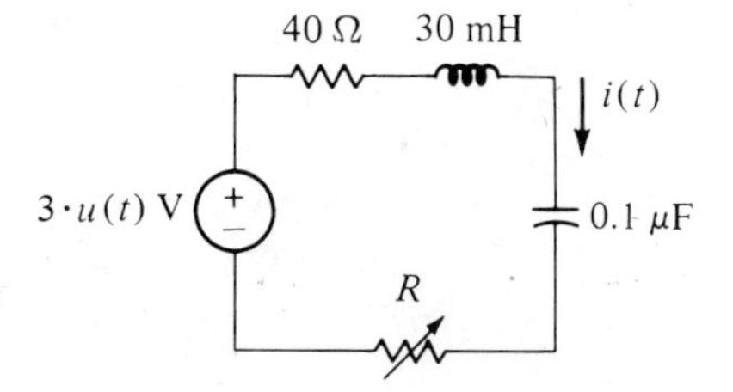

SPICE

*6-2. Determine and graph $v(t)$.

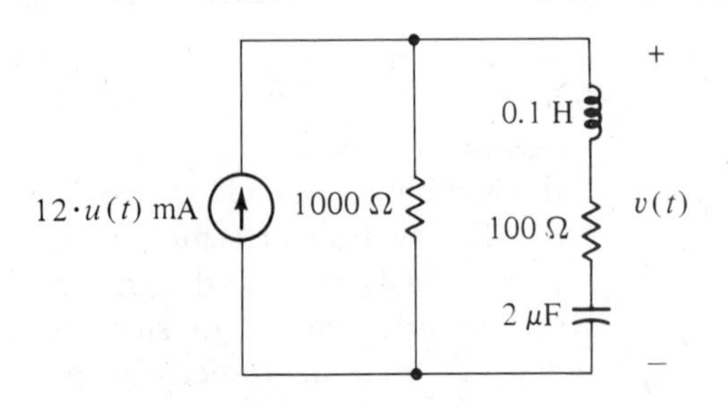

SPICE

*6-3. Determine and graph $i(t)$.

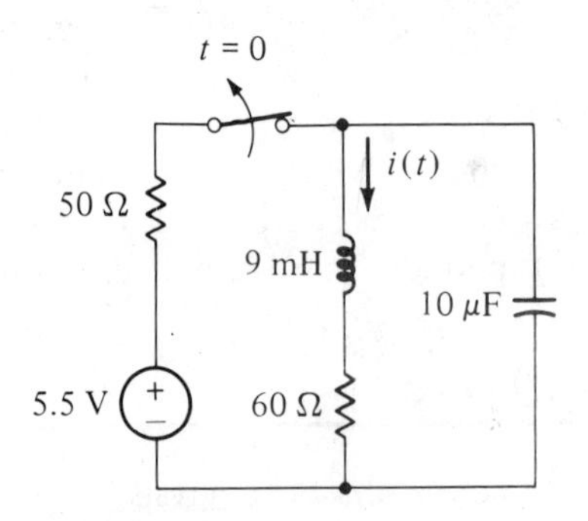

SPICE

*6-4. Determine and graph $v(t)$.

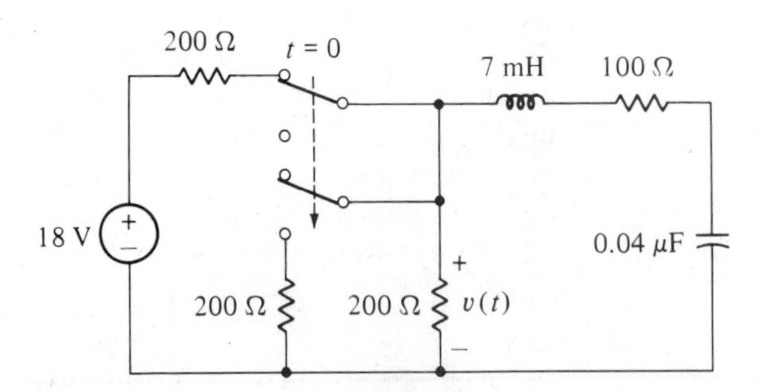

SPICE

*6-5. Determine $v(t)$ for the circuit shown in Figure 6.5 (Example 6.4) if the 20-Ω internal resistance of the inductor is neglected.

6-6. a. Assume that the switch has been closed for a long time and is opened at $t = 0$. Write the expression for $V_C(s)$.

*b. Determine $v_C(t)$ for $t > 0$.

c. Calculate the maximum value of $v_C(t)$, and use the result to calculate the maximum voltage across the switch contacts. What important consideration in illustrated here regarding the use of switches in R–L–C circuits?

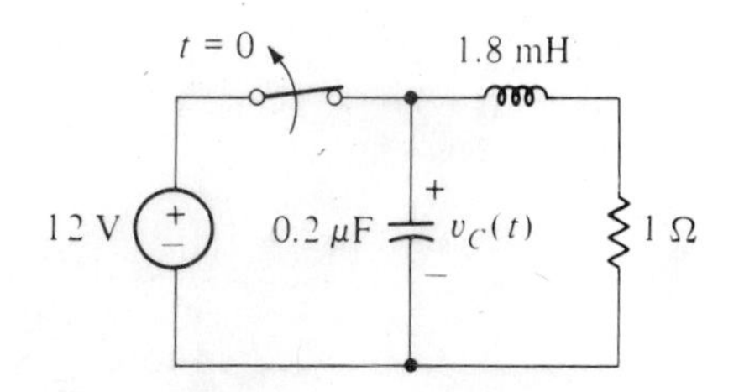

6-7. Use the initial and final value theorems to find the initial and final values, respectively, of voltage or current.

*a. $I(s) = \dfrac{0.1s + 1250}{s^2 + 12\,500s + 25 \times 10^6}$ A · sec.

b. $V(s) = \dfrac{6s^2 + 120 \times 10^3 s + 12 \times 10^9}{s(s^2 + 16\,000s + 2 \times 10^9)}$ V · sec.

*c. $V(s) = \dfrac{8s^2 + 2400}{s(s^2 + 60s + 200)}$ V · sec.

d. $I(s) = \dfrac{30s^2 + 200s + 16\,000}{s^2(s^2 + 30s + 800)}$ A · sec.

SPICE

***6-8.** Determine and graph $i(t)$.

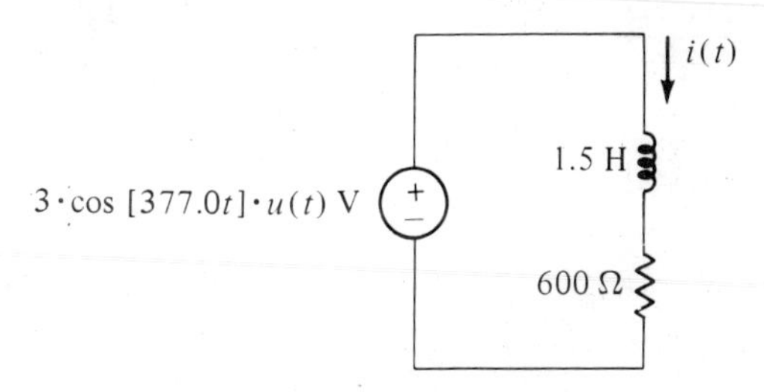

SPICE

6-9. *a. Write the expression for $I(s)$ in terms of R.

b. Determine $i(t)$ for $t > 0$ when $R = 250\ \Omega$, $R = 2.5\ \text{k}\Omega$, and $R = 25\ \text{k}\Omega$. Graph $i(t)$ in each case.

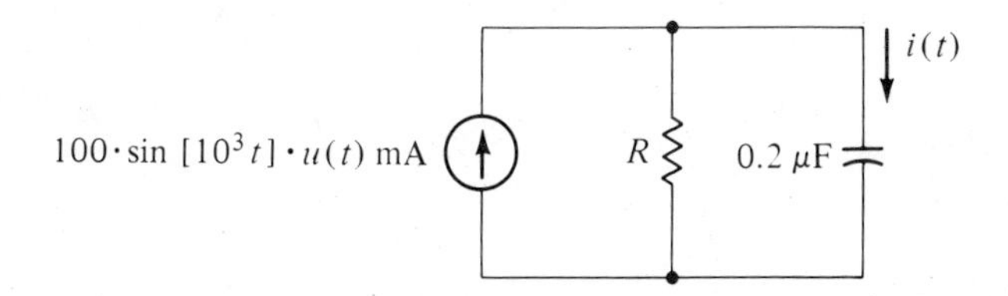

SPICE

***6-10.** Determine and graph $v(t)$. Hint: Show by use of a trignometric identity and Pairs VII and VIII that

$$\mathscr{L}\{\cos[\omega t + \theta] \cdot u(t)\} = \frac{s \cdot \cos\theta - \theta \cdot \sin\theta}{s^2 + \omega^2}$$

500 Ω
1200 pF
$8 \cdot \cos[10^6 t + 60°] \cdot u(t)$ V
$v(t)$
500 Ω

***6-11.** Write the expression for $V_R(s)$ for a series R–C circuit which consists of a 10-kΩ resistor, a 0.02-μF capacitor, and a periodic pulsed voltage source. The pulse amplitude is V_A for an interval T_1 and V_B for an interval T_2 (the period of the waveform is $T_1 + T_2$).

SPICE **6-12.** Use the results of Problem 6-11 to determine and graph $v_R(t)$ for each of the following cases:

V_A (V)	V_B (V)	T_1 (msec)	T_2 (msec)
15	0	1.2	1.2
*15	0	0.3	1.2
15	0	0.3	0.3
*15	5	0.3	1.2

SPICE ***6-13.** Determine and graph $v_R(t)$.

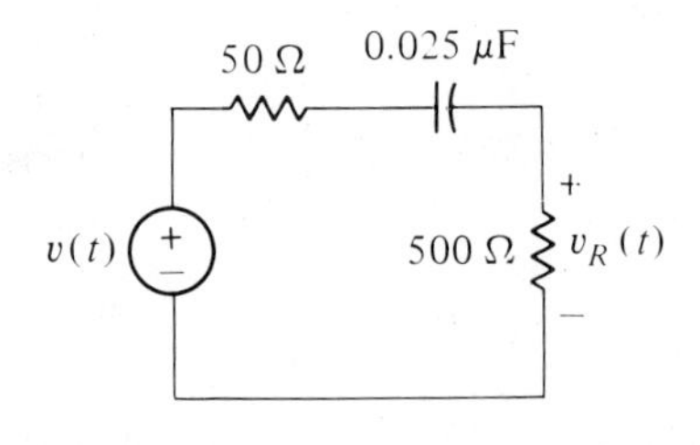

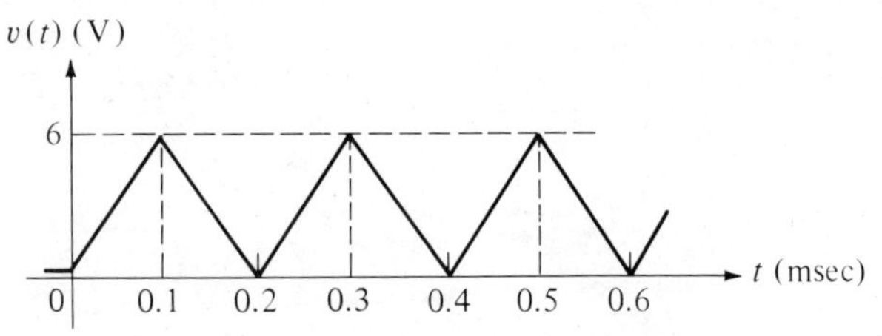

6-14. Verify Equation (6.15) for the circuit of Example 6.4 for arbitrary excitation and source resistance.

6-15. Verify Equations (6.17a) and (6.17b) by simultaneous solution of the equations for $f_1(T_1)$ and $f_2(T_2)$.

7 Transform Circuit Analysis Theorems and Techniques

7.1 INTRODUCTION

We stated in the introduction to Chapter 5 that one advantage to Laplace transformation of a circuit is that it allows us to apply principles and theorems from dc analysis to circuits containing energy storage elements and time-varying sources as well as resistors and dc sources. In Chapters 5 and 6, we applied the Kirchhoff laws to transform circuit analysis and began to use the immediate consequences—element combination and voltage and current division—in simple circuits. In this chapter we will consider more complex circuits and apply source conversion, node analysis, mesh analysis, superposition, and Thevenin's and Norton's theorems to transform circuit analysis.

7.2 Source Conversion

A **practical voltage source** consists of an ideal voltage source in *series* with a network of linear passive elements (resistors, capacitors, and inductors) [Figure 7.1(a)], and a **practical current source** consists of an ideal current source in *parallel* with a network of linear passive elements [Figure 7.1(b)]. **Source conversion** is a technique for converting a practical voltage source to an equivalent practical current source, and vice versa. The reader should review the technique in Chapter 1 for conversion of practical dc sources.

In general, ideal sources may have any time dependence, and passive linear networks may consist of any combination of resistance, capacitance, and inductance. If the Laplace transform exists for the ideal source, Laplace transform methods can be used to obtain the equivalent practical source. Consider a practical voltage source in which $V_S(s)$ is the Laplace transform of the ideal voltage source, and $Z_S(s)$ is the Laplace transform impedance of the

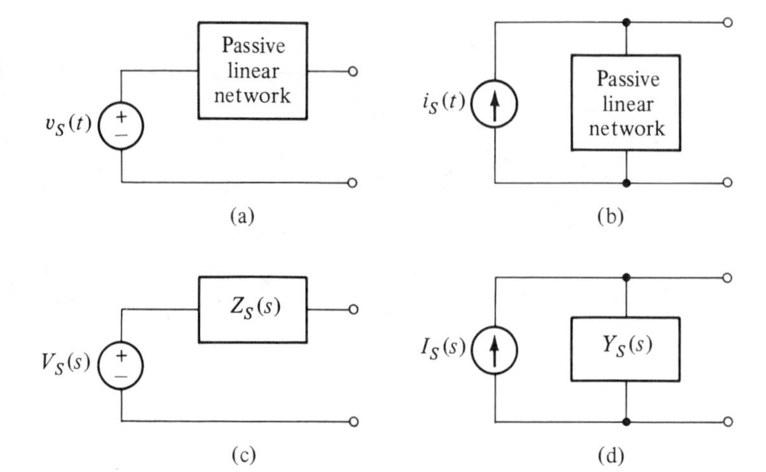

Figure 7.1
Practical sources and source conversion.
(a) Practical voltage source.
(b) Practical current source.
(c) Transformed practical voltage source.
(d) Transformed practical current source.

series linear passive network [Figure 7.1(c)]. The Laplace transform of the equivalent practical current source consists of an ideal current source $I_S(s)$ in parallel with a passive linear network whose admittance is $Y_S(s)$ (Figure 7.1(d)), where

$$I_S(s) = \frac{V_S(s)}{Z_S(s)} \quad \text{and} \quad Y_S(s) = \frac{1}{Z_S(s)}$$

Similarly, the practical voltage source equivalent to a practical current source with $I_S(s)$ in parallel with $Y_S(s)$ consists of $V_S(s)$ in series with $Z_S(s)$, where

$$V_S(s) = \frac{I_S(s)}{Y_S(s)} \quad \text{and} \quad Z_S(s) = \frac{1}{Y_S(s)}$$

Example 7.1

Problem:

Find the equivalent practical current source for a ramp generator with a 50-Ω internal resistance and a 10-μF blocking capacitor [Figure 7.2(a)]. The ramp slope is 50 kV/sec.

Solution:

1. Determine $v_S(t)$:

$$v_S(t) = 50 \times 10^3 t \cdot u(t) \text{ V}$$

2. Use Pair IIIa to determine $V_S(s)$:

$$V_S(s) = 50 \times 10^3/s^2 \text{ V} \cdot \text{sec}$$

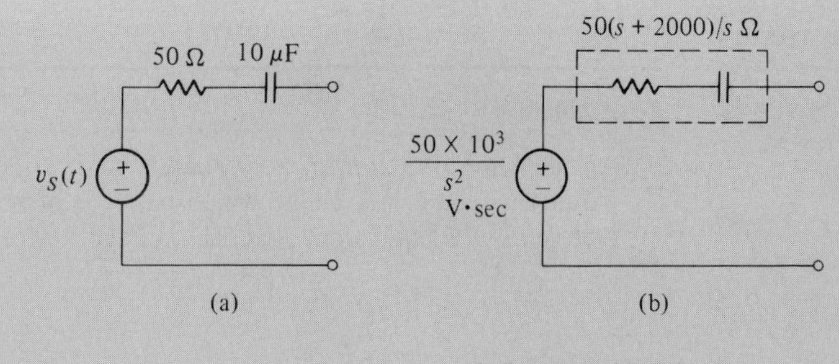

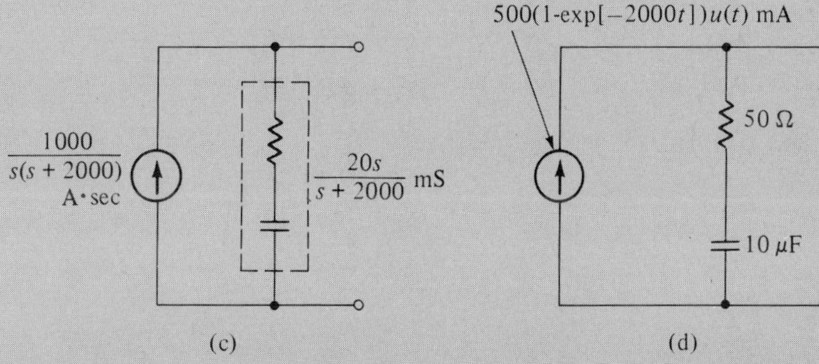

Figure 7.2
Example 7.1.
(a) Practical voltage source.
(b) Transformed practical voltage source.
(c) Equivalent transformed current source.
(d) Equivalent practical current source.

2. Determine $Z_S(s)$:

$$Z_S(s) = 50 + \frac{0.1 \times 10^6}{s} = \frac{50(s + 2000)}{s}\ \Omega$$

The transformed practical voltage source is known in Figure 7.2(b).

3. Calculate $I_S(s)$ for the equivalent current source Figure 7.2(c):

$$I_S(s) = \frac{V_S(s)}{Z_S(s)} = \frac{50 \times 10^3/s^2}{50(s + 200)/s} = \frac{10^6}{s(s + 2000)}\ \text{mA} \cdot \text{sec}$$

$$Y_S(s) = \frac{1}{Z_S(s)} = \frac{20s}{s + 2000}\ \text{mS}$$

4. Use Pair Vb to determine the inverse Laplace transformation of $I_S(s)$:

$$i_S(t) = 500(1 - \exp[-2000t])u(t)\ \text{mA}$$

The equivalent practical current source is shown in Figure 7.2(d). Notice that the passive linear network in parallel with the ideal current source is the same network that was in series with the ideal voltage source.

Exercise 7.1

Problem:

An ideal step-function current source, $i_S(t) = 50 \cdot u(t)$ mA, is connected in parallel with a 2-mH choke which has 6 Ω of wire resistance [Figure 7.3(a)]. Determine the equivalent practical voltage source.

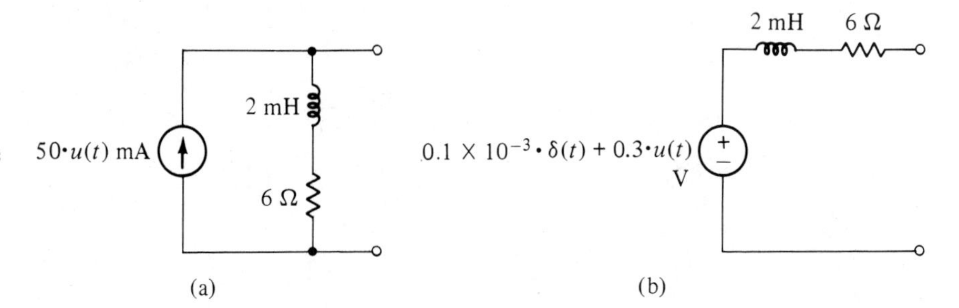

Figure 7.3
Exercise 7.1.
(a) Practical current source.
(b) Equivalent voltage source.

Answer:

$$I_S(s) = 0.050/s \text{ A} \cdot \text{sec}$$

$$Y_S(s) = \frac{500}{s + 3000} \text{ S}$$

$$V_S(s) = 0.1 \times 10^{-3}\left(1 + \frac{3000}{s}\right) \text{V} \cdot \text{sec}$$

$$v_S(t) = 0.1 \times 10^{-3} \cdot \delta(t) + 0.3 \cdot u(t) \text{ V}$$

See Figure 7.3(b).

7.3 Node Analysis

The reader will recall from dc analysis that **node analysis** is a technique involving systematic application of Kirchhoff's current law (KCL) at each node of a circuit. When the node currents are expressed in terms of voltage and conductance between nodes, the result is a system of linear algebraic equations which may be solved to obtain the node voltages. Since we have already established that the Kirchhoff laws apply to transform circuit analysis, and their application results in algebraic equations, node analysis is an obvious extension of the transform circuit analysis techniques.

Node analysis of a transformed circuit is performed in essentially the same way as node analysis of a dc circuit:

1. Identify and label every node, and select one node as reference (ground or common).
2. Convert any practical voltage sources to practical current sources. If ideal voltage sources are present between nodes, it will be necessary to write an auxiliary equation.
3. Determine the Laplace transform admittance $Y(s)$ between adjacent nodes.
4. Apply KCL at each node, other than the reference node, and express the node currents in terms of node voltages and admittances.
5. Solve the resulting system of algebraic equations for the node voltages.

Example 7.2

SPICE

Problem:

Use node analysis to determine the voltage at node A of the circuit shown in Figure 7.4(a).

Solution:

1. Obtain the transformed equivalent circuit [Figure 7.4(b)]. Note that all sources are current sources, and the admittance has been determined for each passive network.
2. Apply KCL at node A and at node B:[1]

$$I_1 + I_2 = I_{S1}$$
$$-I_2 + I_3 = -I_{S2}$$

$$Y_1 V_A + Y_2(V_A - V_B) = I_{S1}$$
$$Y_2(V_B - V_A) + Y_3 V_B = -I_{S2}$$

$$(Y_1 + Y_2)V_A - Y_2 V_B = I_{S1}$$
$$-Y_2 V_A + (Y_2 + Y_3)V_B = -I_{S2}$$

[1] When performing node and mesh analysis, it is usually more convenient to work with symbols until ready to obtain a solution. Although the author prefers to emphasize the s-dependence of the transformed quantities through the use of functional notation [e.g., $Y(s)$, $I(s)$, etc.], a concession will be made to brevity, and the use of functional notation will be suspended when writing systems of node and mesh equations.

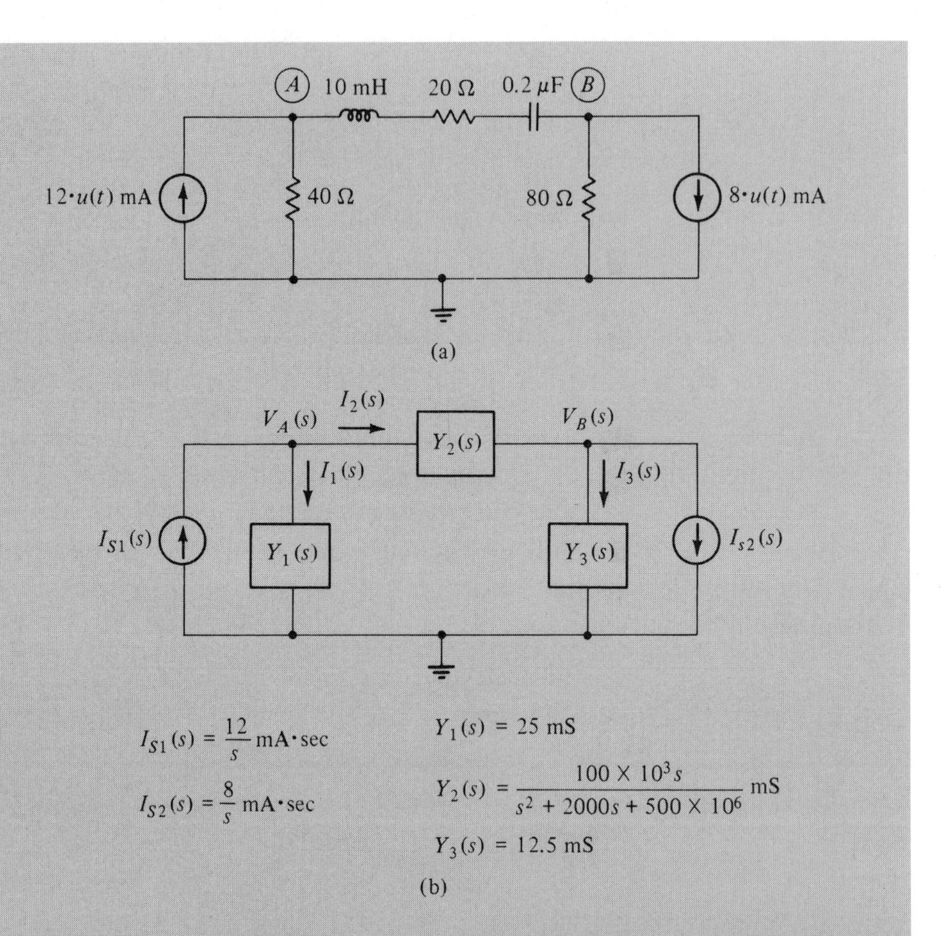

Figure 7.4
Example 7.2.

3. Solve the equations obtained in step 2 for $V_A(s)$. From the node B equation,

$$V_B = -\frac{I_{S2}}{Y_2 + Y_3} + \frac{Y_2 V_A}{Y_2 + Y_3}$$

4. Substitute $V_B(s)$ into the node A equation and solve for $V_A(s)$:

$$(Y_1 + Y_2)V_A + \frac{Y_2 I_{S2}}{Y_2 + Y_3} - \frac{Y_2^2 V_A}{Y_2 + Y_3} = I_{S1}$$

$$V_A(s) = \frac{Y_2(I_{S1} - I_{S2}) + Y_3 I_{S1}}{Y_1 Y_2 + Y_1 Y_3 + Y_2 Y_3} \text{ V} \cdot \text{sec}$$

5. Calculate $V_A(s)$:

$$Y_2(I_{S1} - I_{S2}) + Y_3 I_{S1} = \frac{100 \times 10^3 s(12/s - 8/s)}{s^2 + 2000s + 500 \times 10^6} + 12.5\frac{12}{s}$$

$$= \frac{150(s^2 + 4667s + 500 \times 10^6)}{s(s^2 + 2000s + 500 \times 10^6)} \text{ V} \cdot \text{sec}$$

$$Y_1Y_2 + Y_1Y_3 + Y_2Y_3 = \frac{(25)(100 \times 10^3 s)}{s^2 + 2000s + 500 \times 10^6} + (25)(12.5) + \frac{(12.5)(100 \times 10^3 s)}{s^2 + 2000s + 500 \times 10^6}$$

$$= \frac{312.5(s^2 + 14 \times 10^3 s + 500 \times 10^6)}{s^2 + 2000s + 500 \times 10^6} \text{ S}^2$$

$$V_A(s) = \frac{0.48(s^2 + 4667s + 500 \times 10^6)}{s(s^2 + 14 \times 10^3 s + 500 \times 10^6)} \text{ V} \cdot \text{sec}$$

6. Determine $v(t)$ by inverse Laplace transformation:

$$V_A(s) = \frac{K_1}{s} + \frac{\mathbf{K}_2}{s + 7000 - j21.24 \times 10^3} + \frac{\mathbf{K}_2^*}{s + 7000 + j21.24 \times 10^3}$$

$$v_A(t) = \left(K_1 + 2|\mathbf{K}_2| \cdot \exp[-7000t] \cdot \cos[21.40 \times 10^3 t + \theta]\right) u(t) \text{ V}$$

$$K_1 = s \cdot V_A(s)\big|_{s=0} = \left.\frac{0.48(s^2 + 4667s + 500 \times 10^6)}{s^2 + 14 \times 10^3 s + 500 \times 10^6}\right|_{s=0} = 0.48$$

$$\mathbf{K}_2 = (s + 7000 - j21.24 \times 10^3) V_S(s)\big|_{s=-7000+j21.24\times 10^3}$$

$$= \left.\frac{0.48(s^2 + 4667s + 500 \times 10^6)}{s(s + 7000 + j21.24 \times 10^3)}\right|_{s=-7000+j21.24\times 10^3}$$

$$= 0.1055 \cdot \exp[-j270.0°]$$

$$v_A(t) = (0.48 + 0.2110 \cdot \exp[-7000t] \cdot \cos[21.40 \times 10^3 t - 270.0°]) \cdot u(t)$$

$$= (0.48 - 0.2110 \cdot \exp[-7000t] \cdot \sin[21.40 \times 10^3 t]) \cdot u(t) \text{ V}$$

If $v_B(t)$ is also required, $V_A(s)$ can be substituted into either of the node equations, and the equation solved for $V_B(s)$; $v_B(t)$ can then be obtained by inverse Laplace transformation of $V_B(s)$. Alternatively, the entire procedure used to obtain $v_A(t)$ may be used to obtain $v_B(t)$; this is often no more difficult than using substitution, as will be seen in Example 7.4.

Example 7.3

Problem:

Determine the Laplace transform of the capacitor voltage in the circuit shown in Figure 7.5(a).

Solution:

1. Obtain the transformed equivalent circuit [Figure 7.5(b)]. Note that the practical voltage source between node A and ground has been converted to a practical current source, but the ideal source between node B and ground cannot be converted.
2. Apply KCL at node A:

$$I_{1A} + I_{1B} + I_2 = I_{S1} + I_{S2}$$
$$(Y_{1A} + Y_{1B})V_A + Y_2(V_A - V_B) = I_{S1} + I_{S2}$$
$$(Y_{1A} + Y_{1B} + Y_2)V_A - Y_2V_B = I_{S1} + I_{S2}$$

3. $V_B(s)$ is determined by the ideal voltage source connected between node B and ground:

$$V_B(s) = 2/s \text{ V} \cdot \text{sec}$$

   This equation constitutes the auxiliary equation for this system of equations.
4. Substitute $V_B(s)$ and the expressions for $Y_{1A}(s)$, $Y_{1B}(s)$, $Y_2(s)$, $I_{S1}(s)$, and $I_{S2}(s)$ into the node A equation, and solve for $V_A(s)$.

$$(20 + 4 \times 10^{-3}s + 20)V_A - 20\frac{2}{s} = \frac{40 \times 10^3}{s^2} + \frac{200 \times 10^3}{s^2 + 25 \times 10^6}$$

$$4 \times 10^{-3}(s + 10 \times 10^3)V_A = \frac{40}{s} + \frac{40 \times 10^3}{s^2} + \frac{200 \times 10^3}{s^2 + 25 \times 10^6}$$

Figure 7.5
Example 7.3.

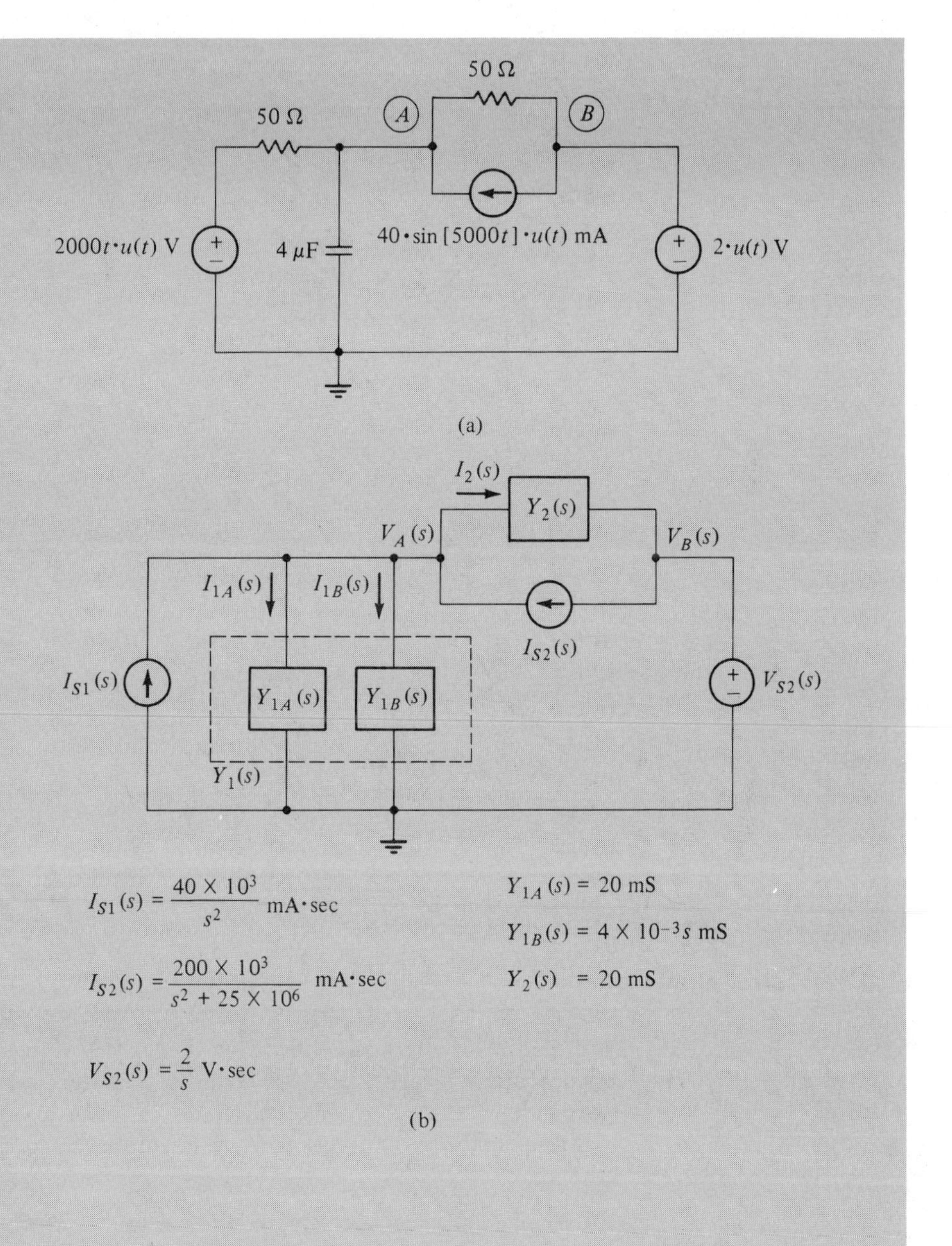

$$V_A(s) = \frac{10 \times 10^3(s^3 + 6000 \times s^2 + 25 \times 10^6 s + 25 \times 10^9)}{s^2(s + 10 \times 10^3)(s^2 + 25 \times 10^6)} \text{ V} \cdot \text{sec}$$

Inverse transformation of $V_A(s)$ is left as an exercise (Problem 7-3).

Exercise 7.2

SPICE

Problem:

Use node analysis to determine $v_R(t)$ for the circuit shown in Figure 7.6.

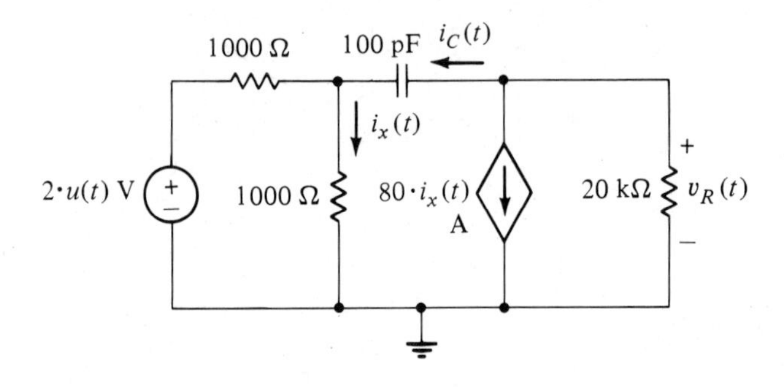

Figure 7.6 Exercise 7.2.

Answer:

$$V_R(s) = \frac{24.38 \times 10^{-3}(s - 800 \times 10^6)}{s(s + 12.19 \times 10^3)} \text{ V} \cdot \text{sec}$$

$$v_R(t) \simeq -1600(1 - \exp[-12.19 \times 10^3 t])u(t) \text{ V}$$

7.4 Mesh Analysis

Mesh analysis is a technique involving Kirchhoff's voltage law (KVL) in which voltages around each mesh are expressed in terms of mesh current and impedance.

Mesh analysis of a transformed circuit is performed in the following manner.

1. Identify and label every mesh, and select a reference direction (clockwise or counterclockwise) for mesh currents.
2. Convert any practical current sources to practical voltage sources. If ideal current sources are present it will be necessary to write an auxiliary equation.
3. Determine the Laplace transform impedance $Z(s)$ for each branch.

4. Apply KVL around each mesh, and express the voltage drops in terms of mesh currents and impedances.
5. Solve the resulting system of algebraic equations for the mesh currents.

Example 7.4

SPICE

Problem:

Use mesh analysis to determine the inductor current in the circuit shown in Figure 7.7(a).

Figure 7.7
Example 7.4.

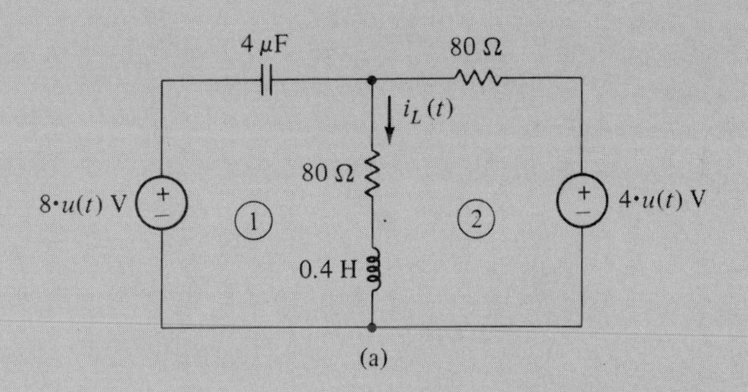

(a)

$V_A(s)$, $V_C(s)$, $Z_A(s)$, $Z_C(s)$, $I_L(s)$, $V_{S1}(s)$, $V_B(s)$, $Z_B(s)$, $I_1(s)$, $I_2(s)$, $V_{S2}(s)$

$$V_{S1}(s) = \frac{8}{s}\ \text{V}\cdot\text{sec} \qquad Z_A(s) = \frac{250 \times 10^3}{s}\ \Omega$$

$$V_{S2}(s) = \frac{4}{s}\ \text{V}\cdot\text{sec} \qquad Z_B(s) = 0.4\,(s + 200)\ \Omega$$

$$Z_C(s) = 80\ \Omega$$

(b)

Solution:

1. Obtain the transformed equivalent circuit [Figure 7.7(b)]. Note that all sources are voltage sources, and the impedance has been determined for each passive network.
2. Apply KVL in mesh 1 and in mesh 2:

$$V_A + V_B = V_{S1}$$
$$-V_B + V_C = -V_{S2}$$

$$Z_A I_1 + Z_B(I_1 - I_2) = V_{S1}$$
$$Z_B(I_2 - I_1) + Z_C I_2 = -V_{S2}$$

$$(Z_A + Z_B)I_1 - Z_B I_2 = V_{S1}$$
$$-Z_B I_1 + (Z_B + Z_C)I_2 = -V_{S2}$$

3. Solve the resulting system of equations for $I_1(s)$ and $I_2(s)$. From the mesh 2 equation,

$$I_2 = -\frac{V_{S2}}{Z_B + Z_C} + \frac{Z_B I_1}{Z_B + Z_C}$$

Substitute $I_2(s)$ into the mesh 1 equation and solve for $I_1(s)$:

$$(Z_A + Z_B)I_1 + \frac{Z_B V_{S2}}{Z_B + Z_C} - \frac{Z_B^2 I_1}{Z_B + Z_C} = V_{S1}$$

$$I_1(s) = \frac{(Z_B + Z_C)V_{S1} - Z_B V_{S2}}{Z_A Z_B + Z_A Z_C + Z_B Z_C} \text{ V} \cdot \text{sec}$$

Also, from the mesh 2 equation,

$$I_1 = \frac{V_{S2} + (Z_B + Z_C)I_2}{Z_B}$$

Substitute $I_1(s)$ into the mesh 1 equation and solve for $I_2(s)$:

$$\frac{(Z_A + Z_B)[V_{S2} + (Z_B + Z_C)I_2]}{Z_B} - Z_B I_2 = V_{S1}$$

$$I_2(s) = \frac{Z_B V_{S1} - (Z_A + Z_B)V_{S2}}{Z_A Z_B + Z_A Z_C + Z_B Z_C} \text{ V} \cdot \text{sec}$$

4. Calculate $I_L(s)$:

$$I_L(s) = I_1(s) - I_2(s)$$

$$= \frac{[(Z_B + Z_C)V_{S1} - Z_B V_{S2}] - [Z_B V_{S1} - (Z_A + Z_B)V_{S2}]}{Z_A Z_B + Z_A Z_C + Z_B Z_C}$$

$$= \frac{Z_C V_{S1} + Z_A V_{S2}}{Z_A Z_B + Z_A Z_C + Z_B Z_C}$$

$$Z_A Z_B + Z_A Z_C + Z_B Z_C = (0.4s + 80)\frac{250 \times 10^3}{s} + 80\frac{250 \times 10^3}{s} + 80(0.4s + 80)$$

$$= 32\frac{s^2 + 3325s + 1.25 \times 10^6}{s}\ \Omega^2$$

$$I_L(s) = \frac{80(8/s) + (250 \times 10^3/s)(4/s)}{32(s^2 + 3325s + 1.25 \times 10^6)/s}\ \text{A} \cdot \text{sec}$$

$$= \frac{20s + 31.25 \times 10^3}{s(s^2 + 3325s + 1.25 \times 10^6)}\ \text{A} \cdot \text{sec}$$

5. Determine $i_L(t)$ by inverse Laplace transformation:

$$I_L(s) = \frac{K_1}{s} + \frac{K_2}{s + 2893} + \frac{K_3}{s + 432.1}$$

$$i_L(t) = (K_1 + K_2 \cdot \exp[-2893t] + K_3 \cdot \exp[-432.1t])u(t)\ \text{A}$$

$$K_1 = s \cdot I_L(s)|_{s=0} = \left.\frac{20s + 31.25 \times 10^3}{s^2 + 3325s + 1.25 \times 10^6}\right|_{s=0} = 25 \times 10^{-3}$$

$$K_2 = (s + 2893)I_L(s)|_{s=-2893} = \left.\frac{20s + 31.25 \times 10^3}{s(s + 432.1)}\right|_{s=-2893} = -3.738 \times 10^{-3}$$

$$K_3 = (s + 432.1)I_L(s)|_{s=-432.1} = \left.\frac{20s + 31.25 \times 10^3}{s(s + 2893)}\right|_{s=-432.1} = -21.26 \times 10^{-3}$$

$$i_L(t) = (25 - 3.738 \cdot \exp[-2893t] - 21.26 \cdot \exp[-432.1t])u(t)\ \text{mA}$$

Example 7.5

Problem:

Use mesh analysis to determine the current $I_s(s)$ in the independent voltage source for the circuit shown in Figure 7.8(a).

Figure 7.8
Example 7.5.

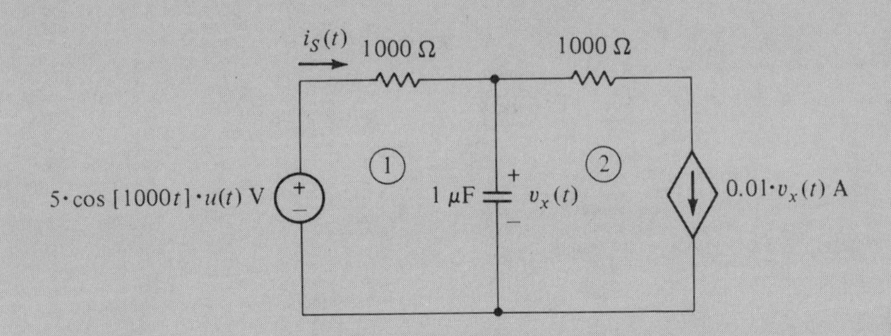

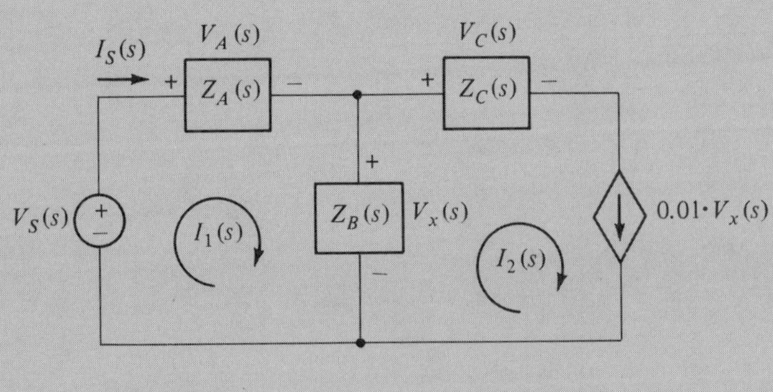

$$V_S(s) = \frac{5s}{s^2 + 10^6} \text{ V·sec}$$

$$Z_A(s) = 1000\ \Omega$$

$$Z_B(s) = \frac{10^6}{s}\ \Omega$$

$$Z_C(s) = 1000\ \Omega$$

Solution:

1. Obtain the transformed equivalent circuit [Figure 7.8(b)]. Note that the ideal current source in mesh 2 cannot be converted to a voltage source.

2. Apply KVL in mesh 1:

$$V_A + V_X = V_S$$

$$Z_A I_1 + Z_B(I_1 - I_2) = V_S$$
$$(Z_A + Z_B)I_1 - Z_B I_2 = V_S$$

3. $I_2(s)$ is determined by the ideal current source in mesh 2:

$$I_2 = 0.01V_x = 0.01Z_B(I_1 - I_2)$$

This equation constitutes the auxiliary equation for this circuit.

4. Solve the auxiliary equation for $I_2(s)$ in terms of $I_1(s)$, and substitute the result into the mesh 1 equation to obtain $I_1(s)$:

$$I_2 = \frac{0.01Z_B}{1 + 0.01Z_B} I_1$$

$$(Z_A + Z_B)I_1 - Z_B \frac{0.01Z_B}{1 + 0.01Z_B} I_1 = V_S$$

$$(Z_A + Z_B)(1 + 0.01Z_B)I_1 - 0.01Z_B^2 I_1 = V_S(1 + 0.01Z_B)$$

$$I_s(s) = I_1 = \frac{V_S(1 + 0.01Z_B)}{(Z_A + Z_B)(1 + 0.01Z_B) - 0.01Z_B^2}$$

$$= \frac{\dfrac{5s}{s^2 + 10^6}\left(1 + 0.01\dfrac{10^6}{s}\right)}{\left(1000 + \dfrac{10^6}{s}\right)\left(1 + 0.01\dfrac{10^6}{s}\right) - 0.01\dfrac{10^{12}}{s^2}}$$

$$= \frac{5s(s + 10 \times 10^3)}{(s + 11 \times 10^3)(s^2 + 10^6)} \text{ mA} \cdot \text{sec}$$

Inverse transformation of $I_S(s)$ is left as an exercise (Problem 7-6).

Exercise 7.3

SPICE

Problem:

Use mesh analysis to determine $i_C(t)$ for the circuit of Exercise 7.2 (Figure 7.6).

Answer:

$$I_C(s) = \frac{-1.951 \times 10^{-3}}{s + 12.19 \times 10^3} \text{ A} \cdot \text{sec}$$

$$i_C(t) = -1.951 \cdot \exp[-12.19 \times 10^3 t] \cdot u(t) \text{ mA}$$

7.5 Superposition

The **superposition principle** is used to determine the response of a circuit excited by multiple independent sources. In applying the superposition principle, as was illustrated in Chapter 1 (Section 1.6), the multiple-source circuit is decomposed into simpler circuits, each containing exactly one of the independent sources. Independent voltage sources are removed from the circuit by replacing them with short circuits (0 V), and independent current sources are removed from the circuit by replacing them with open circuits (0 A). The s-domain response and, by inverse Laplace transformations, the time-domain response of each of the simpler circuits is determined separately. The time-domain results are added algebraically, according to the superposition principle, to determine the total response to all sources.

In theory, circuit decomposition may be accomplished either before or after transformation; however, circuit transformation may introduce additional independent sources due to initial capacitor voltage or inductor current (Chapter 5). For this reason, the time-domain circuit should be transformed, and the transformed circuit decomposed. The independent sources that arise in the transformed circuit because of capacitors or inductors with stored energy at switching time make superposition a particularly valuable technique in transform circuit analysis. In the first place, a circuit which contains only one time-domain source will have multiple s-domain sources if the capacitors or inductors have stored energy at switching time (Example 7.7). Secondly, decomposition of the transformed circuit provides a clear indication of which portion of the response is due to initial conditions and which portion is due to the source. This type of information is very useful to the circuit designer in producing a desired response or in suppressing an undesirable response.

Example 7.6

Problem:

Use the superposition principle to determine $V_A(s)$ for the circuit of Figure 7.5(a) (Example 7.3).

Figure 7.9
Example 7.6.

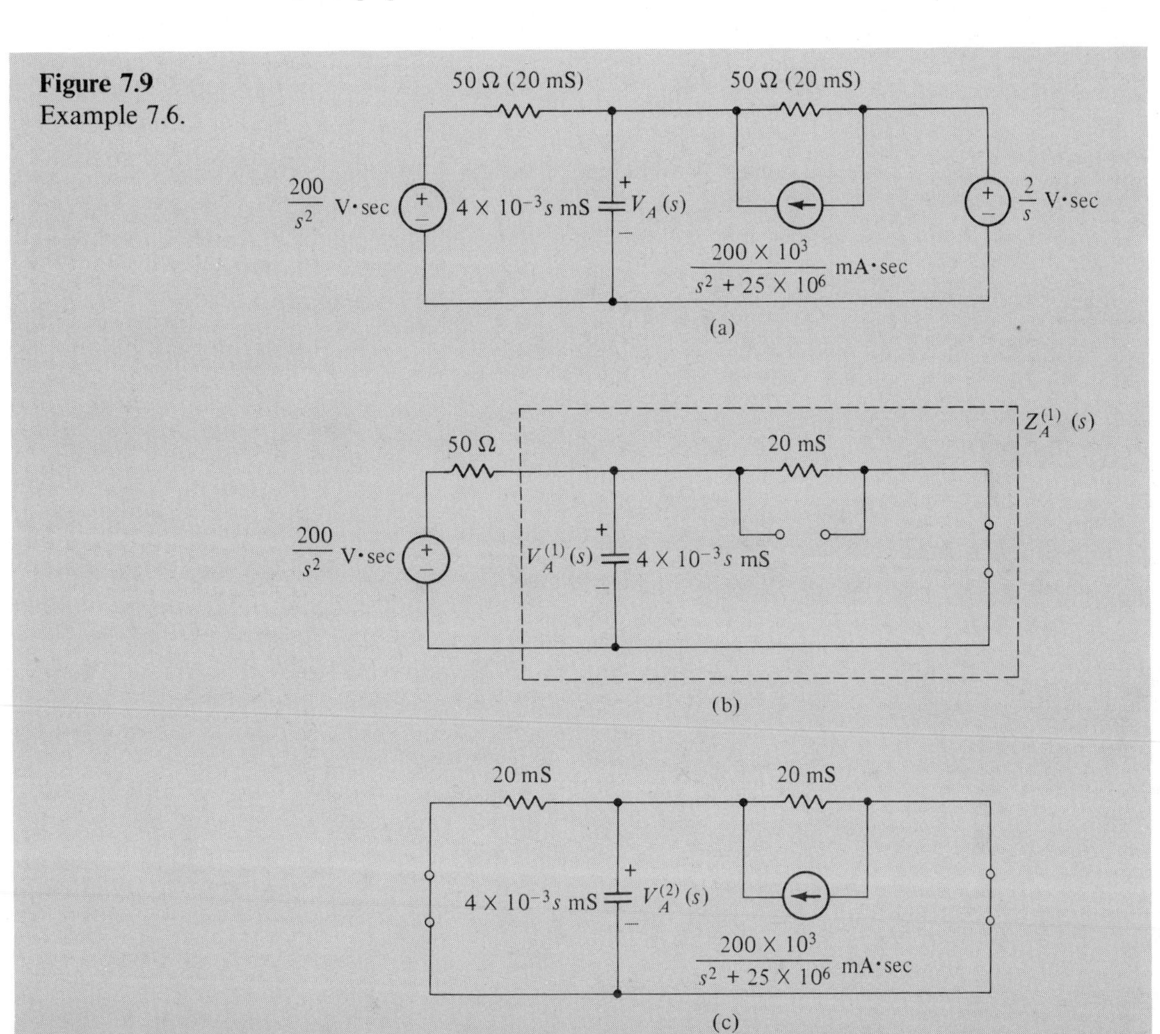

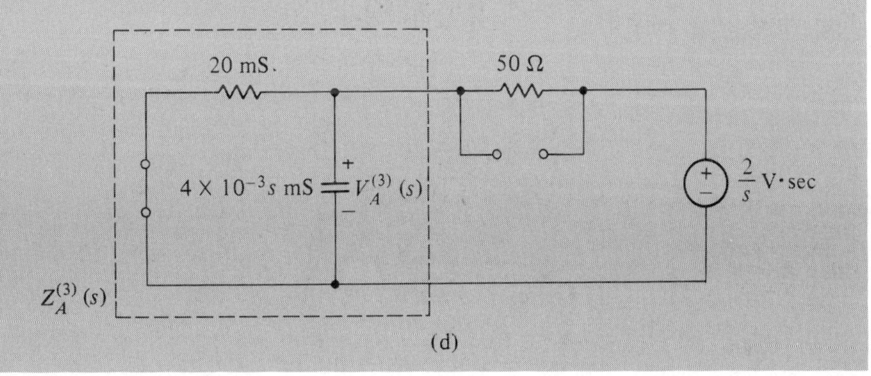

Solution:

1. Transform the circuit as shown in Figure 7.9(a).
2. Decompose the circuit into three simpler circuits, each containing exactly one independent source, as shown in Figure 7.9(b), (c), and (d). Note that voltage sources are replaced by short circuits and current circuits are replaced by open circuits.
3. Determine $V_A^{(1)}(s)$ in the circuit of Figure 7.9(b):

$$Y_A^{(1)} = 4 \times 10^{-3}s + 20 = 4 \times 10^{-3}(s + 5000) \text{ mS}$$

$$Z_A^{(1)} = \frac{1}{Y_A^{(1)}} = \frac{250 \times 10^3}{s + 5000} \ \Omega$$

Use the voltage division rule to calculate $V_A^{(1)}(s)$:

$$V_A^{(1)} = \frac{2000}{s^2} \cdot \frac{250 \times 10^3/(s + 5000)}{50 + 250 \times 10^3/(s + 5000)}$$

$$= \frac{10 \times 10^6}{s^2(s + 10 \times 10^3)} \text{ V} \cdot \text{sec}$$

4. Determine $V_A^{(2)}(s)$ in the circuit of Figure 7.9(c):

$$V_A^{(2)} = \frac{200 \times 10^3/(s^2 + 25 \times 10^6)}{20 + 4 \times 10^{-3}s + 20}$$

$$= \frac{50 \times 10^6}{(s + 10 \times 10^3)(s^2 + 25 \times 10^6)} \text{ V} \cdot \text{sec}$$

5. Determine $V_A^{(3)}(s)$ in the circuit of Figure 7.9(d):

$$Y_A^{(3)} = 4 \times 10^{-3}s + 20 = 4 \times 10^{-3}(s + 5000) \text{ mS}$$

$$Z_A^{(3)} = \frac{1}{Y_A^{(3)}} = \frac{250 \times 10^3}{s + 5000} \ \Omega$$

Use the voltage division rule to calculate $V_A^{(3)}(s)$:

$$V_A^{(3)} = \frac{2}{s} \frac{250 \times 10^3/(s + 5000)}{50 + 250 \times 10^3/(s + 5000)}$$

$$= \frac{10 \times 10^3}{s(s + 10 \times 10^3)} \text{ V} \cdot \text{sec}$$

6. Apply the superposition principle to determine $V_A(s)$:

$$V_A = V_A^{(1)} + V_A^{(2)} + V_A^{(3)}$$

$$= \frac{10 \times 10^6}{s^2(s + 10 \times 10^3)} + \frac{50 \times 10^6}{(s + 10 \times 10^3)(s^2 + 25 \times 10^6)} + \frac{10 \times 10^3}{s(s + 10 \times 10^3)}$$

$$= \frac{10 \times 10^3(s^3 + 6000s^2 + 25 \times 10^6 s + 25 \times 10^9)}{s^2(s + 10 \times 10^3)(s^2 + 25 \times 10^6)} \text{ V} \cdot \text{sec}$$

Example 7.7

SPICE

Problem:

Use the superposition principle to determine the s-domain response current $I(s)$ for the circuit of Figure 7.10(a).

Solution:

1. Determine initial conditions for C_1 and C_2 (Chapter 3):

$$v_1(0^-) = 8 \text{ V}, \qquad v_2(0^-) = 4 \text{ V}$$

2. Transform the circuit as shown in Figure 7.10(b). This particular transformation was chosen because the series model for C_1 combines easily with the series resistor, and the parallel model for C_2 combines easily with the parallel resistor.
3. Decompose the circuit into two simpler circuits, each containing exactly one independent source, as shown in Figure 7.10(c) and (d).
4. Determine $I^{(1)}(s)$ from the circuit shown in Figure 7.10(c). By the voltage division rule,

$$V^{(1)} = \frac{8}{s} \cdot \frac{Z_2}{Z_1 + Z_2} = \frac{8}{s} \cdot \frac{\dfrac{250 \times 10^3}{s + 5000}}{50\dfrac{s + 5000}{s} + \dfrac{250 \times 10^3}{s + 5000}}$$

$$= \frac{40 \times 10^3}{s^2 + 15 \times 10^3 s + 25 \times 10^6} \text{ V} \cdot \text{sec}$$

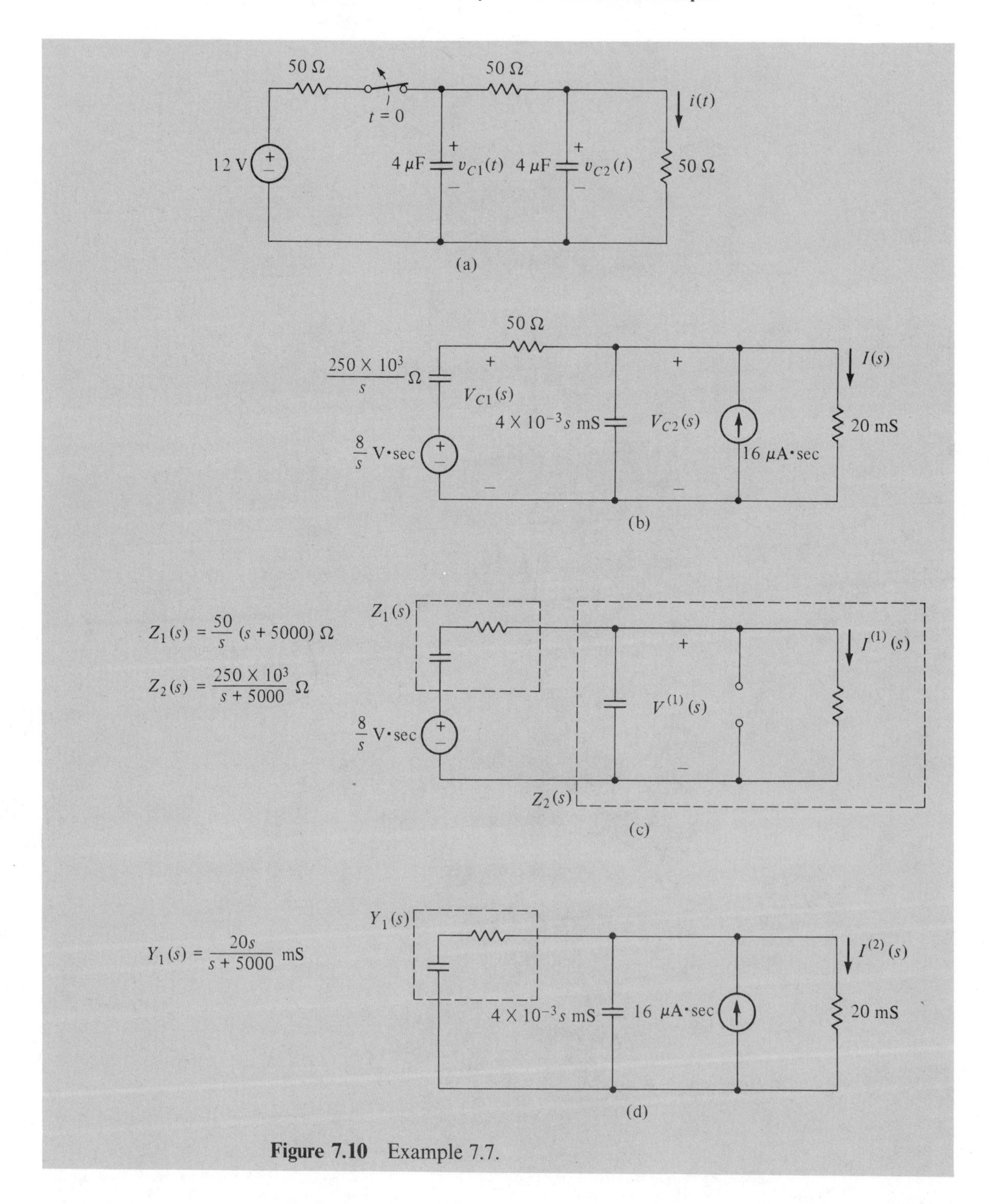

Figure 7.10 Example 7.7.

By Ohm's law,

$$I^{(1)} = \frac{V^{(1)}}{50} = \frac{800}{s^2 + 15 \times 10^3 s + 25 \times 10^6}$$

$$= \frac{800}{(s + 1910)(s + 13090)} \text{ A} \cdot \text{sec}$$

5. Use Pair XI to determine $i^{(1)}(t)$:

$$i^{(1)}(t) = \frac{800}{-1910 + 13.09 \times 10^3}(\exp[-13.09 \times 10^3 t] - \exp[-1910t])u(t)$$

$$= 71.56(\exp[-13.09 \times 10^3 t] - \exp[-1910t])u(t) \text{ mA}$$

6. Use the current division rule to determine $I^{(2)}(s)$ from the circuit shown in Figure 7.10(d):

$$I^{(2)} = 16 \times 10^{-6} \frac{20}{Y_1 + 4 \times 10^{-3}s + 20}$$

$$= \frac{80 \times 10^{-3}(s + 5000)}{s^2 + 15 \times 10^3 s + 25 \times 10^6} \text{ A} \cdot \text{sec}$$

7. Use partial fraction decomposition and Pair IV to determine $i^{(2)}(t)$:

$$I^{(2)}(s) = \frac{80 \times 10^{-3}(s + 5000)}{(s + 1910)(s + 13.09 \times 10^3)}$$

$$= \frac{K_1}{s + 1910} + \frac{K_2}{s + 13.09 \times 10^3} \text{ A} \cdot \text{sec}$$

$$i^{(2)}(t) = (K_1 \cdot \exp[-1910t] + K_2 \cdot \exp[-13.09 \times 10^3 t])u(t) \text{ A}$$

where $K_1 = 22.11 \times 10^{-3}$ and $K_2 = 57.89 \times 10^{-3}$; thus

$$i^{(2)}(t) = (22.11 \cdot \exp[-1910t] + 57.89 \cdot \exp[-13.09 \times 10^3 t])u(t) \text{ mA}$$

8. Apply the superposition principle to obtain $i(t)$:

$$i(t) = i^{(1)}(t) + i^{(2)}(t)$$

$$= (129.5 \cdot \exp[-13.09 \times 10^3 t] - 49.45 \cdot \exp[-1910t])u(t) \text{ mA}$$

Note in Example 7.7 that the single time-domain voltage source manifested itself as two s-domain sources due to the charges on two separate capacitors. The response current could have been obtained by mesh analysis, but application of the superposition principle made it clear what portion of the response was due to C_1 (step 5) and what portion was due to C_2 (step 7).

Exercise 7.4

SPICE

Problem:

Use superposition to determine $v(t)$ for the circuit shown in Figure 7.11.

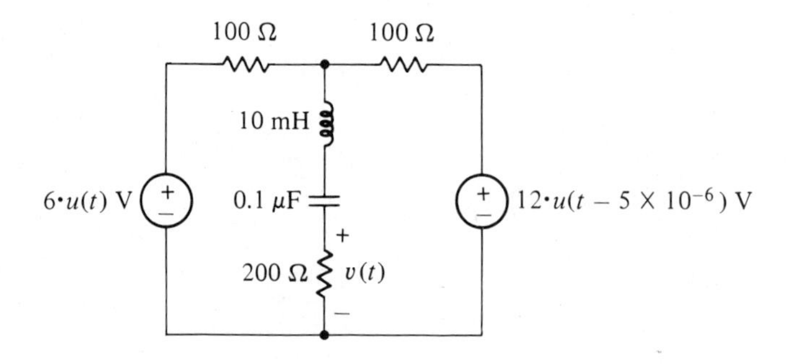

Figure 7.11
Exercise 7.4

Answer:

$$V^{(1)}(s) = \frac{60 \times 10^3}{s^2 + 25 \times 10^3 s + 10^9} \text{ V} \cdot \text{sec}$$

$$V^{(2)}(s) = \frac{120 \times 10^3 \cdot \exp[-5 \times 10^{-6} s]}{s^2 + 25 \times 10^3 s + 10^9} \text{ V} \cdot \text{sec}$$

$$v^{(1)}(t) = 2.066 \cdot \exp[-12.5 \times 10^3 t] \cdot \sin[29.05 \times 10^3 t] \cdot u(t) \text{ V}$$

$$v^{(2)}(t) = 4.131 \cdot \exp\left[-12.5 \times 10^3 (t - 5 \times 10^{-6})\right] \cdot \sin\left[29.05 \times 10^3 (t - 5 \times 10^{-6})\right] \cdot u(t - 5 \times 10^{-6}) \text{ V}$$

7.6 Thevenin's and Norton's Theorems

Thevenin's theorem, as applied to a source network with dc sources and linear resistors (Chapter 1), guarantees the existence of an equivalent source network consisting of a single independent dc source in series with a linear resistor. Equivalent, in this sense, means that both source networks provide the same terminal voltage and deliver the same current to a load network. Norton's theorem is the dual of Thevenin's theorem, and guarantees the existence of an equivalent source network consisting of a single independent current source in parallel with a linear resistor. In practice, Norton's equivalent source network is a source conversion of Thevenin's equivalent source network, and vice versa.

Thevenin's and Norton's theorems can be extended to include all linear source networks. Thevenin's equivalent source network consists of a single independent source (which may be time-dependent) in series with an R–L–C network, and Norton's equivalent source network consists of a current source in parallel with an R–L–C network. The criteria for applying Thevenin's or Norton's theorem to a source network are given in Chapter 1 (Section 1.7). In particular, the source network must be linear, and the control variables for dependent sources must also be in the source network.

When implemented in the s-domain, Thevenin's equivalent source network consists of an s-domain voltage source $V_{\text{Th}}(s)$ in series with an impedance $Z_{\text{Th}}(s)$, and Norton's equivalent source network consists of an s-domain current source $I_{\text{No}}(s)$ in parallel with an admittance $Y_{\text{No}}(s)$. In general,

$$V_{\text{Th}}(s) = V_{\text{oc}}(s)$$

$$I_{\text{No}}(s) = I_{\text{sc}}(s)$$

$$Z_{\text{Th}}(s) = \frac{V_{\text{oc}}(s)}{I_{\text{sc}}(s)}$$

$$Y_{\text{No}}(s) = \frac{I_{\text{sc}}(s)}{V_{\text{oc}}(s)} = \frac{1}{Z_{\text{Th}}(s)}$$

The four cases outlined in Chapter 1 for determining the equivalent source network for a dc source network can be extended to transformed source networks.

Case I Passive Linear Elements and Independent Sources (Example 7.8) For Thevenin's equivalent source, determine $V_{\text{oc}}(s)$ only; for Norton's equivalent source, determine $I_{\text{sc}}(s)$ only. To determine $Z_{\text{Th}}(s)$ or $Y_{\text{No}}(s)$, remove all sources from the source network; replace voltage sources by short circuits and current sources by open circuits. The deactivated source network now consists only of passive linear elements, and $Z_{\text{Th}}(s)$ or $Y_{\text{No}}(s)$ can be calculated at the load terminals.

Case II Passive Linear Elements and Practical Independent Sources (Example 7.9)
Thevenin's or Norton's equivalent source can be determined analytically by repeated source conversions, and combination of series and parallel elements, without directly determining $V_{oc}(s)$ or $I_{sc}(s)$. The equivalent source (Thevenin's or Norton's) will be apparent when all possible source conversions and element combinations have been accomplished.

Case III Passive Linear Elements and Dependent and Independent Sources (Example 7.10)
Determine both $V_{oc}(s)$ and $I_{sc}(s)$.

Case IV Passive Linear Elements and Dependent Sources Only (Example 7.11)
In this case $V_{oc}(s)$ and $I_{sc}(s)$ are both zero, and the value of $Z_{Th}(s)$ [or $Y_{No}(s)$] is indeterminate. To resolve the situation, place a unit (1-V or 1-A) test source at the load terminals. If a 1-A source is used, the voltage rise *in the direction of the test source current* at the load terminals is equal to $Z_{Th}(s)$; if a 1-V source is used, the current *from the source* into the more positive load terminal is equal to $Y_{No}(s)$.

Example 7.8

Problem:

Determine Thevenin's equivalent and Norton's equivalent for the source network shown in Figure 7.12(a).

Solution:

1. Transform the source network [Figure 7.12(b)], and calculate the open-circuit voltage at the load terminals:

$$V_{Th}(s) = V_{oc}(s) = \frac{40}{s^2}(0.1s + 100) = \frac{4}{s^2}(s + 1000)\ \text{V} \cdot \text{sec}$$

2. Deactivate the current source and determine the impedance of the source network [Figure 7.12(c)]:

$$Z_{Th}(s) = 100 + 0.1s + 100 = 0.1(s + 2000)\ \Omega$$

Thevenin's equivalent source network is shown in Figure 7.12(d).

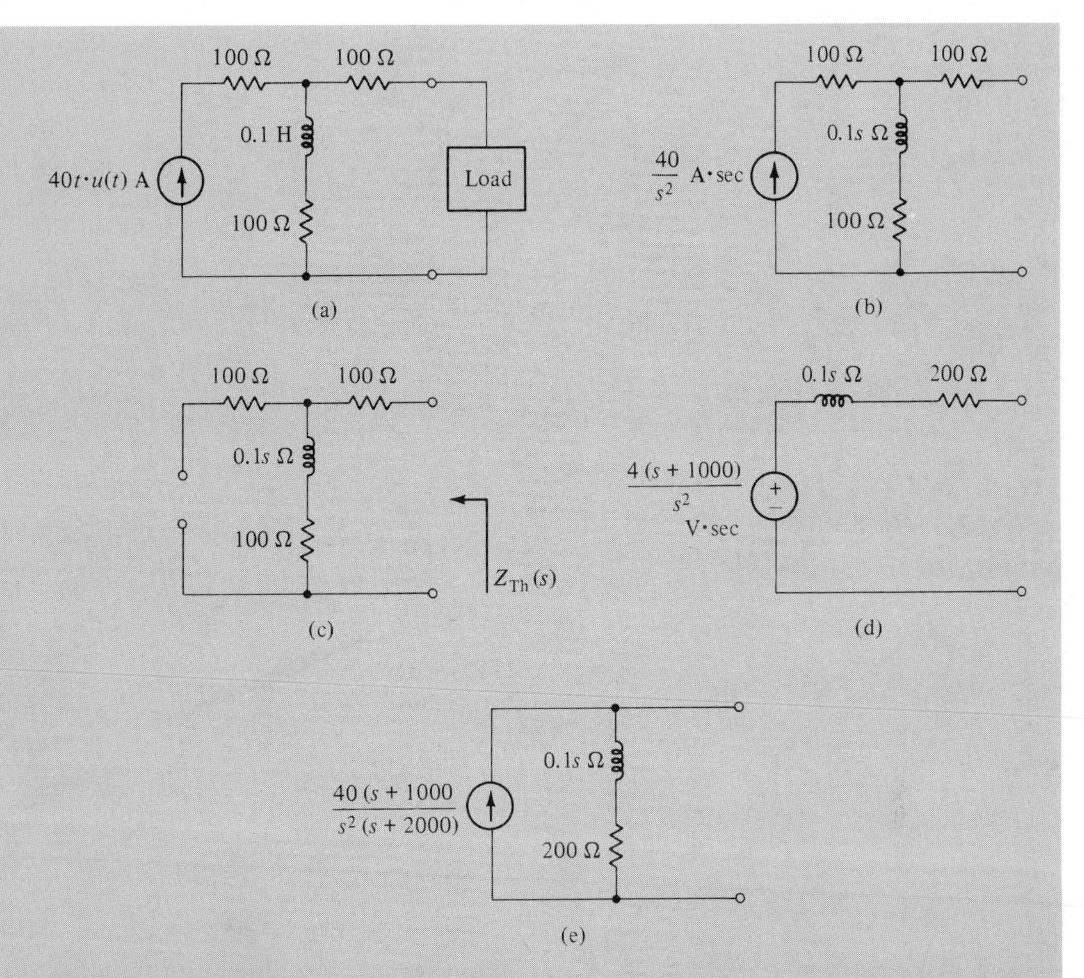

Figure 7.12 Example 7.8.

3. Obtain Norton's equivalent source network [Figure 7.12(e)] by a source conversion of Thevenin's equivalent source network:

$$I_{\text{No}}(s) = \frac{V_{\text{Th}}(s)}{Z_{\text{Th}}(s)} = \frac{(4/s^2)(s+1000)}{0.1(s+2000)}$$

$$= \frac{40(s+1000)}{s^2(s+2000)} \text{ A} \cdot \text{sec}$$

$$Y_{\text{No}}(s) = \frac{1}{Z_{\text{Th}}(s)} = \frac{10}{s+2000} \text{ S}$$

Example 7.9

Problem:

Determine Thevenin's equivalent and Norton's equivalent for the transformed source network shown in Figure 7.13(a).

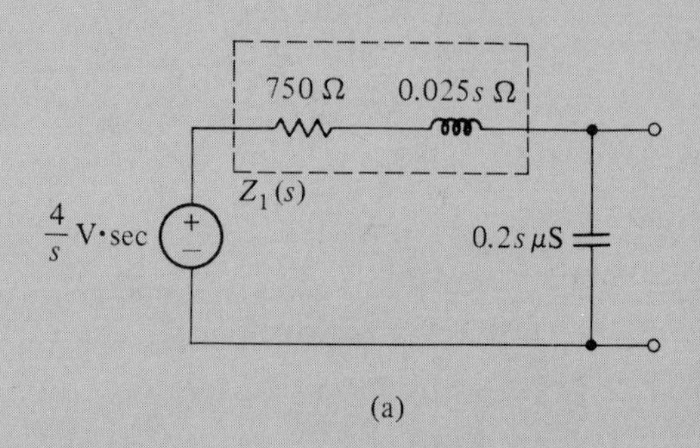

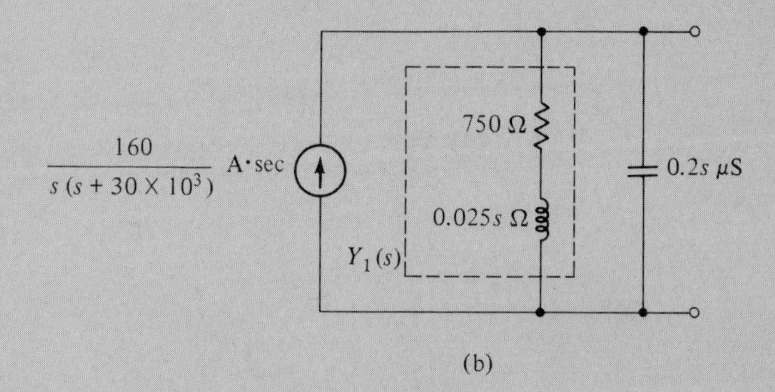

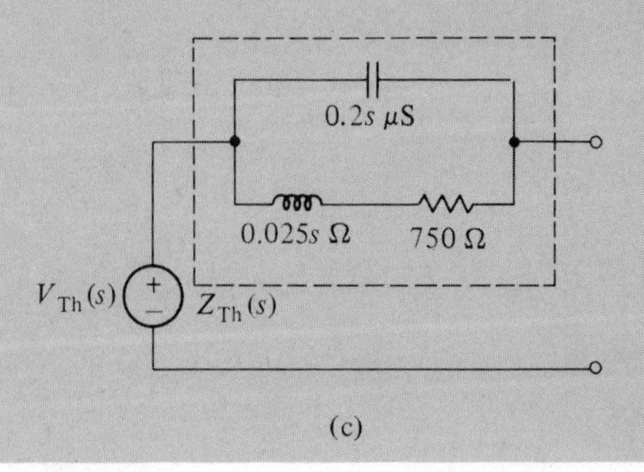

Figure 7.13
Example 7.9.

Solution:

1. Perform a source conversion of the voltage source:

$$Z_1(s) = 0.025s + 750 = 0.025(s + 30 \times 10^3)\ \Omega$$

$$I(s) = \frac{4/s}{Z_1(s)} = \frac{160}{s(s + 30 \times 10^3)}\ \text{A} \cdot \text{sec}$$

The network obtained following the source conversion [Figure 7.13(b)] is Norton's equivalent source network—that is, a single independent current source in parallel with a linear passive network.

2. Perform a source conversion of Norton's equivalent source network to obtain Thevenin's equivalent source network [Figure 7.13(c)]:

$$Y_{\text{No}}(s) = Y_1(s) + 0.2 \times 10^{-6}s$$

$$= \frac{0.2 \times 10^{-6}(s^2 + 30 \times 10^3 s + 200 \times 10^6)}{s + 30 \times 10^3}\ \text{S}$$

$$Z_{\text{Th}}(s) = \frac{1}{Y_{\text{No}}(s)} = \frac{5 \times 10^6(s + 30 \times 10^3)}{s^2 + 30 \times 10^3 s + 200 \times 10^6}\ \Omega$$

$$V_{\text{Th}}(s) = I_{\text{No}}(s) Z_{\text{Th}}(s)$$

$$= \frac{800 \times 10^6}{s(s^2 + 30 \times 10^3 s + 200 \times 10^6)}\ \text{V} \cdot \text{sec}$$

Example 7.10

Problem:

Determine the Thevenin's and Norton's equivalent of the circuit of Figure 7.6 (Exercise 7.2) as a source network. Assume that the load would be connected across the 20-kΩ resistor, which is part of the source network.

Solution:

1. $V_R(s)$, as determined in Exercise 7.2, is the open-circuit load terminal voltage. Therefore,

$$V_{\text{Th}}(s) = V_R(s) = \frac{24.38 \times 10^{-3}(s - 800 \times 10^6)}{s(s + 12.19 \times 10^3)}\ \text{V} \cdot \text{sec}$$

Figure 7.14
Example 7.10.

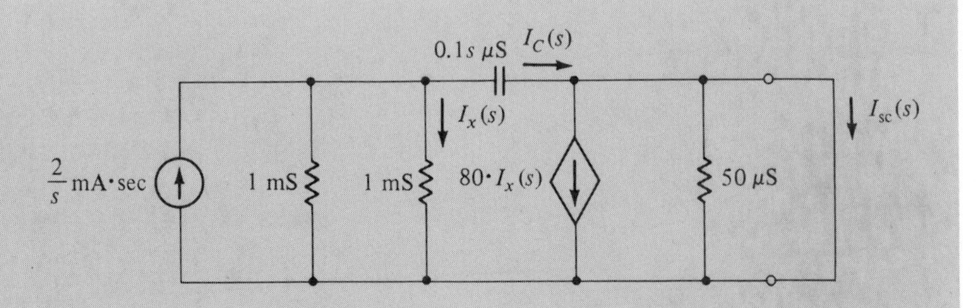

2. Determine $I_{sc}(s)$ from the circuit shown in Figure 7.14:

$$I_{No}(s) = I_{sc}(s) = I_C(s) - 80 \cdot I_x(s)$$

Use the current division rule to determine $I_C(s)$ and $I_x(s)$:

$$I_C(s) = \frac{2}{s} \cdot \frac{100 \times 10^{-9}s}{1 + 1 + 100 \times 10^{-9}s} = \frac{2}{s + 20 \times 10^6} \text{ mA} \cdot \text{sec}$$

$$I_x(s) = \frac{2}{s} \cdot \frac{1}{1 + 1 + 100 \times 10^{-9}s} = \frac{20 \times 10^6}{s(s + 20 \times 10^6)} \text{ mA} \cdot \text{sec}$$

Calculate $I_{No}(s)$:

$$I_{No}(s) = \frac{2}{s + 20 \times 10^6} - \frac{80(20 \times 10^6)}{s(s + 20 \times 10^6)} = \frac{2(s - 800 \times 10^6)}{s(s + 20 \times 10^6)} \text{ mA} \cdot \text{sec}$$

3. Calculate $Z_{Th}(s)$ and $Y_{No}(s)$:

$$Z_{Th}(s) = \frac{\dfrac{24.38 \times 10^{-3}(s - 800 \times 10^6)}{s(s + 12.19 \times 10^3)}}{\dfrac{2 \times 10^{-3}(s - 800 \times 10^6)}{s(s + 20 \times 10^6)}} = \frac{12.19(s + 20 \times 10^6)}{s + 12.19 \times 10^3} \ \Omega$$

$$Y_{No}(s) = \frac{1}{Z_{Th}(s)} = \frac{82.05 \times 10^3(s + 12.19 \times 10^3)}{s + 20 \times 10^6} \text{ mS}$$

Example 7.11

Problem:

Determine Norton's equivalent of the source network shown in Figure 7.15(a).

Solution:

1. Because the source network contains no independent sources, the open-circuit voltage and short-circuit current are both zero. Therefore,

$$I_{\text{No}}(s) = 0$$
$$V_{\text{Th}}(s) = 0$$

and $Y_{\text{No}}(s)$ cannot be determined from $I_{\text{No}}(s)/V_{\text{Th}}(s)$.

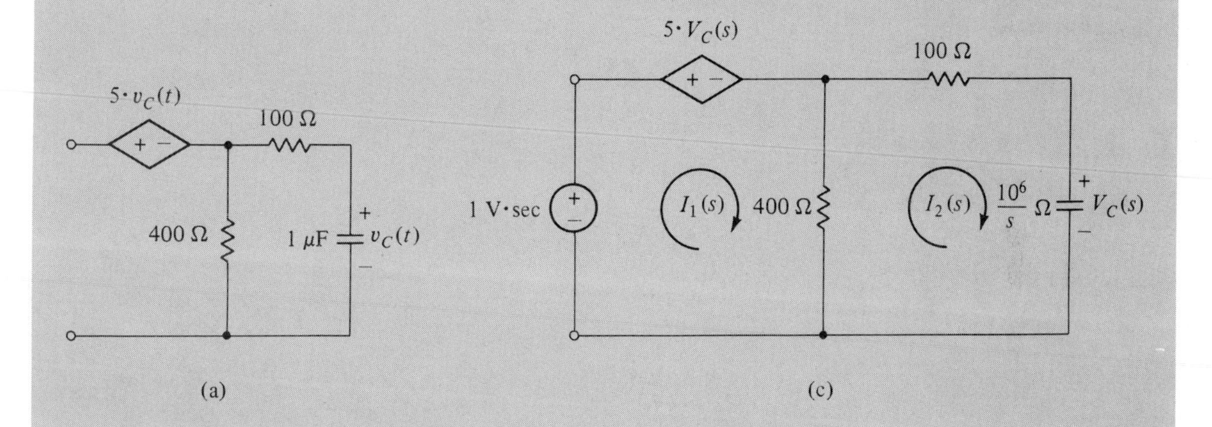

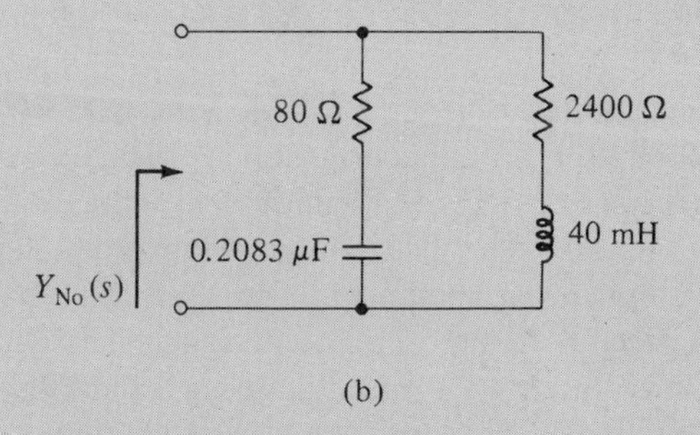

Figure 7.15 Example 7.11.

2. Determine $Y_{No}(s)$ by connecting a 1-V · sec source at the load terminals [Figure 7.15(b)]. $Y_{No}(s)$ is equal to the s-domain current entering the positive load terminal.
3. Solve for the mesh current $I_1(s)$, which is equivalent to the current entering the more positive load terminal:

Mesh 1: $$400I_1 \qquad -400I_2 = 1 - 5V_C$$

Mesh 2: $$-400I_1 + \left(500 + \frac{10^6}{s}\right)I_2 = 0$$

Auxiliary: $$V_C = \frac{10^6}{s}I_2$$

$$400sI_1 - (400s - 5 \times 10^6)I_2 = s$$

$$-400sI_1 + (500s + 10^6)I_2 = 0$$

$$I_1 = \frac{12.5 \times 10^{-3}(s + 2000)}{s + 60 \times 10^3} \text{ A} \cdot \text{sec}$$

4. Calculate $Y_{No}(s)$:

$$Y_{No}(s) = \frac{I_1(s)}{1 \text{ V} \cdot \text{sec}} = \frac{12.5 \times 10^{-3}(s + 2000)}{s + 60 \times 10^3} \text{ S}$$

From these results, the original source network can be represented as an admittance (or an impedance). In this case, the admittance can be realized by the passive linear network shown in Figure 7.15(c). The reader should verify that this network has an admittance equal to $Y_{No}(s)$ (Problem 7-13).

It is interesting to note in Example 7.11 that the original network was composed of resistors, a capacitor, and a voltage-controlled voltage source, whereas the *equivalent* network contained an inductor, as well as resistors and a capacitor, but no controlled source. One must then conclude that it is possible to eliminate inductance from a circuit by the proper use of a controlled source. Electronic circuit designers have made good use of this observation to replace bulky, and often costly, inductors by controlled sources implemented with small, inexpensive integrated circuits (Chapter 8). Minimizing the use of inductors has played an important role in miniaturizing electronic circuits.

Exercise 7.5

Problem:

Determine Thevenin's equivalent source network and Norton's equivalent source network for the circuit shown in Figure 7.16.

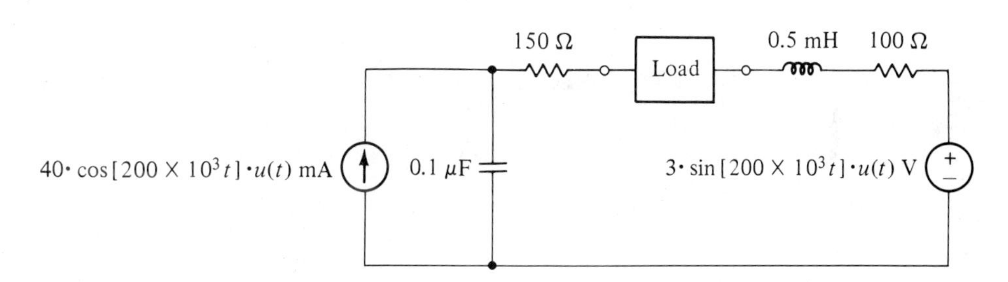

Figure 7.16 Exercise 7.5.

Answer:

$$V_{\mathrm{Th}}(s) = \frac{10^6}{s^2 + 40 \times 10^9} \ \mathrm{V \cdot sec}$$

$$Z_{\mathrm{Th}}(s) = 0.5 \times 10^{-3} \frac{s^2 + 500 \times 10^3 s + 20 \times 10^9}{s} \ \Omega$$

$$I_{\mathrm{No}}(s) = \frac{2 \times 10^9 s}{(s^2 + 40 \times 10^9)(s^2 + 500 \times 10^3 s + 20 \times 10^9)} \ \mathrm{A \cdot sec}$$

$$Y_{\mathrm{No}}(s) = \frac{2000 s}{s^2 + 500 \times 10^3 s + 20 \times 10^9} \ \mathrm{S}$$

7.7 Computer Simulations

The solution of the simultaneous s-domain equations resulting from node or mesh analysis is a significant algebra problem. The presence of controlled sources complicates the algebra even further, since these sources cannot be disabled and the circuit must be analyzed as a multisource circuit. Computer simulations save a significant amount of analytical work when dealing with these complex circuits.

The use of SPICE to perform node and mesh analysis is illustrated in SPICE 7.1 (Exercise 7.2) and SPICE 7.2. (Example 7.5), respectively. Both of these SPICE examples contain linear controlled current sources; examples containing linear controlled voltage sources will be provided in later chapters. The element record format for controlled current sources may be found in Appendix D.

```
*******11/07/88 ********  SPICE 2G.6    3/15/83 ********20:47:08*****
SPICE 7.1
****     INPUT LISTING                        TEMPERATURE =   27.000 DEG C
*******************************************************************************
 R1 1 2 1K
 R2 4 0 1K
 R3 3 0 20K
 C 2 3 100P
 V 1 0 PULSE(0 2 1F 1F)
 F 3 0 VAMX 80
 VAMX 2 4
 .TRAN 20U 400U
 .PLOT TRAN V(2) V(3)
 .END
*******11/07/88 ********  SPICE 2G.6    3/15/83 ********20:47:08*****
SPICE 7.1
****     INITIAL TRANSIENT SOLUTION       TEMPERATURE =   27.000 DEG C
*******************************************************************************
  NODE   VOLTAGE     NODE   VOLTAGE     NODE   VOLTAGE     NODE   VOLTAGE
 (  1)     .0000    (  2)     .0000    (  3)     .0000    (  4)     .0000
     VOLTAGE SOURCE CURRENTS
     NAME       CURRENT
     V          0.000D+00
     VAMX       0.000D+00
     TOTAL POWER DISSIPATION   0.00D+00  WATTS
*******11/07/88 ********  SPICE 2G.6    3/15/83 ********20:47:08*****
SPICE 7.1
****     OPERATING POINT INFORMATION      TEMPERATURE =   27.000 DEG C
*******************************************************************************

**** CURRENT-CONTROLLED CURRENT SOURCES
              F
 I-SOURCE  0.00E+00


*******11/07/88 ********  SPICE 2G.6    3/15/83 ********20:47:08*****
SPICE 7.1
****     TRANSIENT ANALYSIS                   TEMPERATURE =   27.000 DEG C
*******************************************************************************
LEGEND:
 *: V(2)
 +: V(3)
```

```
     TIME       V(2)
(*)------------- -5.000D-01    0.000D+00     5.000D-01     1.000D+00  1.500D+00
                 - - - - - - - - - - - - - - - - - - - - - - - - - - - - - - - -
(+)------------- -2.000D+03   -1.500D+03    -1.000D+03    -5.000D+02  0.000D+00
                 - - - - - - - - - - - - - - - - - - - - - - - - - - - - - - - -
 0.000D+00  0.000D+00 .            *             .             .             +
 2.000D-05  2.349D-01 .            .      *      .             .        +    .
 4.000D-05  4.004D-01 .            .         *   .        +    .             .
 6.000D-05  5.303D-01 .            .             .*   +        .             .
 8.000D-05  6.319D-01 .            .             +   *         .             .
 1.000D-04  7.116D-01 .            .         +   .     *       .             .
 1.200D-04  7.740D-01 .            .       +     .       *     .             .
 1.400D-04  8.230D-01 .            .    +        .        *    .             .
 1.600D-04  8.613D-01 .            .   +         .         *   .             .
 1.800D-04  8.913D-01 .            . +           .          *  .             .
 2.000D-04  9.148D-01 .            .+            .           * .             .
 2.200D-04  9.333D-01 .            +             .           * .             .
 2.400D-04  9.477D-01 .            +             .            *.             .
 2.600D-04  9.590D-01 .           +.             .            *.             .
 2.800D-04  9.679D-01 .           +.             .            *.             .
 3.000D-04  9.749D-01 .          + .             .            *.             .
 3.200D-04  9.803D-01 .          + .             .            *.             .
 3.400D-04  9.846D-01 .          + .             .             *             .
 3.600D-04  9.879D-01 .          + .             .             *             .
 3.800D-04  9.905D-01 .          + .             .             *             .
 4.000D-04  9.926D-01 .          + .             .             *             .
                 - - - - - - - - - - - - - - - - - - - - - - - - - - - - - - - -

         JOB CONCLUDED
```

```
*******11/07/88 ********  SPICE 2G.6    3/15/83 ********20:59:02*****
SPICE 7.2
****      INPUT LISTING                     TEMPERATURE =   27.000 DEG C
****************************************************************************
 R1 5 2 1K
 R2 4 0 1K
 R3 7 0 20K
 C 6 3 100P
 V 1 0 PULSE(0 2 1F 1F)
 F 3 0 VAMX 80
 VAM1 1 5
 VAM2 2 6
 VAM3 3 7
 VAMX 2 4
 .TRAN 20U 400U
 .PLOT TRAN I(VAM1) I(VAM2) I(VAM3)
 .END
*******11/07/88 ********  SPICE 2G.6    3/15/83 ********20:59:03*****
SPICE 7.2
****      INITIAL TRANSIENT SOLUTION        TEMPERATURE =   27.000 DEG C
****************************************************************************
  NODE   VOLTAGE     NODE   VOLTAGE     NODE   VOLTAGE     NODE   VOLTAGE
 (  1)     .0000    (  2)     .0000    (  3)     .0000    (  4)     .0000
 (  5)     .0000    (  6)     .0000    (  7)     .0000
```

```
     VOLTAGE SOURCE CURRENTS
     NAME        CURRENT
     V          0.000D+00
     VAM1       0.000D+00
     VAM2       0.000D+00
     VAM3       0.000D+00
     VAMX       0.000D+00
     TOTAL POWER DISSIPATION   0.00D+00  WATTS
*******11/07/88 ********  SPICE 2G.6    3/15/83 ********20:59:03*****
SPICE 7.2
****     OPERATING POINT INFORMATION      TEMPERATURE =   27.000 DEG C
*******************************************************************************
**** CURRENT-CONTROLLED CURRENT SOURCES
            F
 I-SOURCE  0.00E+00



*******11/07/88 ********  SPICE 2G.6    3/15/83 ********20:59:03*****
SPICE 7.2
****     TRANSIENT ANALYSIS               TEMPERATURE =   27.000 DEG C
*******************************************************************************
LEGEND:
 *: I(VAM1)
 +: I(VAM2)
 =: I(VAM3)

    TIME       I(VAM1)
(*+)-----------  0.000D+00     5.000D-04      1.000D-03      1.500D-03  2.000D-03
                        - - - - - - - - - - - - - - - - - - - - - - - - - - - - -
(=)------------ -8.000D-02    -6.000D-02     -4.000D-02     -2.000D-02  0.000D+00
                        - - - - - - - - - - - - - - - - - - - - - - - - - - - - -
 0.000D+00   0.000D+00 X             .              .             .            =
 2.000D-05   1.765D-03 .             .              .             .+=     *    .
 4.000D-05   1.600D-03 .             .              .      X      .  *         .
 6.000D-05   1.470D-03 .             .            +=.            *.            .
 8.000D-05   1.368D-03 .             .       X      .          *  .            .
 1.000D-04   1.288D-03 .             . +=           .        *    .            .
 1.200D-04   1.226D-03 .            X.              .      *      .            .
 1.400D-04   1.177D-03 .         X   .              .     *       .            .
 1.600D-04   1.139D-03 .       X     .              .    *        .            .
 1.800D-04   1.109D-03 .     X       .              .   *         .            .
 2.000D-04   1.085D-03 .    X        .              .  *          .            .
 2.200D-04   1.067D-03 .   X         .              .  *          .            .
 2.400D-04   1.052D-03 .  X          .              .*            .            .
 2.600D-04   1.041D-03 . X           .              .*            .            .
 2.800D-04   1.032D-03 . X           .              .*            .            .
 3.000D-04   1.025D-03 .X            .              .*            .            .
 3.200D-04   1.020D-03 .X            .              .*            .            .
 3.400D-04   1.015D-03 .X            .              *             .            .
 3.600D-04   1.012D-03 .X            .              *             .            .
 3.800D-04   1.009D-03 .X            .              *             .            .
 4.000D-04   1.007D-03 X             .              *             .            .
                        - - - - - - - - - - - - - - - - - - - - - - - - - - - - -

                 JOB CONCLUDED
```

7.8 SUMMARY

Transform circuit analysis extends the techniques and theorems associated with circuits containing only dc sources and resistors to circuits which also contain time-varying sources and energy storage elements. The increased complexity of the algebra seems a small price to pay when we consider the alternative of formulating and obtaining direct solutions to the initial value problems (consisting of time-dependent differential equations and initial conditions) for these relatively complex circuits.

The reader should recognize and appreciate the experimental, as well as the analytical, implications of the superposition principle and Thevenin's (and Norton's) theorem. In practical design or analysis problems, experimental determination of the behavior of a multisource circuit by using only one source at a time is often enlightening. Similarly, laboratory determination of Thevenin's or Norton's equivalent of a complicated source network by observing open-circuit voltage and short-circuit current (BE CAREFUL!) can be simpler, and certainly more interesting, than mathematical analysis.

7.9 Terms

dual
mesh analysis
node analysis
Norton's theorem
practical voltage source
practical current source
source conversion
Superposition principle
Thevenin's theorem

Problems

7-1. Transform each time-domain source; then perform a source conversion in the *s*-domain. Next perform an inverse transformation, and sketch and label the resulting time-domain source.

***a.**

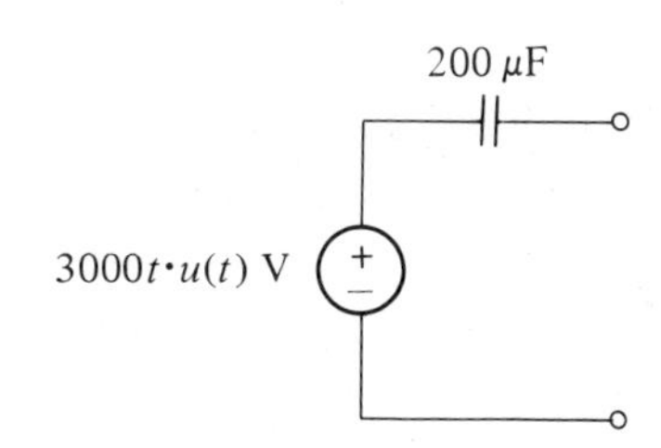

b.

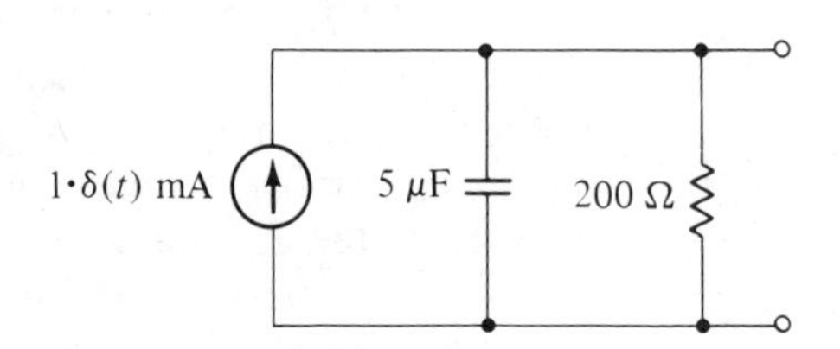

***7-2.** Transform the circuit for node analysis. Express series or parallel passive elements as a single admittance where possible. Write the s-domain node equations for nodes A, B, and C.

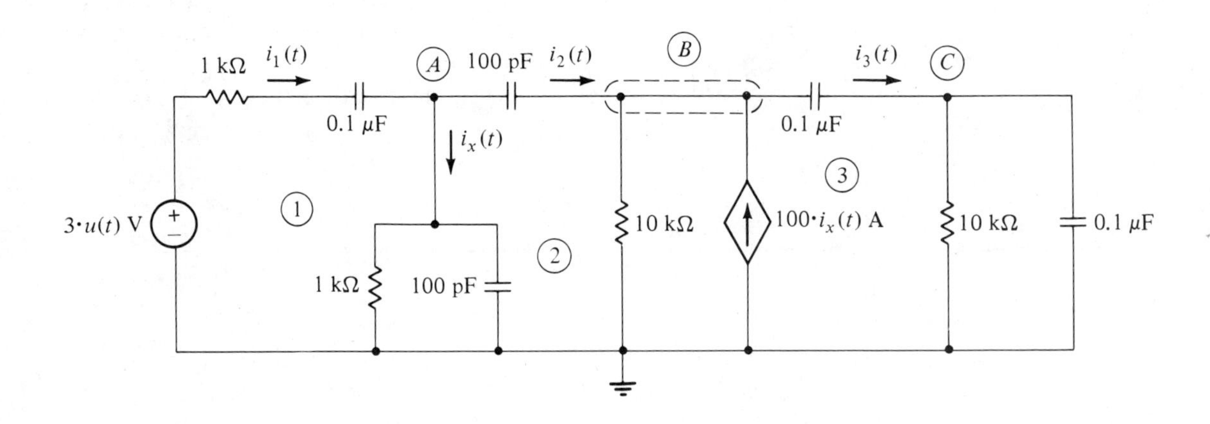

SPICE ***7-3.** Determine $v_A(t)$ for Example 7.3.

SPICE ***7-4.** Use s-domain node analysis to determine $v_C(t)$.

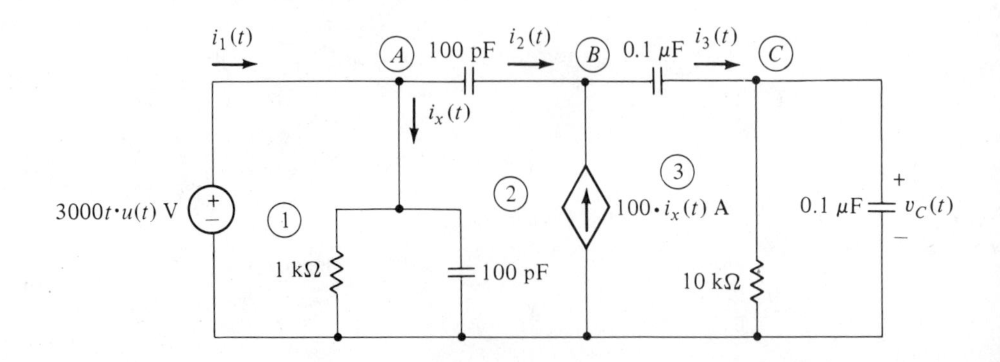

***7-5.** Transform the circuit of Problem 7-2 for mesh analysis. Express series or parallel passive elements as a single impedance where possible. Write the s-domain mesh equations for meshes 1, 2, and 3.

SPICE ***7-6.** Determine $i_S(t)$ for Example 7.5.

SPICE ***7-7.** Use s-domain mesh analysis to determine $I_3(s)$ for the circuit of Problem 7-4.

SPICE ***7-8.** Use s-domain superposition to determine $v(t)$.

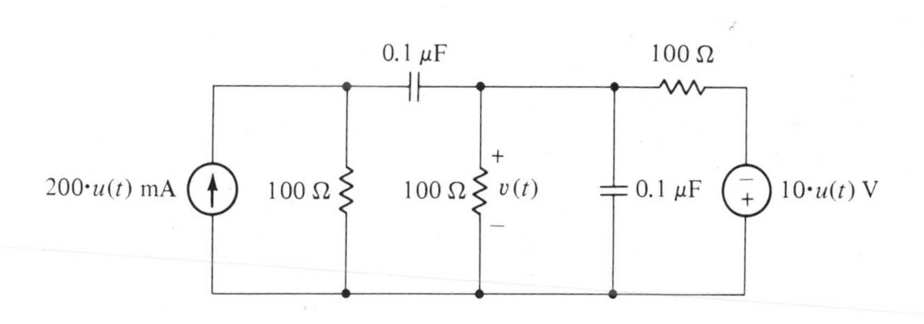

***7-9.** Use s-domain superposition to determine the component of inductor current due to the sinusoidal voltage source alone.

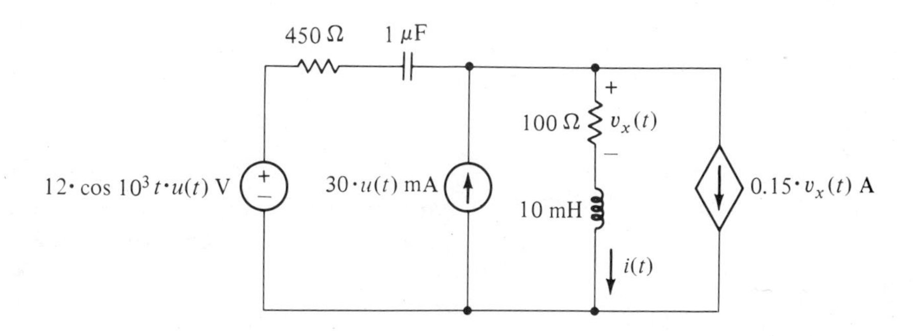

***7-10.** Determine the s-domain Thevenin's equivalent source network by repeated source conversions and passive element combinations.

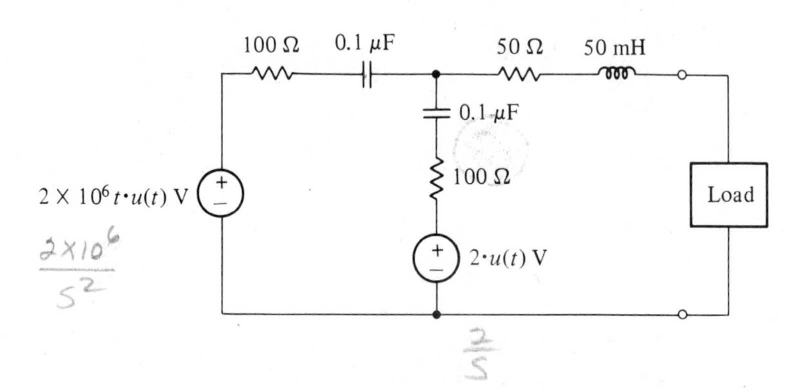

7-11. Disable the sources and confirm $Z_{Th}(s)$ for the circuit of Problem 7-10.

***7-12.** Determine Norton's equivalent source network. Use node analysis to calculate $V_{oc}(s)$ and mesh analysis to calculate $I_{sc}(s)$.

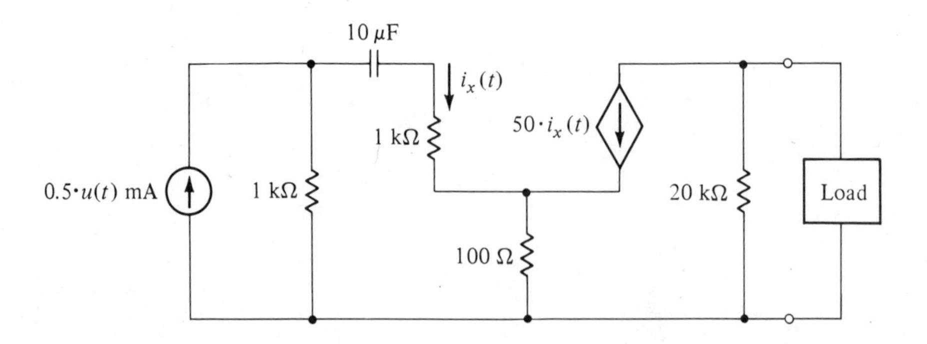

7-13. Confirm that the network of Figure 7.15(c) has the admittance given as $Y_{No}(s)$ in Example 7.11. This will require careful calculation to four or more significant figures. Be sure to factor both numerator and denominator.

***7-14.** Determine $Z_{Th}(s)$, and sketch the network of passive elements which can be used to represent $Z_{Th}(s)$. Remove the controlled source, and compare the remaining passive network with the network used to represent $Z_{Th}(s)$. Explain the effect of the controlled source in terms of the comparison.

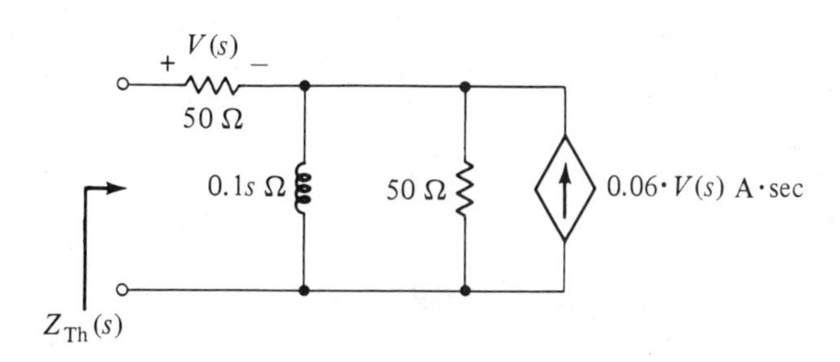

8 Two-Port Network Analysis

8.1 INTRODUCTION

A network is often categorized by the number n of accessible pairs of terminals, called **ports**, as an n-**port network**. An n-port network has at least $n + 1$, and possibly as many as $2n$, accessible terminals, so that a one-port network has at least two accessible terminals, a two-port network has at least three, and possibly four, accessible terminals, and so on. A number of common electrical and electronic devices, as simple as a transformer or a transistor, or as complex as an operational amplifier (OPAMP) or a multi-pole filter, may be treated as two-port networks.

Thus far we have restricted ourselves to one-port analysis (even though more ports may have been accessible)—that is, we have analyzed the voltage between two terminals, or the current in a branch connected between the two terminals. If the network contained only passive linear elements and linear dependent sources, we were able to determine an impedance or admittance which characterized the voltage–current relationship at the port. If the network contained independent sources, as well as passive linear elements and linear dependent sources, we were able employ Thevenin's [or Norton's] Theorem to determine an equivalent network consisting of $V_{Th}(s)$ [or $I_{No}(s)$] and $Z_{Th}(s)$ [or $Y_{No}(s)$]. In one-port analysis, if the network is a load, the port is called a **driving point** or an **input port**; if the network is a source, the port is called the **load terminals** or an **output port**.

In this section, we will consider two-port networks [Figure 8.1(a)] which occur between a source network and a load network [Figure 8.1(b)]. In a given situation, the two-port network may be considered as part of the source [Figure 8.1(c)] or as part of the load [Figure 8.1(d)]; however, when a designer wishes to study the load response to various sources while studying the source behavior under various loads, it is convenient to consider the two-port network as a separate entity.

Figure 8.1
Configuration of two-port networks.
(a) Two-port networks.
(b) Occurrence of a two-port network.
(c) Two-port network associated with source network.
(d) Two-port network associated with load network.

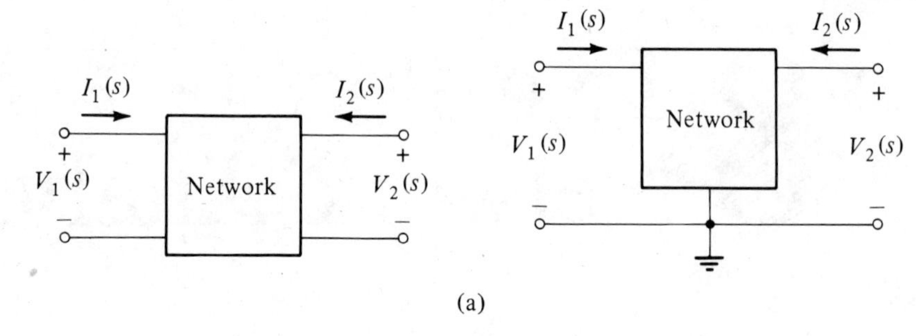

(a)

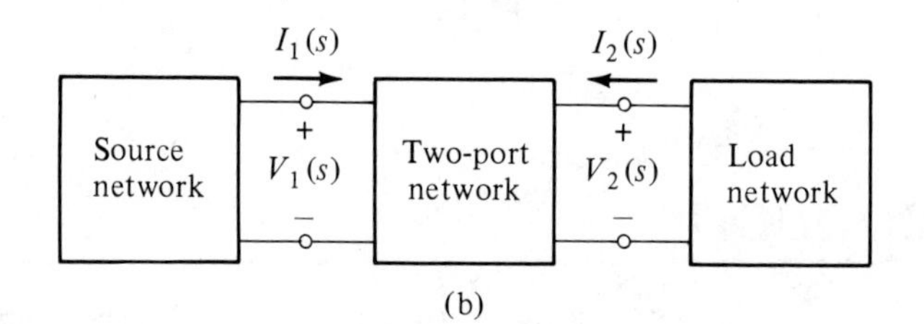

(b)

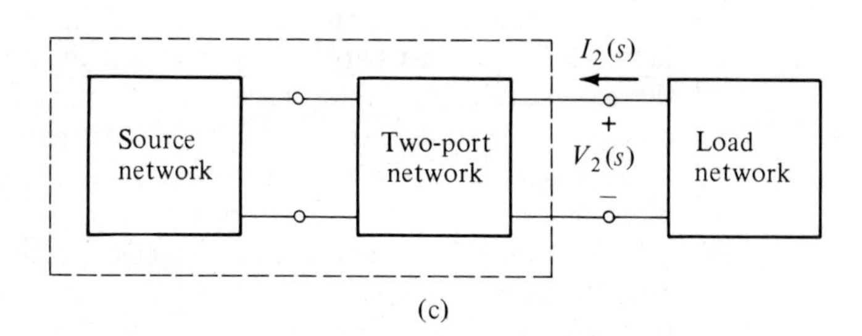

(c)

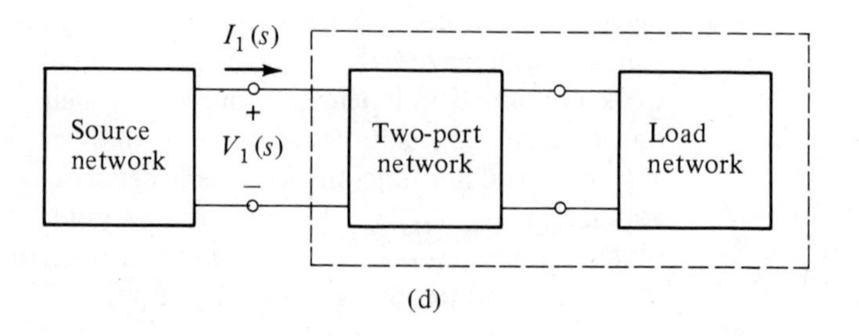

(d)

8.2 Two-Port Network Parameters

Two equations are required to express the relationship of current and voltage at the input port [$V_1(s)$ and $I_1(s)$] to the voltage and current at the output port [$V_2(s)$ and $I_2(s)$],[1] and there are six ways to write these two equations:

$$V_1(s) = z_{11}(s)I_1(s) + z_{12}(s)I_2(s) \tag{8.1a}$$
$$V_2(s) = z_{21}(s)I_1(s) + z_{22}(s)I_2(s) \tag{8.1b}$$

$$I_1(s) = y_{11}(s)V_1(s) + y_{12}(s)V_2(s) \tag{8.2a}$$
$$I_2(s) = y_{21}(s)V_1(s) + y_{22}(s)V_2(s) \tag{8.2b}$$

$$V_1(s) = a_{11}(s)V_2(s) - a_{12}(s)I_2(s) \tag{8.3a}$$
$$I_1(s) = a_{21}(s)V_2(s) - a_{22}(s)I_2(s) \tag{8.3b}$$

$$V_2(s) = b_{11}(s)V_1(s) - b_{12}(s)I_1(s) \tag{8.4a}$$
$$I_2(s) = b_{11}(s)V_1(s) - b_{22}(s)I_1(s) \tag{8.4b}$$

$$V_1(s) = h_{11}(s)I_1(s) + h_{12}(s)V_2(s) \tag{8.5a}$$
$$I_2(s) = h_{21}(s)I_1(s) + h_{22}(s)V_2(s) \tag{8.5b}$$

$$I_1(s) = g_{11}(s)V_1(s) + g_{12}(s)I_2(s) \tag{8.6a}$$
$$V_2(s) = g_{21}(s)V_1(s) + g_{22}(s)I_2(s) \tag{8.6b}$$

The coefficients represent properties of the network, and are commonly called **network parameters** or **two-port parameters**. Each network parameter is, in general, a function of s, although it is quite possible for a parameter to have a constant value. In a resistor network, for example, all parameters are constants. The derivation of each of the two-port parameters is given in Table 8.1.

The parameters z_{ij} in Equations (8.1a) and (8.1b) are called **impedance parameters** (and have the dimensions of ohms) because they relate voltage in the left member of the equation to currents in the right member. Similarly, the parameters y_{ij} in Equations (8.2a) and (8.2b) are called **admittance parameters** (and have the dimensions of siemens) because they relate current in the right member of the equation to voltages in the right member. The impedance and admittance parameters, sometimes called collectively **immittance**[2] parameters,

---

[1] It is conventional to define the output current I_2 as the current *into* the output port.

[2] The term immittance, derived from impedance or admittance, is also used as a general term for the s-domain function which relates excitation and response at the port of a one-port network. If the excitation is current, the function is an impedance function with the dimensions of ohms, and if the excitation is voltage, the function is an admittance function with the dimensions of siemens.

Table 8.1
Two-Port Network Parameters

Impedance parameters			
$z_{11} = \left.\frac{V_1}{I_1}\right\|_{I_2=0}$ Ω	$z_{12} = \left.\frac{V_1}{I_2}\right\|_{I_1=0}$ Ω	$z_{21} = \left.\frac{V_2}{I_1}\right\|_{I_2=0}$ Ω	$z_{22} = \left.\frac{V_2}{I_2}\right\|_{I_1=0}$ Ω
Admittance parameters			
$y_{11} = \left.\frac{I_1}{V_1}\right\|_{V_2=0}$ S	$y_{12} = \left.\frac{I_1}{V_2}\right\|_{V_1=0}$ S	$y_{21} = \left.\frac{I_2}{V_1}\right\|_{V_2=0}$ S	$y_{22} = \left.\frac{I_2}{V_2}\right\|_{V_1=0}$ S
Transmission parameters			
$a_{11} = \left.\frac{V_1}{V_2}\right\|_{I_2=0}$	$a_{12} = \left.\frac{V_1}{I_2}\right\|_{V_2=0}$ Ω	$a_{21} = \left.\frac{I_1}{V_2}\right\|_{I_2=0}$ S	$a_{22} = \left.\frac{I_1}{I_2}\right\|_{V_2=0}$
$b_{11} = \left.\frac{V_2}{V_1}\right\|_{I_1=0}$	$b_{12} = \left.\frac{V_2}{I_1}\right\|_{V_1=0}$ Ω	$b_{21} = \left.\frac{I_2}{V_1}\right\|_{I_1=0}$ S	$b_{22} = \left.\frac{I_2}{I_1}\right\|_{V_1=0}$
Hybrid parameters			
$h_{11} = \left.\frac{V_1}{I_1}\right\|_{V_2=0}$ Ω	$h_{12} = \left.\frac{V_1}{V_2}\right\|_{I_1=0}$	$h_{21} = \left.\frac{I_2}{I_1}\right\|_{V_2=0}$	$h_{22} = \left.\frac{I_2}{V_2}\right\|_{I_1=0}$ S
$g_{11} = \left.\frac{I_1}{V_1}\right\|_{I_2=0}$ S	$g_{12} = \left.\frac{I_1}{I_2}\right\|_{V_1=0}$	$g_{21} = \left.\frac{V_2}{V_1}\right\|_{I_2=0}$	$g_{22} = \left.\frac{V_2}{I_2}\right\|_{V_1=0}$ Ω

are useful in the analysis and design of power distribution networks, filters, and impedance-matching networks.

The parameters a_{ij} and b_{ij} in Equations (8.3a) through (8.4b) are collectively called **transmission parameters** because they relate voltage *or* current at one port to voltage *and* current at the other port. For obvious reasons, the parameters h_{ij} and g_{ij} in Equations (8.5a) through (8.6b) are called **cross** or **hybrid parameters**. As can be seen in Table 8.1, different transmission parameters have different dimensions, and the same is true for hybrid parameters. The h-parameters are used extensively in analysis and design of circuits containing transistors which operate in a linear mode. Standard z-, y-, h-, and g-parameter models are given in Figure 8.2.

Careful consideration of Table 8.1 will provide a greater understanding of network parameters. For example,

$$z_{11} = \left.\frac{V_1}{I_1}\right|_{I_2=0}$$

means that z_{11} is the impedance (ratio of voltage to current) at the input port when there is no load connected to the output port ($I_2 = 0$), and

$$z_{21} = \left.\frac{V_2}{I_1}\right|_{I_2=0}$$

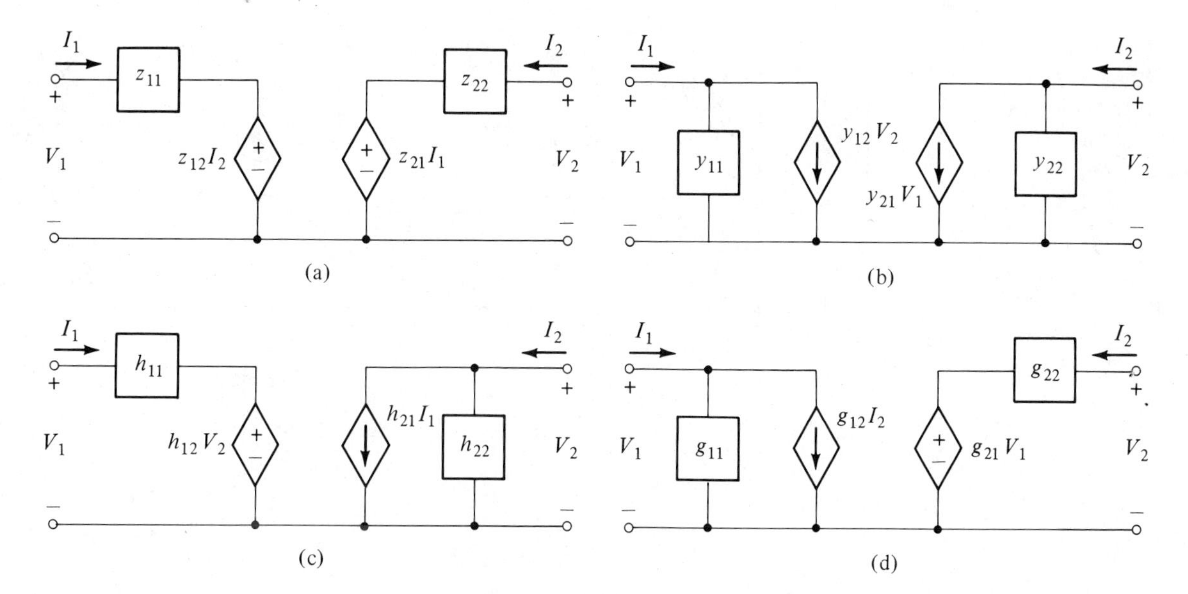

Figure 8.2 Two-port network models. (a) z-parameter model. (b) y-parameter model. (c) h-parameter model. (d) g-parameter model.

relates the no-load output voltage to the input current.

$$z_{22} = \left.\frac{V_2}{I_2}\right|_{I_1=0}$$

means that z_{22} is the impedance at the output port when the input port is open ($I_1 = 0$), and

$$z_{12} = \left.\frac{V_1}{I_2}\right|_{I_1=0}$$

relates output current to voltage across the input port when the input port is open. z_{11} and z_{22} are sometimes called **driving-point impedances**, and z_{21} and z_{12} are sometimes called **transfer impedances**.

Calculation of Network Parameters

Any set of network parameters can be determined by measuring or calculating three of the four terminal quantities (V_1, I_1, V_2, and I_2) under two different conditions. For example, to determine the y-parameters (Figure 8.3), it is necessary to measure or calculate V_1, I_1, and I_2 with the output port shorted and a source connected to the input port, then measure or calculate I_1, V_2, and

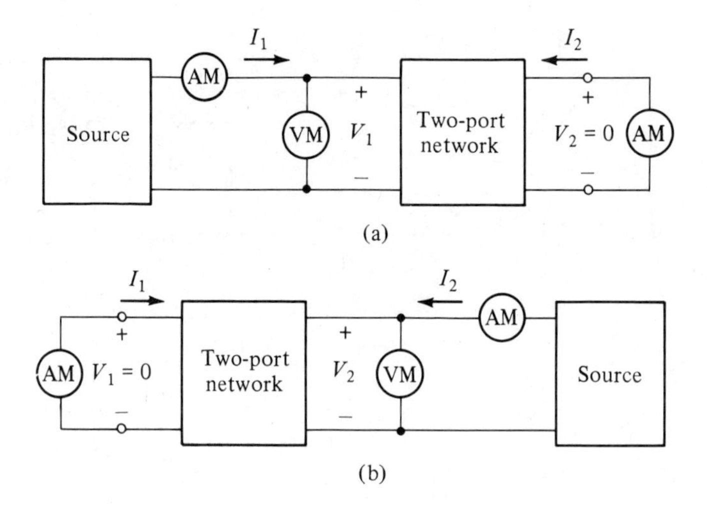

Figure 8.3 Measurement of y-parameters.

I_2 with the input port shorted and a source connected to the output port. From the first three measurements,

$$y_{11} = I_1/V_1 \quad \text{and} \quad y_{21} = I_2/V_1$$

and from the next three measurements,

$$y_{12} = I_1/V_2 \quad \text{and} \quad y_{22} = I_2/V_2$$

Example 8.1

Problem:

Determine the z-parameters for the passive linear network shown in Figure 8.4(a). Sketch and label the standard z-parameter model for the network.

Solution:

1. Connect a 1-mA · sec source to the input port [Figure 8.4(b)], and use node analysis to calculate $V_1(s)$ and $V_2(s)$ with the output port open:

$$\text{Input node:} \quad (2 \times 10^{-3}s + 1)V_1 \qquad - V_2 = 1$$

$$\text{Output node:} \quad - V_1 + (10^{-3}s + 1)V_2 = 0$$

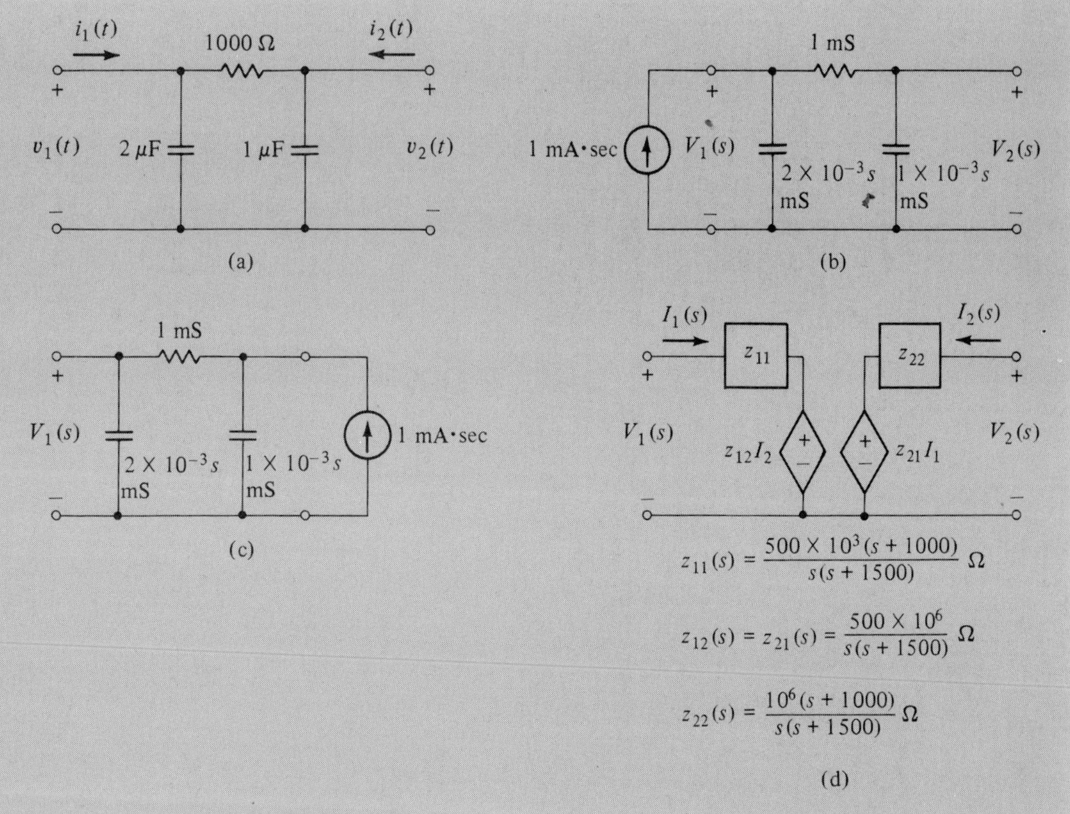

Figure 8.4 Example 8.1.

From the output node equation,

$$V_2 = \frac{1000}{s + 1000} V_1$$

Substitute $V_2(s)$ into the input node equation, and solve for $V_1(s)$:

$$(2 \times 10^{-3}s + 1)V_1 - \frac{1000}{s + 1000} V_1 = 1$$

$$2 \times 10^{-3}(s + 500)(s + 1000)V_1 - 1000V_1 = s + 1000$$

$$V_1 = \frac{500(s + 1000)}{s(s + 1500)} \text{ V} \cdot \text{sec}$$

Substitute $V_1(s)$ into the output node equation, and solve for $V_2(s)$:

$$V_2 = \frac{1000}{s + 1000} \cdot \frac{500(s + 1000)}{s(s + 1500)} = \frac{500 \times 10^3}{s(s + 1500)} \text{ V} \cdot \text{sec}$$

2. Use $V_1(s)$, $I_1(s)$, and $V_2(s)$ to calculate $z_{11}(s)$ and $z_{21}(s)$:

$$z_{11} = \frac{V_1}{I_1} = \frac{\dfrac{500(s + 1000)}{s(s + 1500)}}{10^{-3}} = \frac{500 \times 10^3(s + 1000)}{s(s + 1500)} \ \Omega$$

$$z_{21} = \frac{V_2}{I_1} = \frac{\dfrac{500 \times 10^3}{s(s + 1500)}}{10^{-3}} = \frac{500 \times 10^6}{s(s + 1500)} \ \Omega$$

3. Connect a 1-mA · sec source to the output port [Figure 8.4(c)], and use node analysis to calculate $V_1(s)$ and $V_2(s)$ with the input port open:

Input node: $(2 \times 10^{-3}s + 1)V_1 - V_2 = 0$

Output node: $-V_1 + (10^{-3}s + 1)V_2 = 1$

From the input node equation,

$$V_1 = \frac{500}{s + 500} V_2$$

Substitute $V_1(s)$ into the output node equation, and solve for $V_2(s)$:

$$-\frac{500}{s + 500} V_2 + (10^{-3}s + 1)V_2 = 1$$

$$-500V_2 + (s + 500)(10^{-3}s + 1)V_2 = s + 500$$

$$V_2 = \frac{1000(s + 500)}{s(s + 1500)} \text{ V} \cdot \text{sec}$$

Substitute $V_2(s)$ into the input node equation, and solve for $V_1(s)$:

$$V_1 = \frac{500}{s + 500} \cdot \frac{1000(s + 500)}{s(s + 1500)} = \frac{500 \times 10^3}{s(s + 1500)} \text{ V} \cdot \text{sec}$$

4. Use $V_1(s)$, $V_2(s)$, and $I_2(s)$ to calculate $z_{22}(s)$ and $z_{12}(s)$:

$$z_{22} = \frac{V_2}{I_2} = \frac{\dfrac{1000(s+500)}{s(s+1500)}}{10^{-3}} = \frac{10^6(s+500)}{s(s+1500)}\ \Omega$$

$$z_{12} = \frac{V_1}{I_2} = \frac{\dfrac{500 \times 10^3}{s(s+1500)}}{10^{-3}} = \frac{500 \times 10^6}{s(s+1500)}\ \Omega$$

The z-parameter model is shown in Figure 8.4(d).

In Example 8.1, $z_{12}(s)$ is the same as $z_{21}(s)$ because a network containing only passive linear elements, which are all **bilateral** elements, is itself bilateral —that is, its transfer impedance is the same in either direction. Conversely, any network for which the transfer impedance is the same in either direction $[z_{12}(s) = z_{21}(s)]$ is a bilateral network. Also in Example 8.1, $V_2(s)$ with a 1-mA · sec source at the input port (step 1) was the same as $V_1(s)$ with a 1-mA · sec source at the output port (step 3). Either of these observations is sufficient to say that the network is a **reciprocal network.**[3]

Conversion of Network Parameters

The six sets of network parameters are obviously related, so it is possible to express one set of parameters in terms of another set. For example, if the z-parameters are known and the h-parameters are desired, Equations (8.1a) and (8.1b) can be rearranged so that the variables are in the same relative positions as in Equations (8.5a) and (8.5b). This is accomplished by solving Equation (8.1b) for $I_2(s)$:

$$I_2 = -\frac{z_{12}}{z_{22}} I_2 + \frac{1}{z_{22}} V_2 \tag{8.7}$$

Equation (8.7) now corresponds to Equation (8.5b), so

$$h_{21} = -z_{12}/z_{22} \quad \text{and} \quad h_{22} = 1/z_{22}$$

[3] The distinguishing characteristic, and indeed the definition, of a reciprocal network is as follows: A source connected in parallel with, or in series with, branch A will produce a voltage V across or a current I through branch B of the same network. If the source is moved to branch B, the same voltage V or current I will be produced in branch A. This observation is sometimes called the **reciprocity theorem**.

Substituting Equation (8.7) into Equation (8.1a) yields

$$V_1 = \frac{z_{11}z_{22} - z_{12}z_{21}}{z_{22}} I_1 + \frac{z_{12}}{z_{22}} V_2 \tag{8.8}$$

which corresponds to Equation (8.5a); therefore,

$$h_{11} = \frac{z_{11}z_{22} - z_{12}z_{21}}{z_{22}} \quad \text{and} \quad h_{12} = \frac{z_{12}}{z_{22}}$$

Table 8.2 summarizes the conversions between network parameters.

Table 8.2 Coversion of Two-Port Network Parameters

Parameter Desired	In Terms of z_{ij}	y_{ij}	a_{ij}	b_{ij}	h_{ij}	g_{ij}
z_{11}	z_{11}	$\frac{y_{22}}{\Delta y}$	$\frac{a_{11}}{a_{21}}$	$\frac{b_{22}}{b_{21}}$	$\frac{\Delta h}{h_{22}}$	$\frac{1}{g_{11}}$
z_{12}	z_{12}	$-\frac{y_{21}}{\Delta y}$	$\frac{\Delta a}{a_{21}}$	$\frac{1}{b_{21}}$	$\frac{h_{12}}{h_{22}}$	$-\frac{g_{12}}{g_{11}}$
z_{21}	z_{21}	$-\frac{y_{21}}{\Delta y}$	$\frac{1}{a_{21}}$	$\frac{\Delta b}{b_{21}}$	$-\frac{h_{21}}{h_{22}}$	$\frac{g_{21}}{g_{11}}$
z_{22}	z_{22}	$\frac{y_{11}}{\Delta y}$	$\frac{a_{22}}{a_{21}}$	$\frac{b_{11}}{b_{21}}$	$\frac{1}{h_{22}}$	$\frac{\Delta g}{g_{11}}$
y_{11}	$\frac{z_{22}}{\Delta z}$	y_{11}	$\frac{a_{22}}{a_{12}}$	$\frac{b_{11}}{b_{12}}$	$\frac{1}{h_{11}}$	$\frac{\Delta g}{g_{22}}$
y_{12}	$-\frac{z_{12}}{\Delta z}$	y_{12}	$-\frac{\Delta a}{a_{12}}$	$-\frac{1}{b_{12}}$	$-\frac{h_{12}}{h_{11}}$	$\frac{g_{12}}{g_{22}}$
y_{21}	$-\frac{z_{21}}{\Delta z}$	y_{21}	$-\frac{1}{a_{12}}$	$-\frac{\Delta b}{b_{12}}$	$\frac{h_{21}}{h_{11}}$	$-\frac{g_{21}}{g_{22}}$
y_{22}	$\frac{z_{11}}{\Delta z}$	y_{22}	$\frac{a_{11}}{a_{12}}$	$\frac{b_{22}}{b_{12}}$	$\frac{\Delta h}{h_{11}}$	$\frac{1}{g_{22}}$
a_{11}	$\frac{z_{11}}{z_{21}}$	$-\frac{y_{22}}{y_{21}}$	a_{11}	$\frac{b_{22}}{\Delta b}$	$-\frac{\Delta h}{h_{21}}$	$\frac{1}{g_{21}}$
a_{12}	$\frac{\Delta z}{z_{21}}$	$-\frac{1}{y_{21}}$	a_{12}	$\frac{b_{12}}{\Delta b}$	$-\frac{h_{11}}{h_{21}}$	$\frac{g_{22}}{g_{21}}$
a_{21}	$\frac{1}{z_{21}}$	$-\frac{\Delta y}{y_{21}}$	a_{21}	$\frac{b_{21}}{\Delta b}$	$-\frac{h_{22}}{h_{21}}$	$\frac{g_{11}}{g_{21}}$
a_{22}	$\frac{z_{22}}{z_{21}}$	$-\frac{y_{11}}{y_{21}}$	a_{22}	$\frac{b_{11}}{\Delta b}$	$-\frac{1}{h_{21}}$	$\frac{\Delta g}{g_{21}}$

$\Delta z = z_{11}z_{22} - z_{12}z_{21}$ $\Delta a = a_{11}a_{22} - a_{12}a_{21}$ $\Delta h = h_{11}h_{22} - h_{12}h_{21}$
$\Delta y = y_{11}y_{22} - y_{12}y_{21}$ $\Delta b = b_{11}b_{22} - b_{12}b_{21}$ $\Delta g = g_{11}g_{22} - g_{12}g_{21}$

Table 8.2 Continued

b_{11}	$\frac{z_{22}}{z_{12}}$	$-\frac{y_{11}}{y_{12}}$	$\frac{a_{22}}{\Delta a}$	b_{11}	$\frac{1}{h_{12}}$	$-\frac{\Delta g}{g_{12}}$
b_{12}	$\frac{\Delta z}{z_{12}}$	$-\frac{1}{y_{12}}$	$\frac{a_{12}}{\Delta a}$	b_{12}	$\frac{h_{11}}{h_{12}}$	$-\frac{g_{22}}{g_{12}}$
b_{21}	$\frac{1}{z_{12}}$	$-\frac{\Delta y}{y_{12}}$	$\frac{a_{21}}{\Delta a}$	b_{21}	$\frac{h_{22}}{h_{12}}$	$-\frac{g_{11}}{g_{12}}$
b_{22}	$\frac{z_{11}}{z_{12}}$	$-\frac{y_{22}}{y_{12}}$	$\frac{a_{11}}{\Delta a}$	b_{22}	$\frac{\Delta h}{h_{12}}$	$-\frac{1}{g_{12}}$
h_{11}	$\frac{\Delta z}{z_{22}}$	$\frac{1}{y_{11}}$	$\frac{a_{12}}{a_{22}}$	$\frac{b_{12}}{b_{11}}$	h_{11}	$\frac{g_{22}}{\Delta g}$
h_{12}	$\frac{z_{12}}{z_{22}}$	$-\frac{y_{12}}{y_{11}}$	$\frac{\Delta a}{a_{22}}$	$\frac{1}{b_{11}}$	h_{12}	$-\frac{g_{12}}{\Delta g}$
h_{21}	$-\frac{z_{21}}{z_{22}}$	$\frac{y_{21}}{y_{11}}$	$-\frac{1}{a_{22}}$	$-\frac{\Delta b}{b_{11}}$	h_{21}	$-\frac{g_{21}}{\Delta g}$
h_{22}	$\frac{1}{z_{22}}$	$\frac{\Delta y}{y_{11}}$	$\frac{a_{21}}{a_{22}}$	$\frac{b_{21}}{b_{11}}$	h_{22}	$\frac{g_{11}}{\Delta g}$
g_{11}	$\frac{1}{z_{11}}$	$\frac{\Delta y}{y_{22}}$	$\frac{a_{21}}{a_{11}}$	$\frac{b_{21}}{b_{22}}$	$\frac{h_{22}}{\Delta h}$	g_{11}
g_{12}	$-\frac{z_{12}}{z_{11}}$	$\frac{y_{12}}{y_{22}}$	$-\frac{\Delta a}{a_{11}}$	$-\frac{1}{b_{22}}$	$-\frac{h_{12}}{\Delta h}$	g_{12}
g_{21}	$\frac{z_{21}}{z_{11}}$	$-\frac{y_{21}}{y_{22}}$	$\frac{1}{a_{11}}$	$\frac{\Delta b}{b_{22}}$	$-\frac{h_{21}}{\Delta h}$	g_{21}
g_{22}	$\frac{\Delta z}{z_{11}}$	$\frac{1}{y_{22}}$	$\frac{a_{12}}{a_{11}}$	$\frac{b_{12}}{b_{22}}$	$\frac{h_{11}}{\Delta h}$	g_{22}

$\Delta z = z_{11}z_{22} - z_{12}z_{21}$ $\Delta a = a_{11}a_{22} - a_{12}a_{21}$ $\Delta h = h_{11}h_{22} - h_{12}h_{21}$

$\Delta y = y_{11}y_{22} - y_{12}y_{21}$ $\Delta b = b_{11}b_{22} - b_{12}b_{21}$ $\Delta g = g_{11}g_{22} - g_{12}g_{21}$

Exercise 8.1

Problem:

The following measurements were made for a two-port resistive network:

Output port shorted: $V_1 = 0.1$ V, $I_1 = 85$ μA, $I_2 = 7$ mA.

Input port open: $V_1 = 10$ mV, $V_2 = 50$ V, $I_2 = 2$ mA.

Calculate the h-parameters, and convert the h-parameters to y-parameters. Sketch the standard h-parameter and y-parameter models for the network.

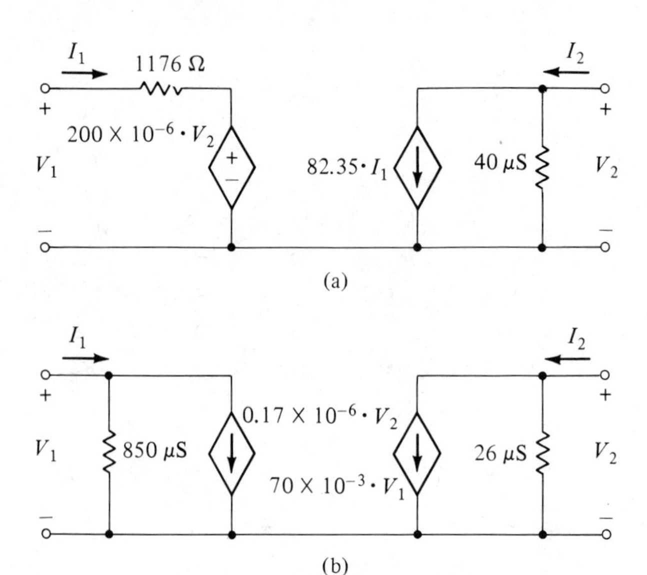

Figure 8.5
Exercise 8.1.
(a) *h*-parameter model.
(b) *y*-parameter model.

Answer:

$$h_{11} = 1176\ \Omega, \qquad h_{21} = 82.35$$

$$h_{12} = 200 \times 10^{-6}, \qquad h_{22} = 40\ \mu\text{S}$$

$$y_{11} = 850\ \mu\text{S}, \qquad y_{21} = 70\ m\text{S}$$

$$y_{12} = -0.17\ \mu\text{S}, \qquad y_{22} = 26\ \mu\text{S}$$

Refer to Figure 8.5 for the network models.

8.3 Transfer Functions

When all that is required in a network analysis problem is to express the relationship between one output variable [$V_2(s)$ or $I_2(s)$] and one input variable [$V_1(s)$ or $I_1(s)$], the use of network parameters introduces undue complication. In these instances, the use of a **transfer function**

$$H(s) = \frac{F_{\text{OUT}}(s)}{F_{\text{IN}}(s)} \tag{8.9}$$

where $F_{IN}(s)$ is the excitation and $F_{OUT}(s)$ is the response, is more appropriate. Four possible transfer functions for a two-port network are

$$H(s) = \frac{V_2(s)}{V_1(s)} \tag{8.10a}$$

$$H(s) = \frac{V_2(s)}{I_1(s)} \tag{8.10b}$$

$$H(s) = \frac{I_2(s)}{V_1(s)} \tag{8.10c}$$

$$H(s) = \frac{I_2(s)}{I_1(s)} \tag{8.10d}$$

Transfer functions which relate like quantities [Equations (8.10a) and (8.10d)] are dimensionless; however, transfer functions which relate unlike quantities have dimensions such as ohms [Equation (8.10b)], siemens [Equation (8.10c)], etc. Equations (8.10a) through (8.10d) are more properly called **forward transfer functions**, since the reciprocal relationships, called **reverse transfer functions**, may also be useful. The term transfer function is used here to imply forward transfer function; the term reverse transfer function will be used explicitly when required.

Conventionally, transfer functions involving output voltage [Equations (8.10a) and (8.10b)] are specified for open-circuit conditions, and transfer functions involving output current [Equations (8.10c) and (8.10d)] are specified for short-circuit conditions. Any deviation from this convention indicates that the analyst has combined the two-port network with a particular load [Figure 8.2(d)] which is neither an open circuit nor a short circuit. The use of a transfer function can still be meaningful if the load conditions are specified; but if the load is changed, a new transfer function must be derived.

Deriving a Transfer Function by Circuit Analysis

The transfer functions for a particular two-port network can be derived by straightforward application of s-domain circuit analysis techniques. In deriving the transfer function, all initial conditions within the two-port network must be zero—that is, no capacitor or inductor may have stored energy.

Example 8.2

Problem:

Derive transfer functions corresponding to Equations (8.10a) through (8.10d) for the network of Figure 8.4 (Example 8.1).

Solution:

1. Determine $H(s) = V_2(s)/V_1(s)$. With the output terminals open, use voltage division to calculate $V_2(s)$ due to $V_1(s)$:

$$V_2(s) = V_1(s) \cdot \frac{10^6/s}{1000 + 10^6/s} = V_1(s) \cdot \frac{1000}{s + 1000}$$

Therefore,

$$H(s) = \frac{V_2(s)}{V_1(s)} = \frac{1000}{s + 1000}$$

2. Determine $H(s) = V_2(s)/I_1(s)$. With the output terminals open, use current division to calculate the current $I_C(s)$ through the 1-μF capacitor due to $I_1(s)$:

$$I_C(s) = I_1(s)\frac{\dfrac{1}{1000 + 10^6/s}}{2 \times 10^{-6}s + \dfrac{1}{1000 + 10^6/s}} = I_1(s)\frac{500}{s + 1500}$$

$$H(s) = \frac{V_2(s)}{I_1(s)} = \frac{(10^6/s) \cdot I_C(s)}{I_1(s)} = \frac{500 \times 10^6}{s(s + 1500)}\ \Omega$$

3. Determine $H(s) = I_2(s)/V_1(s)$. With the output terminals shorted,

$$I_2(s) = -\frac{V_1(s)}{1000}$$

Therefore,

$$H(s) = \frac{I_2(s)}{V_1(s)} = -0.001$$

4. Determine $H(s) = I_2(s)/I_1(s)$. With the output terminals shorted, use current division to calculate $I_2(s)$ due to $I_1(s)$:

$$I_2(s) = -I_1(s) \cdot \frac{10^{-3}}{2 \times 10^{-6}s + 10^{-3}} = -I_1(s) \cdot \frac{500}{s + 500}$$

Therefore,

$$H(s) = \frac{I_2(s)}{I_1(s)} = -\frac{500}{s + 500}$$

Table 8.3
Calculation of Transfer Functions from Network Parameters

Transfer Function	*z*-Parameters	*y*-Parameters	*a*-Parameters	*b*-Parameters	*h*-Parameters	*g*-Parameters
$\frac{V_2}{V_1} =$	$\frac{z_{21}}{z_{11}}$	$-\frac{y_{21}}{y_{22}}$	a_{11}	$\frac{\Delta b}{b_{22}}$	$-\frac{h_{21}}{\Delta h}$	g_{21}
$\frac{V_2}{I_1} =$	z_{21}	$-\frac{y_{21}}{\Delta y}$	a_{21}	$\frac{\Delta b}{b_{21}}$	$-\frac{h_{21}}{h_{22}}$	$\frac{g_{21}}{g_{11}}$
$\frac{I_2}{V_1} =$	$-\frac{z_{21}}{\Delta z}$	y_{21}	$-\frac{1}{a_{12}}$	$-\frac{\Delta b}{b_{12}}$	$\frac{h_{21}}{h_{11}}$	$-\frac{g_{21}}{g_{22}}$
$\frac{I_2}{I_1} =$	$-\frac{z_{21}}{z_{22}}$	$\frac{y_{21}}{y_{11}}$	$-\frac{1}{a_{22}}$	$\frac{\Delta b}{b_{11}}$	h_{21}	$-\frac{g_{21}}{\Delta g}$

$\Delta z = z_{11}z_{22} - z_{12}z_{21}$ $\qquad$ $\Delta h = h_{11}h_{22} - h_{12}h_{21}$

$\Delta y = y_{11}y_{22} - y_{12}y_{21}$ $\qquad$ $\Delta g = g_{11}g_{21} - g_{12}g_{21}$

$\Delta b = b_{11}b_{22} - b_{12}b_{21}$

Deriving a Transfer Function from Network Parameters

If any set of network parameters is known, the transfer functions may be calculated from those parameters without additional network analysis. For example, if the condition $I_2(s) = 0$ (output port open) is enforced on Equations (8.1a) and (8.1b), then

$$V_1(s) = z_{11}(s)I_1(s) \tag{8.11a}$$
$$V_2(s) = z_{21}(s)I_1(s) \tag{8.11b}$$

The ratio of Equation (8.11b) to (8.11a) yields the transfer function

$$H(s) = \frac{V_2(s)}{V_1(s)} = \frac{z_{21}(s)}{z_{11}(s)}$$

Table 8.3 summarizes the calculation of transfer functions from network parameters.

Exercise 8.2

Problem:

Use the *z*-parameters obtained in Example 8.1 to confirm the transfer functions obtained in Example 8.2.

Partial Fraction Decomposition of a Transfer Function

The transfer function is very useful in studying the response $F_{OUT}(s)$ at the output terminals to a particular input excitation $F_{IN}(s)$. This application of the transfer function requires rearranging Equation (8.10) so that

$$F_{OUT}(s) = H(s) \cdot F_{IN}(s) \tag{8.12}$$

$F_{OUT}(s)$ can be called the **response function.**

For linear two-port networks, $H(s)$ is a rational function—that is, $H(s)$ is the ratio of two polynomials in s. For the types of sources most often encountered in circuit analysis, $F_{IN}(s)$ is also a rational function. Under these conditions, $H(s) \cdot F_{IN}(s)$ is a rational function, which means that the response function $F_{OUT}(s)$ can be written as a sum of partial fractions with one term for each pole of $H(s)$ and one term for each pole of $F_{IN}(s)$. Those partial fractions associated with a pole of $H(s)$ will determine the transient (or natural) component of the output response, while those associated with a pole of $F_{IN}(s)$ will determine the steady-state (or forced) component of the output response.

Example 8.3

Problem:

The transfer function for the two-port network shown in Figure 8.6 is

$$H(s) = \frac{V_2(s)}{V_1(s)} = \frac{100}{s + 100}$$

Use $H(s)$ to determine $v_2(t)$ when

$$v_1(t) = 500t \cdot u(t) \text{ V}$$

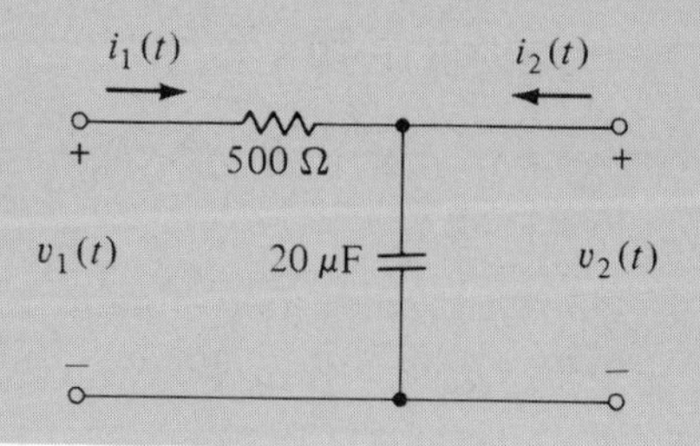

Figure 8.6
Example 8.3.

Solution:

1. Rearrange $H(s)$ to obtain the response function $V_2(s)$:

$$V_2(s) = \frac{100}{s+100} \cdot V_1(s)$$
$$= \frac{100}{s+100} \cdot \frac{500}{s^2} = \frac{50 \times 10^3}{s^2(s+100)} \text{ V} \cdot \text{sec}$$

2. Perform a partial decomposition of $V_2(s)$:

$$V_2(s) = \frac{K_1}{s^2} + \frac{K_2}{s} + \frac{K_3}{s+100} \text{ V} \cdot \text{sec}$$

3. Use Pairs II, IIIa, and IV to obtain $v_2(t)$:

$$v_2(t) = (K_1 t + K_2 + K_3 \cdot \exp[-100t]) \cdot u(t) \text{ V}$$

where

$$K_1 = s^2 \cdot V_2(s)\big|_{s=0} = 500$$

$$K_2 = \frac{d}{ds}[s^2 \cdot V_2(s)]\bigg|_{s=0} = -5$$

$$K_3 = (s+100)V_2(s)\big|_{s=-100} = 5$$

Therefore,

$$v_2(t) = 5(100t - 1 + \exp[-100t])u(t) \text{ V}$$

Clearly the transient component of the response $(5 \cdot \exp[-100t])$ is due to the pole of $H(s)$, while the steady-state response $[5(100t - 1)]$ is due to the poles of the excitation function.

If the excitation is a unit impulse function

$$f_{\text{IN}}(t) = \delta(t)$$

then from Pair I,

$$F_{\text{IN}}(s) = 1$$

and from Equation (8.12),

$$F_{\text{OUT}}(s) = H(s)$$

By inverse transformation,

$$f_{OUT}(t) = h(t)$$

For this reason, $h(t)$ is called the **impulse response.**[4]

Example 8.4

Problem:

Use the transfer function from Example 8.3 to determine $v_2(t)$ for an excitation

$$v_1(t) = \delta(t) \text{ V}$$

Solution:

Rearrange $H(s)$ to obtain the response function $V_2(s)$:

$$V_2(s) = \frac{100}{s+100} \cdot V_1(s) = \frac{100}{s+100} \cdot 1 = \frac{100}{s+100} \text{ V} \cdot \text{sec}$$

From Pair IV,

$$v_2(t) = 100 \cdot \exp[-100t] \cdot u(t) \text{ V}$$

In this example, $v_2(t)$ is identical to $h(t)$, and is the impulse response.

Recall that the steady-state response depends upon the excitation—in particular, upon the pole (or poles) of the excitation function $F_{IN}(s)$. If a pole of the excitation function equals a root of the *numerator* of $H(s)$, then the pole of $F_{IN}(s)$ is canceled by the operation $H(s) \cdot F_{IN}(s)$, with the result that the steady-state response is zero.

The roots of the numerator of a function are called the **zeros** of the function, because the value of the function will be zero for those values of the independent variable. Accordingly, using a source with a pole of $F_{IN}(s)$ equal to a zero of $H(s)$ is referred to as **exciting the network at a zero** (Example 8.5).

Exciting the network at a pole (Example 8.6) also causes the steady-state response to be zero, because the excitation is also a transient. Excitation at a pole does, however, change the nature of the transient response, depending on which pole is excited.

---

[4] The reader is cautioned that $f_{OUT}(t)$ is not the product of $h(t)$ and $f_{IN}(t)$, so the impulse response $h(t)$ cannot be used to determine $f_{OUT}(t)$, *except when the excitation is a unit impulse.*

Example 8.5

Problem:

The transfer function for the two-port network shown in Figure 8.7 is

$$H(s) = \frac{I_2(s)}{I_1(s)} = \frac{-s(s+1500)}{(s+500)(s+1000)}$$

Use $H(s)$ to determine $i_2(t)$ for

$$i_1(t) = u(t) \text{ A}$$

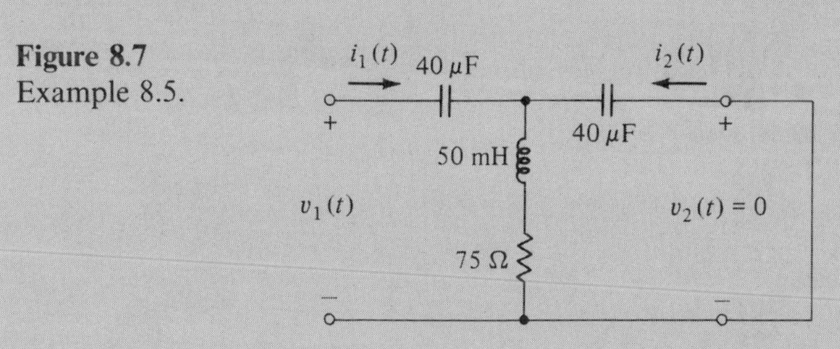

Figure 8.7 Example 8.5.

Solution:

$$I_2(s) = H(s) \cdot I_1(s) = \frac{-s(s+1500)}{(s+1000)(s+500)} \cdot \frac{1}{s}$$

$$= \frac{-(s+1500)}{(s+500)(s+1000)} \text{ A} \cdot \text{sec}$$

[Note that the pole of $I_1(s)$ has been canceled.]

Therefore,

$$I_2(s) = \frac{K_1}{s+500} + \frac{K_2}{s+1000} \text{ A} \cdot \text{sec}$$

From Pair IV,

$$i_2(t) = (K_1 \cdot \exp[-500t] + K_2 \cdot \exp[-1000t])u(t) \text{ A}$$

where

$$K_1 = (s + 500) I_2(s)|_{s=-500} = -2$$
$$K_2 = (s + 1000) I_2(s)|_{s=-1000} = 1$$

Therefore,

$$i_2(t) = -(2 \cdot \exp[-500t] - \exp[-1000t])u(t) \text{ A}$$

Clearly, the steady-state value of $i_2(t)$ is zero due to cancellation of the pole of the excitation $I_1(s)$.

Example 8.6

Problem:

Use the transfer function from Example 8.5 to determine $i_2(t)$ for

$$i_1(t) = \exp[-500t] \cdot u(t) \text{ A}$$

Solution:

$$I_2(s) = H(s) \cdot I_1(s) = \frac{-s(s + 1500)}{(s + 1000)(s + 500)} \cdot \frac{1}{s + 500}$$
$$= \frac{-s(s + 1500)}{(s + 500)^2(s + 1000)} \text{ A} \cdot \text{sec}$$

Therefore,

$$I_2(s) = \frac{K_1}{(s + 500)^2} + \frac{K_2}{s + 500} + \frac{K_3}{s + 1000} \text{ A} \cdot \text{sec}$$

From Pairs VIa and IV,

$$i_2(t) = \{(K_1 t + K_2) \cdot \exp[-500t] + K_2 \cdot \exp[-1000t]\}u(t) \text{ A}$$

where

$$K_1 = (s + 500)^2 I_2(s)|_{s=-500} = 1000$$
$$K_2 = \frac{d}{ds}\left[(s + 500)^2 I_2(s)\right]\Big|_{s=-500} = -3$$
$$K_3 = (s + 1000) I_2(s)|_{s=-1000} = 2$$

Therefore,

$$i_2(t) = \{(1000t - 3)\exp[-500t] + 2 \cdot \exp[-1000t]\}\, u(t) \text{ A}$$

The steady-state value of $i_2(t)$ is zero, as in Example 8.5, but the term $(1000t - 3)\exp[-500t]$ alters the nature of the transient response.

8.4 OPAMP Networks

In this section, we will employ transform circuit analysis techniques to develop a general approach to the analysis of circuits containing **operational amplifiers** (OPAMPS). The discussions that follow assume that the reader is familiar with OPAMPS and their applications in electronic circuits.

The utilization of transform techniques necessarily restricts us to circuit applications in which the OPAMP remains within its linear operating region. We will not, therefore, consider many OPAMP applications such as comparator, clipper, multivibrator, etc., in which the OPAMP becomes **saturated** during normal operation. Further, we will treat the OPAMP only as an ideal three-terminal device.

The Ideal OPAMP

An **ideal** OPAMP [Figure 8.8(a)] has these characteristics:

1. The output voltage v_o depends only upon the **differential input voltage** v_d, and the OPAMP **gain** A:

$$v_o = A(v^+ - v^-) = Av_d$$

2. The OPAMP gain is constant, but is essentially infinite, so for values of v_o within the linear operating range,

$$v_d \simeq 0 \quad \text{and} \quad v_+ \simeq v_-$$

3. The impedance between the **inverting** $(-)$ and **noninverting** $(+)$ input terminals is purely resistive and essentially infinite, so the input current is zero.
4. The output (Thevenin's) impedance is also purely resistive, but is essentially zero, so the output voltage is independent of load.

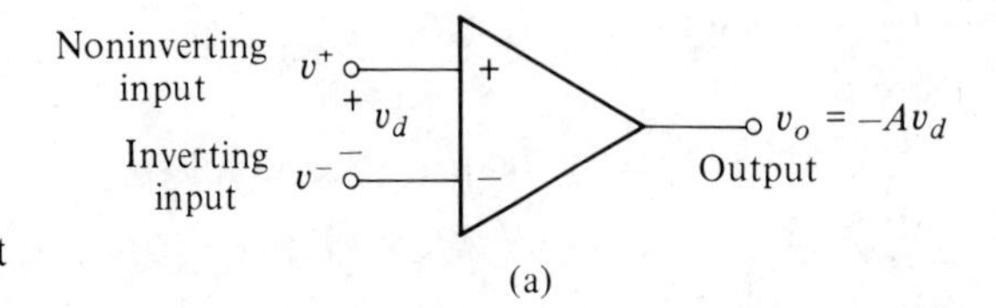

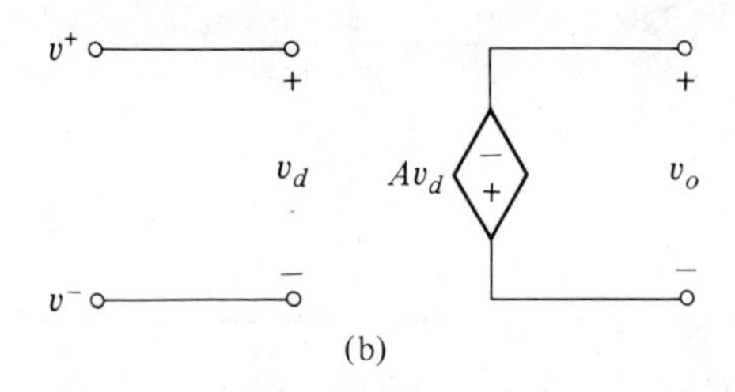

Figure 8.8
Operational amplifier (OPAMP).
(a) OPAMP symbol.
(b) Ideal OPAMP circuit model.

These characteristics of the ideal OPAMP lead to the circuit model of Figure 8.8(b).

In considering the OPAMP as an ideal device, the practical considerations of OPAMP design are suppressed; in particular:

1. The **common-mode rejection ratio** (CMRR) is infinite, which means that v_o is exactly zero when v^+ is exactly equal to v^-.
2. The **slew rate** is infinite, which means that the operation of the OPAMP is independent of the rate of change (or the frequency) of the input.

OPAMP Networks

Restriction of OPAMP operation to the linear region means that the OPAMP circuit must have a **feedback network**—that is, a path between the output terminal and one of the differential input terminals—to prevent saturation of the output voltage v_o. The presence of feedback does not, of itself, prevent saturation, because an OPAMP circuit with feedback can still be driven into saturation. However, through proper design of the feedback network *and* proper choice of the range of the differential input voltage, the OPAMP can be maintained within its linear operating region. Without feedback, and under the assumption of infinite gain, the OPAMP will saturate with the smallest differential input voltage.

Accordingly, two general OPAMP networks, each incorporating exactly one OPAMP, may be considered (Figure 8.9). The network of Figure 8.9(a) will be referred to as an **inverting configuration**, and the configuration of Figure 8.9(b) as a **noninverting configuration**. Laplace transformation of the linear networks

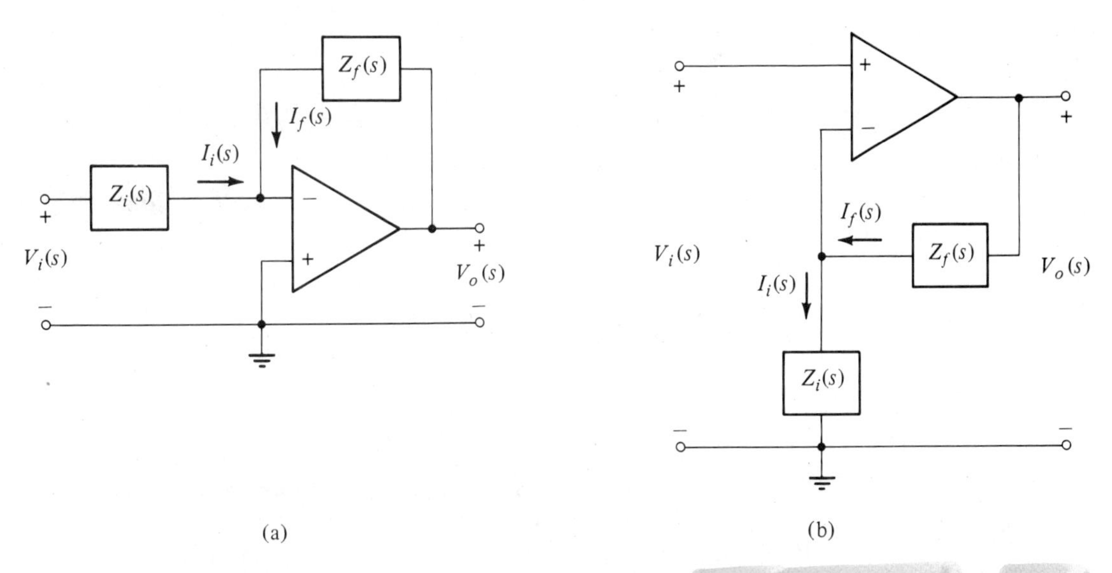

Figure 8.9 General OPAMP networks. (a) Inverting configuration. (b) Noninverting configuration.

external to the OPAMP, and of network voltages and currents, is entirely appropriate, since the OPAMP is restricted to linear operation.

Application of Kirchhoff's current law (KCL) at the inverting input terminal in Figure 8.9(a) yields

$$I_f + I_i = 0 \tag{8.13}$$

since the current entering the inverting terminal is essentially zero for an ideal OPAMP. Further, for an ideal OPAMP, the differential input voltage is essentially zero, so if the noninverting input terminal is at ground potential, the inverting terminal is also at ground potential.[5] Accordingly,

$$I_f = V_o/Z_f \tag{8.14}$$

$$I_i = V_i/Z_i \tag{8.15}$$

Substitution of Equations (8.14) and (8.15) into Equation (8.13) yields

$$\frac{V_o}{Z_f} + \frac{V_i}{Z_i} = 0$$

[5]This is the **virtual ground** concept used by OPAMP circuit designers.

or

$$V_o = -\frac{Z_f}{Z_i} V_i \tag{8.16a}$$

From Equation (8.16a), the voltage transfer function for the inverting configuration is

$$\frac{V_0}{V_i} = -\frac{Z_f}{Z_i} \tag{8.16b}$$

KCL at the inverting input terminal in Figure 8.9(b) yields

$$I_f = I_i \tag{8.17}$$

Since the differential input voltage is essentially zero, if the voltage at the noninverting terminal is $V_i(s)$, then the voltage at the inverting terminal is also $V_i(s)$. Accordingly, Equation (8.15) also applies to the noninverting configuration, but

$$I_f = \frac{V_o - V_i}{Z_f} \tag{8.18}$$

Substitution of Equations (8.15) and (8.18) into (8.17) yields

$$\frac{V_o - V_i}{Z_f} = \frac{V_i}{Z_i}$$

or

$$V_o = \frac{Z_i + Z_f}{Z_i} V_i \tag{8.19a}$$

From Equation (8.19a), the voltage transfer function for the noninverting configuration is

$$\frac{V_o}{V_i} = \frac{Z_i + Z_f}{Z_i} \tag{8.19b}$$

Example 8.7

SPICE

Problem:

In the circuit of Figure 8.10(a), the input voltage is a single positive pulse generated by

$$v_i(t) = 3[u(t) - u(t-2)] \text{ V}$$

Find and graph $v_o(t)$ for $0 < t < 3$ sec.

Figure 8.10
Example 8.7.

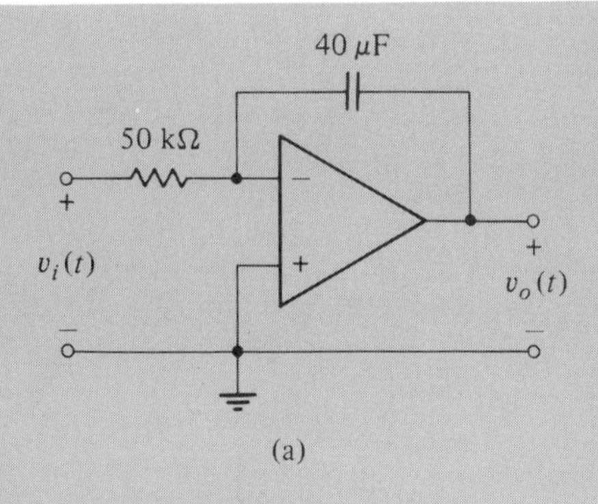

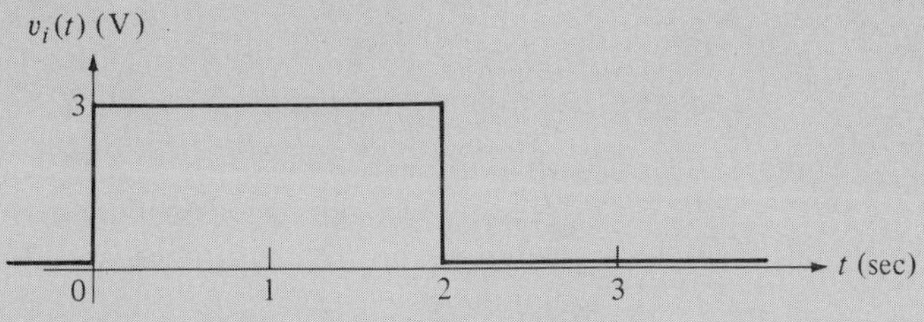

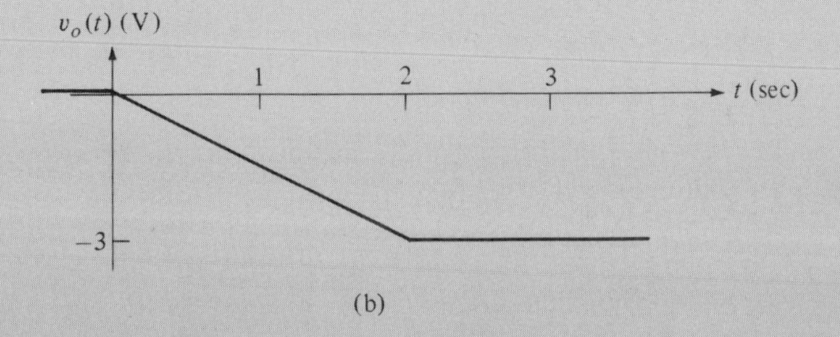

Solution:

1. Transform the input and feedback elements, and find the Laplace transform of the input voltage:

$$Z_i(s) = 50 \times 10^3\ \Omega$$

$$Z_f(s) = \frac{1}{40 \times 10^{-6}s} = \frac{25 \times 10^3}{s}\ \Omega$$

From Pair II and the time-domain shifting property,

$$V_i(s) = \frac{3}{s} - \frac{3}{s} \cdot \exp[-2s] = \frac{3}{s}(1 - \exp[-2s])\ \text{V} \cdot \text{sec}$$

2. Use Equation (8.16) to calculate $V_o(s)$:

$$V_o(s) = -\frac{25 \times 10^3/s}{50 \times 10^3} \cdot \frac{3(1 - \exp[-2s])}{s}$$
$$= -\frac{1.5(1 - \exp[-2s])}{s^2} \text{ V} \cdot \text{sec}$$

3. From Pair IIIa and the Time-Domain Shifting Property,

$$v_o(t) = -1.5t \cdot u(t) + 1.5(t - 2)u(t - 2) \text{ V}$$

The graph of $v_o(t)$ is given in Figure 8.10(b).

The reader probably recognized the configuration of the network in Figure 8.10 as that of an **integrator**.[6] When proper substitutions are made for $Z_i(s)$ and $Z_f(s)$ in Equation (8.16a),

$$V_0(s) = -\frac{0.5}{s} \cdot V_i(s)$$

By inverse Laplace transformation using the linearity and integration properties,

$$v_o(t) = -0.5 \int_0^{t_1} v_i(t)\, dt$$

This general result confirms the specific result obtained in Example 8.7—the output is the integral of the input, scaled by 0.5 and inverted. The scale factor is the reciprocal of the product RC.

Exercise 8.3

Problem:

Use transform circuit analysis to show that if R and C are interchanged in Figure 8.10, the resulting network is a **differentiator**.

[6] The circuit is actually an inverting scaler–integrator, because the sign of the output is the opposite of the sign of the input, and the output is scaled by the factor $1/(RC)$.

Exercise 8.4

Problem:

Show that $v_o(t) = v_i(t)$ for both OPAMP networks shown in Figure 8.11. These configurations are called **follower networks**.

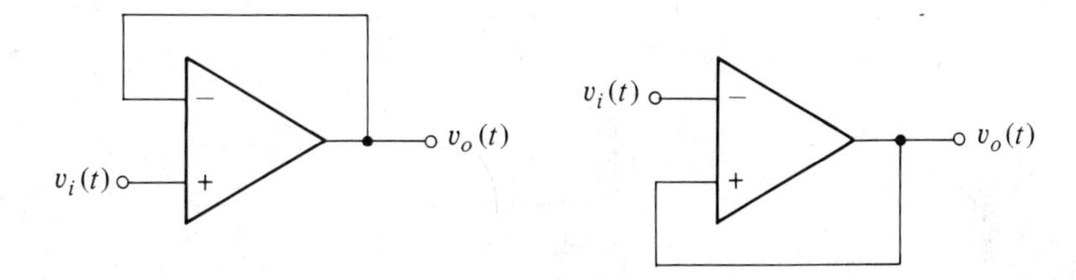

Figure 8.11
Exercise 8.4.

Impedance-Converting Networks

In **impedance conversion** applications, a two-terminal, one-port network (or a single element) is incorporated into an OPAMP network in such a way that the input impedance of the OPAMP network is related to the impedance of the two-terminal network. Impedance converters are characterized by the presence of two feedback networks—one between output and inverting input terminal, and another between output and noninverting input terminal—and a third network between the inverting input terminal and ground [Figure 8.12(a)]. In some cases another OPAMP is placed in one of the feedback paths (Example 8.9).

Referring to Figure 8.12(a), KCL at the inverting input terminal requires that

$$I_1 = I_2 \tag{8.20}$$

since essentially no current enters the OPAMP input terminals. The voltage at the inverting input terminal is the same as $V_i(s)$, because the differental input voltage is essentially zero. As a result, Equation (8.20) can be expressed in terms of input and output voltages as

$$\frac{V_i}{Z_1} = \frac{V_o - V_i}{Z_2}$$

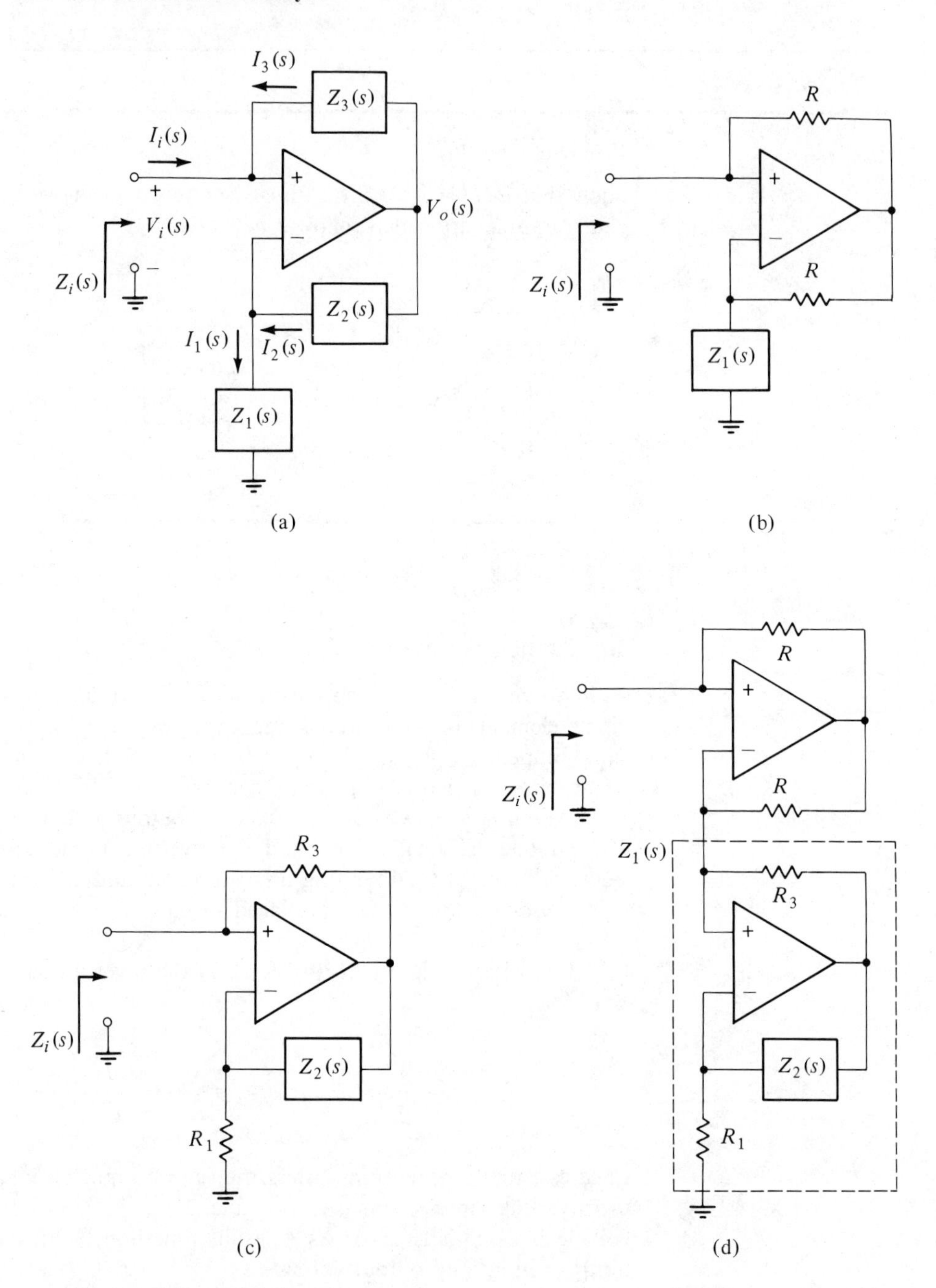

Figure 8.12 Impedance-converting networks. (a) General impedance converter. (b) Negative impedance converter. (c) Negative gyrator. (d) Positive gyrator.

which can be solved to obtain $V_o(s)$:

$$V_o = \left(1 + \frac{Z_2}{Z_1}\right) V_i \tag{8.21}$$

KCL at the noninverting terminal yields

$$I_i + I_3 = 0$$

which can be expressed in terms of voltages as

$$I_i = -\frac{V_o - V_i}{Z_3} \tag{8.22}$$

Substitution of Equation (8.21) into (8.22) yields

$$I_i = -V_i \frac{Z_2}{Z_1 Z_3}$$

or

$$Z_i = \frac{V_i}{Z_i} = -\frac{Z_1 Z_3}{Z_2} \tag{8.23}$$

Equation (8.23) provides some interesting results. For example, if $Z_2(s) = Z_3(s) = R$ as in Figure 8.12(b), then

$$Z_i(s) = -Z_1(s) \tag{8.24}$$

This network is called a **negative-impedance converter**. One useful application results from making $Z_1(s)$ a resistor (R_1). The OPAMP network input then appears to be a **negative resistance**, $-R_1$.

Another possibility is to have $Z_1(s) = R_1$ and $Z_3(s) = R_3$ [Figure 8.12(c)]. The result is that

$$Z_i(s) = -R_1 R_3 \cdot \frac{1}{Z_2(s)} \tag{8.25}$$

For this network, called a **negative gyrator**, the input impedance is the **negative reciprocal** of Z_2, scaled by a constant ($R_1 R_3$).

A **positive gyrator** is obtained by connecting the input of a negative gyrator to the inverting terminal of a negative impedance converter as $Z_1(s)$

[Figure 8.12(d)], so

$$Z_1(s) = -\frac{R_1R_3}{Z_2(s)}$$

The input impedance of the negative impedance converter is

$$Z_i(s) = \frac{R_1R_3}{Z_2(s)} \tag{8.26}$$

The most useful application of the positive gyrator is illustrated in Example 8.8, which follows.

Example 8.8

Problem:

Determine the input impedance $Z_i(s)$ at the noninverting terminal of OPAMP A in the network of Figure 8.13.

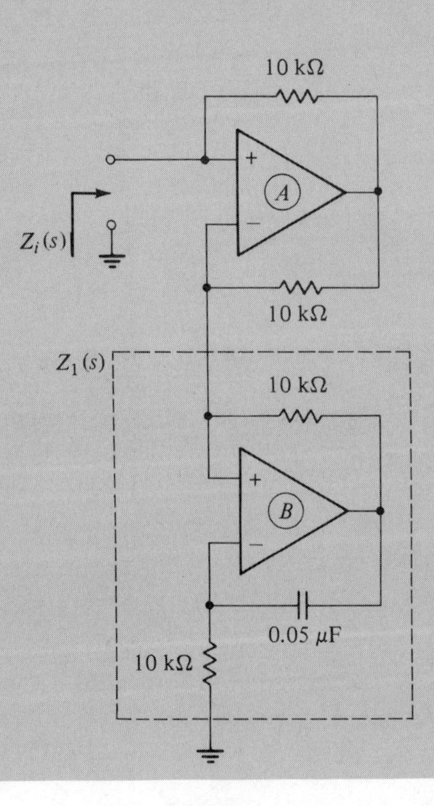

Figure 8.13
Example 8.8.

Solution:

1. The input impedance to OPAMP B, which is a negative gyrator, constitutes $Z_1(s)$. So, from Equation (8.25),

$$Z_1 = -(10 \times 10^3)(10 \times 10^3)\frac{1}{1/(0.05 \times 10^{-6}s)} = -5s\ \Omega$$

2. The OPAMP A network is a negative impedance converter, so from Equation (8.24),

$$Z_i = -Z_1 = 5s\ \Omega,$$

which is the impedance of a 5-H inductor.

Special emphasis must be given to the result obtained in Example 8.8. A network consisting of a few relatively small, lightweight, and electrically linear components is providing the same impedance as a rather large, heavy inductor which would probably have significant nonlinearities. The importance of this category of networks in the miniaturization and weight reduction of electronic equipment cannot be overstated.

Example 8.9

Problem:

Show that the OPAMP network of Figure 8.14 is an **impedance magnifier**—that is, show that

$$Z_i(s) = K \cdot Z(s) \quad \text{and} \quad K > 1$$

Solution:

1. OPAMP A is a follower; therefore,

$$V_{oA} = V_{iA}$$

2. R_1 and R_2 form a voltage divider, so

$$V_{iB} = V_{oA}\frac{R_2}{R_1 + R_2} = V_{iA}\frac{R_2}{R_1 + R_2}$$

3. OPAMP B is also a follower; therefore,

$$V_{oB} = V_{iB} = V_{iA}\frac{R_2}{R_1 + R_2}$$

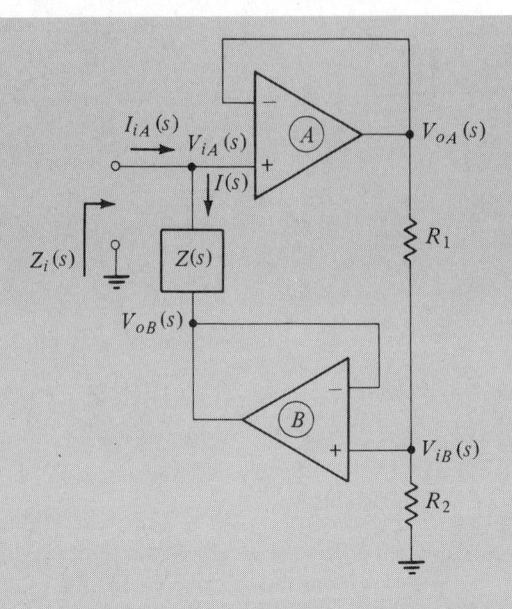

Figure 8.14
Example 8.9.

4. Since the current entering the input terminal of OPAMP A is essentially zero, $I_{iA}(s)$ is equal to the current through $Z(s)$:

$$I_{iA} = \frac{V_{iA} - V_{oB}}{Z} = \frac{V_{iA} - V_{iA}R_2/(R_1 + R_2)}{Z} = V_{iA}\frac{R_1}{(R_1 + R_2)Z}$$

5. Calculate $Z_i(s)$:

$$Z_i = \frac{V_{iA}}{I_{iA}} = \frac{R_1 + R_2}{R_1}Z = \left(1 + \frac{R_2}{R_1}\right)Z$$

Since $1 + R_2/R_1 > 1$,

$$Z_i(s) = K \cdot Z(s), \qquad \text{where} \quad K > 1$$

Exercise 8.5

Problem:

Use transform circuit analysis to show that the OPAMP network in Figure 8.15 is an **impedance reducer**—that is, $Z_i(s) = K \cdot Z(s)$, where $K < 1$.

Figure 8.15
Exercise 8.5.

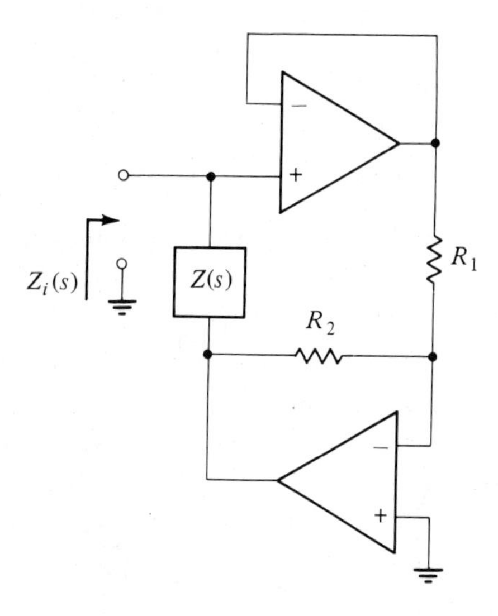

OPAMP Networks with Two Inputs

The model shown in Figure 8.16 represents a useful class of OPAMP networks which have two inputs. To begin analysis of this network, apply KCL at the inverting terminal to obtain

$$I_{i1} + I_f = 0,$$

Figure 8.16
Two-port OPAMP network.

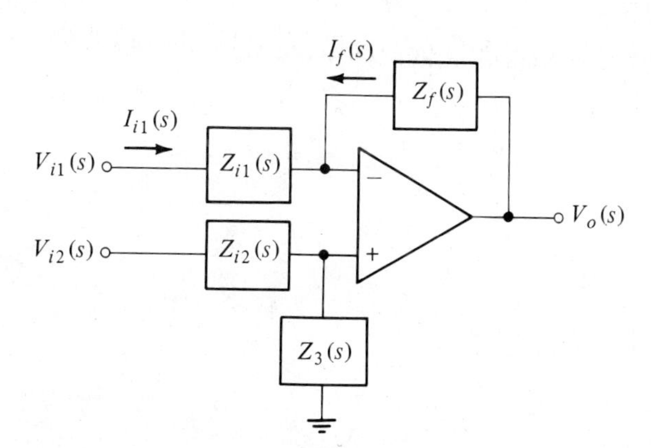

or, expressed in terms of voltage,

$$\frac{V_{i1} - V^-}{Z_{i1}} + \frac{V_o - V^-}{Z_f} = 0 \tag{8.27}$$

By the voltage division rule,

$$V^+ = V_{i2}\frac{Z_3}{Z_{i2} + Z_3} \tag{8.28}$$

Because the differential input voltage is essentially zero,

$$V^- = V^+$$

so Equation (8.28) may be substituted into Equation (8.27), and Equation (8.27) rearranged, to obtain

$$V_o = \frac{Z_3(Z_{i1} + Z_f)}{Z_{i1}(Z_{i2} + Z_3)}V_{i2} - \frac{Z_f}{Z_{i1}}V_{i1} \tag{8.29}$$

Example 8.10

Problem:

Determine $v_o(t)$ for the two-input OPAMP network shown in Figure 8.17, in which $RC = 1$.

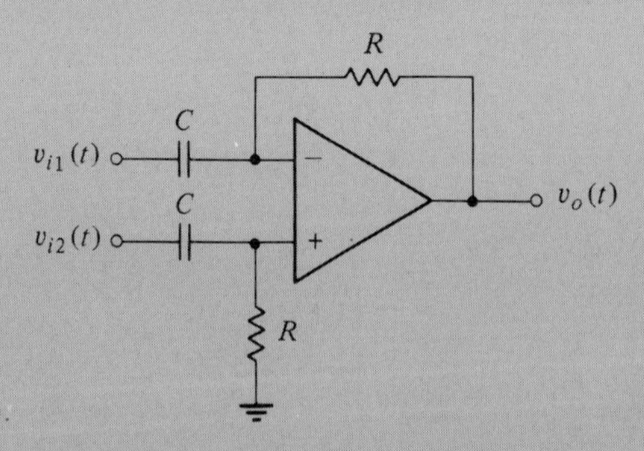

Figure 8.17
Example 8.10.

Solution:

1. Transform the network elements:

$$Z_{i1}(s) = Z_{i2}(s) = 1/Cs\ \Omega,$$
$$Z_f(s) = Z_3(s) = R\ \Omega.$$

2. Use Equation (8.29) to calculate $V_o(s)$:

$$V_o = \frac{R\left(\frac{1}{Cs} + R\right)}{\frac{1}{Cs}\left(\frac{1}{Cs} + R\right)} V_{i2} - \frac{R}{1/Cs} V_{i1} = s \cdot (V_{i2} - V_{i1})$$

Multiplication by s in the s-domain is equivalent to differentiation with respect to t in the time domain (differentiation property). The OPAMP network is therefore a **differentiator** which provides an output

$$v_o(t) = \frac{d}{dt}[v_{i2}(t) - v_{i1}(t)]$$

One subtle modification of this network is to ground input 1 [$v_{i1}(t) = 0$], with the result that

$$v_o(t) = \frac{d}{dt} v_{i2}(t)$$

This network is a *positive* differentiator, in contrast with the *negative* differentiator in Exercise 8.3. Exercise 8.6 provides the opportunity to analyze a positive integrator, which may be compared with the negative integrator in Example 8.7.

Exercise 8.6

Problem:

Use transform circuit analysis to show that if the resistors and capacitors are interchanged in Figure 8.17, the resulting network is a positive integrator.

8.5 Magnetically Coupled Networks

In this section we will apply transform circuit analysis to networks in which **magnetic coupling** is present. Magnetic coupling exists in a network when a portion of the magnetic field associated with one inductor **links** the coils of another inductor and induces a voltage in that inductor. In general, the two inductors are not connected electrically, although they may have a common

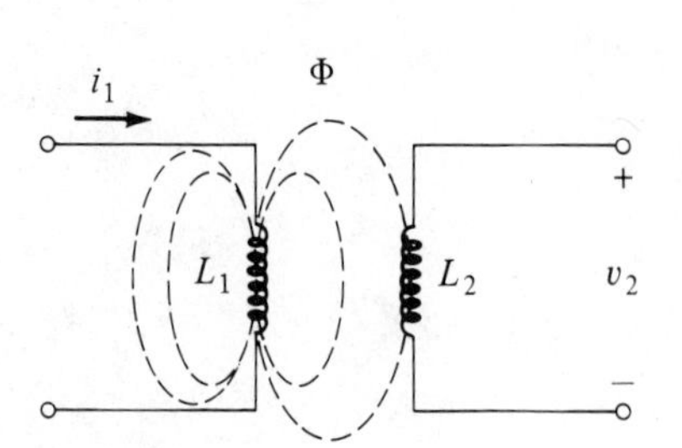

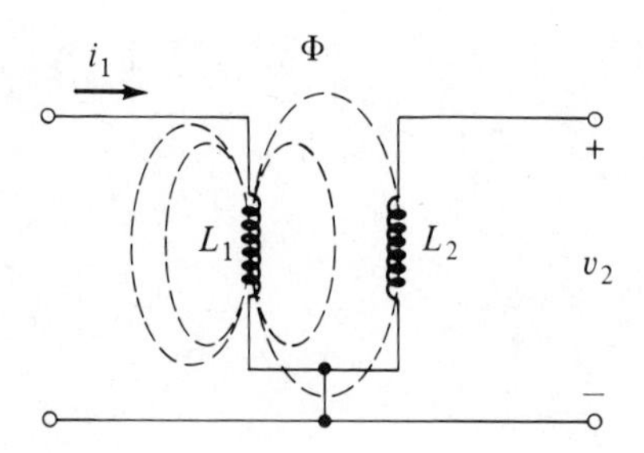

Figure 8.18
Magnetic coupling.

terminal (Figure 8.18). An important objective in this section is the development of electrical network models, in which magnetic coupling does not exist, to represent magnetically coupled networks.

Magnetic coupling may exist unintentionally in a network, in which case it is a problem to be dealt with, possibly by changing the physical placement of the inductors concerned or providing magnetic shielding for one or the other of them. Our interest is the intentional use of two or more inductors, often in the same physical package, in such a way as to exploit the magnetic coupling that exists between them. Such an arrangement of inductors is commonly called a **transformer**.

In most applications, the transformer is excited by a source connected to one inductor, called the **primary**, while a load is connected to the other inductor, which is called the **secondary**. For the same transformer, the inductor which is the primary in one application may be the secondary in another.

Mutual Inductance

The more correct term for the property of inductance, which was defined and quantified in Chapter 3, is **self-inductance**, because this property relates the voltage induced in an inductor to the current in the same inductor. Equation (3.5) is repeated here for convenience:

$$v(t) = L\frac{d}{dt}i(t) \tag{3.5}$$

The relationship between voltage in one inductor and current in another (Figure 8.19) is given by

$$v_2(t) = M\frac{d}{dt}i_1(t) \tag{8.30a}$$

$$v_1(t) = M\frac{d}{dt}i_2(t) \tag{8.30b}$$

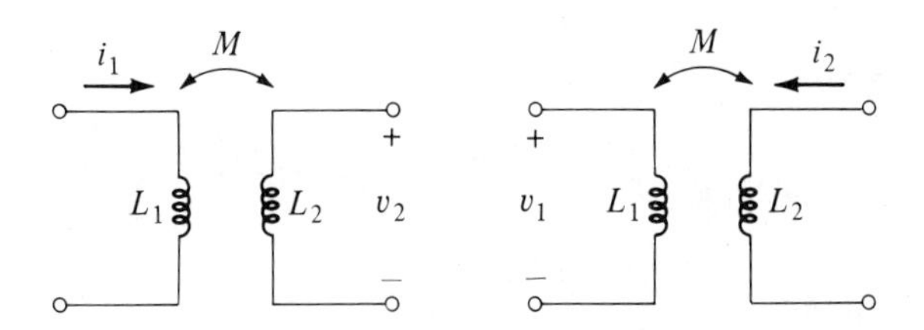

Figure 8.19
Mutual inductance.

where M is called the **mutual inductance**, or **coefficient of mutual inductance**. Self-inductance (L) and mutual inductance (M) obviously have the same dimensions—V · sec/A, or henries. At this point in the discussion, the induced voltage is an open-circuit voltage—that is, Equation (8.30a) assumes $i_2(t) = 0$, and Equation (8.30b) assumes $i_1(t) = 0$. This restriction will soon be removed.

The value of the mutual inductance is related to L_1 and L_2 by

$$M \leq \sqrt{L_1 L_2} \tag{8.31}$$

and is therefore the same in Equation (8.30b) as it is in Equation (8.30a).[7] A more useful way of expressing Equation (8.31) is

$$M = k\sqrt{L_1 L_2} \tag{8.32}$$

where

$$0 < k \leq = 1$$

The physical construction of a transformer determines k, which is called the **coefficient of coupling**. Transformers with coils widely separated on a core with **low permeability** have small values of k, and are said to be **loosely coupled**; transformers with coils closely spaced on a core with **high permeability** have values of k approaching unity, and are said to be **tightly coupled**. The remainder of this section will be devoted to so-called **linear transformers** in which the core material is nonmagnetic (low permeability), and for which k is less than 1. Linear transformers have a wide variety of applications in communications circuits and in pulse-forming networks. In Chapter 10, we will briefly consider **ideal transformers** in which the core is highly permeable, ferromagnetic material, which allows us to assume that k is unity.

[7]The proof of Equation (8.31) is derived from energy considerations.

Figure 8.20
Transformer dot convention.

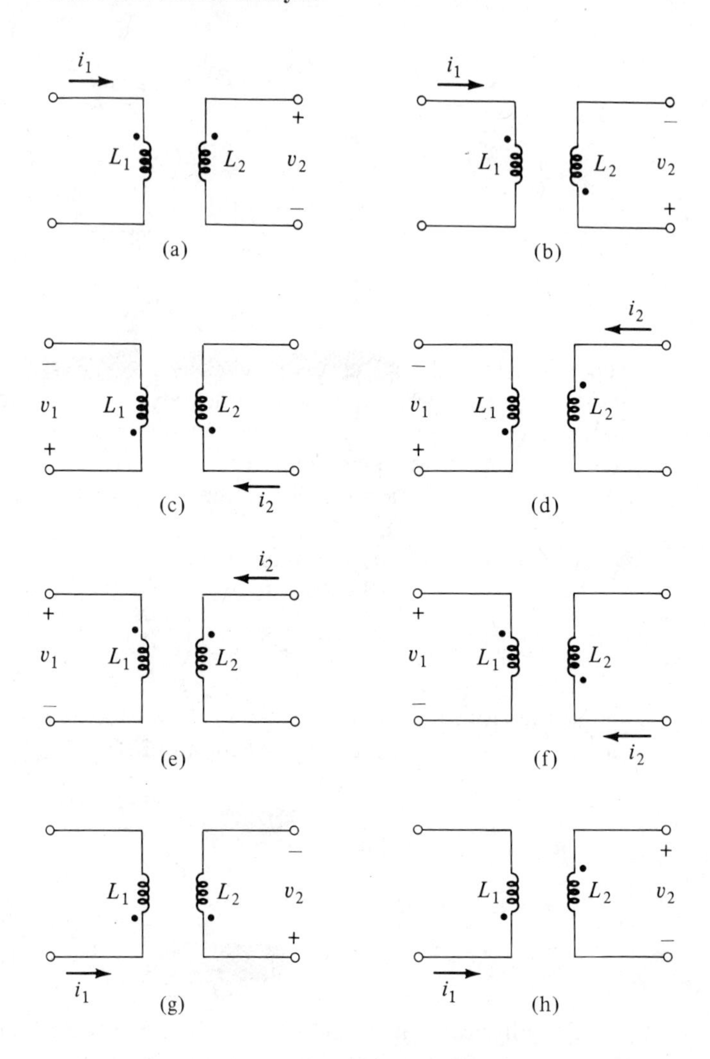

Polarity of Induced Voltage

For the discussion that follows, the transformer will be considered as a four-terminal, two-port device (Figure 8.20), with a test current entering one inductor and an open circuit at the terminals of the other inductor. The current i_1 entering the upper terminal of L_1 produces a magnetic field. If current entering the upper terminal of L_2 produces a magnetic field in the *same* direction, the polarity of v_2 is shown in Figure 8.20(a). If current entering the upper terminal of L_2 produces a magnetic field in the *opposite* direction, the polarity of v_2 is shown in Figure 8.20(b).

The polarity of the induced voltage depends on the direction in which one inductor is wound with respect to the other inductor.[8] The information regarding direction of winding is conveyed in the transformer symbol by placing a **dot** on one terminal of each coil. Current entering the **dotted terminal** of one coil will induce an open-circuit voltage in the second coil which is positive at the dotted terminal of that coil. The reader is encouraged to verify all cases in Figure 8.20.

Controlled Source Network Model of a Transformer

If a source is connected to L_1 and a load connected to L_2, the voltage at each port has two components—one due to the self-inductance and one due to mutual inductance:

$$v_{\text{port}} = v_{\text{self}} + v_{\text{mutual}}$$

The component due to self-inductance, given by Equation (3.5), is always positive because of the passive sign convention; the component due to mutual inductance, given by Equation (8.30a) or (8.30b), may be positve or negative depending on the dot convention. The network description of the time-domain transformer model in Figure 8.21(a) results in **coupled differential equations**, which are sometimes called the **linear transformer equations**:

$$v_1(t) = L_1 \frac{d}{dt} i_1(t) + M \frac{d}{dt} i_2(t) \tag{8.33a}$$

$$v_2(t) = M \frac{d}{dt} i_1(t) + L_1 \frac{d}{dt} i_2(t) \tag{8.33b}$$

In the equations above, M is allowed to *carry its sign* according to the dot convention. The dot convention, and its effect on the sign of M, accounts for the polarity of the induced voltage, and in no way implies that mutual inductance is a negative quantity.

Laplace transformaton of the linear transformer equations yields

$$V_1(s) = L_1[s \cdot I_1(s) - i_1(0)] + M[s \cdot I_2(s) - i_2(0)]$$
$$V_2(s) = M[s \cdot I_1(s) + i_1(0)] + L_2[s \cdot I_2(s) - i_2(0)]$$

or

$$V_1(s) = L_1 s \cdot I_1(s) + Ms \cdot I_2(s) - [L_1 \cdot i_1(0) + M \cdot i_2(0)] \tag{8.34a}$$

$$V_2(s) = Ms \cdot I_1(s) + L_2 s \cdot I_2(s) - [L_2 \cdot i_2(0) + M \cdot i_1(0)] \tag{8.34b}$$

[8] The direction of coil winding must be provided by the manufacturer or determined by laboratory test or (worst case!) by examining the coils.

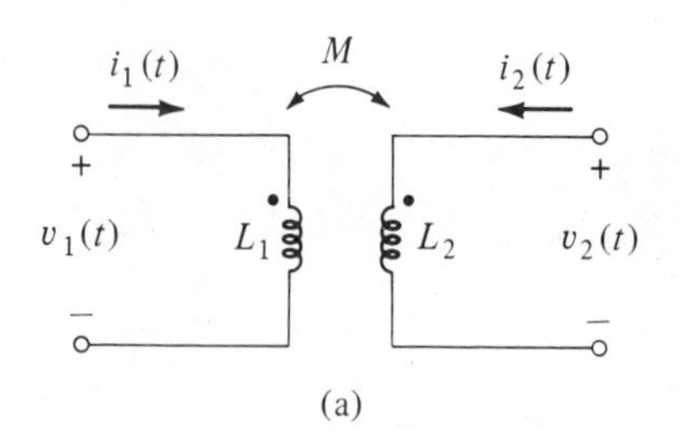

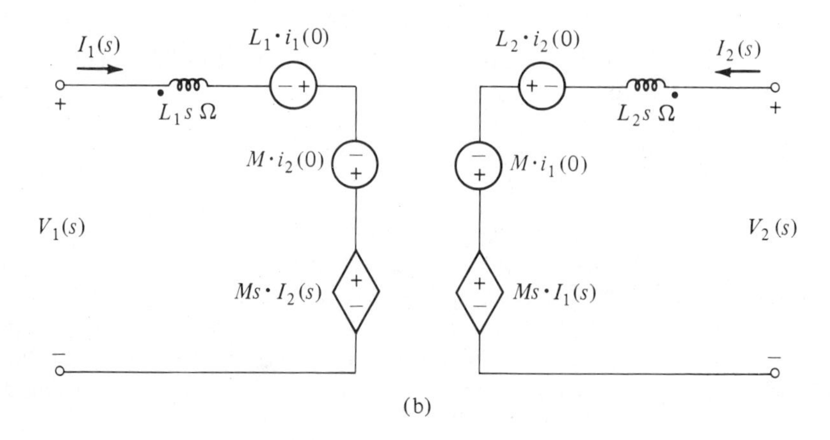

Figure 8.21 Controlled-source model of a transformer.

which leads to the s-domain network model for a transformer shown in Figure 8.21(b). This is a purely electrical model of a transformer with no magnetic coupling between the inductors; magnetic coupling in the actual transformer is accounted for by the presence of sources in the model. With magnetic coupling eliminated from the s-domain model, the usual techniques of transform circuit analysis can be applied.

Care must be taken to establish the correct polarity of the sources used in the model. The model is based on the usual current direction for two-port analysis—that is, positive current entering both ports. If a port current is assumed to have the other direction, the polarity of the independent source which accounts for the initial current in that coil must be reversed in that mesh. Further, the proper sign of M must be used for all sources in the model which depend on M.

Example 8.11

SPICE

Problem:

Assume that the switch has been open for a long time prior to $t = 0$, and determine the load voltage $v_2(t)$ for the circuit shown in Figure 8.22.

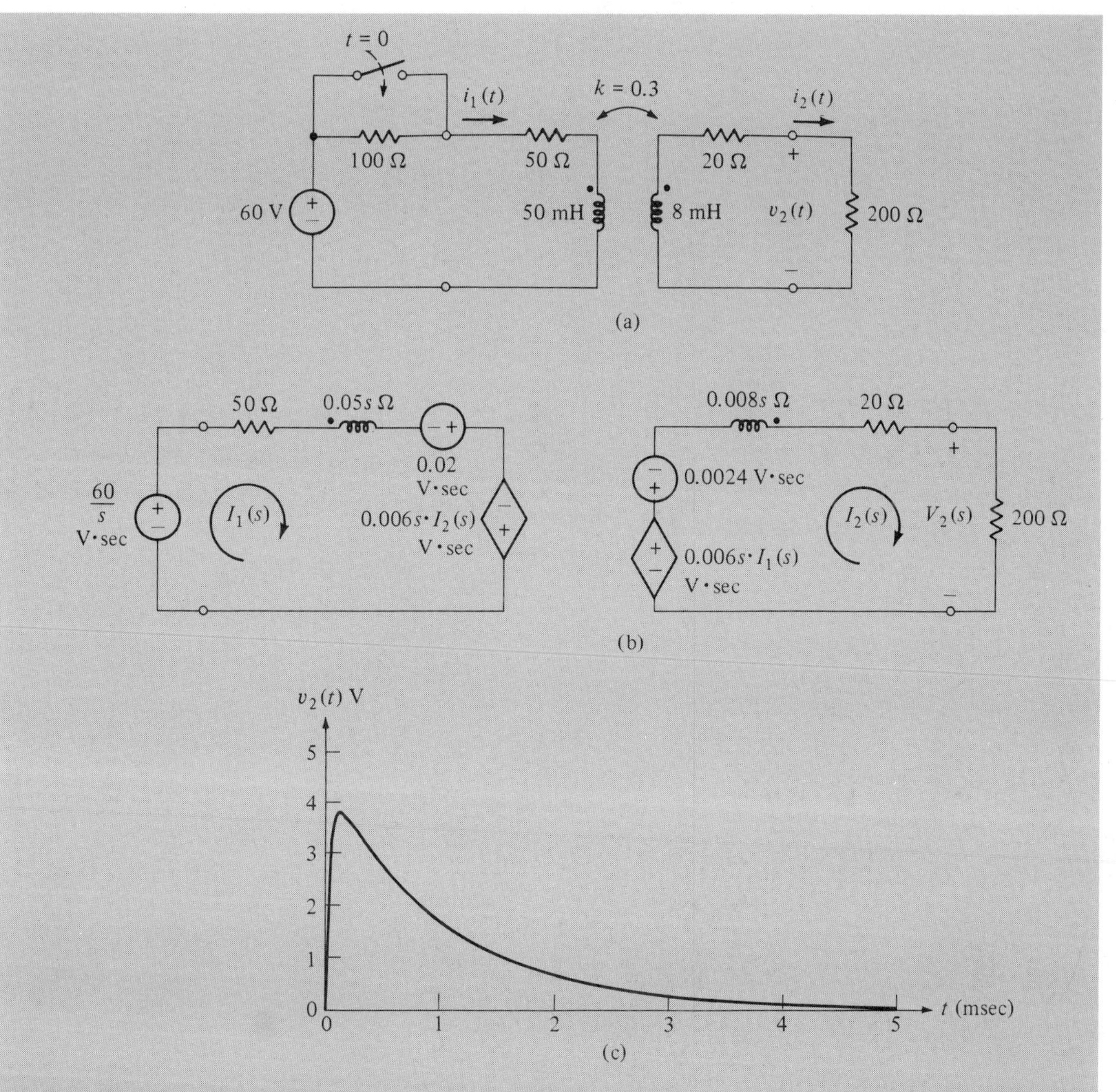

Figure 8.22 Example 8.11.

Solution:

1. Calculate initial conditions for $i_1(t)$ and $i_2(t)$:

$$i_1(0) = i_1(0^-) = 60/150 = 0.4 \text{ A}$$

The primary current is at steady state, so there is no flux linkage, and hence no induced voltage in either primary or secondary; therefore,

$$i_2(0) = i_2(0^-) = 0$$

2. Calculate mutual inductance:

$$M = k\sqrt{L_1 L_2} = 0.3\sqrt{(50 \times 10^{-3})(8 \times 10^{-3})} = 6 \text{ mH}$$

3. The transformed network is shown in Figure 8.22(b). Note that the polarity of the controlled source has been reversed in the primary mesh because of the direction assumed for $i_2(t)$.
4. Write the mesh equations for the transformed circuit:

$$\text{Mesh 1: } (0.05s + 50)\text{I}_1 = 60/s + 0.02 + 0.006sI_2$$

$$\text{Mesh 2: } (0.008s + 20 + 200)I_2 = 0.006sI_1 - 0.0024$$

$$0.05s(s + 1000)I_1 \qquad -0.006s^2 I_2 = 0.02(s + 3000)$$

$$-0.006sI_1 + 0.008(s + 27.5 \times 10^3)I_2 = -0.0024$$

5. Solve the mesh 2 equation for $I_1(s)$:

$$I_1 = \frac{0.4}{s} + \frac{1.333(s + 27.5 \times 10^3)}{s} I_2$$

6. Substitute I_1 into the mesh 1 equation and solve for I_2:

$$0.05(s + 1000)\left[0.4 + 1.333(s + 27.5 \times 10^3)I_2\right] - 0.006s^2 I_2$$

$$= 0.02(s + 3000)$$

$$I_2 = \frac{659.3}{s^2 + 31.32 \times 10^3 s + 30.21 \times 10^6}$$

$$= \frac{659.3}{(s + 996.6)(s + 30.33 \times 10^3)} \text{ A} \cdot \text{sec}$$

7. From Pair XI,

$$i_2(t) = 22.48(\exp[-996.6t] - \exp[-30.22 \times 10^3 t])u(t) \text{ mA}$$

and from Ohm's law,

$$v_2(t) = 4.497(\exp[-996.6t] - \exp[-30.22 \times 10^3 t])u(t) \text{ V}$$

A graph of $v_2(t)$ is given in Figure 8.22(c)

Exercise 8.6

SPICE

Problem:

1. Sketch and label the s-domain model for the network shown in Figure 8.23(a).
2. Write the s-domain mesh equations for the model.
3. Solve the mesh equations and obtain $i_2(t)$.

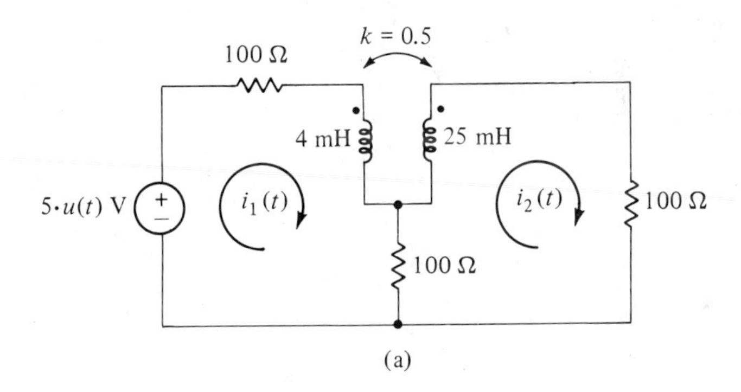

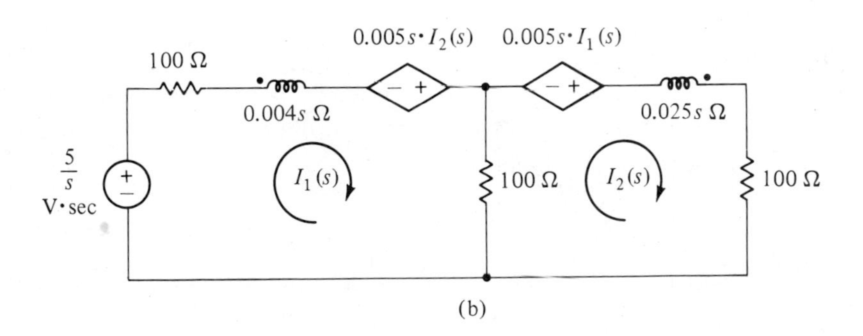

Figure 8.23
Exercise 8.6.

Answers:

1. The s-domain model is given in Figure 8.23(b).
2. $0.004s(s + 50 \times 10^3)I_1 - 0.005s(s + 20 \times 10^3)I_2 = 5$

   $-0.005(s + 20 \times 10^3)I_1 \quad +0.025(s + 8 \times 10^3)I_2 = 0$

63. $i_2(t) = (16.67 - 12.34 \cdot \exp[-7020t]$

   $-4.330 \cdot \exp[-56.98 \times 10^3 t]) \cdot u(t)$ mA

T-Equivalent Network Model of a Transformer

The **T-equivalent** network model (Figure 8.24) is another way of representing a transformer as an electrical network without magnetic coupling; however, the

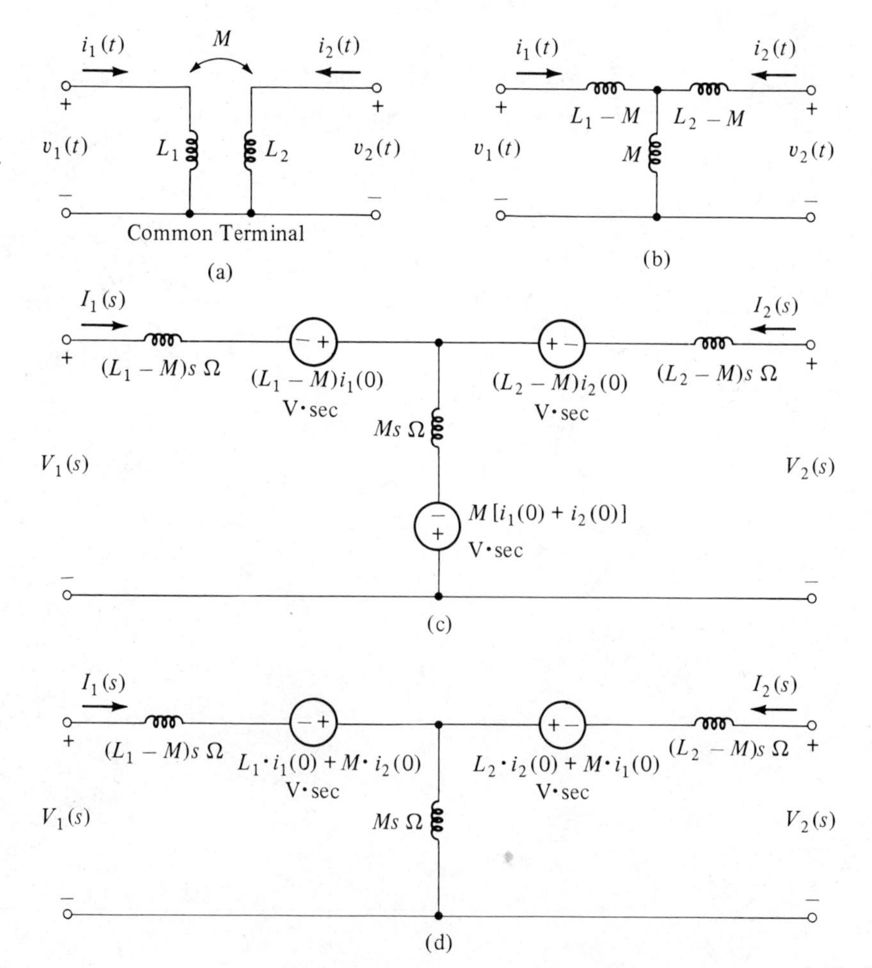

Figure 8.24
T-equivalent of a transformer.
(a) Transformer as a three-terminal device.
(b) Time-domain T-equivalent network.
(c) s-domain T-equivalent network.
(d) Simplified s-domain T-equivalent network.

T-equivalent model is restricted to applications in which the transformer is connected as a three-terminal device [Figure 8.24(a)]. The T-equivalent model, which begins as a time-domain equivalent circuit [Figure 8.24(b)], can be verified by mesh analysis:

Mesh 1:

$$(L_1 - M)\frac{d}{dt}i_1(t) + M\frac{d}{dt}[i_1(t) + i_2(t)] = v_1(t)$$

Mesh 2:

$$M\frac{d}{dt}[i_2(t) + i_1(t)] + (L_2 - M)\frac{d}{dt}i_2(t) = v_2(t)$$

Rearrangement of the mesh equations results in the linear transformer equations [Equations (8.33a) and (8.33b)].

Care must be taken in transforming the T-equivalent network model [Figure 8.24(c)], but since the central node is fictitious and of no interest, an acceptable, simpler model is given in Figure 8.24(d). The reader should confirm that this model yields the *s*-domain linear transformer equations (Exercise 8.7).

A major advantage of the T-equivalent model is that it does not contain controlled sources. The T-equivalent is especially favored in sinusoidal steady-state analysis (Chapter 10) where initial conditions are not considered, and the model is even simpler. Another possible, but less favored, model for the three-terminal transformer is the **Π-equivalent model**, the derivation of which is left as a problem for the reader (Problem 8-14).

Example 8.12

SPICE

Problem:

Use the *s*-domain T-equivalent network model to solve the problem of Exercise 8.6.

Solution:

1. Because of the step-function source, the initial conditions are zero in both meshes.
2. Calculate M:

$$M = 0.5\sqrt{(4 \times 10^{-3})(25 \times 10^{-3})} = 5 \text{ mH}$$

M carries a plus sign unless the "dotted" end of one coil is connected to the "undotted" end of the other coil to form the common terminal.

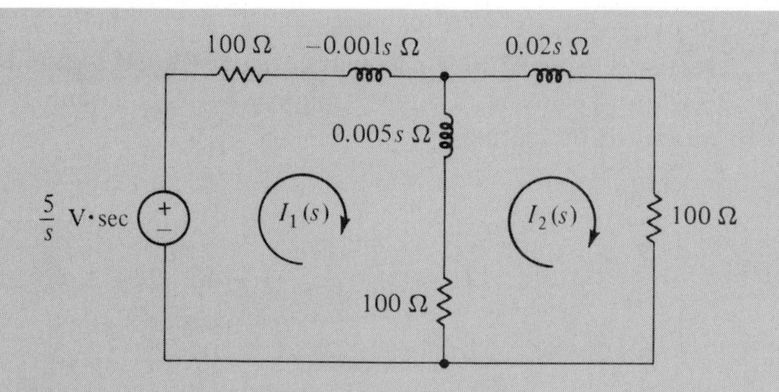

Figure 8.25
Example 8.12.

3. Write the mesh equations for the s-domain network model, which is shown in Figure 8.25:

   Mesh 1:

$$(-0.001s + 0.005s + 100 + 100)I_1 - (0.005s + 100)I_2 = 5/s$$

   Mesh 2:

$$-(0.005s + 100)I_1 + (0.02s + 0.005s + 100 + 100)I_2 = 0$$

   Hence

$$0.004s(s + 50 \times 10^3)I_1 - 0.005s(s + 20 \times 10^3)I_2 = 5$$

$$-0.005(s + 20 \times 10^3)I_1 \quad +0.025(s + 8 \times 10^3)I_2 = 0$$

   It is comforting that these equations agree with the equations obtained in Exercise 8.6!

4. Solve the mesh 2 equation for $I_1(s)$ in terms of $I_2(s)$:

$$I_1 = \frac{5(s + 8 \times 10^3)}{s + 20 \times 10^3} I_2$$

5. Substitute $I_1(s)$ into the mesh-1 equation and solve for $I_2(s)$:

$$0.004s(s + 50 \times 10^3)\frac{5(s + 8 \times 10^3)}{s + 20 \times 10^3} I_2$$

$$- 0.005s(s + 20 \times 10^3)I_2 = 5$$

$$I_2 = \frac{333.3(s + 20 \times 10^3)}{s^2 + 64 \times 10^3 s + 400 \times 10^6} \text{ A} \cdot \text{sec}$$

6. Perform a partial fraction decomposition of $I_2(s)$:

$$I_2 = \frac{K_1}{s} + \frac{K_2}{s + 7020} + \frac{K_3}{s + 56.98 \times 10^3} \text{ A} \cdot \text{sec}$$

where

$$K_1 = 16.67 \times 10^{-3},$$

$$K_2 = -12.34 \times 10^{-3}, \quad \text{and} \quad K_3 = -4.330 \times 10^{-3}$$

7. Use Pairs II and IV to determine $i_2(t)$:

$$i_2(t) = (16.67 - 12.34 \cdot \exp[-7020t] - 4.330 \cdot \exp[-56.98 \times 10^3 t])u(t) \text{ mA}$$

Exercise 8.7

Problem:

Use mesh analysis of the simpified s-domain T-equivalent network model [Figure 8.24(d)] to show that the model conforms to the linear transformer equations [Equations (8.34a) and (8.34b)].

Transfer Functions

The network shown in Figure 8.26 represents a general case for a transformer excited by a source $v_1(s)$ with source impedance $Z_g(s)$ and loaded by $Z_L(s)$.

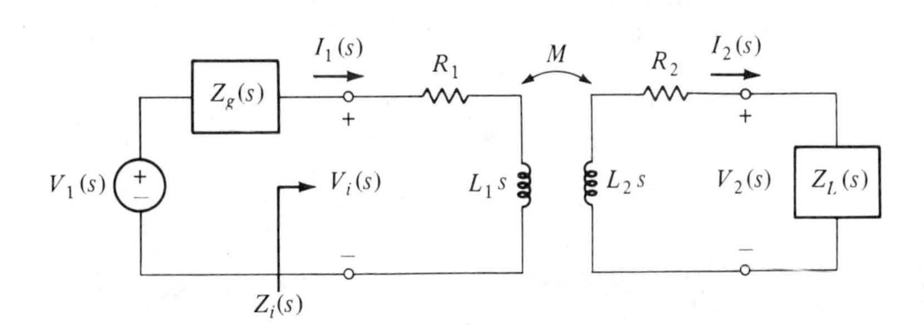

Figure 8.26 General case for a driven, loaded transformer.

Winding resistances, R_1 and R_2, are also considered. For this network,

$$[Z_g(s) + R_1 + L_1 s]\, I_1(s) \qquad\qquad + Ms \cdot I_2(s) = V_1(s) \tag{8.35a}$$

$$Ms \cdot I_1(s) + [Z_L(s) + R_2 + L_2 s]\, I_2(s) = 0 \tag{8.35b}$$

$$Z_L(s) \cdot I_2(s) = V_2(s) \tag{8.35c}$$

As usual, M must carry its sign according to the dot convention.

Equation (8.35b) yields the current transfer function:

$$\frac{I_2(s)}{I_1(s)} = \frac{-Ms}{Z_L(s) + R_2 + L_2 s} \tag{8.36}$$

The voltage transfer function is obtained by substitution (Problem 8-19):

$$\frac{V_2(s)}{V_1(s)} = \frac{-Ms \cdot Z_L(s)}{[Z_g(s) + R_1 + L_1 s][Z_L(s) + R_2 + L_2 s] - M^2 s^2} \tag{8.37}$$

Assigning the incorrect sign to the mutual inductance term will change the sign of the transfer function.

The input impedance, obtained from Equations (8.36) and (8.37) (Problem 8-20), is

$$Z_i(s) = \frac{(R_1 + L_1 s)[Z_L(s) + R_2 + L_2 s] - M^2 s^2}{Z_L(s) + R_2 + L_2 s} \tag{8.38}$$

and is independent of the polarity of the mutual inductance because M is squared in its only occurrence.

Example 8.13

Problem:

Write the voltage transfer function

$$H(s) = \frac{V_2(s)}{V_1(s)}$$

for the magnetically coupled circuit of Figure 8.27.

Figure 8.27
Example 8.13.

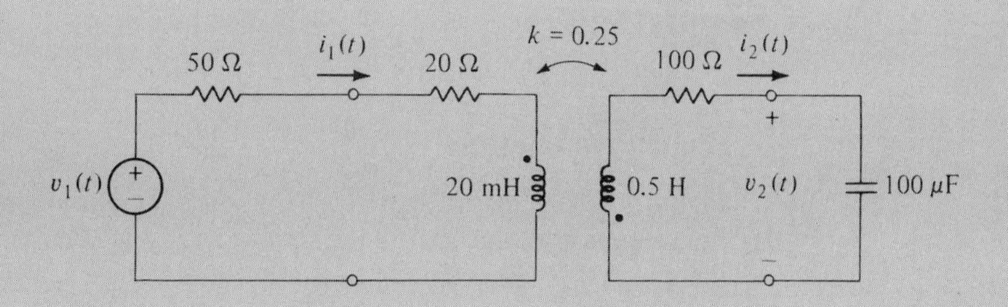

Solution:

1. Use Equation (8.32) to calculate the mutual inductance:

$$M = 0.25\sqrt{(20 \times 10^{-3})(0.5)} = 25 \text{ mH}$$

M carries a plus sign because both $I_1(s)$ and $I_2(s)$ enter dotted terminals of the transformer.

2. Use Equation (8.37) to write $H(s)$:

$$H(s) = \frac{-0.025s(10 \times 10^3/s)}{(50 + 20 + 0.02s)(10 \times 10^3/s + 100 + 0.5s) - 625 \times 10^{-6}s^2}$$

$$= \frac{-26.67 \times 10^3 s}{(s + 3747)(s^2 + 200s + 20 \times 10^3)}$$

Exercise 8.8

Problem:

Obtain the current transfer function

$$H(s) = \frac{I_2(s)}{I_1(s)}$$

for the circuit of Example 8.13.

Answer:

$$H(s) = \frac{-0.05s^2}{s^2 + 200s + 20 \times 10^3}$$

8.6 Computer Simulations

OPAMP Simulation

Figure 8.8(b) suggests a method for SPICE simulation of the ideal OPAMP. In practice a very large resistance is connected between v^+ and v^- to eliminate the dangling nodes, the control constant (A) is set to a very large value, and the positive terminal of the VCVS is grounded. The element record format for a VCVS is given in Appendix D. The control voltage is v_d, and the control constant is the OPAMP gain A. SPICE 8.1 illustrates the use of a VCVS in simulating the OPAMP network of Example 8.7.

SPICE 8.2 (Example 8.8), which also uses the VCVS to simulate OPAMPs, illustrates a method of verifying the input impedance of an impedance converter network. Since

$$v(t) = L\frac{d}{dt}i(t)$$

the 5-V step response to the 1-A/sec input current confirms that the input impedance is that of a 5-H inductor.

When detailed performance data are required for a network incorporating a real OPAMP IC, the VCVS approximation of the OPAMP may not be adequate. In these cases it is necessary to simulate the entire OPAMP circuit, as well as the remainder of the network within which the OPAMP resides. The SPICE feature which provides for **subcircuit** calls can be used to facilitate this type of simulation. Subcircuit libraries which contain the commonly used ICs are available, or can be developed by the user. Some manufacturers of ICs provide SPICE subcircuit data for their products.

Transformer Simulation

SPICE provides the capability to simulate magnetic coupling between inductors, so linear transformers of the type considered in this chapter may be included in SPICE simulations. SPICE 8.3 (Exercise 8.6) and SPICE 8.4 (Example 8.13) illustrate the simulation of mutually coupled inductors. Each inductor must be entered as a separate element with the "dotted" node entered first. The coupling coefficient for each pair of mutually coupled inductors is entered in a separate element record (Appendix D). Thus, three records are required for specification of a transformer with two windings, six records are required for three windings, etc.

```
*******11/09/88 ********  SPICE 2G.6    3/15/83 ********19:25:29*****
SPICE 8.1
****     INPUT LISTING                       TEMPERATURE =   27.000 DEG C
******************************************************************************
 R 1 2 50K
 C 2 3 40U
 V 1 0 PULSE(0 3 1F 1F 1F 2)
 E 0 3 2 0 1E6
 .TRAN 150M 3
 .PLOT TRAN V(3)
 .END
*******11/09/88 ********  SPICE 2G.6    3/15/83 ********19:25:29*****
SPICE 8.1
****     INITIAL TRANSIENT SOLUTION          TEMPERATURE =   27.000 DEG C
******************************************************************************
  NODE   VOLTAGE      NODE   VOLTAGE      NODE   VOLTAGE
 (  1)      .0000    (  2)      .0000    (  3)      .0000
     VOLTAGE SOURCE CURRENTS
     NAME        CURRENT
     V          0.000D+00
     TOTAL POWER DISSIPATION   0.00D+00  WATTS
*******11/09/88 ********  SPICE 2G.6    3/15/83 ********19:25:29*****
SPICE 8.1
****     OPERATING POINT INFORMATION         TEMPERATURE =   27.000 DEG C
******************************************************************************

**** VOLTAGE-CONTROLLED VOLTAGE SOURCES
          E
 V-SOURCE      .000
 I-SOURCE  0.00E+00
*******11/09/88 ********  SPICE 2G.6    3/15/83 ********19:25:30*****
SPICE 8.1
****     TRANSIENT ANALYSIS                  TEMPERATURE =   27.000 DEG C
******************************************************************************

     TIME       V(3)
                  -3.000D+00    -2.000D+00    -1.000D+00     0.000D+00  1.000D+00
                        - - - - - - - - - - - - - - - - - - - - - - - - - - - - - -
 0.000D+00  0.000D+00 .             .             .             *             .
 1.500D-01 -2.250D-01 .             .             .          *  .             .
 3.000D-01 -4.500D-01 .             .             .       *     .             .
 4.500D-01 -6.750D-01 .             .             .    *        .             .
 6.000D-01 -9.000D-01 .             .             .*            .             .
 7.500D-01 -1.125D+00 .             .           * .             .             .
 9.000D-01 -1.350D+00 .             .        *    .             .             .
 1.050D+00 -1.575D+00 .             .     *       .             .             .
 1.200D+00 -1.800D+00 .             .  *          .             .             .
 1.350D+00 -2.025D+00 .             *             .             .             .
 1.500D+00 -2.250D+00 .          *  .             .             .             .
 1.650D+00 -2.475D+00 .      *      .             .             .             .
 1.800D+00 -2.700D+00 .   *         .             .             .             .
 1.950D+00 -2.925D+00 .*            .             .             .             .
 2.100D+00 -3.000D+00 *             .             .             .             .
 2.250D+00 -3.000D+00 *             .             .             .             .
 2.400D+00 -3.000D+00 *             .             .             .             .
 2.550D+00 -3.000D+00 *             .             .             .             .
 2.700D+00 -3.000D+00 *             .             .             .             .
 2.850D+00 -3.000D+00 *             .             .             .             .
 3.000D+00 -3.000D+00 *             .             .             .             .
                        - - - - - - - - - - - - - - - - - - - - - - - - - - - - - -
```

```
*******11/09/88 ********  SPICE 2G.6    3/15/83 ********19:32:10*****
SPICE 8.2
****     INPUT LISTING                      TEMPERATURE =   27.000 DEG C
*******************************************************************************
 R1 1 3 10K
 R2 2 3 10K
 R3 2 4 10K
 R4 5 0 10K
 C 4 5 50N
 I 0 1 PWL(0 0 1 1)
 EA 0 3 1 2 1E6
 EB 0 4 2 5 1E6
 .TRAN 50M 1
 .PLOT TRAN V(1)
 .END
*******11/09/88 ********  SPICE 2G.6    3/15/83 ********19:32:11*****
SPICE 8.2
****     INITIAL TRANSIENT SOLUTION         TEMPERATURE =   27.000 DEG C
*******************************************************************************
  NODE   VOLTAGE     NODE   VOLTAGE     NODE   VOLTAGE     NODE   VOLTAGE
 (  1)     .0000    (  2)     .0000    (  3)     .0000    (  4)     .0000
 (  5)     .0000
    TOTAL POWER DISSIPATION   0.00D+00  WATTS
*******11/09/88 ********  SPICE 2G.6    3/15/83 ********19:32:11*****
SPICE 8.2
****     OPERATING POINT INFORMATION        TEMPERATURE =   27.000 DEG C
*******************************************************************************
**** VOLTAGE-CONTROLLED VOLTAGE SOURCES
            EA          EB
 V-SOURCE       .000        .000
 I-SOURCE  0.00E+00   0.00E+00
*******11/09/88 ********  SPICE 2G.6    3/15/83 ********19:32:11*****
SPICE 8.2
****     TRANSIENT ANALYSIS                 TEMPERATURE =   27.000 DEG C
*******************************************************************************

    TIME        V(1)
                 0.000D+00      2.000D+00      4.000D+00      6.000D+00  8.000D+00
                - - - - - - - - - - - - - - - - - - - - - - - - - - - - - - - - -
 0.000D+00  0.000D+00 *            .              .              .              .
 5.000D-02  5.000D+00 .            .              .      *       .              .
 1.000D-01  5.000D+00 .            .              .      *       .              .
 1.500D-01  5.000D+00 .            .              .      *       .              .
 2.000D-01  5.000D+00 .            .              .      *       .              .
 2.500D-01  5.000D+00 .            .              .      *       .              .
 3.000D-01  5.000D+00 .            .              .      *       .              .
 3.500D-01  5.000D+00 .            .              .      *       .              .
 4.000D-01  5.000D+00 .            .              .      *       .              .
 4.500D-01  5.000D+00 .            .              .      *       .              .
 5.000D-01  5.000D+00 .            .              .      *       .              .
 5.500D-01  5.000D+00 .            .              .      *       .              .
 6.000D-01  5.000D+00 .            .              .      *       .              .
 6.500D-01  5.000D+00 .            .              .      *       .              .
 7.000D-01  5.000D+00 .            .              .      *       .              .
 7.500D-01  5.000D+00 .            .              .      *       .              .
 8.000D-01  5.000D+00 .            .              .      *       .              .
 8.500D-01  5.000D+00 .            .              .      *       .              .
 9.000D-01  5.000D+00 .            .              .      *       .              .
 9.500D-01  5.000D+00 .            .              .      *       .              .
 1.000D+00  5.000D+00 .            .              .      *       .              .
                - - - - - - - - - - - - - - - - - - - - - - - - - - - - - - - - -
```

```
*******11/09/88 ********  SPICE 2G.6    3/15/83 ********21:05:12*****
SPICE 8.3
****     INPUT LISTING                    TEMPERATURE =   27.000 DEG C
*******************************************************************************
 RS 1 2 100
 R 3 0 100
 RL 5 0 100
 LP 2 3 4M
 LS 4 3 25M
 K LP LS .5
 V 1 0 PULSE(0 5 1F 1F)
 VAM 4 5
 .TRAN 50U 1M
 .PLOT TRAN I(VAM)
 .END
*******11/09/88 ********  SPICE 2G.6    3/15/83 ********21:05:12*****
SPICE 8.3
****     INITIAL TRANSIENT SOLUTION       TEMPERATURE =   27.000 DEG C
*******************************************************************************
  NODE   VOLTAGE     NODE   VOLTAGE     NODE   VOLTAGE     NODE   VOLTAGE
 (  1)     .0000    (  2)     .0000    (  3)     .0000    (  4)     .0000
 (  5)     .0000
     VOLTAGE SOURCE CURRENTS
     NAME       CURRENT
     V         0.000D+00
     VAM       0.000D+00
     TOTAL POWER DISSIPATION   0.00D+00  WATTS
*******11/09/88 ********  SPICE 2G.6    3/15/83 ********21:05:12*****
SPICE 8.3
****     TRANSIENT ANALYSIS               TEMPERATURE =   27.000 DEG C
*******************************************************************************
     TIME      I(VAM)
                    0.000D+00     5.000D-03     1.000D-02     1.500D-02  2.000D-02
                         - - - - - - - - - - - - - - - - - - - - - - - - - - - - -
 0.000D+00  0.000D+00 *             .             .             .             .
 5.000D-05  7.776D-03 .             .       *     .             .             .
 1.000D-04  1.053D-02 .             .             .*            .             .
 1.500D-04  1.237D-02 .             .             .       *     .             .
 2.000D-04  1.364D-02 .             .             .          *  .             .
 2.500D-04  1.454D-02 .             .             .             *.            .
 3.000D-04  1.517D-02 .             .             .             *             .
 3.500D-04  1.561D-02 .             .             .             . *           .
 4.000D-04  1.592D-02 .             .             .             .  *          .
 4.500D-04  1.615D-02 .             .             .             .  *          .
 5.000D-04  1.630D-02 .             .             .             .   *         .
 5.500D-04  1.641D-02 .             .             .             .   *         .
 6.000D-04  1.648D-02 .             .             .             .   *         .
 6.500D-04  1.654D-02 .             .             .             .   *         .
 7.000D-04  1.658D-02 .             .             .             .   *         .
 7.500D-04  1.660D-02 .             .             .             .   *         .
 8.000D-04  1.662D-02 .             .             .             .    *        .
 8.500D-04  1.664D-02 .             .             .             .    *        .
 9.000D-04  1.664D-02 .             .             .             .    *        .
 9.500D-04  1.665D-02 .             .             .             .    *        .
 1.000D-03  1.666D-02 .             .             .             .    *        .
                         - - - - - - - - - - - - - - - - - - - - - - - - - - - - -
        JOB CONCLUDED
```

```
*******11/17/88 ********  SPICE 2G.6    3/15/83 ********19:53:36*****
SPICE 8.4
****     INPUT LISTING                      TEMPERATURE =   27.000 DEG C
*****************************************************************************
 R 1 2 50
 RP 2 3 20
 LP 3 0 20M
 LS 0 4 .5
 RS 4 5 100
 C 5 0 100U
 K LP LS .25
 V 1 0 PULSE(0 100 1F 1F)
 .TRAN 3.75M 75M
 .PLOT TRAN V(5)
 .END
*******11/17/88 ********  SPICE 2G.6    3/15/83 ********19:53:36*****
SPICE 8.4
****     INITIAL TRANSIENT SOLUTION         TEMPERATURE =   27.000 DEG C
*****************************************************************************
  NODE   VOLTAGE     NODE  VOLTAGE     NODE   VOLTAGE     NODE   VOLTAGE
 (  1)    .0000     (  2)    .0000    (  3)     .0000    (  4)     .0000
 (  5)    .0000
     VOLTAGE SOURCE CURRENTS
     NAME       CURRENT
     V         0.000D+00
     TOTAL POWER DISSIPATION   0.00D+00  WATTS
*******11/17/88 ********  SPICE 2G.6    3/15/83 ********19:53:36*****
SPICE 8.4
****     TRANSIENT ANALYSIS                 TEMPERATURE =   27.000 DEG C
*****************************************************************************

     TIME       V(5)
                  -3.000D+00     -2.000D+00     -1.000D+00      0.000D+00  1.000D+00
                           - - - - - - - - - - - - - - - - - - - - - - - - - - - - - - -
  0.000D+00  0.000D+00 .              .              .              *              .
  3.750D-03 -1.689D+00 .              .    *         .              .              .
  7.500D-03 -2.293D+00 .          *   .              .              .              .
  1.125D-02 -2.123D+00 .            * .              .              .              .
  1.500D-02 -1.638D+00 .              .     *        .              .              .
  1.875D-02 -1.091D+00 .              .             *.              .              .
  2.250D-02 -6.197D-01 .              .              .     *        .              .
  2.625D-02 -2.790D-01 .              .              .          *   .              .
  3.000D-02 -6.102D-02 .              .              .             *.              .
  3.375D-02  5.211D-02 .              .              .              .*             .
  3.750D-02  9.713D-02 .              .              .              .*             .
  4.125D-02  9.890D-02 .              .              .              .*             .
  4.500D-02  8.106D-02 .              .              .              .*             .
  4.875D-02  5.684D-02 .              .              .              .*             .
  5.250D-02  3.434D-02 .              .              .              *              .
  5.625D-02  1.715D-02 .              .              .              *              .
  6.000D-02  5.569D-03 .              .              .              *              .
  6.375D-02 -8.985D-04 .              .              .              *              .
  6.750D-02 -3.866D-03 .              .              .              *              .
  7.125D-02 -4.482D-03 .              .              .              *              .
  7.500D-02 -3.934D-03 .              .              .              *              .
                           - - - - - - - - - - - - - - - - - - - - - - - - - - - - - - -

        JOB CONCLUDED
```

It is also possible to conduct a purely electrical simulation of a network which includes transformers by replacing any transformer with one of its electrical equivalents (i.e., controlled source model, T-equivalent model, or Π-equivalent model).

8.7 SUMMARY

Two-port networks usually exist between a source network and a load network for the purpose of making a given load function more effectively with a given source. The reader may be familiar with some of the terms used to describe various roles of the two-port network—**signal conditioning, filtering, amplification, attenuation, impedance matching, isolation, wave shaping,** etc.

In this chapter we have not dwelt on the functions of particular two-port networks, but instead introduced two analysis methods which are independent of the role of the two-port network. The method of network parameters, which characterizes any two-port network in terms of one of six standard models, is used to examine all possible relationships between input and output voltage and current. Transfer functions are used to relate one output quantity, either voltage or current, to one input quantity.

Transform network analysis techniques were applied to two important types of two-port devices—the OPAMP and the linear transformer. The reader should take the opportunity in the problem section which follows to practice obtaining network parameters and transfer functions for networks containing these devices.

8.8 Terms

admittance parameters
amplification
attenuation
bilateral
coefficient of coupling
coefficient of mutual inductance
common-mode rejection ratio (CMRR)
coupled differential equations
cross parameters
differential input voltage
differentiator
dot
dotted terminal
driving point
driving-point impedance
exciting the network at a pole
exciting the network at a zero
feedback network
ferromagnetic
filtering
forward transfer functions
gain
hybrid parameters
ideal transformer

immittance
impedance conversion
impedance magnifier
impedance matching
impedance parameters
impedance reducer
impulse response
input port
integrator
inverting configuration
inverting input terminal
isolation
linear transformer
linear transformer equations
load terminals
loosely coupled
magnetic coupling
magnetically coupled network
mutual inductance
n-port network
negative gyrator
negative impedance converter
negative resistance
network parameters
noninverting configuration
noninverting input terminal
nonmagnetic
operational amplifier (OPAMP)
output port
permeability
Π-equivalent network model
port
positive gyrator
primary
reciprocal network
response function
reverse transfer functions
saturated
secondary
self-inductance
signal conditioning
slew rate
T-equivalent network model
tightly coupled
transfer function
transfer impedance
transformer
transmission parameters
two-port parameters
virtual ground
wave shaping
zeros of the function

Problems

***8-1.** **a.** Use transform circuit analysis to derive the z-parameters.
b. Use the z-parameters to derive the y-, g-, and h-parameters.
c. Sketch and label the y-, z-, g-, and h-parameters models.

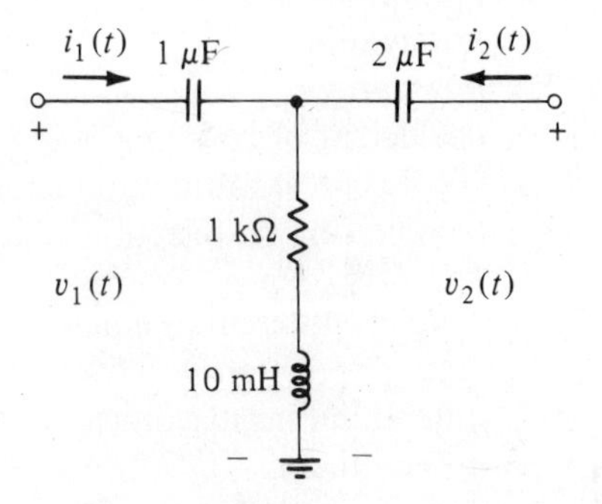

***8-2.** For the two-port network of Problem 8-1,

a. Use transform circuit analysis to derive the transfer functions

$$H(s) = \frac{V_2(s)}{V_1(s)}, \qquad H(s) = \frac{V_2(s)}{I_1(s)}$$

$$H(s) = \frac{I_2(s)}{V_1(s)}, \qquad H(s) = \frac{I_2(s)}{I_1(s)}$$

Assume open-circuit conditions at port 2 where $V_2(s)$ is involved, and short-circuit conditions where $I_2(s)$ is involved.

b. Repeat part a using the appropriate set of two-port parameters as derived in Problem 8-1 for direct calculation of the transfer functions.

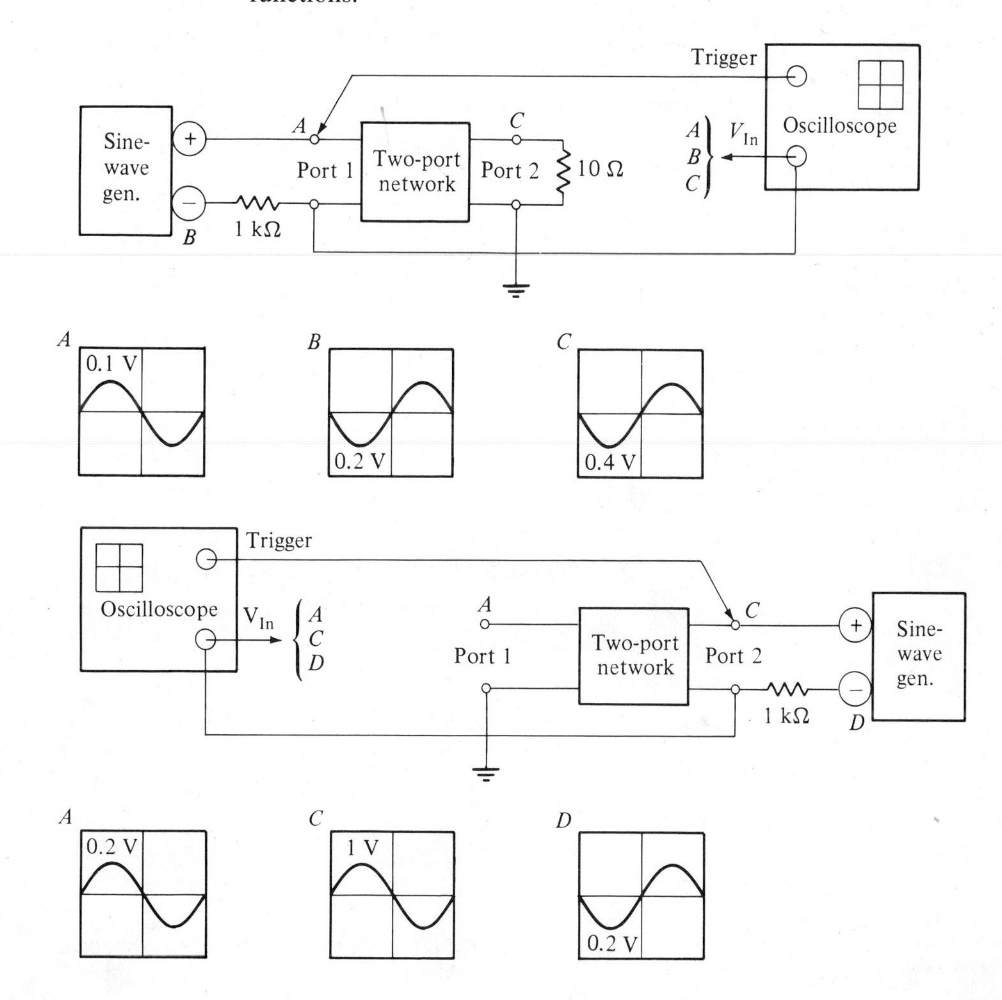

8-3. The two-port network in the test configurations contains no inductors or capacitors.

***a.** Calculate the h-parameters from the experimental results. Assume that the 10-Ω resistor constitutes a short circuit.

***b.** Derive the z-parameters from the h-parameters.

c. Sketch and label the h- and z-parameter models.

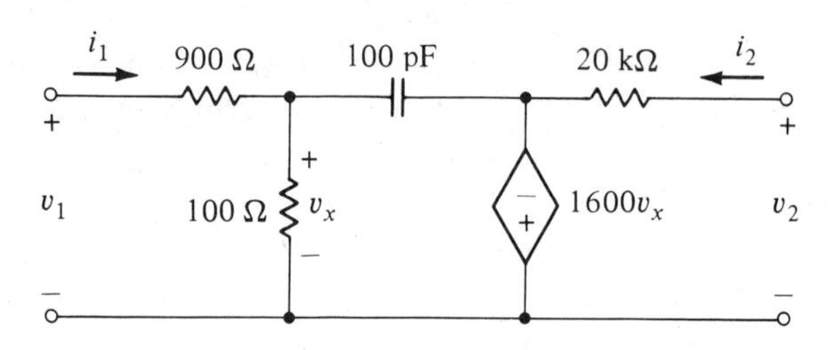

8-4. For the two-port network of Problem 8-3,

***a.** Calculate the transfer function

$$H(s) = \frac{I_2(s)}{I_1(s)}$$

under short-circuit conditions.

b. Calculate the transfer function

$$H(s) = \frac{V_2(s)}{V_1(s)}$$

with a 2.5-kΩ load resistor.

***8-5.** **a.** Use transform circuit analysis to determine the y-parameters.

b. Sketch and label the y-parameter model.

8-6. For the two-port network in Problem 8-5,

a. Use the y-parameters to calculate the z-parameters, and then use the z-parameters to derive the transfer function for open-circuit conditions.

$$H(s) = \frac{V_2(s)}{V_1(s)}$$

***b.** Recalculate $H(s)$ with a 10-μF capacitor across port 2.

8-7. For the inverting OPAMP network configuration of Figure 8.9(a),

a. Prove that

$$z_{11}(s) = Z_i(s), \quad z_{21}(s) = -Z_f(s), \text{ and } z_{12}(s) = z_{22}(s) = 0.$$

b. Sketch and label the z-parameter model.

SPICE

***8-8.** **a.** Find and sketch $v_o(t)$ for $v_i(t) = 25t \cdot u(t)$ V.

b. Repeat part a for $v_i(t) = 1000t^2 \cdot u(t)$ V.

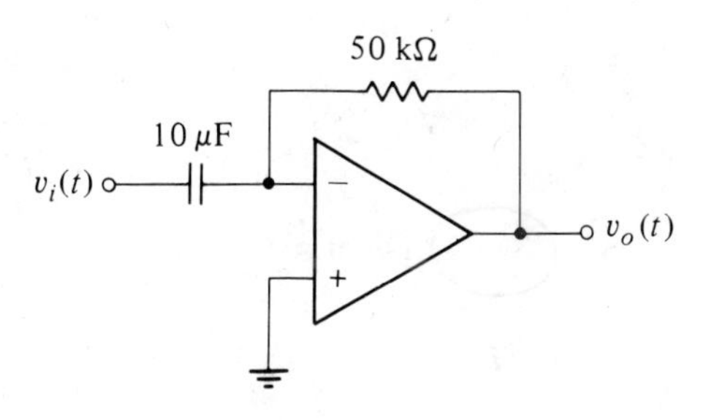

SPICE

8-9. ***a.** Derive the transfer function

$$H(s) = \frac{V_o(s)}{V_i(s)}$$

b. Use $H(s)$ to calculate $V_o(s)$ for $v_i(t) = 5 \cdot u(t)$ V.

***c.** Use $H(s)$ to calculate $V_o(s)$ for $v_i(t) = 240t \cdot u(t)$ V.

d. Use the initial and final value theorems to calculate the initial and steady-state values of $v_o(t)$ for each of part b and part c.

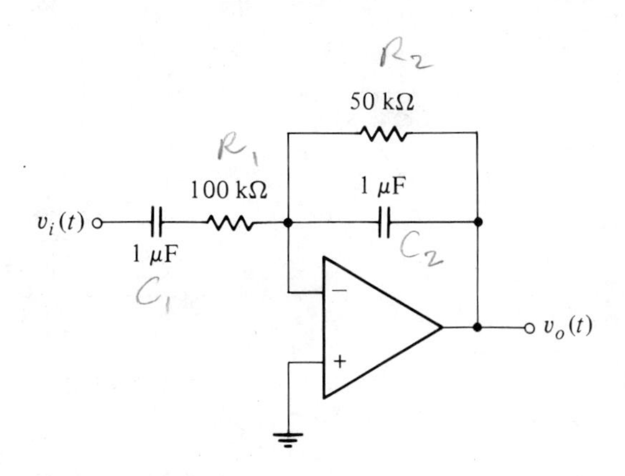

SPICE

8-10. a. Determine $v_o(t)$ for $v_i(t) = 0.5 \cdot \sin[100t] \cdot u(t)$ V.
*b. Repeat part a for $v_i(t) = 10\exp[-100t] \cdot u(t)$ V.

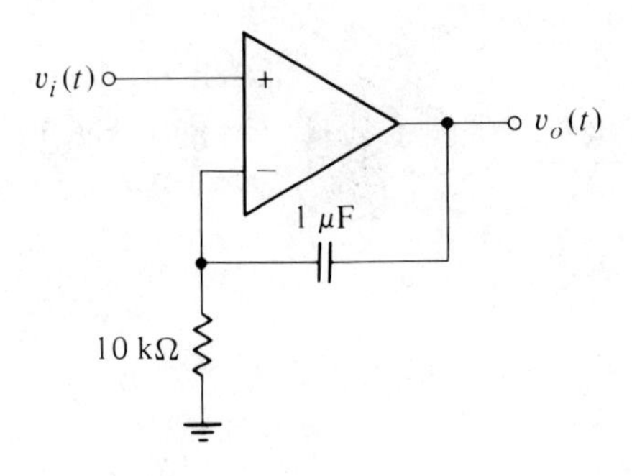

***8-11.** Determine $Z_i(s)$.

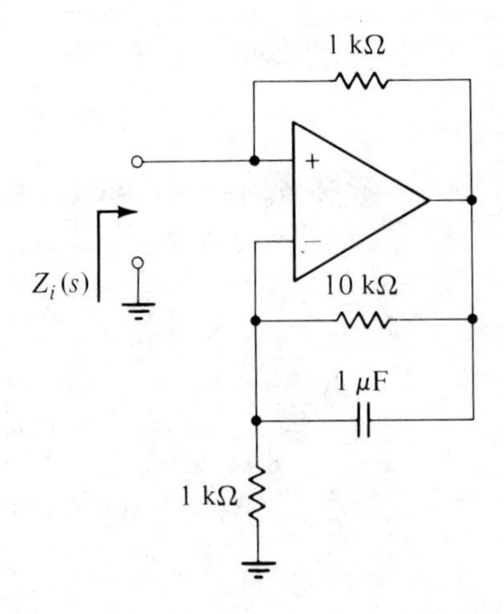

***8-12.** Replace the transformer with a controlled-source model, then find the z-parameters for the network. Assume initial conditions are zero.

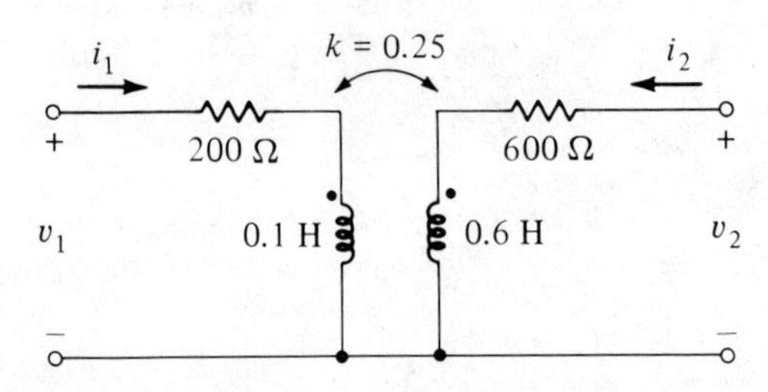

8-13. Repeat Problem 8-12 using the T-equivalent model for the transformer.

8-14. Show that the figure below is a Π-equivalent model for a linear transformer with zero initial conditions.

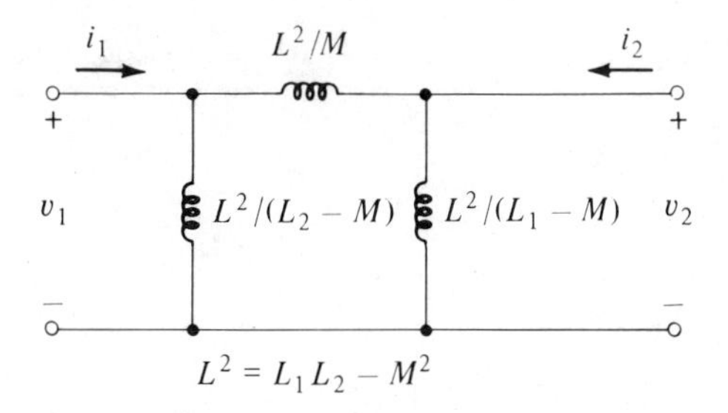

***8-15.** Sketch and label a transformed equivalent two-port network using a controlled-source model for the transformer.

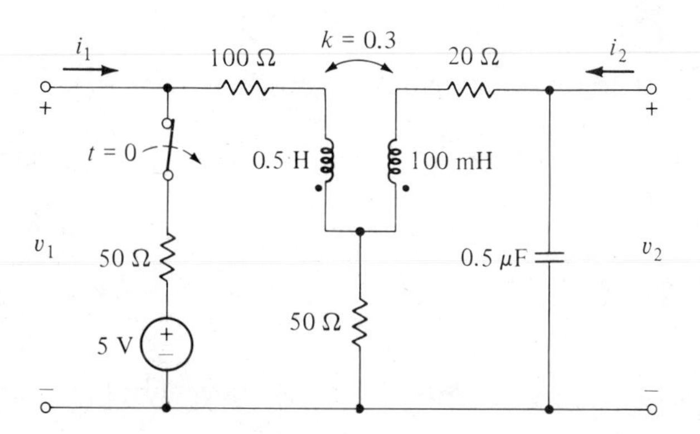

***8-16.** Sketch and label a transformed equivalent two-port network using a T-equivalent model for the transformer.

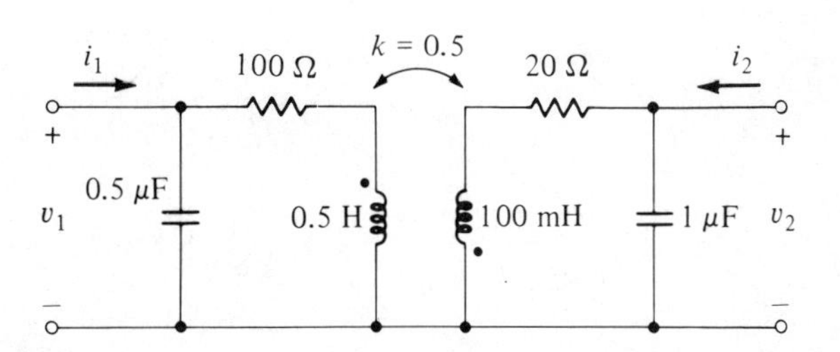

***8-17.** Determine the transfer function

$$H(S) = \frac{V_2(s)}{V_1(s)}$$

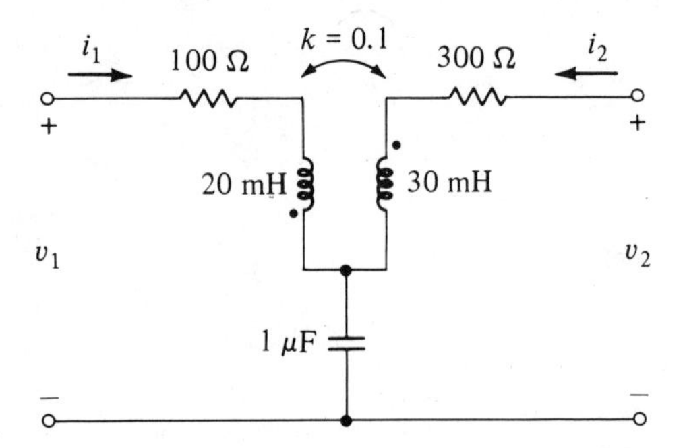

***8-18.** Determine the equivalent inductance between terminals A and B for each of the three configuratoins below.

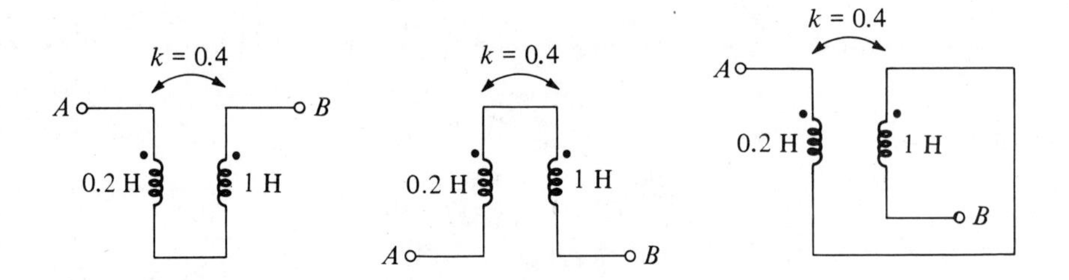

8-19. Verify that Equation (8.37) is obtained from Equations (8.35a), (8.35b), and (8.35c).

8-20. Verify that Equation (8.38) is obtained from Equations (8.35a), (8.35b), and (8.35c). Note that $Z_g(s)$ does not appear in Equation (8.38).

9 Sinusoidal Steady-State Analysis

9.1 INTRODUCTION

The reader has undoubtedly used **sinusoidal steady-state analysis**, or **ac analysis**, as it is often called, during a basic circuit analysis course. The major technique of ac analysis involves **frequency-domain transformation** of capacitors and inductors and the use of complex quantities called **phasors** to represent sinusoidal steady-state functions. In this chapter we will define the phasor and use it, along with the s-domain immittance functions [$Y(s)$ or $Z(s)$], to develop formally the ac analysis technique. This approach will show clearly that ac analysis and use of phasors is but a specialization of the more general technique of transform circuit analysis which we have detailed in the past few chapters.

Immediately after defining the phasor and developing the ac analysis technique, we will then review the use of ac analysis in determining the response of circuits operating at a single, constant frequency. These techniques should be familiar to most readers, since single-frequency operation is emphasized in basic circuit analysis texts and courses.

A major goal in this chapter is to introduce a general technique, commonly called **frequency-response** analysis, in which we examine the effect on steady-state circuit response of varying the frequency of the sinusoidal source. Our investigations in the general area of frequency-response analysis will include characterization of the phenomenon called **resonance**.

In addition to considering the response to a source operating at a single frequency, and the effect of varying the source frequency, we must consider the possibility of more than one source frequency. This situation arises when a given source is operating simultaneously at more than one frequency, or when there are two or more sources operating at different frequencies. In either case, if the multiple frequencies involved are **harmonically related**—that is, all frequencies present are **integer multiples** of the lowest frequency—we can utilize the Fourier analysis techniques presented in Chapter 11. If the frequencies are not harmonically related, we must use phasor techniques to analyze the circuit at each frequency, then *superpose* the results in the time domain.

9.2 The Sinusoidal Steady-State (AC) Analysis Technique

Definition of the Phasor

Consider first the complex number **A** whose magnitude (or absolute value) is A_m and whose polar angle is θ. Now consider the product of **A** and the complex time-domain function $\exp[j\omega t]$,

$$\begin{aligned}\mathbf{A} \cdot \exp[j\omega t]) &= (A_m \cdot \exp[j\theta])(\exp[j\omega t]) \\ &= A_m \cdot \exp[j(\omega t + \theta)]\end{aligned}$$

By Euler's identity,

$$\mathbf{A} \cdot \exp[j\omega t] = A_m(\cos[\omega t + \theta] + j\sin[\omega t + \theta])$$

If the real part is extracted from the right member of the preceding equation, as in

$$A_m \cdot \cos[\omega t + \theta] = \text{Real}\{\mathbf{A} \cdot \exp[j\omega t]\} \tag{9.1}$$

then the *magnitude* (or absolute value) of **A** is the *amplitude*, and the *polar angle* of **A** is the *displacement* (*or phase*) *angle*, of the sinusoid $A_m \cdot \cos[\omega t + \theta]$.

A complex number whose magnitude equals the amplitude, and whose polar angle equals the phase angle, of a sinusoidal function is called a **phasor**. In the foregoing discussion, which establishes the mathematical relationship between a phasor and a sinusoidal time-domain function, the complex number **A** is the **phasor representation** of $A_m \cdot \cos[\omega t + \theta]$.[1]

The use of the phasor to represent a sinusoid is an example of mathematical transformation (though not an integral transform as described in Chapter 4). The first sentence of the preceding paragraph defines the *transformation*, and Equation (9.1) defines the *inverse transformation*. The domain of a phasor is called the **frequency domain**, because the phasor magnitude and polar angle

[1] Some authors, most often those who write basic textbooks in circuit or network analysis, *equate* the phasor to the sinusoid which it represents. Certainly there is no mathematical *equality* between a complex number and a real, time-domain function. As we will soon see, this is tantamount to saying that $F(s) = f(t)$.

depend upon frequency (ω). It will soon become apparent how the frequency domain is related to the s-domain.

Development of the AC Analysis Technique

Development of the ac analysis technique in basic texts and courses is usually based on the intuitively appealing **rotating vector**, the tip of which traces a sinusoid in the time domain [Figure 9.1(b)]. The representation of sinusoidal quantities by vectors leads to the **phasor diagram** [Figure 9.1(c)]. Before electronic computers and scientific calculators became available, the use of the phasor diagram for graphical analysis of sinusoidal steady-state response was an important tool of the working analyst or designer. Phasor diagrams, which are in some respects snapshots or freeze frames of the time domain, are now used primarily to provide greater insight into the nature of and relationship between sinusoidal steady-state responses.

Another approach to developing the ac analysis technique is to use transform circuit analysis and the s-domain immittance function for a one-port, passive linear network. Admittance will be used in this development, but impedance could be used just as well.

Admittance is the ratio of response current to excitation voltage,

$$Y(s) = \frac{I(s)}{V(s)}, \qquad \text{or} \quad I(s) = V(s) \cdot Y(s)$$

If the excitation is a sinusoid,

$$v(t) = V_m \cdot \cos[\omega t + \theta] \cdot u(t) \text{ V} \tag{9.2}$$

then

$$V(s) = \frac{V_m(s \cdot \cos\theta - \omega \cdot \sin\theta)}{s^2 + \omega^2} \text{ V}$$

and

$$I(s) = \frac{V_m(s \cdot \cos\theta - \omega \cdot \sin\theta)}{s^2 + \omega^2} \cdot Y(s) \tag{9.3}$$

The immittance function for a passive, linear network is the ratio of two polynomials, with the degree of the numerator at most one more than the degree of the denominator. Multiplying $Y(s)$ by $V(s)$ will raise the degree of the numerator by one at most, and the degree of the denominator by two, with the result that the denominator of $I(s)$ is at least one greater than the degree

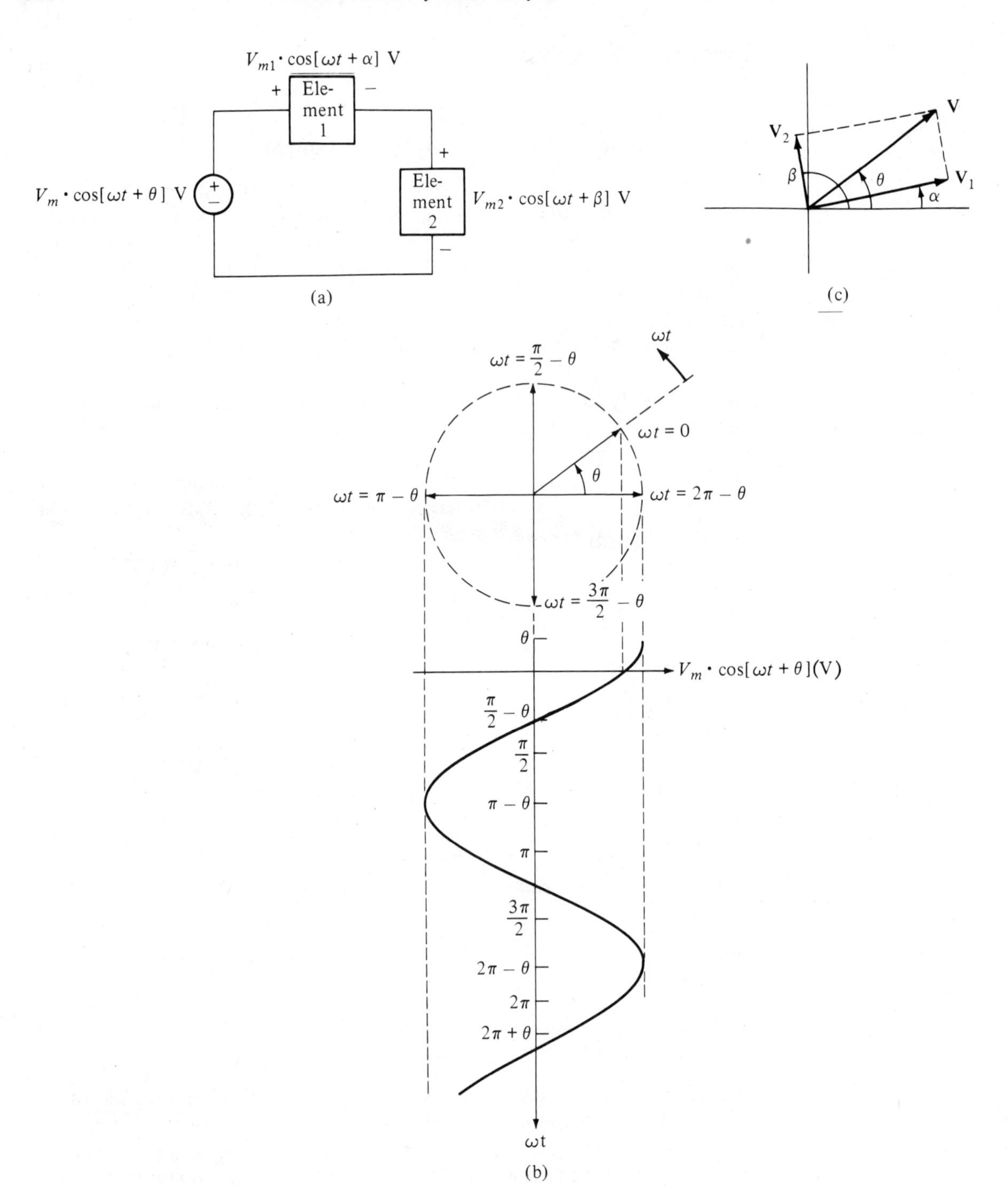

Figure 9.1 Phasor as a rotating vector.

of the numerator.[2] The importance of this observation is that it assures that the response of a passive, linear network to sinusoidal excitation can be decomposed into partial fractions. Therefore,

$$I(s) = \frac{\mathbf{K}}{s - j\omega} + \frac{\mathbf{K}^*}{s + j\omega} + \frac{K_1}{s - p_1} + \frac{K_2}{s - p_2} + \cdots + \frac{K_n}{s - p_n} \text{ A} \cdot \text{sec} \quad (9.4)$$

where the first two terms are due to $V(s)$, and the remaining terms are due to $Y(s)$.

By inverse Laplace transformation of Equation (9.4) using Equation 4.17 and Pair IV,

$$i(t) = (2|\mathbf{K}| \cdot \cos[\omega t + \phi] + K_1 \cdot \exp[p_1 t] + K_2 \cdot \exp[p_2 t] + \cdots + K_n \cdot \exp[p_n t])u(t) \text{ A}$$

Any constant K_i in the decomposition can be real or complex, as can any pole p_i of $Y(s)$, but the real part of every pole is less than zero due to dissipation within the network. As a consequence of $\text{Real}\{p_i\} < 0$, all exponential terms of $i(t)$ vanish for large values of t, and the steady-state value of $i(t)$ is

$$i(t)_{\text{ss}} = 2|\mathbf{K}| \cdot \cos[\omega t + \phi] \text{ A} \quad (9.5)$$

Flashback

$$\lim_{x \to \infty} \exp[-x] = 0$$

Note that the steady-state response differs from the excitation only in amplitude and phase angle, both of which depend on the complex constant **K** in the first term of the partial-fraction decomposition of $I(s)$ [Equation (9.4)]. The steady-state response to sinusoidal excitation is therefore determined by the excitation frequency ω and **K**.

From Equation (4.11),

$$\mathbf{K} = (s - j\omega)I(s)|_{s=j\omega}$$

---

[2] If $\theta = \pi/2$ rad—that is, if $v(t) = V_m \cdot \sin \omega t$ V, then the degree of the numerator is not raised.

Substitution of Equation (9.3) for $I(s)$ yields

$$\begin{aligned}
\mathbf{K} &= (s - j\omega)\frac{V_m(s \cdot \cos\theta - \omega \cdot \sin\theta)}{s^2 + \omega^2} \cdot Y(s)\bigg|_{s=j\omega} \\
&= \frac{V_m(s \cdot \cos\theta - \omega \cdot \sin\theta)}{s + j\omega} \cdot Y(s)\bigg|_{s=j\omega} \\
&= \frac{V_m(j\omega \cdot \cos\theta - \omega \cdot \sin\theta)}{j2\omega} \cdot \mathbf{Y}(j\omega) \\
&= \frac{V_m}{2}(\cos\theta + j \cdot \sin\theta)\mathbf{Y}(j\omega)
\end{aligned}$$

From Euler's identity,

$$\mathbf{K} = \tfrac{1}{2}V_m \cdot \exp[j\theta] \cdot \mathbf{Y}(j\omega)$$

In general, $\mathbf{Y}(j\omega)$ is a complex function

$$\mathbf{Y}(j\omega) = |\mathbf{Y}(j\omega)| \cdot \exp[j \cdot \gamma(j\omega)]$$

where $|\mathbf{Y}(j\omega)|$ is the magnitude and $\gamma(j\omega)$ the polar angle of $\mathbf{Y}(j\omega)$. So

$$\mathbf{K} = \tfrac{1}{2} \cdot V_m|\mathbf{Y}(j\omega)| \cdot \exp[j\{\theta + \gamma(j\omega)\}]$$

The magnitude of $\mathbf{K}$ is

$$|\mathbf{K}| = \tfrac{1}{2}V_m \cdot |\mathbf{Y}(j\omega)| \tag{9.6a}$$

and the polar angle is

$$\phi = \theta + \gamma(j\omega) \tag{9.6b}$$

Substitution of Equations (9.6a) and (9.6b) into Equation (9.5) yields the final result:

$$\begin{aligned}
i(t)_{ss} &= V_m \cdot |\mathbf{Y}(j\omega)| \cdot \cos[\omega t + \theta + \gamma(j\omega)] \\
&= I_m \cdot \cos[\omega t + \phi] \text{ A}
\end{aligned} \tag{9.7}$$

where $I_m = V_m \cdot |\mathbf{Y}(j\omega)|$ and $\phi = \theta + \gamma(j\omega)$.

As previously observed, the steady-state response differs from the excitation only in amplitude and phase angle. From Equation (9.7), the amplitude I_m of the response depends on the amplitude V_m of the excitation and the magnitude $|\mathbf{Y}(j\omega)|$ of the admittance; the phase angle ϕ of the response depends on the phase angle θ of the excitation and the polar angle $\gamma(j\omega)$ of the admittance.

A similar development involving current excitation,

$$i(t) = I_m \cdot \cos[\omega t + \theta] \cdot u(t) \text{ A}$$

results in a response voltage of

$$\begin{aligned} v(t)_{ss} &= I_m \cdot |\mathbf{Z}(j\omega)| \cdot \cos[\omega t + \theta + \zeta(j\omega)] \\ &= V_m \cdot \cos[\omega t + \phi] \text{ A} \end{aligned} \tag{9.8}$$

where $\mathbf{Z}(j\omega) = |\mathbf{Z}(j\omega)| \cdot \exp[\zeta(j\omega)]$, $V_m = I_m \cdot |\mathbf{Z}(j\omega)|$, and $\phi = \theta + \zeta(j\omega)$.

In either development above, it must be recognized that admittance and impedance are reciprocals, so it is possible to write for Equation (9.7)

$$I_m = \frac{V_m}{|\mathbf{Z}(j\omega)|} \quad \text{and} \quad \phi = \theta - \zeta(j\omega)$$

and for Equation (9.8)

$$V_m = \frac{I_m}{|\mathbf{Y}(j\omega)|} \quad \text{and} \quad \phi = \theta - \gamma(j\omega)$$

In computing steady-state response to sinusoidal excitation, the frequency ω can be suppressed from the computation, since only the amplitude and phase angle of the excitation are needed. Equation (9.1a) indicates that the phasor representation of a sinusoid carries amplitude and phase angle, but not frequency information. Moreover, the immittance functions $[Y(s)$ or $Z(s)]$ are evaluated at $s = j\omega$ for computation of the steady-state response to sinusoidal excitation, so in transforming circuit elements, s can be replaced by $j\omega$. The ac analysis technique can now be summarized (Figure 9.2):

1. Transform the passive elements of the network as in Laplace transformation, replacing s by $j\omega$. Either the impedance transformation or the admittance transformation may be used, depending on circuit topology and the requirements of the problem.
2. Represent the sinusoidal sources as phasors. Voltage sources are represented as

$$\mathbf{V} = V_m \angle \theta \text{ V}$$

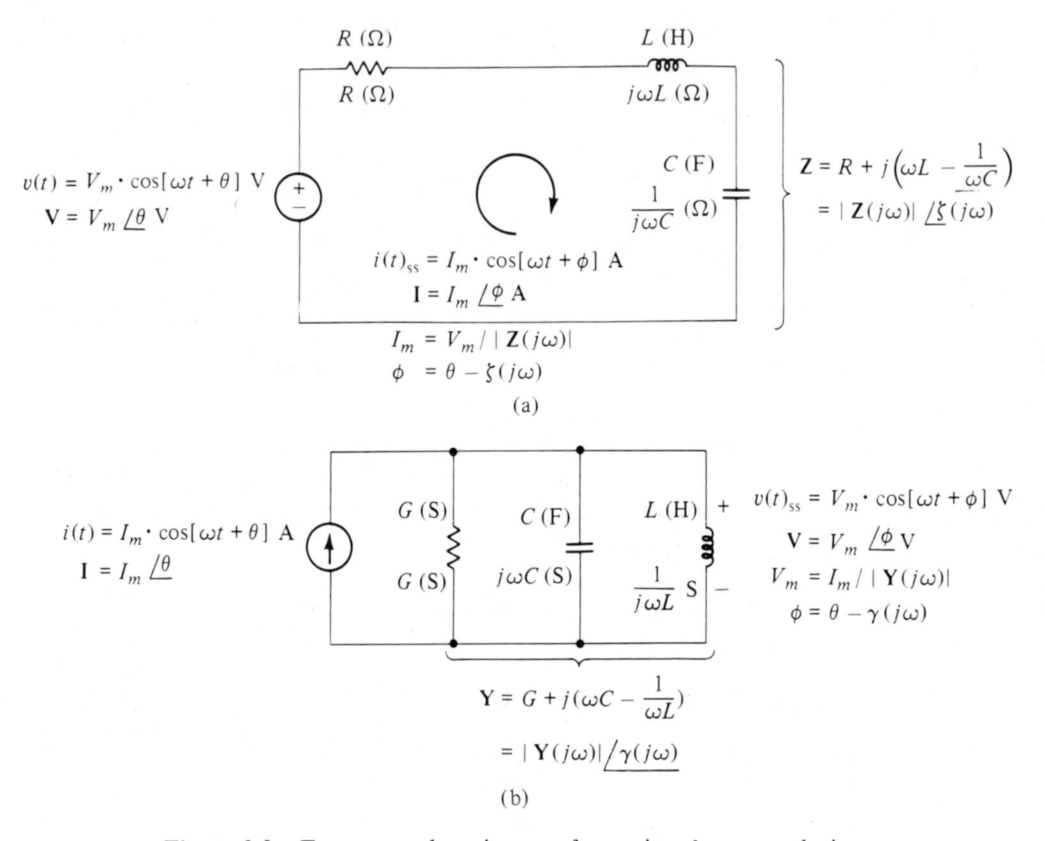

Figure 9.2 Frequency-domain transformation for ac analysis.

where $v(t) = V_m \cdot \cos[\omega t + \theta]$ V, and current sources are represented as

$$\mathbf{I} = I_m \angle \theta \text{ A}$$

where $i(t) = I_m \cdot \cos[\omega t + \theta]$ A.[3]

3. Compute the response voltage or current in phasor form:

$$\mathbf{V} = V_m \angle \phi$$

[3] The use of $\angle\theta$ is standard shorthand notation for $\exp[j\theta]$. Thus

$$V_m\angle\theta = V_m \cdot \exp[j\theta] \quad \text{and} \quad I_m\angle\theta = I_m \cdot \exp[j\theta]$$

or

$$\mathbf{I} = I_m \angle \phi$$

4. Obtain the steady-state response by inverse transformation of the phasor:

$$\begin{aligned} v(t)_{ss} &= \text{Real}\{\mathbf{V} \cdot \exp[j\omega t]\} = \text{Real}\{V_m \cdot \exp[\phi] \cdot \exp[j\omega t]\} \\ &= \text{Real}\{V_m \cdot \exp[j\omega t + \phi]\} \\ &= \text{Real}\{V_m(\cos[\omega t + \phi] + j \cdot \sin[\omega t + \phi]\} \\ &= V_m \cdot \cos[\omega t + \phi] \text{ V} \\ i(t)_{ss} &= \text{Real}\{\mathbf{I} \cdot \exp[j\omega t]\} = \text{Real}\{I_m \cdot \exp[\phi] \cdot \exp[j\omega t]\} \\ &= \text{Real}\{I_m \cdot \exp[j\omega t + \phi]\} \\ &= \text{Real}\{I_m(\cos[\omega t + \phi] + j \cdot \sin[\omega t + \phi]\} \\ &= I_m \cdot \cos[\omega t + \phi] \text{ A} \end{aligned}$$

Any of the methods of transform circuit analysis—combination of elements, voltage and current division, source conversion, node and mesh analysis, superposition, Thevenin's and Norton's theorems, etc.—can be used in step 3 to obtain the response phasor.

Step 4 in the ac analysis technique is stated formally in the interest of completeness, but with experience, the necessary information for expressing the steady-state response (i.e., the magnitude and polar angle of the phasor for amplitude and phase angle, respectively, of the sinusoid) can be extracted from the phasor without formal computation.

If the desired response were computed in the s-domain, the time-domain response would have to be determined by inverse Laplace transformation; the steady-state time-domain response would then only be apparent after eliminating the transient components. By using the ac analysis techniques described above, the desired response phasor, which appears as a function of $j\omega$, is a complex number for any explicit frequency ω. Inverse transformation of the phasor is simply a matter of extracting the information required to express the steady-state response at that value of ω—namely, magnitude and polar angle—directly from the complex number.

Some Particulars of AC Analysis

The frequency-domain impedance of a resistor, inductor, and capacitor in series [Figure 9.3(a)] is

$$\begin{aligned} \mathbf{Z}(j\omega) &= R + j\omega L + \frac{1}{j\omega C} = R + j\left(\omega L - \frac{1}{\omega C}\right) \\ &= R + j(X_L - X_C) = R + jX \end{aligned}$$

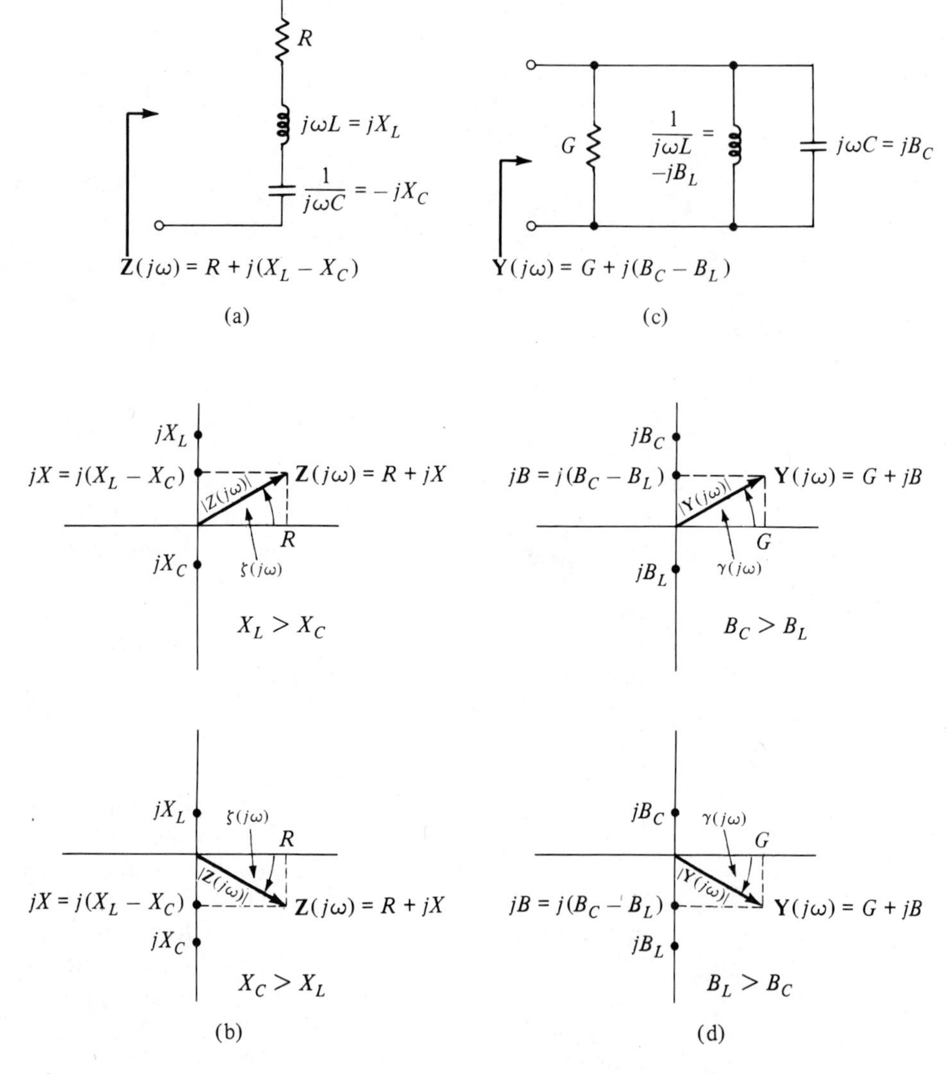

Figure 9.3 Frequency-domain impedance and admittance diagrams.

The imaginary part $X = X_L - X_C$ of $\mathbf{Z}(j\omega)$ is called **reactance**, where X_L is **inductive reactance**, and X_C is **capacitive reactance**. Reactance may be either positive or negative, depending on the relative values of X_L and X_C, but R is a positive number, so $\mathbf{Z}(j\omega)$ always lies in the right half of the complex plane [Figure 9.3(b)].

Similarly, the frequency-domain admittance of a resistor, inductor, and capacitor in parallel [Figure 9.3(c)] is

$$\mathbf{Y}(j\omega) = G + j\omega C + \frac{1}{j\omega L} = G + j\left(\omega C - \frac{1}{\omega L}\right)$$
$$= G + j(B_C - B_L) = G + jB$$

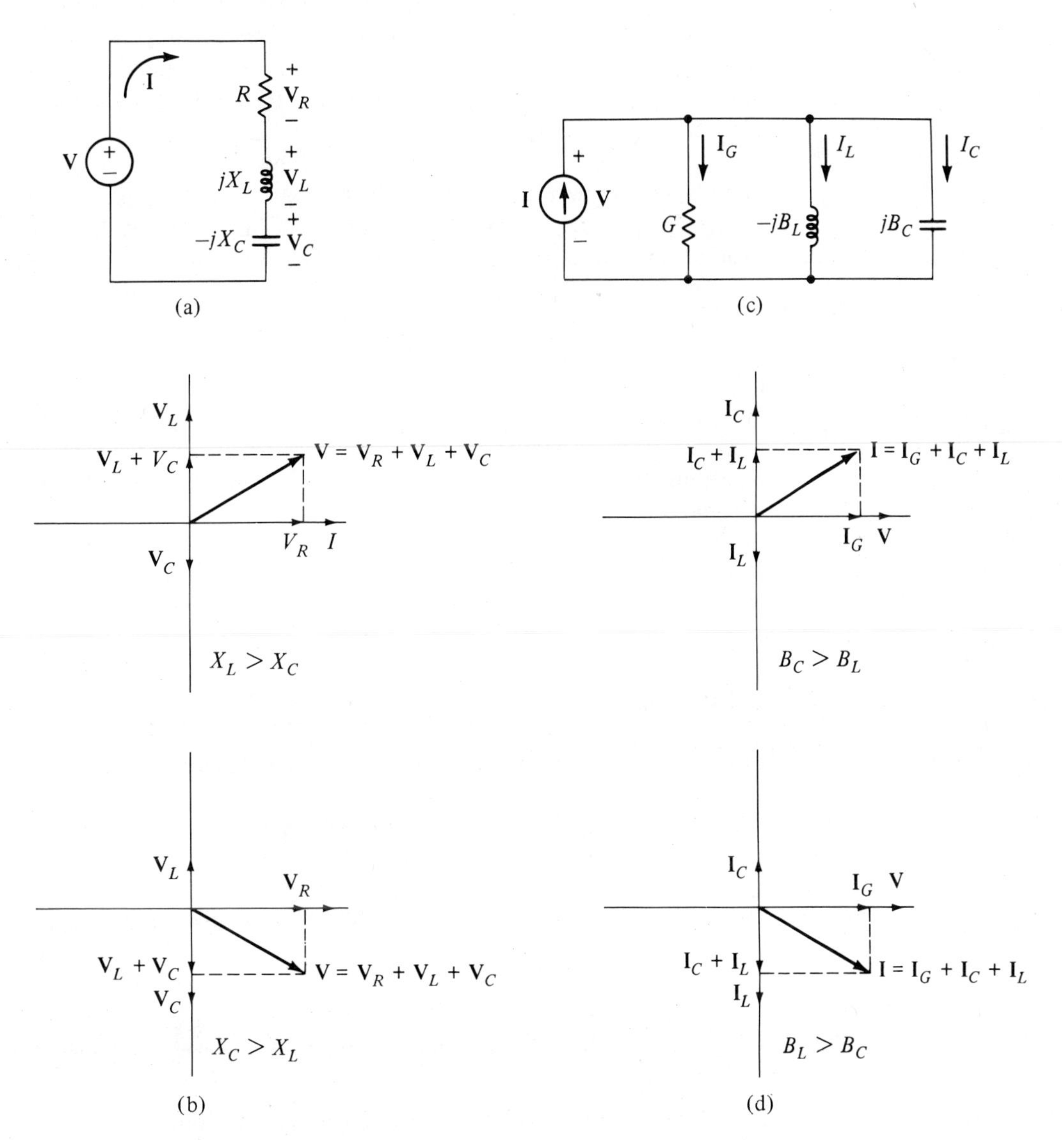

Figure 9.4 Phasor diagrams.

The imaginary part $B = B_C - B_L$ of $\mathbf{Y}(j\omega)$ is called **susceptance**, where B_L is **inductive susceptance**, and B_C **capacitive susceptance**. Susceptance can be either positive or negative, but G is positive, so $\mathbf{Y}(j\omega)$ also lies in the right half of the complex plane [Figure 9.3(d)]. Clearly susceptance and reactance are electrical duals.

Location of $\mathbf{Z}(j\omega)$ and $\mathbf{Y}(j\omega)$ in the right half of the complex plane restricts their polar angles to $\pm\pi/2$ ($\pm 90°$), which means that the response of a passive, linear network leads or lags the excitation by no more than 90°. For a series R–L–C network, the voltage leads the current if $X_L > X_C$, and lags if $X_C > X_L$. For a parallel R–L–C network, current leads voltage if $B_C > B_L$, and lags if $B_L > B_C$. These time-domain relationships are represented in the frequency domain by the phasor diagrams shown in Figure 9.4. The phasor diagrams of Figure 9.4(b) are constructed by superposing the phasors on the impedance diagram so that $\mathbf{V}$ passes through $\mathbf{Z}(j\omega)$, $\mathbf{V}_R$ passes through R, $\mathbf{V}_L$ passes through jX_L, and $\mathbf{V}_C$ passes through jX_C. The length of each phasor is proportional to its magnitude, so $\mathbf{V}$ is the *vector sum* of $\mathbf{V}_R$, $\mathbf{V}_L$, and $\mathbf{V}_C$. The direction of the response current phasor $\mathbf{I}$ coincides with that of $\mathbf{V}_R$ because, as the reader well knows, voltage and current are **in phase** for a resistor. Correspondingly, $\mathbf{I}$ lags $\mathbf{V}_L$ by 90°, and leads $\mathbf{V}_C$ by 90°, which also substantiates the well-known relationship between current and voltage for an inductor and a capacitor, respectively. The phasor diagrams of Figure 9.4(d) are constructed in a dual manner, and a dual set of observations can be made regarding the phase angle relationship between voltage and current in each element.

9.3 Review of Single-Frequency AC Analysis

In the previous section, we discussed a formal basis for sinusoidal steady-state (ac) analysis and the use of phasor techniques, and established the relationship between the frequency domain and the s-domain. Our assumption, as stated earlier, is that the reader has previously practiced single-frequency ac analysis, so we will present only a limited number of examples as a review. Single-frequency ac analysis problems can be found at the end of this chapter if the reader feels the need for practice.

Example 9.1 demonstrates

1. circuit element transformation, including phasor representation of the source,
2. combination of resistive and reactive elements to form a complex impedance or admittance,

3. computations using the relationships $\mathbf{Z}(j\omega) = \mathbf{V}/\mathbf{I}$ and $\mathbf{Y}(j\omega) = \mathbf{I}/\mathbf{V}$, the frequency-domain equivalent of Ohm's law,
4. use of the current division rule, and
5. inverse transformation of the response phasor.

Example 9.2 involves the determination Thevenin's and Norton's equivalent frequency-domain sources, and includes a demonstration of source conversion. Example 9.3 illustrates the application of the superposition principle in the frequency domain, and includes use of the voltage division rule.

Example 9.1

SPICE

Problem:

Determine the steady-state inductor current for the circuit shown in Figure 9.5.

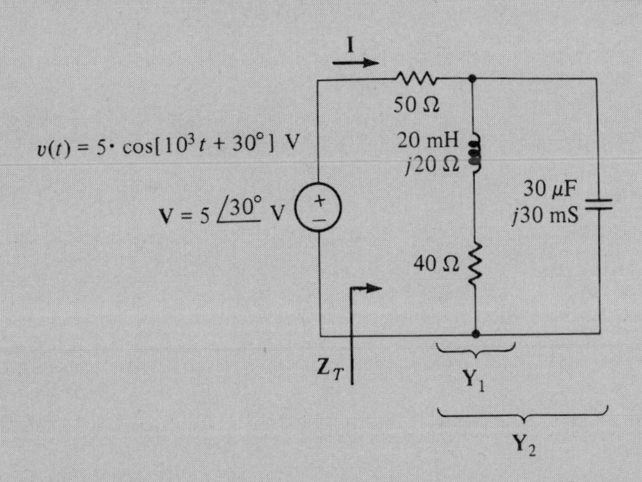

Figure 9.5
Example 9.1.

Solution:

1. Perform the frequency-domain transformation of the circuit elements, as shown on the circuit diagram.
2. Compute the admittance of the R–L branch:

$$\mathbf{Y}_1 = \frac{1}{\mathbf{Z}_1} = \frac{1}{40 + j20\ \Omega} = 22.36 \times 10^{-3} \angle -26.57^\circ = 20 - j10\ \text{mS}$$

3. Compute the impedance of the R–L–C network:

$$\mathbf{Y}_2 = \mathbf{Y}_1 + j30 \text{ mS} = 20 + j20 \text{ mS} = 28.28\angle 45^\circ \text{ mS}$$

$$\mathbf{Z}_2 = 1/\mathbf{Y}_2 = 35.36\angle -45^\circ \ \Omega = 25 - j25 \ \Omega$$

4. Compute the total impedance at the source terminals:

$$\mathbf{Z}_T = \mathbf{Z}_2 + 50 \ \Omega = 75 - j25 \ \Omega = 79.06\angle -18.43^\circ \ \Omega$$

5. Compute the source current phasor:

$$\mathbf{I} = \frac{\mathbf{V}}{\mathbf{Z}_T} = \frac{5\angle 30^\circ}{79.06\angle -18.43^\circ} = 63.24\angle 48.43^\circ \text{ mA}$$

6. Use the current division rule to obtain the inductor current phasor:

Flashback

If $\mathbf{A} = A\angle\alpha$ and $\mathbf{B} = B\angle\beta$,

$$\mathbf{A} \cdot \mathbf{B} = (A \cdot B)\angle \alpha + \beta$$

and

$$\frac{\mathbf{A}}{\mathbf{B}} = \frac{A}{B}\angle \alpha - \beta$$

$$\mathbf{I}_L = \mathbf{I}\frac{\mathbf{Y}_1}{\mathbf{Y}_2} = 63.24\angle 48.43^\circ \ \frac{22.36\angle -26.57^\circ}{28.28\angle 45^\circ}$$

$$= 50\angle -23.14^\circ \text{ mA}$$

7. Obtain the steady-state inductor current by inverse transformation of the phasor:

$$i_L(t)_{ss} = \text{Real}\{\mathbf{I}_L \cdot \exp[j10^3t]\} = \text{Real}\{50 \cdot \exp[j10^3t - 23.14^\circ]\}$$
$$= 50 \cdot \cos[10^3t - 23.14^\circ] \text{ mA}$$

Example 9.2

Problem:

For Example 9.2 (Figure 9.5), assume that the R–L branch is the load network, and find Thevenin's and Norton's equivalents for the remainder of the circuit [Figure 9.6(a)].

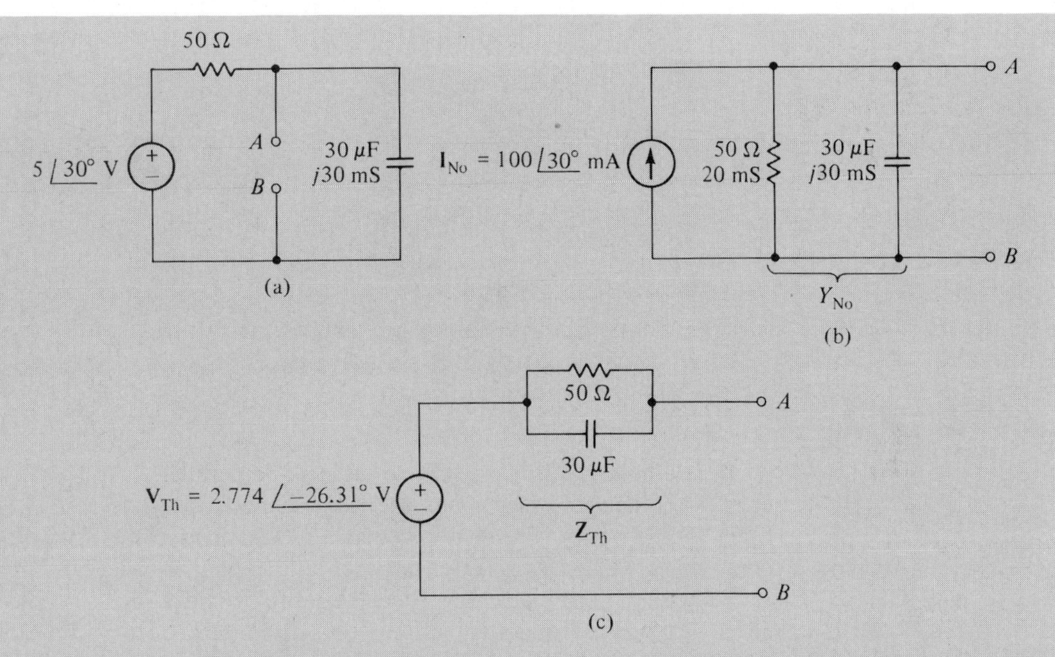

Figure 9.6 Example 9.2.

Solution:

1. Convert the practical voltage source to a practical current source; the resulting circuit is Norton's equivalent source network [Figure 9.6(b)]:

$$I_{No} = \frac{5\angle 30^\circ}{50} = 100\angle 30^\circ \text{ mA}$$

2. Obtain the steady-state Norton's equivalent source current by inverse transformation of the phasor:

$$i_{No}(t)_{ss} = \text{Real}\{I_{No} \cdot \exp[j10^3 t]\} = 100 \cdot \cos[10^3 t + 30^\circ] \text{ mA}$$

3. Compute the admittance and impedance of the parallel R–C network:

$$\mathbf{Y}_{No} = 20 + j30 \text{ mS} = 36.06\angle 56.31^\circ \text{ mS}$$

$$\mathbf{Z}_{Th} = 1/\mathbf{Y}_{No} = 27.74\angle -56.31^\circ \ \Omega = 15.38 - j23.08 \ \Omega$$

4. Compute Thevenin's equivalent source voltage phasor:

$$\mathbf{V}_{Th} = \mathbf{I}_{No}\mathbf{Z}_{Th} = (100\angle 30^\circ)(27.74\angle -56.31^\circ) = 2.774\angle -26.31^\circ \text{ V}$$

5. Obtain the steady-state Thevenin's equivalent source voltage by inverse transformation of the phasor:

$$v_{Th}(t)_{ss} = \text{Real}\{V_{Th} \cdot \exp[j10^3 t]\} = 2.774 \cdot \cos[10^3 t - 26.31^\circ] \text{ V}$$

6. Thevenin's equivalent source network is obtained by placing the parallel R–C network in series with Thevenin's equivalent source [Figure 9.6(c)].

In Example 9.2, it is possible to replace the parallel R–C network with a series R–C network which has the same impedance at 1000 rad/sec. The reader should verify (Problem 9-8) that at 1000 rad/sec a 15.38-Ω resistor in series with a 43.33-μF capacitor has the same impedance as a 50-Ω resistor in parallel with a 30-μF capacitor. It can be useful to make such a replacement if the circuit will be operated at that one discrete frequency, but the series network is not equivalent to the parallel network because the two networks have a different *frequency response*. While the impedance may be the same for both networks at 1000 rad/sec, the impedance will be different for all other frequencies.

Example 9.3

SPICE

Problem:

Use the superposition principle to determine the capacitor voltage in the circuit of Figure 9.7(a).

Solution:

1. Deactivate the current source, and solve for $\mathbf{V}_C^{(1)}$, the component of $\mathbf{V}_C$ due to the voltage source acting alone [Figure 9.7(b)]. Compute the impedance of the parallel R–C network:

$$\mathbf{Y}_1 = 10 + j5 \text{ mS} = 11.18\angle 26.57^\circ \text{ mS}$$

$$\mathbf{Z}_1 = 1/\mathbf{Y}_1 = 89.44\angle -26.57^\circ \ \Omega = 80 - j40 \ \Omega$$

Use the voltage division rule to compute $\mathbf{V}_C^{(1)}$:

$$\mathbf{V}_C^{(1)} = \mathbf{V}_1 \cdot \frac{\mathbf{Z}_1}{\mathbf{Z}_1 + 20 + j140} = 10\angle 0^\circ \frac{89.44\angle -26.57^\circ}{100 + j100} = 2 - j6 \text{ V}$$

Figure 9.7
Example 9.3.

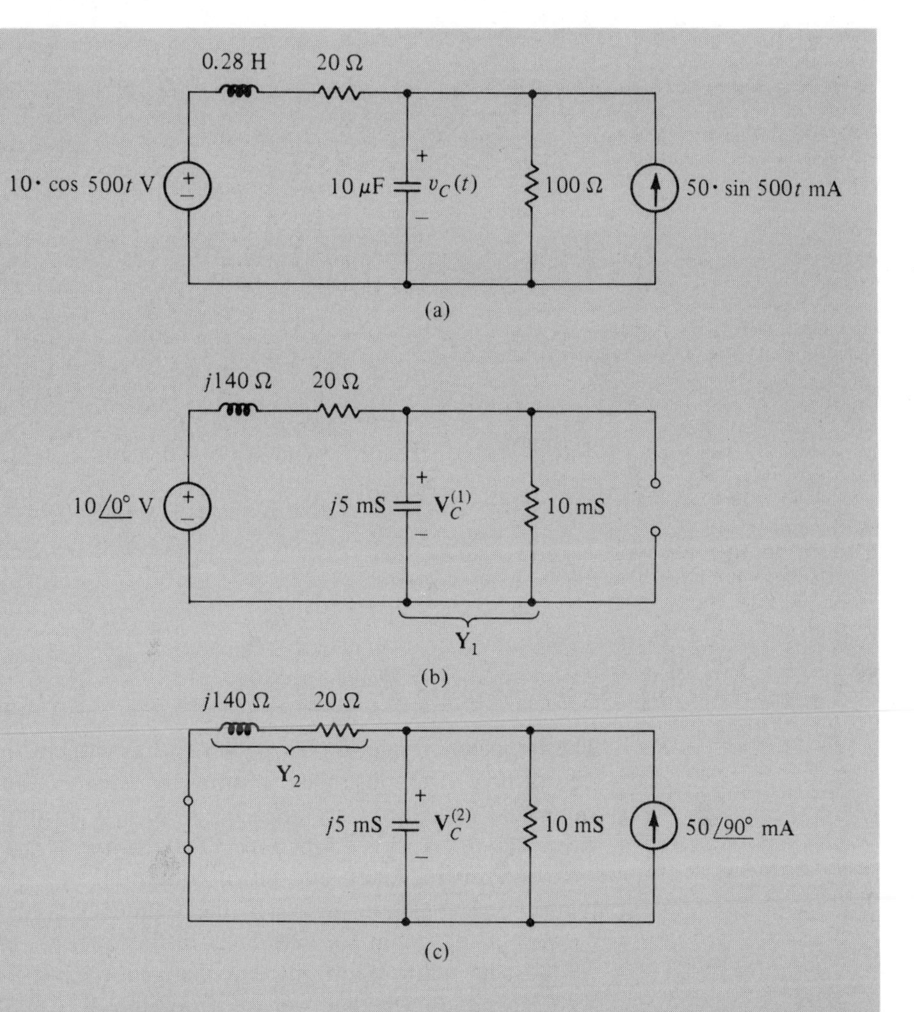

2. Deactivate the voltage source, and solve for $\mathbf{V}_C^{(2)}$, the component of $\mathbf{V}_C$ due to the current source acting alone [Figure 9.7(c)]. Compute the admittance of the R–L–C network:

$$\mathbf{Y}_2 = \frac{1}{20 + j140} = 1 - j7 \text{ mS}$$

$$\mathbf{Y}_T = \mathbf{Y}_2 + 10 + j5 \text{ mS} = 11 - j2 = 11.18\underline{/-10.30^\circ} \text{ mS}$$

Compute $\mathbf{V}_C^{(2)}$:

$$\mathbf{V}_C^{(2)} = \frac{50\angle 90^\circ}{11.18\angle -10.30^\circ} = -0.8000 + j4.400 \text{ V}$$

3. Use the superposition principle to obtain $\mathbf{V}_C$:

$$\mathbf{V}_C = \mathbf{V}_C^{(1)} + \mathbf{V}_C^{(2)} = (2 - j6) + (-0.8000 + j4.400)$$
$$= 1.200 - j1.600 = 2\angle -53.13^\circ \text{ V}$$

4. Obtain $v_C(t)_{ss}$ by inverse transformation of $\mathbf{V}_C$:

$$v_C(t)_{ss} = 2 \cdot \cos[500t - 53.13^\circ] \text{ V}$$

9.4 Frequency Response and Resonance

The **frequency response** of a network is a mathematical or graphical description of how a response phasor varies in *amplitude* and *phase angle* when the sinusoidal excitation is varied in *frequency*. It should be emphasized that frequency response is a function *of the network*. If the excitation is assumed to vary only in frequency, and not in amplitude and phase angle, the excitation phasor is a constant, and only the frequency-domain characteristics of the network can account for variations in the response phasor. Having said that, we must hasten to point out that the frequency response of a given network will depend on whether we are observing a current response or a voltage response, and whether the excitation is a voltage source or a current source. For example, the frequency response of a one-port network excited by a voltage source refers to the current entering the network; the frequency response of the same network excited by a current source refers to the voltage developed between the network terminals. The responses are related, but quite different.

Frequency response may be determined experimentally, analytically, or by computer simulation. Most often, frequency response is displayed graphically, and described verbally in terms of the characteristics of the frequency response graph. Some of the terms used to describe the graphical characteristics of frequency response will be introduced in Chapter 10.

We will see that the mathematical function describing the frequency response of a one-port network differs only by a complex constant from the frequency-domain immittance function of the network. Later we will generalize this observation to include the transfer function between ports of a multiport network.

A word on notation is in order at this point. We will use $\mathbf{I}(j\omega)$ or $\mathbf{V}(j\omega)$ to denote a response phasor, since the amplitude and phase angle of the response phasor vary with frequency. We will use the simpler $\mathbf{I}$ and $\mathbf{V}$ to denote a source phasor, since amplitude and phase angle of the source are held constant for frequency-response analysis.

In this section, we will limit ourselves to analysis and graphical display of the frequency response of two simple, but important, networks which occur frequently in electrical and electronics circuits—the series R–L–C network and the parallel R–L–C network with nonideal elements. The relative simplicity of the networks will allow us to develop mathematical concepts and graphical display techniques without undue complications.

Frequency Response of a One-Port Network

If a one-port network is excited by a voltage source

$$v(t) = V_m \cdot \cos[\omega t + \theta] \text{ V}$$

the steady-state response current is

$$i(t)_{ss} = I_m \cdot \cos[\omega t + \phi] \text{ A}$$

where I_m and ϕ are determined from the phasor calculation

$$\mathbf{I}(j\omega) = \mathbf{V} \cdot \mathbf{Y}(j\omega) = \left(V_m \cdot |\mathbf{Y}(j\omega)|\right)\underline{/\theta + \gamma(j\omega)} \text{ A} \tag{9.9}$$

If V_m and θ are held constant, then $\mathbf{V}$ is a constant, and $\mathbf{I}(j\omega)$ varies directly with $\mathbf{Y}(j\omega)$, so $I_m(j\omega) = V_m \cdot |\mathbf{Y}(j\omega)|$ and $\phi(j\omega) = \theta + \gamma(j\omega)$.

Similarly, the response voltage phasor at the terminals of a one-port network excited by a current source, $I_m \cdot \cos[\omega t + \theta]$ A, is $V_m \cdot \cos[\omega t + \phi]$ V, where $V_m(j\omega)$ and $\phi(j\omega)$ are determined by

$$\mathbf{V}(j\omega) = \mathbf{I} \cdot \mathbf{Z}(j\omega) = \left(I_m \cdot |\mathbf{Z}(j\omega)|\right)\underline{/\theta + \zeta(j\omega)} \text{ V} \tag{9.10}$$

In this case, $\mathbf{I}$ is a constant, and $\mathbf{V}(j\omega)$ varies with directly with $\mathbf{Z}(j\omega)$.

In either case, the amplitude of the response is the magnitude of the immittance multiplied by the constant amplitude of the source, and the phase angle of the response is the polar angle of the immittance plus the constant phase angle θ of the source.

The first observation to be made after the foregoing discussion is that, since immittance is a complex function of ω, frequency response is also a

Figure 9.8
Graphical displays of frequency response.

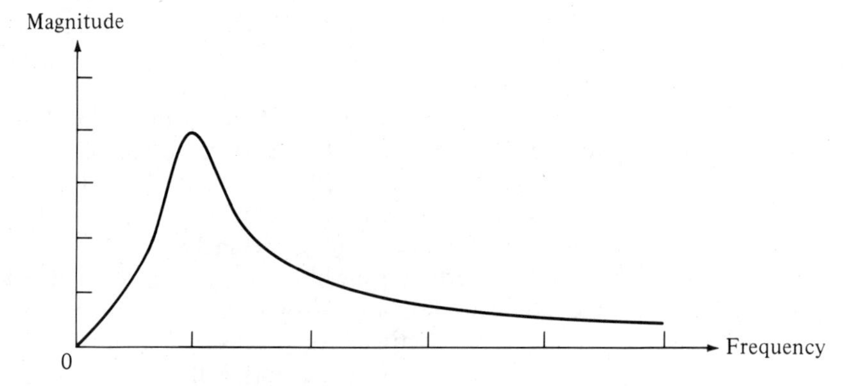

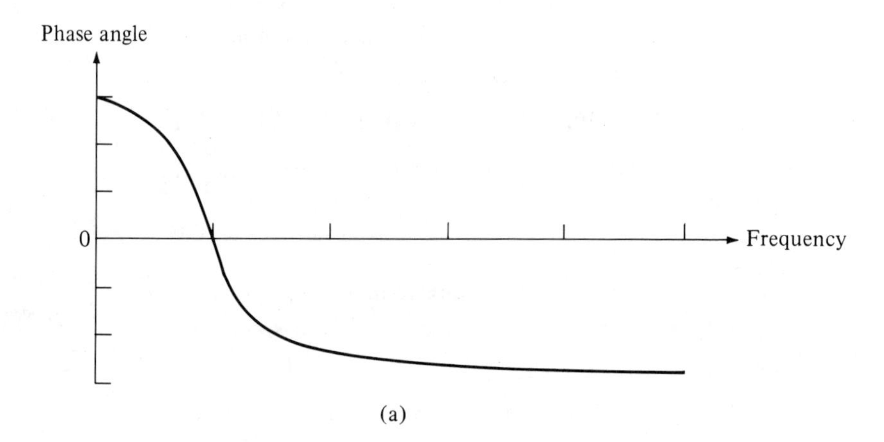

(a)

complex function of ω and must be treated mathematically and displayed graphically as such. Figure 9.8 shows three different ways of displaying the same complex function.[4] The display in Figure 9.8(a), which consists of two graphs (one for amplitude vs. frequency and one for phase angle vs. frequency), is the method of display most widely used in network analysis and design. When plotted on logarithmic or semilogarithmic axes, the display of Figure 9.8(a) is called a **Bode plot**; more will be said later about Bode analysis.

The display in Figure 9.8(b) also consists of two graphs—one for the real part of the response vs. frequency and one for the imaginary part of the response vs. frequency. This method of display is largely reserved for theoretical studies of complex functions. For the display in Figure 9.8(c), the imagi-

[4]The function plotted here is the response current of a network consisting of a 250-Ω resistor, a 0.5-H inductor, and a 2-μF capacitor in series.

Figure 9.8
Continued

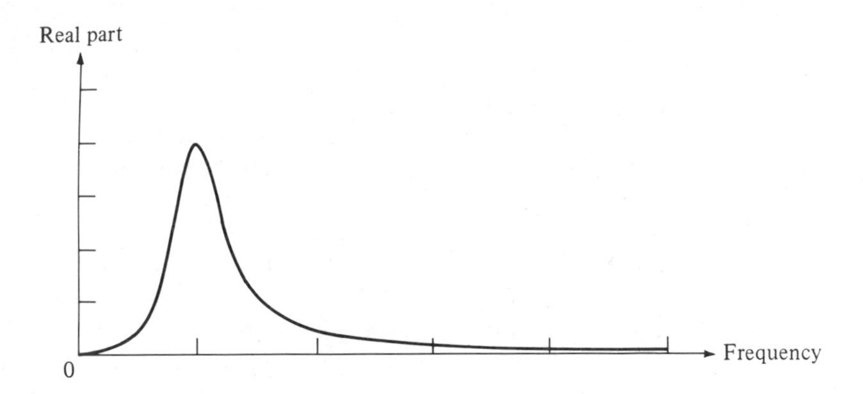

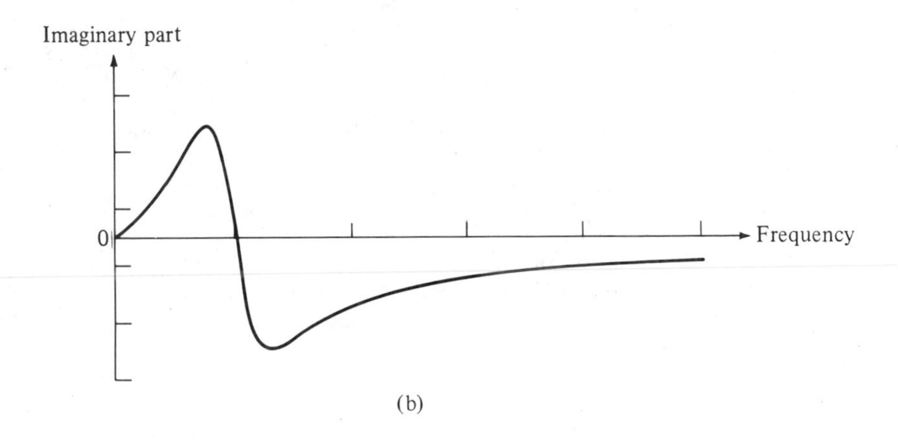

(b)

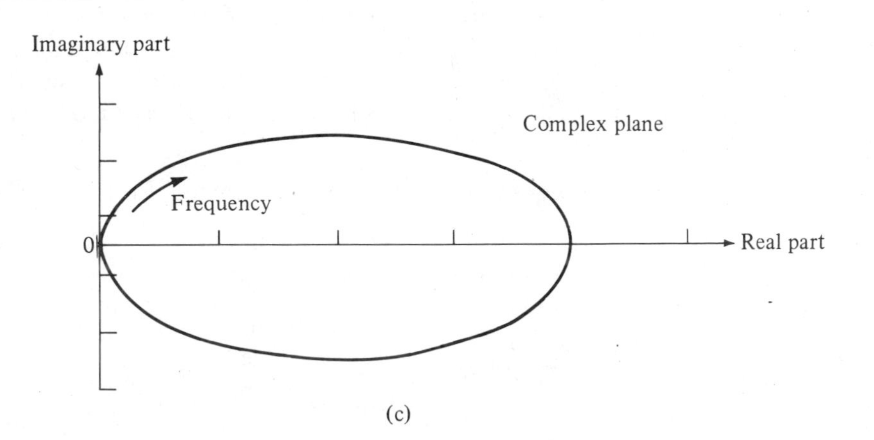

(c)

nary part of the response is plotted on the vertical axis vs. the real part on the horizontal axis. Such a graph is a **parametric** plot, where frequency is the parameter. The term **polar plot** is also applied to Figure 9.8(c), since for any value of the parameter ω the distance along a straight line to the origin is the amplitude of the response, and the angle between the positive real axis and the straight line is the phase angle of the response. Polar plots are used in **Nyquist analysis**, a technique used in feedback control systems, and they also form the basis of the **Smith chart**, a important tool for the microwave technologist.

A second observation to be made regarding frequency response is that when the excitation phasor is $1\angle 0°$, the frequency response of a one-port network is equivalent to the immittance function for the network. Immittance can be regarded as the *normalized* frequency response of the network. For example, if $1\angle 0°$ V is substituted for **V** in Equation (9.9),

$$\mathbf{I}(j\omega) = \mathbf{V} \cdot \mathbf{Y}(j\omega) = \mathbf{Y}(j\omega)$$

and if $1\angle 0°$ A is substituted for **I** in Equation (9.10),

$$\mathbf{V}(j\omega) = \mathbf{I} \cdot \mathbf{Z}(j\omega) = \mathbf{Z}(j\omega)$$

Thus, the normalized frequency response of a one-port network excited by a voltage source is the admittance of the network, and the normalized frequency response of a one-port network excited by a current source [Equation (9.10)] is the impedance of the network. In practice, only the normalized frequency response, and hence only the immittance of the network, is needed for frequency response analysis. If the excitation phasor is different from $1\angle 0°$, the response amplitude is simply multiplied by the amplitude of the source, and the response phase angle displaced by an amount equal to the phase angle of the source (Example 9.4). All subsequent references to frequency response imply normalized frequency response; exceptions are explicitly noted.

Frequency Response of a Series *R–L–C* Network

The current response of a series *R–L–C* network, such as in Figure 9.9(a), when excited by the voltage source $1\angle 0°$ V, is equal to the admittance of the network

$$\mathbf{I}(j\omega) = \mathbf{Y}(j\omega) = \frac{1}{R + j\omega L + 1/(j\omega C)}$$

$$= |\mathbf{Y}(j\omega)| \angle \gamma(j\omega) \tag{9.11a}$$

Figure 9.9
Frequency response of a series $R-L-C$ circuit.

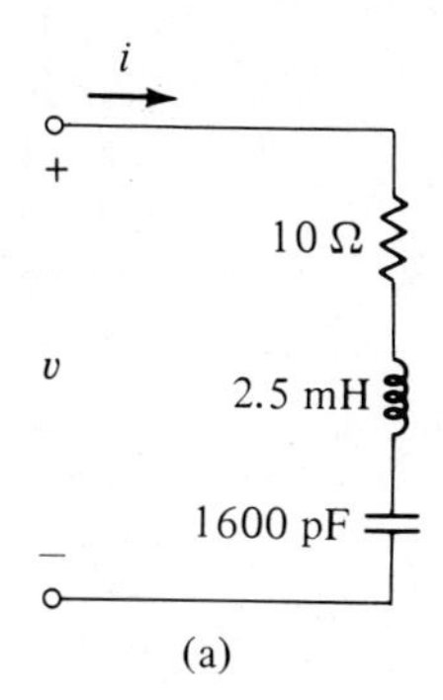

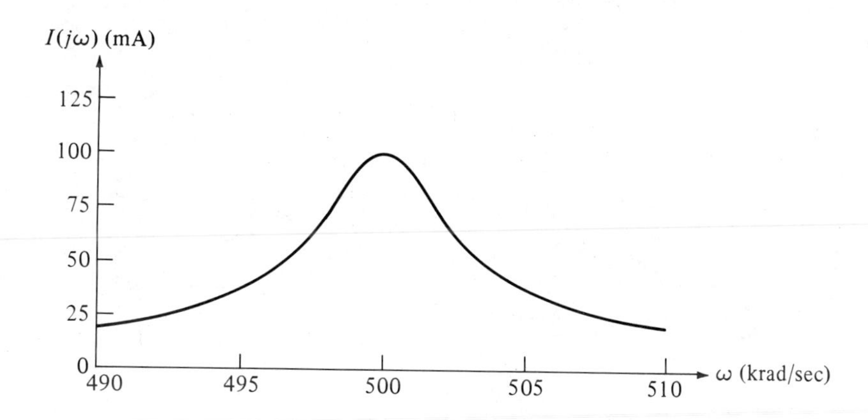

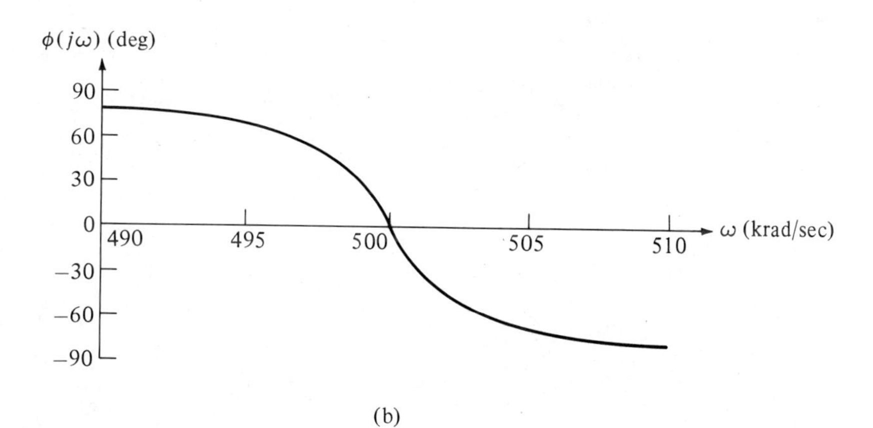

Figure 9.9 Continued

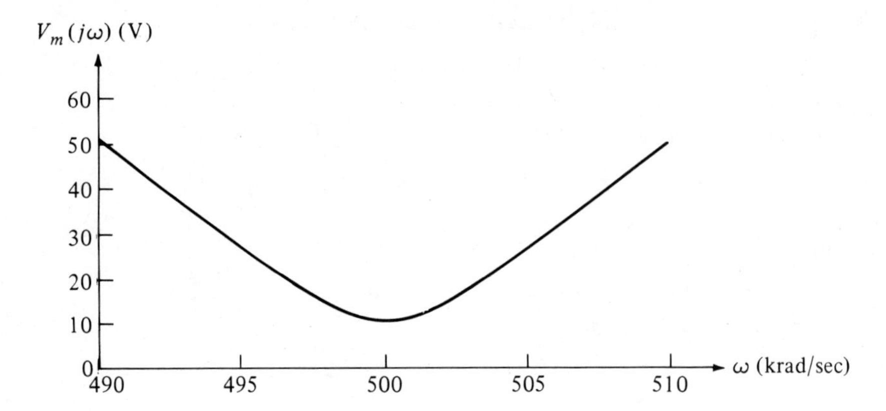

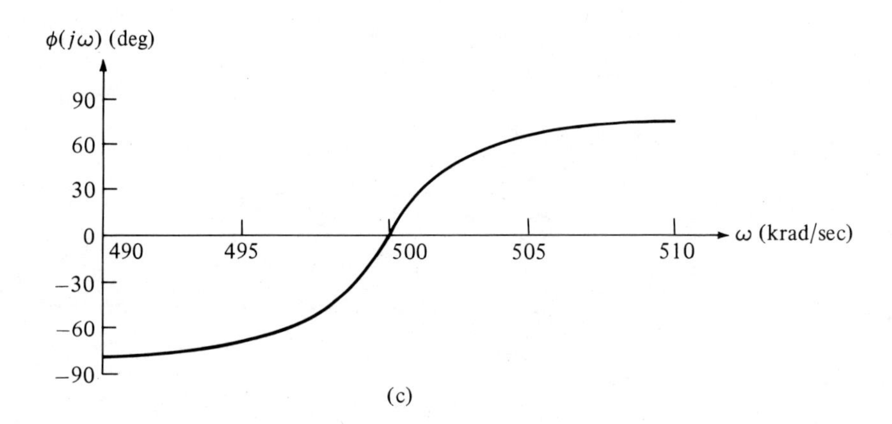

(c)

where

$$I_m(j\omega) = |\mathbf{Y}(j\omega)| = \frac{1}{\sqrt{R^2 + [\omega L - 1/(\omega C)]^2}} \tag{9.11b}$$

and

$$\phi(j\omega) = \gamma(j\omega) = -\tan^{-1}\left[\frac{\omega L - 1/(\omega C)}{R}\right] \tag{9.11c}$$

The impedance, and hence the voltage response to the current source $1\angle 0°$ A, of a series R–L–C network is

$$\mathbf{V}(j\omega) = \mathbf{Z}(j\omega) = R + j\omega L + \frac{1}{j\omega C}$$
$$= |\mathbf{Z}(j\omega)| \angle \zeta(j\omega) \tag{9.12a}$$

where

$$V_m(j\omega) = |\mathbf{Z}(j\omega)| = \sqrt{R^2 + \left(\omega L - \frac{1}{\omega C}\right)^2} \tag{9.12b}$$

and

$$\phi(j\omega) = \zeta(j\omega) = \tan^{-1}\left[\frac{\omega L - 1/(\omega C)}{R}\right] \tag{9.12c}$$

The frequency response of a series R–L–C network excited by a current source is shown in Figure 9.9(b).

Resonance

The phase angle $\phi(j\omega)$ of the response current for a series R–L–C network is a continuous function, negative at low frequencies and positive at high frequencies, and thus must have a value of 0 at some intermediate frequency. Equation (9.11c) reveals that $\gamma(j\omega)$, and hence $\phi(j\omega)$, is 0 when

$$\omega L - \frac{1}{\omega C} = 0, \qquad \text{or} \quad \omega = \frac{1}{\sqrt{LC}} \tag{9.13}$$

This value of ω is called the **resonant frequency**, and is conventionally designated ω_0. The condition which exists in a circuit excited at its resonant frequency is called **resonance**.

At resonance,

$$I_m(j\omega_0) = |\mathbf{Y}(j\omega_0)| = \frac{1}{R} \quad \text{and} \quad \phi(j\omega_0) = \gamma(j\omega_0) = 0°$$

which means that response current is at *maximum* amplitude and in phase with the source voltage. Also, at resonance,

$$V_m(j\omega_0) = |\mathbf{Z}(j\omega_0)| = R \quad \text{and} \quad \phi(j\omega_0) = \zeta(j\omega_0) = 0°$$

which means that response voltage is at a *minimum* amplitude and in phase with the source current.

It is worthwhile noting that maximum (or minimum) amplitude and minimum phase difference do not, in general, occur at the same frequency. More will be said about this later.

At frequencies below resonance, network behavior is dominated by capacitance and the term $1/(\omega C)$, so that at very low frequencies the current amplitude is small and approximated by

$$I_m(j\omega) = |\mathbf{Y}(j\omega)| \simeq \omega C \quad \text{and} \quad \phi(j\omega) = \gamma(j\omega) \simeq \tan^{-1}\left[\frac{1}{\omega RC}\right]$$

The voltage amplitude at very low frequencies is large:

$$V_m(j\omega) = |\mathbf{Z}(j\omega)| \simeq \frac{1}{\omega C} \quad \text{and} \quad \phi(j\omega) = \zeta(j\omega) \simeq -\tan^{-1}\left[\frac{1}{\omega RC}\right]$$

At frequencies above resonance, network behavior is dominated by inductance and the term ωL, so that at very high frequencies the current amplitude is again small:

$$I_m(j\omega) = |\mathbf{Y}(j\omega)| \simeq \frac{1}{\omega L} \quad \text{and} \quad \phi(j\omega) = \gamma(j\omega) \simeq -\tan^{-1}\left[\frac{\omega L}{R}\right]$$

and the voltage amplitude is large:

$$V_m(j\omega) = |\mathbf{Z}(j\omega)| \simeq \omega L \quad \text{and} \quad \phi(j\omega) = \zeta(j\omega) \simeq \tan^{-1}[\omega L/R]$$

Quality Factor and Bandwidth

Quality factor (Q) and **bandwidth** (BW) are descriptive terms, both based on energy considerations, associated with frequency response. Both terms apply to networks of any configuration, and are not restricted to simple series or parallel networks. Quality factor is indicative of the efficiency of a network in storing energy during sinusoidal steady-state operation. In particular, the quality factor is the ratio of the energy stored in the network to the energy dissipated per cycle of operation:

$$Q = \frac{\text{energy stored}}{\text{energy dissipated}/(2\pi)}$$

The reader should be aware that the quality factor of a network varies with operating frequency, and references to the quality factor of a network are usually understood to mean the quality factor at resonance (Q_0).

Bandwidth is the frequency range over which power dissipation is more than one-half of the maximum power dissipation (in cases where the response amplitude has a maximum value) or less than twice the minimum power dissipation (in cases where the response amplitude has a minimum value). For many one-port networks, including the series R–L–C and parallel R–L–C networks, there exist two finite, non-zero frequencies at which the average power dissipation is exactly half the maximum power dissipation, or exactly twice the minimum power dissipation. The smaller of these two frequencies is called the **lower cutoff frequency** (ω_{c1}), and the larger is called the **upper cutoff frequency** (ω_{c2}). The bandwidth of the frequency response of these networks is simply the difference between the upper cutoff frequency and the lower cutoff frequency [Figure 9.10(a)],

$$\text{BW} = \omega_{c2} - \omega_{c1} \tag{9.14a}$$

Some modification of the definition of bandwidth will be made later to cover those cases for which $\omega_{c1} \to 0$ or for which $\omega_{c2} \to \infty$.

Quality factor and bandwidth are obviously related, since both are based on energy considerations. For series R–L–C or parallel R–L–C networks,

$$\text{BW} = \omega_0/Q_0 \tag{9.14b}$$

Quality factors for passive networks may range from less than unity to several hundred; with active networks, quality factors of 1000 and greater are possible.

The frequency response of a high-Q network is said to be **sharp**, referring to the characteristic shape of the graph of response amplitude vs. frequency (Figure 9.11); whereas, the frequency response of a low-Q network is said to be **broad**. In a high-Q network, the slope of the graph of response phase angle vs. frequency, as the phase angle shifts from leading to lagging (or vice versa), is greater than for a low-Q network. A high-Q network is also said to be more **selective** than a low-Q network. Selectivity, a measure of the ability of a network to discriminate in its response to excitation sources of different frequencies, refers to the ratio of the response amplitude at maximum (or minimum) to the response amplitude at some other frequency.

Most textbooks on basic network analysis include a thorough presentation of resonance, quality factor, and bandwidth, with an extensive set of formulas applicable to series and parallel networks. Those formulas are summarized in Table 9.1. For practical reasons stated earlier, the parallel R–L–C circuit is of less interest than the series R–L–C circuit. In the next section, we will see that the winding resistance of a practical inductor complicates the frequency-response analysis for a parallel network, and produces some interesting results with regard to resonance.

Figure 9.10
Example 9.4.

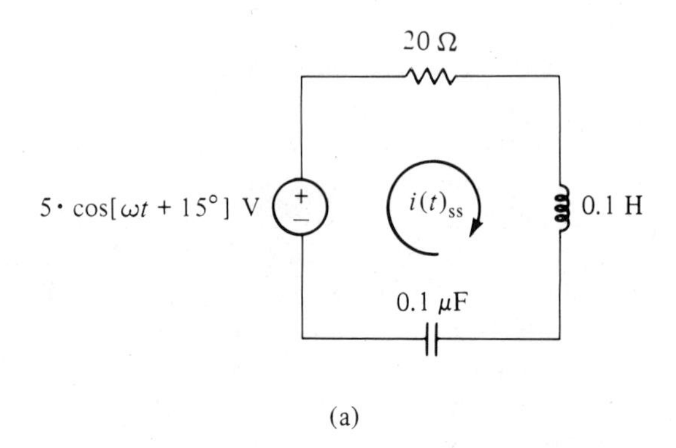

(a)

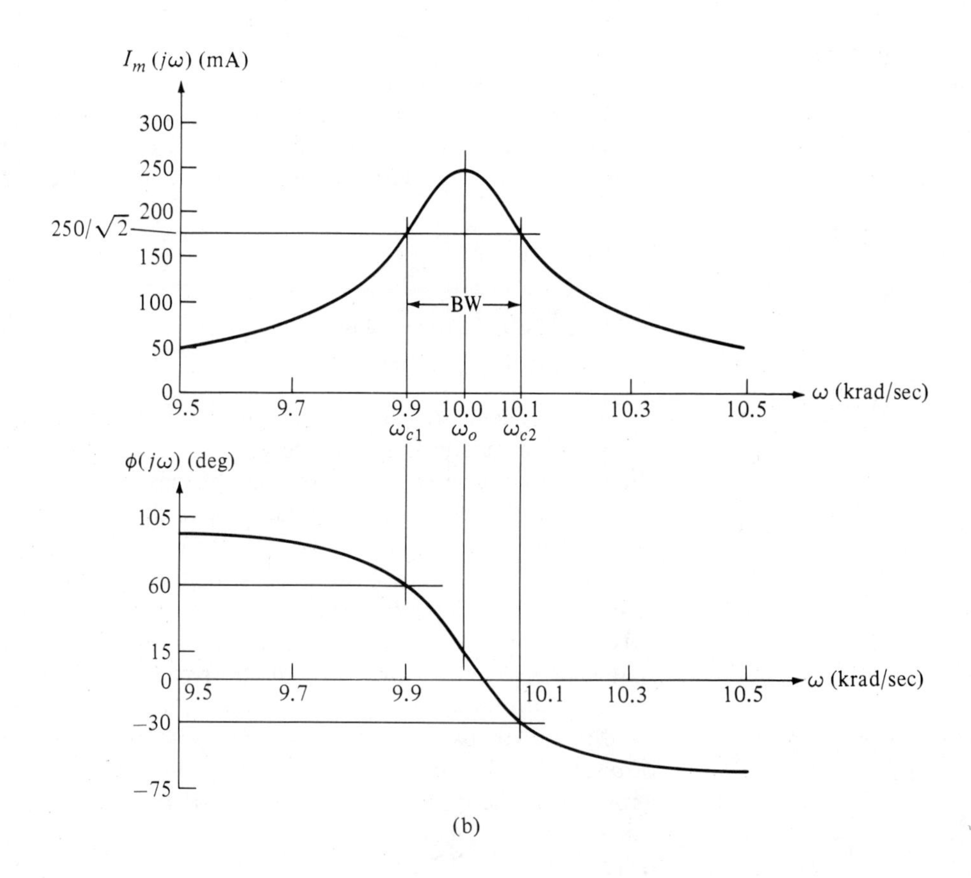

(b)

Figure 9.11
Example 9.5.

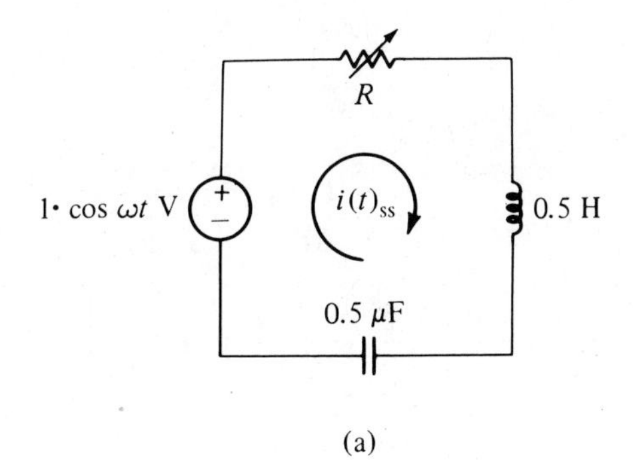

(a)

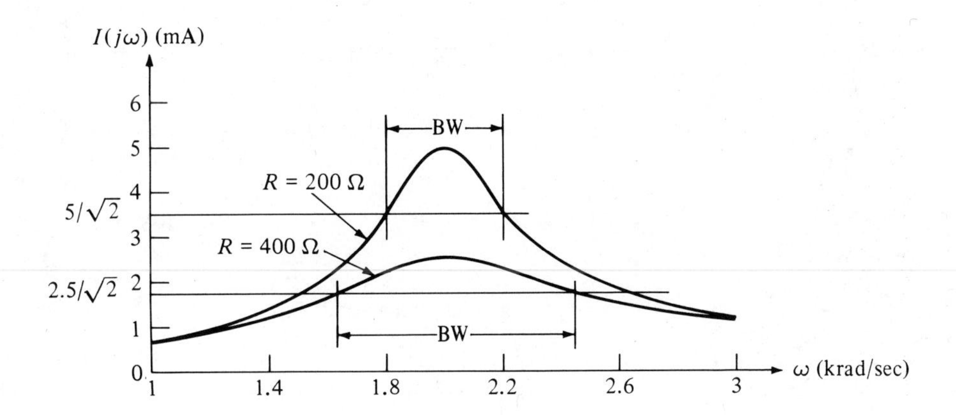

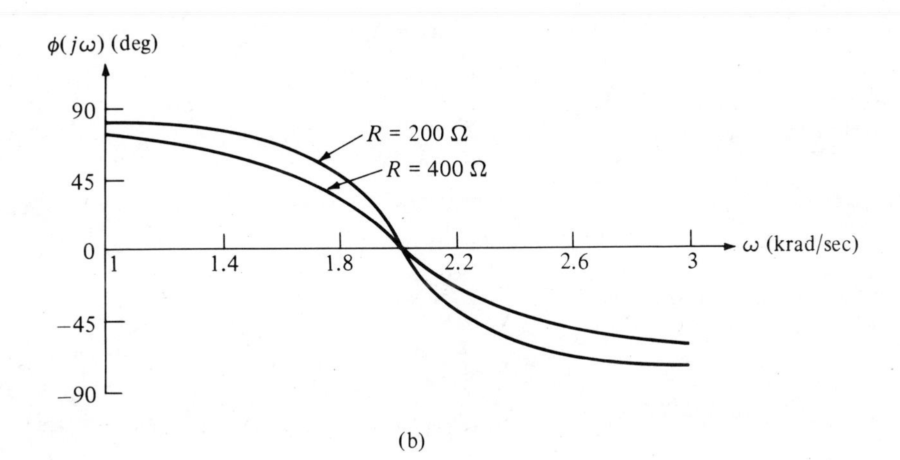

(b)

Table 9.1 Resonance, Quality Factor, and Bandwidth Formulas for Series and Parallel One-Port Networks

Series Network	Parallel Network
$\omega_0 = \dfrac{1}{\sqrt{LC}}$	$\omega_0 = \dfrac{1}{\sqrt{LC}}$
$Q_0 = \dfrac{\omega_0 L}{R} = \dfrac{1}{\omega_0 RC} = \dfrac{1}{R}\sqrt{\dfrac{L}{C}}$	$Q_0 = \dfrac{\omega_0 C}{G} = \dfrac{1}{\omega_0 GL} = \dfrac{1}{G}\sqrt{\dfrac{C}{L}}$
$\omega_{c1} = \dfrac{1}{2}\left[-\dfrac{R}{L} + \sqrt{\left(\dfrac{R}{L}\right)^2 + \dfrac{4}{LC}}\right]$	$\omega_{c1} = \dfrac{1}{2}\left[-\dfrac{G}{C} + \sqrt{\left(\dfrac{G}{C}\right)^2 + \dfrac{4}{LC}}\right]$
$\omega_{c2} = \dfrac{1}{2}\left[+\dfrac{R}{L} + \sqrt{\left(\dfrac{R}{L}\right)^2 + \dfrac{4}{LC}}\right]$	$\omega_{c2} = \dfrac{1}{2}\left[+\dfrac{G}{C} + \sqrt{\left(\dfrac{G}{C}\right)^2 + \dfrac{4}{LC}}\right]$
$\omega_{c1} = \tfrac{1}{2}\omega_0\left(-\dfrac{1}{Q_0} + \sqrt{\left(\dfrac{1}{Q_0}\right)^2 + 4}\right)$	$\omega_{c1} = \tfrac{1}{2}\omega_0\left(-\dfrac{1}{Q_0} + \sqrt{\left(\dfrac{1}{Q_0}\right)^2 + 4}\right)$
$\omega_{c2} = \tfrac{1}{2}\omega_0\left(+\dfrac{1}{Q_0} + \sqrt{\left(\dfrac{1}{Q_0}\right)^2 + 4}\right)$	$\omega_{c2} = \tfrac{1}{2}\omega_0\left(+\dfrac{1}{Q_0} + \sqrt{\left(\dfrac{1}{Q_0}\right)^2 + 4}\right)$
$\mathrm{BW} = \omega_{c2} - \omega_{c1} = \omega_0/Q_0$	$\mathrm{BW} = \omega_{c2} - \omega_{c1} = \omega_0/Q_0$
$\omega_0 = \sqrt{\omega_{c1}\omega_{c2}}$ (geometric mean; all Q_0)	$\omega_0 = \sqrt{\omega_{c1}\omega_{c2}}$ (geometric mean; all Q_0)
$\omega_0 \simeq \tfrac{1}{2}(\omega_{c1} + \omega_{c2})$ (arithmetic mean; $Q_0 > 10$)	$\omega_0 \simeq \tfrac{1}{2}(\omega_{c1} + \omega_{c2})$ (arithmetic mean; $Q_0 > 10$)
For $\omega = \omega_0$	
$X_L = X_C$	$B_C = B_L$
$\lvert \mathbf{Y}(j\omega_0) \rvert = 1/R$	$\lvert \mathbf{Z}(j\omega_0) \rvert = 1/G$
$\lvert \mathbf{Z}(j\omega_0) \rvert = R$	$\lvert \mathbf{Y}(j\omega_0) \rvert = G$
$\gamma(j\omega_0) = 0°$	$\zeta(j\omega_0) = 0°$
$\zeta(j\omega_0) = 0°$	$\gamma(j\omega_0) = 0°$
For excitation $V_m \cdot \cos \omega_0 t$ V,	For excitation $I_m \cdot \cos \omega_0 t$ A,
$\mathbf{I} = \mathbf{V}/R$	$\mathbf{V} = \mathbf{I}/G$
$\mathbf{V}_R = \mathbf{V}$	$\mathbf{I}_G = \mathbf{I}$
$\mathbf{V}_L = Q_0 \cdot V_m \angle 90°$	$\mathbf{I}_C = Q_0 \cdot I_m \angle 90°$
$\mathbf{V}_C = Q_0 \cdot V_m \angle -90°$	$\mathbf{I}_L = Q_0 \cdot I_m \angle -90°$

Example 9.4

SPICE

Problem:

1. Determine the resonant frequency, bandwidth, and quality factor of the series R–L–C network in Figure 9.10(a).
2. Sketch the frequency response $\mathbf{I}(j\omega)$ if the source is

$$v(t) = 5 \cdot \cos[\omega t + 15°] \text{ V}.$$

Solution:

1. Calculate the resonant frequency ω_0:

$$\omega_0 = \frac{1}{\sqrt{LC}} = \frac{1}{\sqrt{(0.1)(0.1 \times 10^{-6})}} = 10 \text{ krad/sec}$$

2. Calculate the quality factor at resonance (Q_0):

$$Q_0 = \frac{1}{R}\sqrt{\frac{L}{C}} = \frac{1}{20}\sqrt{\frac{0.1}{0.1 \times 10^{-6}}} = 50$$

3. Calculate the bandwidth BW:

$$\text{BW} = \omega_0/Q_0 = 10 \times 10^3/50 = 200 \text{ rad/sec}$$

4. Calculate ω_{c1} and ω_{c2}:

$$\omega_{c1} = \tfrac{1}{2}\omega_0\left(-1/Q_0 + \sqrt{(1/Q_0)^2 + 4}\right)$$

$$= \tfrac{1}{2}(10 \times 10^3)\left(-1/50 + \sqrt{(1/50)^2 + 4}\right) = 9.901 \text{ krad/sec}$$

$$\omega_{c2} = \tfrac{1}{2}\omega_0\left(+1/Q_0 + \sqrt{(1/Q_0)^2 + 4}\right)$$

$$= \tfrac{1}{2}(10 \times 10^3)\left(+1/50 + \sqrt{(1/50)^2 + 4}\right) = 10.101 \text{ krad/sec}$$

5. Refer to Figure 9.10(b) for a graph of $\mathbf{I}(j\omega)$. Note that the response current is maximum and in phase with the source voltage at resonant frequency.

Example 9.5

Problem:

For the series R–L–C network in Figure 9.11(a), determine the effect on selectivity of decreasing R from 400 to 200 Ω. Compare the selectivity for a voltage source operating at resonance to one operating 500 rad/sec below resonance.

Solution:

1. Calculate the resonant frequency ω_0:

$$\omega_0 = \frac{1}{\sqrt{LC}} = \frac{1}{\sqrt{(0.5)(0.5 \times 10^{-6})}} = 2 \text{ krad/sec}$$

2. Calculate the amplitude of the response at resonance:

$$R = 200\ \Omega: \qquad I_m(j\omega_0) = 1/200 = 5 \text{ mA}$$

$$R = 400\ \Omega: \qquad I_m(j\omega_0) = 1/400 = 2.5 \text{ mA}$$

3. Calculate the amplitude of the response 500 rad/sec below resonance:

$R = 200\ \Omega$:

$$I_m(j1500) = \frac{1}{\sqrt{200^2 + \left(1500 \times 0.5 - \dfrac{1}{1500 \times 0.5 \times 10^{-6}}\right)^2}}$$

$$= 1.622 \text{ mA}$$

$R = 400\ \Omega$:

$$I_m(j1500) = \frac{1}{\sqrt{400^2 + \left(1500 \times 0.5 - \dfrac{1}{1500 \times 0.5 \times 10^{-6}}\right)^2}}$$

$$= 1.414 \text{ mA}$$

4. The selectivity with $R = 400\ \Omega$ is

$$2.5 : 1.414, \qquad \text{or} \qquad 1.768 : 1$$

The selectivity with $R = 200\ \Omega$ is

$$5:1.622, \quad \text{or} \quad 3.083:1$$

Thus, network selectivity is increased by a factor of 1.744 (3.083/1.768) by halving the resistance.

Frequency Response of an R–L–C Network with Practical Inductor

A practical network consisting of an inductor and a capacitor connected in parallel with a source [Figure 9.12(a)] can be modeled as shown in Figure 9.12(b) and (c). The validity of the model was discussed in Section 6.2. The resistance r_L represents the winding resistance of the inductor, and the conductance G in parallel with the capacitor is attributed to the combined resistance r_C of the capacitor dielectric, the source resistance R_S, and any other resistance R_P across the network terminals:

$$G = \frac{1}{R_S} + \frac{1}{r_C} + \frac{1}{R_P}$$

The normalized frequency response of the network of Figure 9.12(a) is equal to the network impedance:

$$\mathbf{V}(j\omega) = \mathbf{Z}(j\omega) = \frac{1}{G + j\omega C + 1/(r_L + j\omega L)}$$

$$= \frac{r_L + j\omega L}{\left(Gr_L + 1 - \omega^2 LC\right) + j\omega(GL + r_L C)} \tag{9.15a}$$

$$V_m(j\omega) = |\mathbf{Z}(j\omega)| = \sqrt{\frac{r_L^2 + (\omega L)^2}{\left(Gr_L + 1 - \omega^2 LC\right)^2 + \omega^2(GL + r_L C)^2}} \tag{9.15b}$$

$$\phi(j\omega) = \zeta(j\omega) = \tan^{-1}\left[\frac{\omega L}{r_L}\right] - \tan^{-1}\left[\frac{\omega(GL + r_L C)}{Gr_L + 1 - \omega^2 LC}\right] \tag{9.15c}$$

At resonance, $\phi(j\omega_0) = 0$; therefore, from Equation (9.15c),

$$\tan^{-1}\left[\frac{\omega_0 L}{r_L}\right] = \tan^{-1}\left[\frac{\omega_0(GL + r_L C)}{Gr_L + 1 - \omega_0^2 LC}\right]$$

$$\frac{\omega_0 L}{r_L} = \frac{\omega_0(GL + r_L C)}{Gr_L + 1 - \omega_0^2 LC}\cdot$$

Figure 9.12
Practical parallel $R-L-C$ network.

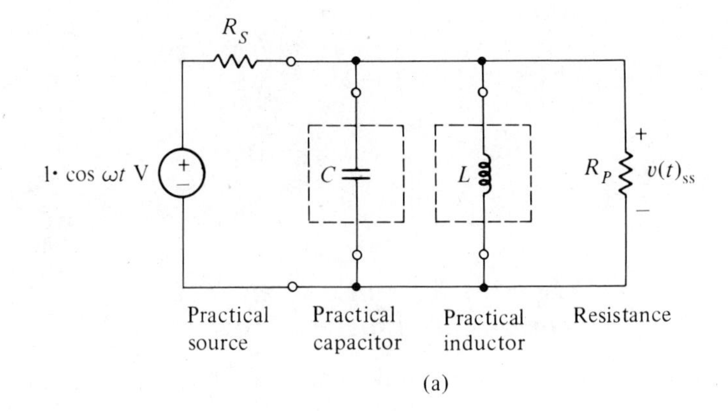

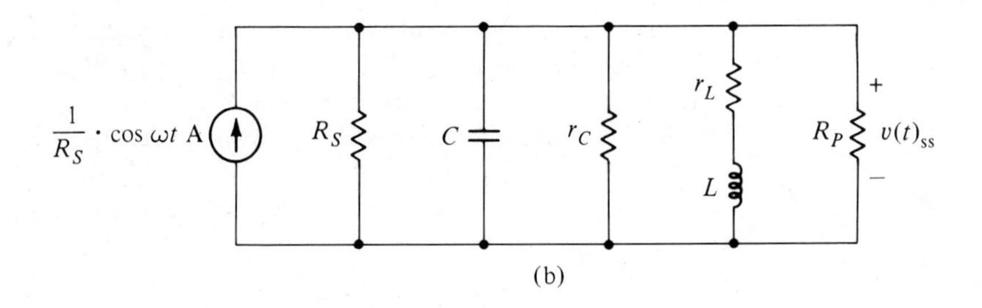

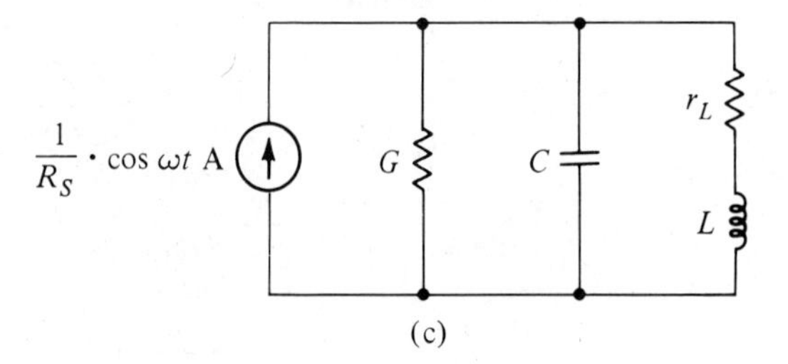

or

$$\omega_0 = \frac{1}{\sqrt{LC}} \sqrt{\left(1 - \frac{r_L^2 C}{L}\right)} \tag{9.16}$$

Equation (9.16) indicates that the resonant frequency of the practical network is lower than the resonant frequency of a network of ideal elements with equivalent capacitance and inductance.

The maximum value of $V_m(j\omega)$, or the maximum response amplitude, is obtained by the straightforward (but tedious) process of setting the derivative of $V_m(j\omega)$ equal to zero and solving for ω (Problem 9-11). The result is

$$\omega_{\text{MAX}} = \frac{1}{\sqrt{LC}} \cdot \sqrt{\sqrt{1 + \frac{2r_L^2 C}{L} + 2Gr_L} - \frac{r_L^2 C}{L}} \tag{9.17}$$

which can be significantly different from ω_0. It is worthwhile to note that G is not a factor in determining the resonant frequency [Equation (9.16)], but does enter into the computation of the frequency for which response amplitude is maximum. Problem 9-12 provides some interesting observations concerning the effect of G on the frequency of maximum response amplitude.

Example 9.6

SPICE

Problem:

For the network of Figure 9.13(a):

1. Determine the resonant frequency.
2. Sketch the amplitude response, $V_m(j\omega)$, for an excitation source $v(t) = 20 \cdot \cos \omega t$ V.
3. Determine the bandwidth and the quality factor.

Solution:

1. Calculate G:

$$G = \frac{1}{2000} + \frac{1}{2000} = 1 \text{ mS}$$

2. Use Equation (9.16) to calculate ω_0:

$$\omega_0 = \sqrt{\frac{1 - (100)^2(0.5 \times 10^{-6})/(20 \times 10^{-3})}{(20 \times 10^{-3})(0.5 \times 10^{-6})}} = 8.660 \text{ krad/sec}$$

3. Use Equation (9.17) to calculate ω_{MAX}:

$$\omega_{\text{MAX}} = \sqrt{\frac{N}{(0.5 \times 10^{-6})(20 \times 10^{-3})}}$$

Figure 9.13
Example 9.6.

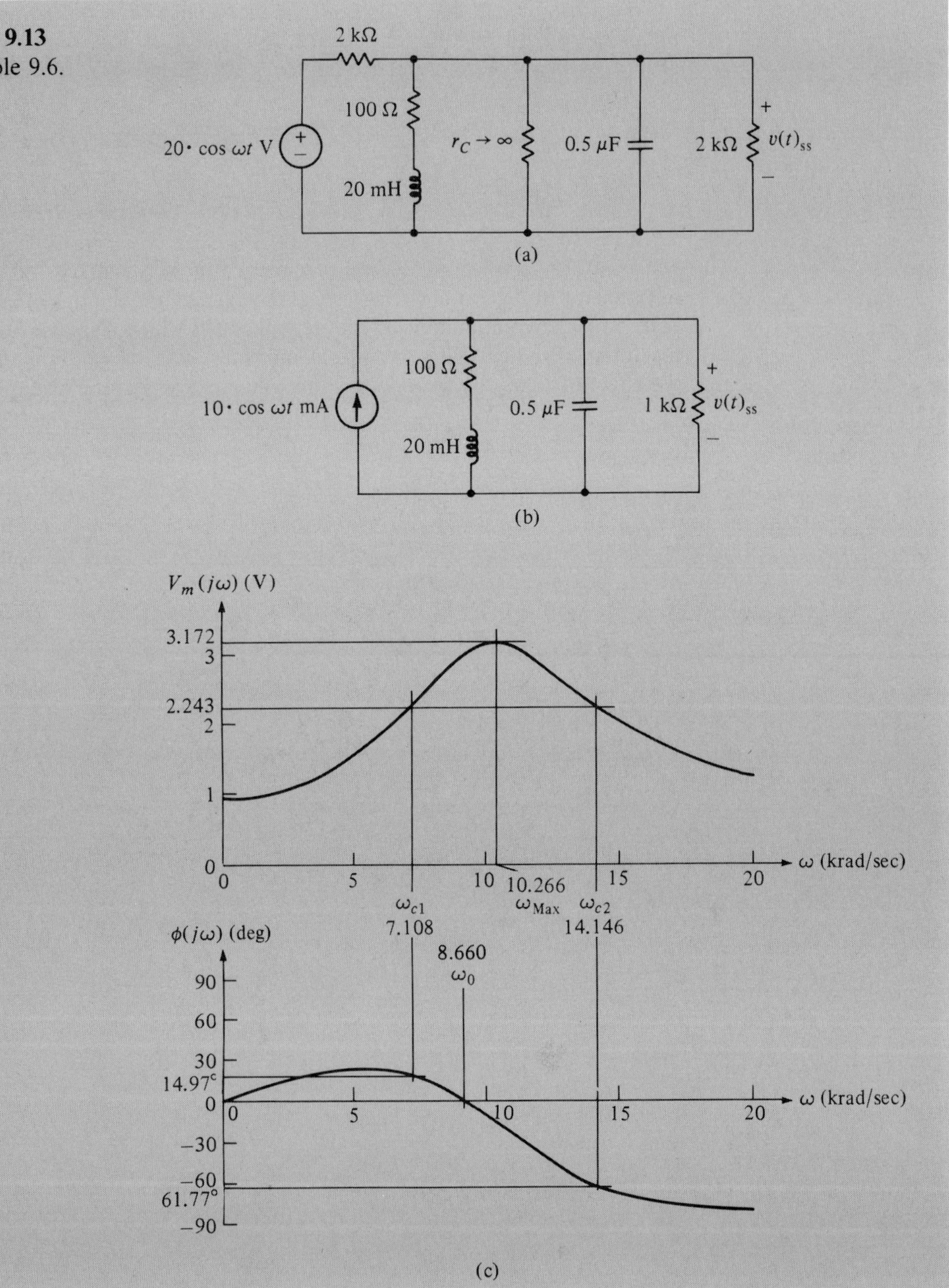

where

$$N = \sqrt{1 + \frac{2(100)^2(0.5 \times 10^{-6})}{20 \times 10^{-3}} + 2(10^{-3})(100)} - \frac{(100)^2(0.5 \times 10^{-6})}{20 \times 10^{-3}}$$

Thus

$$\omega_{\text{MAX}} = 10.266 \text{ krad/sec}$$

4. Use Equation (9.15a) to evaluate $\mathbf{Z}(j\omega_0)$ and $\mathbf{Z}(j\omega_{\text{MAX}})$:

$$\mathbf{Z}(j\omega_0) = \frac{100 + j(8.660 \times 10^3)(20 \times 10^{-3})}{\mathbf{D}}$$

$$\text{Real}\{\mathbf{D}\} = (10^{-3})(100) + 1 - (8.660 \times 10^3)^2(20 \times 10^{-3})(0.5 \times 10^{-6})$$

$$\text{Imag}\{\mathbf{D}\} = 8.660 \times 10^3 \times [(10^{-3})(20 \times 10^{-3}) + (100)(0.5 \times 10^{-6})]$$

$$\mathbf{Z}(j\omega_0) = \frac{200\angle 60^\circ}{0.7\angle 60^\circ} = 285.7\angle 0^\circ \ \Omega$$

$$\mathbf{Z}(j\omega_{\text{MAX}}) = \frac{100 + j(10.27 \times 10^3)(20 \times 10^{-3})}{\mathbf{D}}$$

$$\text{Real}\{\mathbf{D}\} = (10^{-3})(100) + 1 - (10.27 \times 10^3)^2(20 \times 10^{-3})(0.5 \times 10^{-6})$$

$$\text{Imag}\{\mathbf{D}\} = 10.27 \times 10^3 \times [(10^{-3})(20 \times 10^{-3}) + (100)(0.5 \times 10^{-6})]$$

$$\mathbf{Z}(j\omega_{\text{MAX}}) = \frac{228.4\angle 64.03^\circ}{0.7201\angle 86.32^\circ} = 317.2\angle -22.29^\circ \ \Omega$$

5. By source conversion, $i(t) = v(t)/2000 = 10 \cdot \cos \omega t$ mA, so $I_m = 10$ mA. Calculate $V_m(j\omega_0)$ and $V_m(j\omega_{\text{MAX}})$:

$$V_m(j\omega_0) = I_m \cdot |\mathbf{Z}(j\omega_0)| = 2.857 \text{ V}$$

$$V_m(j\omega_{\text{MAX}}) = I_m \cdot |\mathbf{Z}(j\omega_{\text{MAX}})| = 3.172 \text{ V}$$

6. At the cutoff (half-power) frequencies,

$$|\mathbf{Z}(j\omega_{c1})| = |\mathbf{Z}(j\omega_{c2})| = \frac{|\mathbf{Z}(j\omega_{\text{MAX}})|}{\sqrt{2}} = 224.3 \ \Omega$$

Set $|\mathbf{Z}(j\omega_c)|$ [Equation (9.15b)] equal to 224.3 Ω, and solve for ω_{c1} and ω_{c2}:

$$\sqrt{\frac{100^2 + \left[\omega_c(20 \times 10^{-3})\right]^2}{|\mathbf{D}|}} = 224.3$$

$$|\mathbf{D}| = \left[(10^{-3})(100) + 1 - \omega_c^2(20 \times 10^{-3})(0.5 \times 10^{-6})\right]^2 + \omega_c^2\left[(10^{-3})(20 \times 10^{-3}) + (100)(0.5 \times 10^{-6})\right]^2$$

Square both members of the equation and cross-multiply to obtain

$$50.31 \times 10^3\left[(1.1 - 10 \times 10^{-9}\omega_c)^2 + 4.9 \times 10^{-9}\omega_c^2\right] = 10 \times 10^3 + 400 \times 10^{-6}\omega_c^2$$

$$\omega_c^4 - 250.6 \times 10^6\omega_c^2 + 10.11 \times 10^{15} = 0$$

$$\omega_c^2 = 50.53 \times 10^6, 200.1 \times 10^6$$

$$\omega_{c1} = 7.108 \text{ krad/sec}$$

$$\omega_{c2} = 14.146 \text{ krad/sec}$$

7. Calculate bandwidth and quality factor:

$$\text{BW} = \omega_{c2} - \omega_{c1} = 14.146 - 7.108 = 7.038 \text{ krad/sec}$$

$$Q \simeq \omega_{\text{MAX}}/\text{BW} = 1.459$$

8. Use Equation (9.15c) to evaluate $\phi(j\omega_{c1})$ and $\phi(j\omega_{c2})$:

$$\phi(j\omega_{c1}) = \tan^{-1}\left[\frac{(7.108 \times 10^3)(20 \times 10^{-3})}{100}\right] - \tan^{-1}\left[\frac{7.108 \times 10^3\left[(10^{-3})(20 \times 10^{-3}) + (100)(0.5 \times 10^{-6})\right]}{(10^{-3})(100) + 1 - (7.108 \times 10^3)^2(20 \times 10^{-3})(0.5 \times 10^{-6})}\right]$$

$$= 54.88° - 39.91° = 14.97°$$

$$\phi(j\omega_{c2}) = \tan^{-1}\left[\frac{(14.146 \times 10^3)(20 \times 10^{-3})}{100}\right] - \tan^{-1}\left[\frac{14.146 \times 10^3((10^{-3})(20 \times 10^{-3}) + (100)(0.5 \times 10^{-6}))}{(10^{-3})(100) + 1 - (14.146 \times 10^3)^2(20 \times 10^{-3})(0.5 \times 10^{-6})}\right]$$

$$= 70.53° - 132.3° = -61.77°$$

9. A graph of the frequency response is given in Figure 9.13(c).

Exercise 9.1

SPICE

Problem:

1. Sketch the response current $\mathbf{I}(j\omega)$ for a series circuit consisting of a 40-Ω resistor, a 2000-pF capacitor, a 500-μH inductor with a 10-Ω winding, and a source

$$v(t) = 2.5 \cdot \cos \omega t \text{ V}$$

Figure 9.14
Exercise 9.1.

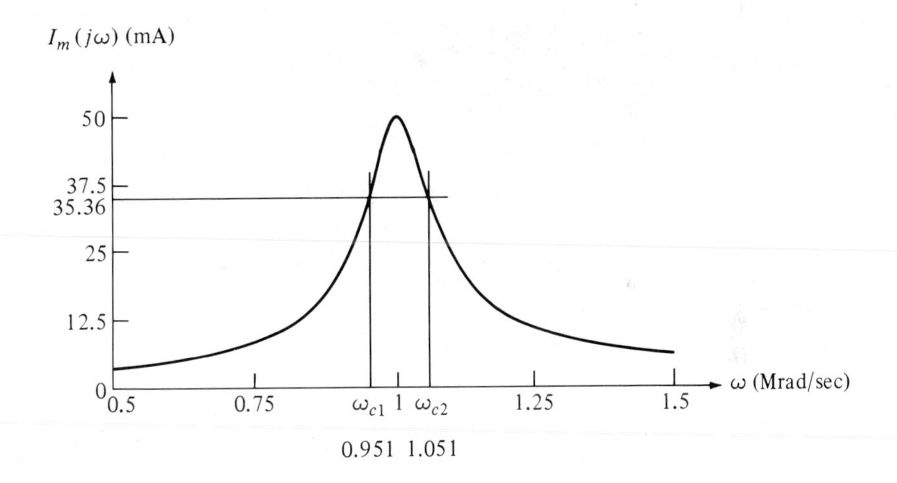

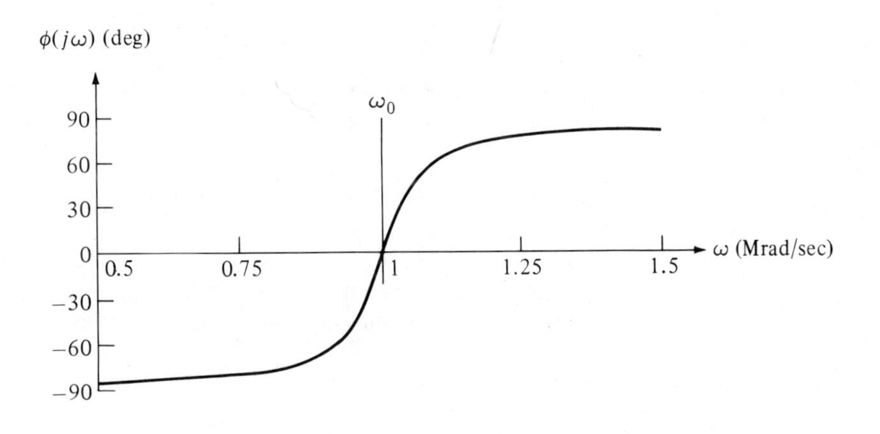

(a)

Figure 9.14
Continued

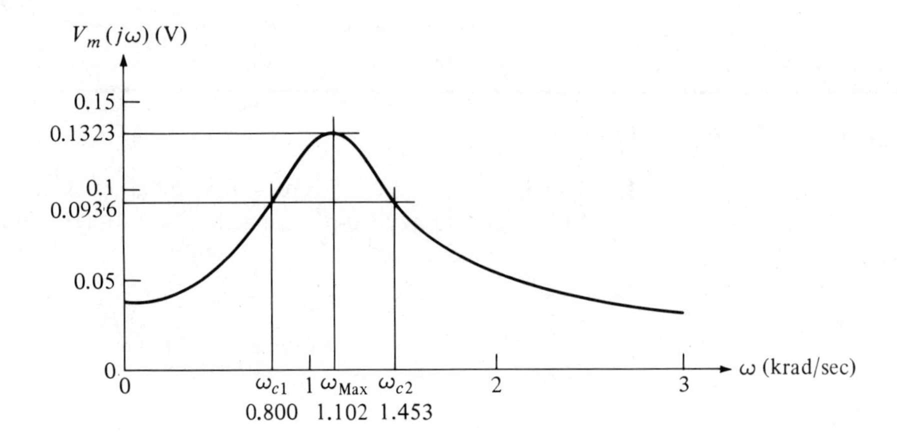

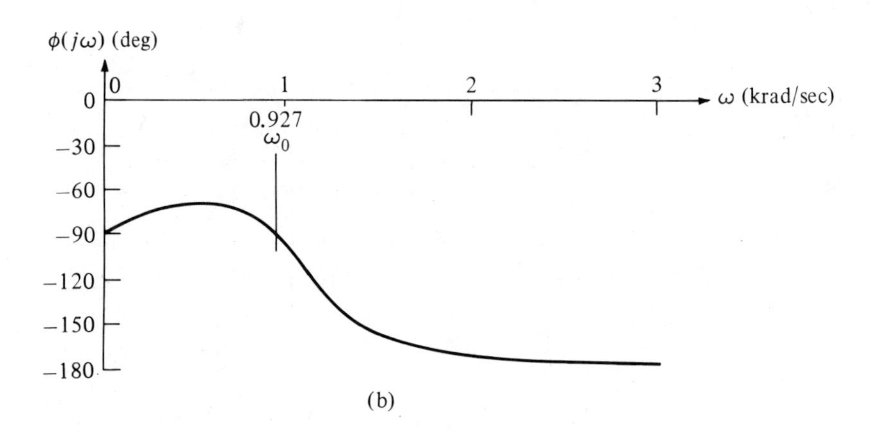

(b)

2. Sketch the response voltage for a parallel network consisting of a 0.2-H inductor and a 4-μF capacitor excited by

$$i(t) = 300 \cdot \sin \omega t \ \mu\text{A}$$

The winding resistance of the inductor is 125 Ω, and the total parallel conductance of the network is 0.1 mS.

Answer:

See Figure 9.14.

9.5 Computer Simulations

Sinusoidal steady-state simulations with SPICE are accomplished by using independent sinusoidal steady-state (ac) sources, specifying the ac analysis mode, and adjusting the output (.PLOT or .PRINT) directives to accommodate phasor notation. Independent ac sources are specified in the same way as dc sources, except that the keyword AC is used instead of DC. The two values following AC are the amplitude and phase angle (in degrees), respectively. The phase angle defaults to zero if only one value is specified.

The ac analysis mode is specified by use of the .AC directive, which replaces the .TRAN record used in all previous SPICE examples. The format of the .AC directive is given in Appendix D. Note that the keyword following .AC (which may be LIN, OCT, or DEC, for linear, octave, or decade, respectively) refers to the scaling of the frequency axis. The number-of-points (NP) value

```
*******11/29/88 ********   SPICE 2G.6    3/15/83 ********16:02:43*****
SPICE 9.1
****     INPUT LISTING                          TEMPERATURE =   27.000 DEG C
******************************************************************************

 RG 2 3 50
 L 3 4 20M
 RL 4 0 40
 C 3 0 30U
 V 1 0 AC 5 30
 VAM 1 2
 .AC LIN 1 159.2 159.2
 .PRINT AC IM(VAM) IP(VAM)
 .END

*******11/29/88 ********   SPICE 2G.6    3/15/83 ********16:02:43*****
SPICE 9.1
****     SMALL SIGNAL BIAS SOLUTION             TEMPERATURE =   27.000 DEG C
******************************************************************************

  NODE   VOLTAGE     NODE   VOLTAGE     NODE   VOLTAGE     NODE   VOLTAGE
 (  1)     .0000    (  2)     .0000    (  3)     .0000    (  4)     .0000
     VOLTAGE SOURCE CURRENTS
     NAME        CURRENT
     V          0.000D+00
     VAM        0.000D+00
     TOTAL POWER DISSIPATION   0.00D+00  WATTS

*******11/29/88 ********   SPICE 2G.6    3/15/83 ********16:02:43*****
SPICE 9.1
****     AC ANALYSIS                            TEMPERATURE =   27.000 DEG C
******************************************************************************

     FREQ        IM(VAM)     IP(VAM)

  1.592E+02     6.325E-02   4.844E+01
```

```
*******11/29/88 ********  SPICE 2G.6    3/15/83 ********16:11:15*****
SPICE 9.2
****     INPUT LISTING                       TEMPERATURE =   27.000 DEG C
*******************************************************************************

 RG 1 2 50
 L 2 3 20M
 RL 3 0 40
 C 2 0 30U
 I 0 1 AC 1 0
 .AC LIN 1 159.2 159.2
 .PRINT AC VM(1) VP(1)
 .END

*******11/29/88 ********  SPICE 2G.6    3/15/83 ********16:11:15*****
SPICE 9.2
****     SMALL SIGNAL BIAS SOLUTION          TEMPERATURE =   27.000 DEG C
*******************************************************************************

  NODE   VOLTAGE     NODE   VOLTAGE      NODE   VOLTAGE
 (  1)     .0000    (  2)     .0000     (  3)     .0000
     TOTAL POWER DISSIPATION   0.00D+00  WATTS

*******11/29/88 ********  SPICE 2G.6    3/15/83 ********16:11:15*****
SPICE 9.2
****     AC ANALYSIS                         TEMPERATURE =   27.000 DEG C
*******************************************************************************

     FREQ        VM(1)       VP(1)

  1.592E+02     7.905E+01  -1.844E+01
```

refers to the total number of data points if LIN is used, or to the number of points per octave or decade if OCT or DEC is used. The beginning frequency (FSTART) and ending frequency (FSTOP) are specified in hertz. Single-frequency analysis is accomplished by assigning the desired frequency value to both FSTART and FSTOP.

The results of ac analysis are complex numbers, so two specifications are required in the output lists for each output voltage or current required. The format options for the output variables are given in Appendix D.

SPICE 9.1 illustrates the use of SPICE for single-frequency analysis (Example 9.1). SPICE 9.2 illustrates the use of SPICE to determine impedance (Example 9.1). The source has been replaced by a $1\angle 0°$-A source, and since $\mathbf{Z} = \mathbf{V}/\mathbf{I}$, $\mathbf{Z}$ is numerically the same as $\mathbf{V}$ when $\mathbf{I} = 1\angle 0°$ A. A similar procedure with a source of $1\angle 0°$ V may be used to obtain admittance. SPICE 9.3 (Example 9.1, part 2) illustrates the use of SPICE for frequency response analysis. Further illustrations of frequency response analysis will be given in Chapter 10.

```
*******11/30/88 ********  SPICE 2G.6    3/15/83 ********11:10:00*****
SPICE 9.3
****    INPUT LISTING                      TEMPERATURE =   27.000 DEG C
*****************************************************************************
 RG 2 3 50
 L 3 4 20M
 RL 4 0 40
 C 3 0 30U
 V 1 0 AC 5 30
 VAM 1 2
 .AC DEC 10 20 2K
 .PLOT AC IM(VAM) IP(VAM)
 .END
*******11/30/88 ********  SPICE 2G.6    3/15/83 ********11:10:00*****
SPICE 9.3
****    SMALL SIGNAL BIAS SOLUTION         TEMPERATURE =   27.000 DEG C
*****************************************************************************
  NODE   VOLTAGE     NODE   VOLTAGE     NODE   VOLTAGE     NODE   VOLTAGE
 (  1)    .0000     (  2)    .0000     (  3)    .0000     (  4)    .0000
     VOLTAGE SOURCE CURRENTS
     NAME       CURRENT
     V          0.000D+00
     VAM        0.000D+00
     TOTAL POWER DISSIPATION   0.00D+00  WATTS
*******11/30/88 ********  SPICE 2G.6    3/15/83 ********11:10:00*****
SPICE 9.3
****    AC ANALYSIS                        TEMPERATURE =   27.000 DEG C
*****************************************************************************
LEGEND:
 *: IM(VAM)
 +: IP(VAM)
     FREQ       IM(VAM)
(*)------------- 5.012D-02     6.310D-02     7.943D-02     1.000D-01  1.259D-01
                 - - - - - - - - - - - - - - - - - - - - - - - - - - - - - - -
(+)------------- 3.000D+01     4.000D+01     5.000D+01     6.000D+01  7.000D+01
                 - - - - - - - - - - - - - - - - - - - - - - - - - - - - - - -
 2.000D+01  5.561D-02 .  +  *        .             .             .           .
 2.518D+01  5.564D-02 .   + *        .             .             .           .
 3.170D+01  5.569D-02 .    +*        .             .             .           .
 3.991D+01  5.578D-02 .     +*       .             .             .           .
 5.024D+01  5.594D-02 .      *+      .             .             .           .
 6.325D+01  5.622D-02 .      *  +    .             .             .           .
 7.962D+01  5.674D-02 .       *    + .             .             .           .
 1.002D+02  5.774D-02 .        *     .   +         .             .           .
 1.262D+02  5.966D-02 .          *   .       +     .             .           .
 1.589D+02  6.321D-02 .             *.           + .             .           .
 2.000D+02  6.909D-02 .              .      *      .+            .           .
 2.518D+02  7.704D-02 .              .            * . +          .           .
 3.170D+02  8.520D-02 .              .             +   *         .           .
 3.991D+02  9.152D-02 .              .        +    .           * .           .
 5.024D+02  9.543D-02 .              .   +         .             *.          .
 6.325D+02  9.757D-02 .              +             .             *.          .
 7.962D+02  9.868D-02 .          +   .             .             *.          .
 1.002D+03  9.926D-02 .        +     .             .              *          .
 1.262D+03  9.958D-02 .      +       .             .              *          .
 1.589D+03  9.975D-02 .    +         .             .              *          .
 2.000D+03  9.985D-02 .   +          .             .              *          .
                 - - - - - - - - - - - - - - - - - - - - - - - - - - - - - - -
```

9.6 SUMMARY

We began this relatively brief chapter by demonstrating the relationship between Laplace transform analysis and sinusoidal steady-state, or ac, analysis. It should now be clear that sinusoidal steady-state analysis, which is often learned intuitively and without a sound theoretical basis, is but a specialization of Laplace transform analysis. Frequency-domain transformation, and the attendant calculations in complex algebra, is a small price to pay for relief from the requirement to integrate and differentiate trignometric functions and obtain solutions to differential equations in the time domain. This is especially true in view of the availability of electronic calculators capable of performing complex arithmetic and making rectangular-to-polar and polar-to-rectangular conversions.

Additional problems in single-frequency ac analysis are available at the end of the chapter for any readers who did not find the examples to be an adequate review. Any reader who is not experienced in phasor notation and complex arithmetic should seek additional examples in any basic circuit analysis text for engineering or engineering technology before attempting to progress to Chapters 10 and 11.

A major effort in this chapter was devoted to introducing the concept of frequency response analysis of one-port networks. The associated concepts of resonance, bandwidth, and quality factor were also explored. This work will serve as a springboard to the frequency-domain analysis of network transfer functions in the next chapter.

9.7 Terms

ac analysis
amplitude
bandwidth
broad frequency response
capacitive reactance
capacitive susceptance
cutoff frequency
frequency
frequency domain
frequency-domain transformation
frequency response
Fourier analysis
harmonically related
in phase
inductive reactance
inductive susceptance
integer multiple
inverse transformation
lag
lead
lower cutoff frequency
Nyquist analysis
parametric plot
phase angle
phasors
phasor diagram
quality factor
reactance

resonance
resonant frequency
rotating vector
selective
sharp frequency response
sinusoidal steady-state analysis
Smith chart
superpose
susceptance
transformation
upper cutoff frequency

Problems

9-1. Transform the sinusoidal functions to phasors.

***a.** $v(t) = 32 \cdot \sin[\omega t - 48°]$ V.

b. $v(t) = 0.2 \cdot \cos[800t + 39°]$ V.

***c.** $i(t) = 2.7 \cdot \cos[2 \times 10^6 t - 64°]$ mA.

d. $v(t) = 1.2 \cdot \sin[377.0(t - 2) + 22°]$ V.

***e.** $i(t) = 636.4 \cdot \cos[377.0t + 107°]$ μA.

9-2. Obtain the sinusoidal function by inverse transformation of the phasor.

a. $\mathbf{I} = 22\angle 110°$ mA; $\omega = 300$ rad/sec.

***b.** $\mathbf{V} = 12\angle -42°$ V; $f = 640$ Hz.

c. $\mathbf{V} = 100\angle 44°$ V; ω not specified.

***d.** $\mathbf{I} = 8\angle 396°$ μA; $\omega = 10^6$ rad/sec.

e. $\mathbf{I} = 5.4\angle -10°$ mA; $f = 60$ Hz.

SPICE

9-3. Construct a phasor diagram for each circuit. Find the unknown phasor.

***a.**

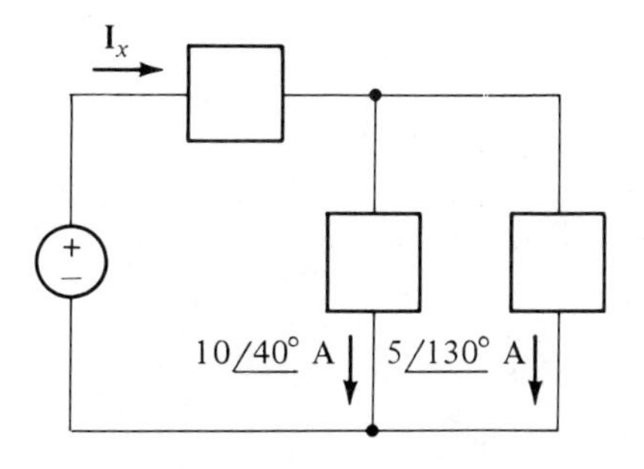

b.

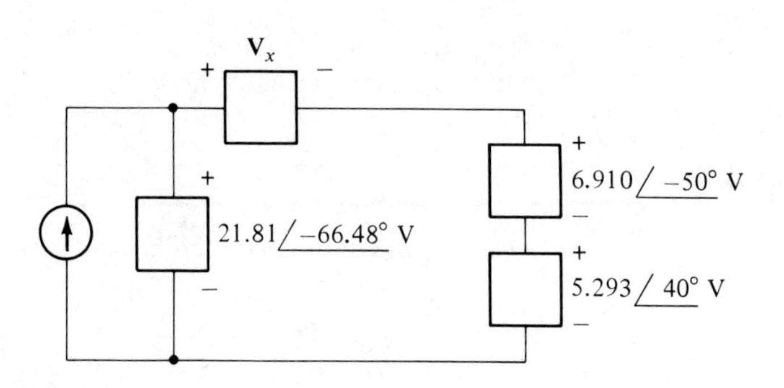

SPICE **9-4.** Find the impedance or admittance, as required, for each network.

a. $\omega = 3$ krad/sec.

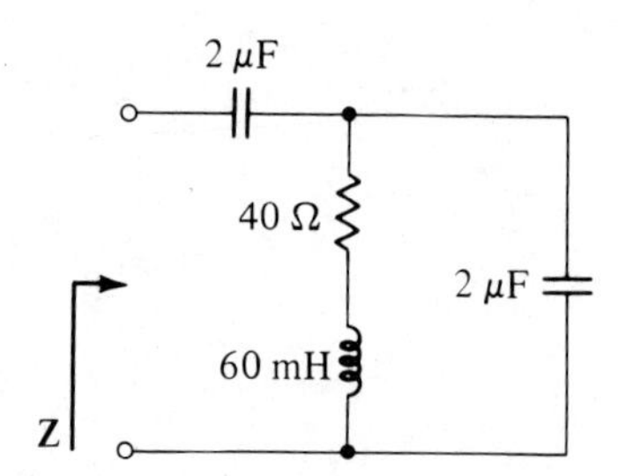

***b.** $f = 50$ Hz.

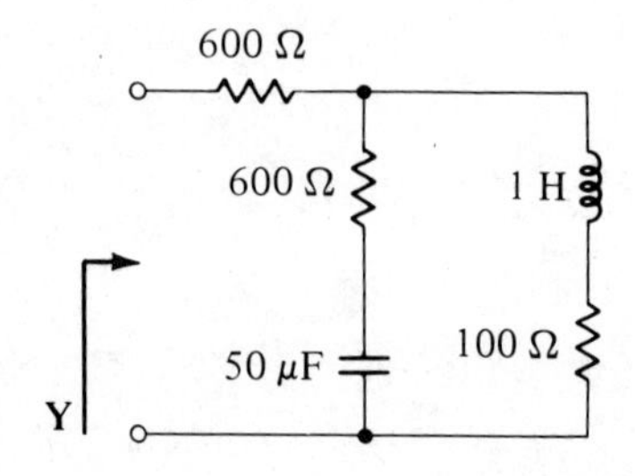

c. $\omega = 377.0$ rad/sec.

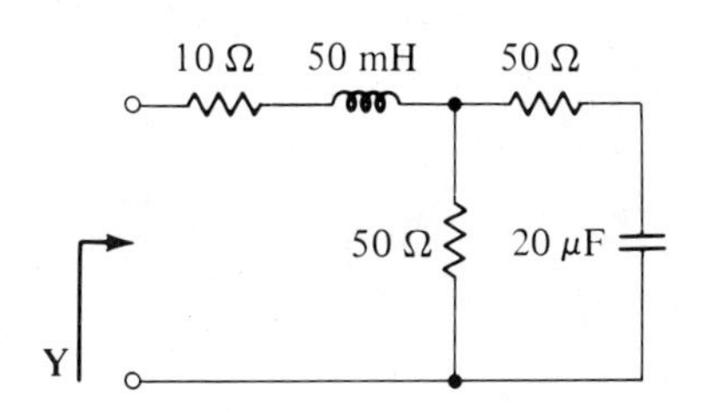

***d.** $\omega = 1$ krad/sec.

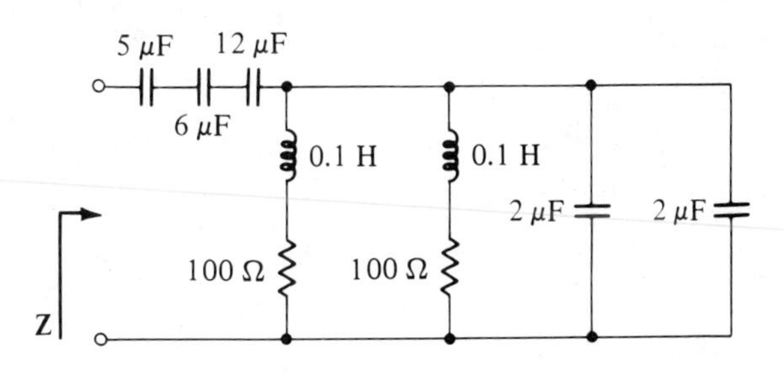

SPICE

9-5. Use phasor analysis and the current division rule to find the steady-state unknown current in each circuit.
***a.** $i(t) = 300 \cdot \sin[600t - 30°]$ mA.

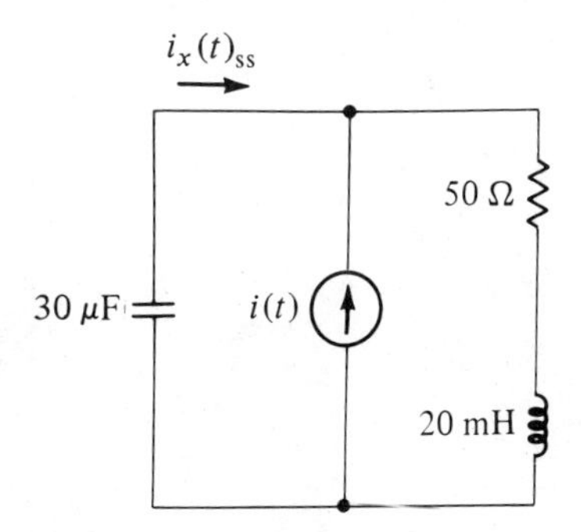

b. $v(t) = 12 \cdot \cos 2 \times 10^3 t$ V.

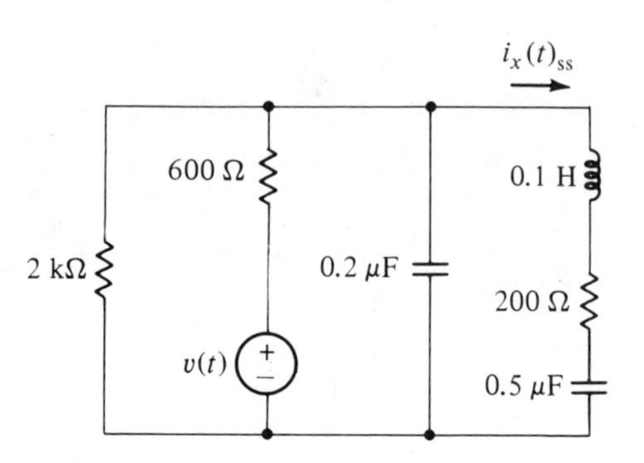

SPICE

9-6. Use phasor analysis and the voltage division rule to find the steady-state unknown voltage in each circuit.

a. $v(t) = 2 \cdot \cos[800t + 40°]$ V.

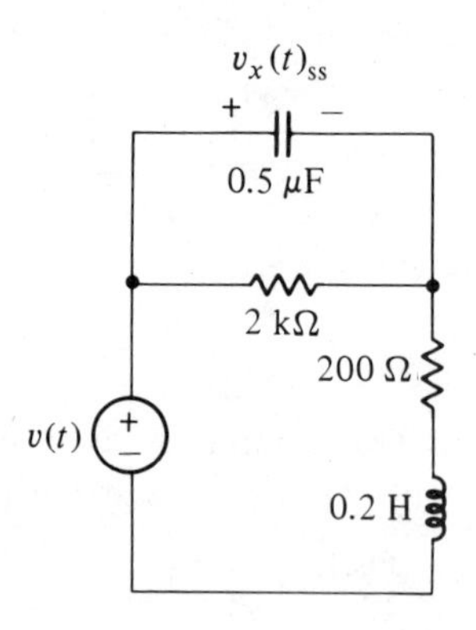

***b.** $i(t) = 300 \cdot \sin[10 \times 10^3 t]$ μA.

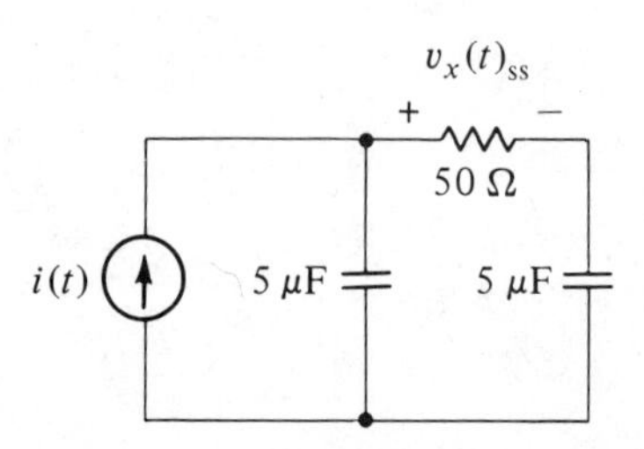

9-7. Find the steady-state Thevenin's and Norton's equivalent for each circuit.

***a.** $v(t) = 5 \cdot \cos[4 \times 10^6 t - 60°]$ V.

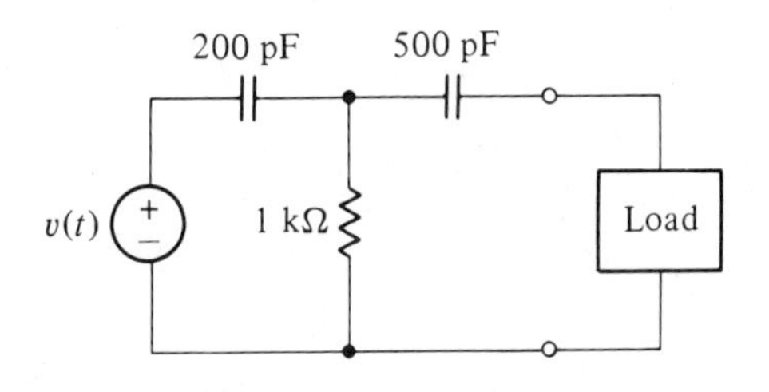

b. $i(t) = 3 \cdot \cos 377.0t$ A, $v(t) = 120 \cdot \sin 377.0t$ V.

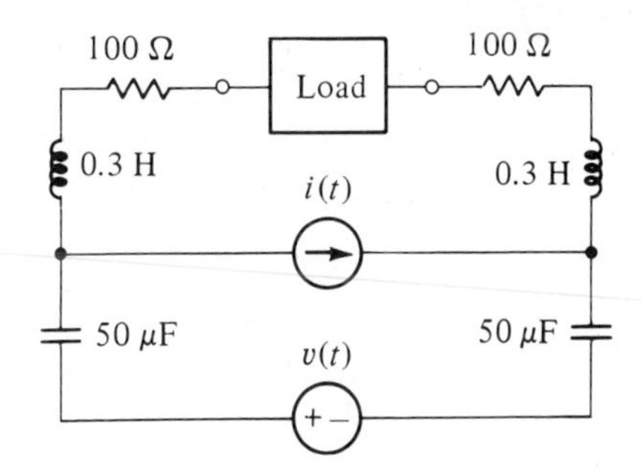

9-8. **a.** Verify that Thevenin's equivalent impedance for Example 9.2 is also realized by a 15.38-Ω resistor in series with a 43.33-μF capacitor.

***b.** Derive a formula for converting a parallel R–C or R–L network to a series R–C or R–L network, respectively, with the same impedance at the given operating frequency.

SPICE

9-9. Use phasor analysis and the superposition principle to determine the unknown voltage or current in each circuit.

***a.** $v(t) = 12 \cdot \cos[10^3 t - 45°]$ V, $i(t) = 20 \cdot \sin 10^3 t$ mA.

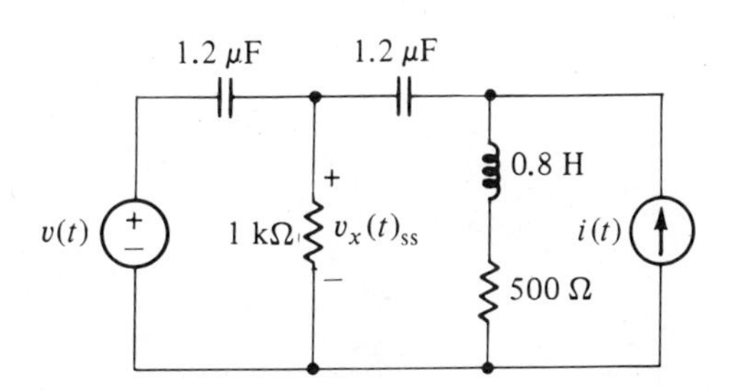

b. $v_1(t) = 100 \cdot \sin[5 \times 10^3 t + 60°]$ V, $v_2(t) = 80 \cdot \cos 5 \times 10^3 t$ V.

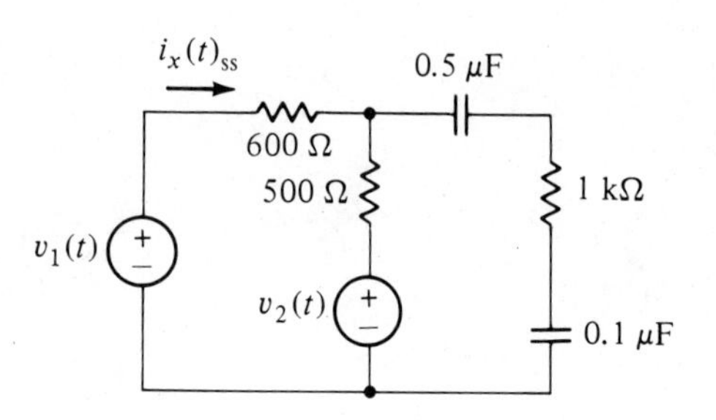

SPICE

9-10. For each network, find ω_0, ω_{MAX} or ω_{MIN}, BW, and Q_0. Sketch the normalized frequency response (amplitude and phase angle) of each network.

a.

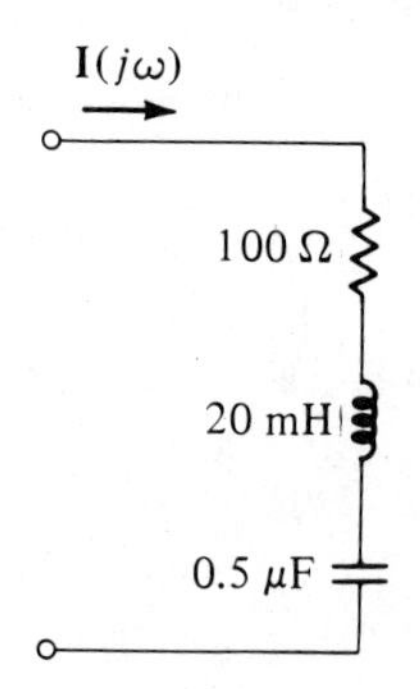

***b.**

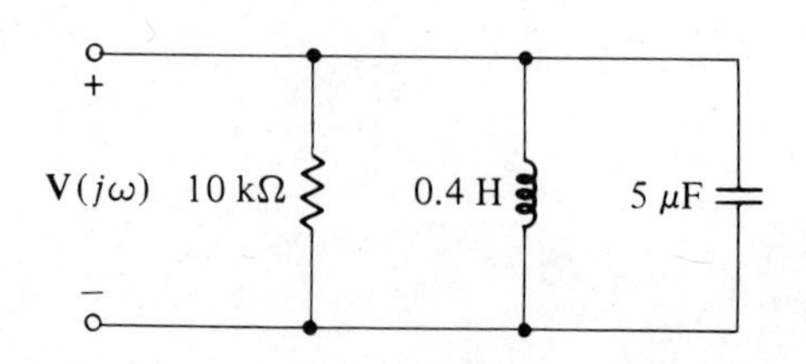

c.

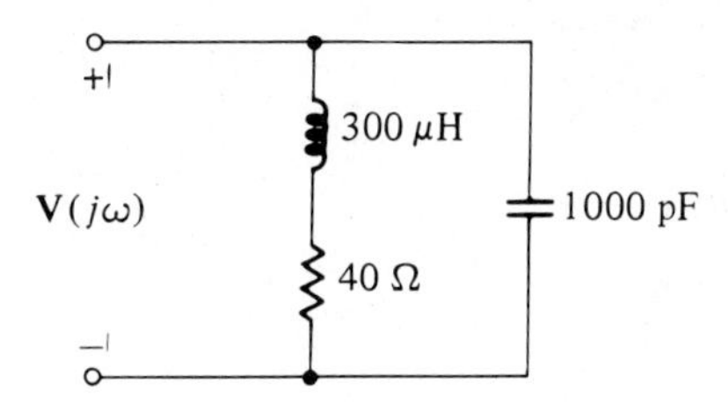

***d.**

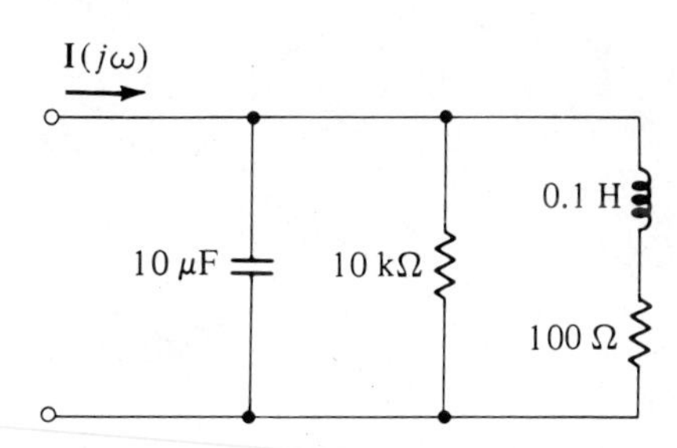

9-11. Verify Equation (9.17) by differentiating Equation (9.15b) and setting the result equal to zero.

9-12. Suppose that the parallel resistor providing the conductance G in Figure 9.12(c) is replaced with an active electronic device designed to provide an adjustable conductance. What is the effect on ω_{MAX}:

a. If $G = r_L^3 C^2/(2L^2)$?

b. If $G = -r_L C/L$?

10 Frequency-Domain Transfer Functions

10.1 INTRODUCTION

We will now turn our attention to sinusoidal steady-state operation of multiport networks. This requires that we again examine the transfer function $H(s)$, which establishes the relationship between the input (excitation) at one port and the output (response) at another port. In this case the excitation is a sinusoid, and the response is considered after all transients have decayed.

Before proceeding to applications of the frequency-domain transfer function, we will introduce the **Bode plot**, a procedure for graphing the frequency-domain transfer function and displaying the frequency response of the network. Several methods for graphing complex functions were illustrated in Chapter 9, one of which was, in fact, a Bode plot.

We will then investigate two significant applications of the frequency-domain transfer function—filter networks and transformers. The applications presented here are only intended to be representative, so we will restrict our discussion of filters to the **Butterworth filter**, and our discussion of transformers will involve primarily the **ideal transformer**.

10.2 Frequency-Domain Transfer Functions

We will begin this discussion by duplicating the process used in Section 9.2 to develop the phasor analysis technique. $H(s)$ was previously defined in Equation (8.9) as

$$H(s) = \frac{F_{OUT}(s)}{F_{IN}(s)}$$

from which

$$F_{OUT}(s) = F_{IN}(s) \cdot H(s)$$

If the excitation is a sinusoid,

$$f_{IN}(t) = (F_{IN})_m \cdot \cos[\omega t + \theta]$$

then

$$F_{IN}(s) = \frac{(F_{IN})_m(s \cdot \cos\theta - \omega \cdot \sin\theta)}{s^2 + \omega^2}$$

$$F_{OUT}(s) = \frac{(F_{IN})_m(s \cdot \cos\theta - \omega \cdot \sin\theta)}{s^2 + \omega^2} \cdot H(s)$$

Reasoning identical to that used in Section 9.2 to develop Equations (9.4) through (9.7) leads to the steady-state response in terms of the frequency-domain transfer function

$$f_{OUT}(t)_{ss} = (F_{IN})_m \cdot |\mathbf{H}(j\omega)| \cdot \cos[\omega t + \theta + \eta(j\omega)] \tag{10.1}$$

$\mathbf{H}(j\omega) = |\mathbf{H}(j\omega)|\angle\eta(j\omega)$ is the frequency-domain transfer function and is obtained by substituting $j\omega$ for s everywhere in $H(s)$.

Equation (10.1) confirms that the sinusoidal steady-state output of a multiport network differs only in amplitude and phase angle from the input, so the phasor form may be used,

$$\mathbf{F}_{OUT}(j\omega) = (F_{OUT})_m(j\omega)\angle\phi(j\omega) \tag{10.2}$$

where $(F_{OUT})_m(j\omega) = (F_{IN})_m \cdot |\mathbf{H}(j\omega)|$ and $\phi(j\omega) = \theta + \eta(j\omega)$.

If the input is normalized ($\mathbf{F}_{IN} = 1\angle 0°$), then the steady-state response is identical to $\mathbf{H}(j\omega)$. This observation is equivalent to that made regarding steady-state response and the immittance function for a one-port network, so the normalized frequency response of a multiport network is determined directly from the transfer function. In the procedural aspects of frequency response analysis, the immittance function of a one-port network is comparable to the transfer function of a multiport network, and it is not necessary to distinguish between the two.

$H(s)$ is the ratio of two polynomials:

$$H(s) = K\frac{(s - z_1)(s - z_2) \cdots (s - z_m)}{(s - p_1)(s - p_2) \cdots (s - p_n)} \tag{10.3}$$

Flashback

A polynomial of degree n may be expressed in its *factored form* as

$$P(x) = C_n(x - r_1)(x - r_2)\cdots(x - r_n)$$

where C_n is the coefficient of the highest-order term, and each r_i is a root of the polynomial. [Refer to Equation (4.6).]

The roots of the numerator polynomial are called the **zeros** of the function, and are denoted by z_i. The roots of the denominator polynomial are called the **poles** of the function, and are denoted by p_i. The ratio of the coefficients of the highest-order terms is given by K. The zeros and the poles of $H(s)$ may be real or complex, but complex zeros and poles, if any, must occur in conjugate pairs.

From Equation (10.3), the frequency-domain transfer function is

$$\mathbf{H}(j\omega) = K \cdot \frac{(-z_1 + j\omega)(-z_2 + j\omega)\cdots(-z_m + j\omega)}{(-p_1 + j\omega)(-p_2 + j\omega)\cdots(-p_m + j\omega)} \tag{10.4}$$

The magnitude of $\mathbf{H}(j\omega)$, which is the amplitude of the frequency response, is

$$|\mathbf{H}(j\omega)| = K\frac{\sqrt{z_1^2 + \omega^2}\sqrt{z_2^2 + \omega^2}\cdots\sqrt{z_n^2 + \omega^2}}{\sqrt{p_1^2 + \omega^2}\sqrt{p_2^2 + \omega^2}\cdots\sqrt{p_m^2 + \omega^2}} \tag{10.5}$$

Flashback

Let $\mathbf{C}_1 = a + jb$ and $\mathbf{C}_2 = c + jd$. If $\mathbf{C} = \mathbf{C}_1 \cdot \mathbf{C}_2$, then $|\mathbf{C}| = \sqrt{a^2 + b^2}\sqrt{c^2 + d^2}$; and if $\mathbf{C} = \mathbf{C}_1/\mathbf{C}_2$, then $|\mathbf{C}| = \sqrt{a^2 + b^2}/\sqrt{c^2 + d^2}$.

The phase angle of $\mathbf{H}(j\omega)$, and of the frequency response, is

$$\eta(j\omega) = \tan^{-1}\left[\frac{\omega}{-z_1}\right] + \tan^{-1}\left[\frac{\omega}{-z_2}\right] + \cdots + \tan^{-1}\left[\frac{\omega}{-z_m}\right] - \left(\tan^{-1}\left[\frac{\omega}{-p_1}\right] + \tan^{-1}\left[\frac{\omega}{-p_2}\right] + \cdots + \tan^{-1}\left[\frac{\omega}{-p_n}\right]\right) \tag{10.6}$$

Flashback

Let $\mathbf{C}_1 = a + jb$ and $\mathbf{C}_2 = c + jd$. If $\mathbf{C} = \mathbf{C}_1\mathbf{C}_2$, then $\angle\mathbf{C} = \tan^{-1}[b/a] + \tan^{-1}[d/c]$; and if $\mathbf{C} = \mathbf{C}_1/\mathbf{C}_2$, then $\angle\mathbf{C} = \tan^{-1}[b/a] - \tan^{-1}[d/c]$.

Bode Plots

Before proceeding to applications of the frequency-domain transfer function, it is necessary to formalize a procedure for graphing the frequency-domain transfer function and displaying the frequency response of the network. Several methods for graphing complex functions were illustrated in Figure 9.8, among which was the Bode plot. A Bode plot consists of two x–y graphs—one graph of the magnitude of the function vs. frequency, and another of the phase angle vs. frequency.

Conventionally, Bode plots are **semi-logarithmic**—that is, the axis of the independent variable (frequency) is logarithmic, and the axis of the dependent variable (magnitude or phase angle) is linear. In practice, the frequency may be in rad/sec (ω) or in hertz (f), the magnitude may be an absolute value ($|\mathbf{H}(j\omega)|$) or a relative value in **decibels** (dB), and the phase angle may be in radians or degrees. The term Bode plot is derived from **Bode analysis**, a technique which uses this type of plot to detect the potential for instability in a feedback control system. While system stability is not a primary focus of this discussion, Bode plots of network transfer functions will nevertheless be used to display frequency response.

The Decibel

Logarithmic scaling serves to compress data and makes it possible to display a wider range of data on a manageable display surface. The decibel not only provides logarithmic scaling of a quantity, but also can be used to compare the quantity with a specified **reference datum**. The **bel** was originally used to relate the power output of a system to the power input to the system. The reference datum is the input power, and

$$\text{No. of bels} = \log[p_o/p_i]$$

A decibel (dB) is simply 1/10 of a bel:

$$\text{No. of dB} = 10 \cdot \log[p_o/p_i]$$

The decibel can be used for other than power ratios, if its meaning and application is well understood and properly applied. For example,

$$\frac{p_o}{p_i} = \frac{v_o^2/R_L}{v_i^2/R_i} = \frac{i_o^2 R_L}{i_i^2 R_i}$$

where R_L and R_i are the load and input resistances, respectively. In situations where the input resistance is equal to the load resistance,

$$p_o/p_i = (v_o/v_i)^2 = (i_o/i_i)^2$$

and the ratio in decibels can be expressed as

$$\text{No. of dB} = 20 \cdot \log[v_o/v_i] = 20 \cdot \log[i_o/i_i]$$

The reader should note well that **power transfer** in decibels can be expressed in terms of input and output voltages or currents only if the input resistance is the same as the load resistance.

In the Bode plots which follow, and in many other network analysis situations, the decibel is used to express the magnitude of **voltage transfer** or **current transfer**, without regard to input and load resistances. The decibel values so obtained must *not* be used to express the power transfer of the network, and especially must not be used in comparing the power transfer of different networks.

If Equation (10.5) is expressed in decibels,

$$\begin{aligned} H(j\omega)_{\text{dB}} &= 20 \cdot \log|\mathbf{H}(j\omega)| \\ &= 20 \cdot \log K \\ &\quad + 20\left(\log\sqrt{z_1^2 + \omega^2} + \log\sqrt{z_2^2 + \omega^2} + \cdots + \log\sqrt{z_m^2 + \omega^2}\right) \\ &\quad - 20\left(\log\sqrt{p_1^2 + \omega^2} + \log\sqrt{p_2^2 + \omega^2} + \cdots + \log\sqrt{p_n^2 + \omega^2}\right) \\ &= 20 \cdot \log K \\ &\quad + 10(\log[z_1^2 + \omega^2] + \log[z_2^2 + \omega^2] + \cdots + \log[z_m^2 + \omega^2]) \\ &\quad - 10(\log[p_1^2 + \omega^2] + \log[p_2^2 + \omega^2] + \cdots + \log[p_n^2 + \omega^2]) \end{aligned} \tag{10.7}$$

Flashback

$$\log ab = \log a + \log b$$

$$\log \frac{a}{b} = \log a - \log b$$

$$\log \sqrt{a} = \frac{\log a}{2}$$

Example 10.1

SPICE

Problem:

Write the frequency-domain transfer function, $\mathbf{V}_o/\mathbf{V}_i$, for the passive two-port network in Figure 10.1, and construct a Bode plot of the frequency response.

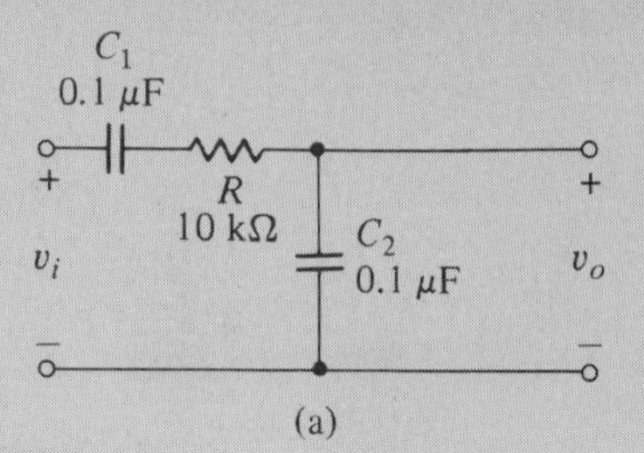

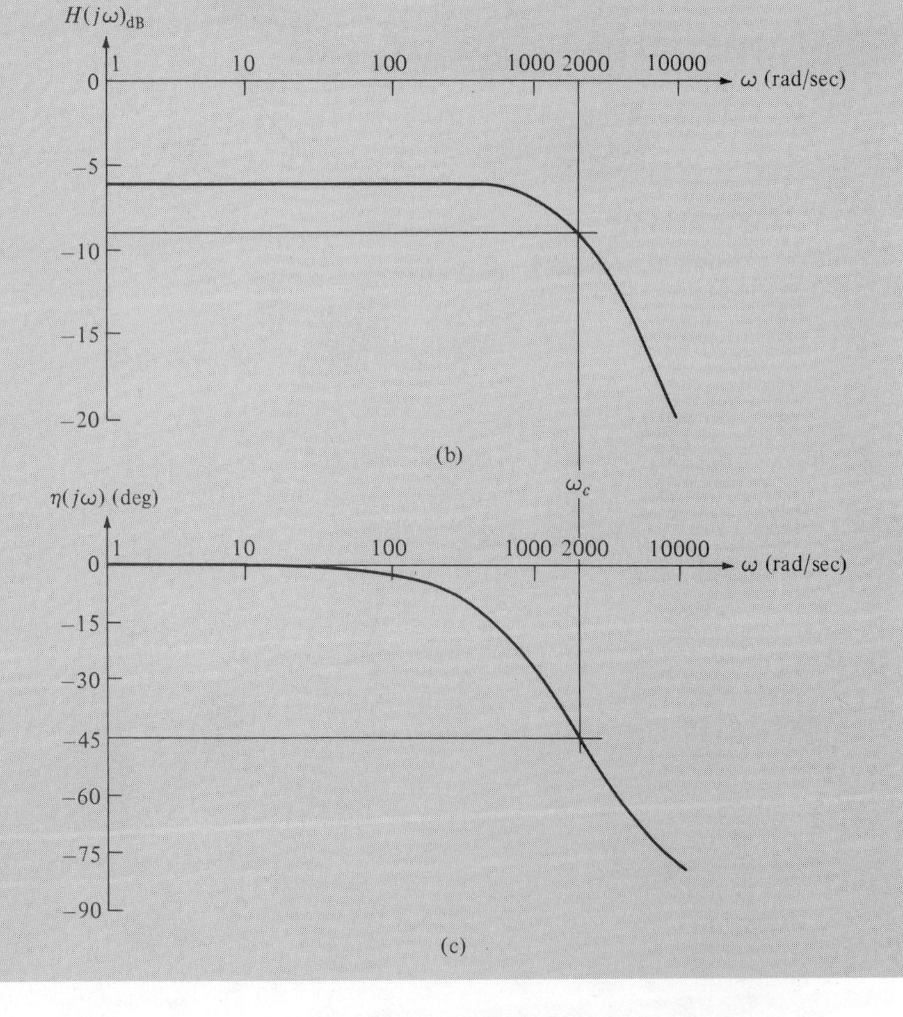

Figure 10.1
Example 10.1.
(a) Network.
(b) Magnitude plot.
(c) Phase angle plot.

Solution:

1. Use the voltage division rule to obtain the voltage at the output port of the network in Figure 10.1(a):

$$\mathbf{V}_o = \mathbf{V}_i \frac{1/(j\omega C_2)}{R + 1/(j\omega C_1) + 1/(j\omega C_2)}$$

$$= \mathbf{V}_i \frac{C_1}{C_1 + C_2 + j\omega R C_1 C_2}$$

$$= \mathbf{V}_i \frac{1/(RC_2)}{(C_1 + C_2)/(RC_1C_2) + j\omega}$$

2. Divide both members of the equation by $\mathbf{V}_i$ and substitute circuit parameters to obtain the required transfer function:

$$\mathbf{H}(j\omega) = \frac{\mathbf{V}_o}{\mathbf{V}_i} = \frac{1000}{2000 + j\omega}$$

3. Use Equation (10.7) to obtain $H(j\omega)_{\mathrm{dB}}$:

$$H(j\omega)_{\mathrm{dB}} = 20 \cdot \log 1000 - 10 \cdot \log[4 \times 10^6 + \omega^2]$$

$$= 60 - 10 \cdot \log[4 \times 10^6 + \omega^2]$$

4. Use Equation (10.6) to obtain $\eta(j\omega)$:

$$\eta(j\omega) = -\tan^{-1}[\omega/2000]$$

5. The Bode plot is shown in Figure 10.1(b) and (c)

Example 10.2

SPICE

Problem:

Write the frequency-domain transfer function $\mathbf{V}_o/\mathbf{V}_i$ for the OPAMP network in Figure 10.2(a), and construct a Bode plot of the frequency response.

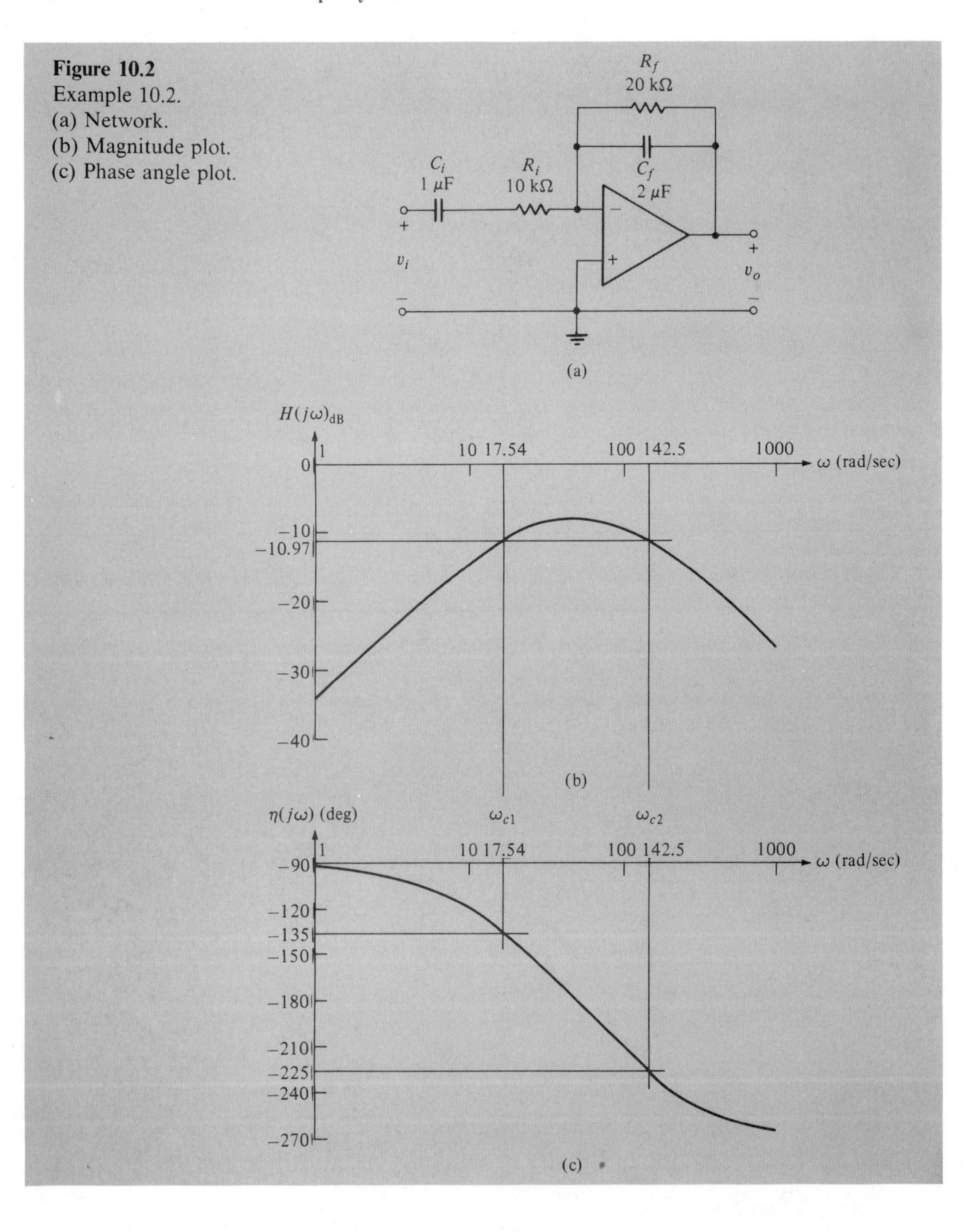

Figure 10.2
Example 10.2.
(a) Network.
(b) Magnitude plot.
(c) Phase angle plot.

Solution:

1. Use Equation (8.16b) to obtain the network transfer function:

$$\mathbf{H}(j\omega) = \frac{\mathbf{V}_o}{\mathbf{V}_i} = -\frac{\mathbf{Z}_f(j\omega)}{\mathbf{Z}_i(j\omega)} = -\frac{1/(G_f + j\omega C_f)}{R_i + 1/(j\omega C_i)}$$

$$= -\frac{j\omega C_i}{(G_f + j\omega C_f)(1 + j\omega R_i C_i)}$$

$$= -\frac{1}{C_f R_i} \cdot \frac{j\omega}{[1/(R_f C_f) + j\omega][1/(R_i C_i) + j\omega]}$$

$$= -50 \frac{j\omega}{(25 + j\omega)(100 + j\omega)}$$

2. Use Equation (10.7) to obtain $H(j\omega)_{\text{dB}}$:

$$H(j\omega)_{\text{dB}} = 20 \cdot \log 50 + 10(\log[\omega^2] - \log[625 + \omega^2] - \log[10 \times 10^3 + \omega^2])$$

$$= 33.98 + 10(2 \cdot \log \omega - \log[625 + \omega^2] - \log[10 \times 10^3 + \omega^2])$$

3. Use Equation (10.6) to obtain $\eta(j\omega)$:

$$\eta(j\omega) = -180° + \tan^{-1}[\omega/0] - \tan^{-1}[\omega/25] - \tan^{-1}[\omega/100]$$

$$= -90° - \tan^{-1}[\omega/25] - \tan^{-1}[\omega/100]$$

4. The Bode plot is shown in Figure 10.2(b) and (c).

Frequency-Response Characteristics

Frequency-domain transfer functions, and the multiport networks from which they are derived, are often described as **low-pass, high-pass, band-pass,** or **band-stop** (Figure 10.3). These descriptions refer to the frequency range for which the magnitude of the transfer function is greater than $1/\sqrt{2}$ (or 0.7071) times the maximum magnitude,

$$|\mathbf{H}(j\omega)| > \frac{|\mathbf{H}(j\omega)|_{\text{MAX}}}{\sqrt{2}}$$

or in the case of the band-stop network, less than $1/\sqrt{2}$ times the maximum magnitude. The limits of a **pass band** or a **stop band** are designated as the cutoff frequencies (ω_{c1} and ω_{c2}). For a low-pass network, $\omega_{c1} \to 0$, and for a

Figure 10.3
Frequency-response characteristics.
(a) Low-pass characteristic.
(b) High-pass characteristic.
(c) Band-pass characteristic.
(d) Band-stop characteristic.

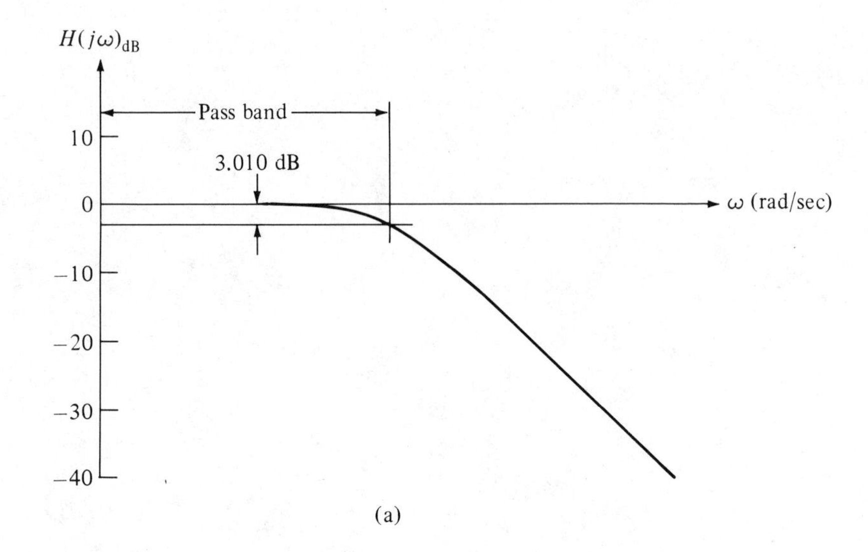

(a)

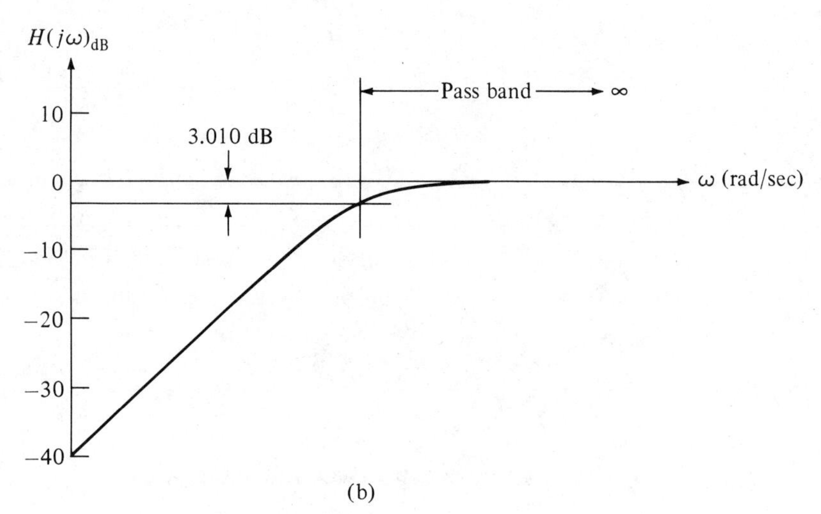

(b)

high-pass network, $\omega_{c2} \to \infty$. Band-stop networks are also called **band-elimination** or **band-reject** networks.

A network for which the magnitude of the transfer function is never less than $1/\sqrt{2}$ times the maximum magnitude is said to be an **all-pass** network. All-pass networks are generally used to produce a desired phase angle vs. frequency characteristic, without affecting magnitude. Only an active network can have such a frequency response.

Figure 10.3
Continued

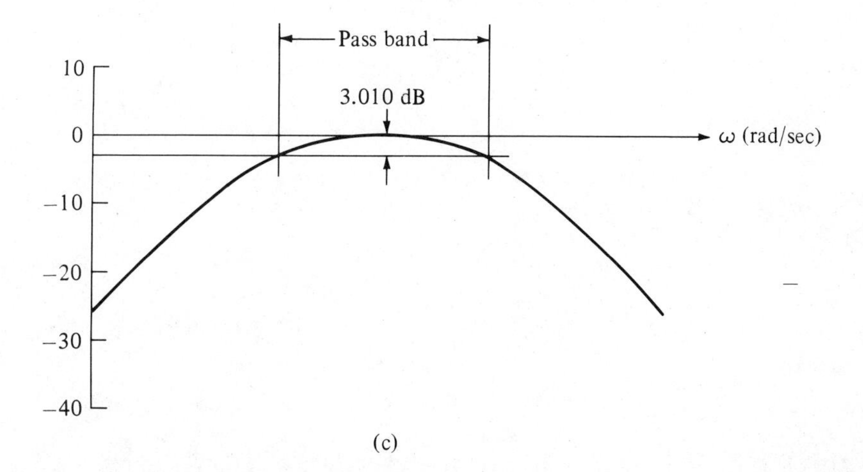

(c)

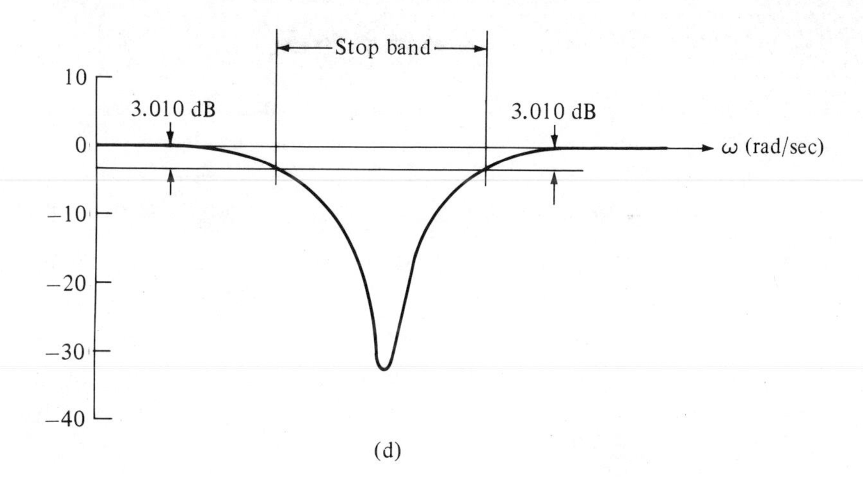

(d)

When expressed in decibels, the magnitude of a transfer function at the cutoff frequencies is

$$H(j\omega_c)_{\text{dB}} = 20 \cdot \log \frac{|\mathbf{H}(j\omega)|_{\text{MAX}}}{\sqrt{2}}$$

$$= 20 \cdot \log|\mathbf{H}(j\omega)|_{\text{MAX}} - 20 \cdot \log\sqrt{2}$$

The value of the first term is simply the maximum magnitude in decibels, and the second term is -3.010 dB, so

$$H(j\omega_c)_{\text{dB}} = (H(j\omega)_{\text{MAX}})_{\text{dB}} - 3.010 \text{ dB} \tag{10.8}$$

Example 10.3

Problem:

Categorize the networks of Examples 10.1 and 10.2 according to frequency-response characteristics.

Solution:

1. From the graph of $H(j\omega)_{\text{dB}}$ vs. ω in Figure 10.1(b), the network has a low-pass characteristic with $\omega_c = 2$ krad/sec.
2. From the graph of $H(j\omega)_{\text{dB}}$ vs. ω in Figure 10.2(b), the network has a band-pass characteristics with $\omega_{c1} \simeq 17.54$ rad/sec and $\omega_{c2} \simeq 142.5$ rad/sec.

Exercise 10.1

SPICE

Problem:

1. Write the transfer function $\mathbf{H}(j\omega) = \mathbf{V}_o/\mathbf{V}_i$ for the network shown in Figure 10.4(a).
2. Construct a Bode plot of $\mathbf{H}(j\omega)$.
3. From the Bode plot, characterize the frequency response of the network.

Answers:

1. $\mathbf{H}(j\omega) = \dfrac{10 \times 10^3 + j\omega}{50 \times 10^3 + j\omega}$.

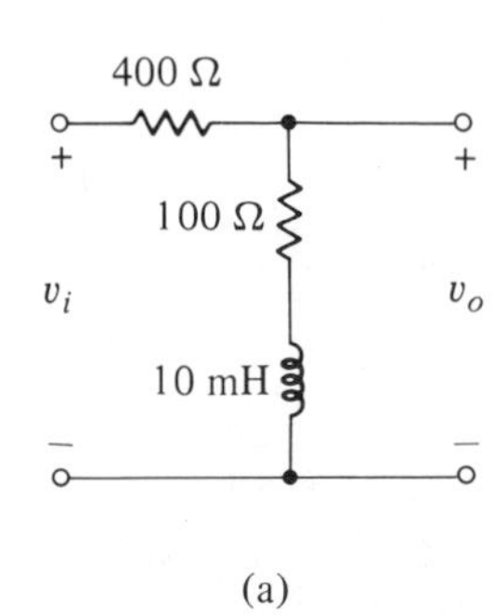

Figure 10.4
Exercise 10.1.
(a) Network.
(b) Magnitude plot.
(c) Phase angle plot.

Figure 10.4
Continued

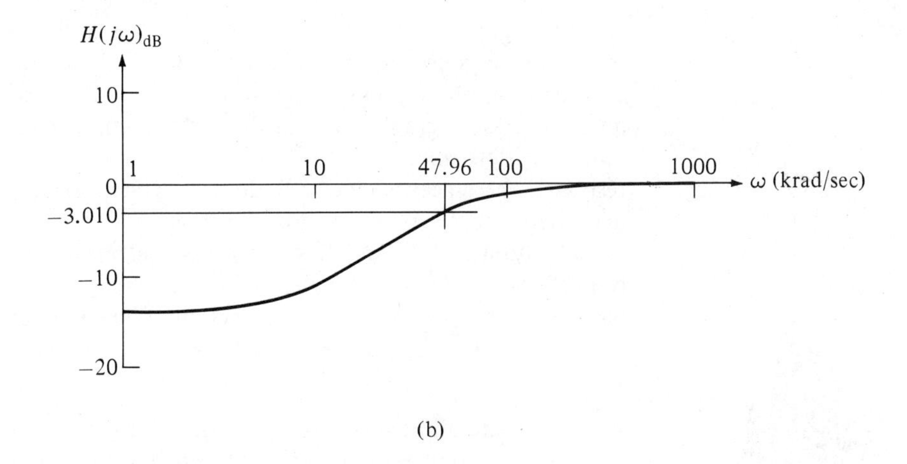

(b)

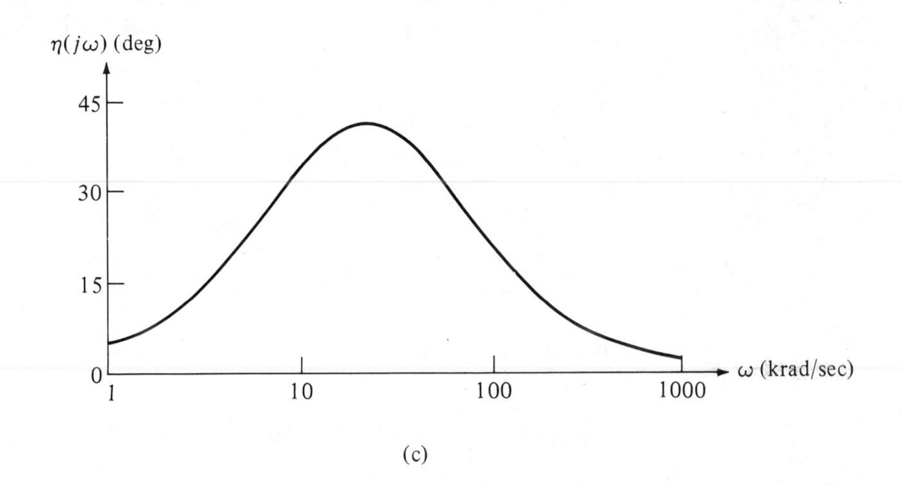

(c)

2. See Figure 10.4(b) and (c).
3. High-pass.

10.3 Piecewise-Linear Bode Plot

Equations (10.5), (10.6), and (10.7) are convenient for graphing frequency response if the transfer function is simple, as in the examples and exercises in the previous section, with one or two zeros and/or one of two poles, or if a computer graphics program is available. The author generated the original frequency response graphs in this text using these equations with a popular computer graphics program. However, when it is necessary to prepare frequency-response graphs without the use of computer graphics, or more importantly, to determine the effect of certain network components on frequency response, the **piecewise-linear** Bode plot, which uses straight lines to approximate frequency response, is a useful tool.

Equations (10.6) and (10.7) express the phase angle and the magnitude, respectively, of $\mathbf{H}(j\omega)$ as a sum of terms, each of which is associated with a particular zero or pole of the function, plus one term in Equation (10.7) associated with the constant K. To construct a piecewise-linear Bode plot of $\mathbf{H}(j\omega)$,

1. construct a piecewise-linear Bode plot for each term separately to approximate the contribution of that term to the frequency response of $\mathbf{H}(j\omega)$;
2. graphically combine the frequency response plots for the individual terms to obtain the piecewise-linear Bode plot for $\mathbf{H}(j\omega)$.

Frequency Response Contribution of *K*

From Equation (10.7), the contribution of K to $H(j\omega)_{\text{dB}}$ is

$$H_K(j\omega)_{\text{dB}} = 20 \cdot \log|K| \tag{10.9}$$

K does not contribute to the phase angle $\eta(j\omega)$ unless the sign of K is negative, in which case the contribution is a constant $-180°$. The frequency response contribution of K is shown in Figure 10.5.

Magnitude Contribution of a Real Pole Term

From Equation (10.7), the contribution of a real pole term to $H(j\omega)_{\text{dB}}$ is

$$H_p(j\omega)_{\text{dB}} = -10 \cdot \log[p^2 + \omega^2] \tag{10.10}$$

As frequency decreases, ω becomes less significant in comparison with $|p|$, so

$$H_p(j\omega)_{\text{dB}} \to -10 \cdot \log p^2 = -20 \cdot \log|p| \quad \text{as} \quad \omega \to 0 \tag{10.11}$$

Equation (10.11), which represents a horizontal line [Figure 10.6(a)], is the **low-frequency asymptote** of $H_p(j\omega)_{\text{dB}}$. An asymptote is the straight line which a function approaches for limiting values of the independent variable. In this case, ω is approaching zero as a limit, hence the name low-frequency asymptote.

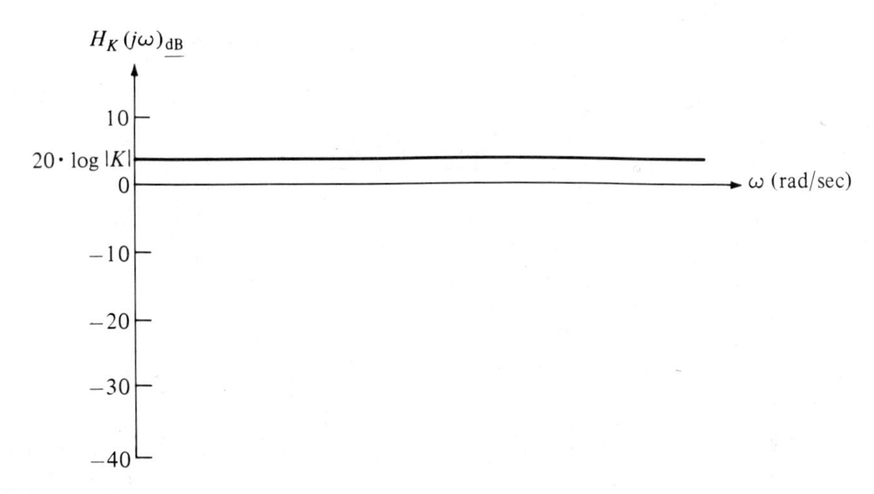

Figure 10.5
Contribution of K to the Bode plot of $\mathbf{H}(j\omega)$.

As frequency increases, $|p|$ becomes less and less significant in comparison with ω, so

$$H_p(j\omega)_{dB} \to -10 \cdot \log \omega^2 = -20 \cdot \log \omega \qquad \text{as} \quad \omega \to \infty \qquad (10.12)$$

Equation (10.12), which represents a straight line with a slope of -20 on a semi-logarithmic graph [Figure 10.6(a)], is the **high-frequency asymptote** of $H_p(j\omega)_{dB}$. Because the logarithmic axis is scaled in **decades** $(\ldots, 0.01, 0.1, 1, 10, 100, \ldots)$, the slope is specified in decibels/decade (dB/dec).

The low-frequency and high-frequency asymptotes intersect [Figure 10.6(a)] to create a piecewise-linear approximation to the frequency response of $H_p(j\omega)_{dB}$, which is seen to have a low-pass characteristic. At the point of intersection,

$$H_p(j\omega)_{dB} = -20 \cdot \log|p|$$

The frequency at the point of intersection can be determined by equating the low- and high-frequency asymptotes [Equations (10.11) and (10.12)]:

$$-20 \cdot \log|p| = -20 \cdot \log \omega_p$$

or

$$\omega_p = |p| \qquad (10.13)$$

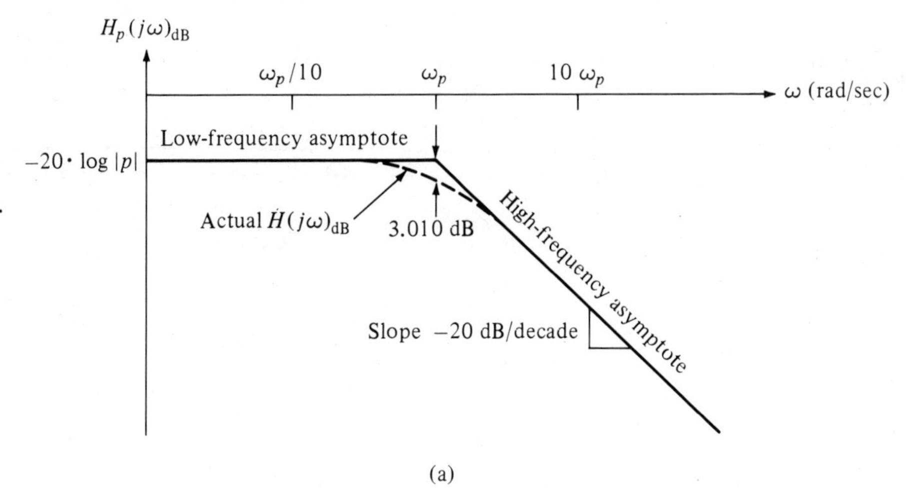

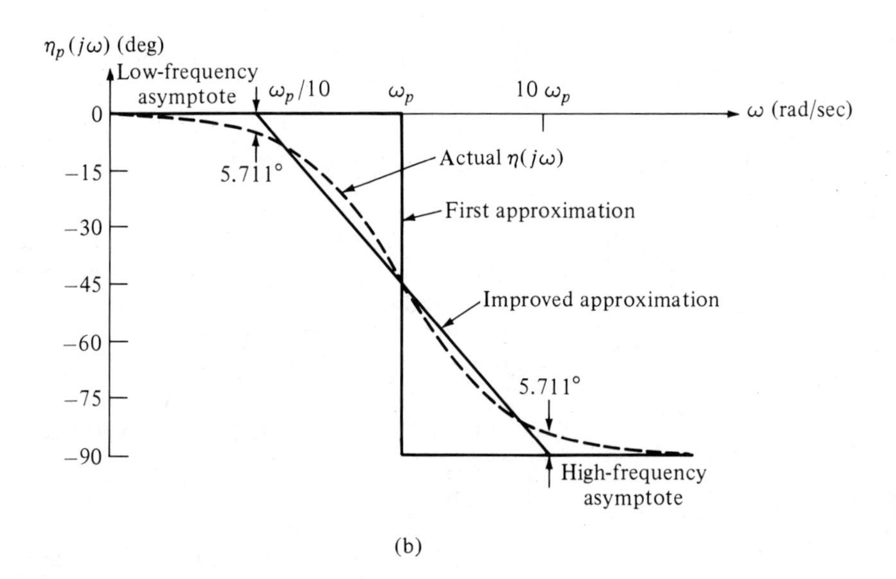

Figure 10.6 Contribution of a real pole term to the Bode plot of $H(j\omega)$. (a) Magnitude, $H(j\omega)_{dB}$. (b) Phase angle, $\eta(j\omega)$.

The maximum magnitude of $\mathbf{H}_p(j\omega)$ occurs as $\omega \to 0$, and so is equal to the low-frequency asymptote

$$\left(H_p(j\omega)_{MAX}\right)_{dB} = -20 \cdot \log|p| \tag{10.14}$$

From Equation 10.10, the magnitude of $\mathbf{H}_p(j\omega)$ at ω_p is

$$H_p(j\omega_p)_{dB} = -10 \cdot \log\left[p^2 + \omega_p^2\right]$$

From Equations (10.13) and (10.14),

$$
\begin{aligned}
H_p(j\omega_p)_{\mathrm{dB}} &= -10 \cdot \log[p^2 + p^2] = -10 \cdot \log 2p^2 \\
&= -10 \cdot \log 2 - 20 \cdot \log|p| \\
&= -3.010 - 20 \cdot \log|p| \\
&= \left(H_p(j\omega)_{\mathrm{MAX}}\right)_{\mathrm{dB}} - 3.010 \qquad (10.15)
\end{aligned}
$$

Equation (10.15) is identical to Equation (10.8), which means that ω_p is also the cutoff frequency for $\mathbf{H}_p(j\omega)$. In Bode analysis, the cutoff frequency for each term of $\mathbf{H}(j\omega)$ is called a **corner frequency** or a **breakpoint**.

In Figure 10.6(a), the actual magnitude of $\mathbf{H}_p(j\omega)$ (curved, dashed line) is smaller at all frequencies than the approximate magnitude (straight, solid lines), with the largest difference occurring at the corner frequency, where the approximate magnitude is $-20 \cdot \log|p|$, or $(H_p(j\omega)_{\mathrm{MAX}})_{\mathrm{dB}}$. This observation, together with Equation (10.15), indicates that the piecewise-linear approximation is at most about 3 dB greater than the actual magnitude.

Phase Angle Contribution of a Real Pole Term

From Equation (10.6), the contribution of a real pole term to $\eta(j\omega)$ is

$$\eta_p(j\omega) = -\tan^{-1}\left[\frac{\omega}{-p}\right] \qquad (10.16)$$

The low-frequency asymptote depends on whether p is positive or negative:

$$\eta_p(j\omega) \to -\tan^{-1}\left[\frac{0}{-p}\right] = 0° \qquad \text{for} \quad p < 0 \qquad (10.17a)$$

$$\eta_p(j\omega) \to -\tan^{-1}\left[\frac{0}{-p}\right] = -180° \qquad \text{for} \quad p > 0 \qquad (10.17b)$$

The high-frequency asymptote is

$$\eta_p(j\omega) \to -\tan^{-1}\infty = -90° \qquad (10.18)$$

Positive real poles do not occur in the transfer functions for passive linear networks, but can occur in the transfer functions for nonlinear or active networks.

The asymptotes of $\eta_p(j\omega)$ do not intersect [Figure 10.6(b)], so they must be connected by another straight line in order to form an approximation to the phase angle response. An evaluation of the actual phase angle at the corner frequency yields

$$\eta_p(j\omega_c) = \tan^{-1}\left[\frac{|p|}{-p}\right] = -45° \qquad \text{for} \quad p < 0$$

$$\eta_p(j\omega_c) = \tan^{-1}\left[\frac{|p|}{-p}\right] = -135° \qquad \text{for} \quad p > 0$$

Thus, the exact value of $\eta_p(j\omega)$ is midway between its asymptotes at the corner frequency. One way to approximate the phase angle response is to connect the asymptotes with a vertical line at the corner frequency [Figure 10.6(b)], but this results in errors approaching 45° at frequencies just below or just above the corner frequency. A better approximation results from connecting a point on the low-frequency asymptote, one decade below the corner frequency ($\omega_c/10$), with a point on the high-frequency asymptote, one decade above the corner frequency ($10\omega_c$). This line, which has a slope of 45°/decade, passes through the actual value of $\eta_p(j\omega)$ at the corner frequency. The greatest error in the approximation occurs one decade above and below the corner frequency, and is less than 6° (Problem 10-3).

Example 10.4

SPICE

Problem:

Construct a piecewise-linear Bode plot of the voltage transfer function for the network in Figure 10.7(a).

Solution:

1. Use the voltage division rule to obtain the transfer function:

$$\mathbf{H}(j\omega) = \frac{800}{800 + 200 + j0.2\omega}$$

$$= 4000\frac{1}{5000 + j\omega} = K\frac{1}{-p + j\omega}$$

2. Calculate the contribution of K to $\mathbf{H}(j\omega)$:

$$H_K(j\omega)_{\text{dB}} = 20 \cdot \log 4000 \simeq 72.04 \text{ dB}$$

3. Calculate the contribution of $1/(-p + j\omega)$ to $\mathbf{H}(j\omega)$. The low-frequency asymptotes are

$$H_p(j\omega)_{\text{dB}} = -20 \cdot \log 5000 \simeq -73.98 \text{ dB}$$

$$\eta_p(j\omega) = 0, \qquad \text{since} \quad p < 0$$

Figure 10.7
Example 10.4.
(a) Network.
(b) Magnitude plot.
(c) Phase angle plot.

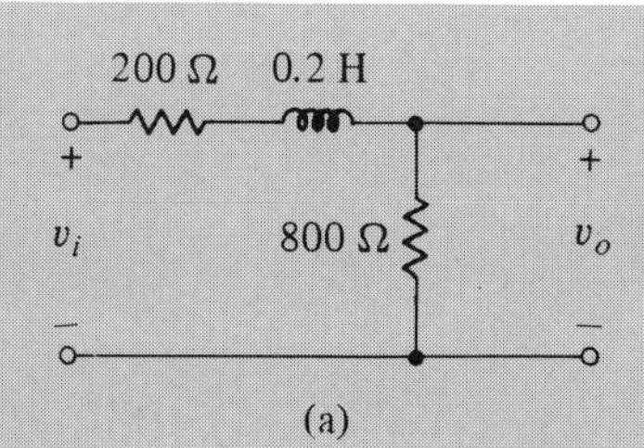

(a)

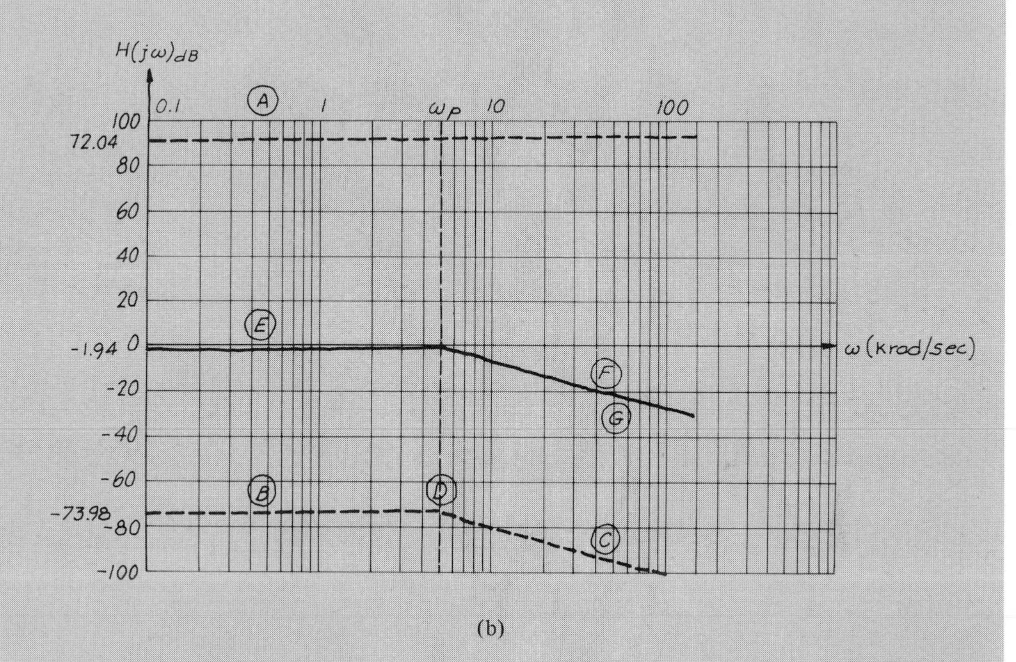

(b)

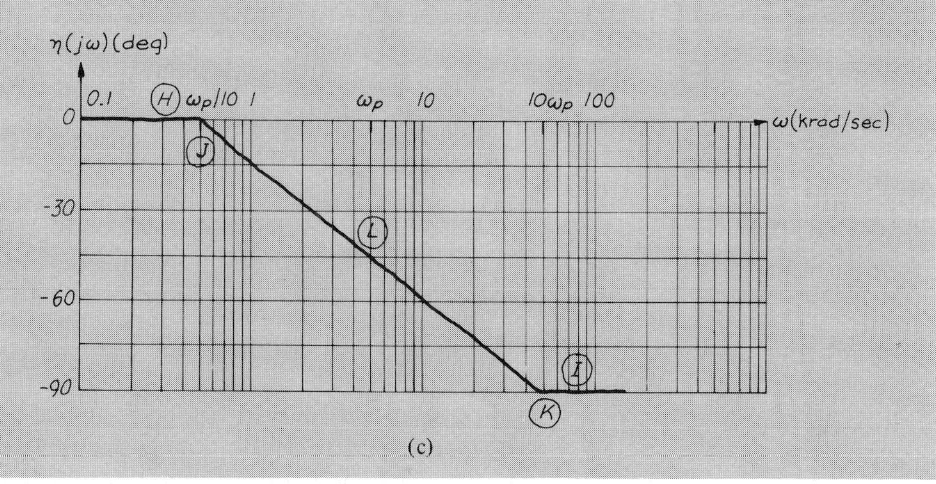

(c)

The high-frequency asymptotes are

$$H_p(j\omega)_{\mathrm{dB}} = -20 \cdot \log \omega$$

$$\eta_p(j\omega) = -90°$$

4. Construct the piecewise-linear magnitude graph for the Bode plot on a *three-cycle* semi-logarithmic graph sheet, as shown in Figure 10.7(b).

   Plot $H_K(j\omega)_{\mathrm{dB}}$, which is a horizontal line at 72.04 dB (A) in the magnitude plot.

   Plot the low-frequency asymptote of $H_p(j\omega)_{\mathrm{dB}}$, which is a horizontal line at -73.98 dB (B).

   Plot the high-frequency asymptote, which is a line with a slope of -20 dB/dec (C) intersecting the low-frequency asymptote at $\omega_p = |p| = 5$ krad/sec (D).

   Combine the plotted lines graphically to form a piecewise-linear graph of $H(j\omega)_{\mathrm{dB}}$.

   For frequencies below 5 krad/sec, the graph of $H(j\omega)_{\mathrm{dB}}$ is a horizontal line E at -1.94 dB ($-73.98 + 72.04$). Above 5 krad/sec, the graph of $H(j\omega)_{\mathrm{dB}}$ is a straight line F which parallels the graph of $H_p(j\omega)_{\mathrm{dB}}$, because $H_K(j\omega)_{\mathrm{dB}}$ exhibits no change with frequency. At 50 krad/sec (G), $H(j\omega)_{\mathrm{dB}}$ is -21.94 dB, or 20 dB below the value at the corner frequency.

5. Construct the piecewise-linear graph of $\eta(j\omega)$, as shown in Figure 10.7(c).

   The constant does not contribute to $\eta(j\omega)$.

   Plot the low-frequency asymptote of $\eta_p(j\omega)$, which is a horizontal line at 0° (H).

   Plot the high-frequency asymptote, which is a horizontal line at $-90°$ (I).

   Connect a point J on the low-frequency asymptote at 0.5 krad/sec ($\omega_p/10$) to a point K on the high-frequency asymptote at 50 krad/sec ($10\omega_p$).

   Note that the connecting line L passes through $-45°$ at 5 krad/sec.

Frequency Response Contribution of a Real Zero Term

All terms in Equations (10.6) and (10.7) associated with zeros of $\mathbf{H}(j\omega)$ have positive signs, while those associated with poles have negative signs. Thus, the Bode plot for a real zero term (Figure 10.8) amounts to an inversion of the

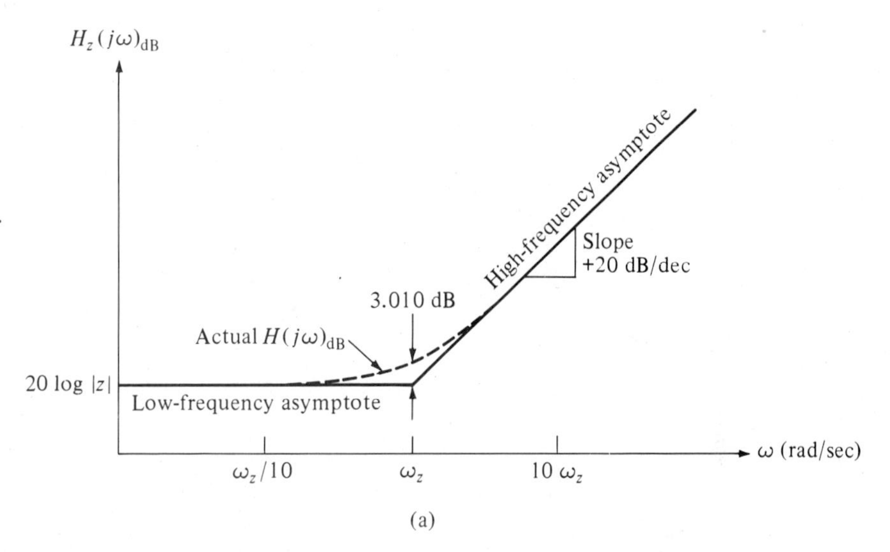

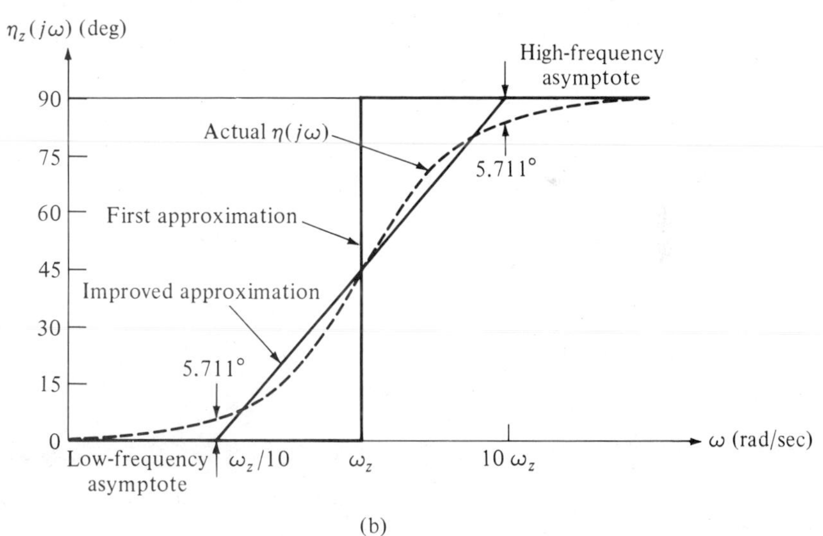

Figure 10.8 Contribution of a real zero term to the Bode plot of $\mathbf{H}(j\omega)$. (a) Magnitude, $H(j\omega)_{dB}$. (b) Phase angle, $\eta(j\omega)$.

Bode plot for a real pole term, and the appropriate equations are given in Table 10.1 without further comment.

Example 10.5 SPICE

Problem:

Construct a piecewise-linear Bode plot of voltage transfer function for the network in Figure 10.9(a).

Figure 10.9
Example 10.5.
(a) Network.
(b) Magnitude plot.
(c) Phase angle plot.

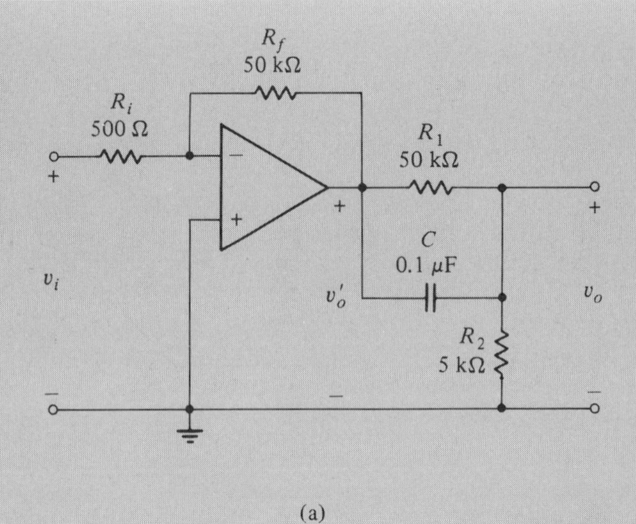

(a)

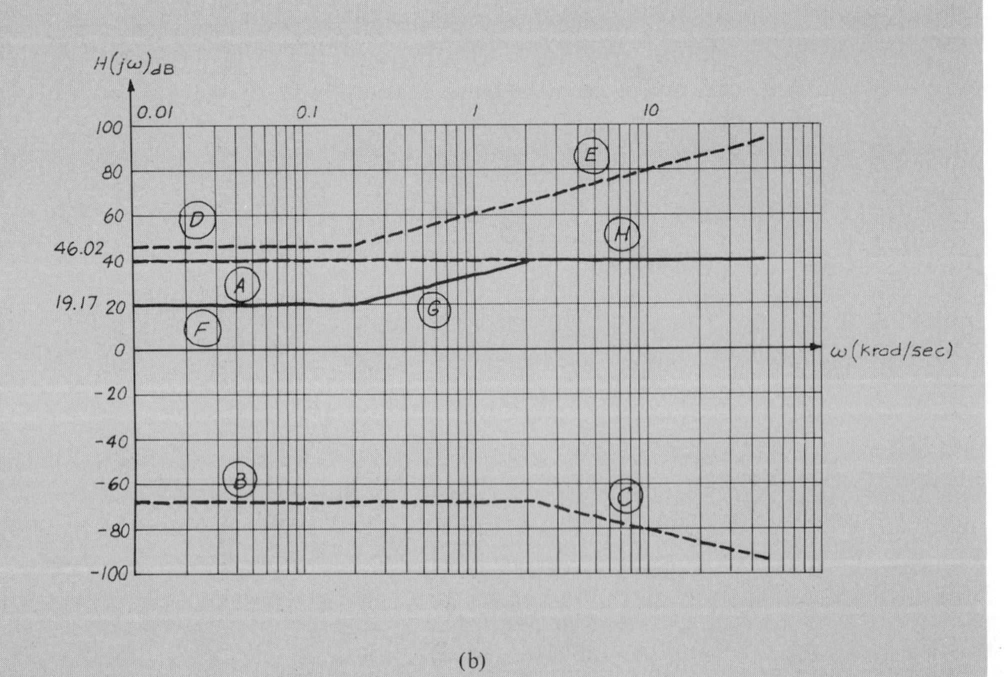

(b)

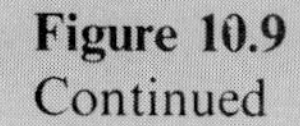

Figure 10.9 Continued

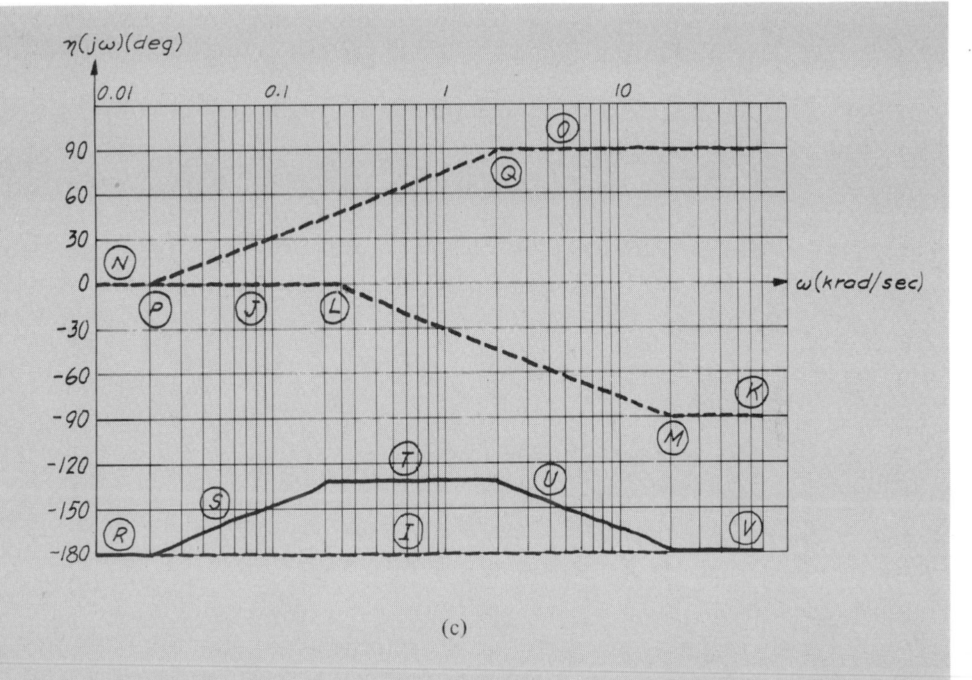

(c)

Solution:

1. Use the voltage division rule to calculate the transfer function $\mathbf{V}_o/\mathbf{V}_o'$ for the passive network:

$$\frac{\mathbf{V}_o}{\mathbf{V}_o'} = \frac{R_2}{R_2 + 1/(G_1 + j\omega C)}$$

$$= \frac{G_1/C + j\omega}{(G_1 + G_2)/C + j\omega} = \frac{200 + j\omega}{2200 + j\omega}$$

2. Use Equation (8.16b) to calculate the transfer function $\mathbf{V}_o'/\mathbf{V}_i$ for the OPAMP network:

$$\frac{\mathbf{V}_o'}{\mathbf{V}_i} = -\frac{R_f}{R_i} = -100$$

3. If the OPAMP output resistance is sufficiently small, the network transfer function is

$$\mathbf{H}(j\omega) = \frac{\mathbf{V}_o}{\mathbf{V}_i} = \frac{\mathbf{V}_o'}{\mathbf{V}_i} \cdot \frac{\mathbf{V}_o}{\mathbf{V}_o'} = -100\frac{200 + j\omega}{2200 + j\omega} = K\frac{-z + j\omega}{-p + j\omega}$$

4. Calculate the contribution of K to $\mathbf{H}(j\omega)$:

$$H_K(j\omega)_{\text{dB}} = 20 \cdot \log 100 = 40 \text{ dB}$$

5. Calculate the contribution of $1/(-p + j\omega)$ to $\mathbf{H}(j\omega)$. The low-frequency asymptotes are

$$H_p(j\omega)_{\text{dB}} = -20 \cdot \log 2200 \simeq -66.85 \text{ dB}$$
$$\eta_p(j\omega) = 0, \qquad \text{since} \quad p < 0$$

The high-frequency asymptotes are

$$H_p(j\omega)_{\text{dB}} = -20 \cdot \log \omega$$
$$\eta_p(j\omega) = -90°$$

6. Calculate the contribution of $-z + j\omega$ to $\mathbf{H}(j\omega)$. The low-frequency asymptotes are

$$H_z(j\omega)_{\text{dB}} = 20 \cdot \log 200 \simeq 46.02 \text{ dB}$$
$$\eta_z(j\omega) = 0, \qquad \text{since} \quad z < 0$$

The high-frequency asymptotes are

$$H_z(j\omega)_{\text{dB}} = 20 \cdot \log \omega$$
$$\eta_z(j\omega) = 90°$$

7. Construct the piecewise-linear magnitude graph, as shown in Figure 10.9(b):

Plot $H_K(j\omega)_{\text{dB}}$ (A).

Plot the asymptotes for $H_p(j\omega)_{\text{dB}}$ (B and C). The corner frequency is $\omega_p = 2.2$ krad/sec.

Plot the asymptotes for $H_z(j\omega)_{\text{dB}}$ (D and E). The corner frequency is $\omega_z = 0.2$ krad/sec.

Combine the plotted lines graphically to form a piecewise-linear graph of $H(j\omega)_{\text{dB}}$ (F, G, and H).

8. Construct the piecewise-linear graph of $\eta(j\omega)$, as shown in Figure 10.9(c):

Plot a horizontal line through $-180°$ (I) to represent the phase angle associated with the negative sign preceding K.

Plot the asymptotes of $\eta_p(j\omega)$ (J and K).

Connect a point L on the low-frequency asymptote at 0.22 krad/sec ($\omega_p/10$) to a point M on the high-frequency asymptote at 22 krad/sec ($10\omega_p$).

Plot the asymptotes of $\eta_z(j\omega)$ (N and O).

Connect a point P on the low-frequency asymptote at 0.02 krad/sec ($\omega_z/10$) to a point Q on the high-frequency asymptote at 2 krad/sec ($10\omega_z$).

Combine the plotted lines graphically to form a piecewise-linear graph of $\eta(j\omega)$ (R, S, T, U, and V).

Magnitude Contribution of a Complex Pole

Complex poles of $\mathbf{H}(j\omega)$ occur, if at all, in conjugate pairs. A complex pole of $\mathbf{H}(j\omega)$ must be considered together with its conjugate in order to determine the effect of the pole on the frequency response of $\mathbf{H}(j\omega)$. From Equation (4.16), the contribution to $H(s)$ of a conjugate pair of complex poles is

$$H_{\text{cp}}(s) = \frac{1}{(s-\mathbf{p})(s-\mathbf{p}^*)} = \frac{1}{(s+\alpha-j\omega_d)(s+\alpha+j\omega_d)}$$

and from Equation (10.4),

$$\mathbf{H}_{\text{cp}}(j\omega) = \frac{1}{(-\alpha+j\omega_d+j\omega)(-\alpha-j\omega_d+j\omega)}$$

$$= \frac{1}{\left(\omega_0^2-\omega^2\right)+j2\alpha\omega} \tag{10.19}$$

where

$$\omega_0^2 = \alpha^2 + \omega_d^2$$

If there are no other poles or zeros of $\mathbf{H}(j\omega)$, then ω_0 is clearly the resonant frequency of the network, but if there are other poles or zeros, this significance cannot be assigned to ω_0. Another point to be made is that ω_0 is greater than $|\alpha|$; otherwise, the poles would be real (see Section 4.9).

The magnitude of $\mathbf{H}_{\text{cp}}(j\omega)$ is

$$H_{\text{cp}}(j\omega) = \frac{1}{\sqrt{\left(\omega_0^2-\omega^2\right)^2+4\alpha^2\omega^2}} \tag{10.20a}$$

Figure 10.10
Contribution of a complex pole term to the magnitude graph of the Bode plot of $\mathbf{H}(j\omega)$.

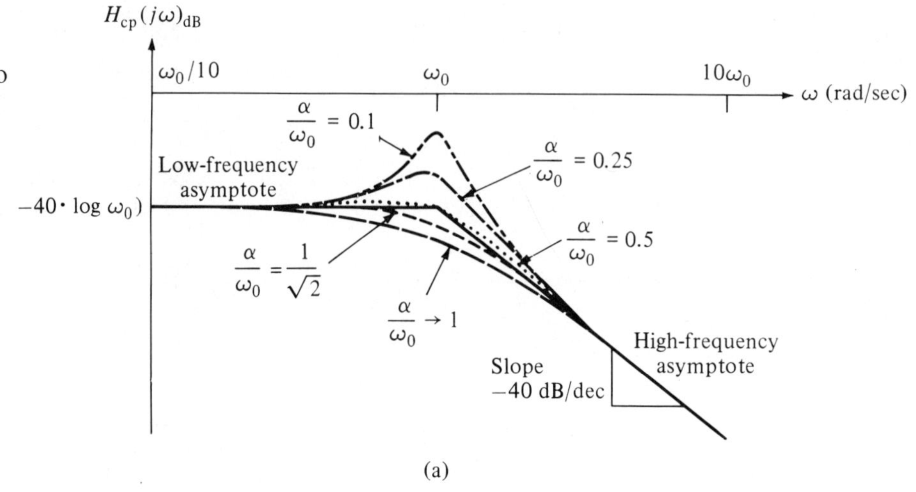

(a)

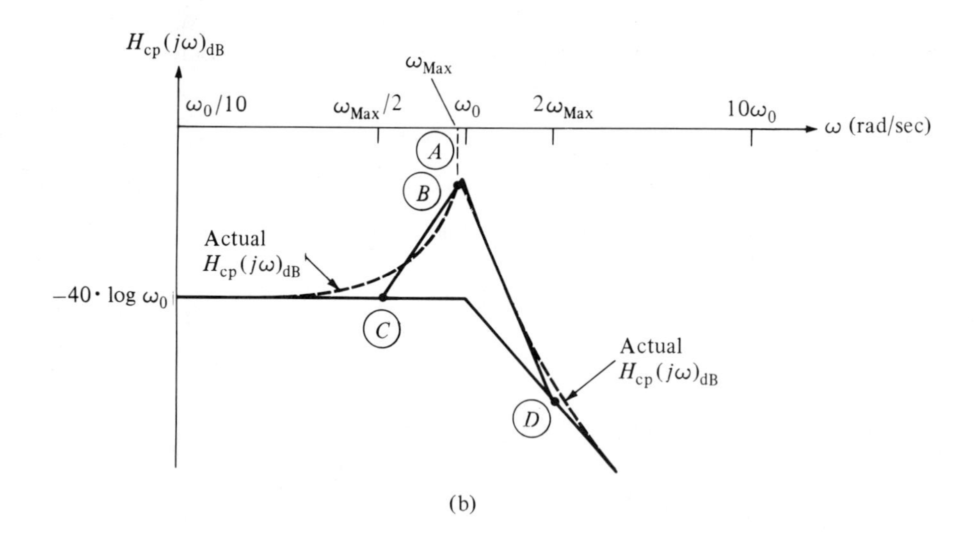

(b)

or

$$H_{cp}(j\omega)_{dB} = -10\log\left[\left(\omega_0^2 - \omega^2\right)^2 + 4\alpha^2\omega^2\right] \tag{10.20b}$$

The *low-frequency* asymptote of $H_{cp}(j\omega)_{dB}$ [Figure 10.10(a)] is a horizontal line,

$$H_{cp}(j\omega)_{dB} \to -10 \cdot \log \omega_0^4 = -40 \cdot \log \omega_0 \qquad \text{as} \quad \omega \to 0 \tag{10.21}$$

and the *high-frequency* asymptote of $H_{cp}(j\omega)_{dB}$ is a straight line with a slope of -40 dB/dec,

$$H_{cp}(j\omega)_{dB} \to -10 \cdot \log \omega^4 = -40 \cdot \log \omega \qquad \text{as} \quad \omega \to \infty \tag{10.22}$$

The low-frequency and high-frequency asymptotes intersect at

$$\omega_{cp} = \omega_0 \tag{10.23}$$

so ω_0 is a corner frequency of the Bode plot.

Difference Analysis for a Complex Pole Term

The difference between the actual graph of $H_{cp}(j\omega)_{dB}$ and the piecewise-linear graph formed by the asymptotes depends on the quantity α/ω_0, which is known as the **damping ratio.**[1] Recall that the only restriction on α and ω_0 is that $\omega_0 > |\alpha|$, so the absolute value of the damping ratio can range from zero to unity (exclusive of 1), or $0 \le |\alpha/\omega_0| < 1$.

If $1/\sqrt{2} \le |\alpha/\omega_0| < 1$, then $H_{cp}(j\omega)_{dB}$ decreases **monotonically** (without peaking) from its maximum value at zero frequency. $H_{cp}(j\omega)_{dB}$ is graphed in Figure 10.10(a) for $|\alpha/\omega_0| = 1/\sqrt{2}$ (or 0.7071) and for $|\alpha/\omega_0| \to 1$. All other cases which are also monotonic lie between these two graphs. The greatest difference between the actual graph of $H_{cp}(j\omega)_{dB}$ and the approximation formed by the asymptotes occurs at the corner frequency ω_0, where

$$\begin{aligned} H_{cp}(j\omega_0)_{dB} &= -10 \cdot \log\left[\left(\omega_0^2 - \omega_0^2\right)^2 + 4\alpha^2\omega_0^2\right] \\ &= -40 \cdot \log \omega_0 - 10 \cdot \log\left[4(\alpha/\omega_0)^2\right] \end{aligned} \tag{10.24}$$

The amount of the difference at the corner frequency is $-10 \cdot \log[4(\alpha/\omega_0)^2]$, or approximately 3 to 6 dB (Table 10.2).

If $0 < |\alpha/\omega_0| < 1/\sqrt{2}$, then $H_{cp}(j\omega)_{dB}$ increases to a peak at a frequency below the corner frequency [Figure 10.10(a)],

$$\begin{aligned} \omega_{MAX} &= \sqrt{\omega_0^2 - 2\alpha^2} \\ &= \omega_0\sqrt{1 - 2(\alpha/\omega_0)^2} \end{aligned} \tag{10.25}$$

Equation (10.25) is found by equating the derivative of Equation (10.20a) to zero (Problem 10-5).

---

[1] In control system theory, the symbol ζ (zeta) is used for damping ratio, and ω_n is used for ω_0, so $\alpha = \zeta\omega_n$.

When Equation (10.25) is substituted into Equation (10.20b),

$$\begin{aligned}
\left(H_{cp}(j\omega)_{dB}\right)_{MAX} \\
&= -10 \cdot \log\left[\left(\omega_0^2 - \left(\omega_0^2 - 2\alpha^2\right)\right)^2 + 4\alpha^2\left(\omega_0^2 - 2\alpha^2\right)\right] \\
&= -40 \cdot \log \omega_0 - 10 \cdot \log\left[4(\alpha/\omega_0)^2\left(1 - (\alpha/\omega_0)^2\right)\right] \qquad (10.26)
\end{aligned}$$

The difference between $(H_{cp}(j\omega)_{dB})_{MAX}$ and the piecewise-linear approximation is $-10 \cdot \log[4(\alpha/\omega_0)^2(1 - (\alpha/\omega_0)^2)]$, which is given in Table 10.2 for several values of the damping ratio. Because of these large differences in the vicinity of the peak, it is often necessary to provide some corrections to the piecewise-linear approximation. One straightforward method, demonstrated in Figure 10.10(b), is outlined below.

1. Calculate and plot ω_{MAX} (A) from Equation (10.25) or Table 10.2.
2. Calculate and plot the peak value (B) from Equation (10.26) or Table 10.2.
3. Connect B with a straight line to a point C on the low-frequency asymptote one **octave**[2] below ω_{MAX}, or $\omega_{MAX}/2$.
4. Connect B with a straight line to a point D one octave above ω_{MAX}, or $2\omega_{MAX}$. Point D will lie on the high-frequency asymptote, because the difference associated with a peak occurring more than an octave below the corner frequency is too small to warrant correction.

This technique may be inadequate when a high degree of accuracy is required and the damping ratio is very small. In such cases, the reader is advised to abandon the piecewise-linear plotting techniques altogether.

Phase Angle Contribution of Complex Conjugate Poles

From Equation (10.19), the phase angle of $\mathbf{H}_{cp}(j\omega)$ is

$$\eta_{cp}(j\omega) = -\tan^{-1}\left[\frac{2\alpha\omega}{\omega_0^2 - \omega^2}\right] \qquad (10.27)$$

The low-frequency asymptote depends on whether α is positive or negative:

$$\eta_{cp}(j\omega) = 0°, \qquad \text{for} \quad \alpha > 0 \qquad (10.28a)$$

$$\eta_{cp}(j\omega) = 180°, \qquad \text{for} \quad \alpha < 0 \qquad (10.28b)$$

---

[2] An octave is a frequency range for which the upper frequency is twice the lower frequency.

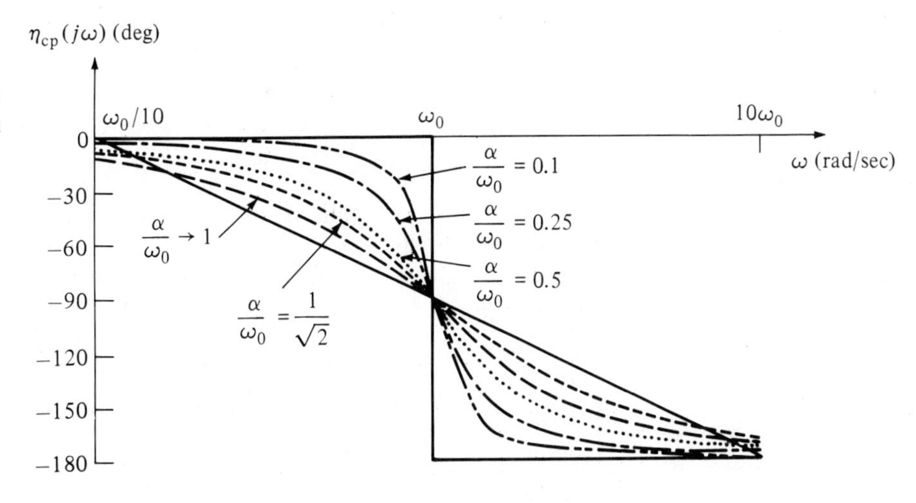

Figure 10.11 Contribution of a complex pole term to the phase angle graph of the Bode plot of $\mathbf{H}(j\omega)$.

The high-frequency asymptote is

$$\eta_{cp}(j\omega) = -180^\circ \quad \text{for} \quad \alpha > 0 \tag{10.29a}$$

$$\eta_{cp}(j\omega) = 0^\circ \quad \text{for} \quad \alpha < 0 \tag{10.29b}$$

The reader should recall that a complex pole is defined [Equation (4.16)] as $\mathbf{p} = -\alpha + j\omega_d$, and a negative value for α results in a positive real part for the pole. Negative values for α do not occur in the transfer functions for passive linear networks, but they can occur in the transfer functions for nonlinear or active networks.

The asymptotes of $\eta_p(j\omega)$ do not intersect (Figure 10.11), so they must be connected by another straight line in order to form an approximation to the phase angle response. An evaluation of the actual phase angle at the corner frequency yields

$$\eta_{cp}(j\omega_0) = \begin{cases} -90^\circ & \text{for} \quad \alpha > 0 \\ 90^\circ & \text{for} \quad \alpha < 0 \end{cases}$$

Thus, the exact value of $\eta_{cp}(j\omega)$ is midway between its asymptotes at the corner frequency. Figure 10.11 shows that the actual phase shift in the vicinity of the corner frequency is highly dependent on the damping ratio. A straight line with a slope of -90°/dec, and passing through -90°, provides a good approximation of the phase angle response for high damping ratios, but a vertical line connecting the asymptotes at the corner frequency is a better for low damping ratios. Greater accuracy in approximating the phase angle may be obtained by plotting the data points given in the "phase angle graph" section of Table 10.2, and connecting those points with straight lines.

Frequency Response Contribution of a Complex Zero

The Bode plot for a pair of complex conjugate zeros (Figure 10.12) is an inversion of the Bode plot for a complex pole term, and the appropriate equations are given in Table 10.1 without further comment.

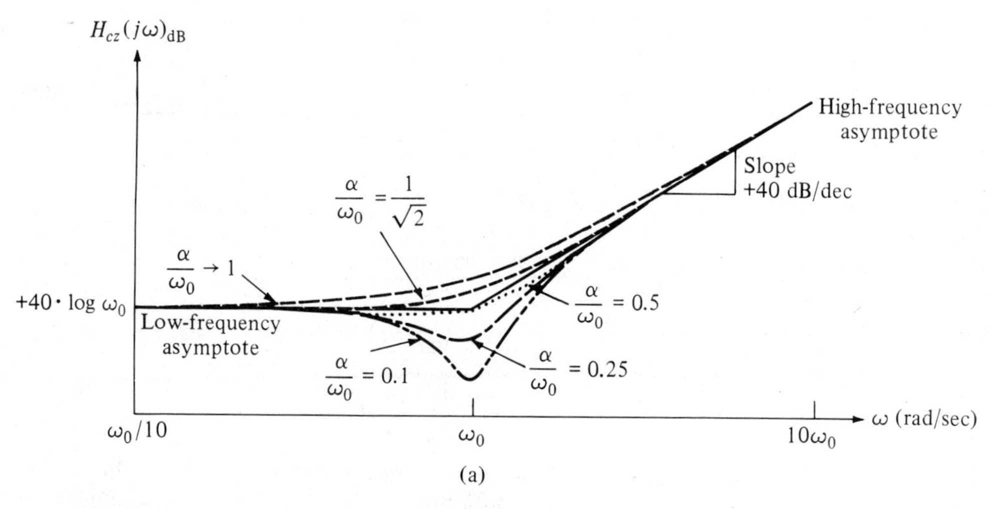

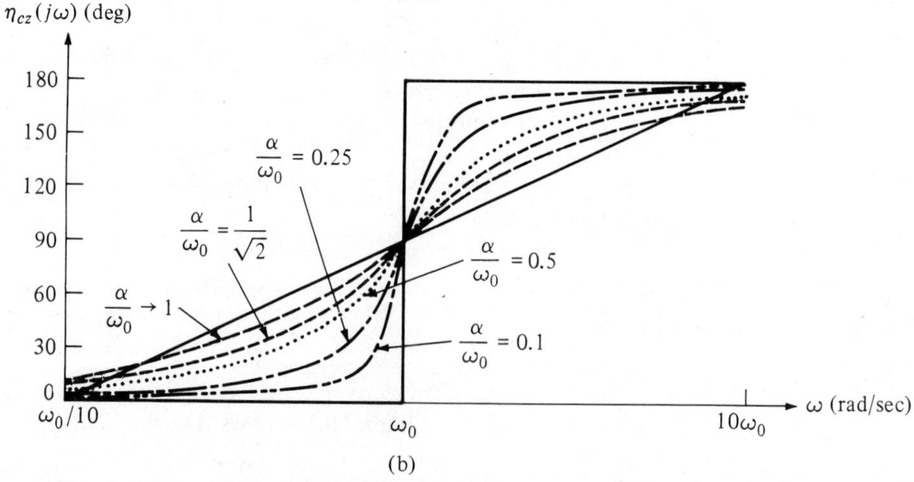

Figure 10.12 Contribution of a complex zero term to the Bode plot of $H(j\omega)$. (a) Magnitude, $\mathbf{H}(j\omega)_{dB}$. (b) Phase angle, $\eta(j\omega)$.

Example 10.6 **Problem:**

SPICE

Construct a piecewise-linear Bode plot of the voltage transfer function for the network in Figure 10.13(a).

Figure 10.13
Example 10.6.
(a) Network.
(b) Magnitude plot.
(c) Phase angle plot.

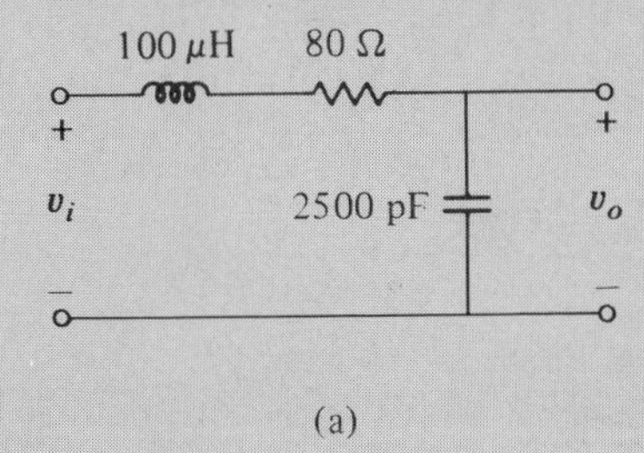

(a)

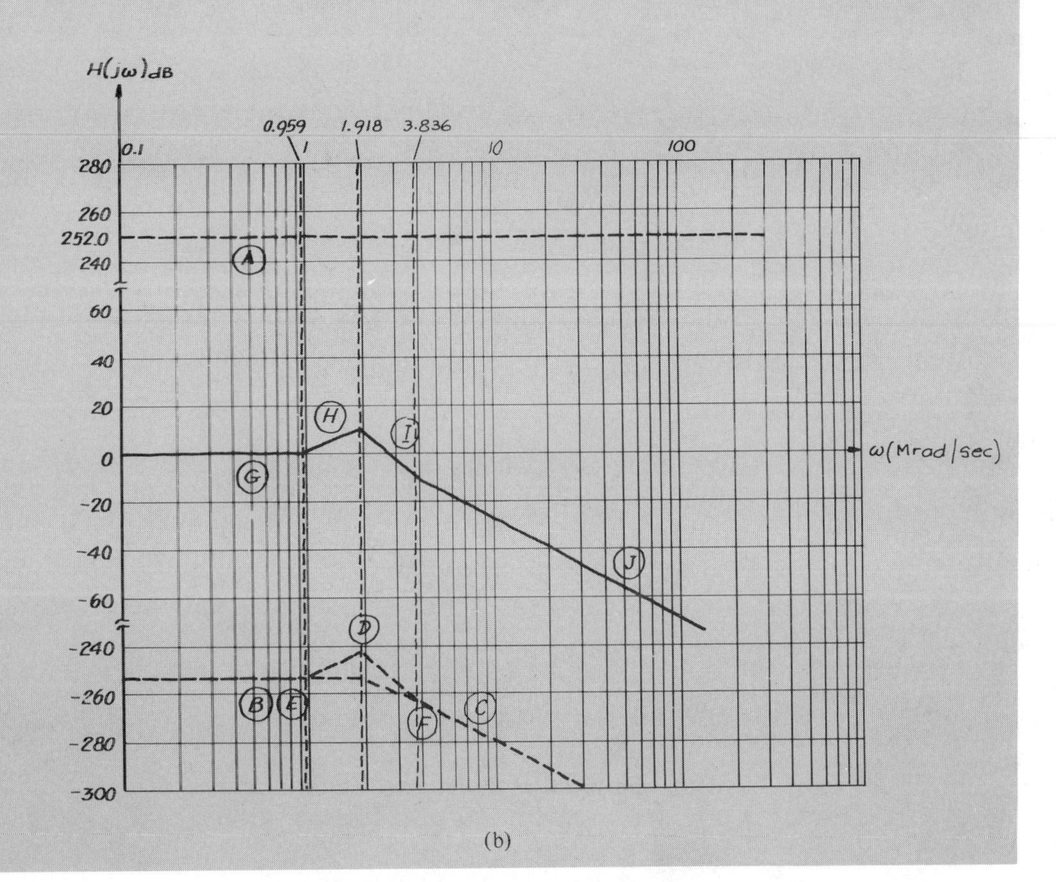

(b)

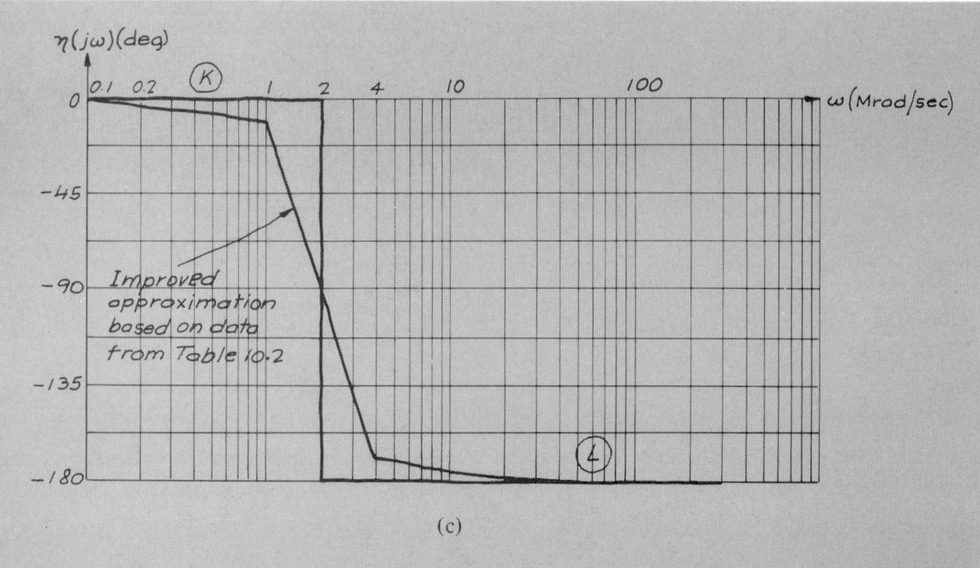

(c)

Figure 10.13 Continued

Solution:

1. Use the Voltage Division Rule to determine $V_o(s)$:

$$V_o(s) = V_i(s) \cdot \frac{400 \times 10^6/s}{100 \times 10^{-6}s + 80 + 400 \times 10^6/s}$$

$$= V_i(s) \cdot \frac{4 \times 10^{12}}{s^2 + 800 \times 10^3 s + 4 \times 10^{12}}$$

2. Express $H(s)$ with the denominator factored:

$$H(s) = \frac{V_o(s)}{V_i(s)} = \frac{4 \times 10^{12}}{s^2 + 800 \times 10^3 s + 4 \times 10^{12}}$$

$$= 4 \times 10^{12} \frac{1}{(s + 400 \times 10^3 - j1.960 \times 10^6)(s + 400 \times 10^3 + j1.960 \times 10^6)}$$

3. Since $H(s)$ contains complex conjugate poles, use the result from above to express $\mathbf{H}(j\omega)$ in the form of Equation (10.19). From step 2,

$$K = 4 \times 10^{12}, \quad \alpha = 400 \times 10^3 \text{ sec}^{-1}, \text{ and } \omega_d = 1.960 \text{ Mrad/sec}$$

therefore,

$$\omega_0 = \sqrt{\alpha^2 + \omega_d^2} = 2 \times 10^6 \text{ rad/sec}$$

and

$$\mathbf{H}(j\omega) = 4 \times 10^{12} \frac{1}{(4 \times 10^{12} - \omega^2) + j800 \times 10^3 \omega}$$

4. Use Equation (10.9) to calculate $H_K(j\omega)_{\text{dB}}$:

$$H_K(j\omega)_{\text{dB}} = 20 \log 4 \times 10^{12} = 252.0 \text{ dB}$$

5. Use Equations (10.21) and (10.22) to calculate the asymptotes of $H_{\text{cp}}(j\omega)_{\text{dB}}$:

$$H_{\text{cp}}(j\omega)_{\text{dB}} \rightarrow \begin{cases} -40 \cdot \log 2 \times 10^6 = -252.0 \text{ dB} & \text{as} \quad \omega \rightarrow 0 \\ -40 \cdot \log \omega & \text{as} \quad \omega \rightarrow \infty \end{cases}$$

6. Use Equations (10.25) and (10.26) to calculate the peak value of $H_{\text{cp}}(j\omega)_{\text{dB}}$:

$$\alpha/\omega_0 = 400 \times 10^3 / 2 \times 10^6 = 0.2$$

so

$$\omega_{\text{MAX}} = 2 \times 10^6 \sqrt{1 - 2(0.2)^2} = 1.918 \times 10^6 \text{ rad/sec}$$

and

$$\begin{aligned} H_{\text{cp}}(j\omega_{\text{MAX}})_{\text{dB}} &= -252.0 - 10 \cdot \log\left[4(0.2)^2 \sqrt{1 - (0.2)^2}\right] \\ &= -244.0 \text{ dB} \end{aligned}$$

7. Construct the piecewise-linear magnitude graph, as shown in Figure 10.13(b):
   Plot $H_K(j\omega)_{\text{dB}}$ (A).
   Plot the asymptotes for $H_{\text{cp}}(j\omega)_{\text{dB}}$ (B and C). The corner frequency is $\omega_0 = 2$ Mrad/sec.

Plot $H_{cp}(j\omega_{MAX})_{dB}$ (D), and connect D to a point E one octave below ω_{MAX} on the low-frequency asymptote, and to a point F one octave above ω_{MAX} on the high-frequency asymptote.

Combine the plotted lines graphically to form a piecewise-linear graph of $H(j\omega)_{dB}$ (G, H, I, and J).

8. Construct the piecewise-linear graph of $\eta(j\omega)$, as shown in Figure 10.13(b):

   Plot the asymptotes of $\eta_p(j\omega)$ (K and L).

   Connect the low-frequency asymptote to the high-frequency asymptote with a vertical line at ω_0.

Example 10.7

SPICE

Problem:

Construct a piecewise-linear Bode plot of the voltage transfer function of the network in Figure 10.14(a).

Solution:

1. Use Equation (8.16b) to determine $H(s)$ with the denominator factored:

$$H(s) = -\frac{Z_f(s)}{Z_i(s)} = -\frac{R_f}{Ls + R_i + 1/(Cs)}$$

$$= -\frac{R_f}{L} \cdot \frac{s}{s^2 + (R_i/L)s + 1/(LC)}$$

$$= -200 \times 10^3 \frac{s}{s^2 + 16 \times 10^3 s + 100 \times 10^6}$$

$$= -200 \times 10^3 \frac{s}{(s + 8 \times 10^3 - j6 \times 10^3)(s + 8 \times 10^3 + j6 \times 10^3)}$$

2. Use the result above to express $\mathbf{H}(j\omega)$ in the form of Equation (10.19). From step 1,

$$K = -200 \times 10^3, \qquad \alpha = 8000/\text{sec}, \quad \text{and} \quad \omega_d = 6\ \text{krad/sec}$$

Therefore,

$$\omega_0 = \sqrt{\alpha^2 + \omega_d^2} = 10\ \text{krad/sec}$$

Figure 10.14
Example 10.7.
(a) Network.
(b) Magnitude plot.
(c) Phase angle plot.

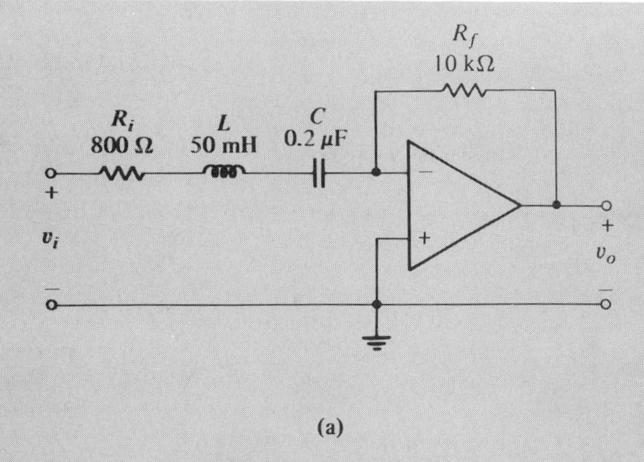

(a)

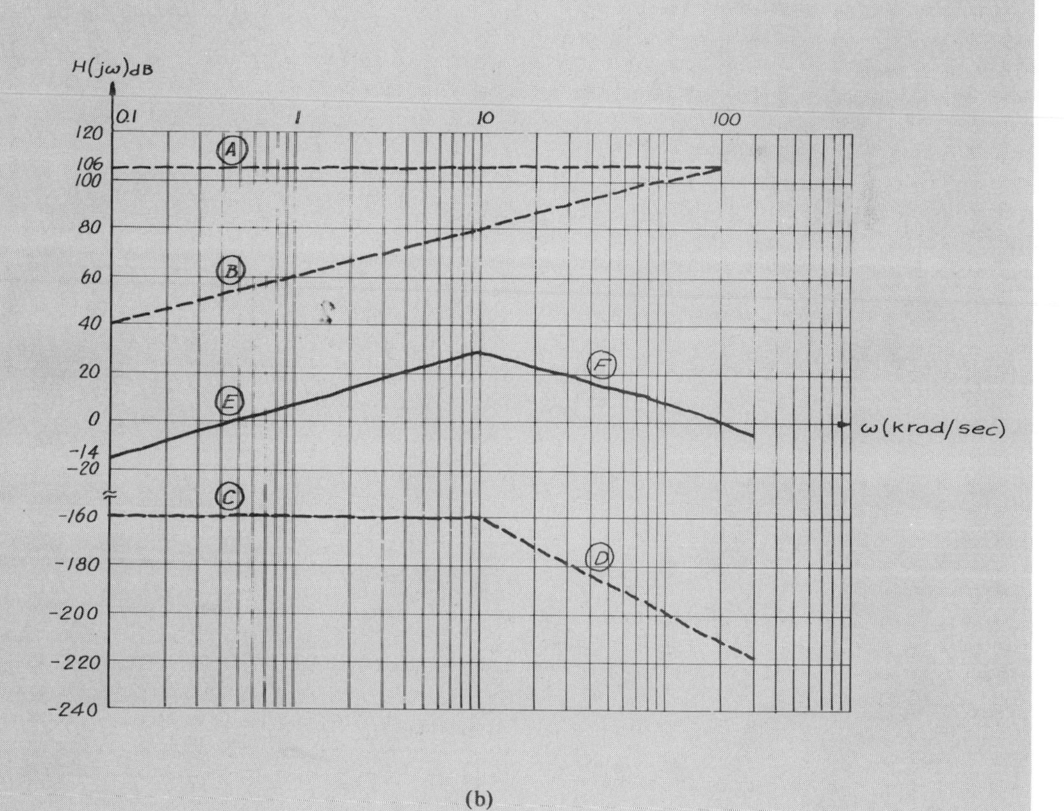

(b)

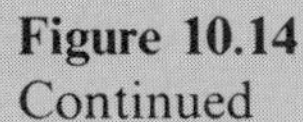

Figure 10.14
Continued

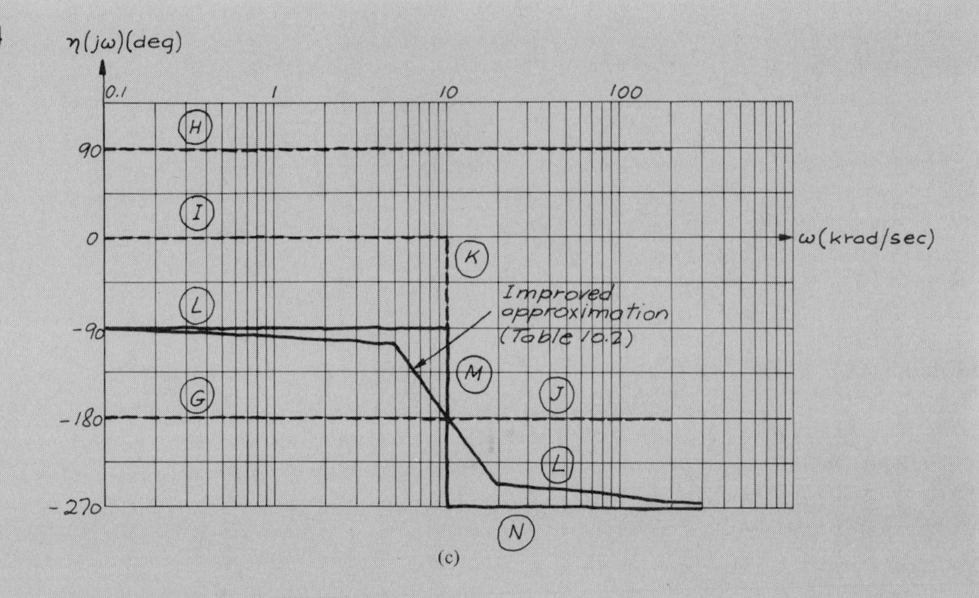

(c)

and

$$\mathbf{H}(j\omega) = -200 \times 10^3 \frac{j\omega}{(100 \times 10^6 - \omega^2) + j16 \times 10^3\omega}$$

3. Use Equation (10.9) to calculate $H_K(j\omega)_{\text{dB}}$:

$$H_K(j\omega)_{\text{dB}} = 20 \cdot \log[200 \times 10^3] = 106.0 \text{ dB}$$

4. Use the formulas from Table 10.1 to calculate the asymptotes of $H_z(j\omega)_{\text{dB}}$. Note that the corner frequency is $\omega_z = 0$, so there is no low-frequency asymptote:

$$H_z(j\omega)_{\text{dB}} \to 20 \cdot \log \omega \qquad \text{as} \quad \omega \to \infty$$

5. Use Equations (10.21) and (10.22) to calculate the asymptotes of $H_{\text{cp}}(j\omega)_{\text{dB}}$:

$$H_{\text{cp}}(j\omega)_{\text{dB}} \to \begin{cases} -40 \cdot \log[10 \times 10^3] = -160.0 \text{ dB} & \text{as} \quad \omega \to 0 \\ -40 \cdot \log \omega & \text{as} \quad \omega \to \infty \end{cases}$$

6. Calculate the damping ratio:

$$\frac{\alpha}{\omega_0} = \frac{8 \times 10^3}{10 \times 10^3} = 0.8$$

Since the damping ratio exceeds $1/\sqrt{2}$ (or 0.7071), the magnitude graph is monotonic, and no correction for peaking is required.

7. Construct the piecewise-linear magnitude graph, as shown in Figure 10.14(b):

   Plot $H_K(j\omega)_{\text{dB}}$ (A).

   Plot the high-frequency asymptote of $H_z(j\omega)$ (B). In this case, $\omega = 100$ rad/sec is the smallest value on the frequency axis, and $20 \cdot \log 100 = 40$, so the high frequency asymptote is plotted to pass through 40 dB at $\omega = 100$ rad/sec.

   Plot the asymptotes for $H_{\text{cp}}(j\omega)_{\text{dB}}$ (C and D). The corner frequency is $\omega_0 = 10$ krad/sec.

   Combine the plotted lines graphically to form a piecewise-linear graph of $H(j\omega)_{\text{dB}}$ (E and F).

8. Construct the piecewise-linear graph of $\eta(j\omega)$, as shown in Figure 10.14(b):

   Plot a horizontal line through $-180°$ (G) to represent the phase angle associated with the negative sign preceding K.

   Plot the asymptote of η_z (H), a constant 90°.

   Plot the asymptotes of $\eta_{\text{cp}}(j\omega)$ (I and J), and connect the low-frequency asymptote to the high-frequency asymptote with a vertical line K at ω_0.

   Combine the plotted lines graphically to form a piecewise-linear graph of $\eta(j\omega)$ (L, M, and N).

Exercise 10.2

Problem:

Construct a piecewise-linear Bode plot of $\mathbf{H}(j\omega)$ for the transfer function

$$H(s) = \frac{10s + 30}{s(s^2 + 50s + 900)}$$

Answer:

See Figure 10.15.

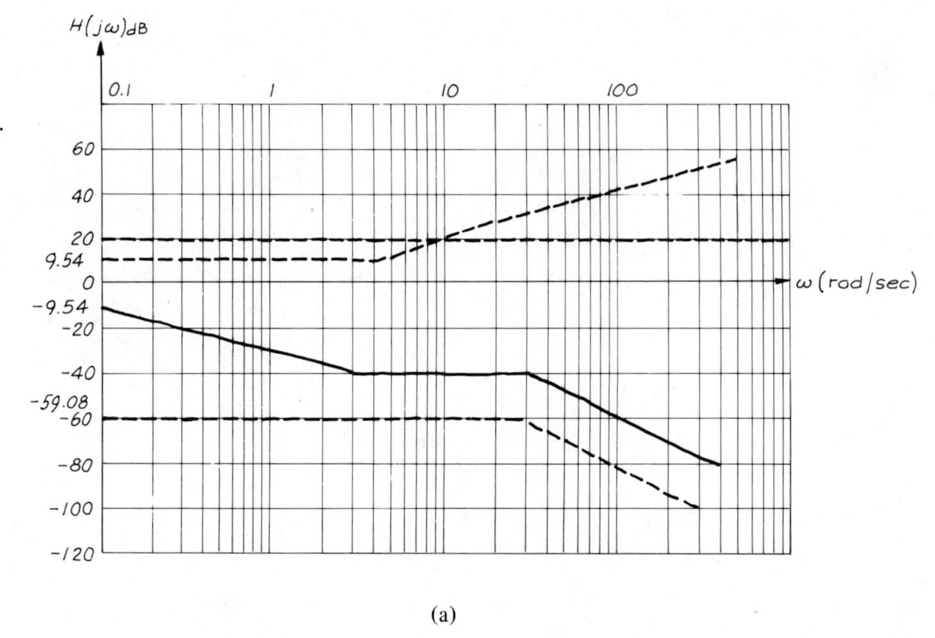

(a)

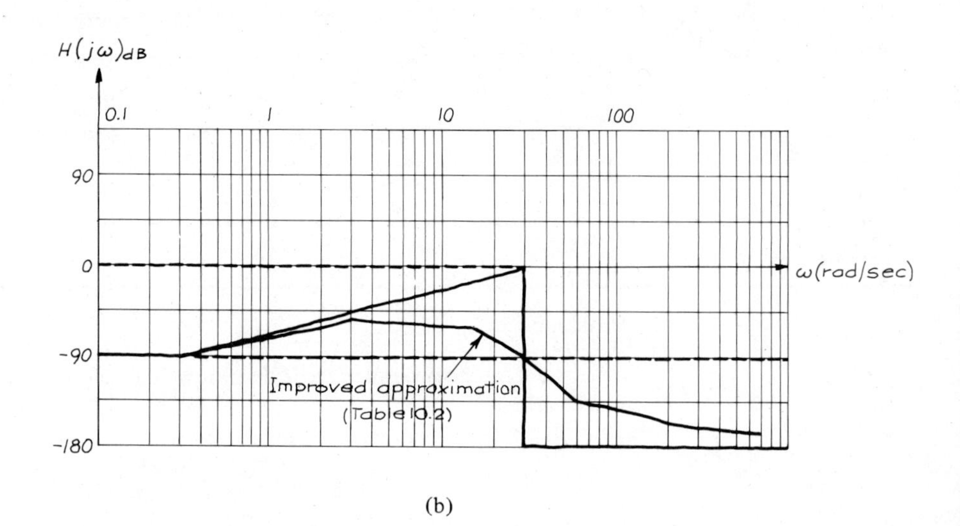

(b)

Figure 10.15
Exercise 10.2.
(a) Magnitude plot.
(b) Phase angle plot.

Table 10.1 Formulas for Piecewise-Linear Bode Plot

Constant term	$\mathbf{H}_K(j\omega) = K$
Magnitude:	$\lvert\mathbf{H}_K(j\omega)\rvert = K$
Magnitude in dB:	$H_K(j\omega)_{\text{dB}} = 20 \cdot \log\lvert K\rvert$
Phase angle:	$\eta_K(j\omega) = \begin{cases} 0^\circ & \text{if } K > 0 \\ -180^\circ & \text{if } K < 0 \end{cases}$
Real pole term	$\mathbf{H}_p(j\omega) = 1/(-p + j\omega)$
Magnitude:	$\lvert\mathbf{H}_p(j\omega)\rvert = 1/\sqrt{p^2 + \omega^2}$
Magnitude in dB:	$H_p(j\omega)_{\text{dB}} = -10 \cdot \log[p^2 + \omega^2]$
Low-frequency asymptote:	$H_p(j\omega)_{\text{dB}} \to -20 \cdot \log\lvert p\rvert$
High-frequency asymptote:	$H_p(j\omega)_{\text{dB}} \to -20 \cdot \log \omega$
Corner frequency:	$\omega_p = \lvert p\rvert$
Maximum magnitude:	$(H_p(j\omega)_{\text{MAX}})_{\text{dB}} = -20 \cdot \log\lvert p\rvert$ at $\omega = 0$
Correction:	-3.010 dB at $\omega = \omega_p$
Phase angle:	$\eta_p(j\omega) = -\tan^{-1}[\omega/(-p)]$
Low-frequency asymptote:	$\eta_p(j\omega) \to \begin{cases} 0^\circ & \text{for } p < 0 \\ -180^\circ & \text{for } p > 0 \end{cases}$
High-frequency asymptote:	$\eta_p(j\omega) \to -90^\circ$
Correction:	Connect $\omega_p/10$ on low-frequency asymptote with $10\omega_p$ on high-frequency asymptote
Real zero term	$\mathbf{H}_z(j\omega) = -z + j\omega$
Magnitude:	$\lvert\mathbf{H}_z(j\omega)\rvert = \sqrt{z^2 + \omega^2}$
Magnitude in dB:	$H_z(j\omega)_{\text{dB}} = 10 \cdot \log[z^2 + \omega^2]$
Low-frequency asymptote:	$H_z(j\omega)_{\text{dB}} \to 20 \cdot \log\lvert z\rvert$
High-frequency asymptote:	$H_z(j\omega)_{\text{dB}} \to 20 \cdot \log \omega$
Corner frequency:	$\omega_z = \lvert z\rvert$
Minimum magnitude:	$(H_z(j\omega)_{\text{MIN}})_{\text{dB}} = 20 \cdot \log\lvert z\rvert$ at $\omega = 0$
Correction:	3.010 dB at $\omega = \omega_z$
Phase angle	$\eta_z(j\omega) = \tan^{-1}[\omega/(-z)]$
Low-frequency asymptote:	$\eta_z(j\omega) \to \begin{cases} 0^\circ & \text{for } z < 0 \\ 180^\circ & \text{for } z > 0 \end{cases}$
High-frequency asymptote:	$\eta_z(j\omega) \to 90^\circ$
Correction:	Connect $\omega_z/10$ on low-frequency asymptote with $10\omega_z$ on high-frequency asymptote

continued

Table 10.1 Continued

Complex pole term	$\mathbf{H}_{cp}(j\omega) = 1/[(-\alpha + j\omega_d + j\omega)(-\alpha - j\omega_d + j\omega)]$ $= 1/[(\omega_0^2 - \omega^2) + j2\alpha\omega]$, where $\omega_0^2 = \alpha^2 + \omega_d^2$
Magnitude:	$\lvert H_{cp}(j\omega)\rvert = 1/\sqrt{(\omega_0^2 - \omega^2)^2 + 4\alpha^2\omega^2}$
Magnitude in dB:	$H_{cp}(j\omega)_{dB} = -10 \cdot \log[(\omega_0^2 - \omega^2)^2 + 4\alpha^2\omega^2]$
Low-frequency asymptote:	$H_{cp}(j\omega)_{dB} \to -40 \cdot \log \omega_0$
High-frequency asymptote:	$H_{cp}(j\omega)_{dB} \to -40 \cdot \log \omega$
Corner frequency:	$\omega_{cp} = \omega_0$
Maximum magnitude:	$(H_{cp}(j\omega)_{MAX})_{dB} = -40 \cdot \log \omega_0$ at $\omega = 0$ for $\alpha/\omega_0 < 1/\sqrt{2}$ $(H_{cp}(j\omega)_{MAX})_{dB} = -40 \cdot \log \omega_0$ $-10 \cdot \log[4(\alpha/\omega_0)^2(1 - (\alpha/\omega_0)^2)]$ at $\omega = \omega_0\sqrt{1 - 2(\alpha/\omega_0)^2}$ for $\alpha/\omega_0 > 1/\sqrt{2}$
Correction:	See Table 10.2
Phase angle	$\eta_{cp}(j\omega) = -\tan^{-1}[2\alpha\omega/(\omega_0^2 - \omega^2)]$
Low-frequency asymptote:	$\eta_{cp}(j\omega) \to \begin{cases} 0° & \text{for } \alpha > 0 \\ 180° & \text{for } \alpha < 0 \end{cases}$
High-frequency asymptote:	$\eta_{cp}(j\omega) \to \begin{cases} -180^* & \text{for } \alpha > 0 \\ 0° & \text{for } \alpha < 0 \end{cases}$
Correction:	See comments in text
Complex zero term	$\mathbf{H}_{cz}(j\omega) = (-\alpha + j\omega_d + j\omega)(-\alpha - j\omega_d + j\omega)$ $= (\omega_0^2 - \omega)^2 + j2\alpha\omega$, where $\omega_0^2 = \alpha^2 + \omega_d^2$
Magnitude:	$\lvert \mathbf{H}_{cz}(j\omega)\rvert = \sqrt{(\omega_0^2 - \omega^2)^2 + 4\alpha^2\omega^2}$
Magnitude in dB:	$H_{cz}(j\omega)_{dB} = 10 \cdot \log[(\omega_0^2 - \omega^2)^2 + 4\alpha^2\omega^2]$
Low-frequency asymptote:	$H_{cz}(j\omega)_{dB} \to 40 \cdot \log \omega_0$
High-frequency asymptote:	$H_{cz}(j\omega)_{dB} \to 40 \cdot \log \omega$
Corner frequency:	$\omega_{cz} = \omega_0$
Minimum magnitude:	$(H_{cz}(j\omega)_{MIN})_{dB} = 40 \cdot \log \omega_0$ at $\omega = 0$ for $\alpha/\omega_0 < 1/\sqrt{2}$ $(H_{cz}(j\omega)_{MIN})_{dB} = 40 \cdot \log[\omega_0]$ $+10 \cdot \log[4(\alpha/\omega_0)^2(1 - (\alpha/\omega_0)^2)]$ at $\omega = \omega_0\sqrt{1 - 2(\alpha/\omega_0)^2}$ for $\alpha/\omega_0 > 1/\sqrt{2}$
Correction:	See Table 10.2
Phase angle:	$\eta_{cz}(j\omega) = \tan^{-1}[2\alpha\omega/(\omega_0^2 - \omega^2)]$
Low-frequency asymptote:	$\eta_{cz}(j\omega) \to \begin{cases} 0° & \text{for } \alpha > 0 \\ -180° & \text{for } \alpha < 0 \end{cases}$
High-frequency asymptote:	$\eta_{cz}(j\omega) \to \begin{cases} 180° & \text{for } \alpha > 0 \\ 0° & \text{for } \alpha < 0 \end{cases}$
Correction:	See comments in text

Table 10.2
Corrections to Piecewise-Linear Bode Plot for Complex Pole or Complex Zero Terms

Magnitude graph

Corrections in dB relative to piecewise-linear graph for the complex pole term; reverse sign for complex zero term.

$\lvert\alpha / \omega_0\rvert$	$\omega_{MAX/MIN}$	$\omega_0 / 2$	$\omega_{MAX/MIN}$	ω_0	$2\omega_0$
$\to 1$	0	−1.938	NA	−6.021	−1.938
0.9	0	−1.375	NA	−5.105	−1.375
0.8	0	−0.801	NA	−4.082	−0.801
$1/\sqrt{2}$	0	−0.263	NA	−3.010	−0.263
0.7	$0.1414\omega_0$	−0.222	+0.002	−2.923	−0.222
0.6	$0.5292\omega_0$	+0.350	+0.355	−1.584	+0.350
0.5	$0.7071\omega_0$	+0.902	+1.249	0	+0.902
0.4	$0.8246\omega_0$	+1.412	+2.695	+1.938	+1.412
0.3	$0.9055\omega_0$	+1.854	+4.847	+4.437	+1.854
0.2	$0.9592\omega_0$	+2.200	+8.136	+7.959	+2.200
0.1	$0.9899\omega_0$	+2.422	+14.023	+13.979	+2.422
0.05	$0.9975\omega_0$	+2.480	+20.011	+20.000	+2.480
0.01	$0.9999\omega_0$	+2.498	+33.980	+33.979	+2.498
0.005	$1.0000\omega_0$	+2.499	+40.000	+40.000	+2.499
0.001	$1.0000\omega_0$	+2.499	+53.979	+53.979	+2.499

Phase angle graph

Actual value in degrees for complex pole term ($\alpha > 0$); reverse sign for complex zero term.

$\lvert\alpha / \omega_0\rvert$	$\omega_0 / 10$	$\omega_0 / 2$	ω_0	$2\omega_0$	$10\omega_0$
$\to 1$	−11.42	−53.13	−90	−126.87	−168.58
0.9	−10.30	−50.19	−90	−129.81	−169.70
0.8	−9.18	−46.84	−90	−133.15	−170.82
$1/\sqrt{2}$	−8.13	−43.31	−90	−136.69	−171.87
0.7	−8.05	−43.03	−90	−136.97	−171.95
0.6	−6.91	−38.66	−90	−141.34	−173.09
0.5	−5.77	−33.69	−90	−146.31	−174.23
0.4	−4.62	−28.07	−90	−151.93	−175.38
0.3	−3.47	−21.80	−90	−158.20	−176.53
0.2	−2.31	−14.93	−90	−165.07	−177.69
0.1	−1.16	−7.59	−90	−172.41	−178.84
0.05	−0.58	−3.81	−90	−176.19	−179.42
0.01	−0.11	−0.76	−90	−179.24	−179.88
0.005	−0.06	−0.38	−90	−179.62	−179.94
0.001	−0.01	−0.08	−90	−179.92	−179.99

10.4 Passive-Filter Applications

Perhaps the widest application of the frequency-domain transfer function is in **modern filter theory**.[3] A filter is a multiport network which is capable of altering the input signal to produce a desired output signal. Described in frequency-domain terms, a filter is a frequency-selective network which responds differently to signals of different frequency. In this sense, every

[3] The use of computers in filter design has caused modern filter theory to eclipse classical filter theory, which makes use of tabulated parameters, as the method of choice in practical filter design.

multiport network is a filter, because even simple resistor networks respond differently to high-frequency signals than to low-frequency signals.

Filter networks are so widespread and so varied, and so much has been written about filter analysis and design, that we cannot hope to do anything more than introduce the application of the frequency-domain transfer function to the simplest of filter networks. We will therefore limit ourselves to **low-order passive** filters.

The order of a filter refers to the number of poles in the denominator of the transfer function. The frequency response of a high-order filter changes more sharply from pass band to stop band than that of a low-order filter, but the filter network is generally more complex. A passive filter incorporates only energy sinks (resistance, inductance, and capacitance), so the filter attenuates input signals at all frequencies, but some more than others. An active filter incorporates active electronic devices capable of selectively supplying energy to the network, thereby amplifying some signals while allowing others to be attenuated. Active devices, particularly OPAMP networks, can also be used in filters to simulate the impedance, especially inductive reactance, required for particular frequency responses.

Filter network design, in modern filter theory, begins with the design of a low-pass **prototype** filter with a cutoff frequency of 1 rad/sec and terminated in a 1-Ω resistive load. Filter networks with any other frequency response or load resistance are realized by a combination of **frequency mapping, frequency scaling**, and **impedance scaling**.

Low-Pass Prototype Filters

A network for which the magnitude of the transfer function is

$$|H(j\omega)|_{\text{LP}} = \frac{1}{\sqrt{1 + \omega^{2n}}}, \qquad n = 1, 2, 3, \ldots \tag{10.30}$$

is called a prototype **Butterworth filter**. The prototype filter has a low-pass characteristic (Figure 10.16) with a cutoff frequency of 1 rad/sec. The frequency response of the Butterworth design is said to be **maximally flat** because the magnitude of $\mathbf{H}(j\omega)$ is monotonic, varying gradually within the pass band, then much more steeply within the transition, or **rolloff**, from pass band to stop band. Rolloff becomes steeper as the number of poles, n, increases.

The principal advantage of the Butterworth filter is the flatness of the amplitude response within the pass band. The principal disadvantage is that the rolloff is not as steep as that of some other filter designs of the same complexity (number of poles).

Equation (10.30) arises from the transfer function,

$$H(s)_{\text{LP}} = \frac{1}{D(s)} \tag{10.31}$$

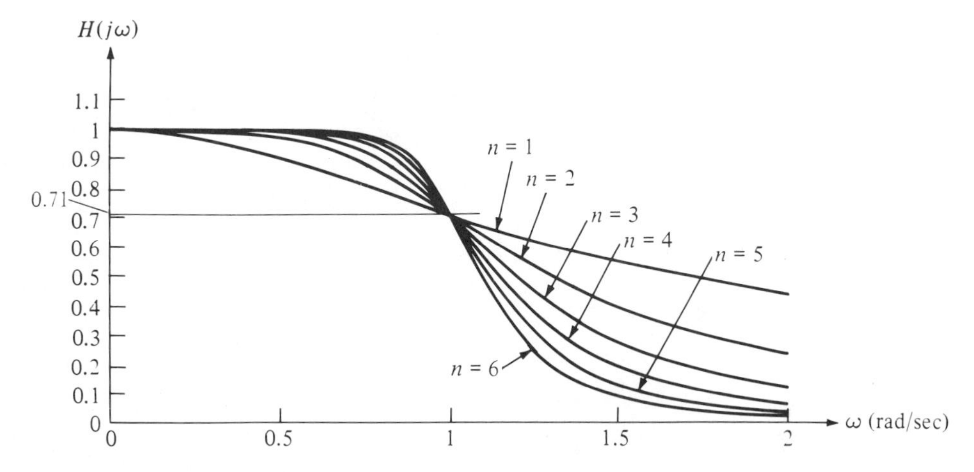

Figure 10.16 Amplitude response of a prototype Butterworth filter.

Table 10.3 Coefficients of the Butterworth Polynomial

n	a_5	a_4	a_3	a_2	a_1
2					1.414
3				2.000	2.000
4			2.613	3.414	2.613
5		3.236	5.236	5.236	3.236
6	3.864	7.464	9.142	7.464	3.864

where $D(s)$ is the **Butterworth polynomial** of degree n,

$$D(s) = s^n + a_{n-1}s^{n-1} + \cdots + a_1 s + 1 \qquad (10.32)$$

The **Butterworth coefficients** a_i, which are given in Table 10.3 for polynomials of low degree, are widely available in design handbooks and in textbooks devoted to filter design. $D(s)$, and from it $H(s)$, will be used later in designing the filter network.

Example 10.8

Problem:

Determine $H(s)$ and $|\mathbf{H}(j\omega)|$ for a third-order Butterworth low-pass prototype filter.

Solution:

1. Determine $D(s)$ from Table 10.3:

$$D(s) = s^3 + 2s^2 + 2s + 1$$

so from Equation (10.31),

$$H(s)_{\mathrm{LP}} = \frac{1}{s^3 + 2s^2 + 2s + 1}$$

2. Use $H(s)_{\mathrm{LP}}$ to obtain $\mathbf{H}_{\mathrm{LP}}(j\omega)$ and its magnitude, $|\mathbf{H}_{\mathrm{LP}}(j\omega)|$:

$$\mathbf{H}_{\mathrm{LP}}(j\omega) = \frac{1}{(j\omega)^3 + 2(j\omega)^2 + 2(j\omega) + 1}$$

$$= \frac{1}{(1 - 2\omega^2) + j\omega(2 - \omega^2)}$$

$$|\mathbf{H}_{\mathrm{LP}}(j\omega)| = \frac{1}{\sqrt{(1 - 2\omega^2)^2 - \omega^2(2 - \omega^2)^2}} = \frac{1}{\sqrt{1 + \omega^6}}$$

The Butterworth prototype is an example of an **all-pole** filter. The **Chebyshev filter**, another all-pole design, has a frequency response characteristic with a steeper rolloff than that of the Butterworth filter, but **ripples** are present in the pass band. The Chebyshev filter is sometimes called an **equiripple** filter because the ripples have the same peak-to-peak amplitude throughout the pass band. Information on design and applications of the Chebyshev filter can be found in any text on modern filter theory.

Filter Realization

Passive low-pass filters are usually realized by an L–C ladder network, which takes on one of the forms shown in Figure 10.17. The configuration of Figure 10.17(a), with an odd number of elements, is used to realize odd-order filters, and the configuration of Figure 10.17(b), with an even number of elements, is used to realize even-order filters. In either case, the arrangement of elements provides low impedance between input and output through the series inductors at low frequencies, and high impedance at high frequencies. The shunt

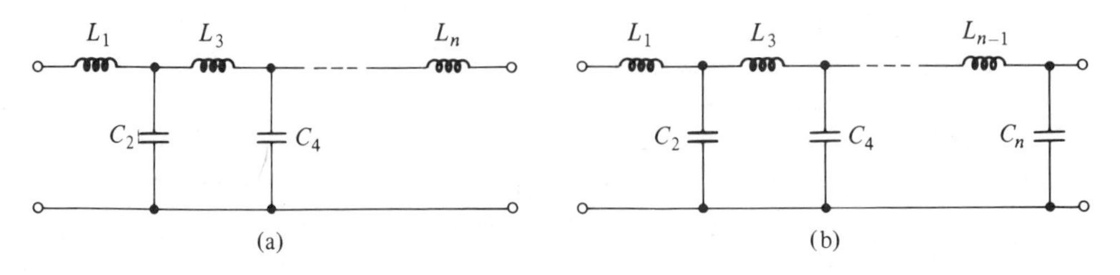

Figure 10.17 L–C ladder networks for passive low-pass filters.

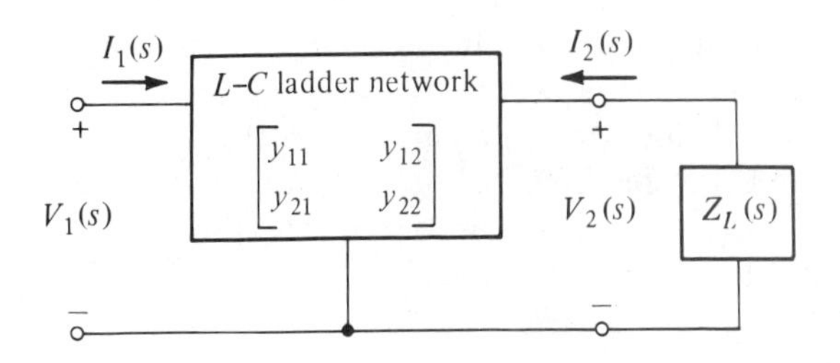

Figure 10.18 Terminated ladder network.

capacitors reinforce this characteristic by providing low conductance to the common terminals at low frequencies and high conductance at high frequencies. For initial design considerations, the capacitors and inductors are assumed to be **lossless**—that is, the capacitors have no dielectric conductance and the inductors have no winding resistance. This discussion of filter realization will be limited to these lossless L–C ladder networks.

When terminated by $Z_L(s)$, the entire circuit consisting of an L–C ladder network filter and a load (Figure 10.18) must have the transfer function, $V_2(s)/V_1(s)$, which produces the specified low-pass characteristic. The L–C ladder network can be characterized by its network admittance parameters [Equations (8.2a) and (8.2b)],

$$I_1 = y_{11}V_1 + y_{12}V_2 \tag{10.33a}$$

$$I_2 = y_{21}V_1 + y_{22}V_2 \tag{10.33b}$$

When the filter network is terminated by $Z_L(s)$,

$$I_2 = -V_2Z_L$$

so Equation (10.33b) becomes

$$-V_2Z_L = y_{21}V_1 + y_{22}V_2$$

or

$$\frac{V_2}{V_1} = \frac{-y_{21}}{Z_L + y_{22}} \tag{10.34}$$

For a prototype design, $Z_L = 1\ \Omega$, so

$$\frac{V_2}{V_1} = \frac{-y_{21}}{1 + y_{22}} \tag{10.35}$$

The expressions for y_{21} and y_{22} can be obtained by rearranging $H(s)$ for the

Butterworth prototype low-pass filter in the form of Equation (10.35). The problem in filter network design is to select inductors and capacitors so that the L–C ladder network parameters are y_{21} and y_{22}, as determined by the specified transfer function. It is important to note that the foregoing procedures neglected source impedance. The presence of source impedance further complicates the design procedure.

Example 10.9

Problem:

Design the L–C ladder network for a fourth-order Butterworth low-pass prototype filter.

Solution:

1. Use Table 10.3 and Equation (10.31) to determine $H(s)$:

$$H_{LP}(s) = \frac{1}{s^4 + 2.613s^3 + 3.414s^2 + 2.613s + 1}$$

2. Group the terms of $D(s)$ according to odd or even degree:

$$D(s) = (2.613s^3 + 2.613s) + (s^4 + 3.414s^2 + 1)$$

Therefore,

$$H(s)_{LP} = \frac{1}{(2.613s^3 + 2.613s) + (s^4 + 3.414s^2 + 1)}$$

3. Divide the numerator and denominator of $H(s)$ by the group of *odd-degree* terms:

$$H_{LP}(s) = \frac{\dfrac{1}{2.613s^3 + 2.613s}}{1 + \dfrac{s^4 + 3.414s^2 + 1}{2.613s^3 + 2.613s}}$$

By comparing $H(s)_{LP}$ with Equation (10.35),

$$y_{21} = -\frac{1}{2.613s^3 + 2.613s}$$

$$y_{22} = \frac{s^4 + 3.414s^2 + 1}{2.613s^3 + 2.613s}$$

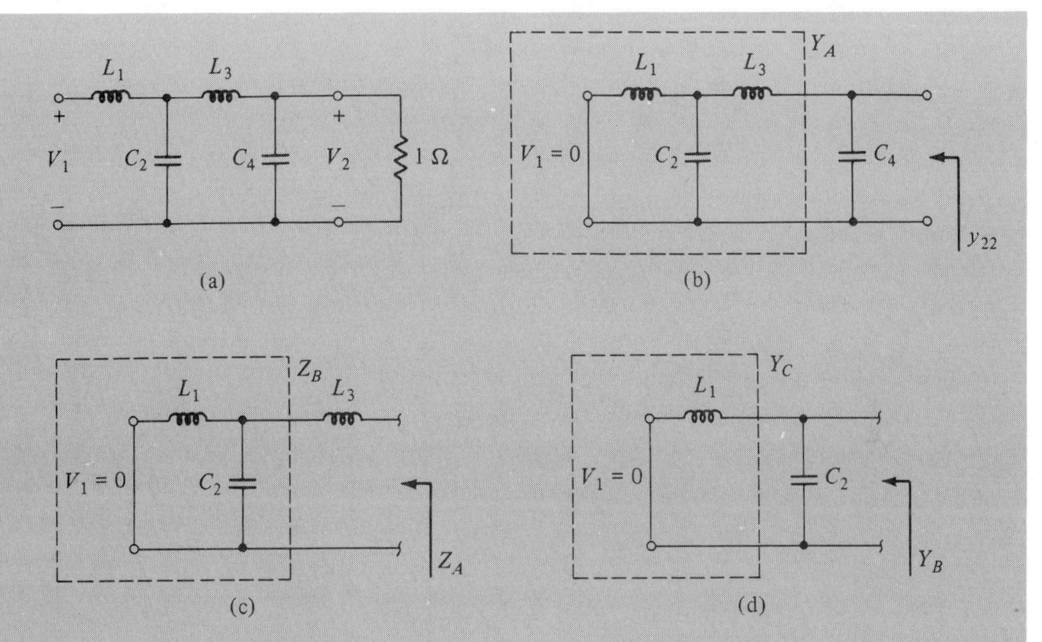

Figure 10.19 Example 10.9.

4. The network will have the configuration of Figure 10.19(a), with two inductors and two capacitors, because the order of the filter is four.
5. Select values of L and C to realize y_{22}. From Equation (8.2b), $y_{22} = I_2/V_2$ when $V_1 = 0$, so y_{22} is the admittance looking into the L–C ladder from the output terminals with the input terminals shorted [Figure 10.19(b)]:

$$y_{22} = C_4 s + Y_A = \frac{s^4 + 3.414s^2 + 1}{2.613s^3 + 2.613s}$$

By long division,

$$y_{22} = 0.383s + \frac{2.414s^2 + 1}{2.613s^3 + 2.613s}$$

so

$$C_4 = 0.383 \text{ F} \quad \text{and} \quad Y_A = \frac{2.414s^2 + 1}{2.613s^3 + 2.613s}$$

6. From Figure 10.19(b),

$$Z_A = \frac{1}{Y_A} = \frac{2.613s^3 + 2.613s}{2.414s^2 + 1} = L_3 s + Z_B$$

By long division,

$$Z_A = 1.082s + \frac{1.531s}{2.414s^2 + 1}$$

so

$$L_3 = 1.082 \text{ H} \quad \text{and} \quad Z_B = \frac{1.531s}{2.414s^2 + 1}$$

7. From Figure 10.19(c),

$$Y_B = \frac{1}{Z_B} = \frac{2.414s^2 + 1}{1.531s} = C_2 s + Y_C$$

By long division,

$$Y_B = 1.577s + \frac{1}{1.531s}$$

so

$$C_2 = 1.577 \text{ F} \quad \text{and} \quad Y_C = 1/1.531s$$

Clearly, Y_C is the admittance of L_1, so

$$L_1 = 1.531 \text{ H}$$

8. The reader should confirm that the required y_{21} in step 3 is obtained from this network (Problem 10-9).

Frequency Mapping

The procedure for designing any filter using modern filter theory is to design a low-pass prototype, then use **frequency mapping** to obtain the desired frequency response characteristic. In frequency mapping, s is replaced by a mapped value everywhere in the low-pass transfer function [Equation (10.31) or (10.30)]. The mathematical basis for frequency mapping is not of interest here, so the results only are presented in Table 10.4.

Table 10.4
Frequency Mapping of Transfer Functions

	s-Domain	Frequency Domain
Low-pass to high-pass	$s \to \frac{1}{s}$	$\omega \to \frac{1}{\omega}$
Low-pass to band-pass	$s \to \frac{1 + s^2}{\mathrm{BW} \cdot s}$	$\omega \to \frac{\lvert 1 - \omega^2 \rvert}{\mathrm{BW} \cdot \omega}$
Low-pass to band-stop	$s \to \frac{\mathrm{BW} \cdot s}{1 + s^2}$	$\omega \to \frac{\mathrm{BW} \cdot \omega}{\lvert 1 - \omega^2 \rvert}$

The cutoff frequency for a frequency-mapped high-pass transfer function, or the center frequency for a frequency-mapped band-pass or band-stop transfer function, is the same as the cutoff frequency for the low-pass prototype, namely 1 rad/sec (Figure 10.20). In effect, the bandwidth (BW) is expressed as a percentage of the center frequency, so a bandwidth specified as 0.2 for a prototype band-pass or band-stop filter will result in a bandwidth of 20 Hz for the corresponding filter with a center frequency of 100 Hz.

If the phase angle response is required, it is necessary to frequency-map Equation (10.31) according to the s-domain column of Table 10.4,

$$H(s)_{\mathrm{LP}} \to H(s)_{\mathrm{HP,BP,BS}}$$

and then obtain the complex frequency-domain transfer function $\mathbf{H}(j\omega)_{\mathrm{HP,BP,BS}}$, and from that $\eta(j\omega)_{\mathrm{HP,BP,BS}}$. This is a tedious, but manageable, process which will not be demonstrated here. If only amplitude response

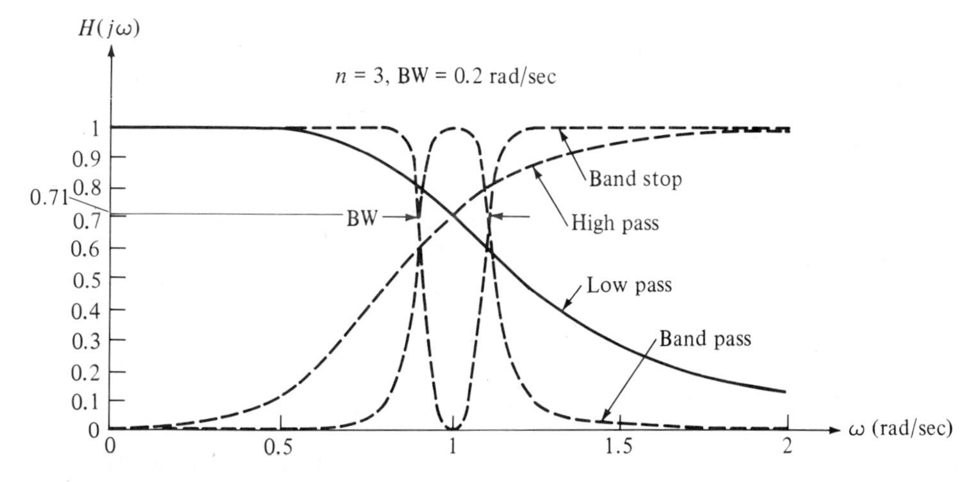

Figure 10.20 Mapped frequency responses.

is required, as is often the case, it is only necessary to frequency-map Equation (10.30):

$$|\mathbf{H}(j\omega)|_{\mathrm{HP}} = \frac{1}{\sqrt{1 + (1/\omega)^{2n}}}$$

$$= \frac{\omega^n}{\sqrt{1 + \omega^{2n}}}, \qquad n = 1, 2, 3, \ldots \tag{10.36a}$$

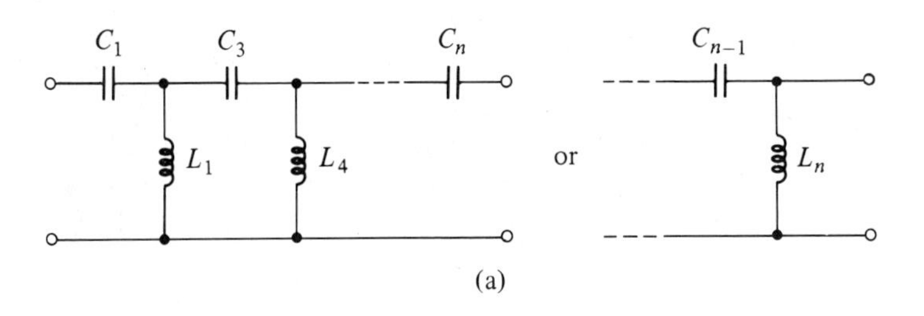

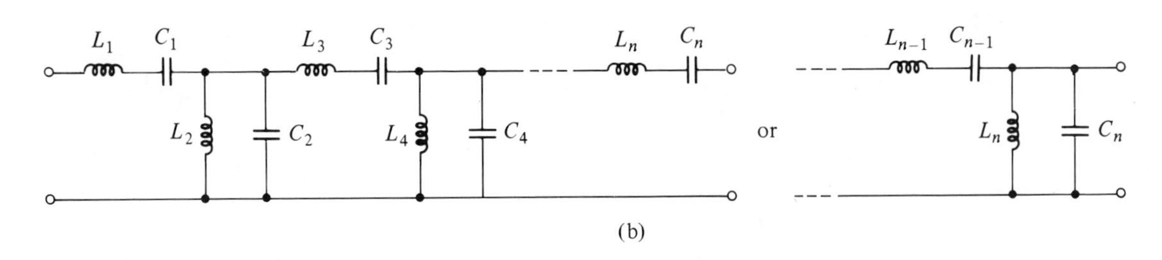

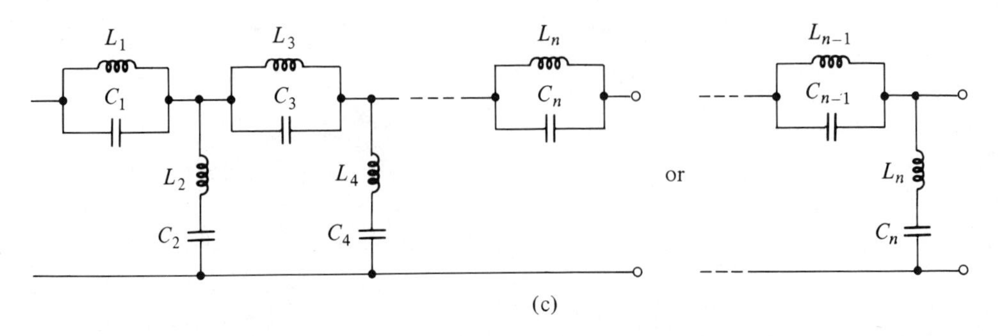

Figure 10.21 Mapped filter networks. (a) High-pass. (b) Band-pass. (c) Band-stop.

$$
|\mathbf{H}(j\omega)|_{\mathrm{BP}} = \frac{1}{\sqrt{1 + \left(|1 - \omega^2|/(\mathrm{BW}\cdot\omega)\right)^{2n}}}
= \frac{(\mathrm{BW}\cdot\omega)^n}{\sqrt{(\mathrm{BW}\cdot\omega)^{2n} + |1-\omega^2|^{2n}}}, \qquad n = 1, 2, 3, \ldots \qquad (10.36\mathrm{b})
$$

$$
|\mathbf{H}(j\omega)|_{\mathrm{BS}} = \frac{1}{\sqrt{1 + \left(\mathrm{BW}\cdot\omega/|1 - \omega^2|\right)^{2n}}}
= \frac{|1-\omega^2|^n}{\sqrt{(\mathrm{BW}\cdot\omega)^{2n} + |1-\omega^2|^{2n}}}, \qquad n = 1, 2, 3, \ldots \qquad (10.36\mathrm{c})
$$

The highest degree term in the denominator of a mapped band-pass or band-stop frequency-domain transfer function [Equation (10.36b) or (10.36c)] is twice that of the low-pass transfer function (Equation 10.30). This means that the s-domain transfer function must have twice as many poles, and the L–C ladder network must have twice as many elements (Figure 10.21).

Table 10.5
Frequency Mapping of Network Elements

	Inductor	Capacitor
Low-pass	L_{LP} (series inductor)	C_{LP} (series capacitor)
High-pass	$C_{\mathrm{HP}} = \frac{1}{L_{\mathrm{LP}}}$ (series capacitor)	$L_{\mathrm{HP}} = \frac{1}{C_{\mathrm{LP}}}$ (series inductor)
Band-pass	$L_{\mathrm{BP}} = \frac{L_{\mathrm{LP}}}{\mathrm{BW}}$ in series with $C_{\mathrm{BP}} = \frac{\mathrm{BW}}{L_{\mathrm{LP}}}$	$L_{\mathrm{BP}} = \frac{\mathrm{BW}}{C_{\mathrm{LP}}}$ in parallel with $C_{\mathrm{BP}} = \frac{C_{\mathrm{LP}}}{\mathrm{BW}}$
Band-stop	$L_{\mathrm{BS}} = \mathrm{BW}\cdot L_{\mathrm{LP}}$ in parallel with $C_{\mathrm{BS}} = \frac{1}{\mathrm{BW}\cdot L_{\mathrm{LP}}}$	$L_{\mathrm{BS}} = \frac{1}{\mathrm{BW}\cdot C_{\mathrm{LP}}}$ in series with $C_{\mathrm{BS}} = \mathrm{BW}\cdot C_{\mathrm{LP}}$

Frequency mapping also applies to the elements of the low-pass prototype network. When the inductors and capacitors of the low-pass prototype are replaced according to the scheme shown in Table 10.5, the resulting network will have the desired frequency response characteristic.

Example 10.10

SPICE

Problem:

Use the prototype low-pass filter from Example 10.9 to design

a. A high-pass prototype filter.

b. A band-pass prototype filter with a bandwidth of 0.2.

c. A band-stop prototype filter with a bandwidth of 0.1.

Graph the amplitude response of each prototype filter.

Solution:

1. Use the low-pass prototype from Figure 10.19 and the element mapping from Table 10.5 to obtain the high-pass prototype (Figure 10.22(a)).

$$(C_1)_{\text{HP}} = 1/L_1 = 0.653 \text{ F}, \qquad (L_2)_{\text{HP}} = 1/C_2 = 0.634 \text{ H}$$

$$(C_3)_{\text{HP}} = 1/L_3 = 0.924 \text{ F}, \qquad (L_4)_{\text{HP}} = 1/C_4 = 2.611 \text{ H}$$

2. Use Equation (10.36a) to obtain $H(j\omega)_{\text{HP}}$:

$$|\mathbf{H}(j\omega)|_{\text{HP}} = \frac{\omega^4}{\sqrt{1+\omega^8}}$$

3. Repeat step 1 for the band-pass prototype [Figure 10.22(b)]:

$$(L_1)_{\text{BP}} = L_1/\text{BW} = 7.655 \text{ H}, \qquad (C_1)_{\text{BP}} = \text{BW}/L_1 = 0.131 \text{ F}$$

$$(L_2)_{\text{BP}} = \text{BW}/C_2 = 0.127 \text{ H}, \qquad (C_2)_{\text{BP}} = C_2/\text{BW} = 7.885 \text{ F}$$

$$(L_3)_{\text{BP}} = L_3/\text{BW} = 5.410 \text{ H}, \qquad (C_3)_{\text{BP}} = \text{BW}/L_3 = 0.185 \text{ F}$$

$$(L_4)_{\text{BP}} = \text{BW}/C_4 = 0.522 \text{ H}, \qquad (C_4)_{\text{BP}} = C_4/\text{BW} = 1.915 \text{ F}$$

4. Use Equation (10.37a) to obtain $H(j\omega)_{\text{BP}}$:

$$|\mathbf{H}(j\omega)|_{\text{BP}} = \frac{(0.2\omega)^4}{\sqrt{(0.2\omega)^8 + (1-\omega^2)^8}}$$

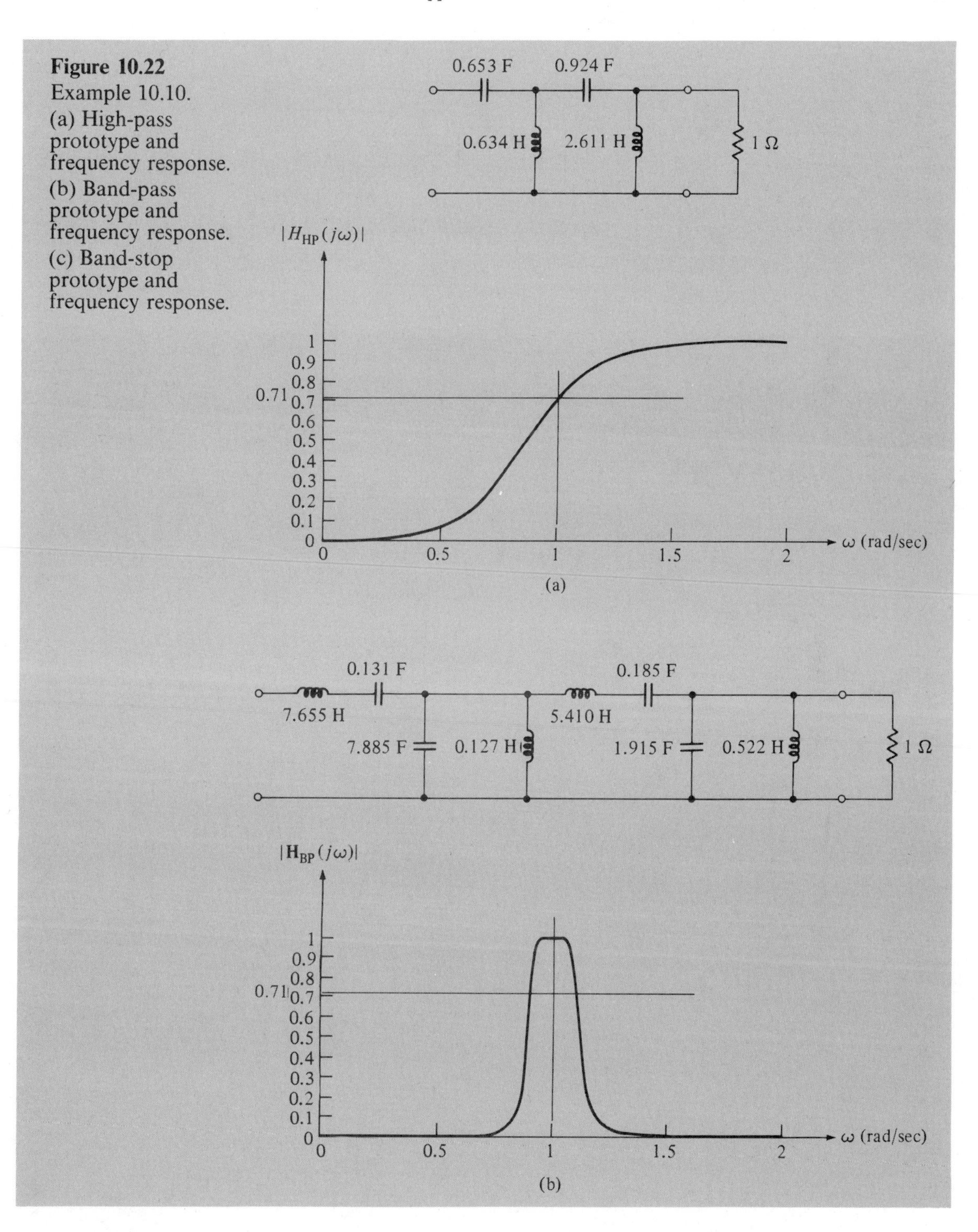

Figure 10.22
Example 10.10.
(a) High-pass prototype and frequency response.
(b) Band-pass prototype and frequency response.
(c) Band-stop prototype and frequency response.

Figure 10.22
Continued

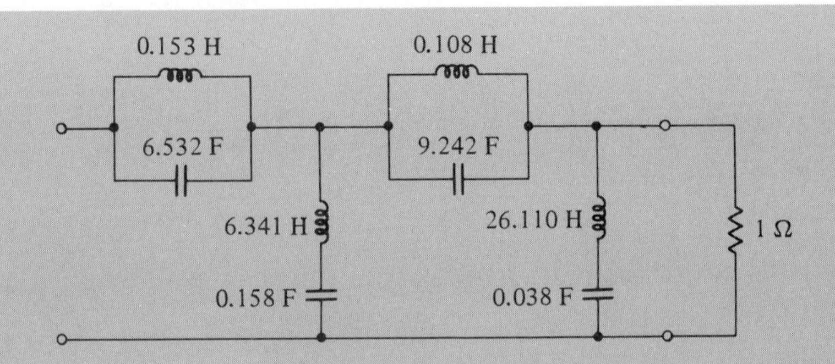

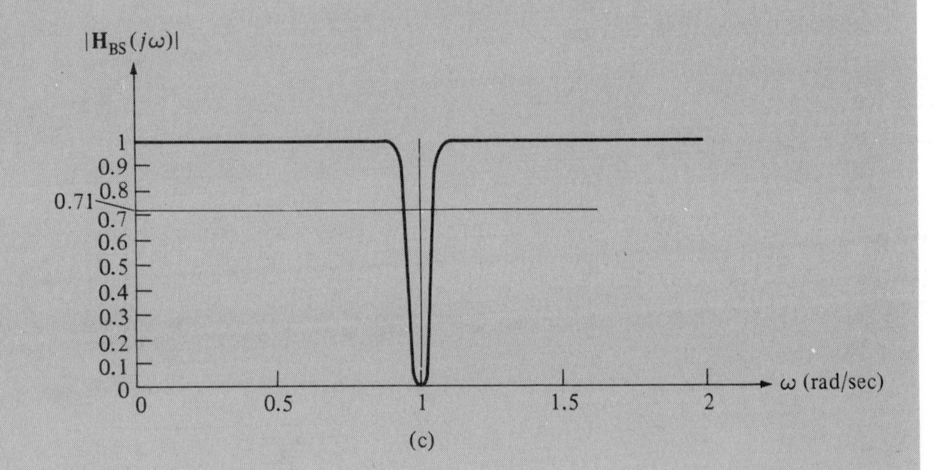

(c)

5. Repeat step 1 for the band-stop prototype [Figure 10.22(c)].

$(C_1)_{BS} = 1/(BW \cdot L_1) = 6.532$ F $\quad (L_1)_{BS} = BW \cdot L_1 = 0.153$ H

$(C_2)_{BS} = BW \cdot C_2 = 0.158$ F $\quad (L_2)_{BS} = 1/(BW \cdot C_2) = 6.341$ H

$(C_3)_{BS} = 1/(BW \cdot L_3) = 9.242$ F $\quad (L_3)_{BS} = BW \cdot L_3 = 0.108$ H

$(C_4)_{BS} = BW \cdot C_4 = 0.038$ F $\quad (L_4)_{BS} = 1/(BW \cdot C_4) = 26.110$ H

6. Use Equation (10.37c) to obtain $H(j\omega)_{BS}$:

$$|\mathbf{H}(j\omega)|_{BS} = \frac{(1-\omega^2)^4}{\sqrt{(0.1\omega)^8 + (1-\omega^2)^8}}$$

Frequency Scaling and Impedance Scaling

Once the prototype filter has been designed, the inductor and capacitor values must be **scaled** for operation at the specified cutoff or center frequency ω_c with the specified load resistance R_L. Scaling simply means that the values of L and C from the prototype are multiplied by a constant, called a **scale factor**. The scale factor actually consists of two scale factors—one due to frequency scaling (k_ω) and one due to impedance scaling (k_z).

The frequency scale factor is the ratio of the desired cutoff or center frequency, ω_c, to the cutoff or center frequency of the prototype, which is always 1 rad/sec, so

$$k_\omega = \omega_c/1 = \omega_c \tag{10.37}$$

The impedance scale factor is the ratio of the actual load resistance, R_L, to the load resistance of the prototype, which is always 1 Ω, so

$$k_z = R_L/1 = R_L \tag{10.38}$$

For an inductor, the scale factor is

$$k_L = k_z/k_\omega = R_L/\omega_c \tag{10.39a}$$

and for a capacitor it is

$$k_C = \frac{1}{k_z k_\omega} = \frac{1}{R_L \omega_c} = \frac{G_L}{\omega_c} \tag{10.39b}$$

The values of all inductors (L') and capacitors (C') in a network scaled to operate at ω_c with a load R_L are calculated by

$$L' = k_L L \tag{10.40a}$$

$$C' = k_C C \tag{10.40b}$$

where L and C are the values obtained for the corresponding elements in the prototype network.

Example 10.11 **Problem:**

Scale the high-pass prototype network from Example 10.10 for a cutoff frequency of 3 kHz when operating with a 5-kΩ resistive load.

Solution:

1. Use Equations (10.40a) and (10.40b) to obtain k_L and k_C:

$$k_L = \frac{5000}{2\pi \times 3000} = 0.2653$$

$$k_C = \frac{1}{5000 \times 2\pi \times 3000} = 10.61 \times 10^{-9}$$

2. Scale the elements of the high-pass network of Figure 10.22(a):

$$C_1 = 0.653k_C = 6.929 \text{ nF}, \qquad L_2 = 0.634k_L = 0.168 \text{ H}$$

$$C_3 = 0.924k_C = 9.804 \text{ nF}, \qquad L_4 = 2.611k_L = 0.693 \text{ H}$$

Exercise 10.3

Problem:

Use modern filter theory to design a second-order band-pass filter with a center frequency of 100 kHz, bandwidth of 5 kHz, and a resistive load of 5 kΩ.

Answer:

See Figure 10.23.

$$|\mathbf{H}(j\omega)|_{\text{BP}} = \frac{(0.05\omega)^2}{\sqrt{(0.05\omega)^4 + (1 - \omega^2)^4}}$$

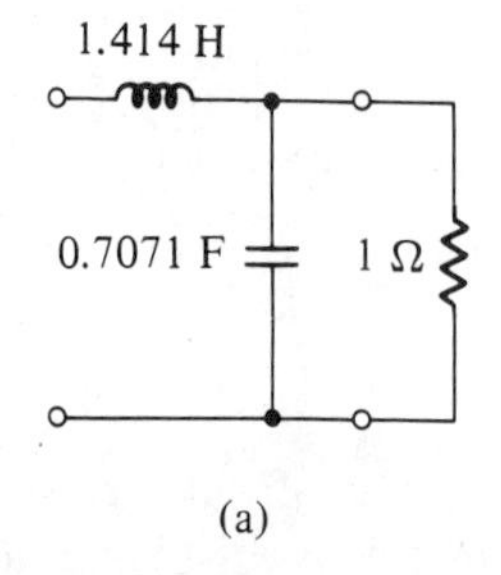

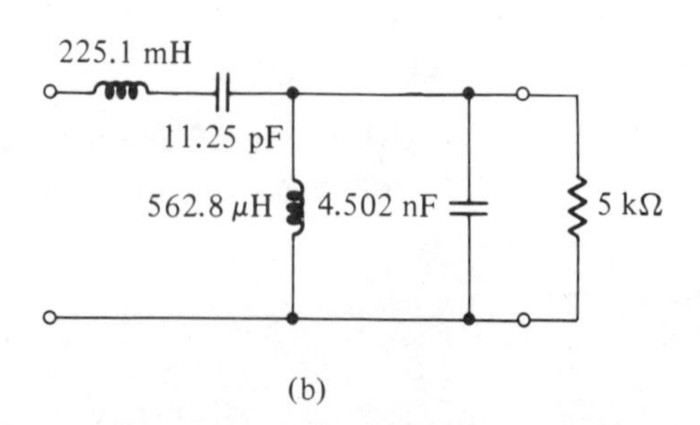

Figure 10.23 Exercise 10.3. (a) Low-pass prototype. (b) Band-pass filter ($f_c = 100$ kHz, BW = 5 kHz).

10.5 Transformer Applications

Application of frequency-domain transfer functions to networks containing transformers is a straightforward extension of the techniques introduced in Chapter 8. All of the equations and relationships presented in Chapter 8 for the linear transformer, and circuits in which it is used, can be extended to the frequency domain by ignoring initial conditions and replacing s with $j\omega$.

We will first consider sinusoidal steady-state operation of the linear transformer with its nonmagnetic core, and then we will examine briefly the **ideal transformer** in which the coils are wound on highly permeable ferromagnetic material.

Frequency-Domain Transfer Functions for a Linear Transformer

The frequency-domain current transfer function, voltage transfer function, and input impedance are obtained by frequency-domain transformation of Equations (8.36), (8.37), and (8.38), respectively:

$$\frac{\mathbf{I}_2(j\omega)}{\mathbf{I}_1(j\omega)} = \frac{-j\omega M}{\mathbf{Z}_L(j\omega) + R_2 + j\omega L_2} \tag{10.41a}$$

$$\frac{\mathbf{V}_2(j\omega)}{\mathbf{V}_1(j\omega)} = \frac{-j\omega M \cdot \mathbf{Z}_L(j\omega)}{[\mathbf{Z}_g(j\omega) + R_1 + j\omega L_1][\mathbf{Z}_L(j\omega) + R_2 + j\omega L_2] + \omega^2 M^2} \tag{10.41b}$$

$$\mathbf{Z}_i(j\omega) = \frac{(R_1 + j\omega L_1)[\mathbf{Z}_L(j\omega) + R_2 + j\omega L_2] + \omega^2 M^2}{\mathbf{Z}_L(j\omega) + R_2 + j\omega L_2} \tag{10.42}$$

In all cases, M must carry its sign according to the dot convention. Assigning incorrect polarity to M will not affect the magnitude of the transfer functions, but will produce a phase reversal (180° error). Impedance is independent of the polarity of the mutual inductance because M is squared in its only occurrence in $\mathbf{Z}_i(s)$.

Example 10.12 **Problem:**

a. Write the frequency-domain current transfer function

$$\mathbf{H}(j\omega) = \frac{\mathbf{I}_2(j\omega)}{\mathbf{I}_1(j\omega)}$$

for the circuit of Exercise 8.8.

b. Construct a piecewise-linear Bode plot of the frequency-domain transfer function.

Solution:

1. $H(s)$ has been calculated in Exercise 8.8:

$$H(s) = -0.05\frac{s^2}{(s + 100 - j100)(s + 100 + j100)}$$

2. Transform $H(s)$ to obtain $\mathbf{H}(j\omega)$:

$$\mathbf{H}(j\omega) = -0.05\frac{(j\omega)^2}{(j\omega + 100 - j100)(j\omega + 100 + j100)}$$

$$\mathbf{H}(j\omega) = 0.05\frac{\omega^2}{(20 \times 10^3 - \omega^2) + j200\omega}$$

$$|\mathbf{H}(j\omega)| = 0.05\frac{\omega^2}{\sqrt{(20 \times 10^3 - \omega^2)^2 + (200\omega)^2}}$$

$$\begin{aligned} H(j\omega)_{\text{dB}} &\simeq 20 \cdot \log 0.05 + 20 \cdot \log \omega^2 \\ &\quad - 20 \cdot \log\sqrt{(20 \times 10^3 - \omega^2)^2 + (200\omega)^2} \\ &= -26.02 \text{ dB} + 40 \cdot \log \omega \\ &\quad - 10 \cdot \log\left[(20 \times 10^3 - \omega^2)^2 + (200\omega)^2\right] \end{aligned}$$

$$\eta(j\omega) = -\tan^{-1}\frac{200\omega}{20 \times 10^3 - \omega^2}$$

3. Construct the piecewise-linear Bode plot. From step 2, the corner frequencies are

$$\omega_z = 0 \quad \text{and} \quad \omega_{\text{cp}} = \sqrt{20 \times 10^3} \simeq 141.4 \text{ rad/sec}$$

The damping ratio for the complex-pole term is

$$|\alpha/\omega_0| = 100/141.4 \simeq 0.7071$$

The Bode plot is shown in Figure 10.24.

Figure 10.24
Example 10.12.
(a) Magnitude plot.
(b) Phase angle plot.

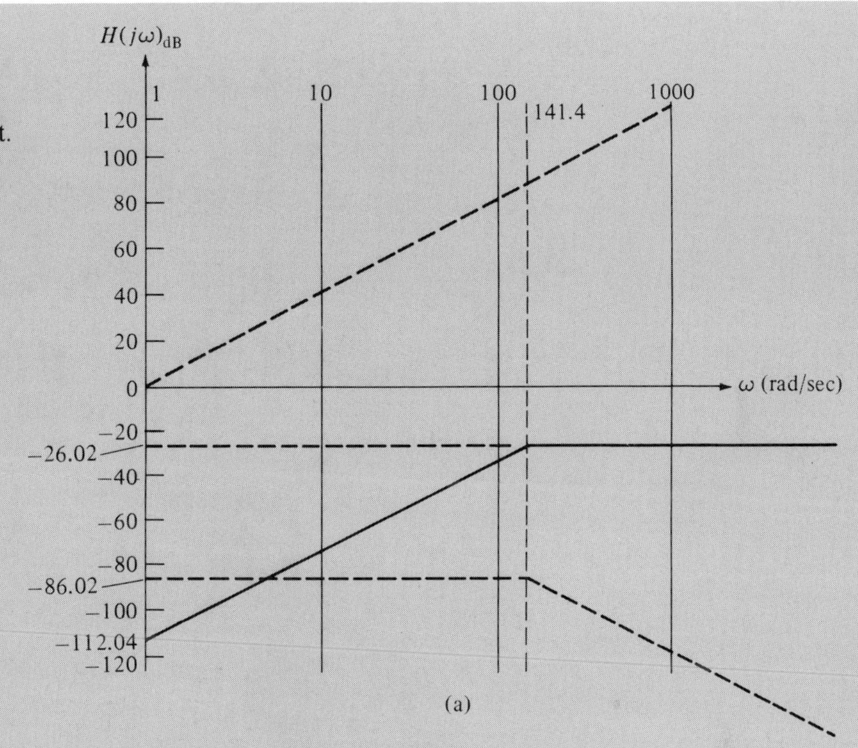

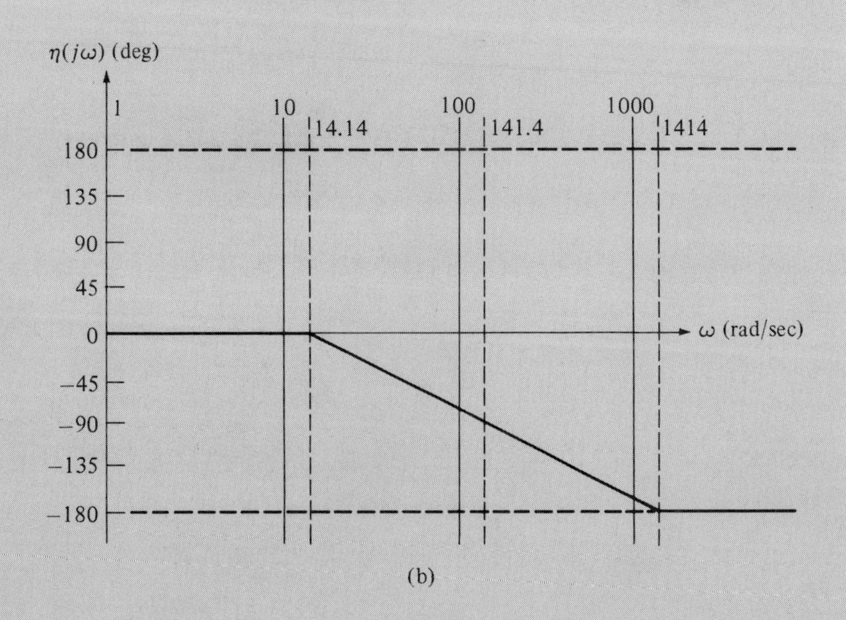

Example 10.13

Problem:

a. Determine an equivalent electrical circuit for the magnetically-coupled circuit of Figure 10.25(a).

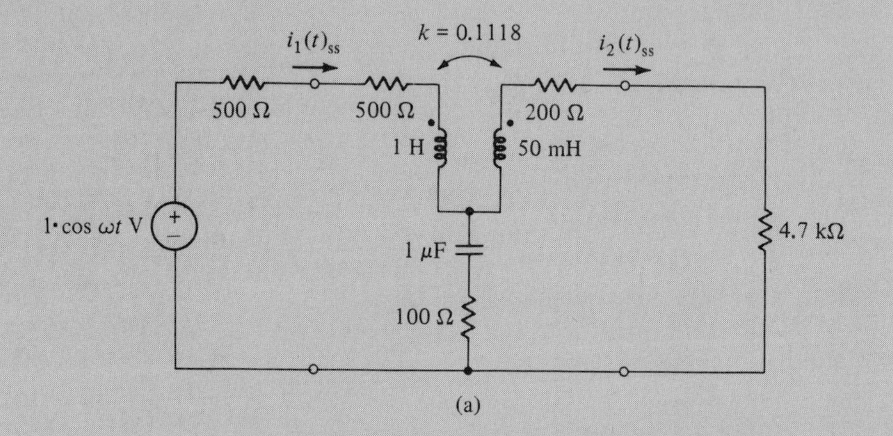

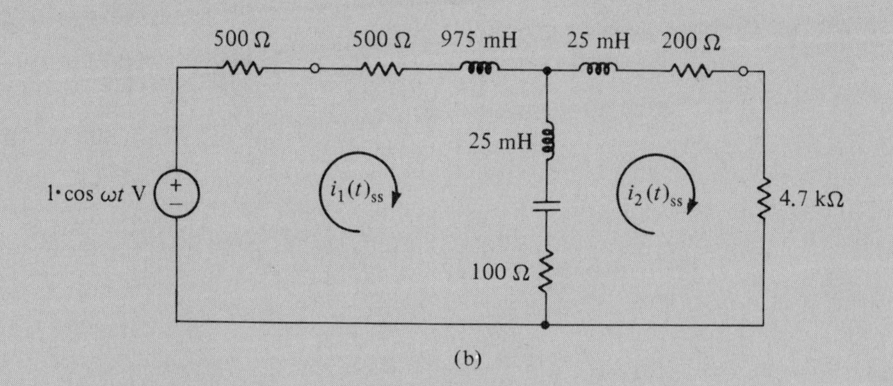

Figure 10.25 Example 10.13. (a) Magnetically coupled circuit. (b) Electrical equivalent.

b. Find the transfer function

$$\mathbf{H}(j\omega) = \frac{\mathbf{V}_2(j\omega)}{\mathbf{V}_1(j\omega)}$$

Solution:

1. Use Equation (8.32) to calculate M:

$$M = 0.1118\sqrt{(1)(0.05)} = 25 \text{ mH}$$

2. Replace the transformer with its T-equivalent [Figure 10.25(b)]:

$$L_1 - M = 975 \text{ mH} \quad \text{and} \quad L_2 - M = 25 \text{ mH}$$

3. Write the mesh equations for the equivalent circuit:

Mesh 1: $\quad Z_{11}I_1 - Z_{12}I_2 = V_1$

Mesh 2: $\quad -Z_{21}I_1 + Z_{22}I_2 = 0$

$Z_{11}(s)$ is the self-impedance of mesh 1:

$$Z_{11} = 0.975s + 0.025s + 500 + 500 + 100 + \frac{1}{10^{-6}s}$$

$$= \frac{s^2 + 1100s + 10^6}{s}$$

$Z_{22}(s)$ is the self-impedance of mesh 2:

$$Z_{22} = 0.025s + 0.025s + 100 + 200 + 4700 + \frac{1}{10^{-6}s}$$

$$= 0.05\frac{s^2 + 100 \times 10^3 s + 20 \times 10^6}{s}$$

$Z_{12}(s)$ is the mutual impedance between mesh 1 and mesh 2:

$$Z_{12} = Z_{21} = 0.025s + 100 + \frac{1}{10^{-6}s}$$

$$= 0.025\frac{s^2 + 4000s + 40 \times 10^6}{s}$$

4. Solve the mesh-2 equation for $I_1(s)$ in terms of $I_2(s)$, substitute the result the mesh-1 equation, and solve for $I_2(s)$ in terms of $V_1(s)$:

Mesh 2: $$I_1 = \frac{Z_{22}}{Z_{21}}I_2$$

Mesh 1: $$\frac{Z_{11}Z_{22}}{Z_{21}}I_2 - Z_{12}I_2 = V_1$$

$$I_2 = \frac{Z_{12}}{Z_{11}Z_{22} - Z_{12}^2}V_1$$

5. Express $V_2(s)$ in terms of $I_2(s)$, then solve for $H(s)$:

$$V_2 = 4700I_2$$
$$= \frac{4700Z_{12}}{Z_{11}Z_{22} - Z_{12}^2}V_1$$
$$H(s) = \frac{V_2}{V_1} = \frac{4700Z_{12}}{Z_{11}Z_{22} - Z_{12}^2}$$
$$= \frac{4700[0.025(s^2 + 4000s + 40 \times 10^6)/s]}{D_1(s)}$$

where

$$D_1(s) = \frac{s^2 + 1100s + 10^6}{s} \cdot \frac{0.05(s^2 + 100 \times 10^3 s + 20 \times 10^6)}{s} - \left(0.025\frac{s^2 + 4000s + 40 \times 10^6}{s}\right)^2$$

$$H(s) = 2350\frac{s(s^2 + 4000s + 40 \times 10^6)}{D_2(s)}$$

where

$$D_2(s) = (s^2 + 1100s + 10^6)(s^2 + 100 \times 10^3 s + 20 \times 10^6) - 0.0125(s^2 + 4000s + 40 \times 10^6)^2$$

Thus

$$H(s) = 2380\frac{s^2 + 4000s + 40 \times 10^6}{s^3 + 102.3 \times 10^3 s^2 + 131.4 \times 10^6 s + 119.5 \times 10^9}$$
$$= \frac{2380(s + 2000 - j6000)(s + 2000 + j6000)}{(s + 644.6 - j876.1)(s + 644.6 + j876.1)(s + 101.1 \times 10^3)}$$

7. Substitute $j\omega$ for s to obtain $\mathbf{H}(j\omega)$:

$$\mathbf{H}(j\omega) = \frac{2380(2000 - j6000 + j\omega)(2000 + j6000 + j\omega)}{(644.6 - j876.1 + j\omega)(644.6 + j876.1 + j\omega)(101.1 \times 10^3 + j\omega)}$$

The Ideal Transformer

The concept of an ideal transformer is based on three approximations:

1. The coefficient of coupling is near unity ($k \simeq 1$).
2. The power dissipated in the core or coils is much less than the power transferred.
3. The reactance of each coil is much greater than the impedance of all other elements in the mesh in which it resides.

These approximations can be realized in a properly designed transformer with a ferromagnetic core common to both coils (Figure 10.26). These devices are commonly referred to as **iron-core** transformers. The ferromagnetic core confines the magnetic field to the core itself and causes all of the flux produced to link all turns of both coils.

The self-inductance of a coil depends on its physical characteristics, the number of turns in the coil, and the flux linkages. When all the flux links all turns of a coil, the self-inductance is proportional to the square of the number

Figure 10.26
Transformers with ferromagnetic cores.

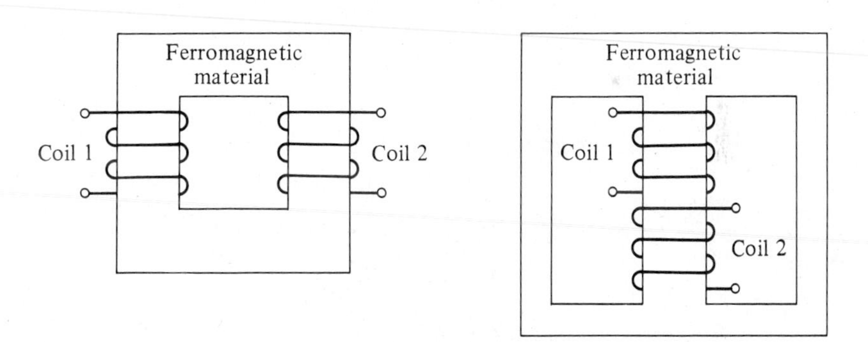

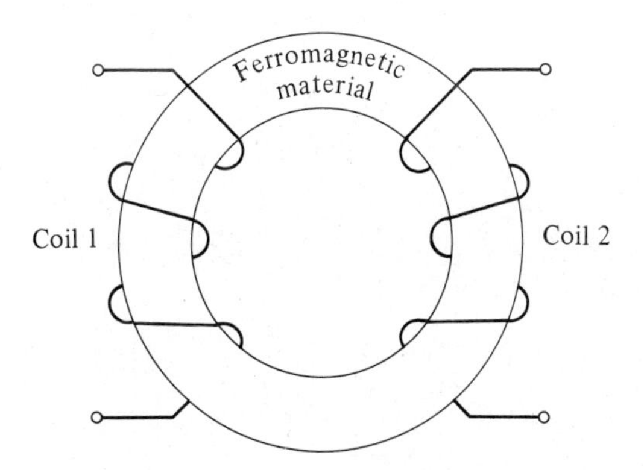

of turns:

$$L = K \cdot N^2 \tag{10.43}$$

It follows that if both primary and secondary coils have the same core, then

$$L_1 = K \cdot N_1^2$$
$$L_2 = K \cdot N_2^2$$

and

$$\frac{L_2}{L_1} = \frac{N_2^2}{N_1^2} = a^2 \tag{10.44}$$

where a is the **turns ratio.**

A network consisting of an ideal transformer driven by a practical source and loaded by $\mathbf{Z}_L$ is shown in Figure 10.27. The notation $1{:}a$ is the usual way of indicating that $N_2/N_1 = a$; two coils with parallel lines between is an accepted symbol for an ideal transformer. Because there are no power losses in the ideal transformer, there is no resistance in the transformer model. The unity coefficient of coupling means that

$$M = \sqrt{L_1 L_2} \tag{10.45}$$

Sinusoidal steady-state analysis of the ideal transformer yields results similar to that of linear transformer analysis, except that the coil resistance terms are removed from Equations (10.41a), (10.41ab), and (10.42):

$$\frac{\mathbf{I}_2(j\omega)}{\mathbf{I}_1(j\omega)} = \frac{-j\omega\mathbf{M}}{\mathbf{Z}_L(i\omega) + j\omega L_2} \tag{10.46}$$

$$\frac{\mathbf{V}_2(j\omega)}{\mathbf{V}_1(j\omega)} = \frac{-j\omega\mathbf{M} \cdot \mathbf{Z}_L(j\omega)}{[\mathbf{Z}_g(j\omega) + j\omega L_1][\mathbf{Z}_L(j\omega) + j\omega L_2] + \omega^2 M^2} \tag{10.47}$$

$$\mathbf{Z}_i(j\omega) = \frac{j\omega L_1[\mathbf{Z}_L(j\omega) + j\omega L_2] + \omega^2 M^2}{\mathbf{Z}_L(j\omega) + j\omega L_2} \tag{10.48}$$

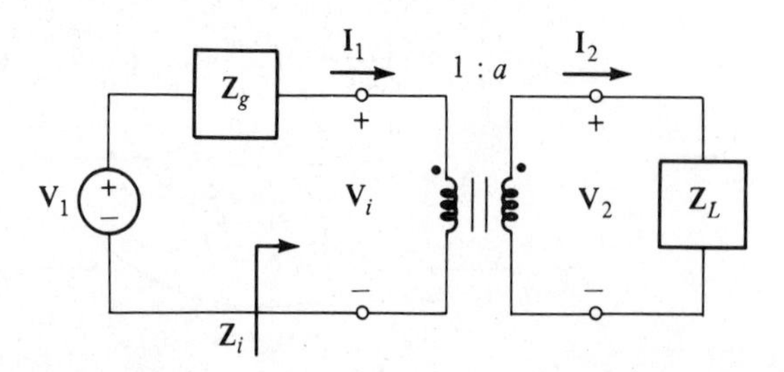

Figure 10.27 Practical network incorporating an ideal transformer.

From Equations (10.44) and (10.45),

$$M = \sqrt{L_1 L_2} = L_2/a$$

and from the ideal transformer approximations,

$$\omega L_2 \gg Z_L$$

so Equation (10.46) becomes

$$\frac{\mathbf{I}_2}{\mathbf{I}_1} = -\frac{j\omega L_2/a}{\mathbf{Z}_L + j\omega L_2} \simeq -\frac{j\omega L_2/a}{j\omega L_2} = -\frac{1}{a} \tag{10.49}$$

The significance of Equation (10.49) is that the current transfer function is a constant, equal to the reciprocal of the turns ratio, and independent of frequency and load. This result is somewhat deceiving, though, because in practice a given transformer design cannot realize the ideal transformer approximations over a wide frequency range, so Equation (10.49) is limited to the specified operating range of the transformer. Furthermore, Equation (10.49) is only valid if the secondary coil reactance is much larger than the load impedance. For values of load impedance of the same order of magnitude as the secondary coil reactance, the ideal transformer approximations are not realized, and Equation (10.49) produces inaccurate results.

From the primary terminals of the transformer in Figure 10.27, Equation (10.47) becomes

$$\frac{\mathbf{V}_2(j\omega)}{\mathbf{V}_i(j\omega)} = \frac{-j\omega M \cdot \mathbf{Z}_L(j\omega)}{j\omega L_1[\mathbf{Z}_L(j\omega) + j\omega L_2] + \omega^2 M^2}$$

From Equations (10.44) and (10.45),

$$L_2 = a^2 L_1$$

and

$$M = \sqrt{L_1 L_2} = aL_1$$

so

$$\frac{\mathbf{V}_2(j\omega)}{\mathbf{V}_i(j\omega)} = \frac{-j\omega a L_1 \cdot \mathbf{Z}_L(j\omega)}{j\omega L_1 \mathbf{Z}_L(j\omega) - \omega^2 a^2 L_1 + \omega^2 a^2 L_1} = -a \tag{10.50}$$

The significance of the result is that the voltage transfer function is a constant, equal to the turns ratio, and independent of frequency and of source and load impedances. Again, accurate results depend upon operation within the specified frequency range of the transformer, and upon the primary coil reactance being much larger than the source impedance.

Equations (10.49) and (10.50) were derived for the current direction indicated in Figure 10.27 and with the sign of M positive. Under these conditions, the secondary voltage is 180° out of phase with the primary voltage, and the secondary current is 180° out of phase with the primary current, as indicated by the minus sign in each equation. If the sign of M is negative, secondary voltage is in phase with primary voltage, and secondary current is in phase with primary current. The phase relationship between voltage and current for the same coil depends upon the polar angle of the load impedance.

The widest application of iron-core transformers is in energy distribution systems and equipment power supplies where the operating frequency is constant. Since the core and coil losses are small, the coil reactances are large, and the primary and secondary voltages and currents are either in phase or 180° out of phase, it follows that the apparent power delivered to the transformer is the same as the apparent power delivered to the load,

$$V_i I_1 = V_2 I_2$$

For this reason, transformers for energy distribution systems are usually rated in volt-amperes (VA or kVA), according to their apparent power capabilities.

The final observation regarding the ideal transformer comes from rearranging Equations (10.50) and (10.49) so that

$$\mathbf{V}_i = -\mathbf{V}_2/a$$
$$\mathbf{I}_1 = -a\mathbf{I}_2$$

The ratio of $\mathbf{V}_i$ to $\mathbf{I}_1$ is the input impedance at the primary terminals of the transformer in Figure 10.27, so by substitution

$$\mathbf{Z}_i = \frac{\mathbf{V}_i}{\mathbf{I}_1} = \frac{-\mathbf{V}_2/a}{-a\mathbf{I}_2} = -\frac{1}{a^2}\cdot\frac{\mathbf{V}_2}{\mathbf{I}_2} = \frac{\mathbf{Z}_L}{a^2} \tag{10.51}$$

The input impedance thus differs from the load impedance by a constant coefficient, so an ideal transformer serves as an impedance-scaling device. This leads to the second most common use of the iron-core transformer—impedance matching in audio power amplifiers. Iron-core audio-frequency transformers are designed to operate over a frequency range of about three decades (20 Hz to 20 kHz). Within the specified frequency range, the normalized frequency response at the transformer input will be the same as that of the load itself.

Equation (10.51) is only valid if $Z_L \ll \omega L_2$. This limitation becomes more apparent if Equation (10.51) is derived by applying the ideal transformer approximations directly to Equation (10.48) (Problem 10-19).

Example 10.14

Problem:

An audio-frequency iron-core transformer is used to match an audio source to an 8-Ω resistive speaker (Figure 10.28). The source specifications are:

Amplitude: 12 V.

Frequency range: 100 Hz to 10 kHz.

The transformer specifications are:

Primary: 8H.

Secondary: 1.28 H.

a. Verify .that the coil reactances are sufficiently large for the ideal transformer approximations.

b. Calculate the amplitude of the speaker current.

Solution:

1. Calculate the reactance of each coil and the source impedance at 100 Hz and at 10 kHz:

$$X_{L1}(j2\pi \times 100) = 2\pi(100)(8) = 5027\ \Omega$$
$$X_{L1}(j2\pi \times 10 \times 10^3) = 2\pi(10 \times 10^3)(8) = 502.7\ \text{k}\Omega$$
$$X_{L2}(j2\pi \times 100) = 2\pi(100)(1.28) = 804.2\ \Omega$$
$$X_{L2}(j2\pi \times 10 \times 10^3) = 2\pi(10 \times 10^3)(1.28) = 80.42\ \text{k}\Omega$$

$$|\mathbf{Z}_g(j2\pi \times 100)| = \sqrt{(50)^2 + \left(\frac{1}{2\pi(100)(100 \times 10^{-6})}\right)^2}$$
$$= 52.47\ \Omega$$

$$|\mathbf{Z}_g(j2\pi \times 10 \times 10^3)| = \sqrt{(50)^2 + \left(\frac{1}{2\pi(10 \times 10^3)(100 \times 10^{-6})}\right)^2}$$
$$= 50\ \Omega$$

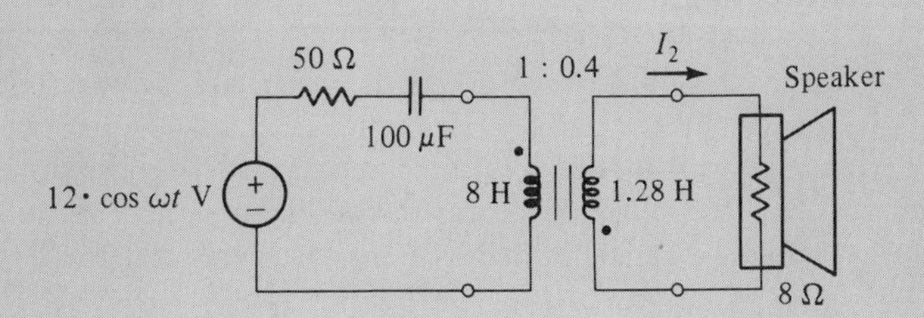

Figure 10.28
Example 10.14.

Thus, at a minimum, X_{L1} is nearly two orders of magnitude larger than $|\mathbf{Z}_g|$, and X_{L2} is more than 100 times larger than $|\mathbf{Z}_L|$. These results confirm that the transformer conforms to the ideal transformer approximation for coil reactances.

2. Calculate the turns ratio:

$$a = \sqrt{L_2/L_1} = 0.4$$

3. Calculate the amplitude of the speaker current:

$$I_1 = \frac{12}{|\mathbf{Z}_g| + |\mathbf{Z}_i|}$$

where

$$|\mathbf{Z}_i| = |\mathbf{Z}_L|/a^2 = 8/0.16 = 50\ \Omega$$

and

$$|\mathbf{Z}_g| \simeq 50\ \Omega$$

throughout the frequency range, so

$$I_1 \simeq 120 \text{ mA}$$

and

$$I_2 = I_1/a = 300 \text{ mA}$$

10.6 Computer Simulations

SPICE 10.1 illustrates frequency response analysis of a two-port network containing an OPAMP (Example 10.2). SPICE 10.2 confirms the band-pass response of the passive filter network in Exercise 10.3. SPICE 10.3 illustrates simulation of an ideal transformer (Example 10.14); note that the coefficient of coupling (k) is specified as 1 for the ideal transformer.

```
*******11/28/88 ********  SPICE 2G.6    3/15/83 ********19:44:32*****
SPICE 10.1
****     INPUT LISTING                     TEMPERATURE =   27.000 DEG C
***************************************************************************
 RI 2 3 10K
 RF 3 4 20K
 CI 1 2 1U
 CF 3 4 2U
 V 1 0 AC 1 0
 E 0 4 3 0 1E6
 .AC DEC 7 .15 150
 .PLOT AC VDB(4) VP(4)
 .END
*******11/28/88 ********  SPICE 2G.6    3/15/83 ********19:44:32*****
SPICE 10.1
****     SMALL SIGNAL BIAS SOLUTION        TEMPERATURE =   27.000 DEG C
***************************************************************************
  NODE   VOLTAGE     NODE   VOLTAGE     NODE   VOLTAGE     NODE   VOLTAGE
 (   1)     .0000    (   2)     .0000    (   3)     .0000    (   4)     .0000
     VOLTAGE SOURCE CURRENTS
     NAME       CURRENT
     V        0.000D+00
     TOTAL POWER DISSIPATION   0.00D+00  WATTS
*******11/28/88 ********  SPICE 2G.6    3/15/83 ********19:44:32*****
SPICE 10.1
****     OPERATING POINT INFORMATION       TEMPERATURE =   27.000 DEG C
***************************************************************************
**** VOLTAGE-CONTROLLED VOLTAGE SOURCES
           E
 V-SOURCE      .000
 I-SOURCE  0.00E+00


*******11/28/88 ********  SPICE 2G.6    3/15/83 ********19:44:32*****
SPICE 10.1
****     AC ANALYSIS                       TEMPERATURE =   27.000 DEG C
***************************************************************************
LEGEND:
 *: VDB(4)
 +: VP(4)

     FREQ      VDB(4)
(*)------------- -4.000D+01    -3.000D+01    -2.000D+01    -1.000D+01  0.000D+00
                      - - - - - - - - - - - - - - - - - - - - - - - - - - - - -
(+)------------- -2.000D+02    -1.000D+02     0.000D+00     1.000D+02  2.000D+02
                      - - - - - - - - - - - - - - - - - - - - - - - - - - - - -
 1.500D-01 -3.450D+01 .       *    .+            .             .             .
 2.084D-01 -3.165D+01 .           * .+           .             .             .
 2.896D-01 -2.880D+01 .             .+*          .             .             .
 4.024D-01 -2.597D+01 .             +     *      .             .             .
 5.591D-01 -2.316D+01 .             +        *   .             .             .
 7.769D-01 -2.038D+01 .            +.          *.             .             .
 1.080D+00 -1.768D+01 .            +.            .   *         .             .
 1.500D+00 -1.511D+01 .           + .            .       *     .             .
 2.084D+00 -1.276D+01 .         +   .            .          *  .             .
 2.896D+00 -1.077D+01 .        +    .            .            *.             .
 4.024D+00 -9.251D+00 .      +      .            .             .*            .
 5.591D+00 -8.305D+00 .    +        .            .             . *           .
 7.769D+00 -7.960D+00 .  +          .            .             .  *          .
 1.080D+01 -8.218D+00 .             .            .             . *          +.
 1.500D+01 -9.077D+00 .             .            .             .*     +      .
 2.084D+01 -1.051D+01 .             .            .            *.    +        .
 2.896D+01 -1.245D+01 .             .            .          *  .   +         .
 4.024D+01 -1.475D+01 .             .            .       *     . +           .
 5.591D+01 -1.729D+01 .             .            .    *        .+            .
 7.769D+01 -1.998D+01 .             .            *             .+            .
 1.080D+02 -2.275D+01 .             .        *   .             +             .
 1.500D+02 -2.556D+01 .             .     *      .             +             .
                      - - - - - - - - - - - - - - - - - - - - - - - - - - - - -
       JOB CONCLUDED
```

```
*******11/28/88 ********  SPICE 2G.6    3/15/83 ********20:34:50*****
SPICE 10.2
****     INPUT LISTING                      TEMPERATURE =   27.000 DEG C
***********************************************************************
 R 3 0 5K
 L1 1 2 225.1M
 L2 3 0 562.8U
 C1 2 3 11.25P
 C2 3 0 4.502N
 V 1 0 AC 1 0
 .AC DEC 10 10K 1MEG
 .PLOT AC VM(3) VP(3)
 .END
*******11/28/88 ********  SPICE 2G.6    3/15/83 ********20:34:50*****
SPICE 10.2
****     SMALL SIGNAL BIAS SOLUTION         TEMPERATURE =   27.000 DEG C
***********************************************************************
  NODE   VOLTAGE     NODE   VOLTAGE     NODE   VOLTAGE
 (  1)     .0000    (  2)     .0000    (  3)     .0000
     VOLTAGE SOURCE CURRENTS
     NAME        CURRENT
     V          0.000D+00
     TOTAL POWER DISSIPATION   0.00D+00  WATTS
*******11/28/88 ********  SPICE 2G.6    3/15/83 ********20:34:50*****
SPICE 10.2
****     AC ANALYSIS                        TEMPERATURE =   27.000 DEG C
***********************************************************************
LEGEND:
 *: VM(3)
 +: VP(3)

     FREQ       VM(3)
(*)------------- 1.000D-06     1.000D-04     1.000D-02     1.000D+00  1.000D+02
                 - - - - - - - - - - - - - - - - - - - - - - - - - - - - - - -
(+)------------- -2.000D+02   -1.000D+02     0.000D+00     1.000D+02  2.000D+02
                 - - - - - - - - - - - - - - - - - - - - - - - - - - - - - - -
 1.000D+04  2.550D-05 .         *   .             .             .          +  .
 1.259D+04  4.090D-05 .          *  .             .             .          +  .
 1.585D+04  6.606D-05 .            *.             .             .          +  .
 1.995D+04  1.079D-04 .             *             .             .          +  .
 2.512D+04  1.797D-04 .             . *           .             .          +  .
 3.162D+04  3.086D-04 .             .  *          .             .          +  .
 3.981D+04  5.594D-04 .             .    *        .             .          +  .
 5.012D+04  1.120D-03 .             .      *      .             .          +  .
 6.310D+04  2.747D-03 .             .         *   .             .          +  .
 7.943D+04  1.158D-02 .             .             *             .         +   .
 1.000D+05  9.999D-01 .             .             +             *             .
 1.259D+05  1.158D-02 .   +         .             *             .             .
 1.585D+05  2.747D-03 .  +          .         *   .             .             .
 1.995D+05  1.120D-03 .  +          .      *      .             .             .
 2.512D+05  5.594D-04 .  +          .    *        .             .             .
 3.162D+05  3.086D-04 .  +          .  *          .             .             .
 3.981D+05  1.797D-04 .  +          . *           .             .             .
 5.012D+05  1.079D-04 .  +          *             .             .             .
 6.310D+05  6.606D-05 .  +         *.             .             .             .
 7.943D+05  4.090D-05 .  +        * .             .             .             .
 1.000D+06  2.550D-05 .  +       *  .             .             .             .
                 - - - - - - - - - - - - - - - - - - - - - - - - - - - - - - -

        JOB CONCLUDED
```

```
*******11/28/88 ********  SPICE 2G.6    3/15/83 ********20:42:10*****
SPICE 10.3
****     INPUT LISTING                      TEMPERATURE =   27.000 DEG C
***********************************************************************
RG 1 2 50
RL 5 0 8
C 2 3 100U
L1 3 0 8
L2 0 4 1.28
K L1 L2 1
V 1 0 AC 12 0
VAM 4 5
.AC DEC 7 50 50K
.PLOT AC I(VAM)
.END
*******11/28/88 ********  SPICE 2G.6    3/15/83 ********20:42:10*****
SPICE 10.3
****     SMALL SIGNAL BIAS SOLUTION         TEMPERATURE =   27.000 DEG C
***********************************************************************
 NODE   VOLTAGE     NODE   VOLTAGE     NODE   VOLTAGE     NODE   VOLTAGE
(  1)     .0000    (  2)     .0000    (  3)     .0000    (  4)     .0000
(  5)     .0000
    VOLTAGE SOURCE CURRENTS
    NAME       CURRENT
    V         0.000D+00
    VAM       0.000D+00
    TOTAL POWER DISSIPATION   0.00D+00  WATTS
*******11/28/88 ********  SPICE 2G.6    3/15/83 ********20:42:20*****
SPICE 10.3
****     AC ANALYSIS                        TEMPERATURE =   27.000 DEG C
***********************************************************************
    FREQ       I(VAM)
                  2.630D-01     2.754D-01     2.884D-01     3.020D-01  3.162D-01
                     - - - - - - - - - - - - - - - - - - - - - - - - - - - - - -
5.000D+01  2.867D-01 .             .          * .             .             .
6.947D+01  2.929D-01 .             .            .      *      .             .
9.653D+01  2.963D-01 .             .            .         *   .             .
1.341D+02  2.981D-01 .             .            .           * .             .
1.864D+02  2.990D-01 .             .            .           * .             .
2.590D+02  2.995D-01 .             .            .           * .             .
3.598D+02  2.998D-01 .             .            .            *.             .
5.000D+02  2.999D-01 .             .            .            *.             .
6.947D+02  3.000D-01 .             .            .            *.             .
9.653D+02  3.000D-01 .             .            .            *.             .
1.341D+03  3.000D-01 .             .            .            *.             .
1.864D+03  3.000D-01 .             .            .            *.             .
2.590D+03  2.999D-01 .             .            .            *.             .
3.598D+03  2.998D-01 .             .            .            *.             .
5.000D+03  2.997D-01 .             .            .            *.             .
6.947D+03  2.993D-01 .             .            .           * .             .
9.653D+03  2.986D-01 .             .            .           * .             .
1.341D+04  2.974D-01 .             .            .         *   .             .
1.864D+04  2.949D-01 .             .            .       *     .             .
2.590D+04  2.904D-01 .             .            . *           .             .
3.598D+04  2.821D-01 .             .       *    .             .             .
5.000D+04  2.681D-01 .     *       .            .             .             .
                     - - - - - - - - - - - - - - - - - - - - - - - - - - - - - -
        JOB CONCLUDED
```

10.7 SUMMARY

In this chapter we have investigated the use of the frequency-domain transfer function to describe steady-state sinusoidal operation of two-port networks. Graphical display of the transfer function, principally through the use of the Bode plot, is an important tool in characterizing two-port networks for purposes of analysis and design. Since frequency-domain transfer functions are complex functions, a Bode plot consists of two frequency-dependent graphs—one of the magnitude and one of the polar angle of the transfer function. Even though Bode plots are easily generated by computer graphics programs, manual graphic techniques involving piecewise-linear approximations of the transfer function are still very useful. Piecewise-linear graphs of the transfer function provide much insight into the operation of the network and the effect of network components on frequency response.

We found important applications of the frequency-domain transfer function in filter networks. Filter theory encompasses such a wide array of published and practiced techniques that we only examined one category of filters—the Butterworth filter—as an example of the use of the transfer function in filter design.

We saw that the use of the frequency-domain transfer function, together with the ideal-transformer approximations, led to a rigorous basis for the well-known ideal transformer terminal relationships. Although the transfer function itself reduces to a constant in ideal-transformer applications, understanding the approximations which were made and the conditions which were enforced to achieve this simplified result will facilitate more informed use of ideal transformers in circuit design.

10.8 Terms

all-pass
all-pole filter
band-elimination
band-reject
band-stop
bel
Bode analysis
Bode plot
breakpoint
Butterworth coefficients
Butterworth filter
Butterworth polynomial
Chebyshev filter
corner frequency
current transfer
damping ratio
decade
decibel
equiripple
filters
frequency mapping
frequency scaling
high-frequency asymptote
high-pass
ideal transformer
impedance scaling
iron-core transformer
logarithmic axis
low-frequency asymptote
low-pass

maximally flat
modern filter theory
monotonic function
pass band
passive filter
piecewise-linear Bode plot
polar plot
poles of a function
power transfer
prototype filter
reference datum
rolloff
semi-logarithmic axis
stop band
turns ratio
voltage transfer
zeros of a function

Problems

10-1. Determine the frequency-domain transfer function

$$\mathbf{H}(j\omega) = \mathbf{V}_o/\mathbf{V}_i$$

for each two-port network:

***a.**

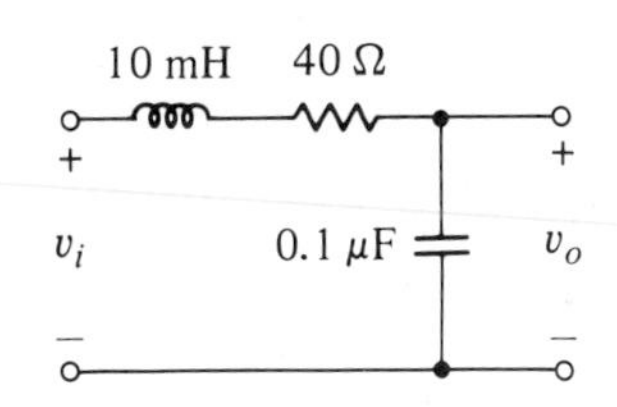

***b.**

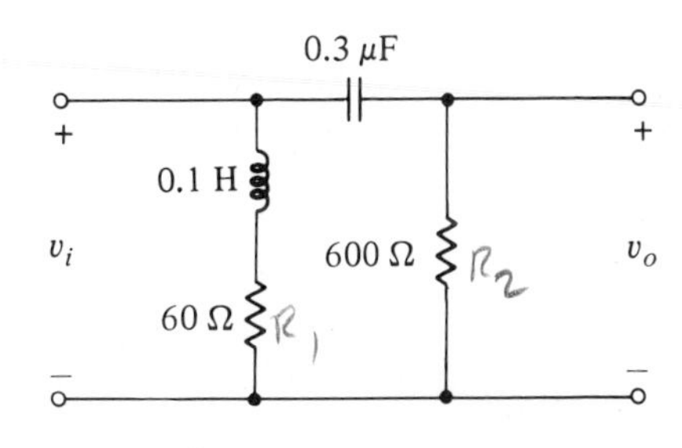

***c.**

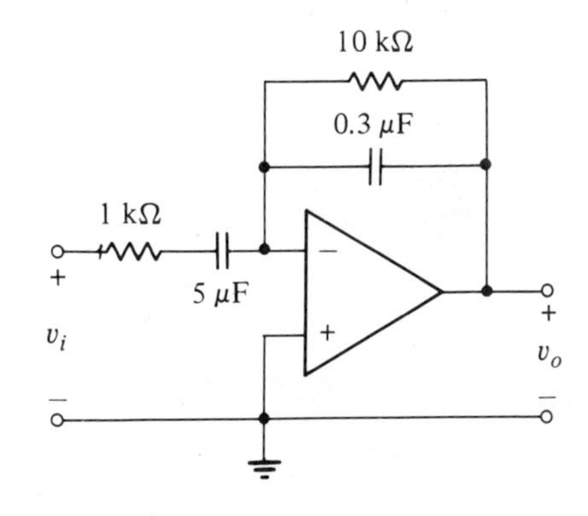

***10-2.** Construct a piecewise-linear Bode plot for each transfer function in Problem 10-1.

10-3. Show that the piecewise-linear graph of $\eta(j\omega)$ for a real-pole or real-zero term differs from the actual value of $\eta(j\omega)$ by 5.711° at a frequency one decade below or one decade above corner frequency.

10-4. Construct a piecewise-linear Bode plot for each transfer function:

***a.** $H(s) = 10^3 \dfrac{s}{s^2 + 115s + 1500}$

b. $H(s) = 4 \times 10^6 \dfrac{s + 10 \times 10^3}{(s + 1000)(s^2 + 4800s + 16 \times 10^6)}$

***c.** $H(s) = 100 \dfrac{s + 20}{(s + 10)(s + 200)^2(s + 1000)}$

d. $H(s) = -500 \dfrac{(s + 1000)^2}{s^3(s^2 + 40s + 10 \times 10^3)}$

***e.** $H(s) = 2500 \dfrac{s^2 + 100s + 10 \times 10^3}{s^2(s + 1000)(s + 5000)}$

10-5. Verify Equation (10.25) by differentiating Equation (10.20a) with respect to ω, setting the result equal to zero, and solving for ω.

***10-6.** Design a third-order high-pass Butterworth filter network which has a 100-kHz cutoff frequency and is terminated in a 600-Ω resistance.

***10-7.** Design a fourth-order band-stop Butterworth filter network which has a center frequency of 2 MHz, has a stop band which is 10% of center frequency, and is terminated in a 5-kΩ resistance.

SPICE

***10-8.** Determine the voltage transfer function

$$H(s) = \frac{V_2(s)}{V_1(s)}$$

realized by the network of Problem 10-6; then prepare a piecewise-linear Bode plot of $H(j\omega)_{\text{dB}}$.

SPICE

10-9. Use SPICE to prepare a Bode plot for the voltage transfer realized by the network of Problem 10-7.

10-10. Verify that

$$y_{21} = -\frac{1}{2.613s^3 + 2.613s}$$

is realized by the network obtained in Example 10.9.

***10-11.** Determine the no-load voltage transfer function $\mathbf{H}(j\omega)$ for Problem 8-12.

SPICE

***10-12.** Determine the no-load voltage transfer function $\mathbf{H}(j\omega)$ for Problem 8-17.

SPICE ***10-13.** Construct a piecewise-linear Bode plot for the transfer function in Problem 10-11.

***10-14.** Construct a piecewise-linear Bode plot of the transfer function in Example 10.13.

SPICE ***10-15.** Construct a piecewise-linear Bode plot of the voltage transfer function.

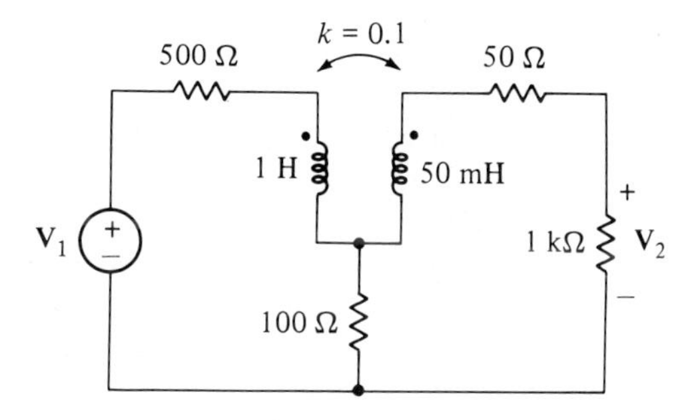

SPICE ***10-16.** Determine the amplitude of each of the following: line current, secondary voltage, load capacitor current, load inductor current. Determine the magnitude of the impedance Z_i seen by the source.

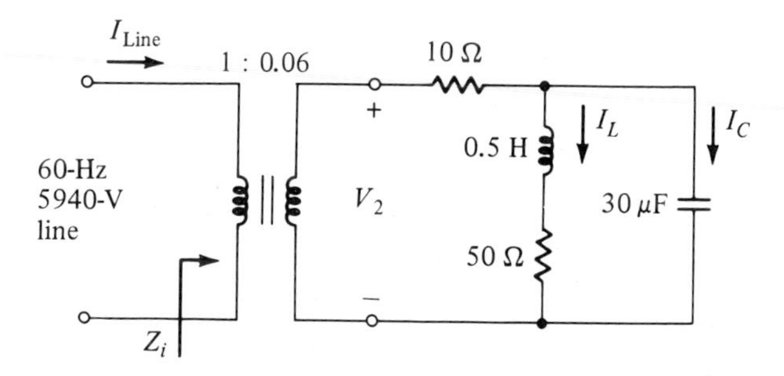

SPICE ***10-17.** Determine the source current amplitude I_m for (a) a 0.5-Ω resistive load, and (b) a 2000-μF capacitive load.

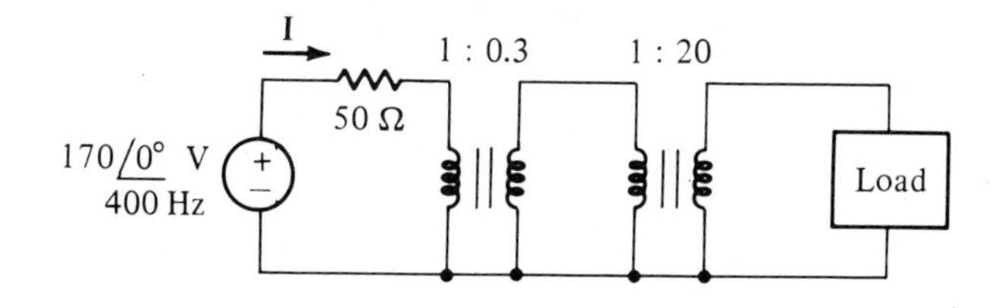

SPICE

***10-18.** A public address system amplifier (PA) with a purely resistive output impedance is designed for maximum power transfer to a directly-connected 8-Ω speaker. What is the effect of connecting the capacitor network between amplifier and speaker? Substantiate the conclusion by means of a piecewise-linear Bode plot.

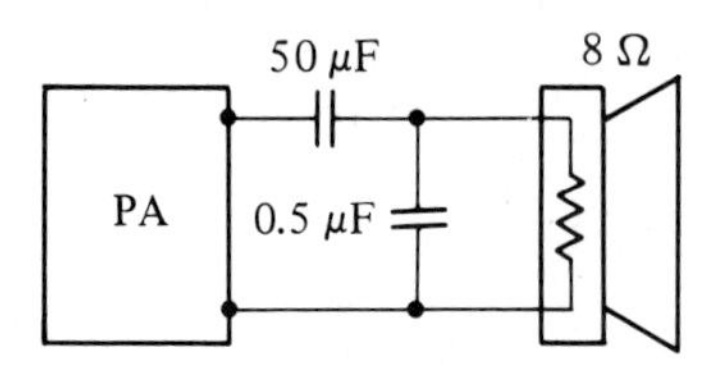

***10-19.** Eliminate the capacitor network in Problem 10-17, and specify the turns ratio for an ideal audio transformer which will permit maximum power transfer from the amplifier to six 8-Ω speakers connected in parallel.

10-20. Derive Equation (10.51) by applying the ideal transformer approximations to Equation (10.48).

11 Fourier Analysis

INTRODUCTION

In Chapter 2, we noted the existence of periodic waveforms (such as pulse trains, sawtooth waveforms, trapezoidal waveforms, etc.) and developed a piecewise technique for expressing the waveform generating function. In Chapter 4, we introduced the periodicity property of the Laplace transform, and subsequently applied the periodicity property to waveform generating functions in various circuit analysis and design problems. Now we come to **Fourier's theorem**, which provides another way of expressing the generating function for a periodic waveform and of analyzing circuits which contain periodic sources.

Fourier's theorem, which we will state mathematically in the next section, observes that any periodic function can be represented by a constant and some combination of **harmonically-related** sinusoids. Thus, Fourier's theorem not only provides a means of dealing with circuits containing periodic sources, it also permits sinusoidal steady-state analysis of circuits which contain nonsinusoidal periodic sources.

We will then extend our discussion of the **Fourier series**, which applies only to periodic waveforms, to the **Fourier transform**, which provides a means of describing nonrecurring (single-event) waveforms. The reader may be familiar with the term **fast Fourier transform**, or FFT, which is used to describe many of the hardware and software techniques of modern signal analysis.

Fourier Series

Fourier's Theorem

A real, periodic function[1] with period T may be expressed as an infinite sum of harmonically related sinusoidal terms,

$$f(t) = \sum_{n=0}^{\infty} (a_n \cdot \cos n\omega_1 t + b_n \cdot \sin n\omega_1 t) \qquad (11.1)$$

where $\omega_1 = 2\pi/T$.

[1] Strictly speaking, Fourier's theorem is restricted to functions having a finite number of discontinuities, all of finite amplitude (i.e., piecewise continuous functions), and a finite number of

Equation (11.1) is known as the **Fourier series**, and a_n and b_n, the amplitudes of the sinusoidal terms, are called the **Fourier coefficients**.

When Equation (11.1) is expanded, the result is

$$
\begin{aligned}
f(t) = {} & a_0 \cdot \cos 0 + b_0 \cdot \sin 0 \\
& + a_1 \cdot \cos \omega_1 t + b_1 \cdot \sin \omega_1 t \\
& + a_2 \cdot \cos 2\omega_1 t + b_2 \cdot \sin 2\omega_1 t + \cdots
\end{aligned}
$$

$$
\begin{aligned}
f(t) = {} & a_0 + a_1 \cdot \cos \omega_1 t + b_1 \cdot \sin \omega_1 t \\
& + a_2 \cdot \cos 2\omega_1 t + b_2 \cdot \sin 2\omega_1 t + \cdots
\end{aligned}
$$

so Equation (11.1) can also be written

$$f(t) = a_0 + \sum_{n=1}^{\infty} (a_n \cdot \cos n\omega_1 t + b_n \cdot \sin n\omega_1 t) \tag{11.2}$$

Note that the lower index on the summation is now $n = 1$.

The frequency ω_1 of the first two sinusoidal components in Equation (11.2) is the same as the frequency of $f(t)$, and is known as the **fundamental frequency**. The frequency $n\omega_1$ of any other sinusoidal component is an integer multiple, or **harmonic**, of the fundamental frequency—in particular, $2\omega_1$ is the **second harmonic**, $3\omega_1$ is the **third harmonic**, etc.

Fourier Coefficients

The first task in using the Fourier series to represent a periodic function is evaluation of the Fourier coefficients, which is accomplished by use of the **Euler formulas** (not to be confused with the Euler identity),

$$a_0 = \frac{1}{T} \int_{t_0}^{t_0+T} f(t)\, dt \tag{11.3a}$$

$$a_n = \frac{2}{T} \int_{t_0}^{t_0+T} f(t) \cdot \cos n\omega_1 t\, dt, \qquad n = 1, 2, \ldots \tag{11.3b}$$

$$b_n = \frac{2}{T} \int_{t_0}^{t_0+T} f(t) \cdot \sin n\omega_1 t\, dt, \qquad n = 1, 2, \ldots \tag{11.3c}$$

It is important to note in Equation (11.3a) that a_0 is simply the average value of the function.

---

extrema. These restrictions are called the Dirichlet conditions; all physically realizable waveforms satisfy the Dirichlet conditions.

Example 11.1

Problem:

Determine the Fourier coefficients for the series which represents the waveform shown in Figure 11.1.

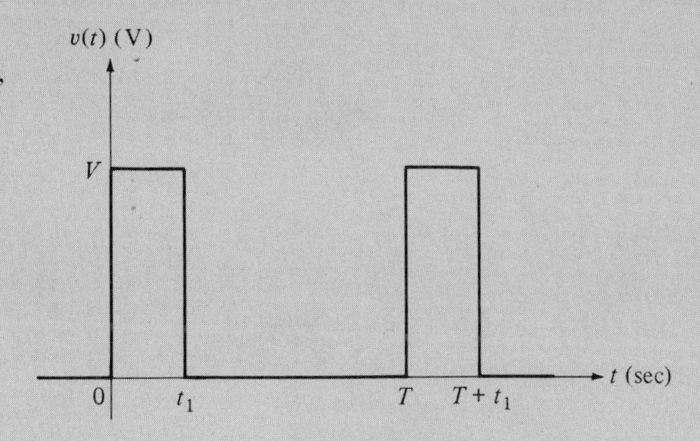

Figure 11.1 Examples 11.1, 11.2, and 11.3.

Solution:

1. Use Equation (11.3a) to determine the expression for a_0. Note that $f(t) = V$ for $0 < t < t_1$, and $f(t) = 0$ for $t_1 < t < T$.

Flashback

$$\int_a^b f(x)\,dx = \int_a^c f(x)\,dx + \int_c^b f(x)\,dx$$

where $a < c < b$

$$a_0 = \frac{1}{T}\left(\int_0^{t_1} V\,dt + \int_{t_1}^{T} 0\,dt\right)$$

$$= \frac{1}{T}Vt\Big|_0^{t_1} = \frac{V \times t_1 - V \times 0}{T} = \frac{Vt_1}{T}$$

2. Use Equations (11.3b) and (11.3c) to evaluate a_n and b_n, respectively. Note that $\omega_1 = 2\pi/T$ rad/sec, so $n\omega_1 = 2n\pi/T$.

Flashback

$$\int \cos ax\,dx = \frac{1}{a} \cdot \sin ax$$

$$\int \sin ax\,dx = -\frac{1}{a} \cdot \cos ax$$

Thus

$$a_n = \frac{2}{T}\int_0^{t_1} V \cdot \cos\frac{2n\pi}{T}t\,dt$$

$$= V\frac{2}{T}\cdot\frac{T}{2n\pi}\cdot\sin\frac{2n\pi}{T}t\bigg|_0^{t_1}$$

$$= \frac{V}{n\pi}\cdot\sin\frac{2n\pi t_1}{T}$$

$$b_n = \frac{2}{T}\int_0^{t_1} V \cdot \sin\frac{2n\pi}{T}t\,dt$$

$$= -V\frac{2}{T}\cdot\frac{T}{2n\pi}\cdot\cos\frac{2n\pi}{T}t\bigg|_0^{t_1}$$

$$= \frac{V}{n\pi}\left(1 - \cos\frac{2n\pi t_1}{T}\right)$$

Alternative Form of the Fourier Series

Since the sinusoidal terms occur in cosine–sine pairs for each frequency, a more useful form of the Fourier series for the purpose of circuit analysis is

$$f(t) = A_0 + \sum_{n=1}^{\infty} (A_m)_n \cdot \cos[n\omega_1 t + \theta_n] \tag{11.4a}$$

where

$$A_0 = a_0 \tag{11.4b}$$

$$(A_m)_n = \sqrt{a_n^2 + b_n^2} \tag{11.4c}$$

$$\theta_n = -\tan^{-1}[b_n/a_n] \tag{11.4d}$$

Flashback

$$a \cdot \cos x + b \cdot \sin x = A_m \cdot \cos[x + \theta]$$

where $A_m = \sqrt{a^2 + b^2}$ and $\theta = -\tan^{-1}[b/a]$

In circuit analysis, A_0 is the **dc component** of the waveform. The collection of all $(A_m)_n$ is the **amplitude spectrum**, and the collection of all θ_n is the **phase spectrum** of $f(t)$; the amplitude spectrum and phase spectrum together are known as the **frequency spectrum** of the waveform. More will be said later about the meaning and application of the frequency spectrum. The use of the subscript m in $(A_m)_n$ emphasizes that the quantity is the amplitude of a sinusoid and is consistent with the notation used previously. Henceforth, V_0 and $(V_m)_n$ will be used with voltage waveforms, and I_0 and $(I_m)_n$ with current waveforms.

Example 11.2

Problem:

Use the results of Example 11.1 to determine the frequency spectrum of the waveform shown in Figure 11.2.

Solution:

1. Use the results of step 2 in Example 11.1 and Equation (11.4c) to determine the amplitude spectrum:

$$(V_m)_n = \sqrt{a_n^2 + b_n^2}$$

$$= \sqrt{\left[\frac{V}{n\pi} \cdot \sin \frac{2n\pi t_1}{T}\right]^2 + \left[\frac{V}{n\pi}\left(1 - \cos \frac{2n\pi t_1}{T}\right)\right]^2}$$

$$= \frac{V}{n\pi}\sqrt{2\left(1 - \cos \frac{2n\pi t_1}{T}\right)}$$

Flashback

$$\sin^2 x + \cos^2 x = 1$$

2. Use the results of step 2 in Example 11.1 and Equation (11.4d) to determine the phase spectrum:

$$\theta_n = -\tan^{-1}\left[\frac{\dfrac{V}{n\pi}\left(1 - \cos \dfrac{2n\pi t_1}{T}\right)}{\dfrac{V}{n\pi} \cdot \sin \dfrac{2n\pi t_1}{T}}\right]$$

$$= -\tan^{-1}\left[\frac{1 - \cos[2n\pi t_1/T]}{\sin[2n\pi t_1/T]}\right]$$

Figure 11.2
Example 11.3.
(a) Pulse train with dc component indicated.
(b) Fundamental plus dc component.
(c) Fundamental and dc component, plus second and third harmonics.
(d) Fundamental and dc component, plus second through sixth harmonics.

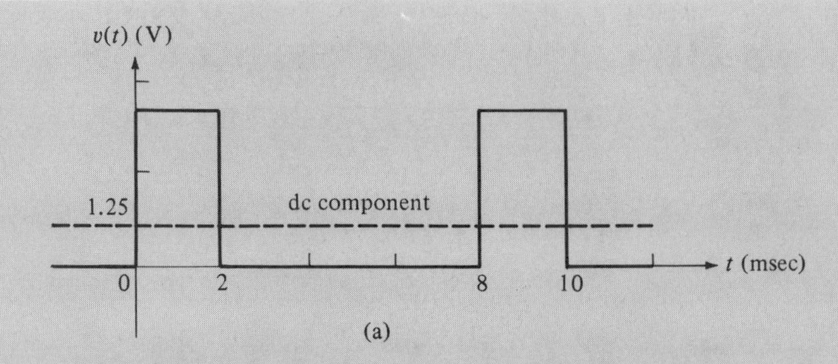

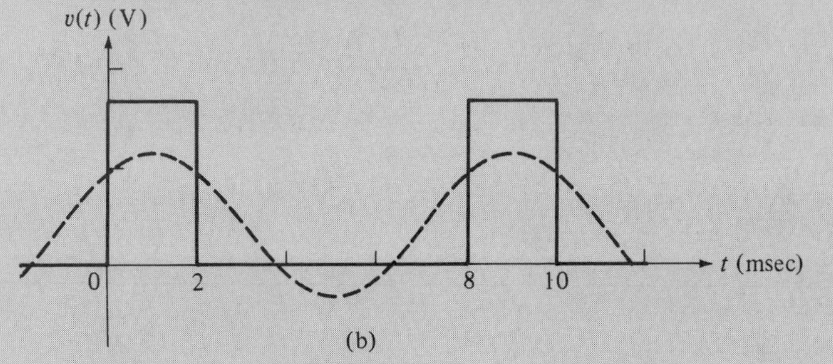

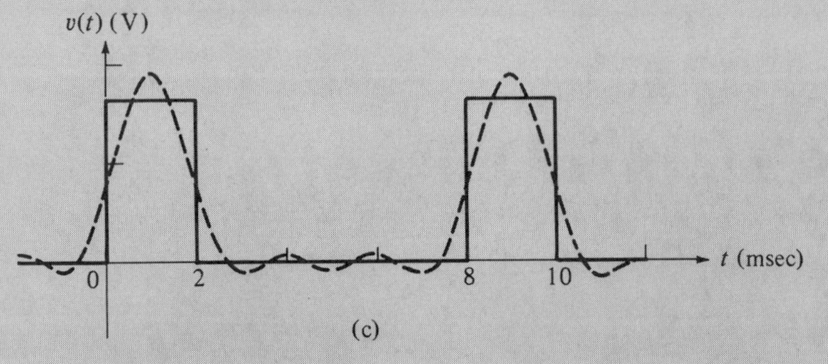

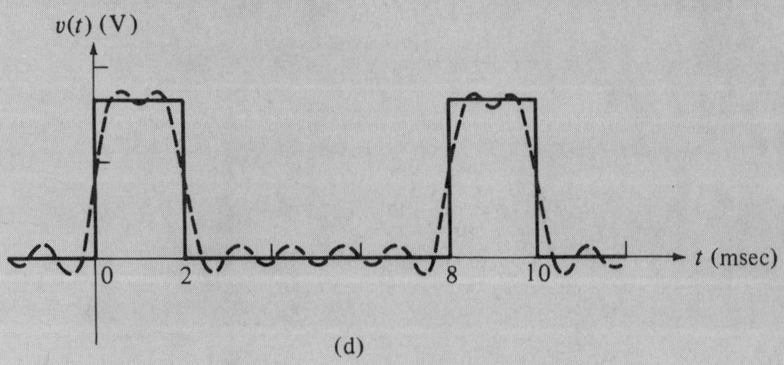

Example 11.3

SPICE

Problem:

Use the results of Examples 11.1 and 11.2 to evaluate the first four components (dc, fundamental, and second and third harmonics) of the Fourier series for the pulse train shown in Figure 11.2(a).

Solution:

1. Determine the dc component from the result of step 1 in Example 11.1 and Equation (11.4b):

$$V_0 = \frac{Vt_1}{T} = \frac{5(0.002)}{0.008} = 1.25 \text{ V}$$

2. Determine $(V_m)_1$, $(V_m)_2$, and $(V_m)_3$ from the result of step 1 in Example 11.2:

$$(V_m)_1 = \frac{V}{\pi}\sqrt{2\left(1 - \cos\frac{2\pi t_1}{T}\right)}$$

$$= \frac{5}{\pi}\sqrt{2\left(1 - \cos\frac{2\pi(0.002)}{0.008}\right)} \simeq 2.251 \text{ V}$$

$$(V_m)_2 = \frac{V}{2\pi}\sqrt{2\left(1 - \cos\frac{4\pi t_1}{T}\right)}$$

$$= \frac{5}{2\pi}\sqrt{2\left(1 - \cos\frac{4\pi(0.002)}{0.008}\right)} \simeq 1.592 \text{ V}$$

$$(V_m)_3 = \frac{V}{3\pi}\sqrt{2\left(1 - \cos\frac{6\pi t_1}{T}\right)}$$

$$= \frac{5}{3\pi}\sqrt{2\left(1 - \cos\frac{6\pi(0.002)}{0.008}\right)} \simeq 0.7503 \text{ V}$$

3. Determine θ_1, θ_2, and θ_3 from the result of step 2 in Example 11.2:

$$\theta_1 = -\tan^{-1}\left[\frac{1 - \cos[2\pi t_1/T]}{\sin[2\pi t_1/T]}\right]$$

$$= -\tan^{-1}\left[\frac{1 - \cos[2\pi(0.002)/0.008]}{\sin[2\pi(0.002)/0.008]}\right]$$

$$= -45°$$

$$\theta_2 = -\tan^{-1}\left[\frac{1 - \cos[4\pi t_1/T]}{\sin[4\pi t_1/T]}\right]$$

$$= -\tan^{-1}\left[\frac{1 - \cos[4\pi(0.002)/0.008]}{\sin[4\pi(0.002)/0.008]}\right]$$

$$= -90°$$

$$\theta_3 = -\tan^{-1}\left[\frac{1 - \cos[6\pi t_1/T]}{\sin[6\pi t_1/T]}\right]$$

$$= -\tan^{-1}\left[\frac{(1 - \cos[6\pi(0.002)/0.008])}{\sin[6\pi(0.002)/0.008]}\right]$$

$$= -135°$$

4. Write the first four components of the Fourier series for $v(t)$:

$$v(t) = 1.25 + 2.251 \cdot \cos[785.4t - 45°]$$

$$+ 1.592 \cdot \cos[1571t - 90°]$$

$$+ 0.7503 \cdot \cos[2356t - 135°] + \cdots$$

According to Fourier's theorem, the *infinite* series [Equation (11.4a)] exactly represents the periodic function $f(t)$. When only a finite number of terms are considered, the Fourier series is said to be **truncated**; the truncated Fourier series only provides an approximation to the periodic function. A graphical analysis of the results of Example 11.3 illustrates the results of truncation. In Figure 11.2(a), only the first term of the Fourier series ($V_0 = 1.25$ V) is considered. Clearly this is not a good approximation of $v(t)$, although it does provide the dc component, or average value, of $v(t)$. The approximation becomes progressively better as the fundamental component and the second and third harmonics are added [Figures 11.2(b) and (c)]. Figure 11.2(d) shows the result of adding the fourth, fifth, and sixth harmonics (Problem 11-3).

Exercise 11.1

SPICE

Problem:

Compute the dc component and the frequency spectrum for the sawtooth waveform shown in Figure 11.3(a).

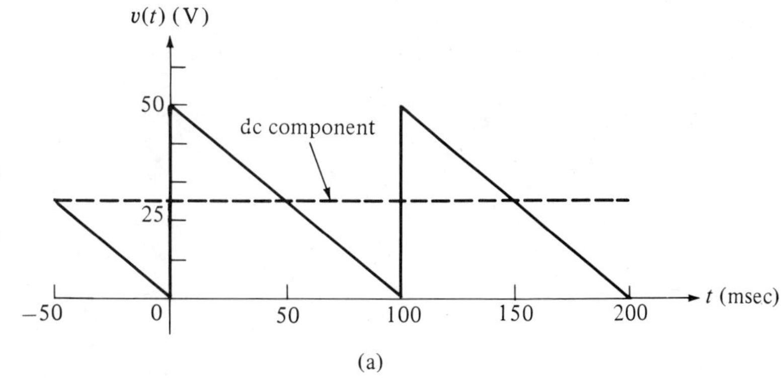

(a)

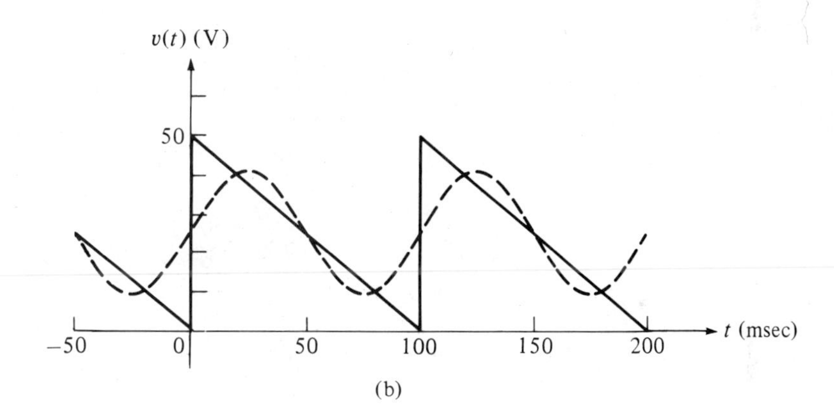

(b)

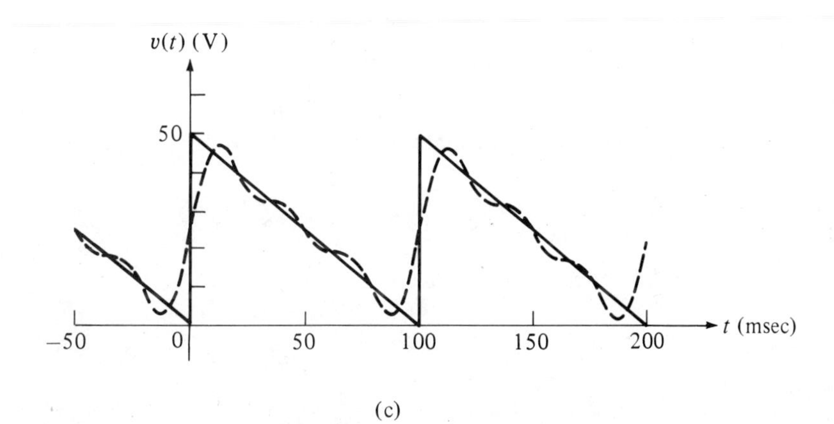

(c)

Figure 11.3
Exercise 11.1.
(a) Sawtooth waveform with dc component indicated.
(b) Fundamental and dc component.
(c) Fundamental and dc component, plus second and third harmonics.

Answer:

$$v(t) = -\frac{V}{T}(t - T) = -500(t - 0.1)\ V$$

for the first period.

$$V_0 = V/2 = 25\ \text{V}$$

$$a_n = 0 \quad \text{and} \quad b_n = \frac{V}{n\pi} = \frac{50}{n\pi}\ \text{V}$$

$$(V_m)_n = \frac{V}{n\pi} = \frac{50}{n\pi}\ \text{V} \quad \text{and} \quad \theta_n = -90^\circ$$

$$\omega_1 = 2\pi/T = 20\pi \simeq 62.83\ \text{rad/sec}$$

$$v(t) = 25 + \frac{50}{n\pi} \sum_{n=1}^{\infty} \sin 20n\pi t\ \text{V}$$

Figure 11.3(d) shows the result of truncating the series after the third harmonic.

Even and Odd Symmetry

Waveform symmetry greatly simplifies the Fourier series, evaluation of the Fourier coefficients, and determination of the frequency spectrum.

Flashback

Symmetry of $f(x)$ (see Figure 11.4):

$f(x)$ is said to have **even symmetry** if

$$f(x) = f(-x)$$

$f(x)$ is said to have **odd symmetry** if

$$f(x) = -f(-x)$$

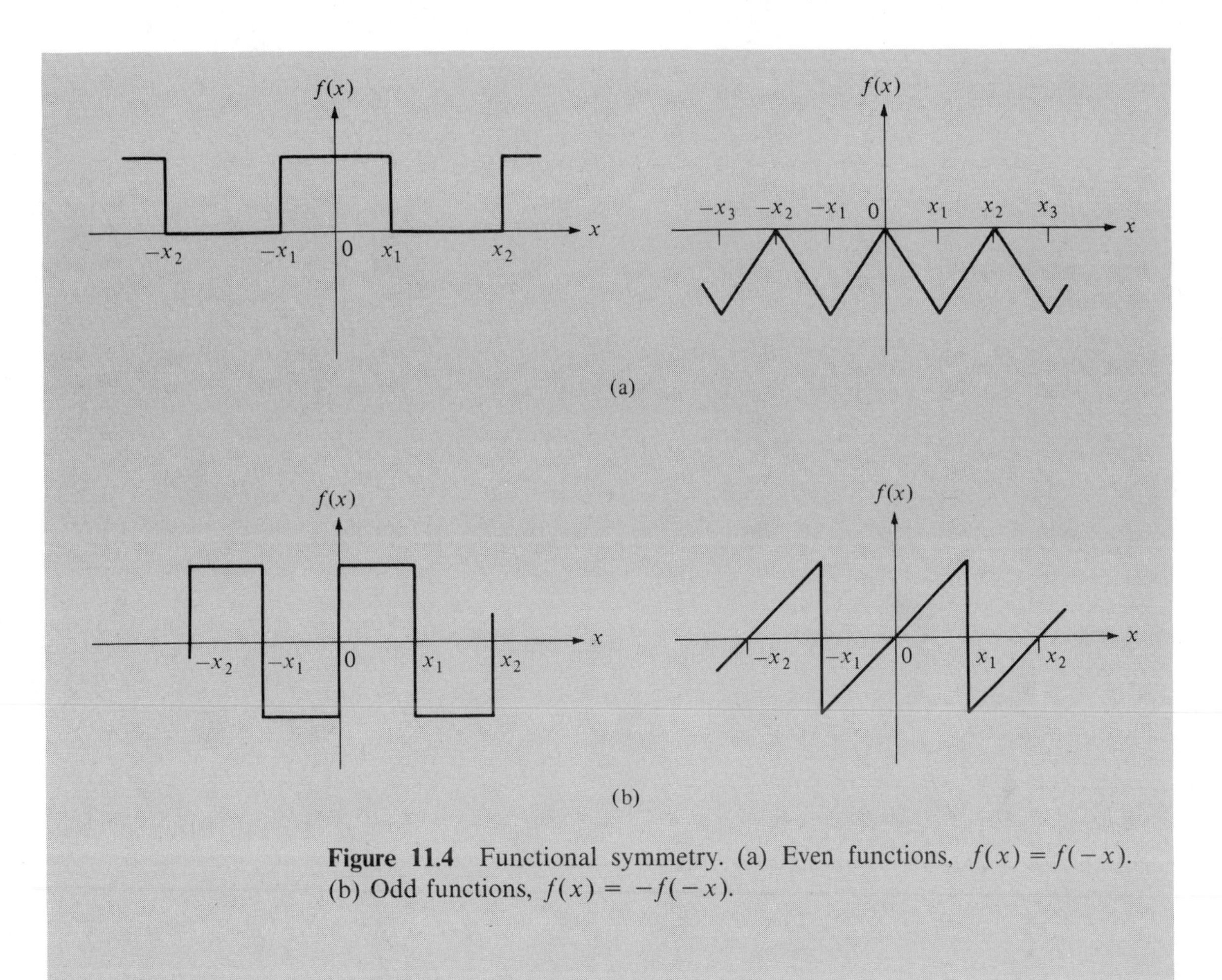

Figure 11.4 Functional symmetry. (a) Even functions, $f(x) = f(-x)$. (b) Odd functions, $f(x) = -f(-x)$.

If $f(t)$ has *even symmetry*, all of the Fourier coefficients b_n vanish from Equation (11.1). Equation (11.4c) becomes $(A_m)_n = |a_n|$, and Equation (11.4d) becomes $\theta_n = 0$ if $a_n > 0$ or $\theta_n = 180°$ if $a_n < 0$. Equation (11.4a) becomes

$$f(t) = A_0 + \sum_{n=1}^{\infty} (A_m)_n \cdot \cos[n\omega_1 t + \theta_n] \qquad \text{(even symmetry)} \quad (11.5)$$

If $f(t)$ has *odd symmetry*, all of the Fourier coefficients a_n (including a_0) vanish from Equation (11.1). In this case, $(A_m)_n = |b_n|$, and $\theta_n = -90°$ if

$b_n > 0$ or $\theta_n = 90°$ if $b_n < 0$, so

$$f(t) = \sum_{n=1}^{\infty} (A_m)_n \cdot \cos[n\omega_1 t + \theta_n] \quad \text{(odd symmetry)} \qquad (11.6)$$

Example 11.4

Problem:

Determine the Fourier series representation of the full-wave rectified sinusoid ($f = 60\text{Hz}$) of Figure 11.5.

Solution:

1. Since the full-wave rectified sinusoid has even symmetry, the Fourier series is determined by Equation (11.5), for which $I_0 = a_0$ and $(I_m)_n = |a_n|$. Use Equation (11.3a) to determine I_0:

$$I_0 = a_0 = \frac{1}{1/120}\int_0^{1/120} 10 \cdot \sin\left[\frac{2\pi}{1/60}t\right] dt$$

$$= -\frac{120(10)}{120\pi} \cdot \cos 120\pi t\Big|_0^{1/120} \simeq 6.366 \text{ mA}$$

2. Use Equation (11.3b) to determine a_n:

$$a_n = \frac{2(10)}{1/120}\int_0^{1/120} 10 \cdot \sin\left[\frac{2\pi}{1/60}t\right] \cdot \cos\left[\frac{2n\pi}{1/120}t\right] dt$$

$$= 2400\int_0^{1/120} \sin 120\pi t \cdot \cos 240 n\pi t \, dt$$

Flashback

$$\int \sin px \cdot \cos qx \, dx$$

$$= -\frac{\cos[(p-q)x]}{2(p-q)} - \frac{\cos[(p+q)x]}{2(p+q)}$$

provided $p^2 \neq q^2$.

Figure 11.5
Example 11.4.
(a) Full-wave rectified sinusoid with dc component indicated.
(b) Fundamental plus dc component.
(c) Fundamental and dc component plus second and third harmonics.

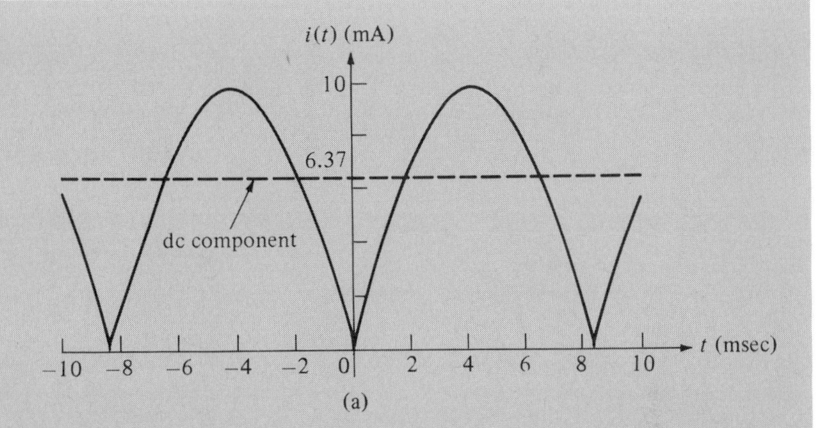

(a)

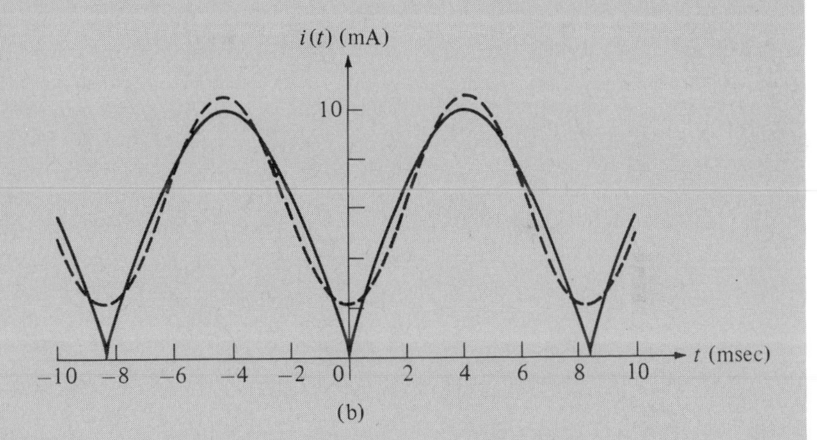

(b)

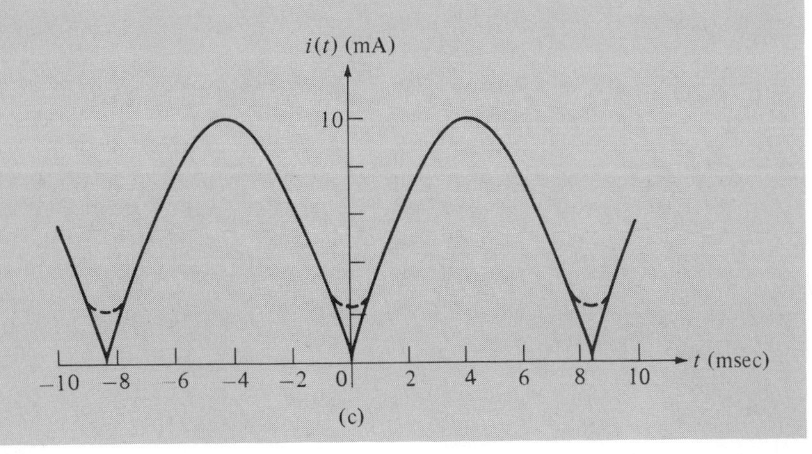

(c)

Thus

$$a_n = -2400\left(\frac{\cos[(120\pi - 240n\pi)t]}{2(120\pi - 240n\pi)} + \frac{\cos[(120\pi + 240n\pi)t]}{2(120\pi + 240n\pi)}\right)\Bigg|_0^{1/120}$$

$$= -10\left[\left(\frac{\cos[(1-2n)\pi]}{(1-2n)\pi} + \frac{\cos[(1+2n)\pi]}{(1+2n)\pi}\right) - \left(\frac{\cos 0}{(1-2n)\pi} + \frac{\cos 0}{(1+2n)\pi}\right)\right]$$

$$= -\frac{10}{\pi}\left[\left(\frac{-1}{1-2n} + \frac{-1}{1+2n}\right) - \left(\frac{1}{1-2n} + \frac{1}{1+2n}\right)\right]$$

$$= \frac{-40}{(4n^2-1)\pi}$$

$$(I_m)_n = |a_n| = \frac{40}{(4n^2-1)\pi} \text{ mA}$$

and $\theta_n = 180°$, because $a_n < 0$ for all n.

3. Write the Fourier series in the form of Equation (11.5):

$$i(t) = \frac{20}{\pi} + \sum_{n=1}^{\infty} \frac{40}{(4n^2-1)\pi} \cdot \cos[n\omega_1 t + 180°]$$

$$= \frac{20}{\pi} - \sum_{n=1}^{\infty} \frac{40}{(4n^2-1)\pi} \cdot \cos n\omega_1 t \text{ mA}$$

$$= 6.366 - 4.244 \cdot \cos 754.0t - 0.8488 \cdot \cos 1508t - \cdots$$

The approximations obtained from the dc component, fundamental, and second and third harmonics are shown in Figure 11.5(b) and (c).

Example 11.5

SPICE

Problem:

Determine the Fourier series representation of the rectangular waveform of Figure 11.6.

Figure 11.6
Example 11.5.
(a) Rectangular waveform with odd symmetry (no dc component).
(b) Fundamental.
(c) Fundamental plus third and fifth harmonics.

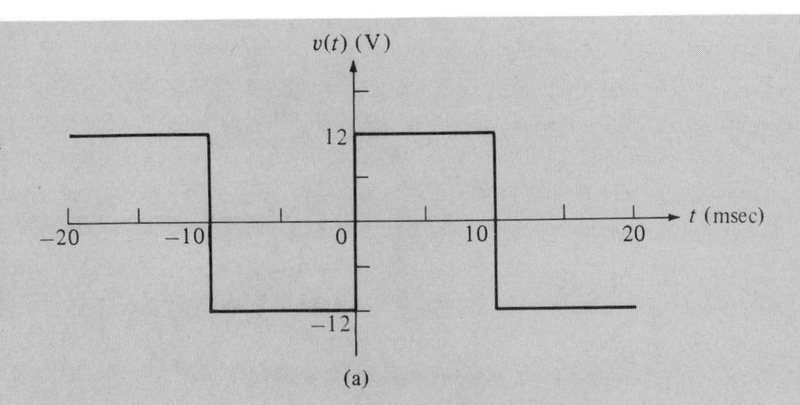

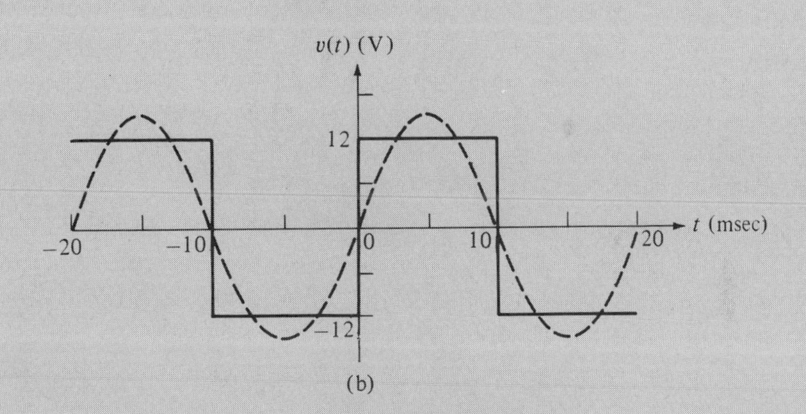

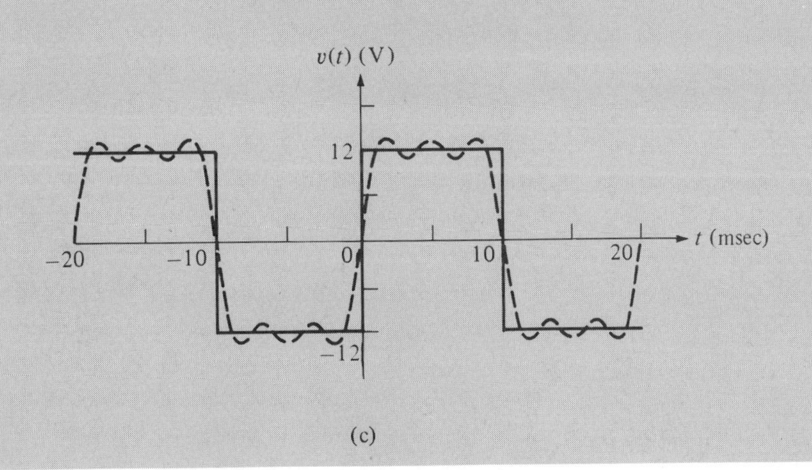

Solution:

1. The rectangular waveform has odd symmetry, so the Fourier series is determined by Equation (11.6). Use Equation (11.3c) to determine b_n:

$$\begin{aligned} b_n &= \frac{2}{0.020}\left(\int_0^{0.010} 12 \cdot \sin \frac{2n\pi}{0.020} t\, dt \right. \\ &\qquad \left. + \int_{0.010}^{0.020} -12 \cdot \sin \frac{2n\pi}{0.020} t\, dt\right) \\ &= \frac{2}{0.020}\left(-\frac{12(0.020)}{2n\pi} \cdot \cos \frac{2n\pi}{0.020} t \Big|_0^{0.010} \right. \\ &\qquad \left. + \frac{12(0.020)}{2\pi n} \cdot \cos \frac{2n\pi}{0.020} t \Big|_{0.010}^{0.020}\right) \\ &= -\frac{12}{n\pi}(2 \cdot \cos n\pi - 2) = \frac{24}{n\pi}(1 - \cos n\pi) \\ &= \frac{48}{n\pi}\ \text{V}, \qquad n = 1, 3, \ldots \end{aligned}$$

Flashback

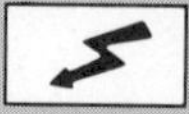

$$\cos n\pi = (-1)^n$$

$$1 - \cos n\pi = \begin{cases} 0 & \text{for} \quad n = 0, 2, 4, \ldots \\ 2 & \text{for} \quad n = 1, 3, 5, \ldots \end{cases}$$

Therefore $(V_m)_n = |b_n| = 48/(n\pi)$ V, and $\theta = -90°$ because $b_n > 0$ for all n.

2. Write the Fourier series in the form of Equation (11.6):

$$\begin{aligned} v(t) &= \frac{48}{\pi} \sum_{n=1,3,5,\ldots}^{\infty} \frac{1}{n} \cdot \cos[n\omega_1 t - 90°] \\ &= \frac{48}{\pi} \sum_{n=1,3,5,\ldots}^{\infty} \frac{1}{n} \cdot \sin n\omega_1 t\ \text{V} \\ &= 15.28 \cdot \sin 314.2t + 5.093 \cdot \sin 942.5t \\ &\quad + 3.056 \cdot \sin 1571t + \cdots \end{aligned}$$

The approximations obtained by using the fundamental and third and fifth harmonics are shown in Figure 11.6(b) and (c).

Exercise 11.2

SPICE

Problem:

Determine the Fourier series for the triangular waveform of Figure 11.7(a).

Answer:

$$\omega_1 = 2\pi/0.300 \simeq 20.94 \text{ rad/sec}$$

$$v(t) = 15 - \frac{120}{\pi^2} \sum_{n=1,3,5,\ldots}^{\infty} \frac{1}{n^2} \cdot \cos n\omega_1 t \text{ V}$$

Figure 11.7
Exercise 11.2.
(a) Triangular waveform with dc component indicated.
(b) Fundamental and dc component.
(c) Fundamental and dc component plus second harmonic.

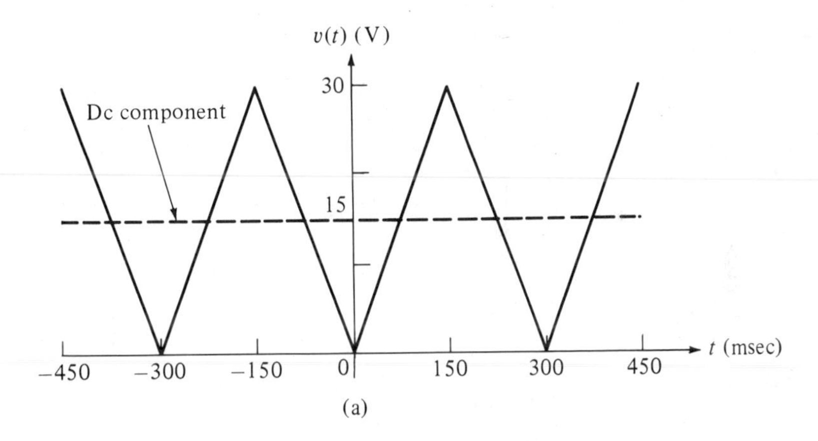

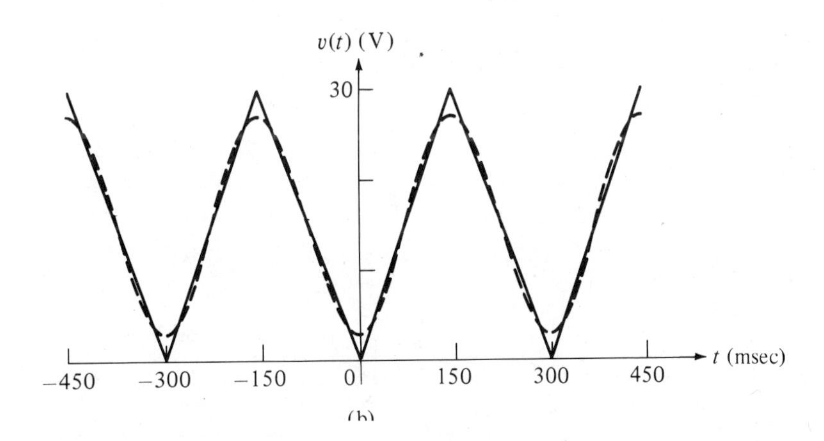

Figure 11.7
Continued

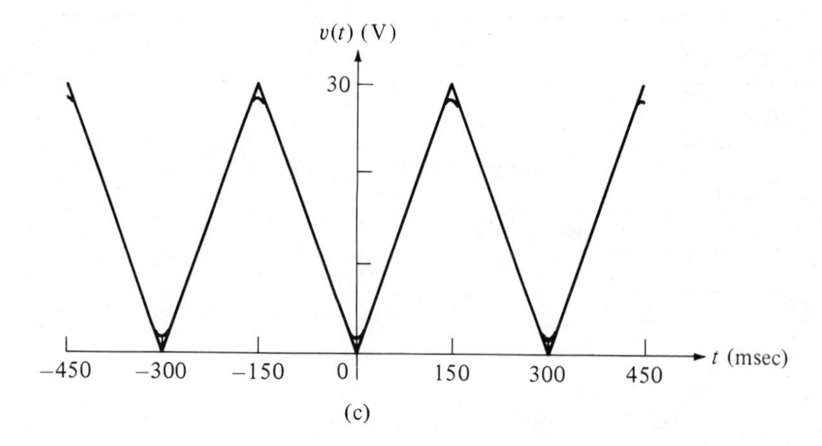

Table 11.1
Fourier Coefficients and Frequency Spectrum

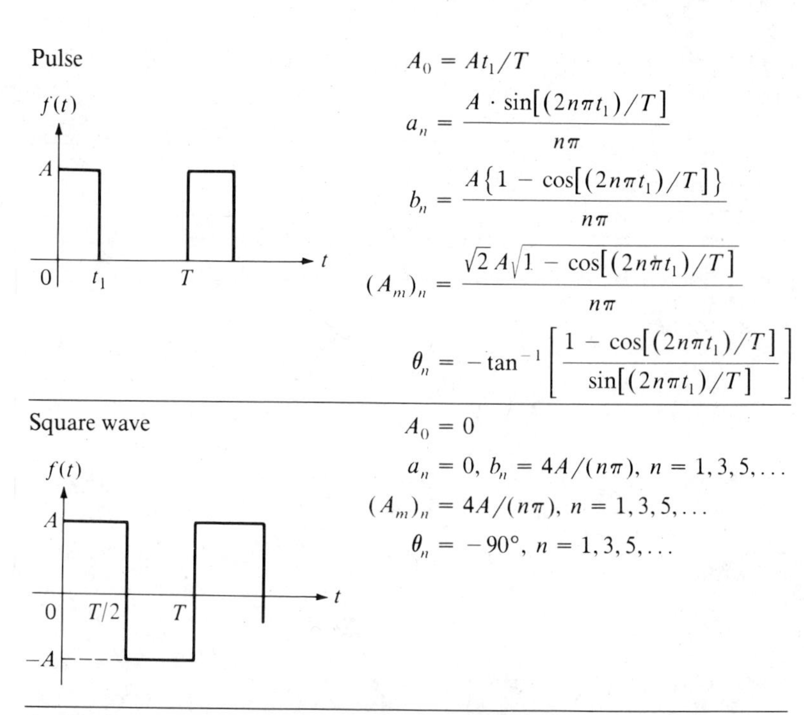

Pulse

$$A_0 = At_1/T$$

$$a_n = \frac{A \cdot \sin[(2n\pi t_1)/T]}{n\pi}$$

$$b_n = \frac{A\{1 - \cos[(2n\pi t_1)/T]\}}{n\pi}$$

$$(A_m)_n = \frac{\sqrt{2}A\sqrt{1 - \cos[(2n\pi t_1)/T]}}{n\pi}$$

$$\theta_n = -\tan^{-1}\left[\frac{1 - \cos[(2n\pi t_1)/T]}{\sin[(2n\pi t_1)/T]}\right]$$

Square wave

$$A_0 = 0$$

$$a_n = 0,\ b_n = 4A/(n\pi),\ n = 1, 3, 5, \ldots$$

$$(A_m)_n = 4A/(n\pi),\ n = 1, 3, 5, \ldots$$

$$\theta_n = -90°,\ n = 1, 3, 5, \ldots$$

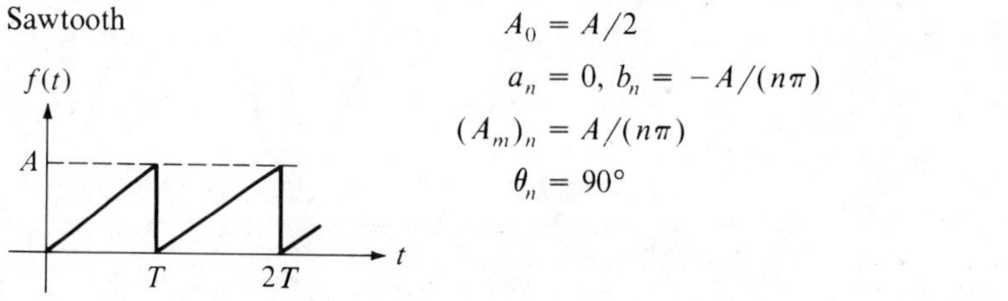

Sawtooth

$$A_0 = A/2$$

$$a_n = 0,\ b_n = -A/(n\pi)$$

$$(A_m)_n = A/(n\pi)$$

$$\theta_n = 90°$$

Table 11.1. Continued

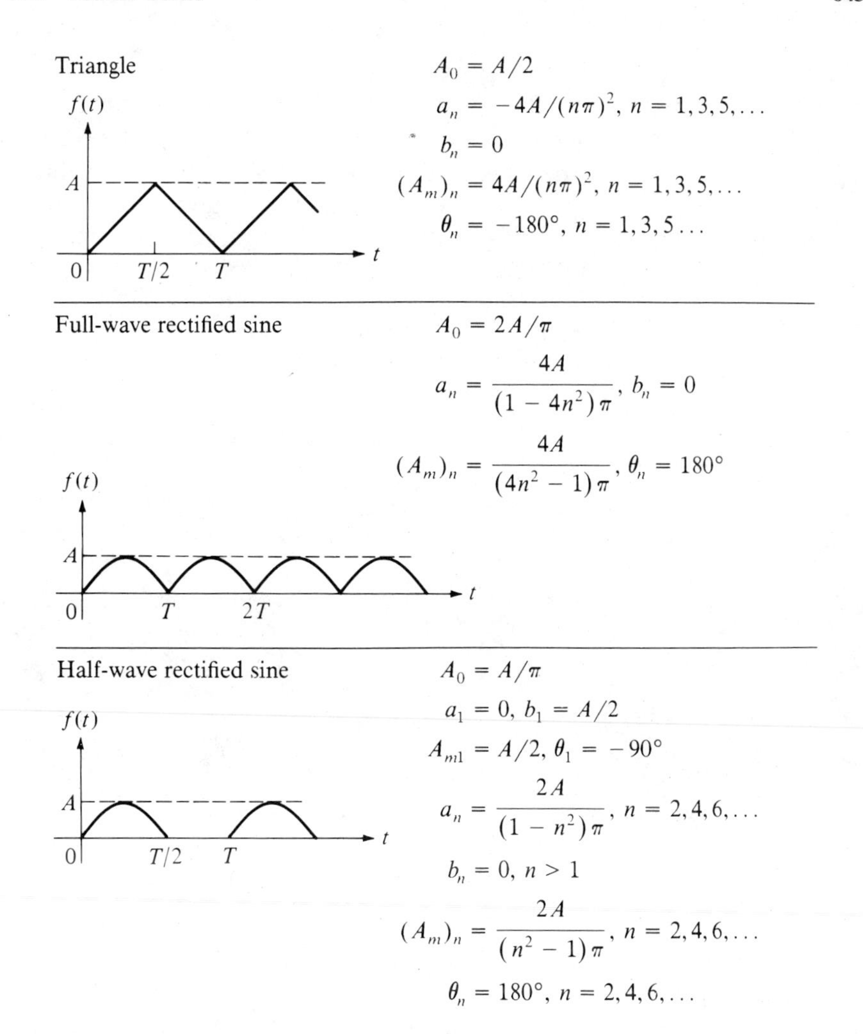

Waveform	Coefficients
Triangle	$A_0 = A/2$ $a_n = -4A/(n\pi)^2,\ n = 1, 3, 5, \ldots$ $b_n = 0$ $(A_m)_n = 4A/(n\pi)^2,\ n = 1, 3, 5, \ldots$ $\theta_n = -180°,\ n = 1, 3, 5 \ldots$
Full-wave rectified sine	$A_0 = 2A/\pi$ $a_n = \dfrac{4A}{(1-4n^2)\pi},\ b_n = 0$ $(A_m)_n = \dfrac{4A}{(4n^2-1)\pi},\ \theta_n = 180°$
Half-wave rectified sine	$A_0 = A/\pi$ $a_1 = 0,\ b_1 = A/2$ $A_{m1} = A/2,\ \theta_1 = -90°$ $a_n = \dfrac{2A}{(1-n^2)\pi},\ n = 2, 4, 6, \ldots$ $b_n = 0,\ n > 1$ $(A_m)_n = \dfrac{2A}{(n^2-1)\pi},\ n = 2, 4, 6, \ldots$ $\theta_n = 180°,\ n = 2, 4, 6, \ldots$

Once the Fourier series representation of a waveform has been determined, adding a **dc offset** to the waveform will change the dc component of the Fourier series by an amount equal to the offset, but will not change the frequency spectrum. This observation is especially useful in dealing with periodic voltage or current waveforms which are **clamped** at some dc level.

Table 11.1 provides a guide for constructing the Fourier series for some frequently encountered waveforms. More extensive tables may be found in advanced network analysis textbooks and in books of standard mathematical tables.

11.3 Circuit-Analysis Applications

At this point the reader may well be concerned about the usefulness of the Fourier series in circuit analysis. Laplace transform circuit analysis, when used in conjunction with piecewise representation of periodic waveforms, appears to be adequate for analysis of circuits with periodic sources. Why then introduce another method of representing periodic waveforms? The answer lies in a closer examination of Equation (11.4a). By invoking Euler's identity, Equation (11.4a) can be written

$$f(t) = A_0 + \sum_{n=1}^{\infty} \text{Real}\{(A_m)_n \cdot \exp[j(n\omega_1 t + \theta_n)]\}$$
$$= A_0 + \sum_{n=1}^{\infty} \text{Real}\{(A_m)_n \cdot \exp[j\theta_n] \cdot \exp[jn\omega_1 t]\} \quad (11.7a)$$

Flashback

Euler's identity:

$$\exp[jx] = \cos x + j \cdot \sin x$$

Therefore, $\cos x = \text{Real}\{\exp[jx]\}$

The term $(A_m)_n \cdot \exp[j\theta_n]$ in Equation (11.7a) is immediately recognized as the phasor

$$\mathbf{A}_n = (A_m)_n \angle \theta_n$$

so

$$f(t) = A_0 + \sum_{n=1}^{\infty} \text{Real}\{\mathbf{A}_n \cdot \exp[jn\omega_1 t]\} \quad (11.7b)$$

It is now clear that use of the Fourier series permits application of ac analysis, already demonstrated to be a transform circuit analysis method, to circuits containing nonsinusoidal sources.

Driving-Point Response to Periodic Excitation

Figure 11.8(a) is an appropriate model for a passive linear network excited by a periodic voltage source for which the Fourier series is

$$v(t) = V_0 + \sum_{n=1}^{\infty} (V_m)_n \cdot \cos[n\omega_1 t + \theta_n]$$

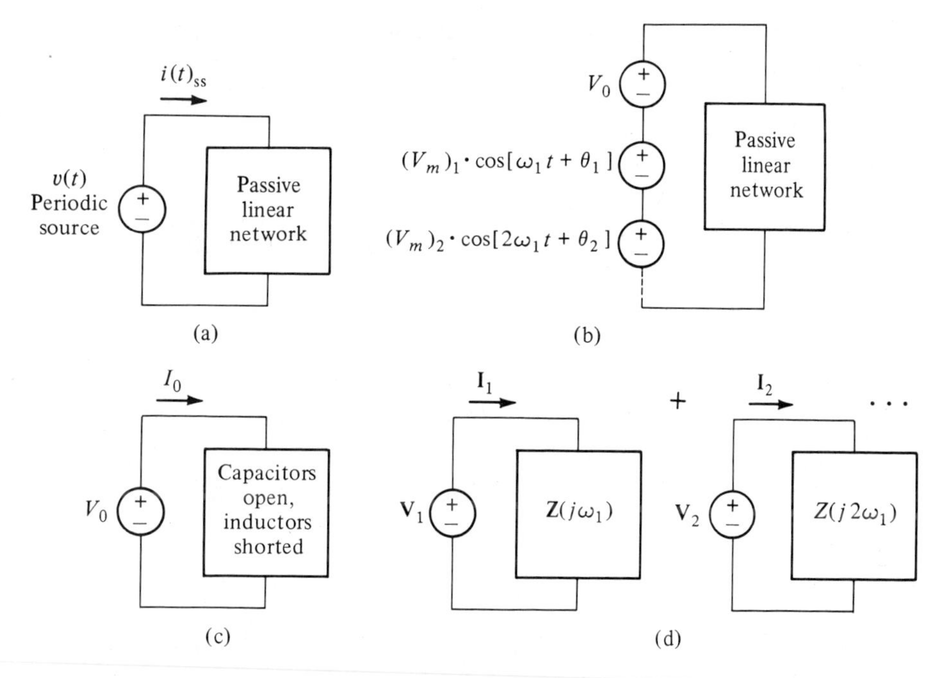

Figure 11.8 Passive linear network excited by periodic source. (a) Passive linear network with periodic excitation. (b) Fourier series representation of periodic source. (c) Analysis of dc component of response. (d) Analysis of steady-state components of response.

The steady-state response current can be obtained by application of the superposition principle if each component of the Fourier series is treated as a separate source [Figure 11.8(b)]. The steady-state response to the dc component of the Fourier series [Figure 11.8(c)] must be determined by time-domain methods (Chapter 3), but the steady-state response due to each sinusoidal component of the Fourier series can be obtained by ac analysis [Figure 11.8(d)]. If the dc component of the Fourier series is zero, then only ac analysis is required.

In Figure 11.8(c), the passive linear network is represented generally by its immittance

$$\mathbf{Z}(j\omega) = |\mathbf{Z}(j\omega)| \underline{/\zeta(j\omega)} \quad \text{or} \quad \mathbf{Y}(j\omega) = |\mathbf{Y}(j\omega)| \underline{/\gamma(j\omega)}$$

because the immittance of the network must be evaluated at the discrete frequency associated with each sinusoidal component of the Fourier series—e.g., $\mathbf{Z}(j\omega_1)$, $\mathbf{Z}(j2\omega_1)$, etc. In this case where the response current to a periodic voltage excitation is required, the response current phasor is

$$\mathbf{I} = \mathbf{I}_1 + \mathbf{I}_2 + \cdots$$

where

$$\mathbf{I}_n = \mathbf{V}_n \cdot \mathbf{Y}(jn\omega_1) \quad \text{or} \quad \mathbf{I}_n = \frac{\mathbf{V}_n}{\mathbf{Z}(jn\omega_1)}$$

Thus,

$$\begin{aligned}
i(t)_{ss} &= I_0 + \sum_{n=1}^{\infty} \text{Real}\{\mathbf{V}_n \cdot \mathbf{Y}(jn\omega_1) \cdot \exp[jn\omega_1 t]\} \\
&= I_0 + \sum_{n=1}^{\infty} \text{Real}\{(V_m)_n \cdot \exp[j\theta_n] \cdot |\mathbf{Y}(jn\omega_1)| \\
&\qquad \cdot \exp[j\gamma(jn\omega_1)] \cdot \exp[jn\omega_1 t]\} \\
&= I_0 + \sum_{n=1}^{\infty} \text{Real}\{(V_m)_n \cdot |\mathbf{Y}(jn\omega_1)| \\
&\qquad \cdot \exp[j(\theta_n + \gamma(jn\omega_1))] \cdot \exp[jn\omega_1 t]\} \\
&= I_0 + \sum_{n=1}^{\infty} (V_m)_n \cdot |Y(jn\omega_1)| \cdot \cos[n\omega_1 t + \theta_n + \gamma(jn\omega_1)] \qquad (11.8a)
\end{aligned}$$

or

$$\begin{aligned}
i(t)_{ss} &= I_0 + \sum_{n=1}^{\infty} \text{Real}\left\{\frac{(V_m)_n}{|\mathbf{Z}(jn\omega_1)|} \cdot \exp[j(\theta_n - \zeta(n\omega_1))] \cdot \exp[jn\omega_1]\right\} \\
&= I_0 + \sum_{n=1}^{\infty} \frac{(V_m)_n}{|\mathbf{Z}(jn\omega_1)|} \cdot \cos[n\omega_1 t + \theta_n - \zeta(jn\omega_1)] \qquad (11.8b)
\end{aligned}$$

The voltage response to excitation by a periodic current source is obtained in a similar manner, with the result

$$\begin{aligned}
v(t)_{ss} &= V_0 + \sum_{n=1}^{\infty} \text{Real}\{(I_m)_n \cdot |\mathbf{Z}(jn\omega_1)| \cdot \exp[j(\theta_n + \zeta(jn\omega_1))] \cdot \exp[jn\omega_1 t]\} \\
&= V_0 + \sum_{n=1}^{\infty} (I_m)_n \cdot |\mathbf{Z}(jn\omega_1)| \cdot \cos[n\omega_1 t + \theta_n + \zeta(jn\omega_1)] \qquad (11.9a)
\end{aligned}$$

or

$$v(t)_{ss} = V_0 + \sum_{n=1}^{\infty} \text{Real}\left\{\frac{(I_m)_n}{|\mathbf{Y}(jn\omega_1)|} \cdot \exp[j(\theta_n - \gamma(n\omega_1))] \cdot \exp[jn\omega_1 t]\right\}$$

$$= V_0 + \sum_{n=1}^{\infty} \frac{(I_m)_n}{|\mathbf{Y}(jn\omega_1)|} \cdot \cos[n\omega_1 t + \theta_n - \gamma(n\omega_1)] \qquad (11.9b)$$

where $(I_m)_n$ and θ_n are the frequency spectrum for the periodic current source.

Example 11.6

SPICE

Problem:

Determine the steady-state response current for the circuit of Figure 11.9.

Solution:

1. Obtain the Fourier series representation of the source voltage from Table 11.1:

$$v(t) = 2.5 + \sum_{n=1,3,5,\ldots}^{\infty} \frac{10}{n\pi} \cdot \cos\left[\frac{2n\pi}{0.100}t - 90^\circ\right]$$

$$= 2.5 + \sum_{n=1,3,5,\ldots}^{\infty} \text{Real}\left\{\frac{10}{n\pi} \cdot \exp[-j90^\circ] \cdot \exp[j20n\pi t]\right\} \text{ V}$$

Note that $V_n = [10/(n\pi)]\underline{/-90^\circ}$ V.

2. Use Ohm's law to calculate the dc response current. Under steady-state conditions, the average capacitor current is zero, so the capacitor can be replaced by an open circuit:

$$I_0 = \frac{2.5}{20 \times 10^3} = 125\ \mu\text{A}$$

3. Calculate $\mathbf{Z}(j\omega)$:

$$\mathbf{Z}(j\omega) = 10 \times 10^3 + \frac{1}{0.1 \times 10^{-3} + j1.5 \times 10^{-6}\omega}$$

$$= 10 \times 10^3 \frac{2 + j0.015\omega}{1 + j0.015\omega}\ \Omega$$

Figure 11.9
Example 11.6.
(a) Network.
(b) Source voltage.
(c) Response current.

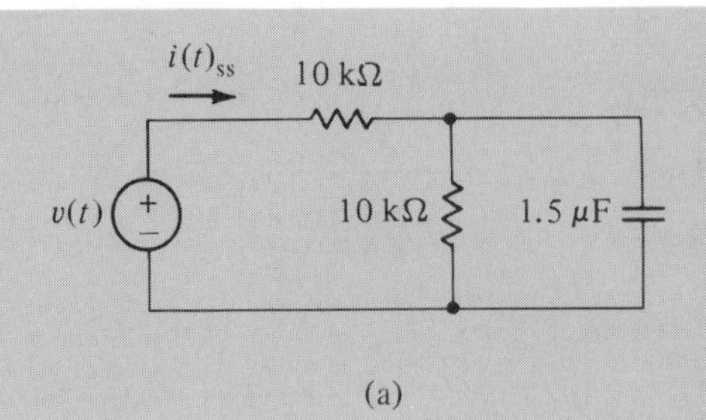

(a)

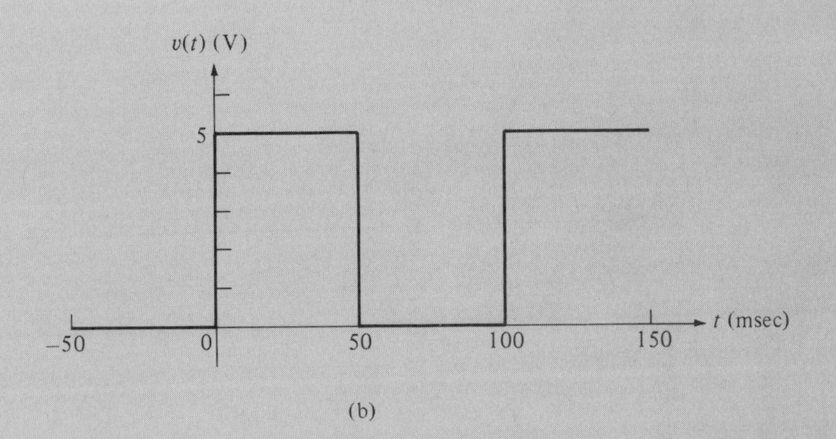

(b)

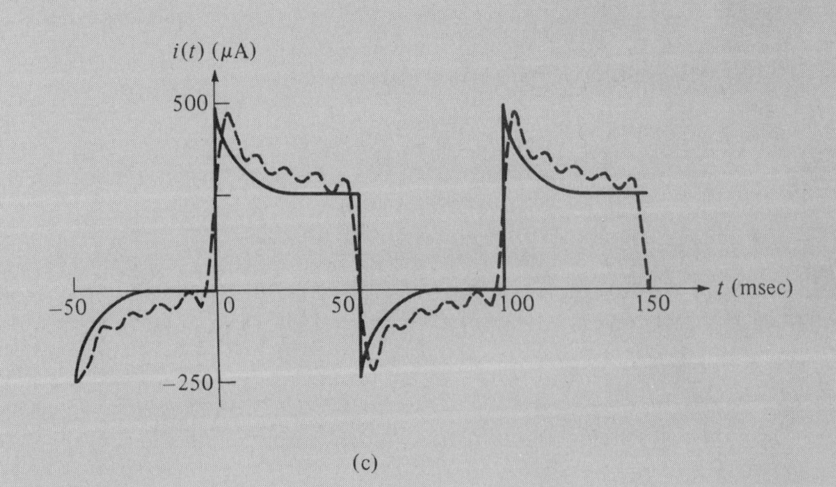

(c)

4. Calculate the phasor $\mathbf{I}_n$ for $n = 1$, 3, and 5:

$$\mathbf{I}_1 = \frac{\mathbf{V}_1}{\mathbf{Z}(j20\pi)} = \frac{(10/\pi)\angle -90^\circ}{16.09 \times 10^3 \angle -18.07^\circ}$$

$$= 197.8\angle -71.93^\circ \ \mu\text{A}$$

$$\mathbf{I}_3 = \frac{\mathbf{V}_3}{\mathbf{Z}(j60\pi)} = \frac{[10/(3\pi)]\angle -90^\circ}{11.55 \times 10^3 \angle -15.80^\circ}$$

$$= 91.88\angle -74.20^\circ \ \mu\text{A}$$

$$\mathbf{I}_5 = \frac{\mathbf{V}_5}{\mathbf{Z}(j100\pi)} = \frac{[10/(5\pi)]\angle -90^\circ}{10.63 \times 10^3 \angle -11.02^\circ}$$

$$= 59.91\angle -78.98^\circ \ \mu\text{A}$$

5. Use Equation (11.8b) to write the Fourier series for $i(t)_{ss}$:

$$i(t)_{ss} = 125 + 197.8 \cdot \cos[62.83t - 71.93^\circ]$$
$$+ 91.88 \cdot \cos[188.5t - 74.20^\circ]$$
$$+ 59.91 \cdot \cos[314.2t - 78.98^\circ] + \cdots \ \mu\text{A}$$

The graph shown in Figure 11.9(c) also contains the fifth and seventh harmonics.

Exercise 11.3

SPICE

Problem:

Use the Fourier series to approximate the solution for Example 6.17. Calculate the dc component, fundamental, and first four harmonics for the steady-state current.

Figure 11.10
Exercise 11.3.

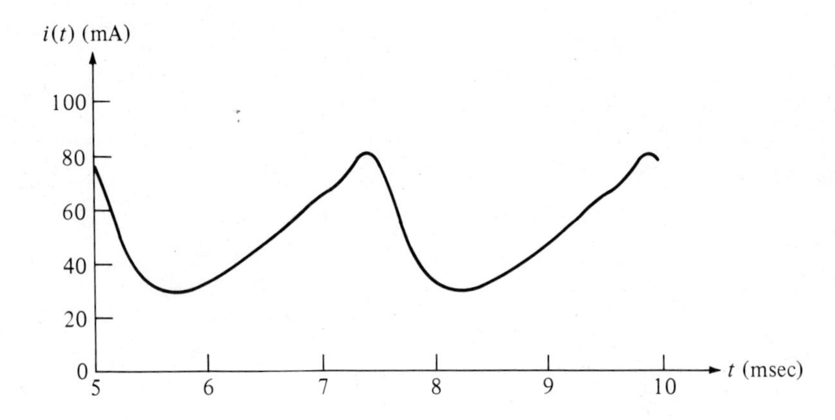

Solution:

$$I_0 = 50 \text{ mA}$$
$$\mathbf{I}_1 = 22.45\angle 44.84^\circ \text{ mA}$$
$$\mathbf{I}_2 = 7.087\angle 26.44^\circ \text{ mA}$$
$$\mathbf{I}_3 = 3.339\angle 18.34^\circ \text{ mA}$$
$$\mathbf{I}_4 = 1.920\angle 13.97^\circ \text{ mA}$$
$$\mathbf{I}_5 = 1.242\angle 11.25^\circ \text{ mA}$$

The graph of the Fourier series approximation for $i(t)_{ss}$ is shown in Figure 11.10.

Output Response to Periodic Input Excitation

The linear two-port network in Figure 11.11(a) is excited by a periodic input voltage

$$v_i(t) = (V_i)_0 + \sum_{n=1}^{\infty} (V_{im})_n \cdot \cos[n\omega_1 t + \theta_n]$$

The steady-state output response can be obtained by application of the superposition principle if each component of the Fourier series is treated as a separate input. As with the driving-point problem, the steady-state response to the dc component of the Fourier series must be determined by time-domain

Figure 11.11
Two-port network with by periodic input.
(a) Linear two-port network with periodic extension.
(b) Dc transfer function.
(c) Sinusoidal steady-state transfer function.

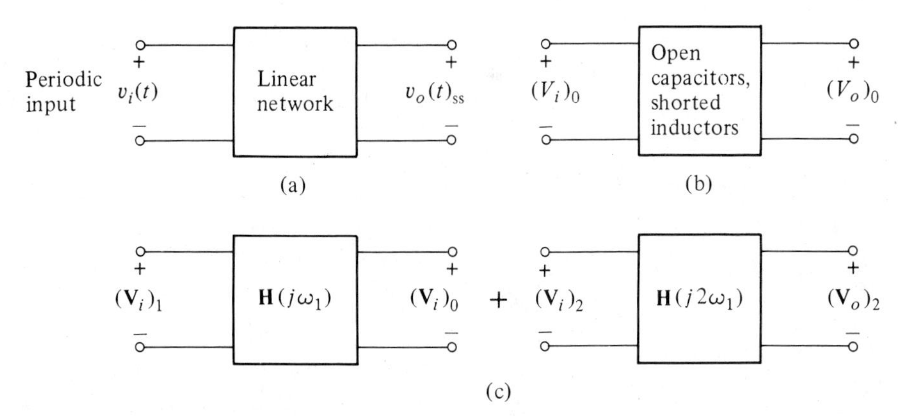

methods, but the steady-state response due to each sinusoidal component of the Fourier series can be obtained by ac analysis.

The two-port network is represented generally by its frequency-domain transfer function

$$\mathbf{H}(j\omega) = |\mathbf{H}(j\omega)|\underline{/\eta(j\omega)}$$

which must be evaluated at the discrete frequency associated with each sinusoidal component of the Fourier series—e.g., $\mathbf{H}(j\omega_1)$, $\mathbf{H}(j2\omega_1)$, etc.

The output voltage phasor is

$$\mathbf{V}_o = (\mathbf{V}_o)_1 + (\mathbf{V}_o)_2 + \cdots$$

where $(\mathbf{V}_o)_n = (\mathbf{V}_i)_n \cdot \mathbf{H}(jn\omega_1)$. Thus, the steady-state output is

$$\begin{aligned}
v_o(t)_{ss} &= (V_o)_0 + \sum_{n=1}^{\infty} \text{Real}\{(\mathbf{V}_i)_n \cdot \mathbf{H}(jn\omega_1) \cdot \exp[jn\omega_1 t]\} \\
&= (V_o)_0 + \sum_{n=1}^{\infty} \text{Real}\{(V_{im})_n \cdot \exp[j\theta_n] \cdot |\mathbf{H}(jn\omega_1)| \\
&\qquad \cdot \exp[j\eta(jn\omega_1)] \cdot \exp[jn\omega_1 t]\} \\
&= (V_o)_0 + \sum_{n=1}^{\infty} \text{Real}\{(V_{im})_n \cdot |\mathbf{H}(jn\omega_1)| \\
&\qquad \cdot \exp[j(\theta_n + \eta(jn\omega_1))] \cdot \exp[jn\omega_1 t]\} \\
&= (V_o)_0 + \sum_{n=1}^{\infty} (V_{im})_n \cdot |\mathbf{H}(jn\omega_1)| \cdot \cos[n\omega_1 t + \theta_n + \eta(jn\omega_1)]
\end{aligned}$$

Similar results are obtained for the output current response to periodic voltage excitation, or the output voltage or current response to periodic current excitation. The form of the equation remains the same as above for all cases; the difference is accounted for by the units of the transfer function.

Example 11.7

SPICE

Problem:

Determine the steady-state output voltage across the capacitor in the network of Example 11.6 (Figure 11.9).

Solution:

1. From Example 11.6,

$$v_i(t) = 2.5 + \sum_{n=1,3,5,\ldots}^{\infty} \text{Real}\left\{\frac{10}{n\pi} \cdot \exp[-j90°] \cdot \exp[n\omega_1 t]\right\} \text{ V}$$

2. Use the voltage division rule to calculate the steady-state dc output voltage. As in Example 11.6, the capacitor is replaced with an open circuit:

$$(V_o)_0 = (V_i)_0 \frac{10 \times 10^3}{10 \times 10^3 + 10 \times 10^3} = 1.25 \text{ V}$$

3. Use the voltage division rule to determine the phasor $\mathbf{V}_o$ in terms of $\mathbf{V}_i$:

$$\mathbf{V}_o = \frac{1/(0.1 \times 10^{-3} + j1.5 \times 10^{-6}\omega)}{10 \times 10^3 + 1/(0.1 \times 10^{-3} + j1.5 \times 10^{-6}\omega)} \cdot \mathbf{V}_i$$

$$= \frac{1}{2 + j0.015\omega} \cdot \mathbf{V}_i$$

Therefore, the frequency-domain transfer function is

$$\mathbf{H}(j\omega) = \frac{\mathbf{V}_o}{\mathbf{V}_i} = \frac{1}{2 + j0.015\omega}$$

4. Calculate the phasor $(\mathbf{V}_o)_n$ for $n = 1$, 3, and 5:

$$(\mathbf{V}_o)_1 = \mathbf{H}(j20\pi)\cdot(\mathbf{V}_i)_1 = \left(\frac{10}{\pi}\angle -90^\circ\right)(0.4523\angle -25.23^\circ)$$
$$= 1.440\angle -115.20^\circ \text{ V}$$
$$(\mathbf{V}_o)_3 = \mathbf{H}(j60\pi)\cdot(\mathbf{V}_i)_3 = \left(\frac{10}{3\pi}\angle -90^\circ\right)(0.2887\angle -54.73^\circ)$$
$$= 0.3064\angle -144.7^\circ \text{ V}$$
$$(\mathbf{V}_o)_5 = \mathbf{H}(j100\pi)\cdot(\mathbf{V}_i)_5 = \left(\frac{10}{5\pi}\angle -90^\circ\right)(0.1953\angle -67.00^\circ)$$
$$= 0.1244\angle -157.0^\circ \text{ V}$$

5. Use Equation (11.10) to write the Fourier series for $v_o(t)_{ss}$:

$$v_o(t)_{ss} = 1.25 + 1.440\cdot\cos[62.83t - 115.2^\circ] + 0.3064\cdot\cos[188.5t - 144.7^\circ] + 0.1244\cdot\cos[314.2t - 157.0^\circ] + \cdots \text{ V}$$

The graph of $v_o(t)_{ss}$, truncated after the seventh harmonic, is shown in Figure 11.12.

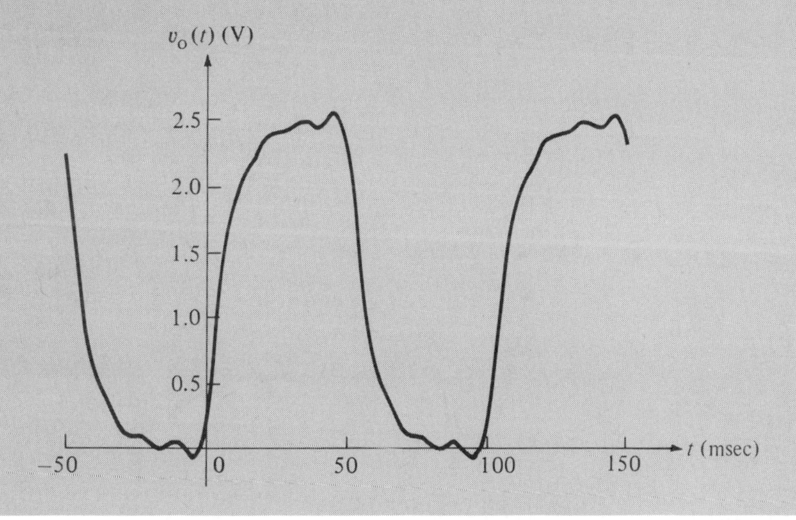

Figure 11.12
Example 11.7.

Exercise 11.4

SPICE

Problem:

Determine the first seven nonzero terms of the Fourier series for the steady-state output voltage of the network shown in Figure 11.13(a). Use the voltage waveform from Example 11.3 as the input.

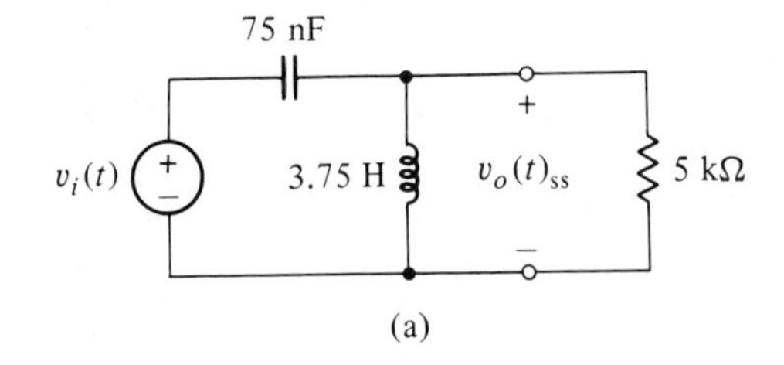

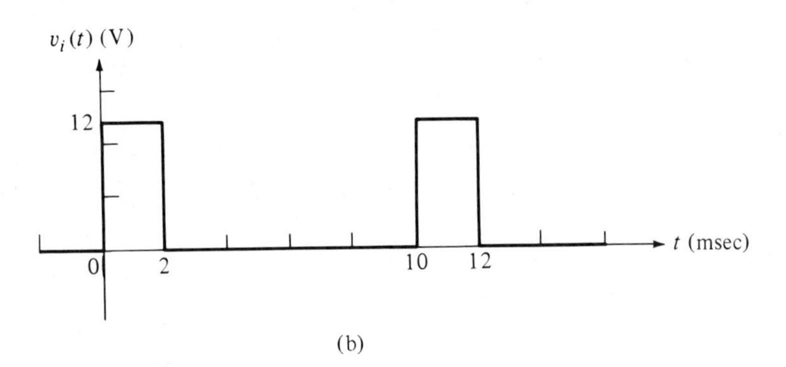

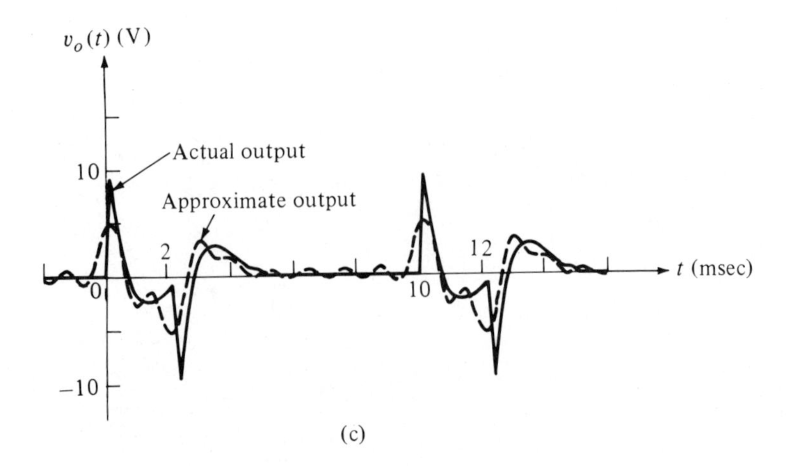

Figure 11.13
Exercise 11.4.
(a) Circuit.
(b) Source voltage.
(c) Response voltage.

Answer:

$$(V_i)_0 = 2.4 \text{ V}$$

$$(V_{im})_n = \frac{12\sqrt{2}}{n\pi}\sqrt{1 - \cos 0.4n\pi}$$

$$\theta_n = -\tan^{-1}\left[\frac{(1 - \cos 0.4n\pi)}{\sin 0.4n\pi}\right]$$

$$\mathbf{H}(j\omega) = -\frac{\omega^2 LC}{(1 - \omega^2 LC) + j\omega L/R}$$

$$\omega_1 = 200\pi \simeq 628.3 \text{ rad/sec}$$

$$\begin{aligned} v_o(t)_{ss} \simeq\ & 0.4955 \cdot \cos[628.3t + 116.1°] \\ & + 1.475 \cdot \cos[1257t + 48.53°] \\ & + 1.712 \cdot \cos[1885t - 17.97°] \\ & + 0.9783 \cdot \cos[2513t - 76.39°] \\ & + 0.7260 \cdot \cos[3770t + 7.331°] \\ & + 1.021 \cdot \cos[4398t - 35.394°] \\ & + 0.8989 \cdot \cos[5027t - 76.31°] + \cdots \end{aligned}$$

The graph in Figure 11.13(b) represents the output voltage truncated after the eighth harmonic. The dc component and the fifth harmonic are both zero.

11.4 Average Power and Effective Values of Periodic Functions

We saw in our review of basic circuit analysis that the instantaneous power absorbed by the network connected between two terminals of a circuit is the product of the instantaneous voltage and current at the terminals. More generally, power is a function of time which depends on the product of the time-domain voltage and current at the terminals:

$$p(t) = v(t) \cdot i(t)$$

The sign of the result will indicate whether the network is absorbing power (positive sign) or supplying power (negative sign) at any instant (Figure 11.14). In a circuit excited by a periodic source, average power is of more practical interest than either instantaneous power or time-varying power. The calculation of average power is based on the mathematical definition of the average

Figure 11.14
Steady-state power dissipation in a circuit excited by a periodic source.
(a) Circuit.
(b) Source voltage.
(c) Response voltage and current.
(d) Power.

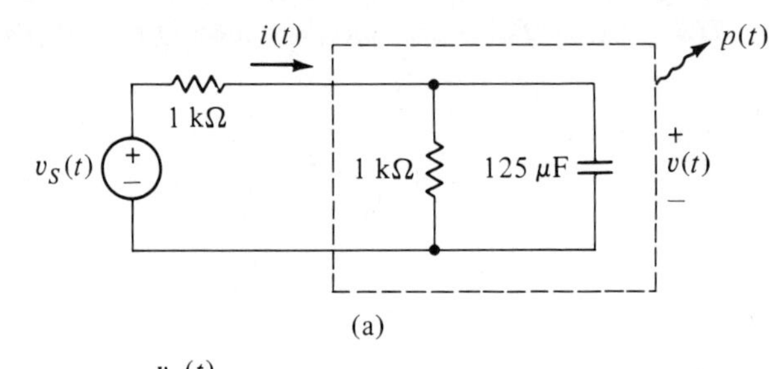

(a)

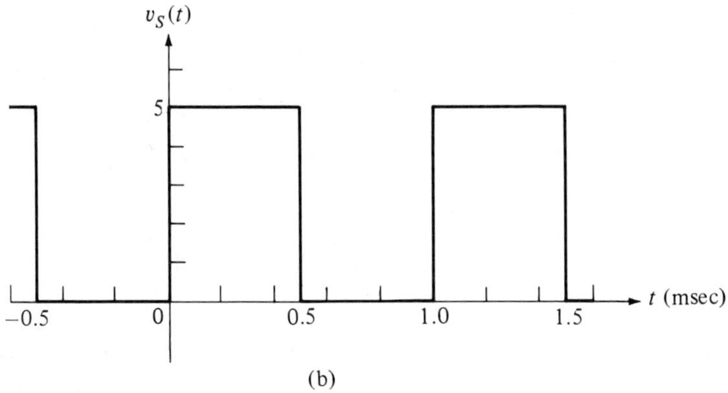

(b)

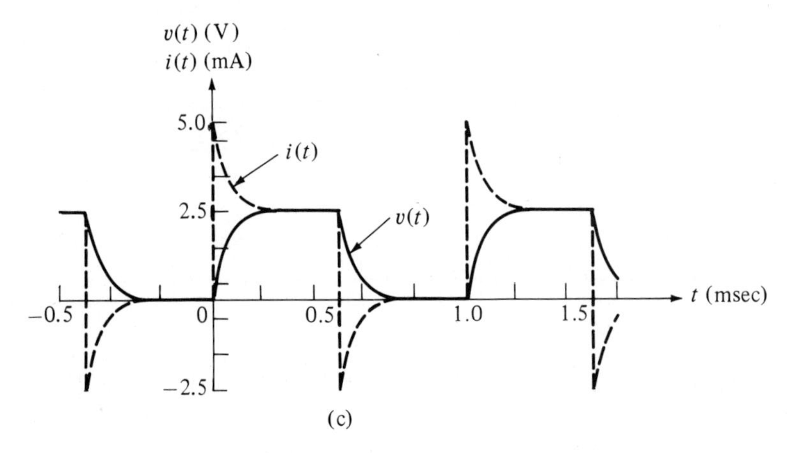

(c)

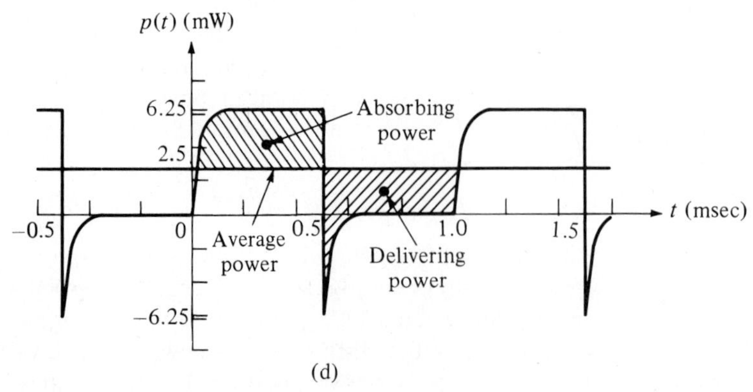

(d)

value of a function, so

$$P = \frac{1}{T}\int_{t_0}^{t_0+T} p(t)\,dt$$

$$= \frac{1}{T}\int_{t_0}^{t_0+T} v(t)\cdot i(t)\,dt \qquad (11.10)$$

The use of Fourier series representations for $v(t)$ and $i(t)$ will allow us to approximate the average power in a circuit excited by any periodic source without performing the integration required by Equation (11.10).

AC Power Calculations

Average power calculations for sinusoidal steady-state operation are usually treated thoroughly in basic circuit analysis courses. Since ac average power calculations were not reviewed in Chapter 9, it is important to do so now. Recall that if the voltage between two terminals of a linear circuit is sinusoidal, then the current is also sinusoidal, differing possibly in amplitude and phase angle, but not frequency. For a linear circuit with only sinusoidal sources operating at the same frequency, the instantaneous power absorbed or supplied by a network is

$$p(t) = (V_m\cdot\cos[\omega t+\theta])(I_m\cdot\cos[\omega t+\phi])$$

and the average power is

$$P = \frac{1}{T}\int_0^{2\pi/\omega} V_m I_m(\cos[\omega t+\theta]\cdot\cos[\omega t+\phi])\,dt$$

$$= \tfrac{1}{2}V_m I_m\cdot\cos[\theta-\phi] \qquad (11.11a)$$

For a passive, linear network, the argument of the cosine function in Equation (11.11a), $\theta-\phi$, is recognized as the polar angle ζ of the network impedance ($\mathbf{Z} = |\mathbf{Z}|\angle\zeta$); therefore [Figure 11.15(a)]

$$P = \tfrac{1}{2}V_m I_m\cdot\cos\zeta \qquad (11.11b)$$

Evaluation of the integral to obtain Equation (11.11a) is left as an exercise (Problem 11-12).

In ac analysis, $\cos[\theta-\phi]$ (or $\cos\zeta$) is sometimes called the network **power factor**. For a purely resistive network, ζ is zero, and the average power [Figure

Figure 11.15
Steady-state power dissipation in a circuit excited by a sinusoidal source.

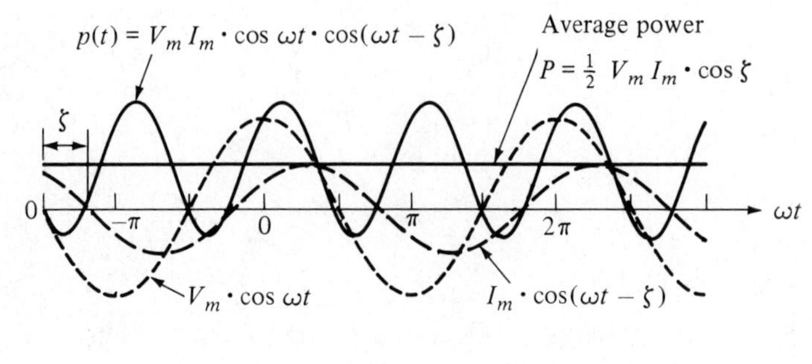

(a)

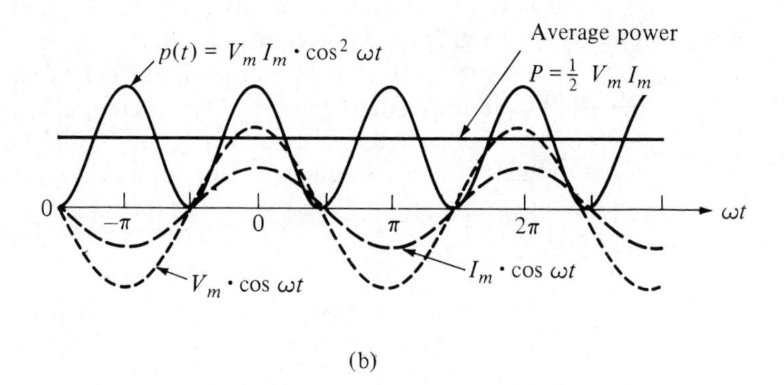

(b)

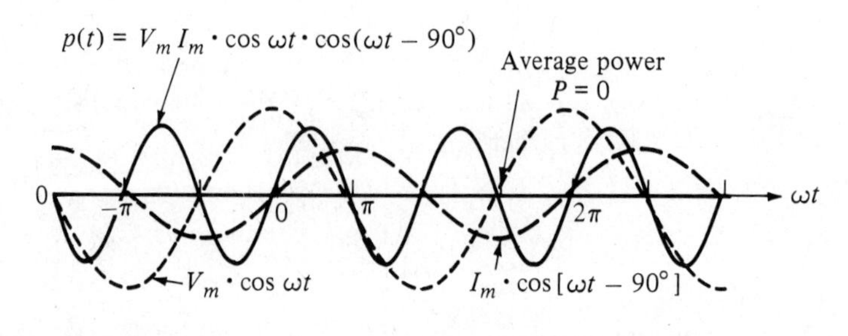

(c)

11.15(b)] is simply

$$P = \tfrac{1}{2}V_m I_m \cdot \cos 0 = \tfrac{1}{2}V_m I_m$$

and for a purely capacitive or purely inductive network, ζ is $-90°$ or $+90°$ [Figure 11.15(c)], and

$$P = \tfrac{1}{2}V_m I_m \cos[\pm 90°] = 0$$

Thus the power factor ranges from zero to unity, and average power ranges from zero to $\frac{1}{2}V_m I_m$, depending on the polar angle of the network impedance.

Average Power Calculations with Periodic Functions

The average power in circuits with periodic sources which are nonsinusoidal or piecewise sinusoidal can be calculated using the Fourier series representations of the terminal voltage and current. If

$$v(t) = V_0 + \sum_{n=1}^{\infty} (V_m)_n \cdot \cos[n\omega_1 t + \theta_n] \text{ V}$$

and

$$i(t) = I_0 + \sum_{n=1}^{\infty} (I_m)_n \cdot \cos[n\omega_1 t + \phi_n] \text{ A}$$

then from Equation (11.9),

$$\begin{aligned} p(t) &= \left(V_0 + \sum_{n=1}^{\infty} (V_m)_n \cdot \cos[n\omega_1 t + \theta_n]\right) \\ &\quad \times \left(I_0 + \sum_{n=1}^{\infty} (I_m)_n \cdot \cos[n\omega_1 t + \phi_n]\right) \\ &= V_0 I_0 + I_0 \sum_{n=1}^{\infty} (V_m)_n \cdot \cos[n\omega_1 t + \theta_n] \\ &\quad + V_0 \sum_{n=1}^{\infty} (I_m)_n \cdot \cos[n\omega_1 t + \phi_n] \\ &\quad + \sum_{p=1}^{\infty} \sum_{q=1}^{\infty} \big((V_m)_p \cdot \cos[p\omega_1 t + \theta_p]\big)\big((I_m)_q \cdot \cos[q\omega_1 t + \phi_q]\big) \end{aligned}$$

Substitution of $p(t)$ into Equation (11.10) to obtain the average power is greatly simplified by some observations about the definite integrals of trigonometric functions.

Flashback

$$\int_0^{2\pi/a} \cos ax\, dx = 0$$

Flashback

$$\int_0^{2\pi/a} \cos pax \cdot \cos qax\, dx = 0 \quad \text{for} \quad p \neq q$$

The first flashback means that

$$I_0 \sum_{n=1}^{\infty} (V_m)_n \cdot \cos[n\omega_1 t + \theta_n] \quad \text{and} \quad V_0 \sum_{n=1}^{\infty} (I_m)_n \cdot \cos[n\omega_1 t + \phi_n]$$

need not be considered in average-power calculations, since the average value of each term in each series is zero. The second flashback means that the average value of a term in the double summation is nonzero only if $p = q$, so in average power calculations, only the terms

$$\sum_{n=1}^{\infty} ((V_m)_n \cdot \cos[n\omega_1 t + \theta_n])((I_m)_n \cdot \cos[n\omega_1 t + \phi_n])$$

need to be considered.

When these observations are incorporated into Equation (11.10),

$$\begin{aligned} P &= \frac{1}{T}\left(\int_0^T V_0 I_0\, dt\right. \\ &\quad \left. + \int_0^T \sum_{n=1}^{\infty} ((V_m)_n \cdot \cos[n\omega_1 t + \theta_n])((I_m)_n \cdot \cos[n\omega_1 t + \phi_n])\, dt\right) \\ &= \frac{1}{T} V_0 I_0 t \Big|_0^T \\ &\quad + \frac{1}{T} \sum_{n=1}^{\infty} \int_0^T (V_m)_n (I_m)_n \cdot \cos[n\omega_1 t + \theta_n] \cdot \cos[n\omega_1 t + \phi_n] \cdot dt \\ &= V_0 I_0 + \sum_{n=1}^{\infty} \tfrac{1}{2}(V_m)_n (I_m)_n \cdot \cos[\theta_n - \phi_n] \end{aligned} \tag{11.12a}$$

The first term represents power due to the dc components of voltage and current, and the second term, which follows directly from Equation (11.11a), represents the summation of power due to the sinusoidal components of the Fourier series for the voltage and for the current.

If the network under consideration is a passive, linear network, $\theta_n - \phi_n$ is the polar angle $\zeta(jn\omega)$ of the network impedance evaluated for each discrete frequency of the Fourier series, so

$$P = V_0 I_0 + \sum_{n=1}^{\infty} \tfrac{1}{2}(V_m)_n (I_m)_n \cdot \cos[\zeta(jn\omega)] \qquad (11.12b)$$

Examination of Equations (11.12a) and (11.12b) reveals the significant fact that only voltage and current at the same frequency interact to supply or absorb power; there is no real power associated with the interaction of voltages and currents at different frequencies.

Example 11.8

Problem:

Use the Fourier series to calculate the approximate average power absorbed by the passive network in Figure 11.16.

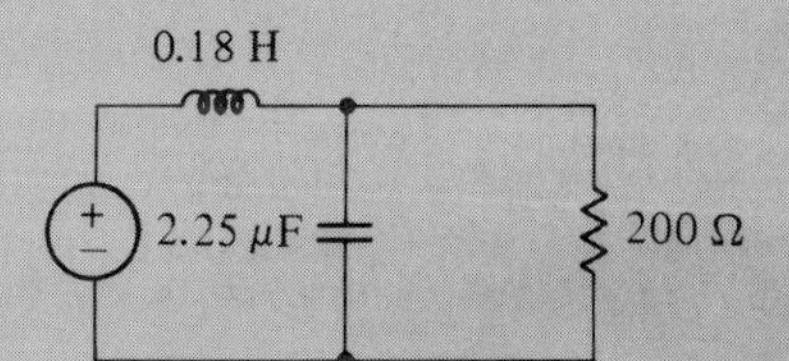

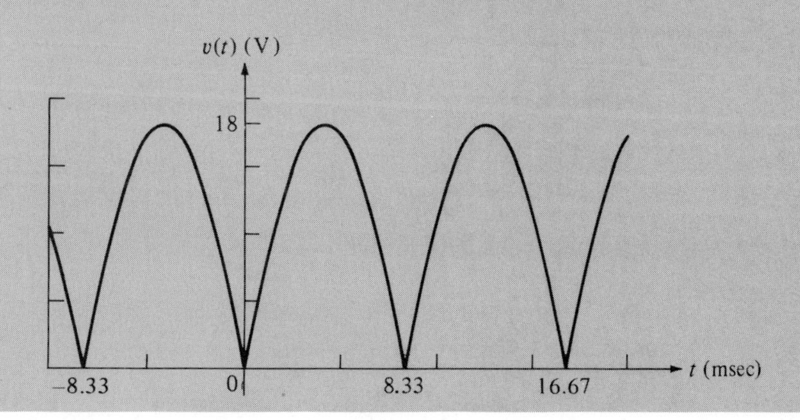

Figure 11.16
Example 11.8.

Solution:

The solution has been organized into tabular form[2] according to the following steps.

1. Use Table 11.1 to obtain the Fourier series for the source voltage (column 1 in the table below):

$$v(t) = \frac{2V_m}{\pi} - \frac{4V_m}{\pi}\sum_{n=1}^{\infty}\frac{1}{4n^2-1}\cos n\omega_1 t$$

$$\omega_1 = 2\pi(120) \simeq 754.0 \text{ rad/sec}$$

2. Determine the network impedance (columns 2 and 3):

$$\mathbf{Z}(j\omega) = j\omega L + \frac{1}{G + j\omega C} = \frac{(1 - \omega^2 LC) + j\omega GL}{G + j\omega C}$$

3. Determine the source current amplitude for each component of $v(t)$ (column 4):

$$(I_m)_n = \frac{(V_m)_n}{|\mathbf{Z}(jn\omega_1)|}$$

4. Use the results of steps 1, 2, and 3 in Equation (11.12b) to calculate the power associated with each component of the Fourier series (column 5).

Column 6 in the table is the cumulative power as successively higher harmonics are considered. It is clear that, in this case, the amount of power absorbed from the harmonic components of the source is insignificant when compared to that absorbed from the dc and fundamental components.

	V VOLTS	*Z* (MAG) OHMS	*Z* (ANG) DEG	*I* mA	*P* mW	Σ*P* mW
dc	11.4592	200.0000	0.0000	57.2958	656.5613	656.5613
FUND	7.6394	194.3504	22.6561	39.3075	138.5576	795.1189
2ND HARM	1.5279	224.9841	52.5064	6.7911	3.1578	798.2767
3RD HARM	0.6548	322.4900	72.2663	2.0305	0.2025	798.4792
4TH HARM	0.3638	452.8577	81.0598	0.8033	0.0227	798.5019
5TH HARM	0.2315	593.3376	85.0135	0.3902	0.0039	798.5058
6TH HARM	0.1603	736.1820	86.9728	0.2177	0.0009	798.5067
7TH HARM	0.1175	879.0052	88.0365	0.1337	0.0003	798.5070

[2] This table is an excellent example of the utility of the electronic spreadsheet, usually considered a tool for business and accounting, in engineering and technology.

Exercise 11.5

Problem:

Calculate the average power dissipated by the 500-Ω resistor in Exercise 11.3.

Answer:

$P \simeq 1.535$ W.

	V VOLTS	Z (MAG) OHMS	Z (ANG) DEG	I mA	P W	ΣP W
dc	25.000	500.000	0.000	50.000	1.250	1.250
FUND.	15.915	708.987	45.152	22.448	0.252	1.502
2ND HARM.	7.958	1122.786	63.556	7.088	0.025	1.527
3RD HARM.	5.305	1588.697	71.656	3.339	0.006	1.533
4TH HARM.	3.979	2071.857	76.035	1.920	0.002	1.534
5TH HARM.	3.183	2562.527	78.748	1.242	0.001	1.535

Effective Values of Periodic Functions

A frequently used characteristic of a periodic waveform is its **effective value**. The effective value of a periodic voltage or current is defined as the dc voltage or current which would deliver the same average power to a resistor. Based on Equation (11.10), the average power delivered to a resistor by any periodic voltage or current is

$$P = \frac{1}{T}\int_{t_0}^{t_0+T} p(t)\,dt$$

where $p(t) = v(t)^2/R$ or $p(t) = R \cdot i(t)^2$, so

$$P = \frac{1}{RT}\int_{t_0}^{t_0+T} v(t)^2\,dt$$

or

$$P = \frac{R}{T}\int_{t_0}^{t_0+T} i(t)^2\,dt$$

The definition of effective value requires that the power calculated above be the same as

$$P = V_{\text{eff}}^2/R \quad \text{or} \quad P = RI_{\text{eff}}^2$$

so

$$\frac{V_{\text{eff}}^2}{R} = \frac{1}{RT}\int_{t_0}^{t_0+T} v(t)^2\,dt$$

or

$$RI_{\text{eff}}^2 = \frac{R}{T}\cdot\int_{t_0}^{t_0+T} i(t)^2\,dt$$

The resistance can be canceled and the equations solved for the effective value, with the results

$$V_{\text{eff}} = \sqrt{\frac{1}{T}\int_{t_0}^{t_0+T} v(t)^2\,dt} \tag{11.13a}$$

$$I_{\text{eff}} = \sqrt{\frac{1}{T}\int_{t_0}^{t_0+T} i(t)^2\,dt} \tag{11.13b}$$

The form of Equations (11.13a) and (11.13b) reveals why effective values are often called **root-mean-squared** (rms) values. The expression under the radical (root) is the average (mean) of the function squared.

Effective values are particularly useful in calculating the average power delivered to a resistive load by a periodic source. In this case, average power calculations can be formulated exactly like dc power calculations—that is,

$$P = V_{\text{eff}} I_{\text{eff}} \tag{11.14a}$$

$$P = V_{\text{eff}}^2/R \tag{11.14b}$$

$$P = RI_{\text{eff}}^2 \tag{11.14c}$$

The effective value of a periodic function must not be confused with the average value or dc component of the function. Voltage or current waveforms which have zero average value (no dc component) are still *effective* in delivering power. The most obvious case is sinusoidal voltage or current. The reader is probably well aware that the effective value of a sinusoidal voltage or current is $V_m/\sqrt{2}$ ($0.7071V_m$) or $I_m/\sqrt{2}$ ($0.7071I_m$), respectively. Calculation of the effective value of a sinusoidal function is done in most basic circuit analysis texts and is left as a review exercise for the reader (Problem 11-15).

The factor $1/\sqrt{2}$ (0.7071) is *not* inherently related to effective value. A common mistake in calculating effective values of nonsinusoidal or piecewise-sinusoidal periodic functions is to divide the peak value by $\sqrt{2}$ (multiply by 0.7071). In general, Equation (11.13a) or (11.13b) must be evaluated for each periodic function by squaring the function, calculating the average of the squared function by integration, then taking the root of the result. The effective value of a periodic function can be approximated, without performing the rms operation and its attendant integration, through the use of the Fourier series representation for the function.

Recall from Equation (11.4a) that

$$f(t) = A_0 + \sum_{n=1}^{\infty} (A_m)_n \cdot \cos[n\omega_1 t - \theta_n]$$

The first step in obtaining the effective value is to square $f(t)$:

$$f(t)^2 = A_0^2 + 2A_0 \sum_{n=1}^{\infty} (A_m)_n \cdot \cos[n\omega_1 t - \theta_n]$$
$$+ \sum_{p=1}^{\infty} \sum_{q=1}^{\infty} \left((A_m)_p \cdot \cos\left[p\omega_1 t + \theta_p\right]\right)\left((A_m)_q \cdot \cos\left[q\omega_1 t + \phi_q\right]\right)$$

The next step is to integrate $f(t)^2$ from t_0 to $t_0 + T$ [Equation (11.13a) or (11.13b)], but the situation here is similar to that encountered in the previous section in obtaining average power. All terms in the single summation and all terms in the double summation for which $p \neq q$ will vanish when integrated and evaluated over one period. As a result, the mean value of $f(t)^2$ is

$$\frac{1}{T}\int_{t_0}^{t_0+T} f(t)^2\, dt = \frac{1}{T}\int_{t_0}^{t_0+T}\left(A_0^2 + \sum_{n=1}^{\infty} (A_m)_n^2 \cos^2(n\omega_1 t + \theta_n)\right) dt$$
$$= A_0^2 + \sum_{n=1}^{\infty}\left(\frac{(A_m)_n}{\sqrt{2}}\right)^2 = A_0^2 + \frac{1}{2}\sum_{n=0}^{\infty} (A_m)_n^2$$

When the root is taken to determine the effective value of $f(t)$,

$$F_{\text{eff}} = \sqrt{A_0^2 + \frac{1}{2}\sum_{n=1}^{\infty} (A_m)_n^2} \qquad (11.15)$$

This result means that the effective value of any periodic function is found by summing the squares of the effective values of its Fourier series components, then taking the square root of the sum. Care must be taken to perform the operations indicated by Equation (11.15) in proper order.

Example 11.9

Problem:

Calculate the approximate effective value of the voltage source in Example 11.8.

Solution:

1. Use the table from Example 11.8 to determine the effective value of each component of the Fourier series for $v(t)$ (column 2).
2. Square the effective value of each component (column 3).
3. Sum the results of step 2 (column 4) and take the square root (column 5). Note in column 5 that the higher harmonics do not contribute significantly to the effective value of $v(t)$.

	V VOLTS	Veff VOLTS	Veff^2 VOLTS^2	ΣVeff^2 VOLTS^2	ΣVeff VOLTS
dc	11.4592	11.4592	131.3123	131.3123	11.4592
FUND	7.6394	5.4019	29.1805	160.4928	12.6686
2ND HARM	1.5279	1.0804	1.1672	161.6600	12.7146
3RD HARM	0.6548	0.4630	0.2144	161.8744	12.7230
4TH HARM	0.3638	0.2572	0.0662	161.9405	12.7256
5TH HARM	0.2315	0.1637	0.0268	161.9673	12.7266
6TH HARM	0.1603	0.1133	0.0128	161.9802	12.7271
7TH HARM	0.1175	0.0831	0.0069	161.9871	12.7274

11.5 Frequency Spectrum

A **spectrum** is simply a collection of related elements—in the case of the Fourier series the related elements are the amplitudes (amplitude spectrum) and the displacement angles (phase spectrum) of the Fourier series components. The amplitude and phase spectra of the Fourier series are **discrete** rather than **continuous** spectra because the elements exist only at discrete frequencies which are multiples of $2\pi/T$ rad/sec. When the elements of the amplitude or phase spectra of a periodic waveform are plotted vs. frequency, as in Figure 11.17, the result is called a **line spectrum**. In this section we will use line spectra in conjunction with the frequency-domain transfer function to help understand the effect of two-port networks—such as filters and coupling networks—on nonsinusoidal or piecewise sinusoidal periodic signals.

Figure 11.17(a) shows the frequency spectrum of a 0-to-1-V, 1-Hz pulse train with a 50% duty cycle. The Fourier series, truncated after the 19th harmonic,, provides a relatively good approximation of the pulse train [Figure 11.17(b)]. Figures 11.18 through 11.21 show the effect of each of four filters on

Figure 11.17 Frequency spectrum of a pulse train. (a) Frequency spectrum. (b) Pulse train and Fourier series approximation.

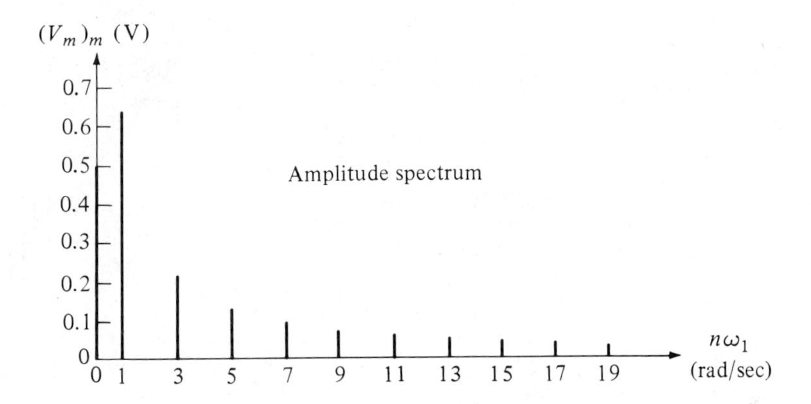

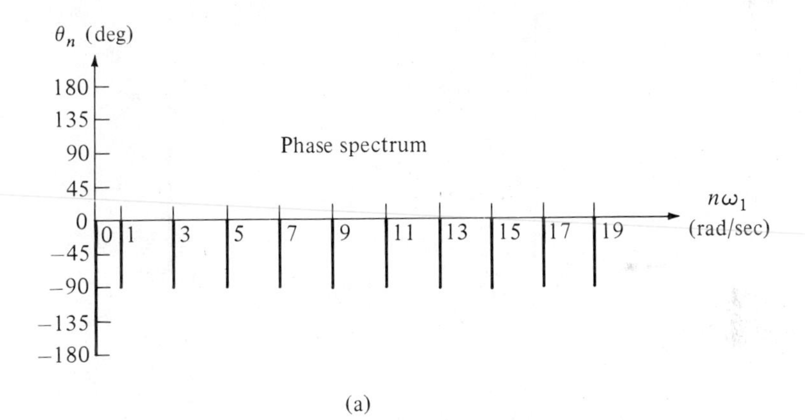

(a)

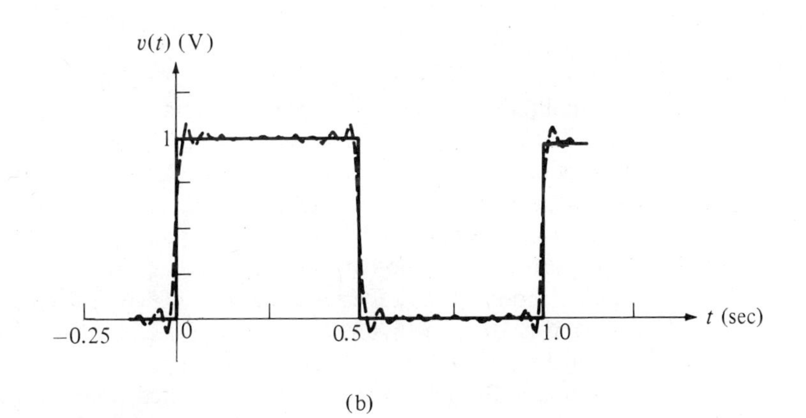

(b)

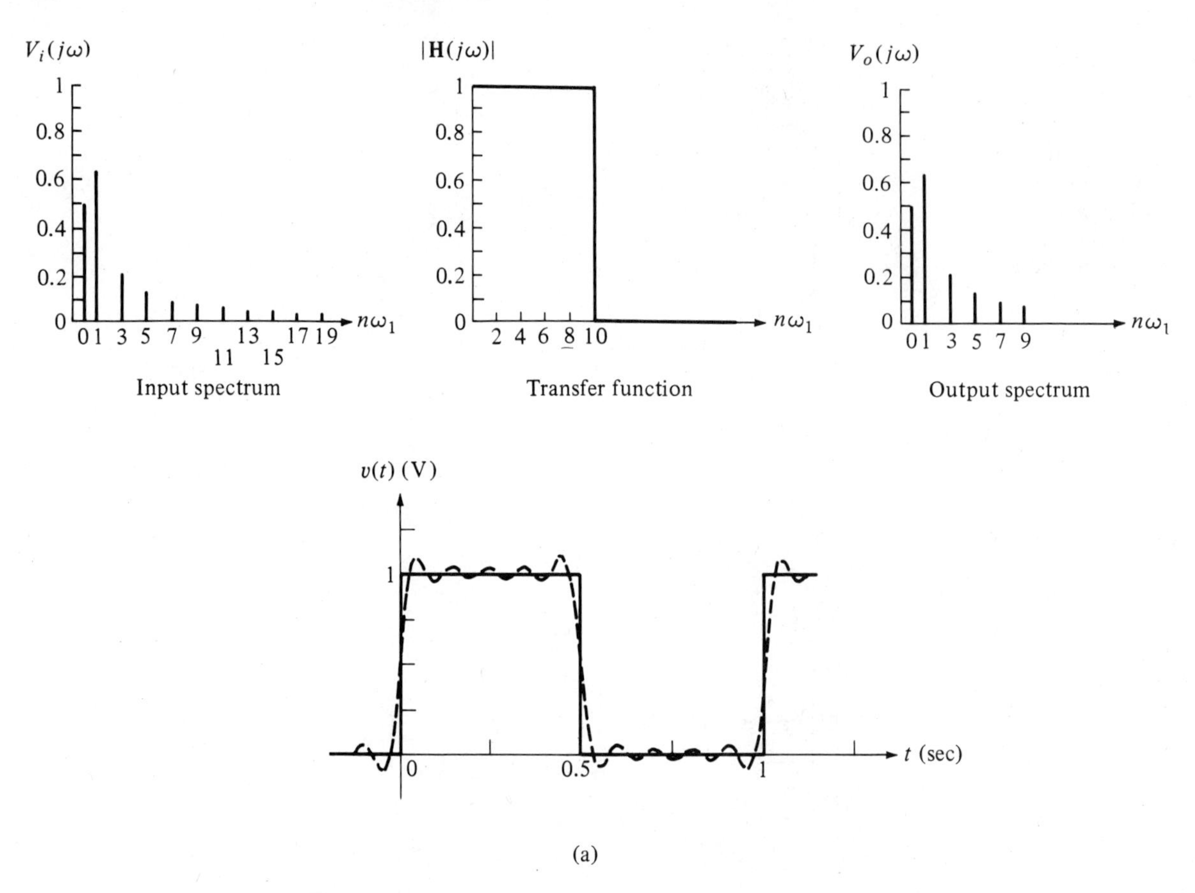

Figure 11.18 Effect of a low-pass filter on a pulse train. (a) Wide pass band, $\omega_c = 10\omega_1$. (b) Narrow pass band, $\omega_c = 6\omega_1$.

the amplitude spectrum of the pulse train and on the shape of the output pulse. The situation is highly idealized, since the filters are considered to have unity gain, infinite rolloff, and zero phase shift within the pass band. Nevertheless, some important generalizations can be made regarding the effect of two-port networks on periodic nonsinusoidal signals.

Figure 11.18(a) and (b) show that a low-pass filter effectively truncates the Fourier series after a certain number of components. The width of the passband is $10\omega_1$ in Figure 11.18(a), so the eleventh and all higher harmonics are blocked; the width of the passband in Figure 11.18(b) is $6\omega_1$, so the seventh and all higher harmonics are blocked. The most evident effect of the low-pass filter on pulse shape is the change in the slope of the rising and falling edges. Clearly the high-frequency components determine rise and fall

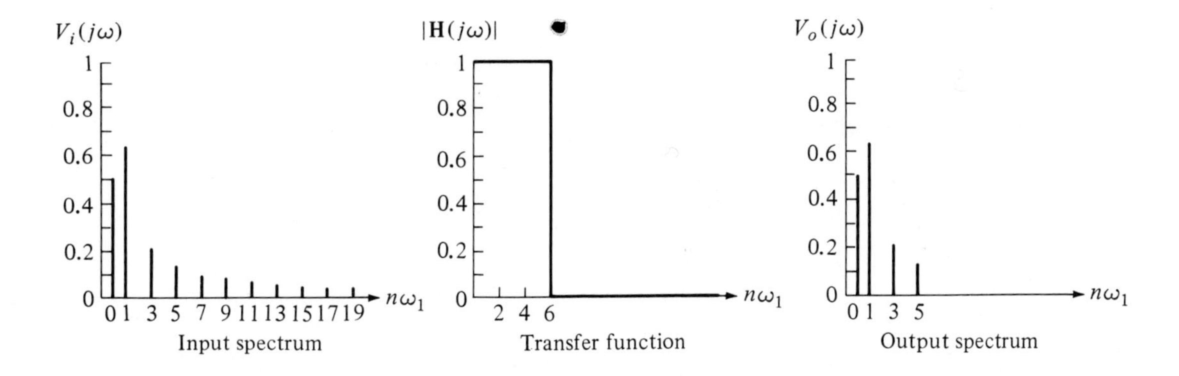

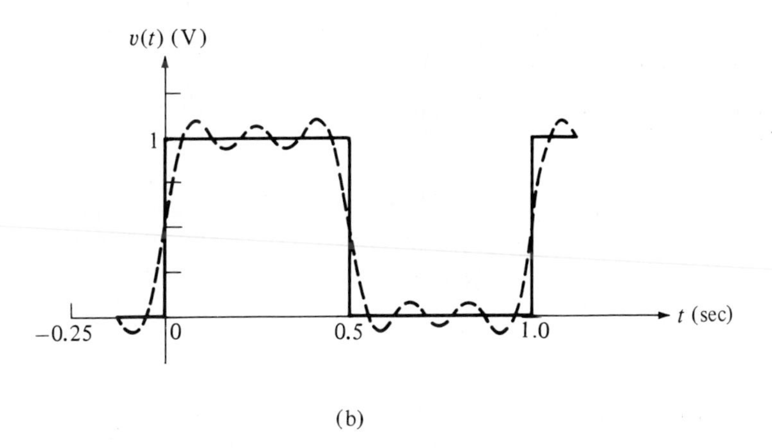

(b)

Figure 11.18 Continued.

times, and the conclusion may be drawn that poor high-frequency response results in **rounding** of the vertical edges of a periodic waveform. A dramatic example can be seen in Figure 11.12, where the simplest of low-pass filters has seriously degraded the pulse shape by virtually eliminating the vertical edges of the pulse.

Figure 11.19(a) and (b) indicate that a high-pass filter displaces the pulse vertically due to loss of the dc component of the Fourier series, and creates **sagging** due to loss of the fundamental component. In Figure 11.19(a), the cutoff frequency has been placed at ω_1, which means that the amplitude of the fundamental has been reduced by 3 dB; in Figure 11.19(b), the more pronounced sagging is due to placing the cutoff frequency below ω_1, effectively eliminating the fundamental. The sagging due to poor low-frequency response

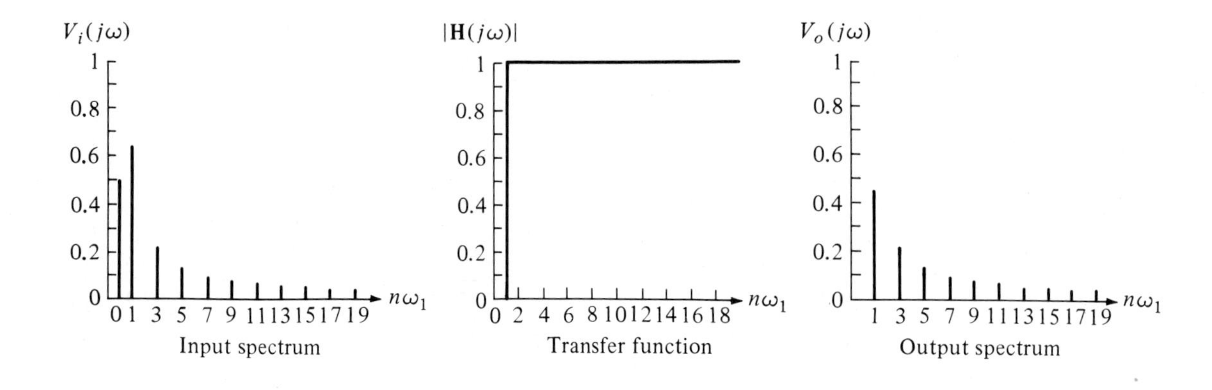

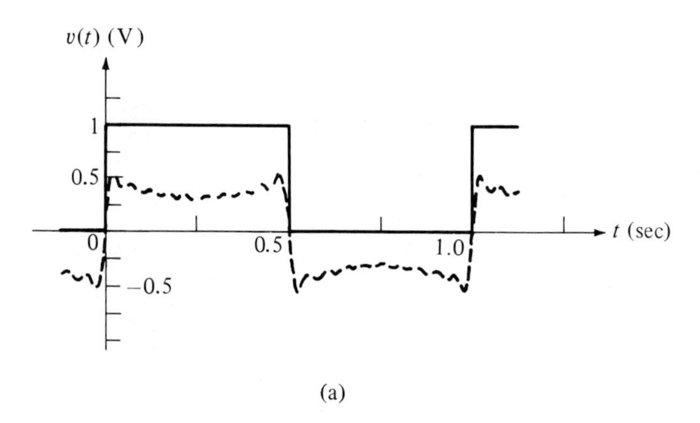

(a)

Figure 11.19 Effect of a high-pass filter on a pulse train. (a) Wide stop band, $\omega_c = \omega_1$. (b) Narrow stop band, $\omega_c = 0.5\omega_1$.

is evident in Figure 11.13, where a simple high-pass filter has dramatically altered the pulse shape.

In Figure 11.20 the fundamental has been rejected, with the result that the pulse sags, but since the dc component is still present, there is no vertical displacement of the pulse. In Figure 11.21, only the third through ninth harmonics are passed, with the result that the pulse not only is displaced vertically and sags, but also has degraded edges.

The effects just noted are referred to as **distortion**; in general, distortion is any variation between the shape of the excitation waveform and the shape of the response waveform. Distortion is the result of amplifying or attenuating one component of the amplitude spectrum more than another, and/or displacing one component of the phase spectrum more than another. When all

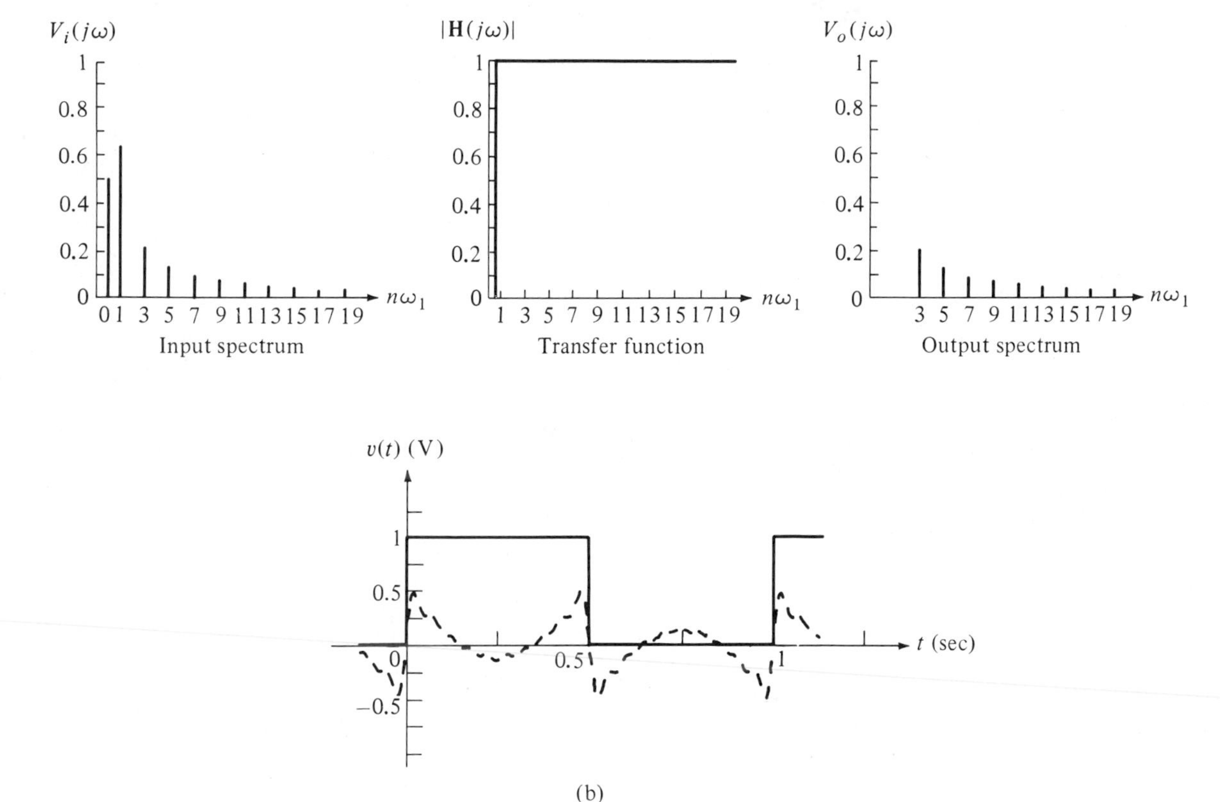

Figure 11.19 Continued.

amplitude components are affected proportionally, the amplitude of the entire response waveform is affected, but it retains the shape of the excitation. Likewise, when all phase components experience an equal angular displacement, the entire response waveform is displaced in time, but its shape is not affected.

The pulse train was chosen to illustrate the effect of filtering on pulse shape because of its importance to digital logic systems. The shunt capacitance at the output of digital circuits and along printed circuit boards and interconnecting cables causes most digital transmission paths to behave as low-pass filters. As a result the sharply rising and falling edges of the pulses generated by the digital circuits are often rounded to the point where they are not suitable for triggering downstream logic circuits. If this effect is recognized by the system designer, it can easily be corrected by the use of circuits (such as

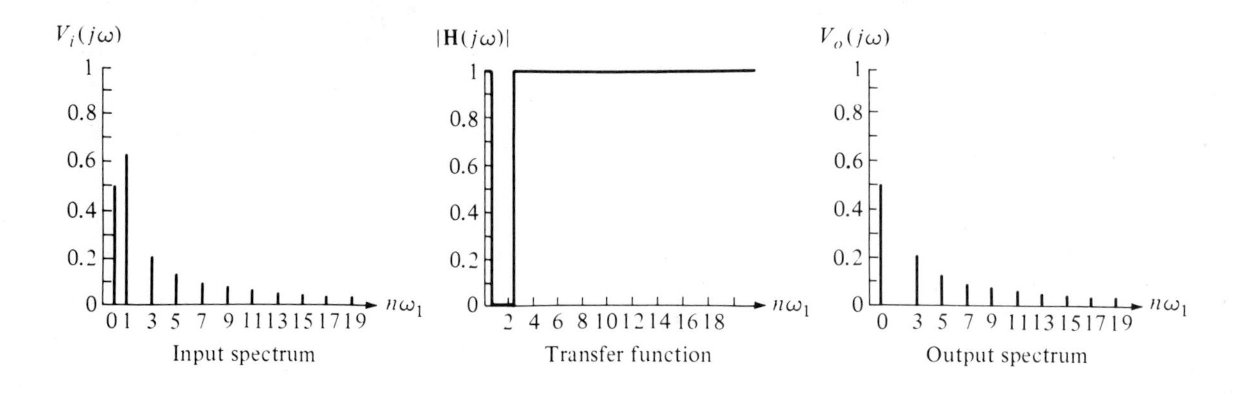

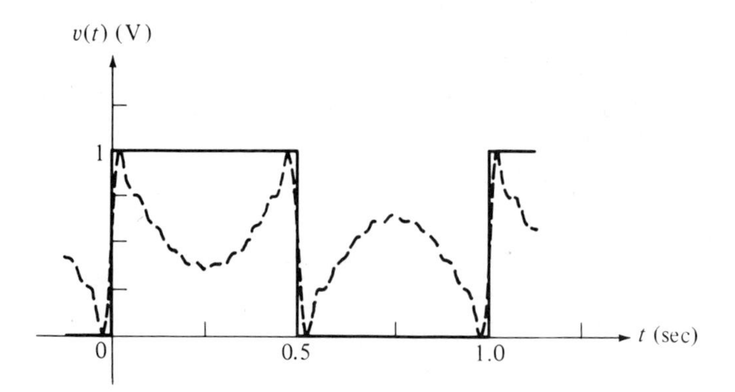

Figure 11.20 Effect of a band-stop filter on a pulse train.

the Schmitt trigger) which restore the sharp edges to the pulse [Figure 11.22(a)].

The requirement to provide capacitive coupling at various points in a digital system, along with the existence of stray inductance, can create high-pass filters along digital transmission paths. Care must be taken in designing digital systems to provide dc restorers (clamps) whenever necessary [Figure 11.22(b)], and especially to prevent the pulse sagging associated with filtering of the fundamental. If a pulse is allowed to sag until it falls below the system low logic level, then as it rises above the low logic level, the system responds as though two pulses, rather than one, had been transmitted [Figure 11.22(c)].

The frequency spectra for several common waveforms are given in Table 11.2; in each case the waveform amplitude is one. Another way of presenting a frequency spectrum is to normalize the components by dividing each by the amplitude of the fundamental; the normalized fundamental then has an

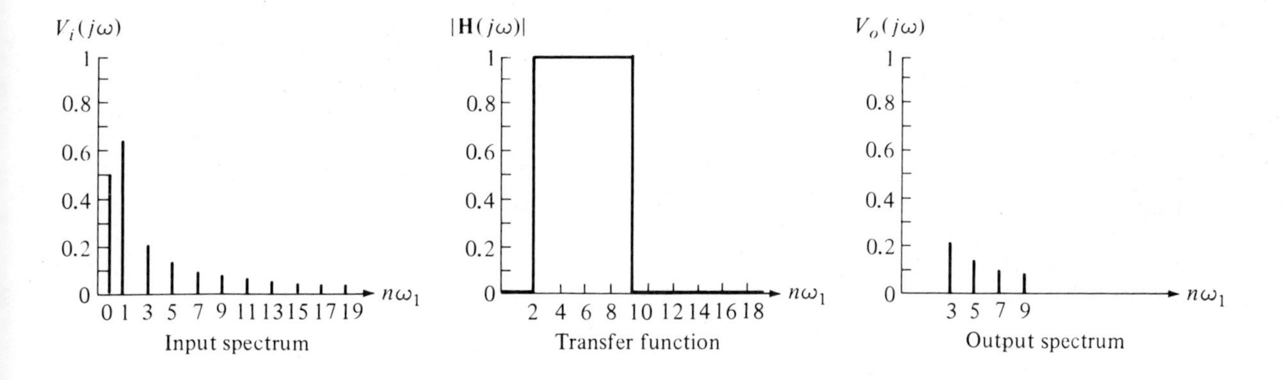

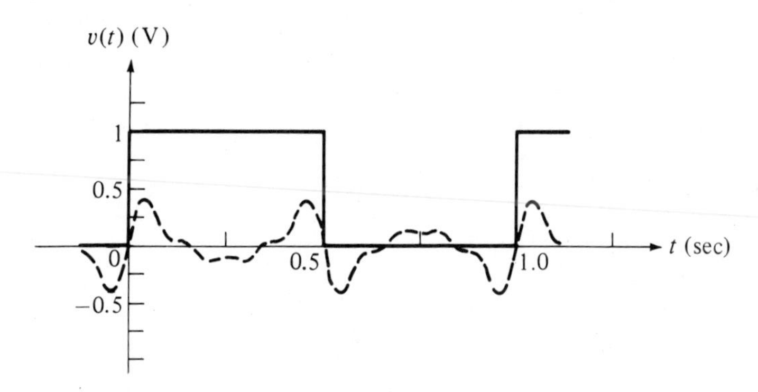

Figure 11.21 Effect of a band-pass filter on a pulse train.

amplitude of one, and the amplitudes of all other components are given as a percentage of the fundamental amplitude.

11.6 Fourier Transforms and Continuous Spectra

In the previous sections of this chapter, we examined the use of the Fourier series to represent a periodic function as a sum of harmonically related sinusoids. With this representation and its associated line spectra, it proved possible to use the techniques of ac analysis and frequency response analysis to determine the response of a linear circuit to a nonsinusoidal periodic excitation. The ideas underlying the Fourier series can be extended, allowing

Figure 11.22
Effects of filtering on a digital pulse.
(a) Restoration of rapid rise and fall time.
(b) Restoration of dc component.
(c) Possible logic error due to sagging.

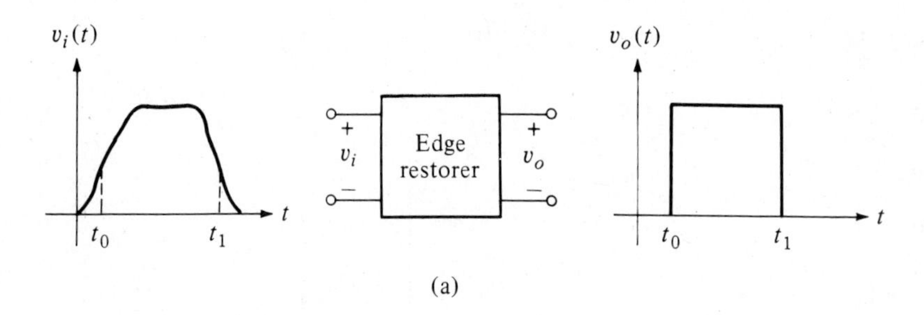

(a)

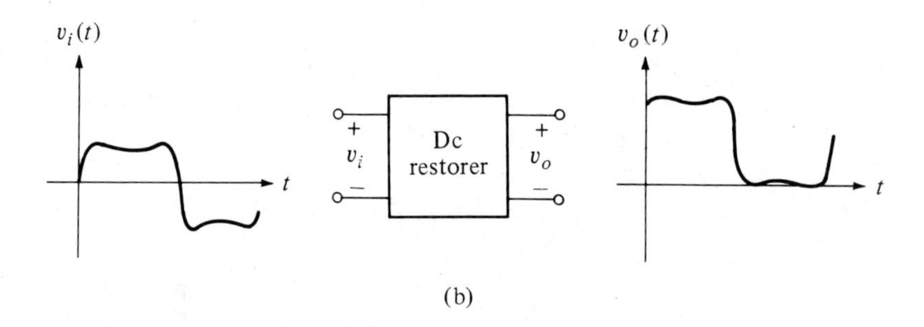

(b)

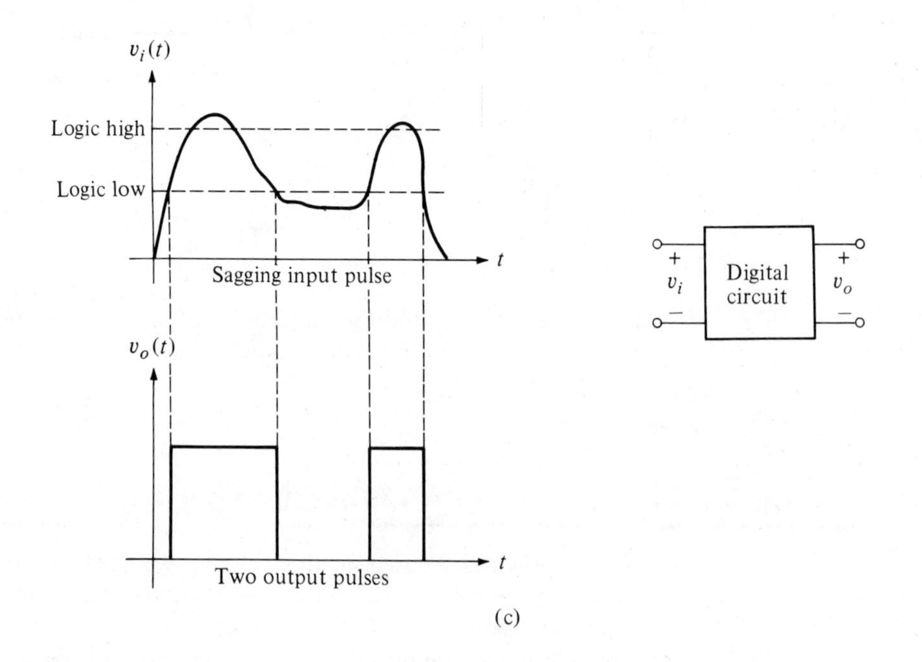

(c)

Table 11.2 Frequency Spectra

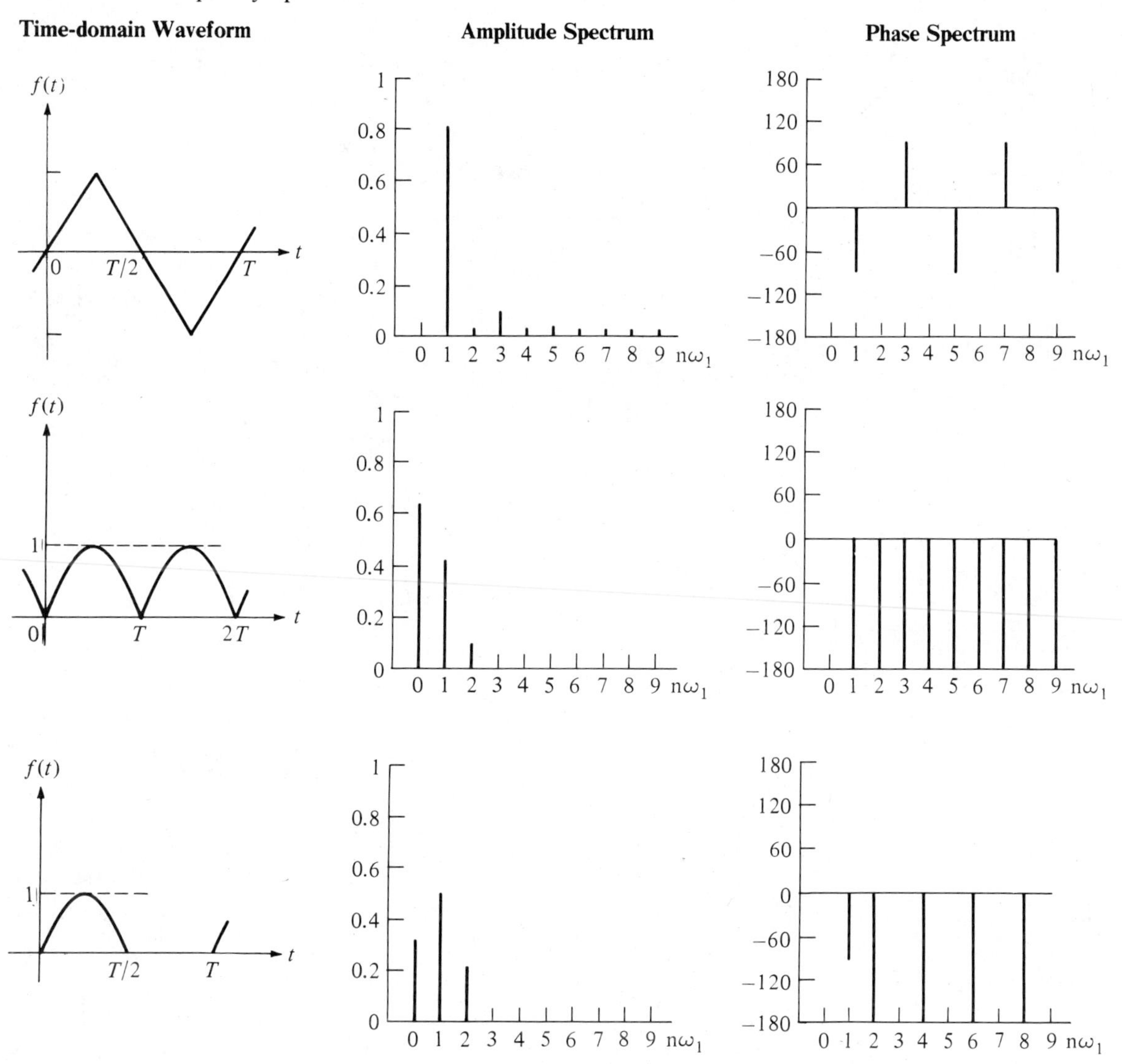

similar treatment of nonperiodic functions. This extension leads us to another useful integral transform, the **Fourier transform**.

A thorough treatment of the Fourier transform is beyond the scope and level of mathematical sophistication of this text, so we will confine ourselves to an intuitive, highly informal presentation of a few fundamentals and a discussion of applications to circuit analysis.

Table 11.2
Continued

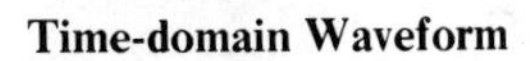

Time-domain Waveform | **Amplitude Spectrum** | **Phase Spectrum**

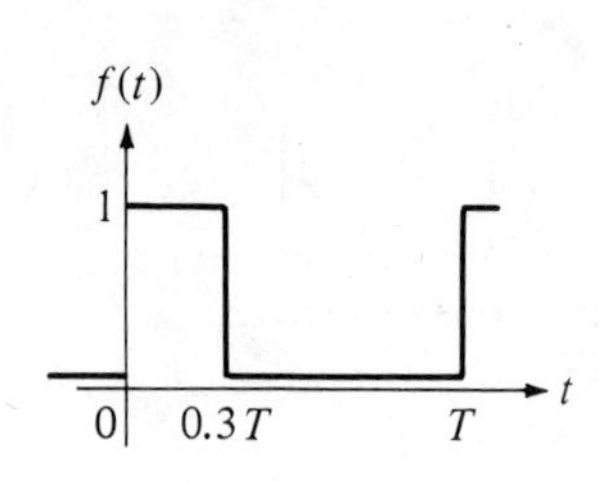

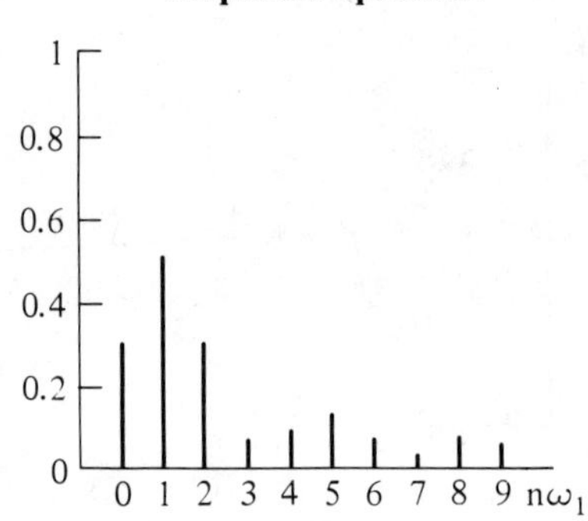

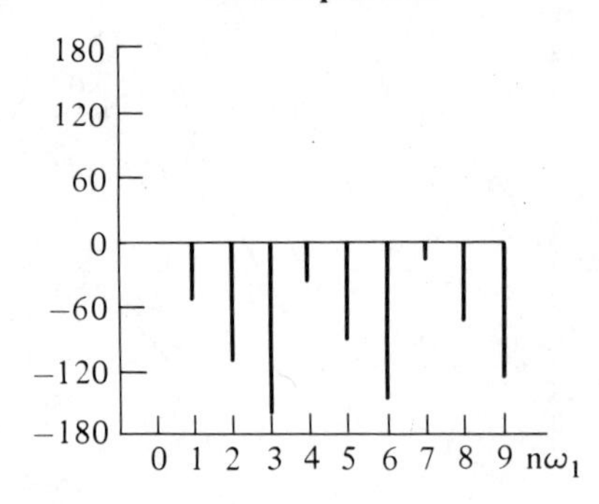

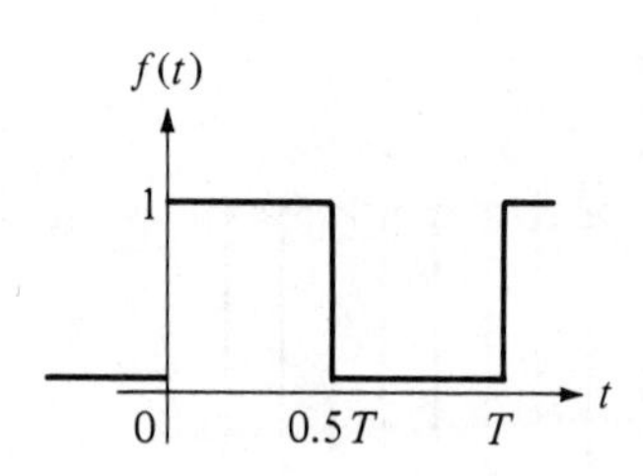

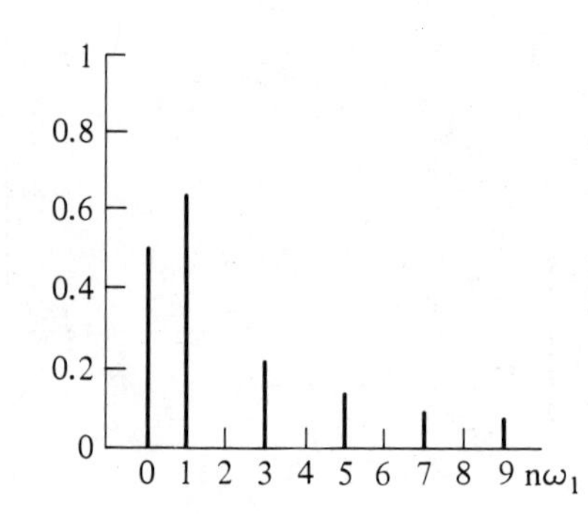

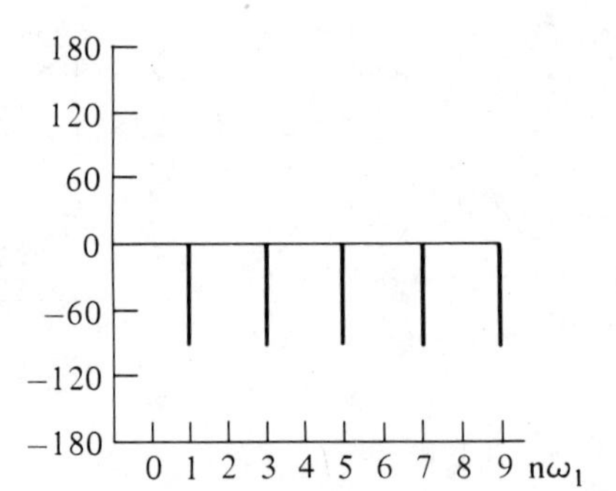

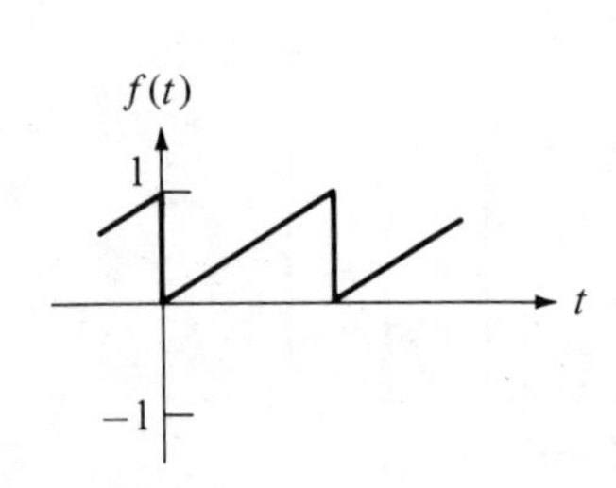

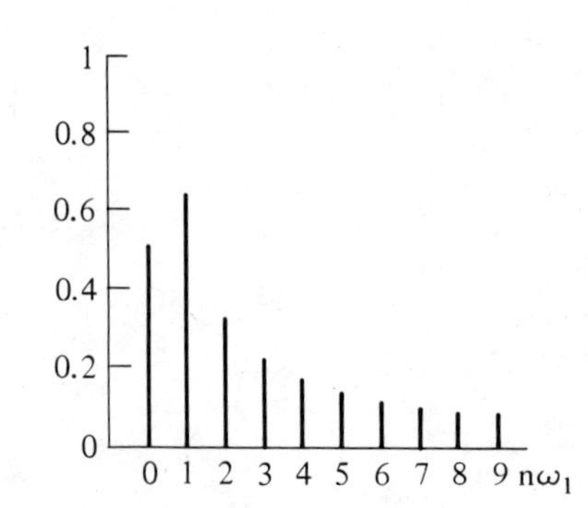

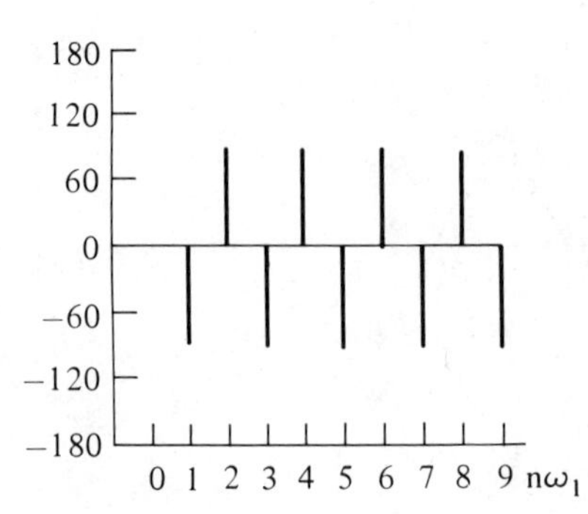

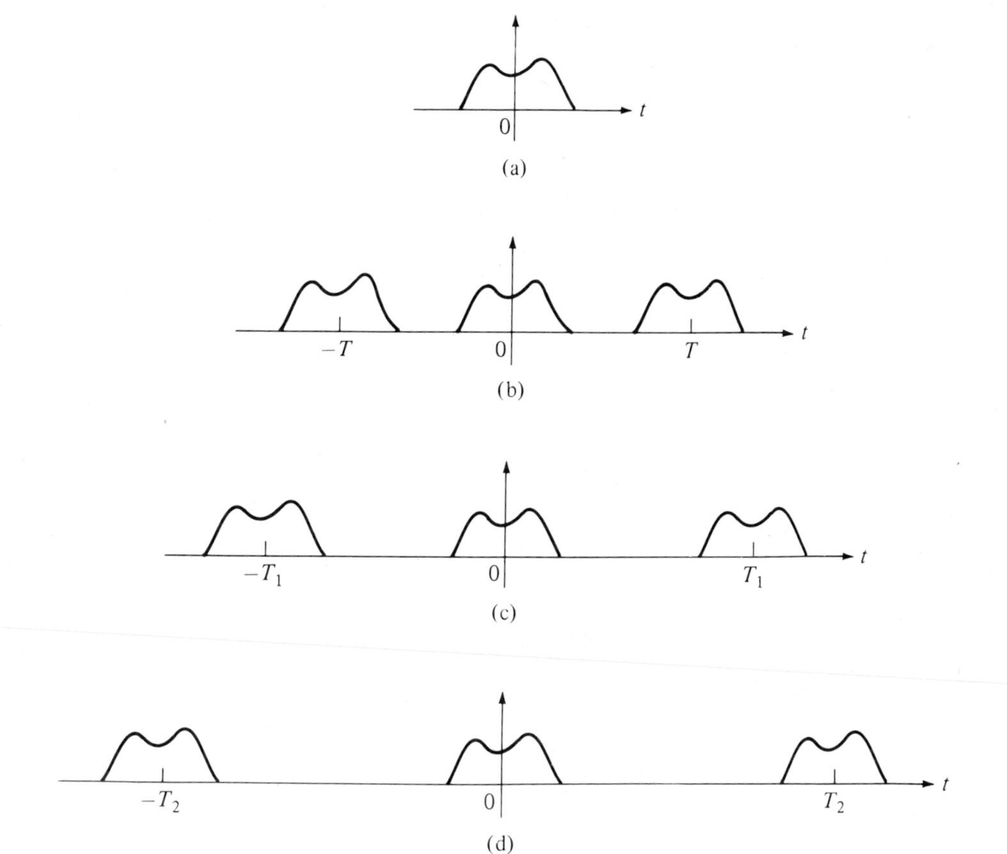

Figure 11.23 Periodic extension of $f(t)$. (a) $f(t)$. (b) $f_T(t)$. (c) $f_{T_1}(t)$, $T_1 > T$. (d) $f_{T_2}(t)$, $T_2 > T_1 > T$.

Nonperiodic Functions and Periodic Extensions

Central to the development of the Fourier transform is the idea that a nonperiodic function is in a sense the *limit* of a periodic function as its period approaches infinity. Figure 11.23(a) shows a function $f(t)$ defined on an interval centered at $t = 0$. Consider a new function $f_T(t)$ constructed by replicating the graph of $f(t)$ every T units along the t-axis, as shown in Figure 11.23(b). This function, called the **periodic extension** of $f(t)$, is a periodic function with period T and defined for all t. It is intuitively clear that as T increases [Figures 11.23(c) and (d)] $f_T(t)$ approaches the aperiodic function $f(t)$. This is formally expressed as:

$$\lim_{T \to \infty} f_T(t) = f(t) \tag{11.16}$$

The important point is that for any value of T, $f_T(t)$ has a Fourier series representation, which suggests that the Fourier series for $f_T(t)$ should be examined as $T \to \infty$.

The Fourier Transform as a Limit of the Fourier Series

To study the Fourier series of $f_T(t)$ as $T \to \infty$, it is convenient to use the **exponential form** of the Fourier series

$$f(t) = \sum_{n=-\infty}^{\infty} \mathbf{C}_n \cdot \exp[jn\omega_1 t] \tag{11.17}$$

where $\mathbf{C}_n$ is the spectrum of complex coefficients for the Fourier series,

$$\mathbf{C}_n = \frac{1}{T}\int_{-T/2}^{T/2} f(t) \cdot \exp[-jn\omega_1 t]\, dt \tag{11.18}$$

Note that the lower limit of the summation index in Equation (11.17) is $-\infty$, rather than 1 as in Equation (11.2). The reader may verify (Problems 11-19 and 11-20) that Equations (11.17) and (11.18) are equivalent to Equations (11.1) through (11.3c).

In the following discussion, it is important to emphasize that the Fourier coefficients $\mathbf{C}_n$ are functions of frequency, so the Fourier coefficients for $f_T(t)$ will be identified as $\mathbf{C}_T(n\omega_1)$. The Fourier series for $f_T(t)$ is

$$f_T(t) = \sum_{n=-\infty}^{\infty} \mathbf{C}_T(n\omega_1) \cdot \exp[jn\omega_1 t] \tag{11.19}$$

where

$$\mathbf{C}_T(n\omega_1) = \frac{1}{T}\int_{-T/2}^{T/2} f_T(t) \cdot \exp[-jn\omega_1 t]\, dt \tag{11.20}$$

The frequency of the fundamental component is $\omega_1 = 2\pi/T$, and all other components have frequencies which are integer multiples of ω_1. If an arbitrary Fourier component of $f_T(t)$ with frequency $\omega = n\omega_1 = 2\pi n/T$ is selected, then the integral in the right member of Equation (11.20) is a function of ω, or

$$\mathbf{C}_T(\omega) = \frac{1}{T}\int_{-T/2}^{T/2} f_T(t) \cdot \exp[-j\omega t]\, dt \tag{11.21}$$

Now, define a new complex function

$$\mathbf{F}_T(\omega) = T \cdot \mathbf{C}_T(\omega) = \int_{-T/2}^{T/2} f_T(t) \cdot \exp[-j\omega t]\, dt \qquad (11.22a)$$

or

$$\mathbf{C}_T(\omega) = \frac{\mathbf{F}_T(\omega)}{T} \qquad (11.22b),$$

so that Equation (11.19) becomes

$$\begin{aligned} f_T(t) &= \sum_{n=-\infty}^{\infty} \frac{\mathbf{F}_T(\omega)}{T} \cdot \exp[j\omega t] \\ &= \frac{1}{2\pi} \sum_{n=-\infty}^{\infty} \frac{2\pi}{T} \cdot \mathbf{F}_T(\omega) \cdot \exp[j\omega t] \end{aligned} \qquad (11.23)$$

Now, let $T \to \infty$, so that $f_T(t) \to f(t)$, and simultaneously let n vary as T varies in order to hold ω ($= 2\pi n/T$) fixed. Then Equation (11.22a) becomes

$$\mathbf{F}(\omega) = \int_{-\infty}^{\infty} f(t) \cdot \exp[-j\omega t]\, dt \qquad (11.24)$$

The separation $\Delta\omega$ between harmonics of the Fourier series is equal to the fundamental frequency:

$$\Delta\omega = \omega_1 = 2\pi/T$$

In the limit as T increases toward infinity, the increment $\Delta\omega$ becomes a differential, so

$$2\pi/T \to d\omega$$

and the sequence of discrete harmonics ($n\omega_1$) becomes a continuous variation in the frequency ω. With these observations incorporated, Equation (11.23) becomes

$$f(t) = \frac{1}{2\pi} \int_{-\infty}^{\infty} \mathbf{F}(\omega) \cdot \exp[j\omega t]\, d\omega \qquad (11.25)$$

Equation (11.24) defines the Fourier transform of $f(t)$, and Equation (11.25) defines the **inverse Fourier transform** of $\mathbf{F}(\omega)$, and is frequently called the **Fourier integral**. In effect, Equation (11.24) *analyzes* a nonperiodic waveform into its frequency spectrum, and Equation (11.25) *synthesizes* a nonperiodic waveform from a continuous spectrum of frequency components.

Table 11.3
Properties of the Fourier Transform

Property	$f(t)$	$\mathbf{F}(\omega)$
Linearity	$k \cdot f(t)$	$k \cdot \mathbf{F}(\omega)$
	$f_1(t) \pm f_2(t)$	$\mathbf{F}_1(\omega) \pm \mathbf{F}_2(\omega)$
Frequency-domain shifting	$\exp[j\omega_0 t] \cdot f(t)$	$\mathbf{F}(\omega - \omega_0)$
Time-domain shifting	$f(t - t_0)$	$\exp[-j\omega t_0] \cdot \mathbf{F}(\omega)$
Differentiation	$\frac{d^n}{dt^n} f(t)$	$(j\omega)^n \cdot \mathbf{F}(\omega)$
Integration	$\int_{-\infty}^{t} f(t)\, dt$	$\frac{\mathbf{F}(\omega)}{\omega}$
Multiplication by t	$t^n \cdot f(t)$	$j^n \cdot \frac{d^n}{d\omega^n} \mathbf{F}(\omega)$

Properties of the Fourier Transform

Many of the properties of the Fourier transform parallel those of the Laplace transform. Some of the more useful properties of the Fourier transform are given without proof in Table 11.3. As in the case of the Laplace transform, the fundamental transform pairs, such as those given in Table 11.4, are obtained by evaluating an integral—in this case, Equation (11.24). Additional Fourier transform pairs may be obtained applying the properties of Table 11.3 to the fundamental transform pairs.

The Fourier transform has an amplitude spectrum and a phase spectrum, similar to the amplitude spectrum and phase spectrum of a Fourier series, except that the spectra are continuous rather than discrete. The amplitude spectrum $|\mathbf{F}(\omega)|$ always has even symmetry, and the phase spectrum always has odd symmetry. The even symmetry of the amplitude spectrum can be observed in Table 11.4.

Application of the Fourier Transform

An examination of Tables 11.3 and 11.4 suggests correctly that Fourier transforms can be used in circuit analysis in much the same way as Laplace transforms—that is, by transforming the elements and transforming the sources. The Laplace transform is generally the preferred integral transform technique in most circuit analysis problems because the Laplace transform permits inclusion of the initial conditions, and the Fourier transform does not. This is easily seen by comparing the differentiation property for the two integral transforms.

The Fourier transform, on the other hand, is often chosen for *waveform analysis* because it resolves the time-domain function into a spectrum of

Table 11.4 Fourier-Transform Pairs

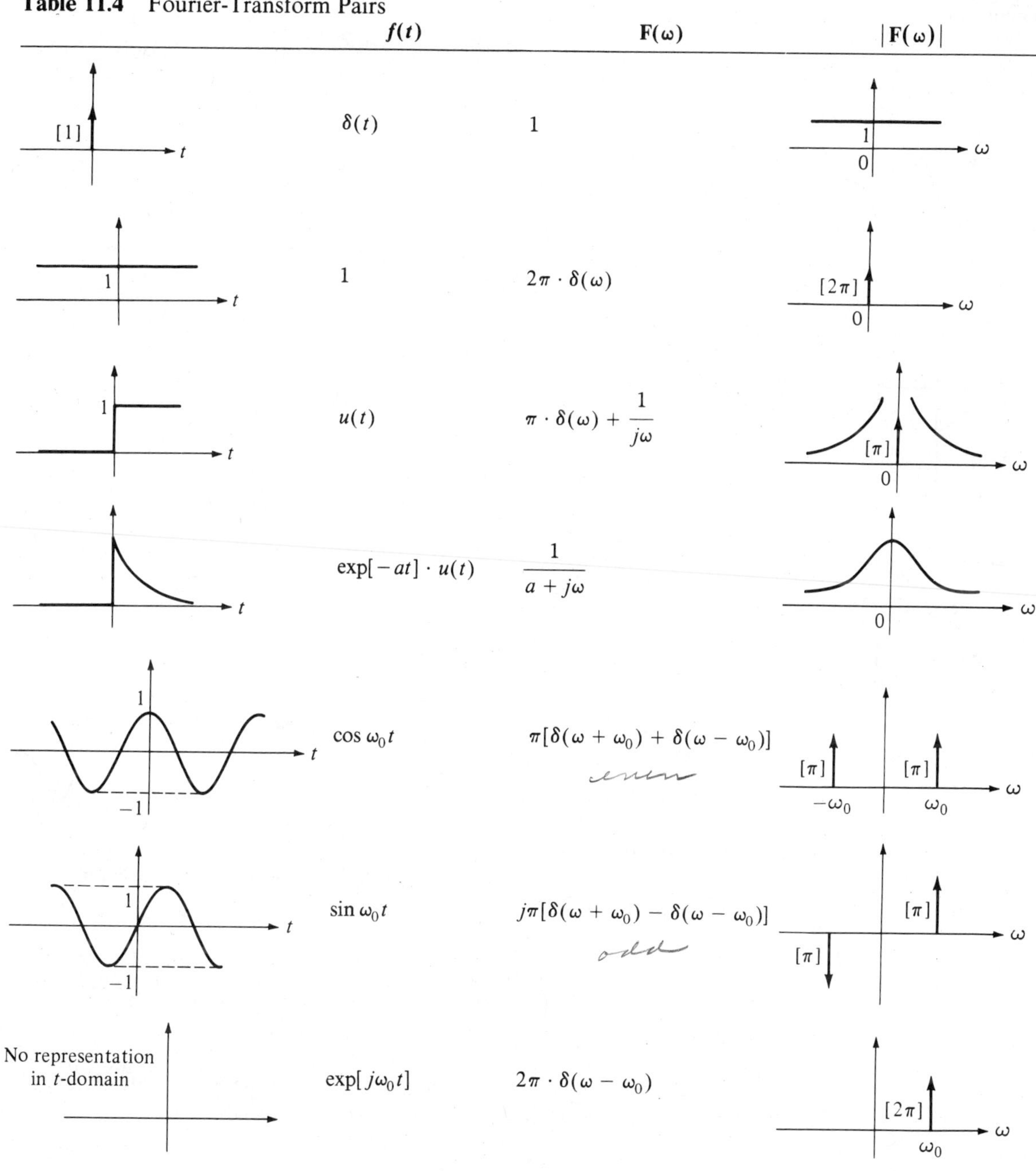

$f(t)$	$\mathbf{F}(\omega)$
$\delta(t)$	1
1	$2\pi \cdot \delta(\omega)$
$u(t)$	$\pi \cdot \delta(\omega) + \frac{1}{j\omega}$
$\exp[-at] \cdot u(t)$	$\frac{1}{a + j\omega}$
$\cos \omega_0 t$	$\pi[\delta(\omega + \omega_0) + \delta(\omega - \omega_0)]$
$\sin \omega_0 t$	$j\pi[\delta(\omega + \omega_0) - \delta(\omega - \omega_0)]$
$\exp[j\omega_0 t]$	$2\pi \cdot \delta(\omega - \omega_0)$

Table 11.4
Continued

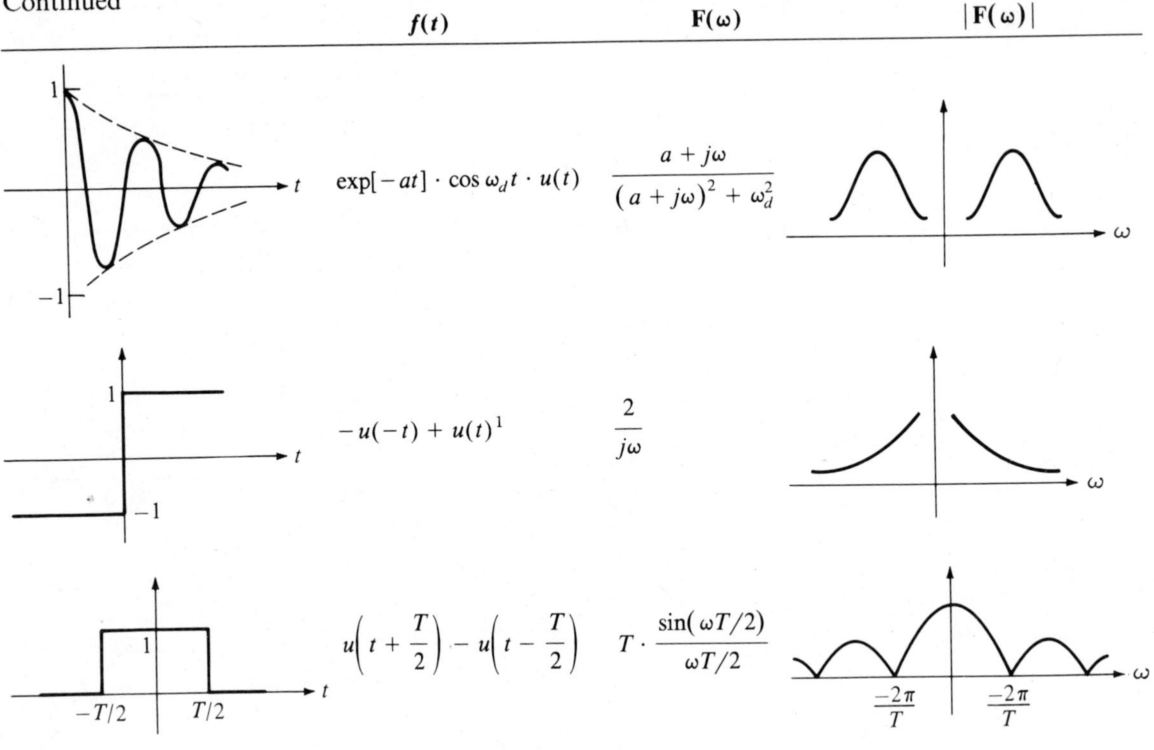

$f(t)$	$\mathbf{F}(\omega)$	$\lvert\mathbf{F}(\omega)\rvert$
$\exp[-at] \cdot \cos \omega_d t \cdot u(t)$	$\dfrac{a + j\omega}{(a + j\omega)^2 + \omega_d^2}$	
$-u(-t) + u(t)$[1]	$\dfrac{2}{j\omega}$	
$u\left(t + \dfrac{T}{2}\right) - u\left(t - \dfrac{T}{2}\right)$	$T \cdot \dfrac{\sin(\omega T/2)}{\omega T/2}$	

[1]This function is called the signum function (sgn t).

phasors, thus providing greater insight into the frequency-domain characteristics of the function. One useful application arises from **Parseval's theorem.**

Parseval's theorem states that

$$\int_{-\infty}^{\infty} f(t)^2\, dt = \frac{1}{2\pi} \int_{-\infty}^{\infty} |\mathbf{F}(\omega)|^2\, d\omega = \frac{1}{\pi} \int_{0}^{\infty} |\mathbf{F}(\omega)|^2\, d\omega \quad (11.26a)$$

where $|\mathbf{F}(\omega)|$ is the amplitude spectrum of the Fourier transform. Parseval's theorem may also be written in terms of cyclic frequency as

$$\int_{-\infty}^{\infty} f(t)^2\, dt = 2 \int_{0}^{\infty} |\mathbf{F}(2\pi f)|^2\, df \quad (11.26b)$$

The power dissipated by a voltage or current waveform is

$$p(t) = G \cdot v(t)^2 = R \cdot i(t)^2$$

or

$$p(t) = v(t)^2 = i(t)^2$$

if the resistance is normalized to 1 Ω. The 1-Ω energy dissipation is

$$W = \int_{t_0}^{t_0+T} v(t)^2\, dt = \int_{t_0}^{t_0+T} i(t)^2\, dt$$

Recall that an aperiodic waveform may be viewed as having an infinite period, so the integral becomes

$$W = \int_{-\infty}^{\infty} v(t)^2\, dt = \int_{-\infty}^{\infty} i(t)^2\, dt$$

Now from Parseval's theorem,

$$W = \frac{1}{\pi}\int_0^{\infty} |\mathbf{V}(\omega)|^2\, d\omega = \frac{1}{\pi}\int_0^{\infty} |\mathbf{I}(\omega)|^2\, d\omega \qquad (11.27a)$$

$$= 2\int_0^{\infty} |\mathbf{V}(2\pi f)|^2\, df = 2\int_0^{\infty} |\mathbf{I}(2\pi f)|^2\, df \qquad (11.27b)$$

where $|\mathbf{V}(\omega)|$ or $|\mathbf{I}(\omega)|$ is the amplitude spectrum of the Fourier transform for the aperiodic voltage or current waveform.

Example 11.11

Problem:

Verify Parseval's theorem for a circuit in which an initially uncharged 10-μF capacitor is charged by a 5-V step source through a 50-Ω resistor.

Solution:

1. The voltage across the resistor is

$$v(t) = 5 \cdot \exp[-t/\tau] \cdot u(t) \text{ V}$$

where

$$\tau = 500\ \mu\text{sec}$$

2. Calculate the power dissipated by the resistor:

$$p(t) = G \cdot v(t)^2 = 0.02(5 \cdot \exp[-t/\tau])^2 u(t)$$
$$= 0.5 \cdot \exp[-4000t] \cdot u(t) \text{ W}$$

Calculate the energy dissipated by the resistor:

$$W = 0.5 \int_0^\infty \exp[-4000t]\, dt = -\frac{0.5}{4000} \exp[-4000t] \Big|_0^\infty$$
$$= 125 \ \mu\text{J}$$

3. Obtain the Fourier transform of $v(t)$ from Table 11.4:

$$\mathbf{V}(\omega) = \frac{5}{2000 + j\omega} \text{ V} \cdot \text{sec}$$

so

$$|\mathbf{V}(\omega)| = \frac{5}{\sqrt{4 \times 10^6 + \omega^2}} \text{ V} \cdot \text{sec}$$

4. Apply Parseval's theorem to obtain the energy dissipated by the resistor:

$$W = \frac{G}{\pi} \int_0^\infty |\mathbf{V}(\omega)|^2\, d\omega = \frac{0.02}{\pi} \int_0^\infty \frac{25}{4 \times 10^6 + \omega^2}\, d\omega$$

Flashback

$$\int \frac{1}{a^2 + x^2}\, dx = \frac{1}{a} \cdot \tan^{-1}\left[\frac{x}{a}\right]$$

Thus

$$W = \frac{250 \times 10^{-6}}{\pi} \tan^{-1}\left(\frac{\omega}{2000}\right)\Big|_0^\infty$$
$$= \frac{250 \times 10^{-6}}{\pi}\left(\frac{\pi}{2} - 0\right) = 125 \ \mu\text{J}$$

—the same result obtained in step 2.

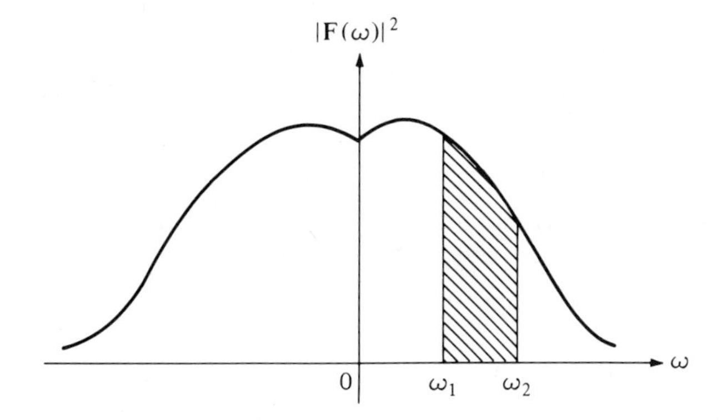

Figure 11.24 Energy associated with a frequency band.

Example 11.11 serves to verify the use of Parseval's theorem for calculating the total energy dissipation associated with a nonperiodic waveform, but it also indicates that there is no particular advantage to the use of Parseval's theorem for this purpose. A more useful application arises from the requirement to determine the energy associated with a particular band of the component frequencies of the waveform.

The integrand of Equation (11.27b), $|\mathbf{V}(2\pi f)|^2\,df$ or $|\mathbf{I}(2\pi f)|^2\,df$, has the units of joules per hertz (J/Hz), which is a measure of **energy density**—that is, the energy associated with $v(t)$ or $i(t)$, respectively, for an infinitesimal band of frequencies. The integral in Equation (11.27b) serves to sum the energy over the entire frequency spectrum. Figure 11.24, indicates that by specifying finite limits for the integral, the energy associated with any particular band of component frequencies can be calculated:

$$W = \frac{1}{\pi}\int_{\omega_1}^{\omega_2} |\mathbf{V}(\omega)|^2\,d\omega = \frac{1}{\pi}\int_{\omega_1}^{\omega_2} |\mathbf{I}(\omega)|^2\,d\omega \tag{11.28}$$

Such calculations are especially useful in determining the ratio of energy available to energy transferred by a two-port network such as a filter.

Example 11.12

Problem:

Determine the energy dissipated in the frequency band 100 kHz to 1 MHz when $i(t) = 12 \cdot u(t)$ mA flows in a 500-Ω resistor.

Solution:

1. Obtain the Fourier transform of $i(t)$ from Table 11.4:

$$\mathbf{I}(\omega) = 0.012\left(\pi \cdot \delta(\omega) + \frac{1}{j\omega}\right) \text{A} \cdot \text{sec}$$

and

$$|\mathbf{I}(\omega)| = 0.012/\omega \text{ A} \cdot \text{sec}$$

except at $\omega = 0$, which is not within the frequency band of interest.

2. Use Equation (11.28) to obtain the required energy dissipation:

$$\omega_1 = 200\pi \times 10^3 \text{ rad/sec} \quad \text{and} \quad \omega_2 = 2\pi \times 10^6 \text{ rad/sec}$$

$$W = 500 \int_{\omega_1}^{\omega_2} \frac{144 \times 10^{-6}}{\omega^2} d\omega$$

$$= 0.072\left(\frac{1}{\omega_1} - \frac{1}{\omega_2}\right) = 103.1 \text{ nJ}$$

Example 11.13

Problem:

The voltage $v(t) = 30 \cdot \exp[-6 \times 10^6 t] \cdot u(t)$ V is applied to the input of an ideal low-pass filter with a cutoff frequency of 1 MHz. Determine the percentage of total 1-Ω energy available at the output.

Solution:

1. Determine $\mathbf{V}(\omega)$ from Table 11.4:

$$\mathbf{V}(\omega) = \frac{30}{6 \times 10^6 + j\omega} \text{ V} \cdot \text{sec}$$

and

$$|\mathbf{V}(\omega)| = \frac{30}{\sqrt{36 \times 10^{12} + \omega^2}} \text{ V} \cdot \text{sec}$$

2. Use Equation (11.27a) to calculate the total 1-Ω energy available at the input:

$$W = \frac{1}{\pi} \int_0^\infty \frac{900}{36 \times 10^{12} + \omega^2} d\omega = \frac{900}{6\pi \times 10^6} \cdot \tan^{-1}\left[\frac{\omega}{6 \times 10^6}\right]\Bigg|_0^\infty$$

$$= 75 \ \mu\text{J}$$

3. Use Equation (11.28) to calculate the 1-Ω energy available at the output:

$$\omega_1 = 0 \quad \text{and} \quad \omega_2 = 2\pi \times 10^6 \text{ rad/sec}$$

$$W = \frac{1}{\pi}\int_0^{\omega_2} \frac{900}{36 \times 10^{12} + \omega^2}\, d\omega = \frac{900}{6\pi \times 10^6} \cdot \tan^{-1}\left[\frac{\omega}{6 \times 10^6}\right]\Bigg|_0^{\omega_2}$$
$$= 38.60\ \mu\text{J}$$

4. Calculate the ratio of the result in step 3 to the result in step 2:

$$\frac{38.60}{75} \times 100\% = 51.47\%$$

Exercise 11.4

Problem:

For the circuit described in Example 11.11, calculate the percentage of the total energy which is dissipated in the frequency band from 5 to 500 Hz.

Answer:

$$W = 78.67\ \mu\text{J}, \qquad 62.91\%$$

Discrete Fourier Transforms and the Fast Fourier Transform

As with the Laplace transform, the usefulness of the Fourier transform is limited in many cases by the difficulty of evaluating the integrals involved. Furthermore, the pervasiveness of discrete functions (e.g. sampled waveforms) in modern circuit analysis suggests the desirability of a transform suited to such functions. Chapter 12 is devoted to a "digitized" version of the Laplace transform, the z-transform, and its application to sampled-data circuits. It is also possible to apply the Fourier transform to discrete functions, and in such a way as to address both of the problems mentioned above. The mathematical details cannot be presented here, but it is worthwhile describing some important consequences of this application.

With a little effort, it is possible to apply the Fourier transform to discrete functions, functions whose argument is an integer variable rather than a continuous variable such as time. The most important change is that the improper integrals involved become infinite series. For a large class of functions of practical interest, these infinite series can be replaced with finite summations, which are more easily evaluated. Moreover, the calculations involved contain important symmetries and recursions, which drastically reduce the amount of computation required. The operation resulting from the manipulations so briefly described here is known as the **fast Fourier transform** (FFT). The basic FFT algorithm is amazingly simple (it can be programmed in a few dozen lines of uncomplicated code), but yields the discrete frequency spectrum of the input function.

With the FFT algorithm in hand, it is possible to apply Fourier analysis to functions whose (continuous) Fourier transform is impractical to compute. The function can be sampled, yielding a discrete function to which the FFT can be applied. If the sampling is done in an appropriate manner, the FFT results will be a good approximation of the Fourier transform of the original function, and can be used in the same fashion (spectrum analysis, energy density, etc.). This utility has made the FFT a fundamental tool of modern signal analysis and processing.

The FFT and related transforms constitute a major field of study in their own right, part of the larger field of digital signal processing. Innumerable versions and variations exist, in both hardware and software implementations. Interested readers may consult the immense literature available.

11.7 Computer Simulations

Fourier analysis by SPICE is based on a transient analysis, so the Fourier analysis directive (.FOUR) must be used in conjunction with a transient analysis directive (.TRAN). In some versions of SPICE, an output directive (.PRINT or .PLOT) is also required, even though .FOUR generates its own output.

Fourier analysis is performed over the interval TSTOP − T to TSTOP, where T = 1/(FUNDAMENTAL FREQ). Accordingly, TSTOP must coincide with the ending of a period of the waveform; otherwise, the phase spectrum will be in error because the waveform is effectively shifted.

SPICE 11.1 (Example 11.3) illustrates spectral analysis of a signal by allowing SPICE to perform Fourier analysis of a voltage waveform applied directly across a 1-Ω resistor. The PWL source specification can be used to extend this technique to virtually any signal waveform. Notice that the output consists of the dc component and first nine harmonics. The phase spectrum is based on sine components rather than cosine components (Equation 11.4a), so 90° must be *subtracted* from the SPICE output to obtain agreement with the phase angles calculated in this chapter.

SPICE 11.2 (Example 11.6) illustrates spectral analysis of a circuit current. To increase the accuracy of the output spectrum, the number of output data

```
*******12/09/88 ********  SPICE 2G.6    3/15/83 ********09:07:00*****
SPICE 11.1
****     INPUT LISTING                        TEMPERATURE =   27.000 DEG C
********************************************************************************
 R 1 0 1
 V 1 0 PULSE(0 5 0 1F 1F 2M 8M)
 .TRAN 40U 8M
 .PRINT TRAN V(1)
 .FOUR 125 V(1)
 .END
*******12/09/88 ********  SPICE 2G.6    3/15/83 ********09:07:01*****
SPICE 11.1
****     INITIAL TRANSIENT SOLUTION           TEMPERATURE =   27.000 DEG C
********************************************************************************
  NODE   VOLTAGE
 (  1)     .0000
     VOLTAGE SOURCE CURRENTS
     NAME       CURRENT
     V         0.000D+00
     TOTAL POWER DISSIPATION   0.00D+00  WATTS
*******12/09/88 ********  SPICE 2G.6    3/15/83 ********09:07:01*****
SPICE 11.1
****     TRANSIENT ANALYSIS                   TEMPERATURE =   27.000 DEG C
********************************************************************************
         (Transient Analysis Table Omitted From Output)

*******12/09/88 ********  SPICE 2G.6    3/15/83 ********09:07:01*****
SPICE 11.1
****     FOURIER ANALYSIS                     TEMPERATURE =   27.000 DEG C
********************************************************************************
 FOURIER COMPONENTS OF TRANSIENT RESPONSE V(1)
DC COMPONENT =    1.244D+00
HARMONIC   FREQUENCY    FOURIER    NORMALIZED    PHASE     NORMALIZED
    NO        (HZ)      COMPONENT   COMPONENT    (DEG)     PHASE (DEG)
     1      1.250D+02   2.242D+00   1.000000    44.328        .000
     2      2.500D+02   1.592D+00    .709951    -1.343     -45.672
     3      3.750D+02   7.593D-01    .338654   -47.015     -91.343
     4      5.000D+02   1.245D-02    .005551   -92.687    -137.015
     5      6.250D+02   4.417D-01    .197018    41.642      -2.687
     6      7.500D+02   5.311D-01    .236901    -4.030     -48.358
     7      8.750D+02   3.309D-01    .147576   -49.701     -94.030
     8      1.000D+03   1.247D-02    .005561   -95.373    -139.701
     9      1.125D+03   2.419D-01    .107908    38.955      -5.373
     TOTAL HARMONIC DISTORTION =      86.437221  PERCENT

         JOB CONCLUDED
```

```
*******12/09/88 ********  SPICE 2G.6    3/15/83 ********11:15:29*****
SPICE 11.2
****     INPUT LISTING                      TEMPERATURE =   27.000 DEG C
*******************************************************************************
 R1 2 3 10K
 R2 3 0 10K
 C 3 0 1.5U
 V 1 0 PULSE(0 5 0 1F 1F 50M 100M)
 VAM 1 2
 .OPTIONS LIMPTS=501
 .TRAN .2M 100M
 .PRINT TRAN I(VAM)
 .FOUR 10 I(VAM)
 .END
*******12/09/88 ********  SPICE 2G.6    3/15/83 ********11:15:29*****
SPICE 11.2
****     INITIAL TRANSIENT SOLUTION         TEMPERATURE =   27.000 DEG C
*******************************************************************************
  NODE   VOLTAGE     NODE   VOLTAGE     NODE   VOLTAGE
 (  1)     .0000    (  2)     .0000    (  3)     .0000
     VOLTAGE SOURCE CURRENTS
     NAME       CURRENT
     V         0.000D+00
     VAM       0.000D+00
     TOTAL POWER DISSIPATION   0.00D+00  WATTS
*******12/09/88 ********  SPICE 2G.6    3/15/83 ********11:15:29*****
SPICE 11.2
****     TRANSIENT ANALYSIS                 TEMPERATURE =   27.000 DEG C
*******************************************************************************
        Transient Analysis Table Omitted From Output

*******12/09/88 ********  SPICE 2G.6    3/15/83 ********11:15:30*****
SPICE 11.2
****     FOURIER ANALYSIS                   TEMPERATURE =   27.000 DEG C
*******************************************************************************
 FOURIER COMPONENTS OF TRANSIENT RESPONSE I(VAM)
DC COMPONENT =    1.250D-04
HARMONIC   FREQUENCY    FOURIER    NORMALIZED     PHASE     NORMALIZED
    NO        (HZ)     COMPONENT    COMPONENT     (DEG)     PHASE (DEG)
     1      1.000D+01  1.972D-04    1.000000     17.571          .000
     2      2.000D+01  3.158D-08     .000160     42.216        24.646
     3      3.000D+01  9.150D-05     .463953     14.663        -2.907
     4      4.000D+01  2.839D-08     .000144     20.579         3.008
     5      5.000D+01  5.967D-05     .302539      9.150        -8.420
     6      6.000D+01  2.938D-08     .000149     10.919        -6.652
     7      7.000D+01  4.383D-05     .222216      5.634       -11.937
     8      8.000D+01  3.234D-08     .000164      5.979       -11.592
     9      9.000D+01  3.450D-05     .174913      3.190       -14.380
     TOTAL HARMONIC DISTORTION =     62.189770  PERCENT

        JOB CONCLUDED
```

points was increased to 501 (from the default of 201) by the .OPTIONS directive.

11.8 SUMMARY

Fourier's theorem guarantees that any periodic waveform may be expressed as an infinite sum of sinusoidal terms, plus perhaps a constant term. The usefulness of this observation is that the powerful techniques of dc analysis and ac analysis, through the use of the superposition theorem, can be brought to bear on a linear network excited by any periodic source.

The Fourier coefficients, taken from the Fourier series which represents a particular waveform, are used to determine the frequency spectrum of the waveform. The frequency spectrum of the excitation waveform, when superposed on the Bode plot of the immittance or transfer function, can provide the designer with valuable insights into causes of distortion in the response waveform.

The Fourier transform extends the concept of spectral analysis to nonperiodic waveforms. One useful application of the Fourier transform to network analysis arises from Parseval's theorem, which provides a means of determining the energy associated with the components of a particular band.

11.9 Terms

amplitude spectrum
clamp
continuous spectrum
dc component
dc offset
Dirichlet conditions
discrete spectrum
distortion
effective value
energy density
Euler formulas
even symmetry
exponential form
fast Fourier transform (FFT)
Fourier coefficients
Fourier integral
Fourier series
Fourier's theorem
Fourier transform
frequency spectrum
fundamental frequency
harmonic
harmonically related
infinite
inverse Fourier transform
line spectrum
odd symmetry
Parseval's theorem
periodic extension
phase spectrum
power factor
root-mean-squared (rms)
rounding
sagging
signum function
spectrum
truncate
waveform symmetry

Problems

Note: An electronic spreadsheet and an associated computer graphics program should be used, if possible, for Problems 11-8 through 11-11, and 11-13 through 11-18.

11-1. Derive formulas for the Fourier coefficients a_0, a_n, and b_n for each waveform below.

***a.**

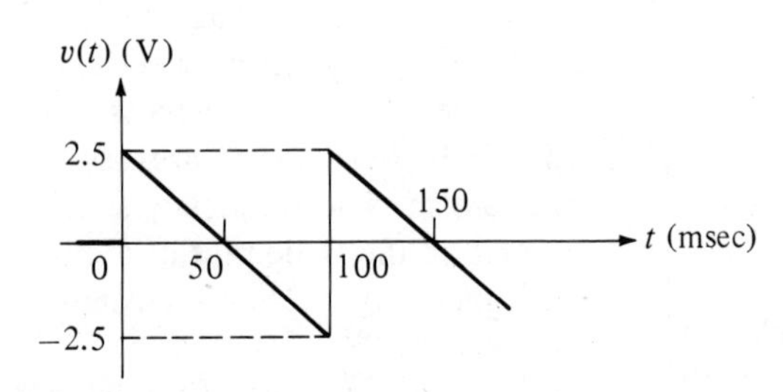

b.

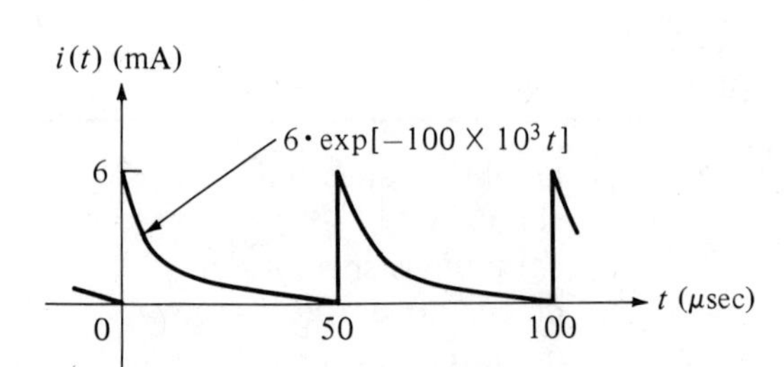

***c.**

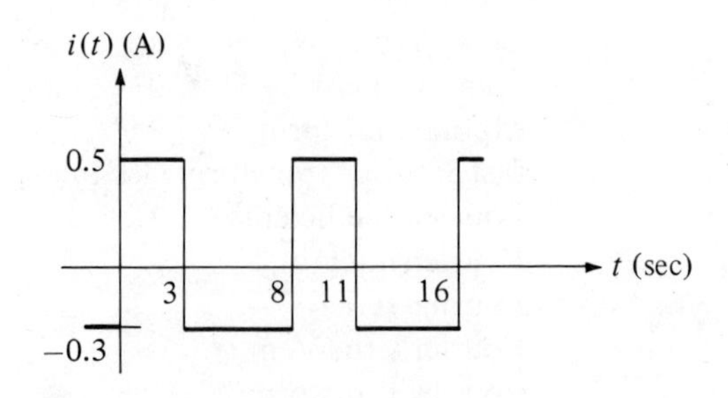

d.

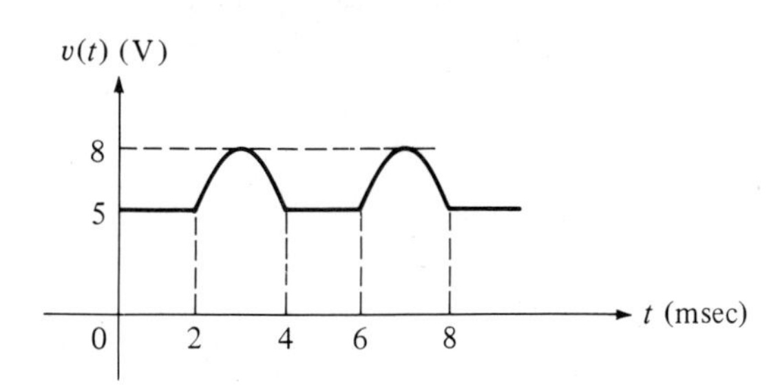

11-2. Derive formulas for the dc component A_0, for the amplitude spectrum $(V_m)_n$ or $(I_m)_n$, and for the phase spectrum θ_n for each waveform in Problem 11-1. [*b, *d]

***11-3.** Evaluate the fourth, fifth, and sixth harmonics of the waveform used for Example 11.3.

***11-4.** Substitute 0.5 for t_1/T in the formulas given in Table 11.1 for a pulse with arbitrary duty cycle to derive the formulas for a pulse with 50% duty cycle.

11-5. Use the formulas derived in Problem 11-4 for a pulse with 50% duty cycle and the observation concerning the effect of dc offset to confirm the formulas given for a square wave.

11-6. Use the Euler formulas to confirm the formulas given in Table 11.1 for the sawtooth waveform.

11-7. Use the Euler formulas to confirm the formulas given in Table 11.1 for the half-wave rectified sinusoid. How are these formulas changed if the waveform is shifted to the right by $T/2$? To the left by $T/2$?

SPICE

11-8. Use the results of Problem 11-2 to calculate the frequency spectrum through the ninth harmonic, and write the truncated Fourier series for each waveform. Graph the waveform obtained from the truncated Fourier series. [*b, *d]

SPICE

11-9. Calculate the frequency spectrum of the response current through the fifth harmonic for each circuit. In each case, write the truncated Fourier series for the response waveform, and graph the waveform obtained from the truncated Fourier series. Refer to Table 11.1 for the frequency spectrum of the source.

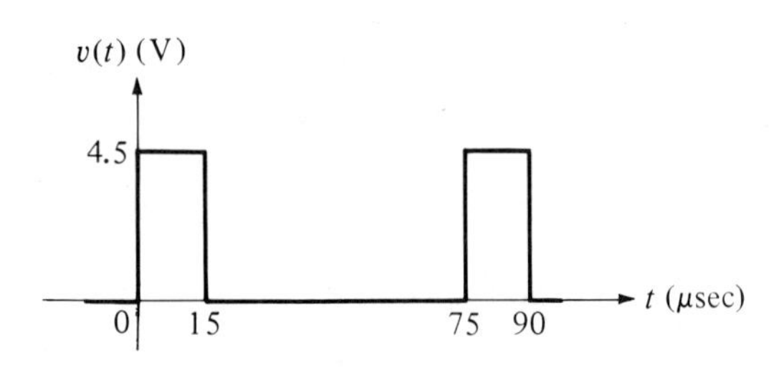

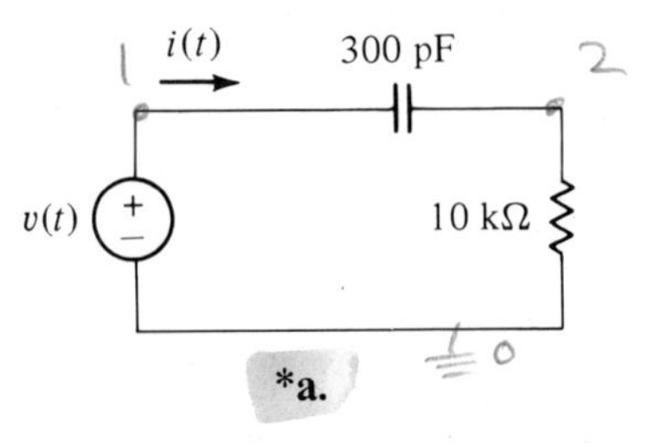

*a.

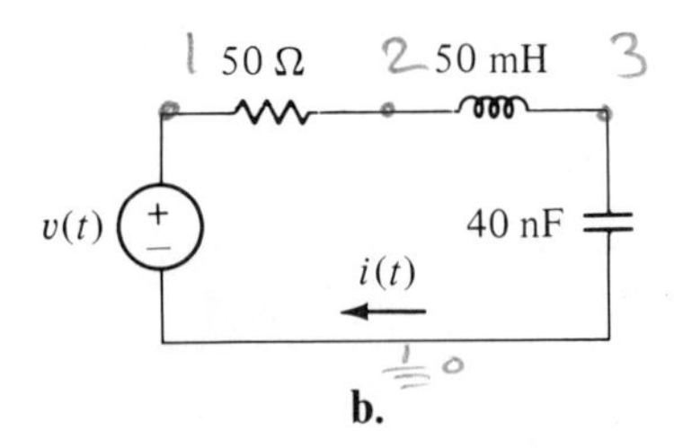

b.

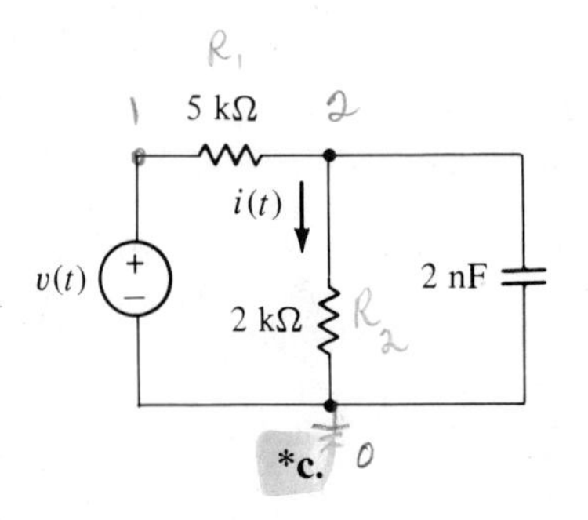

*c.

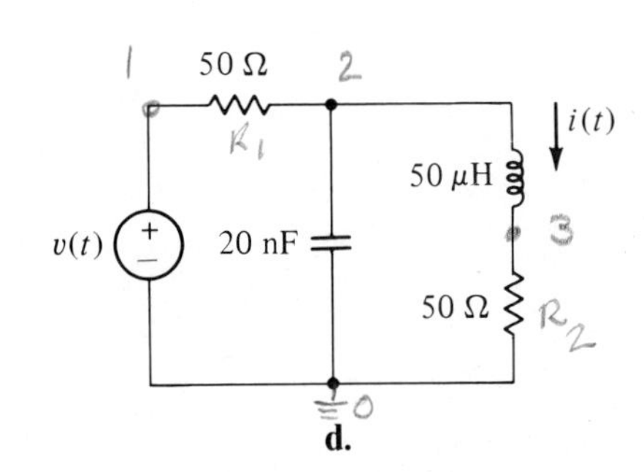

d.

SPICE

11-10. Repeat Problem 11-9 for the response voltage of each network.

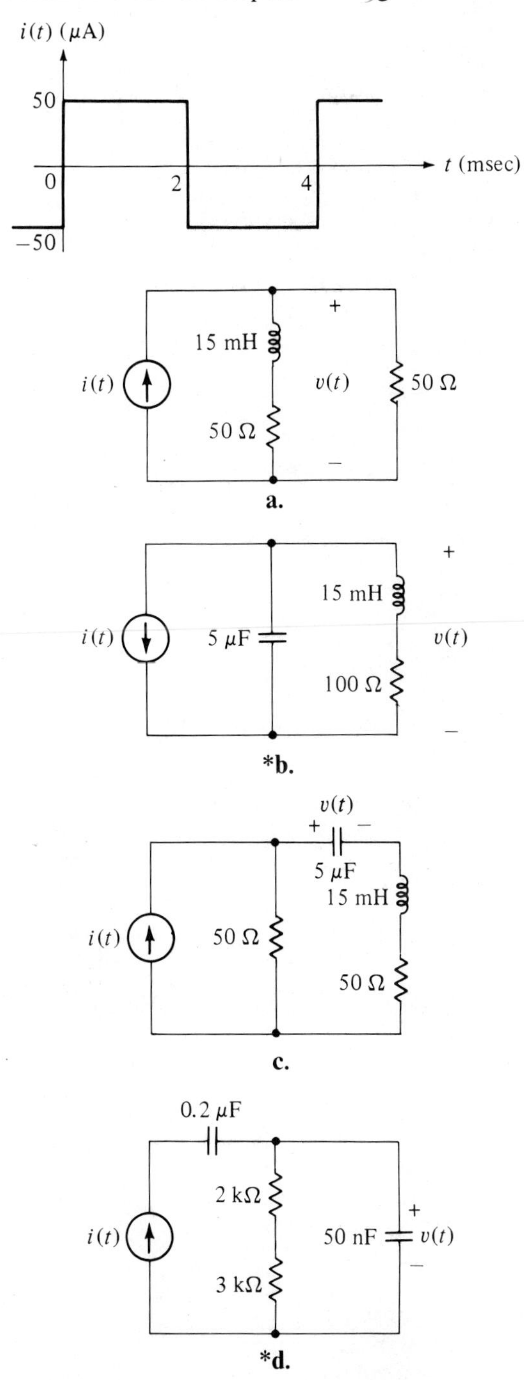

SPICE **11-11.** Repeat Problem 11-9 for the output voltage of each network.

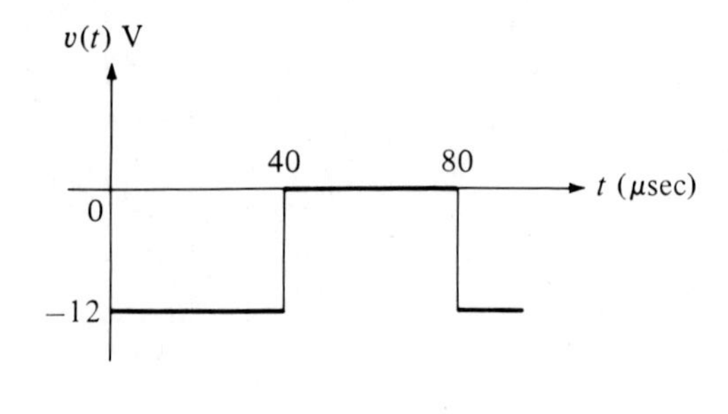

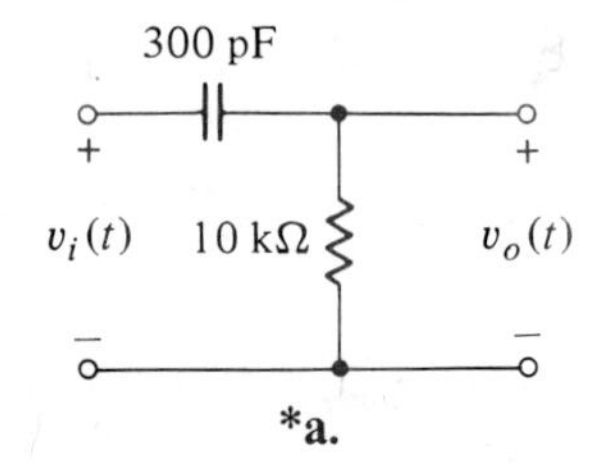

***a.**

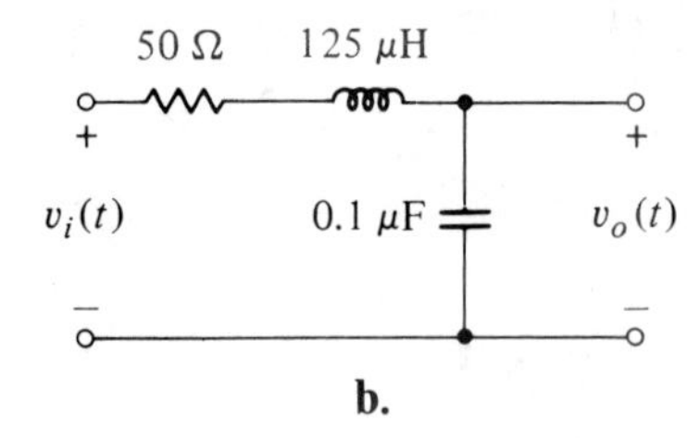

b.

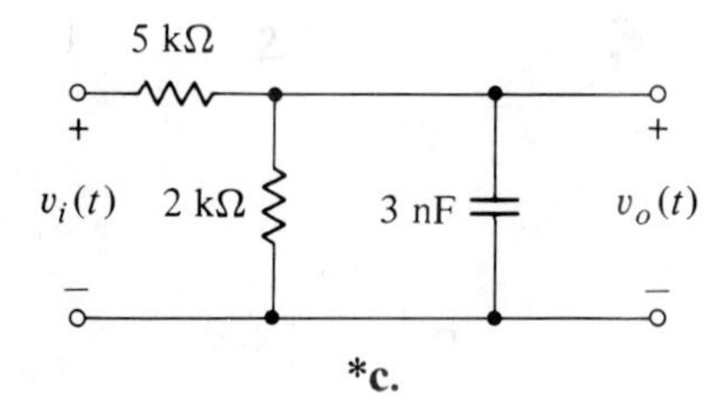

***c.**

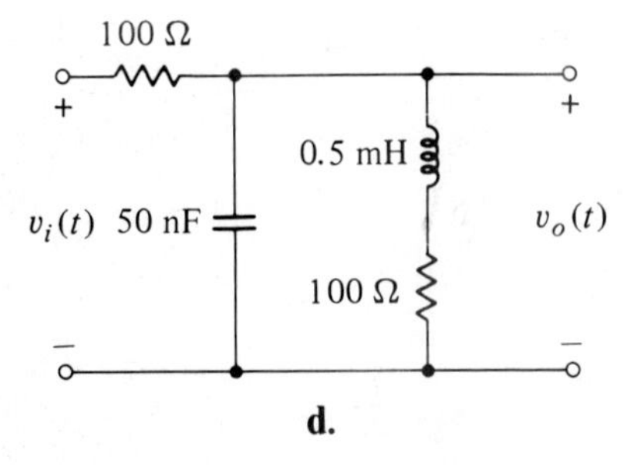

d.

***11-12.** Evaluate the integral in Equation (11.11a). Hint: In the integrand, let $\alpha = \omega t + \theta$ and $\beta = \omega t + \phi$; then use a trignometric identity for $\cos\alpha \cdot \cos\beta$.

11-13. Calculate the approximate power supplied by the source in each network in Problem 11-9. [*a, *c]

11-14. Repeat Problem 11-13 for each circuit in Problem 11-10. [*b, *d]

11-15. Show that the effective, or rms, value of $f(t)$ is $A_m/\sqrt{2}$, where

$$f(t) = A_m \cdot \sin[\omega t + \theta]$$

11-16. Calculate the effective value of each waveform in Problem 11-1. [*b, *d]

11-17. Sketch the line spectra, both amplitude and phase, for each waveform in Problem 11-1. [*a, *c]

***11-18.** Consider each waveform in Problem 11-1 as an input to the ideal filters described below. Sketch the output amplitude spectrum and graph the output waveform in each case. Define an ideal filter as a two-port network having unit gain and zero phase shift in the pass band, with infinite rolloff to zero gain in the stop band.

a. Band reject: $0.1\omega_1$ to $1.9\omega_1$.
b. High pass: Above $4.5\omega_1$.
c. Low pass: Below $3.7\omega_1$.
d. Band pass: $1.9\omega_1$ to $4.5\omega_1$.

11-19. **a.** Verify that $\mathbf{C}_0$ as defined in Equation (11.18) is equal to a_0 as defined in Equation (11.3a).
b. Using the identity $\mathbf{C}_n = (a_n - jb_n)/2$, verify Equation (11.18) for $n = 1, 2, 3, \ldots$.
c. Show that $\mathbf{C}_{-n} = \mathbf{C}_n^*$.

11-20. Use the results of Problem 11-19 to verify that Equation (11.17) is equivalent to Equation (11.2). (Hint: Use Euler's identities to express cosine and sine in terms of complex exponentials.)

***11-21.** Use Parseval's theorem to calculate the energy dissipated by $i(t) = 50 \cdot \exp[-200t] \cdot u(t)$ flowing in a 200-Ω resistor.

***11-22.** Find the 1-Ω energy available at the output of an ideal band-pass filter in response to a 3-V step-function input ($\omega_1 = 200$ rad/sec and $\omega_2 = 1000$ rad/sec). Is it possible to determine what percentage of the 1-Ω energy available at the input is available at the output? Comment.

***11-23.** Calculate the energy dissipated in the frequency range 50 to 2000 Hz if $v(t) = 12(1 - \exp[-5000t])u(t)$ V is applied directly across a 50-Ω resistor.

12 z-Transforms

12.1 INTRODUCTION

Digital control, sometimes called **computer control**, is the preferred method of control for many physical systems and processes. This method of control requires that the continuous input be converted into a sequence of pulses, or **samples**, with duration Δt at intervals of T seconds [Figure 12.1(a)]. In a well-designed system, Δt will be much less than T and any other time considerations within the system, so the sampled data may be represented graphically as in Figure 12.1(b).

The **z-transform**, a specialization of the Laplace transform, has been developed to facilitate analysis of **sampled-data** systems. The z-transform is mainly used in system and process control applications, and is of primary interest to control systems analysts and designers. In this chapter we will introduce the z-transform, show its derivation from and relationship to the Laplace transform, and make some brief comments regarding the electronic circuits used for sampling. We will restrict our applications of the z-transform to a few simple two-port networks, because application to complicated transfer functions is more appropriate to a control systems text. Likewise, we will not become involved with sampling theory and the limitations of data-sampling methods.

12.2 Development of the z-Transform

Ideal sampled data can be represented mathematically by using the unit impulse function: for the first sample, $f(0) \cdot \delta(t)$; for the second sample, $f(T) \cdot \delta(t - T)$; for the third sample, $f(2T) \cdot \delta(t - 2T)$; and so forth, so that the sequence may be expressed as

$$f^*(t) = \sum_{n=0}^{\infty} f(nT) \cdot \delta(t - nT) \tag{12.1}$$

where $f^*(t)$ is sampled data from the continuous function $f(t)$.[1]

[1] The notation for a sampled data function should not be confused with the notation for a complex conjugate.

Figure 12.1
Sampled data.

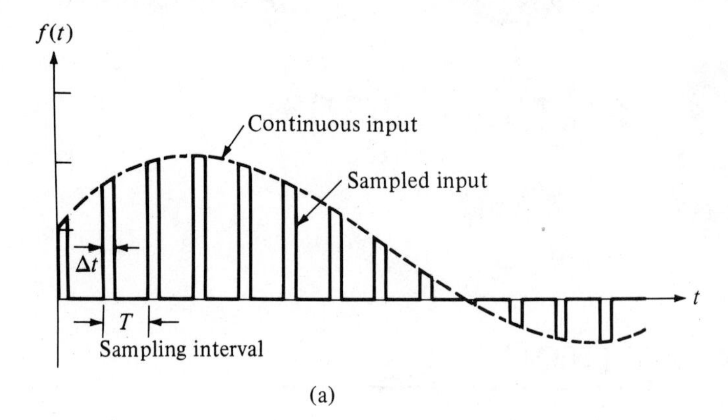

(a)

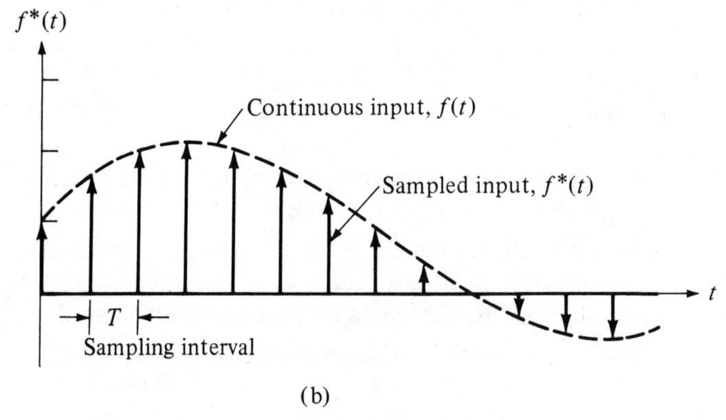

(b)

The Laplace transform of $f^*(t)$ is

$$F^*(s) = \mathscr{L}\{f^*(t)\} = \mathscr{L}\left\{\sum_{n=0}^{\infty} f(nT) \cdot \delta(t - nT)\right\}$$

and by the linearity properties of the Laplace transform,

$$\begin{aligned} F^*(s) &= \mathscr{L}\{f(0) \cdot \delta(t)\} + \mathscr{L}\{f(T) \cdot \delta(t - T)\} \\ &\quad + \mathscr{L}\{f(2T) \cdot \delta(t - 2T)\} + \cdots \\ &= f(0) \cdot \mathscr{L}\{\delta(t)\} + f(T) \cdot \mathscr{L}\{\delta(t - T)\} \\ &\quad + f(2T) \cdot \mathscr{L}\{\delta(t - 2T)\} + \cdots \end{aligned}$$

From Pair I and the time-domain shifting property,

$$F^*(s) = f(0) \cdot 1 + f(T) \cdot \exp[-Ts] + f(2T) \cdot \exp[-2Ts] + \cdots = \sum_{n=0}^{\infty} f(nT) \cdot \exp[-nTs] \tag{12.2}$$

As previously noted, s is a complex variable, so $\exp[-nTs]$ is complex. To obtain the z-transform, the complex variable z is defined as

$$z = \exp[Ts] \tag{12.3a}$$

Note that when s is defined in terms of z, Equation (12.3a) becomes

$$s = \frac{1}{T} \cdot \ln z \tag{12.3b}$$

The z-transform is then defined as

$$F(z) = \mathscr{Z}\{f(t)\} \tag{12.4a}$$

and

$$\mathscr{Z}\{f(t)\} = \mathscr{L}\{f^*(t)\}_{s=(1/T)\cdot \ln z} \tag{12.4b}$$

Therefore,

$$F(z) = F^*(s)_{s=(1/T)\cdot \ln z} = \sum_{n=0}^{\infty} f(nT) \cdot z^{-n} \tag{12.4c}$$

The key to understanding the derivation of the z-transform lies in careful interpretation of Equation (12.4b), which states that the z-transform of the continuous function $f(t)$ is the Laplace transform of the sampled-data function $f^*(t)$, with the change of variable $z = \exp[Ts]$.

Properties of the z-Transform

Some properties of the z-transform will be given here without proof; several are similar to properties of the Laplace transform. Examples of the usefulness of these properties will be given in the next section.

Linearity Properties. z-transformation is a linear operation, so

$$\mathscr{Z}\{k \cdot f(t)\} = k \cdot \mathscr{Z}\{f(t)\} = k \cdot F(z) \tag{12.5a}$$

and

$$\mathscr{Z}\{f_1(t) \pm f_2(t)\} = \mathscr{Z}\{f_1(t)\} \pm \mathscr{Z}\{f_2(t)\} = F_1(z) \pm F_2(z) \quad (12.5b)$$

Time-Domain Shifting Property. Shifting $f(t)$ by one sampling interval requires division of the z-transform of $f(t)$ by z, so

$$\mathscr{Z}\{f(t-T)\} = z^{-1} \cdot \mathscr{Z}\{f(t)\} = z^{-1} \cdot F(z) \quad (12.6a)$$

and

$$\mathscr{Z}\{f(t-nT)\} = z^{-n} \cdot \mathscr{Z}\{f(t)\} = z^{-n} \cdot F(z) \quad (12.6b)$$

Multiplication by exp[$-at$]. Multiplication of $f(t)$ by $\exp[-at]$ requires substitution of $z \cdot \exp[aT]$ for z in the z-transform of $f(t)$. This property corresponds to the s-domain shifting property of the Laplace transform:

$$\mathscr{Z}\{\exp[-at] \cdot f(t)\} = F(z \cdot \exp[aT]) \quad (12.7)$$

Multiplication by t. Multiplication of $f(t)$ by t requires differentiation of the z-transform of $f(t)$ and multiplication of the result by $-Tz$:

$$\mathscr{Z}\{t \cdot f(t)\} = -Tz\frac{d}{dz}F(z) \quad (12.8)$$

These properties are collected in Table 12.1.

Table 12.1
Properties of the z-Transform

$f(t)$	$F(s)$	$F(z)$
	Linearity	
$k \cdot f(t)$	$k \cdot F(s)$	$k \cdot F(z)$
$f_1(t) \pm f_2(t)$	$F_1(s) \pm F_2(s)$	$F_1(z) \pm F_2(z)$
	Time-domain shifting	
$f(t-nT)$	$\exp[-nTs] \cdot F(s)$	$(1/z^n) \cdot F(z)$
	Multiplication by exp[$-$ *at*] *(s-domain shifting)*	
$\exp[-at] \cdot f(t)$	$F(s+a)$	$F(z \cdot \exp[aT])$
	Multiplication by t	
$t \cdot f(t)$		$-Tz\frac{d}{dz}F(z)$

12.3 Obtaining z-Transforms

The z-transform of $f(t)$ can be determined either from Equations (12.4b) or (12.4c). Table 12.2 provides z-transforms for frequently encountered functions. Several of the tabulated transforms are derived within the text, and the remainder are left as exercises (Problems 12-1 through 12-5). Some books of standard mathematical tables and most control systems textbooks contain

Table 12.2 z-Transforms

$f(t)$	$F(s)$	$F(z)$
$\delta(t)$	1	1
$\delta(t-nT)$	$\exp[-nTs]$	z^{-n}
$u(t)$	$1/s$	$\dfrac{z}{z-1}$
$u(t-nT)$	$\dfrac{\exp[-nTs]}{s}$	$\dfrac{z}{z^n(z-1)}$
$t \cdot u(t)$	$\dfrac{1}{s^2}$	$\dfrac{Tz}{(z-1)^2}$
$t^2 \cdot u(t)$	$\dfrac{2}{s^3}$	$\dfrac{T^2 z(z+1)}{(z-1)^3}$
$\exp[-at] \cdot u(t)$	$\dfrac{1}{s+a}$	$\dfrac{z}{z-\exp[-aT]}$
$(1-\exp[-at])u(t)$	$\dfrac{a}{s(s+a)}$	$\dfrac{z(1-\exp[-aT])}{(z-1)(z-\exp[-aT])}$
$t\ \mathrm{D}\exp[-at] \cdot u(t)$	$\dfrac{1}{(s+a)^2}$	$\dfrac{Tz \cdot \exp[-aT]}{(z-\exp[-aT])^2}$
$\dfrac{\exp[-at]-\exp[-bt]}{b-a} \cdot u(t)$	$\dfrac{1}{(s+a)(s+b)}$	$\dfrac{z(\exp[-aT]-\exp[-bT])/(b-a)}{z^2 - z(\exp[-aT]+\exp[-bT]) + \exp[-(a+b)T]}$
$\sin \omega t \cdot u(t)$	$\dfrac{\omega}{s^2+\omega^2}$	$\dfrac{z \sin[\omega T]}{z^2 - 2z\cos\omega T + 1}$
$\cos \omega t \cdot u(t)$	$\dfrac{s}{s^2+\omega^2}$	$\dfrac{z(z-\cos\omega T)}{z^2 - 2z \cdot \cos\omega T + 1}$
$\exp[-\alpha t] \cdot \sin \omega_d t \cdot u(t)$	$\dfrac{\omega_d}{(s+\alpha)^2+\omega_d^2}$	$\dfrac{z \cdot \exp[-\alpha T] \cdot \sin \omega_d T}{z^2 - 2z \cdot \exp[-\alpha T] \cdot \cos \omega_d T + \exp[-2\alpha T]}$
$\exp[-\alpha t] \cdot \cos \omega_d t \cdot u(t)$	$\dfrac{s+\alpha}{(s+\alpha)^2+\omega_d^2}$	$\dfrac{z(z-\exp[-\alpha T] \cdot \cos \omega_d T)}{z^2 - 2z \cdot \exp[-\alpha T] \cdot \cos \omega_d T + \exp[-2\alpha T]}$

more comprehensive tables of z-transforms. Also in this section, a technique is presented for obtaining the z-transform of s-domain transfer functions.

z-Transforms of some Frequently Encountered Functions

Unit Impulse Function. Given

$$f(t) = \delta(t)$$

then

$$f^*(t) = \delta(t)$$

also, because $\delta(t)$ is zero everywhere except at $t = 0$. From Pair I,

$$F^*(s) = 1$$

so from Equation (12.4b),

$$F(z) = \mathscr{Z}\{\delta(t)\} = 1 \tag{12.9}$$

Since s does not appear in $F^*(s)$, then z does not appear in $F(z)$.

Unit Step Function. Given

$$f(t) = u(t)$$

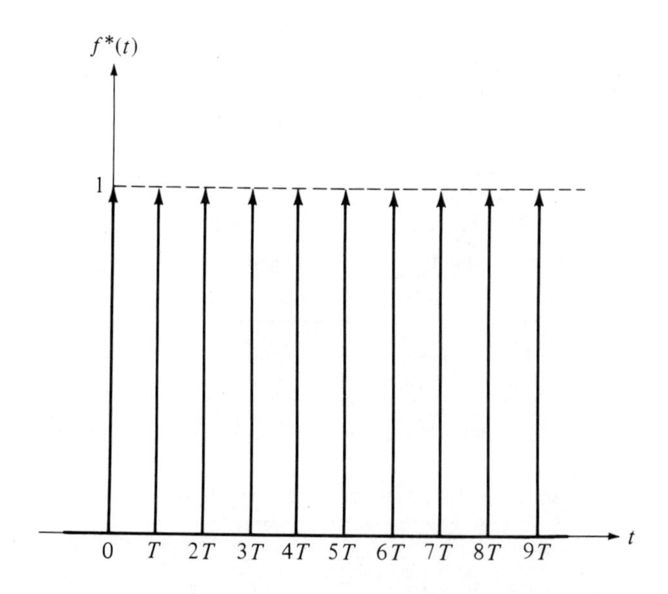

Figure 12.2 Sampled-data unit step function.

then from Figure 12.2,

$$f(nT) = 1$$

for all n. Substitution into Equation (12.4c) yields

$$F(z) = \sum_{n=0}^{\infty} 1 \cdot z^{-n}$$
$$= 1 + \frac{1}{z} + \frac{1}{z^2} + \frac{1}{z^3} + \cdots$$

Flashback

Geometric series:

$$1 + x + x^2 + x^3 + \cdots = \frac{1}{1-x}$$

The resulting series is recognized as the **geometric series**, where $x = 1/z$, so

$$F(z) = \mathscr{Z}\{u(t)\} = \frac{1}{1 - 1/z} = \frac{z}{z-1} \tag{12.10}$$

Ramp Function. Given

$$f(t) = t \cdot u(t)$$

(Figure 12.3), then from the multiplication by t property [Equation (12.8)],

$$F(z) = \mathscr{Z}\{f(t)\} = -Tz\frac{d}{dz}\mathscr{Z}\{u(t)\}$$

From Equation (12.10),

$$F(z) = -Tz\frac{d}{dz}\left(\frac{z}{z-1}\right) = -Tz\frac{-1}{(z-1)^2}$$

Therefore,

$$F(z) = \mathscr{Z}\{t \cdot u(t)\} = \frac{Tz}{(z-1)^2} \tag{12.11}$$

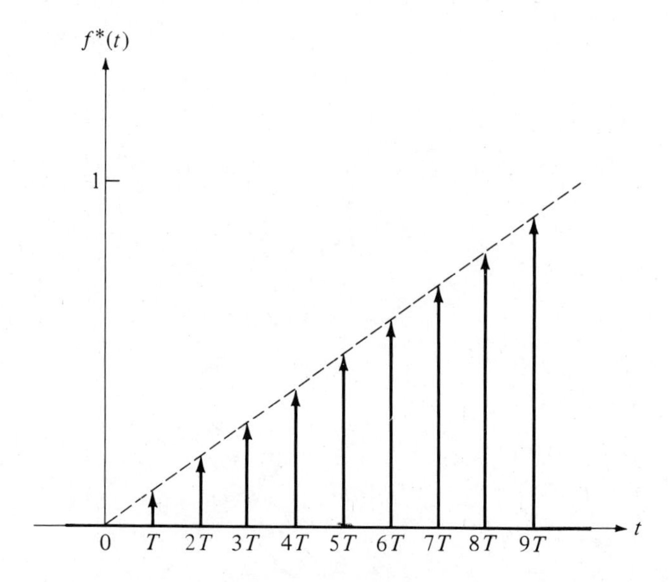

Figure 12.3 Sampled-data ramp function.

Exponential Decay Function. Given

$$f(t) = \exp[-at] \cdot u(t)$$

(Figure 12.4), then from the multiplication by $\exp[-at]$ property [Equation (12.7)]

$$F(z) = \mathscr{Z}\{f(t)\} = F_1(\exp[aT] \cdot z)$$

where $F_1(z) = \mathscr{Z}\{u(t)\} = z/(z-1)$ [from Equation (12.10)]. Therefore,

$$F(z) = \mathscr{Z}\{\exp[-at] \cdot u(t)\} = \frac{z \cdot \exp[aT]}{z \cdot \exp[aT] - 1}$$

$$= \frac{z}{z - \exp[-aT]} \tag{12.12}$$

Sine Function. Given

$$f(t) = \sin \omega t \cdot u(t)$$

(Figure 12.5), then from Euler's identity for $\sin x$,

$$f(t) = \frac{\exp[j\omega t] - \exp[-j\omega t]}{j2}$$

Figure 12.4
Sampled-data exponential decay function.

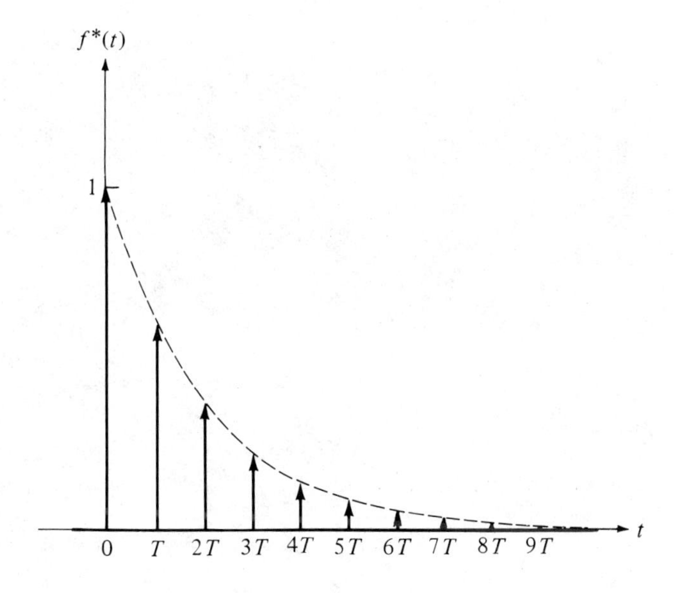

Figure 12.5
Sampled-data sine function.

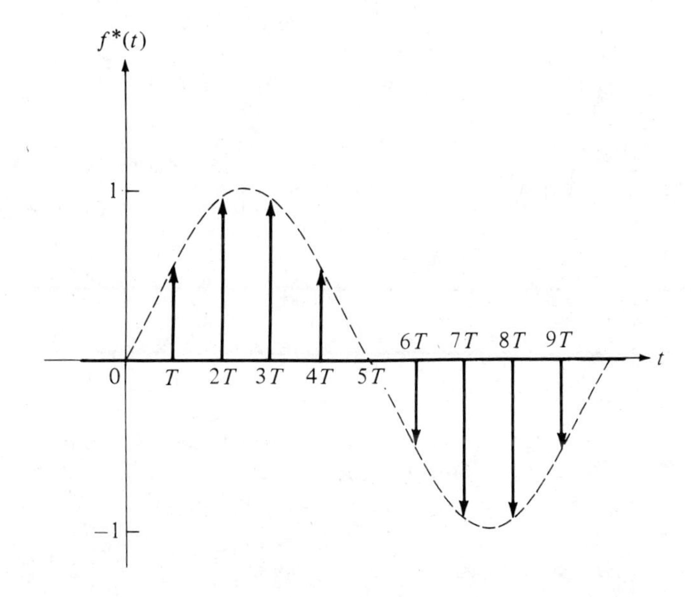

Flashback

Euler identities:

$$\sin x = \frac{\exp[jx] - \exp[-jx]}{j2}$$

$$\cos x = \frac{\exp[jx] + \exp[-jx]}{2}$$

Therefore,

$$\begin{aligned} F(z) &= \frac{1}{j2}(\mathscr{Z}\{\exp[j\omega t]\} - \mathscr{Z}\{\exp[-j\omega t]\}) \\ &= \frac{1}{j2}\left(\frac{z}{z - \exp[j\omega T]} - \frac{z}{z - \exp[-j\omega T]}\right) \\ &= \frac{1}{j2}\frac{z(z - \exp[-j\omega T]) - z(z - \exp[j\omega T])}{(z - \exp[j\omega T])(z - \exp[-j\omega T])} \\ &= \frac{z(\exp[j\omega T] - \exp[-j\omega T])/(j2)}{z^2 - z(\exp[j\omega T] + \exp[-j\omega T]) + 1} \end{aligned}$$

By using the Euler identity for $\sin x$ in the numerator and for $\cos x$ in the denominator,

$$F(z) = \frac{z \cdot \sin \omega T}{z^2 - 2z \cdot \cos \omega T + 1} \tag{12.13}$$

Exercise 12.1

Problem:

Determine the z-transform of the exponential growth function, $(1 - \exp[-at])u(t)$. Hint: Use the linearity property given by Equation (12.5a) along with Equations (12.9) and (12.10).

Answer:

$$F(z) = \frac{z(1 - \exp[-aT])}{(z - 1)(z - \exp[-aT])}$$

z-Transforms of Transfer Functions

If the Laplace transform of a transfer function is known, then the z-transform is obtained directly from the s-domain function. In some cases, a partial fraction decomposition of the s-domain function must be performed before the z-domain function is apparent. Note in the solution for Example 12.1 that the transfer function cannot be completely defined until the sampling interval T for the input function is known.

Example 12.1

Problem:

Find the transfer function $H(z) = V_o(z)/V_i(z)$ for the network of Figure 12.6.

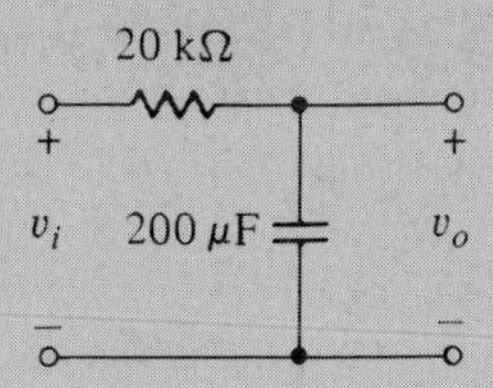

Figure 12.6
Example 12.1.

Solution:

1. Determine the s-domain transfer function:

$$H(s) = \frac{1/(Cs)}{R + 1/(Cs)} = \frac{1/(RC)}{s + 1/(RC)} = \frac{0.25}{s + 0.25}$$

2. Obtain the z-transform of $H(s)$ from Table 12.2:

$$F(s) = \frac{1}{(s+a)} \quad \Rightarrow \quad F(z) = \frac{z}{z - \exp[-aT]}$$

3. Substitute the result of step 2 into the result of step 1:

$$H(z) = \frac{0.25z}{z - \exp[-0.25T]}$$

Example 12.2

Problem:

Find the transfer function $H(z) = V_o(z)/V_i(z)$ for the network of Figure 12.7.

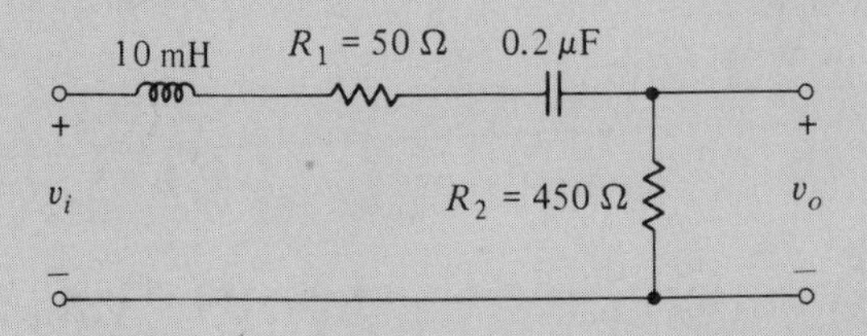

Figure 12.7 Example 12.2.

Solution:

1. Determine the s-domain transfer function:

$$H(s) = \frac{R_2}{Ls + R_1 + R_2 + 1/(Cs)} = \frac{45 \times 10^3 s}{s^2 + 50 \times 10^3 s + 0.5 \times 10^9}$$

2. Perform a partial fraction decomposition of $H(s)$:

$$H(s) = \frac{72.81 \times 10^3}{s + 36.18 \times 10^3} - \frac{27.81 \times 10^3}{s + 13.82 \times 10^3}$$

3. Obtain the z-transform of $H(s)$ from Table 12.2:

$$\begin{aligned} H(z) &= \frac{72.81 \times 10^3 z}{z - \exp[-36.18 \times 10^3 T]} - \frac{27.81 \times 10^3 z}{z - \exp[-13.82 \times 10^3 T]} \\ &= \frac{72.81 \times 10^3 z(z - \exp[-13.82 \times 10^3 T]) - 27.81 \times 10^3 z(z - \exp[-36.18 \times 10^3 T])}{(z - \exp[-36.18 \times 10^3 T])(z - \exp[-13.82 \times 10^3 T])} \\ &= \frac{z(45 \times 10^3 z - 72.81 \times 10^3 \cdot \exp[-13.82 \times 10^3 T] + 27.81 \times 10^3 \cdot \exp[-36.18 \times 10^3 T])}{z^2 - (\exp[-36.18 \times 10^3 T] + \exp[-13.82 \times 10^3 T])z + \exp[-50 \times 10^3 T]} \end{aligned}$$

Exercise 12.2

Problem:

Interchange the resistor and capacitor in Example 12.1 (Figure 12.6) and determine $H(z)$.

Answer:

$$H(s) = \frac{s}{s + 0.25} = 1 - \frac{0.25}{s + 0.25}$$

$$H(z) = 1 - \frac{0.25z}{z - \exp[-0.25T]} = \frac{0.75z - \exp[-0.25T]}{z - \exp[-0.25T]}$$

12.4 Inverse z-Transforms

In obtaining z-transforms, we progress from $f(t)$ via the sampled-data function $f^*(t)$ to $F(z)$, which we call the z-transform of $f(t)$. Recall that $F(z)$ is the Laplace transform of $f^*(t)$. The z-transform is said to be unique, because there is only one $F(z)$ which represents $f(t)$ sampled at a given interval. On the other hand, we see in Figure 12.8 that $f_1(t)$ and $f_2(t)$, clearly distinct time-domain functions, have the same $f^*(t)$, and hence the same $F(z)$. From this we must conclude that the inverse transform of $F(z)$ will yield only $f^*(t)$, and will give no definite indication as to whether $f^*(t)$ was generated by $f_1(t)$ or $f_2(t)$. Lest the picture appear too bleak, we must point out that in most

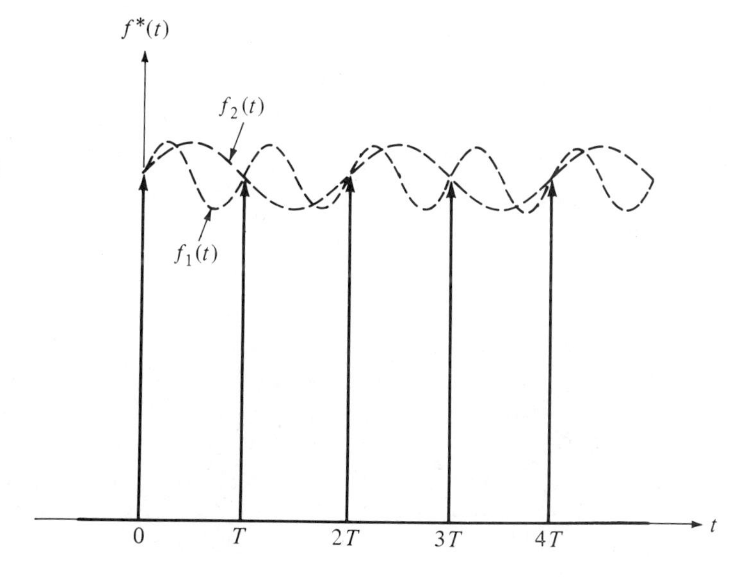

Figure 12.8 Two continuous time-domain functions with the same sampled-data function and the same z-transform.

cases we will be able to identify a particular time-domain function which could have generated $f^*(t)$. From a practical standpoint, a higher sampling rate (smaller intervals) would also alleviate the ambiguity between $f_1(t)$ and $f_2(t)$.

Two straightforward techniques to obtain the inverse of many of the z-transforms encountered in sampled-data system analysis—the **partial fraction method** and the **long division method**—will be presented in this section. A third, more general method—the **residue method**—depends upon the mathematics of complex variables and is beyond the scope of this text.

Partial-Fraction Method

The partial fraction method assumes that the denominator of $F(z)$ can be factored and that $F(z)$ can be decomposed into terms which are then recognized as entries in an available table of z-transforms. The actual decomposition is performed with $F(z)/z$, rather than with $F(z)$ itself, because $F(z)/z$ will always be a proper polynomial function. The algorithm for obtaining an inverse of a z-transform by the partial fraction method is:

1. Decompose $F(z)/z$ into partial fractions.
2. Multiply the partial fractions of $F(z)/z$ by z to obtain a decomposition of $F(z)$.
3. Use an available table of z-transforms to obtain a time-domain function associated with each term of $F(z)$.
4. Combine these results linearly to obtain $f(t)$.
5. Evaluate $f(nT)$ for $n = 0, 1, 2, \ldots$ to obtain $f^*(t)$.

When sampled, the function $f(t)$ yields the $f^*(t)$ whose z-transform is $F(z)$. This $f(t)$ is not unique; there exist other time-domain functions which will yield the same $f^*(t)$.

Example 12.3

Problem:

Determine $f^*(t)$ from a partial fraction decomposition of

$$F(z) = \frac{0.02724z^2 + 0.1812z}{(z-1)(z^2 - 1.038z + 0.2466)}$$

$F(z)$ arises from a sampled-data system with a sampling interval of 0.5 sec.

Solution:

1. Divide $F(z)$ by z to obtain $F(z)/z$, and decompose $F(z)/z$ into partial fractions:

$$\frac{F(z)}{z} = \frac{0.02724z + 0.1812}{(z-1)(z^2 - 1.038z + 0.2466)}$$

$$= \frac{0.02724z + 0.1812}{(z-1)(z-0.6703)(z-0.3679)}$$

$$= \frac{K_1}{z-1} + \frac{K_2}{z-0.6703} + \frac{K_3}{z-0.3679}$$

$$K_1 = (z-1)\frac{F(z)}{z}\bigg|_{z=1} = 1$$

$$K_2 = (z-0.6703)\frac{F(z)}{z}\bigg|_{z=0.6703} = -2$$

$$K_3 = (z-0.3679)\frac{F(z)}{z}\bigg|_{z=0.3679} = 1$$

2. After multiplying $F(z)/z$ by z,

$$F(z) = \frac{z}{z-1} - \frac{2z}{z-0.6703} + \frac{z}{z-0.3679}$$

3. Note that $0.6703 \simeq \exp[-0.4]$ and $0.3679 \simeq \exp[-1]$; therefore,

$$F(z) = \frac{z}{z-1} - \frac{2z}{z-\exp[-0.4]} + \frac{z}{z-\exp[-1]}$$

4. Since $T = 0.5$,

$$F(z) = \frac{z}{z-1} - \frac{2z}{z-\exp[-0.8T]} + \frac{z}{z-\exp[-2T]}$$

5. Obtain $f(t)$ from Table 12.2:

$$f(t) = u(t) - 2\exp[-0.8t]\cdot u(t) + \exp[-2t]\cdot u(t)$$
$$= (1 - 2\exp[-0.8t] + \exp[-2t])u(t)$$

6. Obtain $f^*(t)$ by evaluating $f(nT)$ for $n = 0, 1, 2, \ldots$:

$$
\begin{aligned}
f^*(0) &= 1 - 2 + 1 = 0 \\
f^*(0.5) &= 1 - 1.341 + 0.3679 = 0.02724 \\
f^*(1.0) &= 1 - 0.8987 + 0.1353 = 0.2367 \\
f^*(1.5) &= 1 - 0.6024 + 0.04979 = 0.4474 \\
&\vdots
\end{aligned}
$$

A plot of $f^*(t)$ is given in Figure 12.9.

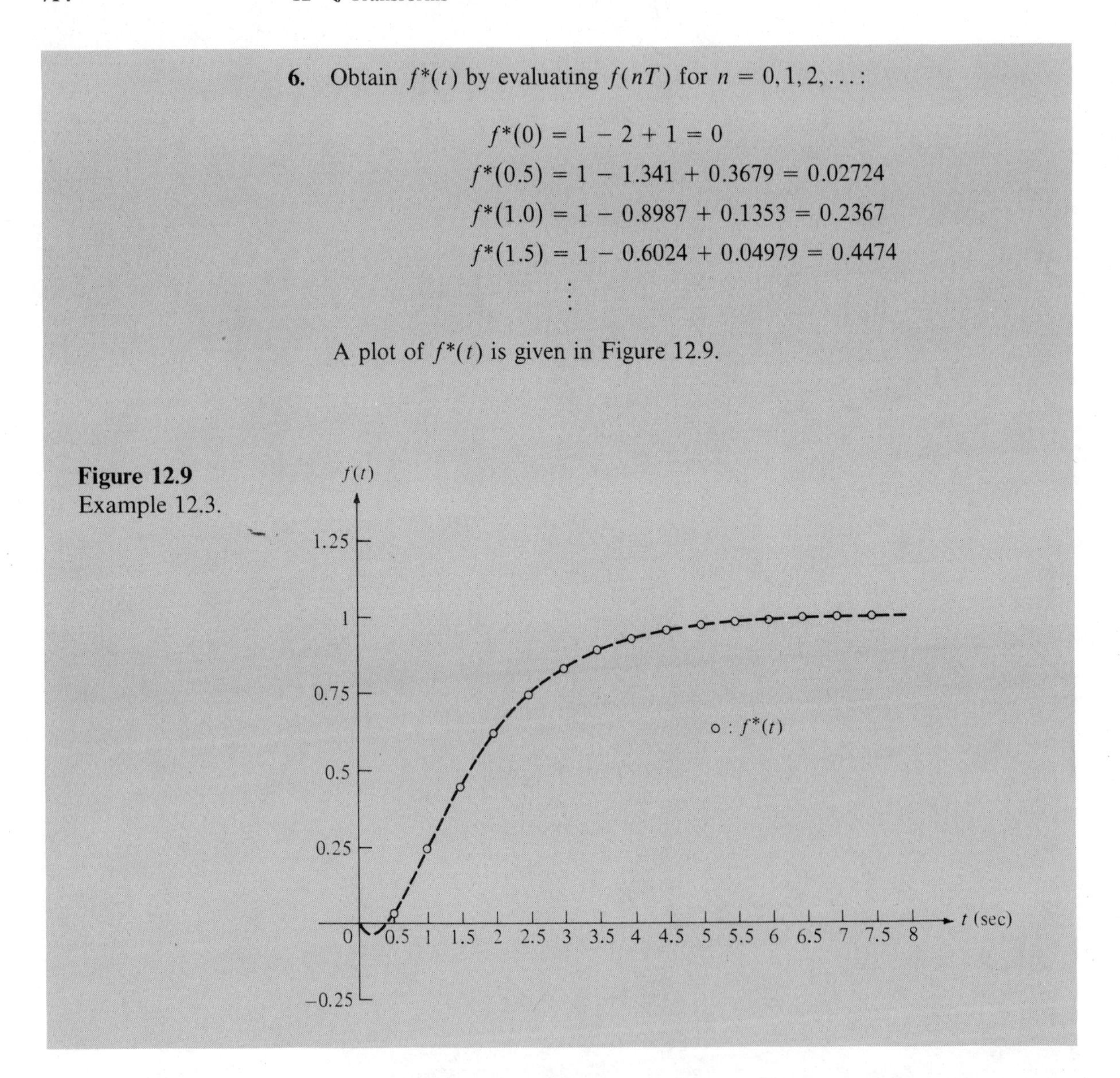

Figure 12.9
Example 12.3.

The partial fraction decomposition method of obtaining inverse z-transforms leads directly to a continuous function $f(t)$ (step 6 in Example 12.3) which could have generated the sampled-data function $f^*(t)$ represented by $F(z)$; $f^*(t)$ is then obtained by evaluation of $f(nT)$ for $n = 0, 1, 2, \ldots$ (step 7 in Example 12.3). The long division method, to be illustrated next, provides the sampled-data function directly, and the continuous function must be inferred, usually from a graph of $f^*(t)$.

Exercise 12.3

Problem:

Determine $v^*(t)$ from a partial fraction decomposition of

$$V(z) = \frac{3z^3 - 5}{z^2(z-1)}$$

$V(z)$ arises from a sampled-data system with a sampling interval of 1 msec.

Answer:

$$V(z) = 5 + \frac{5}{z} + \frac{5}{z^2} - \frac{2z}{z-1}$$

$$\begin{aligned} v(t) &= 5 \cdot \delta(t) + 5 \cdot \delta(t - T) + 5 \cdot \delta(t - 2T) - 2 \cdot u(t) \\ &= 5 \cdot \delta(t) + 5 \cdot \delta(t - 0.001) + 5 \cdot \delta(t - 0.002) - 2 \cdot u(t) \text{ V} \end{aligned}$$

$$\begin{aligned} v^*(0) &= 5 + 0 + 0 - 2 = 3 \\ v^*(0.001) &= 0 + 5 + 0 - 2 = 3 \\ v^*(0.002) &= 0 + 0 + 5 - 2 = 3 \\ v^*(0.003) &= 0 + 0 + 0 - 2 = -2 \\ v^*(0.001n) &= 0 + 0 + 0 - 2 = -2 \qquad \text{for} \quad n \geq 3 \end{aligned}$$

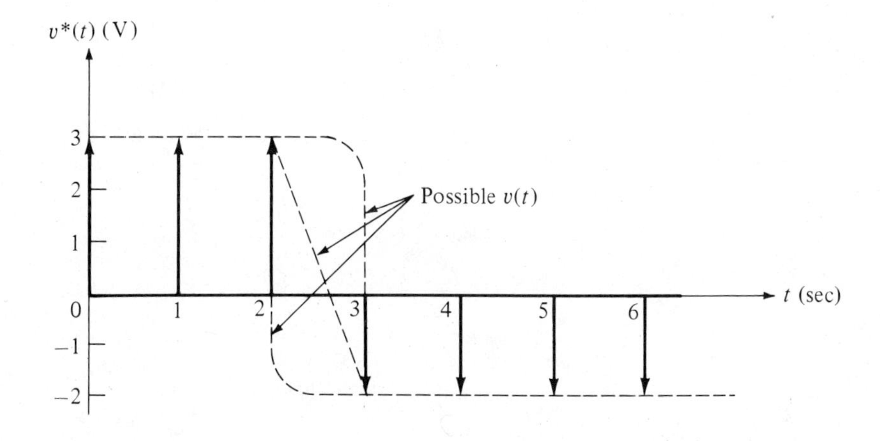

Figure 12.10
Exercise 12.3.

See Figure 12.10 for a graph of $v^*(t)$. Note that $v^*(t)$ could have been generated by any of an infinite number of continuous time-domain functions. Several possibilities are indicated on the graph of $v^*(t)$.

Long Division Method

The long division method for obtaining an inverse z-transform effectively generates the **power series** in powers of $1/z$ for $F(z)$, so that

$$F(z) = a_0 + a_1 z^{-1} + a_2 z^{-2} + \cdots$$
$$= \sum_{n=0}^{\infty} a_n z^{-n}$$

From Table 12.2

$$\mathscr{Z}^{-1}\{z^{-n}\} = \delta(t - nT),$$

so

$$f^*(t) = a_0 \cdot \delta(t) + a_1 \cdot \delta(t - T) + a_2 \cdot \delta(t - 2T) + \cdots$$
$$= \sum_{n=0}^{\infty} a_n \cdot \delta(t - nT)$$

The coefficients a_n in the power series represent $f(nT)$, so

$$f^*(t) = \sum_{n=0}^{\infty} f(nT) \cdot \delta(t - nT)$$

which is precisely the definition of $f^*(t)$ [Equation (12.1)]. Thus, $f^*(t)$ is obtained directly by a power series expansion of $F(z)$.

Example 12.4

Problem:

Determine $f^*(t)$ by long division of $F(z)$ in Example 12.3:

$$F(z) = \frac{0.02724z^2 + 0.1812z}{(z-1)(z^2 - 1.038z + 0.2466)}$$

Solution:

1. Multiply the two factors in the denominator of $F(z)$ in preparation for long division:

$$F(z) = \frac{0.02724z^2 + 0.1812z}{z^3 - 2.038z^2 + 1.285z - 0.2466}$$

2. Divide the numerator of $F(z)$ by the denominator:

$$\begin{array}{r l}
 & 0.02724z^{-1} \\ \cline{2-2}
z^3 - 2.038z^2 + 1.285z - 0.2466 & \big)\, 0.02724z^2 + 0.1812z \\
 & \mp 0.02724z^2 \pm 0.05552z \mp 0.03500 \pm 0.006717z^{-1} \\ \cline{2-2}
 & 0.2367z - 0.03500 + 0.006717z^{-1}
\end{array}$$

Therefore,

$$F(z) = 0.02724z^{-1} + Q_1(z)$$

where

$$Q_1(z) = \frac{0.2367z - 0.03500 + 0.006717z^{-1}}{z^3 - 2.038z^2 + 1.285z - 0.2466}$$

3. Divide the numerator of $Q_1(z)$ by the denominator:

$$\begin{array}{r l}
 & 0.2367z^{-2} \\ \cline{2-2}
z^3 - 2.038z^2 + 1.285z - 0.2466 & \big)\, 0.2367z - 0.03500 + 0.006717z^{-1} \\
 & \mp 0.2367z \pm 0.4824 \mp 0.3042z^{-1} \pm 0.05837z^{-2} \\ \cline{2-2}
 & 0.4474 - 0.2975z^{-1} + 0.05837z^{-2}
\end{array}$$

Therefore,

$$F(z) = 0.02724z^{-1} + 0.2367z^{-2} + Q_2(z)$$

where

$$Q_2(z) = \frac{0.4474 - 0.2975z^{-1} + 0.05837z^{-2}}{z^3 - 2.038z^2 + 1.285z - 0.2466}$$

4. Divide the numerator of $Q_2(z)$ by the denominator:

$$\begin{array}{r l}
 & 0.4474z^{-3} \\ \cline{2-2}
z^3 - 2.038z^2 + 1.285z - 0.2466 & \big)\, 0.4474 - 0.2975z^{-1} + 0.05837z^{-2} \\
 & \mp 0.4474 \pm 0.9118z^{-1} \mp 0.5749z^{-2} \pm 0.1103z^{-3} \\ \cline{2-2}
 & 0.6143z^{-1} - 0.5165z^{-2} + 0.1103z^{-3}
\end{array}$$

Therefore,

$$F(z) = 0.02724z^{-1} + 0.2367z^{-2} + 0.4474z^{-3} + Q_3(z)$$

where

$$Q_3(z) = \frac{0.6143z^{-1} - 0.5165z^{-2} + 0.1103z^{-3}}{z^3 - 2.038z^2 + 1.285z - 0.2466}$$

The long division process can be continued as far as desired, so

$$F(z) = 0.02724z^{-1} + 0.2367z^{-2} + 0.4474z^{-3} + \cdots$$

5. Obtain $f^*(t)$ from Table 12.2:

$$\begin{aligned} f^*(t) &= \mathscr{Z}^{-1}\{0.02724z^{-1}\} + \mathscr{Z}^{-1}\{0.2367z^{-2}\} + \mathscr{Z}^{-1}\{0.4474z^{-3}\} + \cdots \\ &= 0.02724 \cdot \delta(t - T) + 0.2367 \cdot \delta(t - 2T) \\ &\quad + 0.4474 \cdot \delta(t - 3T) + \cdots \end{aligned}$$

6. Evaluate $f^*(nT)$ for $n = 0, 1, 2, \ldots$ ($T = 0.5$ sec):

$$\begin{aligned} f^*(0) &= 0.02724(0) + 0.2367(0) + 0.4474(0) + \cdots = 0 \\ f^*(0.5) &= 0.02724(1) + 0.2367(0) + 0.4474(0) + \cdots = 0.02724 \\ f^*(1.0) &= 0.02724(0) + 0.2367(1) + 0.4474(0) + \cdots = 0.2367 \\ f^*(1.5) &= 0.02724(0) + 0.2367(0) + 0.4474(1) + \cdots = 0.4474 \end{aligned}$$

These results agree with those obtained in Example 12.3 (step 7) by the partial fraction method.

A BASIC program to perform long division and determine the terms of the power series expansion of $F(z)$ is given below. This program is written for GW-BASIC™ (Version 3.2) interpreter, but should execute, with perhaps limited modification, using other versions of BASIC. The reader is encouraged to modify the program to provide more elaborate data-entry cues and to include data-entry error checks and traps.

```
10 CLS:FOR I = 1 TO 10 :PRINT:NEXT I:PRINT "                POLYNOMIAL
LONG DIVISION"
20 FOR I=1 TO 10:PRINT:NEXT I:INPUT "PRESS ANY KEY TO
CONTINUE",A$
30 CLS:PRINT "OUTPUT TO PRINTER?"
40 INPUT "PRESS 'Y' FOR YES; PRESS ANY OTHER KEY FOR NO ",P$
50 IF P$="Y" OR P$="y" THEN P=1 ELSE P=0
```

```
60 CLS:INPUT "ENTER DEGREE OF DIVIDEND (NUMERATOR) ", N1
70 CLS:PRINT "DIVIDEND IS OF DEGREE "N1
80 GOSUB 400:IF A THEN 60
90 N1 = N1 + 1:DIM D1(N1)
100 CLS:PRINT "DIVIDEND COEFFICIENTS":PRINT
110 FOR I = 1 TO N1:PRINT "ENTER COEFFICIENT OF X^"N1 - I:INPUT
D1(I):NEXT I
120 CLS:PRINT "ECHO CHECK DIVIDEND":PRINT
130 FOR I = 1 TO N1:PRINT D1(I)"x^"N1 - I:NEXT I:PRINT
140 GOSUB 400:IF A THEN 100
150 CLS:IF P THEN 160 ELSE 170
160 LPRINT TAB(10) "DIVIDEND":FOR I = 1 TO N1:LPRINT TAB(10)
D1(I)"x^"N1 - I:NEXT I:LPRINT
170 CLS:INPUT "ENTER DEGREE OF DIVISOR (DENOMINATOR) ",N2
180 CLS:PRINT "DIVISOR IS OF DEGREE "N2
190 GOSUB 400:IF A THEN 170
200 N2 = N2 + 1:DIM D2(N2)
210 CLS:PRINT "DIVISOR COEFFICIENTS":PRINT
220 FOR I = 1 TO N2:PRINT "ENTER COEFFICIENT OF x^"N2 - I:INPUT
D2(I):NEXT I
230 CLS:PRINT "ECHO CHECK DIVISOR":PRINT
240 FOR I = 1 TO N2:PRINT D2(I)"x^"N2 - I:NEXT I:PRINT
250 GOSUB 400:IF A THEN 210
260 CLS:IF P THEN 270 ELSE 280
270 LPRINT TAB(10) "DIVISOR":FOR I = 1 TO N2:LPRINT TAB(10)
D2(I)"x^"N2 - I:NEXT I:LPRINT
280 SCALE = D2(1):FOR I = 1 TO N2:D2(I) = D2(I) / SCALE:NEXT I
290 IF N1>N2 THEN DIM D(N1) ELSE DIM D(N2)
300 FOR I = 1 TO N1:D(I) = D1(I):NEXT I
310 E = N1 - N2
320 CLS:PRINT "QUOTIENT":IF P THEN LPRINT TAB(10) "QUOTIENT"
330 PRINT:FOR I = 1 TO 5:DQ = D(1):FOR J = 1 TO N2:D(J) = D(J) -
DQ*D2(J):NEXT J
340 IF N1>N2 THEN N = N1 ELSE N = N2
350 FOR J = 1 TO N - 1:D(J) = D(J + 1):NEXT J:D(N) = 0
360 PRINT DQ / SCALE"x^"E:IF P THEN LPRINT TAB(10) DQ / SCALE"x^"E
370 E = E - 1:NEXT I:PRINT:PRINT:IF P THEN LPRINT:LPRINT
380 PRINT "PRESS ANY KEY FOR FIVE MORE TERMS IN QUOTIENT"
390 INPUT "PRESS 'T' TO TERMINATE ",A$:IF A$ = "T" OR A$ = "t" THEN
END ELSE 330
400 PRINT:PRINT "IF INCORRECT PRESS 'X'; IF CORRECT PRESS ANY
OTHER KEY":PRINT
410 INPUT A$:IF A$ = "X" OR A$ = "x" THEN A = 1 ELSE A = 0
420 RETURN
```

Exercise 12.4

Problem:

Determine $v^*(t)$ by long division of $V(z)$ in Exercise 12.3.

Answer:

$$\begin{aligned} V(z) &= 3 + \frac{3z^2 - 5}{z^3 - z^2} \\ &= 3 + 3z^{-1} + \frac{3z - 5}{z^3 - z^2} \\ &= 3 + 3z^{-1} + 3z^{-2} + \frac{-2}{z^3 - z^2} \\ &= 3 + 3z^{-1} + 3z^{-2} - 2z^{-3} + \frac{-2z^{-1}}{z^3 - z^2} \\ &= 3 + 3z^{-1} + 3z^{-2} - 2z^{-3} + -2z^{-4} - 2z^{-5} - \cdots \end{aligned}$$

$$\begin{aligned} v^*(t) &= 3 \cdot \delta(t) + 3 \cdot \delta(t - T) + 3 \cdot \delta(t - 2T) \\ &\quad -2 \cdot \delta(t - 3T) - 2 \cdot \delta(t - 4T) - 2 \cdot \delta(t - 5T) - \cdots \\ &= 3 \cdot \delta(t) + 3 \cdot \delta(t - 0.001) + 3 \cdot \delta(t - 0.002) \\ &\quad -2 \cdot \delta(t - 0.003) - 2 \cdot \delta(t - 0.004T) \\ &\quad -2 \cdot \delta(t - 0.005) - \cdots \text{ V} \end{aligned}$$

$$\begin{aligned} v^*(0) &= 3 + 0 + 0 - 0 - 0 - 0 - \cdots = 3 \\ v^*(0.001) &= 0 + 3 + 0 - 0 - 0 - 0 - \cdots = 3 \\ v^*(0.002) &= 0 + 0 + 3 - 0 - 0 - 0 - \cdots = 3 \\ v^*(0.003) &= 0 + 0 + 0 - 2 - 0 - 0 - \cdots = -2 \\ v^*(0.004) &= 0 + 0 + 0 - 0 - 2 - 0 - \cdots = -2 \\ v^*(0.005) &= 0 + 0 + 0 - 0 - 0 - 2 - \cdots = -2 \\ v^*(0.001n) &= -2 \qquad \text{for} \quad n \geq 3 \end{aligned}$$

12.5 Sample and Hold

Any system draws energy from its excitation source to provide its output response. The amount of energy drawn from the source is the product of the power supplied by the source and the length of time during which the power is supplied. In the case of sampled-data systems, this relationship is of particular importance and requires special consideration.

Typically, an input data sample is a voltage pulse of very short duration. In order to provide sufficient energy to the circuit during this brief pulse, the source must supply a high level of power, requiring a large current. As the pulse duration decreases (toward the ideal sample of zero duration), the required current increases, and frequently exceeds the capabilities of the excitation source. Directly sampling the excitation source is usually not possible.

Zero-Order Hold

Almost without exception, sampled-data systems incorporate **sample-and-hold** circuits which serve to reduce the current demand on the excitation source during sampling. One such circuit is the **zero-order hold** (ZOH) (Figure 12.11),

Figure 12.11 Zero-order hold (ZOH).
(a) Sampled-data input.
(b) Sampled-data system with ZOH.
(c) Output of ZOH.

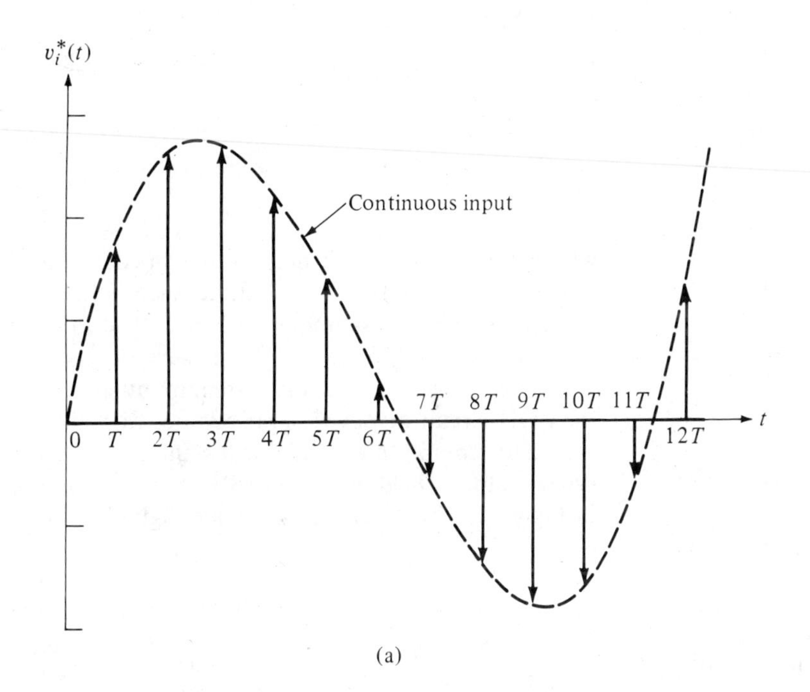

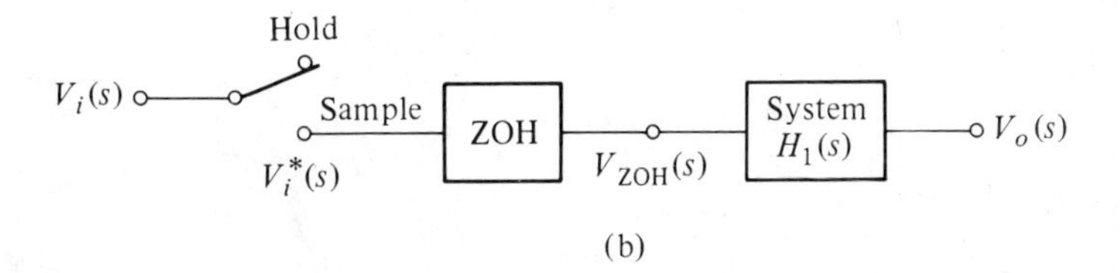

Figure 12.11
Continued

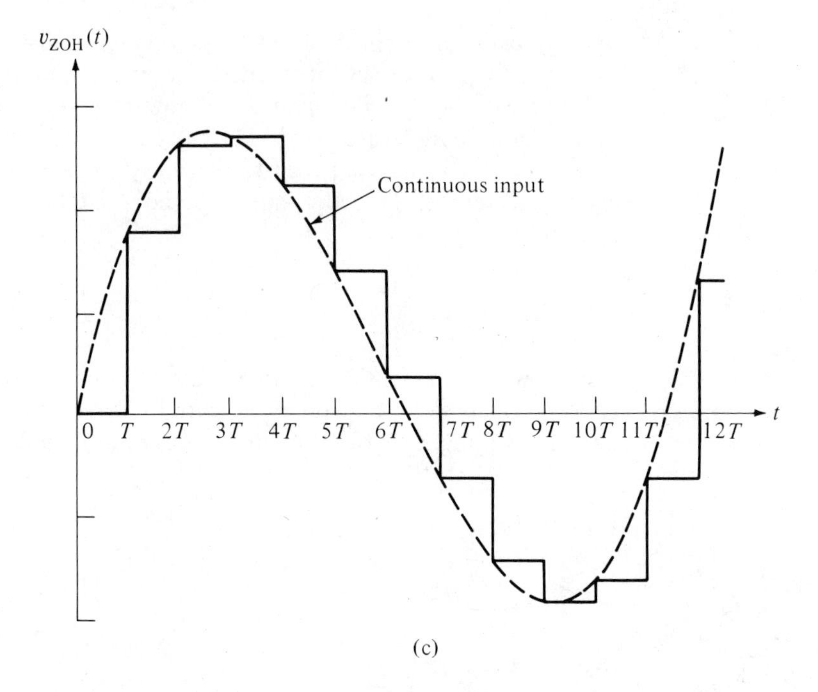

which uses a device with high input impedance to detect the voltage level of the excitation source at sample time, then holds that voltage level as a system input until the next sample is obtained. The ZOH circuit, because of its high input impedance, requires very little current, and hence very little power, from the source. Power to meet the system input energy requirements is delivered by the ZOH circuit dc power supply rather than the excitation source.

ZOH circuits are available as integrated circuits which are known as **sample / track–hold amplifiers** (SHA). A typical SHA configuration is shown in Figure 12.12. When the solid-state digitally controlled **mode switch** is closed

Figure 12.12
Typical zero-order hold circuit.

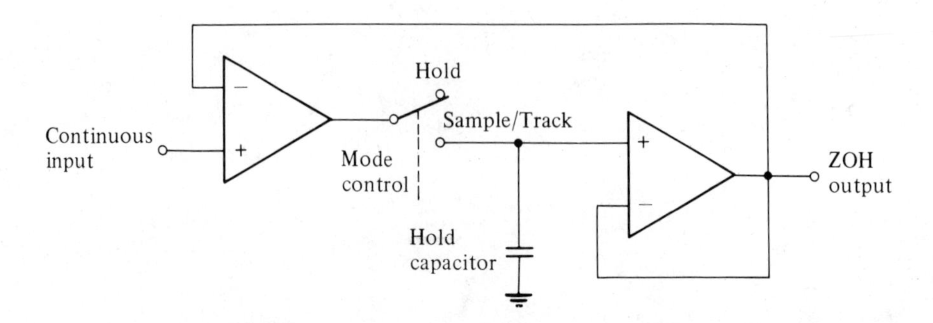

(SAMPLE / TRACK), the input OPAMP functions as a current source to charge the **hold capacitor** to the input voltage. The storage capacitor is small, and the OPAMP output resistance is near zero, so in the sample/track mode the capacitor voltage *tracks* the input voltage with little lag. When the mode switch opens (HOLD), the storage capacitor is isolated and retains its charge until the mode switch returns to the SAMPLE / TRACK position. During the entire cycle, the output OPAMP provides input voltage to the system equal to the hold-capacitor voltage.

Analysis of Networks Containing a ZOH

A ZOH multiplies the sampled value $v_i^*(t)$ of the continuous input voltage $v_i(t)$ by 1, and holds that value until the next sample is available. The ideal response of a ZOH, shown in Figure 12.13, is

$$h_{\mathrm{ZOH}}(t) = u(t) - u(t - T)$$

From Pair II and the time-domain shifting property,

$$H_{\mathrm{ZOH}}(s) = \frac{1}{s} - \frac{1}{s} \cdot \exp(-Ts) = \frac{1 - \exp[-Ts]}{s} \tag{12.14}$$

The ZOH output $v_{\mathrm{ZOH}}(t)$, a piecewise continuous function (Figure 12.11), is the network input.

The response of the ZOH affects the network response, so in the analysis of sampled-data systems, the transfer function of the ZOH must be included in the network transfer function:

$$H(s) = H_{\mathrm{ZOH}}(s) \cdot H_1(s)$$

where $H_1(s)$ is the transfer function of the network without sampling and ZOH. By substituting Equation (12.14), the overall transfer function can be

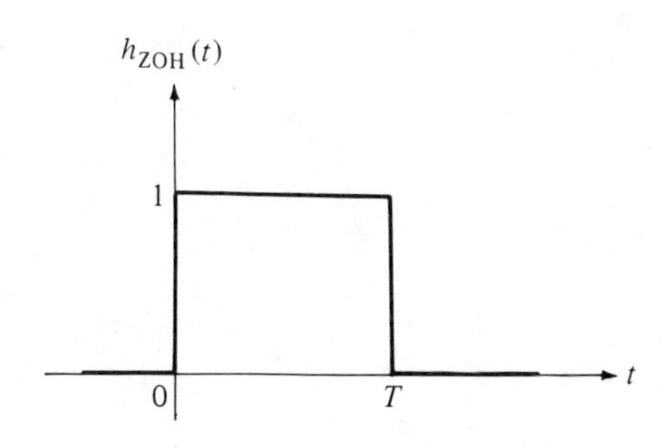

Figure 12.13
Response of a zero-order hold.

written

$$H(s) = (1 - \exp[-Ts])\frac{H_1(s)}{s} \tag{12.15}$$

Equation (12.15) can also be expressed in distributed form as

$$H(s) = \frac{H_1(s)}{s} - \frac{\exp[-Ts] \cdot H_1(s)}{s}$$

The z-transform which corresponds to $H(s)$, found by applying the time-domain shifting property (Table 12.2) to the second term of $H(s)$, is

$$\begin{aligned} H(z) &= H_2(z) - z^{-1} \cdot H_2(z) \\ &= (1 - z^{-1})H_2(z) \end{aligned} \tag{12.16}$$

where $H_2(z)$ is the z-transform which corresponds to $H_1(s)/s$, and $H_1(s)$ is the transfer function of the network without sampling and ZOH.

Example 12.5

Problem:

Use z-transforms to determine the sampled-data output of the network shown in Figure 12.14(a). The input is the triangular waveform shown in Figure 12.14(b),

$$v_i(t) = 5t \cdot u(t) - 10(t-1)u(t-1) + 5(t-2)u(t-2) \text{ V}$$

and the sampler operates at 5 Hz ($T = 0.2$ sec).

Solution:

1. Determine the transfer function $H_1(s)$ of the network without sampling and zero-order hold:

$$H_1(s) = \frac{1/(Cs)}{R + 1/(Cs)} = \frac{4}{s+4}$$

2. Use Equation (12.15) to obtain the network transfer function $H(s)$:

$$H(s) = (1 - \exp[-0.2s])\frac{4}{s(s+4)}$$

Figure 12.14
Example 12.5.

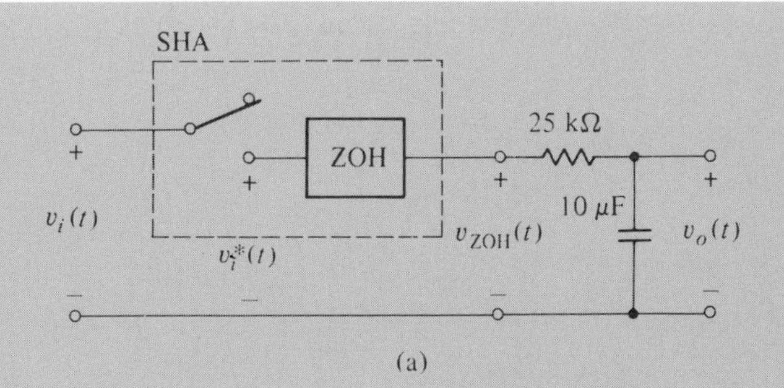

(a)

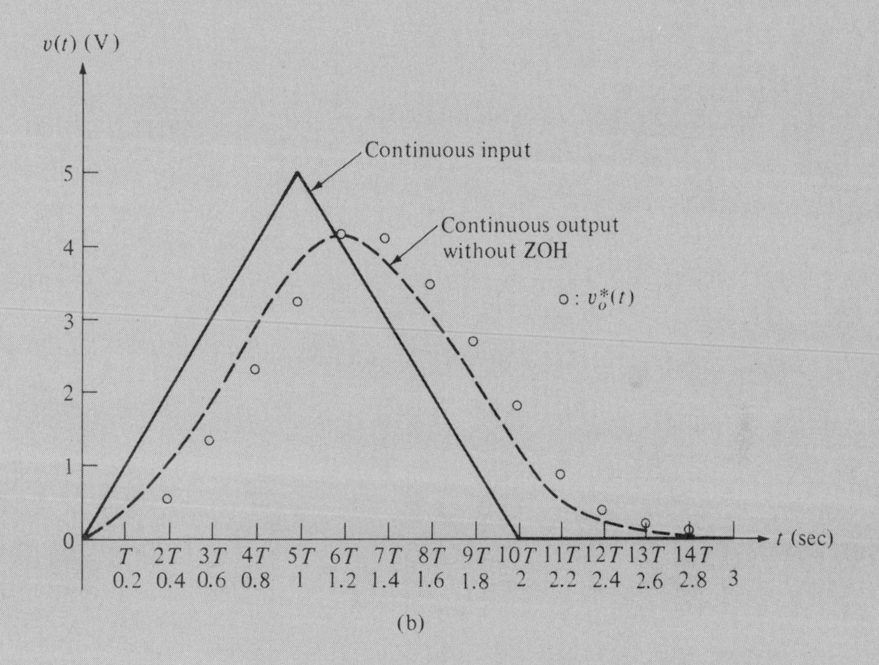

(b)

3. Use Equation (12.16) and Table 12.2 to obtain $H(z)$:

$$H(z) = (1 - z^{-1})\frac{z(1 - \exp[-4T])}{(z-1)(z - \exp[-4T])}$$

$$= \left(\frac{z-1}{z}\right)\left(\frac{z(1 - \exp[-4(0.2)])}{(z-1)\{z - \exp[-4(0.2)]\}}\right)$$

$$= \frac{0.5507}{z - 0.4493}$$

4. Obtain $V_i(z)$ from Table 12.2 and the time-domain shifting property:

$$V_i(z) = \frac{5Tz}{(z-1)^2} - \frac{1}{z^5} \cdot \frac{10Tz}{(z-1)^2} + \frac{1}{z^{10}} \cdot \frac{5Tz}{(z-1)^2}$$

$$= \frac{5(0.2)z \cdot z^{10} - 10(0.2)z \cdot z^5 + 5(0.2)z}{z^{10}(z-1)^2}$$

$$= \frac{z^{10} - 2z^5 + 1}{z^9(z-1)^2}$$

5. Determine $V_o(z)$:

$$V_o(z) = H(z) \cdot V_i(z) = \left(\frac{0.5507}{z - 0.4493}\right)\left(\frac{z^{10} - 2z^5 + 1}{z^9(z-1)^2}\right)$$

$$= \frac{0.5507(z^{10} - 2z^5 + 1)}{z^9(z^3 - 2.4493z^2 + 1.8987z - 0.4493)}$$

6. Determine $v_o^*(t)$ by the long-division method:

$$V_o(z) = 0.5507z^{-2} + 1.3488z^{-3} + 2.2581z^{-4} + 3.2173z^{-5} + \cdots$$

$$v_o^*(t) = 0.5507 \cdot \delta(t - 2T) + 1.3488 \cdot \delta(t - 3T) + 2.2581 \cdot \delta(t - 4T) + 3.2173 \cdot \delta(t - 5T) + \cdots \text{ V}$$

$$v_o^*(0) = 0 \text{ V} \qquad v_o^*(0.6) = 1.3488 \text{ V}$$
$$v_o^*(0.2) = 0 \text{ V} \qquad v_o^*(0.8) = 2.2581 \text{ V}$$
$$v_o^*(0.4) = 0.5507 \text{ V} \qquad v_o^*(1) = 3.2173 \text{ V}$$
$$\vdots$$

Figure 12.14(b) shows the sampled output $v_o^*(nT)$, indicated by the circles, and the continuous output of the network without the SHA, indicated by the dashed line.

Example 12.6

Problem:

Use z-transforms to determine the sampled-data output of the network shown in Figure 12.15(a). The input is the sinusoidal waveform shown in Figure 12.14(b),

$$v_i(t) = \cos 20\pi t \cdot u(t) \text{ V}$$

and the sampler operates at 100 Hz ($T = 10$ msec).

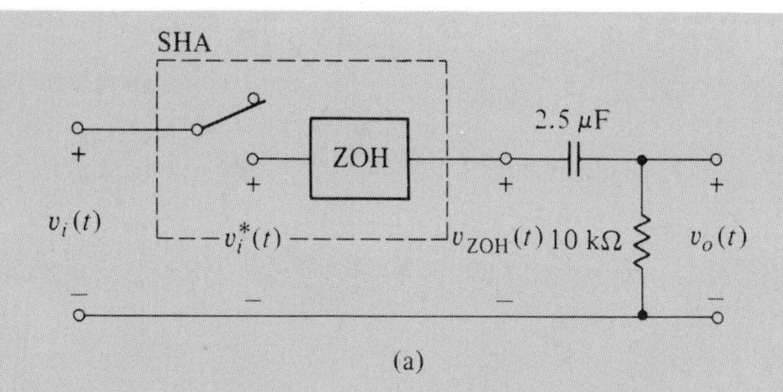

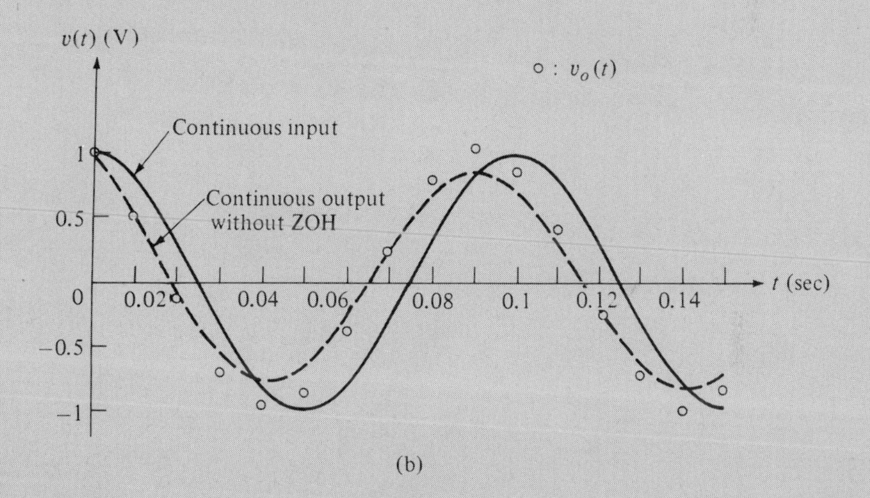

Figure 12.15
Example 12.6.

Solution:

1. Determine the transfer function $H_1(s)$ of the network without sampling and zero-order hold:

$$H_1(s) = \frac{R}{R + 1/(Cs)} = \frac{s}{s + 40}$$

2. Use Equation (12.15) to obtain the network transfer function $H(s)$:

$$H(s) = (1 - \exp[-0.01s])\frac{1}{s + 40}$$

3. Use Equation (12.16) and Table 12.2 to obtain $H(z)$:

$$H(z) = (1 - z^{-1})\frac{z}{z - \exp[-40T]}$$

$$= \left(\frac{z-1}{z}\right)\left(\frac{z}{z - \exp[-40(0.01)]}\right)$$

$$= \frac{z-1}{z - 0.6703}$$

4. Obtain $V_i(z)$ from Table 12.2 and the time-domain shifting property:

$$V_i(z) = \frac{z(z - \cos\omega T)}{z^2 - 2z\cdot\cos\omega T + 1}$$

$$= \frac{z\{z - \cos[20\pi(0.01)]\}}{z^2 - 2z\cdot\cos[20\pi(0.01)] + 1}$$

$$= \frac{z(z - 0.8090)}{z^2 - 1.618z + 1}$$

5. Determine $V_o(z)$:

$$V_o(z) = H(z)\cdot V_i(z) = \frac{z(z-1)(z-0.8090)}{(z-0.6703)(z^2 - 1.618z + 1)}$$

$$= \frac{z(z^2 - 1.809z + 0.8090)}{z^3 - 2.288^2 + 2.085z - 0.6703}$$

6. Determine $v_o^*(t)$ by the long-division method:

$$V_o(z) = 1 + 0.4793z^{-1} - 0.1785z^{-2} - 0.7375z^{-3} - 0.9941z^{-4} - 0.8572z^{-5} - 0.3836z^{-6} + 0.2427z^{-7} + 0.7805z^{-8} + \cdots$$

$$v_o^*(t) = \delta(t) + 0.4793\cdot\delta(t-T) - 0.1785\cdot\delta(t-2T) - 0.7375\cdot\delta(t-3T) - 0.9941\cdot\delta(t-4T) - 0.8572\cdot\delta(t-5T) - 0.3836\cdot\delta(t-6T) + 0.2427\cdot\delta(t-7T) + 0.7805\cdot\delta(t-8T) + \cdots \text{ V}$$

$v_o^*(0) = 1$ V $\quad v_o^*(0.05) = -0.8572$ V
$v_o^*(0.01) = 0.4793$ V $\quad v_o^*(0.06) = -0.3836$ V
$v_o^*(0.02) = -0.1785$ V $\quad v_o^*(0.07) = 0.2427$ V
$v_o^*(0.03) = -0.7375$ V $\quad v_o^*(0.08) = 0.7805$ V
$v_o^*(0.04) = -0.9941$ V $\quad \vdots$

Figure 12.15(b) shows the sampled output $v_o^*(nT)$, indicated by the circles, and the continuous output of the network without the SHA, indicated by the dashed line.

Exercise 12.5

Problem:

Repeat Example 12.5 with the resistor and capacitor interchanged.

Answer:

$$H(z) = \frac{z - 1}{z - 0.4493}$$

$$V_o(z) = \frac{z^{10} - 2z^5 + 1}{z^9(z^2 - 1.449z^2 + 0.4493)}$$

$$= z^{-1} + 1.449z^{-2} + 1.651z^{-3} + 1.742z^{-4} + 1.783z^{-5}$$
$$- 0.1990z^{-6} - 1.089z^{-7} - \cdots$$

$$v_o^*(t) = \delta(t - T) + 1.449 \cdot \delta(t - 2T) + 1.651 \cdot \delta(t - 3T)$$
$$+ 1.742 \cdot \delta(t - 4T) + 1.783 \cdot \delta(t - 5T)$$
$$- 0.1990 \cdot \delta(t - 6T) - 1.089 \cdot \delta(t - 7T) - \cdots \text{ V}$$

$$v_o^*(0) = 0 \qquad v_o^*(0.8) = 1.742 \text{ V}$$
$$v_o^*(0.2) = 1 \text{ V} \qquad v_o^*(1.0) = 1.783 \text{ V}$$
$$v_o^*(0.4) = 1.449 \text{ V} \qquad v_o^*(1.2) = -0.1990 \text{ V}$$
$$v_o^*(0.6) = 1.651 \text{ V} \qquad v_o^*(1.4) = -1.089 \text{ V}$$
$$\vdots$$

Figure 12.16 shows the sampled output $v_o^*(nT)$, indicated by the circles, and the continuous output of the network without the SHA, indicated by the dashed line.

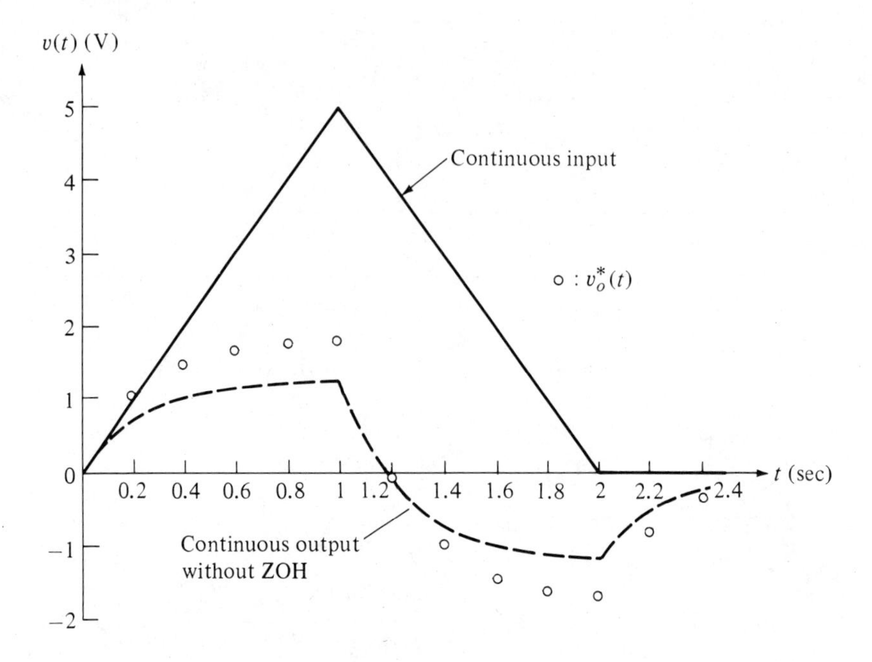

Figure 12.16
Exercise 12.5.

12.6 Second-Order Network with Sample-and-Hold Input

The next three examples illustrate the analysis of an R–L–C circuit with sampled-data input and a zero-order hold. In general the technique is no different than that used in the previous section; however, obtaining $H(z)$ for a second-order network is more complicated. The network used for all three examples and the continuous input is shown in Figure 12.17.

The continuous input is described by

$$v_i(t) = 5000t \cdot u(t) - 5000(t - 0.2 \times 10^{-3})u(t - 0.2 \times 10^{-3}) \text{ V}$$

or

$$V_i(s) = \frac{5000}{s^2} - \frac{5000 \cdot \exp[-0.2 \times 10^{-3}s]}{s^2} \text{ V} \cdot \text{sec}$$

The z-transform of the sampled-data input, which can be determined from

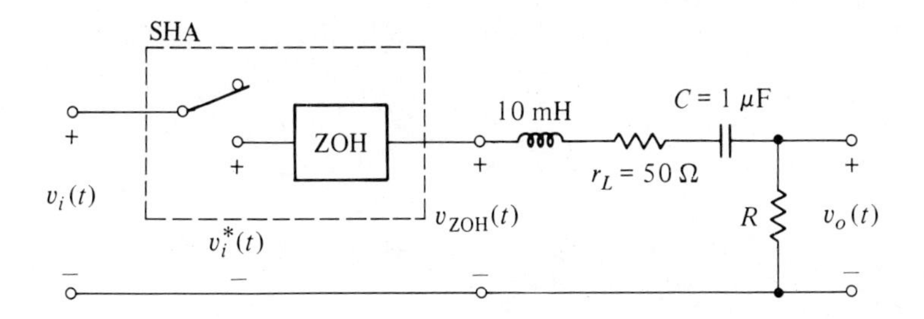

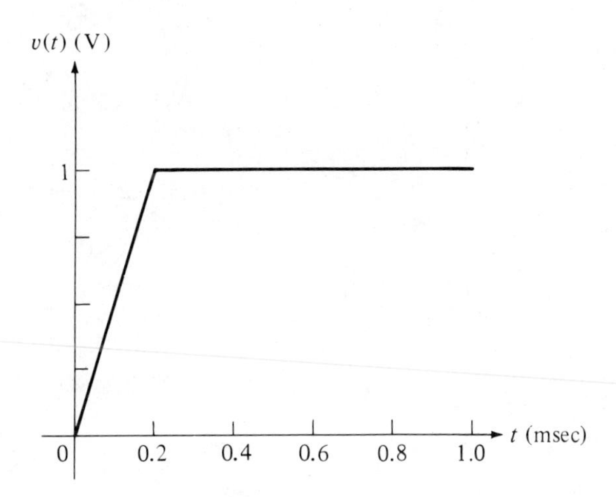

Figure 12.17 Sampled-data system and input for Examples 12.7, 12.8, and 12.9.

Table 12.1 and the time-domain shifting property, is

$$V_i(z) = \frac{5000Tz}{(z-1)^2} - z^{-0.0002/T}\frac{5000Tz}{(z-1)^2}$$

$$= (1 - z^{-0.0002/T})\frac{5000Tz}{(z-1)^2} \tag{12.17}$$

The network transfer function, without zero-order hold, is

$$H_1(s) = \frac{R}{Ls + r_L + R + 1/(Cs)}$$

$$= \frac{(R/L)s}{s^2 + [(r_L + R)/L]s + 1/(LC)}$$

$$= \frac{100Rs}{s^2 + 100(50 + R)s + 100 \times 10^6} \tag{12.18}$$

The critical value of R can be determined from Equation (6.8):

$$R + r_L = 2\sqrt{L/C}$$
$$R_C = 2\sqrt{10 \times 10^{-3}/1 \times 10^{-6}} - 50 = 150\ \Omega$$

Example 12.7

Problem:

Determine the sampled-data output of the network of Figure 12.17 for $R = 450\ \Omega$ and a sampling rate of 20 kHz ($T = 50\ \mu$sec).

Solution:

1. Substitute $t = 50\ \mu$sec into Equation (12.17) to obtain $V_i(z)$:

$$V_i(z) = (1 - z^{-4})\frac{0.25z}{(z-1)^2} = 0.25\frac{z^4 - 1}{z^3(z-1)^2}$$

Note that

$$z^4 - 1 = (z^2 + 1)(z^2 - 1)$$
$$= (z^2 + 1)(z + 1)(z - 1)$$

therefore,

$$V_i(z) = 0.25\frac{z^3 + z^2 + z + 1}{z^3(z-1)}$$

2. Substitute $R = 450\ \Omega$ into Equation (12.18); then use Equation (12.15) to obtain $H(s)$:

$$H_1(s) = \frac{45 \times 10^3 s}{s^2 + 50 \times 10^3 s + 100 \times 10^6} = \frac{45 \times 10^3 s}{(s + 2087)(s + 47\,910)}$$

$$H(s) = (1 - \exp[-50 \times 10^{-6} s])\frac{H_1(s)}{s}$$

$$= (1 - \exp[-50 \times 10^{-6} s])\frac{45 \times 10^3}{(s + 2087)(s + 47\,910)}$$

3. Use Equation (12.16) and Table 12.2 to obtain $H(z)$:

$$H(z) = \left(\frac{z-1}{z}\right)\left(\frac{45 \times 10^3}{47\,910 - 2087} \cdot A(z)\right)$$

where

$$A(z) = \frac{(\exp[-2087T] - \exp[-47\,910T])z}{z^2 - (\exp[-2087T] + \exp[-47910T])z + \exp[-50000T]}$$

Therefore,

$$H(z) = \frac{0.7952(z-1)}{z^2 - 0.9920z + 0.08208}$$

4. Determine $V_o(z)$ from Equation (12.17) and $H(z)$:

$$\begin{aligned} V_o(z) &= H(z) \cdot V_i(z) \\ &= \left(\frac{0.7952(z-1)}{z^2 - 0.9920z + 0.08208}\right)\left(0.25\frac{z^3 + z^2 + z + 1}{z^3(z-1)}\right) \\ &= \frac{0.1988(z^3 + z^2 + z + 1)}{z^3(z^2 - 0.9920z + 0.08208)} \end{aligned}$$

5. Use the long-division method to obtain $v_o^*(t)$:

$$\begin{aligned} V_o(z) = 0.1988z^{-2} &+ 0.3960z^{-3} + 0.5753z^{-4} + 0.7370z^{-5} \\ &+ 0.6839z^{-6} + 0.6180z^{-7} + 0.5569z^{-8} \\ &+ 0.5017z^{-9} + 0.4520z^{-10} + 0.4072z^{-11} + \cdots \end{aligned}$$

$$\begin{aligned} v_o^*(t) = 0.1988 \cdot \delta(t-2T) &+ 0.3960 \cdot \delta(t-3T) \\ &+ 0.5753 \cdot \delta(t-4T) + 0.7370 \cdot \delta(t-5T) \\ &+ 0.6839 \cdot \delta(t-6T) + 0.6180 \cdot \delta(t-7T) \\ &+ 0.5569 \cdot \delta(t-8T) + 0.5017 \cdot \delta(t-9T) \\ &+ 0.4520 \cdot \delta(t-10T) + 0.4072 \cdot \delta(t-11T) + \cdots \text{ V} \end{aligned}$$

$$\begin{aligned} &v_o^*(100 \times 10^{-6}) = 0.1988 \text{ V}, && v_o^*(350 \times 10^{-6}) = 0.6180 \text{ V} \\ &v_o^*(150 \times 10^{-6}) = 0.3960 \text{ V}, && v_o^*(400 \times 10^{-6}) = 0.5569 \text{ V} \\ &v_o^*(200 \times 10^{-6}) = 0.5753 \text{ V}, && v_o^*(450 \times 10^{-6}) = 0.5017 \text{ V} \\ &v_o^*(250 \times 10^{-6}) = 0.7370 \text{ V}, && v_o^*(500 \times 10^{-6}) = 0.4520 \text{ V} \\ &v_o^*(300 \times 10^{-6}) = 0.6839 \text{ V}, && v_o^*(550 \times 10^{-6}) = 0.4072 \text{ V} \end{aligned}$$

The sampled-data output is compared with the continuous output without ZOH in Figure 12.18.

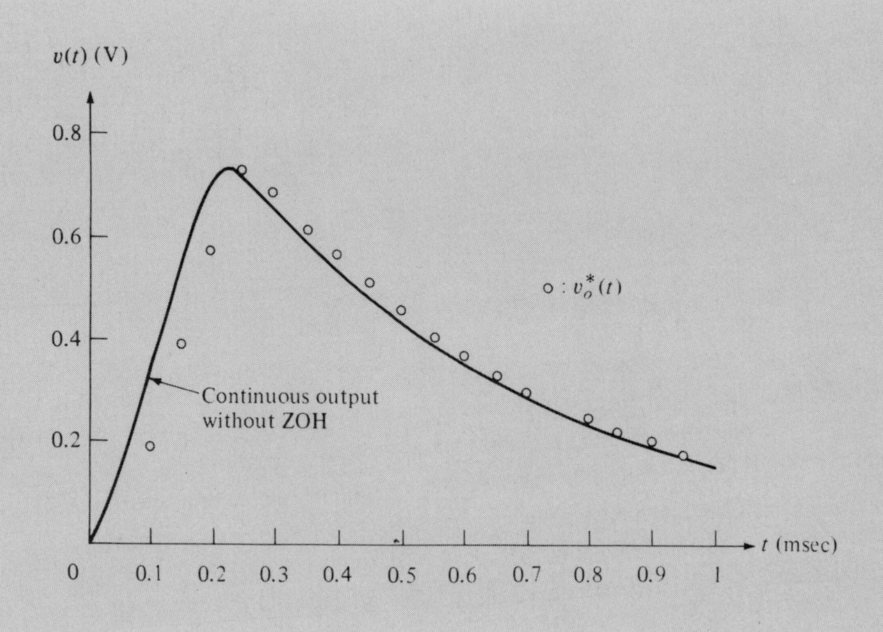

Figure 12.18
Example 12.7.

Example 12.8

Problem:

Determine the sampled-data output of the network of Figure 12.17 for $R = 50\ \Omega$ and the same sampling rate as in Example 12.7.

Solution:

1. Substitute $R = 50\ \Omega$ into Equation (12.18); then use Equation (12.15) to obtain $H(s)$:

$$H_1(s) = \frac{5000s}{s^2 + 10 \times 10^3 s + 100 \times 10^6} = \frac{5000s}{(s + 5000)^2 + (8660)^2}$$

$$H(s) = (1 - \exp[-50 \times 10^{-6}s])\frac{H_1(s)}{s}$$

$$= (1 - \exp[-50 \times 10^{-6}s])\frac{5000}{(s + 5000)^2 + (8660)^2}$$

2. Use Equation (12.16) and Table 12.2 to obtain $H(z)$:

$$H(z) = \left(\frac{z-1}{z}\right)\left(\frac{5000}{8660} \cdot \frac{(\exp[-5000T] \cdot \sin 8660T)z}{z^2 - (2 \cdot \exp[-5000T] \cdot \cos - 8660T)z + \exp[-10000T]}\right)$$

$$= \frac{0.1887(z-1)}{z^2 - 1.414z + 0.6065}$$

3. Determine $V_o(z)$ from Equation (12.17) and $H(z)$:

$$V_o(z) = H(z) \cdot V_i(z)$$

$$= \left(\frac{0.1887(z-1)}{z^2 - 1.414z + 0.6065}\right)\left(0.25\frac{z^3 + z^2 + z + 1}{z^3(z-1)}\right)$$

$$= \frac{0.04717(z^3 + z^2 + z + 1)}{z^3(z^2 - 1.414z + 0.6065)}$$

4. Use the long-division method to obtain $v_o^*(t)$:

$$V_o(z) = 0.04717z^{-2} + 0.1139z^{-3} + 0.1795z^{-4} + 0.2319z^{-5} + 0.2190z^{-6} + 0.1690z^{-7} + 0.1061z^{-8} + 0.04749z^{-9} + 0.002799z^{-10} - 0.02485z^{-11} + \cdots$$

$$v_o^*(t) = 0.04717 \cdot \delta(t-2T) + 0.1139 \cdot \delta(t-3T) + 0.1795 \cdot \delta(t-4T) + 0.2319 \cdot \delta(t-5T) + 0.2190 \cdot \delta(t-6T) + 0.1690 \cdot \delta(t-7T) + 0.1061 \cdot \delta(t-8T) + 0.04749 \cdot \delta(t-9T) + 0.002799 \cdot \delta(t-10T) - 0.02485 \cdot \delta(t-11T) + \cdots \text{ V}$$

$v_o^*(100 \times 10^{-6}) = 0.04717$ V, $v_o^*(350 \times 10^{-6}) = 0.1690$ V

$v_o^*(150 \times 10^{-6}) = 0.1139$ V, $v_o^*(400 \times 10^{-6}) = 0.1061$ V

$v_o^*(200 \times 10^{-6}) = 0.1795$ V, $v_o^*(450 \times 10^{-6}) = 0.04749$ V

$v_o^*(250 \times 10^{-6}) = 0.2319$ V, $v_o^*(500 \times 10^{-6}) = 0.002799$ V

$v_o^*(300 \times 10^{-6}) = 0.2190$ V, $v_o^*(550 \times 10^{-6}) = -0.02485$ V

Figure 12.19
Example 12.8.

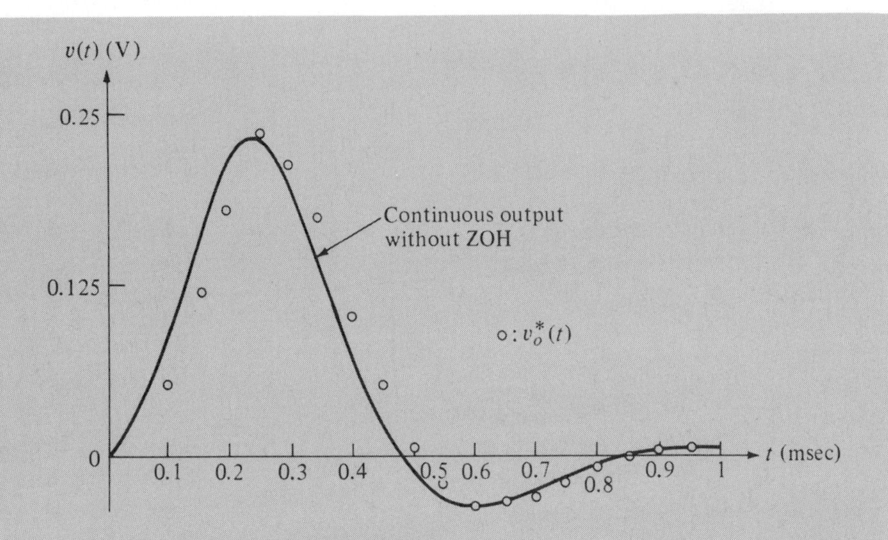

The sampled-data output is compared with the continuous output without ZOH in Figure 12.19.

Example 12.9

Problem:

Determine the sampled-data output of the network of Figure 12.17 for $R = 150\ \Omega$ and the same sampling rate as in Example 12.7.

Solution:

1. Substitute $R = 150\ \Omega$ into Equation (12.18); then use Equation (12.15) to obtain $H(s)$:

$$H_1(s) = \frac{15 \times 10^3 s}{s^2 + 20 \times 10^3 s + 100 \times 10^6}$$

$$= \frac{15 \times 10^3 s}{(s + 10 \times 10^3)^2}$$

$$H(s) = (1 - \exp[-50 \times 10^{-6} s])\frac{H_1(s)}{s}$$

$$= (1 - \exp[-50 \times 10^{-6} s])\frac{15 \times 10^3}{(s + 10 \times 10^3)^2}$$

2. Use Equation (12.16) and Table 12.2 to obtain $H(z)$:

$$H(z) = \left(\frac{z-1}{z}\right)\left(15 \times 10^3 \frac{50 \times 10^{-6}(\exp[-10 \times 10^3 T])z}{(z - \exp[-10 \times 10^3 T])^2}\right)$$

$$= \frac{0.4549(z-1)}{z^2 - 1.213z + 0.3679}$$

3. Determine $V_o(z)$ from Equation (12.17) and $H(z)$:

$$V_o(z) = H(z) \cdot V_i(z)$$

$$= \left(\frac{0.4549(z-1)}{z^2 - 1.213z + 0.3679}\right)\left(0.25\frac{z^3 + z^2 + z + 1}{z^3(z-1)}\right)$$

$$= \frac{0.1137(z^3 + z^2 + z + 1)}{z^3(z^2 - 1.213z + 0.3679)}$$

4. Use the long-division method to obtain $v_o^*(t)$:

$$V_o(z) = 0.1137z^{-2} + 0.2517z^{-3} + 0.3772z^{-4} + 0.4787z^{-5} + 0.4419z^{-6} + 0.3560z^{-7} + 0.2741z^{-8} + 0.2001z^{-9} + 0.1419z^{-10} + 0.09849z^{-11} + \cdots$$

$$v_o^*(t) = 0.1137 \cdot \delta(t - 2T) + 0.2517 \cdot \delta(t - 3T) + 0.3772 \cdot \delta(t - 4T) + 0.4787 \cdot \delta(t - 5T) + 0.4419 \cdot \delta(t - 6T) + 0.3560 \cdot \delta(t - 7T) + 0.2741 \cdot \delta(t - 8T) + 0.2001 \cdot \delta(t - 9T) + 0.1419 \cdot \delta(t - 10T) + 0.09849 \cdot \delta(t - 11T) + \cdots \text{ V}$$

$v_o^*(100 \times 10^{-6}) = 0.1137$ V $v_o^*(350 \times 10^{-6}) = 0.3560$ V

$v_o^*(150 \times 10^{-6}) = 0.2517$ V $v_o^*(400 \times 10^{-6}) = 0.2741$ V

$v_o^*(200 \times 10^{-6}) = 0.3772$ V $v_o^*(450 \times 10^{-6}) = 0.2001$ V

$v_o^*(250 \times 10^{-6}) = 0.4787$ V $v_o^*(500 \times 10^{-6}) = 0.1419$ V

$v_o^*(300 \times 10^{-6}) = 0.4419$ V $v_o^*(550 \times 10^{-6}) = 0.09849$ V

The sampled-data output is compared to the continuous output without ZOH in Figure 12.20.

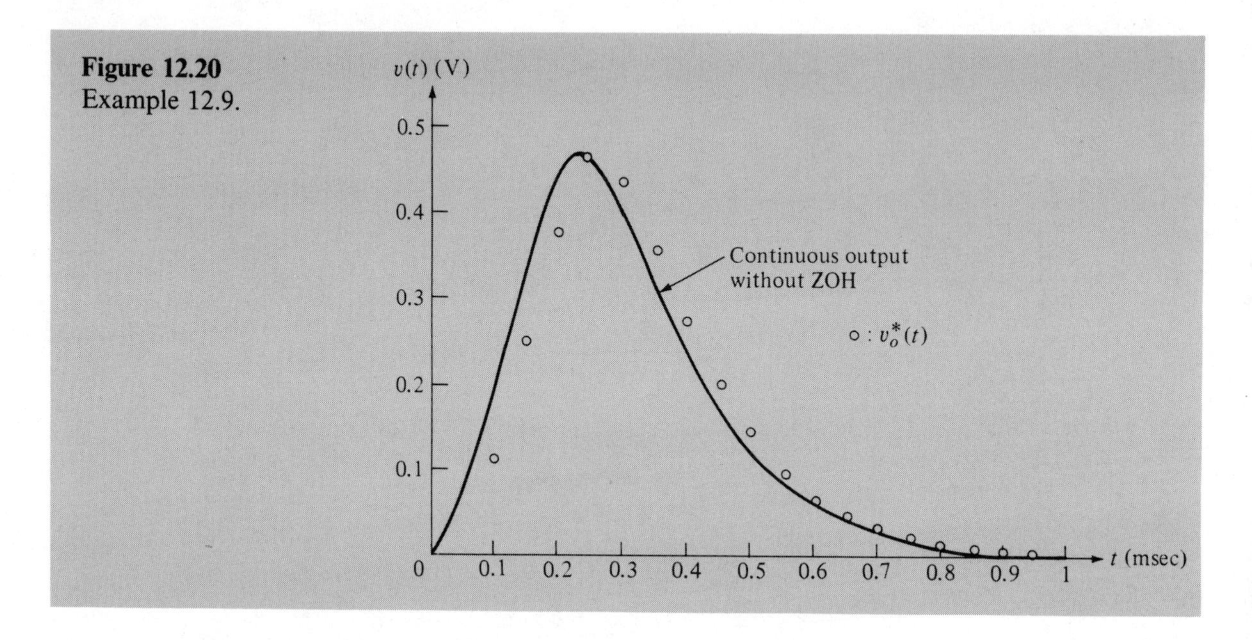

Figure 12.20
Example 12.9.

12.7 SUMMARY

The z-transform is a specialization of the Laplace transform which permits mathematical analysis of sampled-data systems. In this chapter, we have shown the relationship between the Laplace and z-transforms and how z-transforms are obtained from Laplace transforms. An abbreviated table of z-transforms was developed. The z-transform extends to transfer functions, and it is widely used in the analysis and design of digital control systems. The z-transform is also useful in the design of **digital filters**, the term applied to a family of software techniques used in signal processing and analysis. Specific applications of the z-transform to digital filters is beyond the scope of this text.

Two methods for obtaining the inverse z-transform—partial fraction decomposition and long division—were described, and detailed examples were presented for each method. Partial fraction decomposition is more elegant, but it sometimes requires subtle analytical techniques and the use of a table of z-transforms. The long division method, on the other hand, is very straightforward and amenable to computational methods, and does not rely on the use of tables.

Some attention was devoted to data sampling and the effect of sampling on system output. Particular mention was made and examples given of the use of the sample/hold amplifier (SHA) with a zero-order hold (ZOH) circuit. The transfer function for the ZOH was obtained and incorporated into the system transfer function. The reader is referred to a text on control systems for a more complete discussion of sampling theory and techniques, and the effect of sampling on system performance.

12.8 Terms

computer control
digital control
digital filter
geometric series
hold capacitor
long division method
mode switch
partial fraction method
power series
residue method
sample and hold
sampled data
samples
sample/track–hold amplifiers (SHA)
sampling rate
tracks
zero-order hold (ZOH)
z-transform

Problems

12-1. Use the time-domain shifting property of the *z*-transform to show that

a. $\mathscr{Z}\{\delta(t-nT)\} = z^{-n}$.

b. $\mathscr{Z}\{u(t-nT)\} = \dfrac{z}{z^n(z-1)}$.

Compare the results with Table 12.2.

12-2. Use the multiplication by t property of the *z*-transform to determine

$$\mathscr{Z}\{t^2 \cdot u(t)\}$$

Compare the result with Table 12.2.

12-3. Determine the *z*-transform of

$$f(t) = (1 - \exp[-at])u(t)$$

Hint: Use $\mathscr{Z}\{u(t)\}$, $\mathscr{Z}\{\exp[-at] \cdot u(t)\}$, and the linearity property of the *z*-transform. Compare the result with Table 12.2.

12-4. Use the procedure illustrated in Section 12.3 for obtaining the z-transform of the sine function to determine

$$\mathscr{Z}\{\cos \omega t \cdot u(t)\}$$

Compare the result with Table 12.2.

12-5. Use the multiplication by $\exp[-at]$ property of the z-transform to determine:

a. $\mathscr{Z}\{\exp[-\alpha t] \cdot \sin \omega_d t \cdot u(t)\}$.
b. $\mathscr{Z}\{\exp[-\alpha t] \cdot \cos \omega_d t \cdot u(t)\}$.

Compare the results with Table 12.2.

12-6. Use Table 12.2 to determine:

a. $\mathscr{Z}\{3000t \cdot u(t)\}$. Sampling rate: 1.5 kHz.
***b.** $\mathscr{Z}\{12(1 - \exp[-500t])u(t)\}$. Sampling rate: 1 kHz.
c. $\mathscr{Z}\{10 \times 10^3(t - 0.002)u(t - 0.002)\}$. Sampling rate: 2.5 kHz.
***d.** $\mathscr{Z}\{6 \cdot \exp[-300t] \cdot \cos 200t \cdot u(t)\}$. Sampling rate: 2 kHz.

12-7. Find the z-transform of the time-domain function which generates each waveform. (Sampling rate: 1 kHz.)

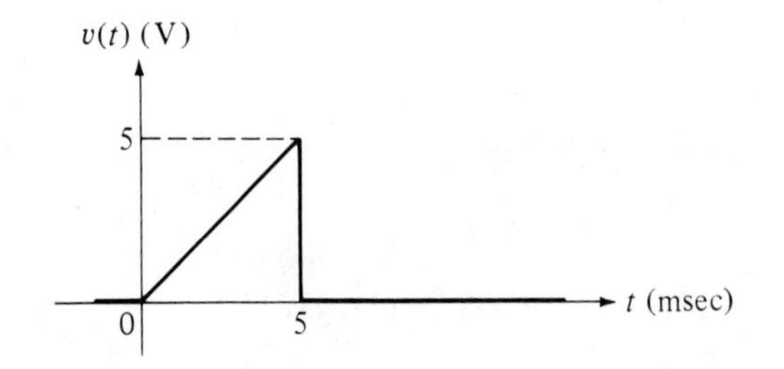

***a.**

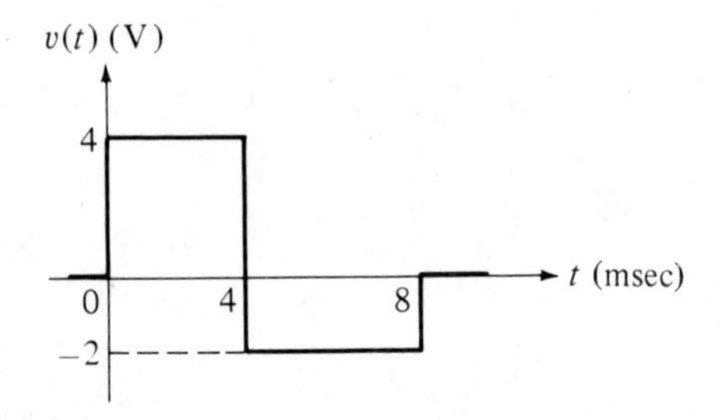

b.

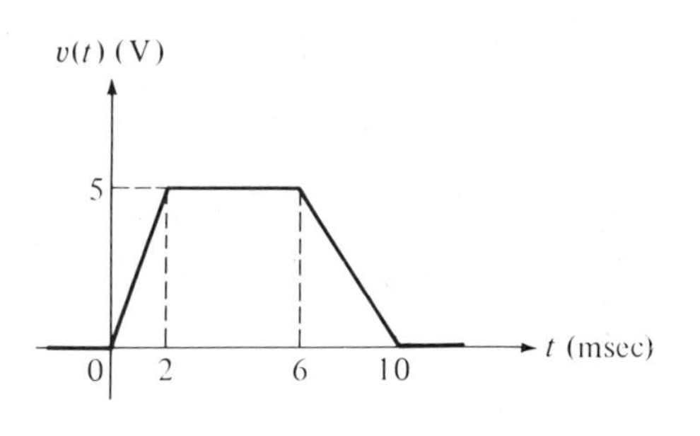

*c.

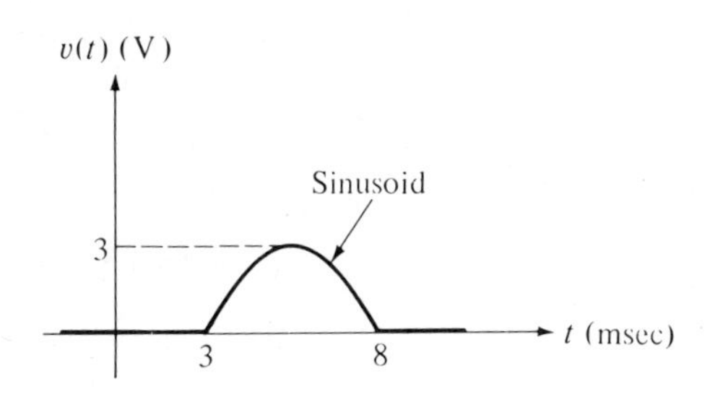

d.

12-8. Find the z-transform of the voltage transfer function for each network.

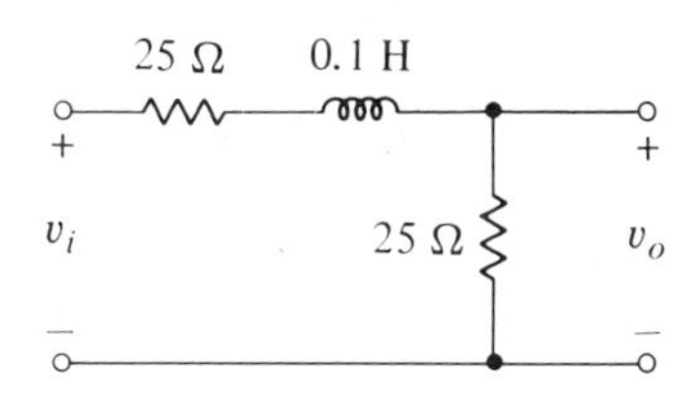

a.

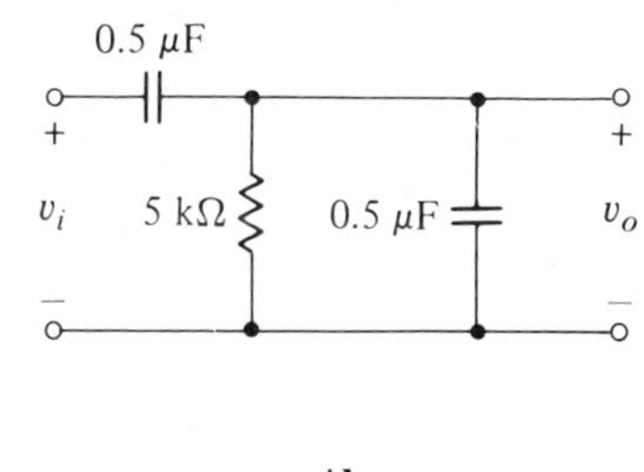

*b.

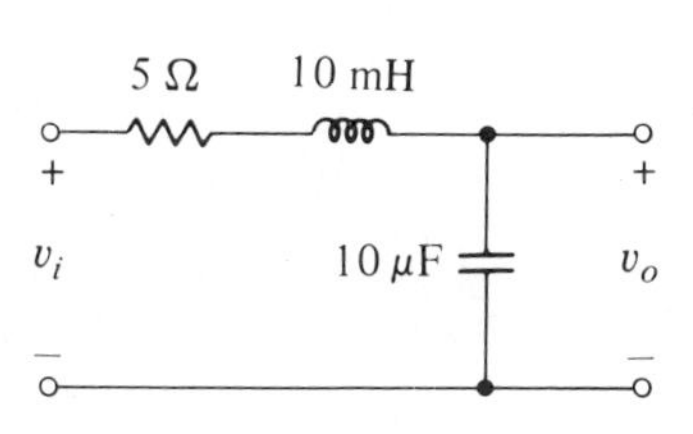

c.

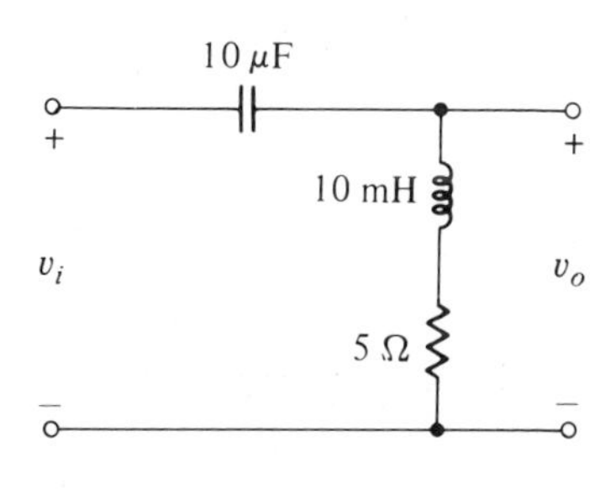

*d.

12-9. Determine the modified voltage transfer function and its z-transform if a SHA with ZOH is connected to the input of each network of Problem 12-8 [***a.**, ***c.**] .

12-10. Use Table 12.2 to determine the inverse z-transform of:

a. $F(z) = \dfrac{0.1813z}{z^2 - 1.8187z + 0.8187}$, $T = 0.2$ msec.

***b.** $F(z) = \dfrac{0.3476z}{z^2 - 0.9932z + 0.2466}$, $T = 0.7$ msec.

Hint: Factor the denominator.

12-11. Use the partial fraction decomposition method to determine the inverse z-transform of

***a.** $V(z) = \dfrac{3z^3 - 2.384z}{z^2 - 1.1511z + 0.5488}$, $T = 1.2$ msec.

***b.** $F(z) = \dfrac{z^3 - 1.4488z^2 + 0.5092z}{(z - 0.9048)(z^2 - 1.5595z + 0.6065)}$, $T = 80$ μsec.

Graph the resulting time-domain functions.

12-12. Repeat Problem 12-11 using the long division method. Graph the resulting time-domain functions and compare with the results of Problem 12-11.

12-13. Use the long division method to determine the inverse z-transform of:

***a.** $F(z) = \dfrac{z(5.9442z - 2.3008)}{z^3 - 2.4253z^2 + 1.9218z - 0.4966}$, $T = 1$ msec.

b. $V(z) = \dfrac{6z(z^2 - 1.1528z + 0.5470)}{z^3 - 2.0757z^2 + 1.6246z - 0.5488}$, $T = 0.5$ msec.

Graph the resulting time-domain functions.

12-14. Graph the sampled-data output of the network of Problem 12-8a with ZOH (Problem 12-9a) for:

a. The triangular input waveform of Problem 12-7a.
***b.** The rectangular input waveform of Problem 12-7b.
c. The trapezoidal input waveform of Problem 12-7c.
***d.** The sinusoidal input waveform of Problem 12-7d.

Use the long division method to obtain the inverse z-transform.

12-15. Repeat Problem 12-14 for the network of Problem 12-8b [***a.**, ***c.**].

12-16. Repeat Problem 12-14 for the network of Problem 12-8c[***b.**, ***d.**].

12-17. Repeat Problem 12-14 for the network of Problem 12-8d [***a.**, ***c.**].

A Solution of Simultaneous Linear Equations

Application of node or mesh analysis to a circuit results in a **system** of linear equations. Although Cramer's rule is widely used to solve such systems, the evaluation of the necessary determinants is a difficult chore, and limits the usefulness of Cramer's rule in systems of large (four or greater) dimension. This appendix presents another technique, **Gaussian elimination**, which is often more efficient for systems of the sizes commonly found in circuit analysis.

Recall from elementary algebra that one method of solving a pair of simultaneous equations is to *add* suitable multiples of the original equations to yield a new equation in which one of the variables is missing, or *eliminated*. This equation is solved for the remaining variable, and this value is *back-substituted* into the original system, and the value of the remaining variable is calculated. Gaussian elimination is merely an extension of this process to systems of larger size, using matrix notation for efficiency.

The systems of equations arising from circuit analysis are usually expressed as the product of a square **coefficient matrix** of conductances (or resistances) and a **vector** of node voltages (or mesh currents) set equal to a vector of source currents (or source voltages). The first step in Gaussian elimination is to form the **augmented matrix**, which is simply the coefficient matrix with the right-hand-side vector (source vector) appended to form the rightmost column. The additional vector *does not* replace any column of the coefficient matrix, as it does in Cramer's rule; the vector is simply attached to the right side of the coefficient matrix. A vertical bar is often drawn to separate the coefficient columns from the right-hand column. An example of a system and its augmented matrix is:

$$\begin{aligned} x + y + z &= 1 \\ 2x - y &= 2 \\ x + 4y + 5z &= -5 \end{aligned} \quad \rightarrow \quad \left[\begin{array}{ccc|c} 1 & 1 & 1 & 1 \\ 2 & -1 & 0 & 2 \\ 1 & 4 & 5 & -5 \end{array}\right]$$

Notice that the augmented matrix is not square, but has one column more than the coefficient matrix. This augmented matrix contains all the information about the original system, so for convenience only the augmented matrix is dealt with; manipulating rows of the augmented matrix is exactly equivalent to manipulating equations in the original system. Such manipulations are at the heart of Gaussian elimination, and are given the name **row operations**.

Once the augmented matrix is formed, the elimination portion of the process can begin. The objective is to modify the augmented matrix so that it represents an *equivalent* system of equations, but one whose solution is apparent. The particular goal is to obtain one equation which contains only one variable, so that the value of that variable is apparent, while another equation contains the variable previously found and *one* other variable, and so on. In essence, the variables will be determined one at a time.

The first question that arises is: How can variables be eliminated from an equation? The answer is unsurprising: Force their coefficients to be zero. Use row operations on the augmented matrix (equivalent to manipulating the original system of equations) to introduce zero coefficients in convenient places in the matrix.

The natural second question is: What are convenient places for these zero coefficients? This question is probably best answered by example. Consider the augmented matrix below:

$$\left[\begin{array}{ccc|c} 1 & 1 & 1 & 1 \\ 0 & 3 & 2 & 0 \\ 0 & 0 & 2 & -6 \end{array}\right]$$

It is a simple matter to solve this system. The third row of the matrix represents the equation

$$2z = -6$$

which is easily solved to yield

$$z = -3$$

The second row of the matrix represents the equation

$$3y + 2z = 0$$

but the value of z is known to be -3, so the equation may be written as

$$3y - 6 = 0$$

yielding $y = 2$.

Now the first row of the matrix, representing

$$x + y + z = 1$$

is really

$$x - 1 = 1$$

yielding

$$x = 2$$

and the solution is complete. The augmented matrix

$$\begin{array}{ccc|c} 0 & 0 & 1 & 1 \\ 0 & 1 & 1 & 0 \\ 1 & 1 & 1 & -1 \end{array}$$

can be approached in a similar manner, starting with the first row and working down, to yield the solution $(-1, -1, 1)$.

Notice the form of these augmented matrices. In each, the coefficient portion contains a triangular region of zeros. A little consideration of the systems of equations represented by such matrices reveals that a matrix containing such a triangle of zeros satisfies the condition mentioned above: one of the equations contains only one variable, while each successive equation introduces one new variable. Any augmented matrix whose coefficient matrix contains a triangle of zeros can be solved by the technique of back-substitution used above. Now the steps in Gaussian elimination are clear:

1. Form the system's augmented matrix.
2. Use row operations to *reduce* the augmented matrix to one containing a triangular region of zeros.
3. Calculate the value of each variable by back-substitution.

The key step is clearly step 2, and leads to a discussion of row operations. It is helpful to remember that these row operations are simply the matrix form of simple algebraic manipulation of equations.

Certainly it makes no difference in what order the equations of a system are written, and that order can be freely changed at any time. The corresponding row operation is:

Operation I

Any two rows of a matrix can be interchanged.

Notation: $R_a \leftrightarrow R_b$.

Example:

$$\begin{array}{ccc|c} 1 & 2 & 3 & 10 \\ 4 & 5 & 6 & 20 \\ 7 & 8 & 9 & 30 \end{array} \rightarrow \begin{array}{ccc|c} 1 & 2 & 3 & 10 \\ 7 & 8 & 9 & 30 \\ 4 & 5 & 6 & 20 \end{array} \qquad R_2 \leftrightarrow R_3$$

Now is the proper time to mention an important, but easily overlooked, property of row operations: They work on *rows*, whole rows, *not* parts of rows. A common mistake in applying Operation I is to interchange the coefficient portion of two rows, but forget to interchange the right-hand elements. This mistake causes the new matrix to represent a different system of equations, and it will be only coincidence if it yields the correct solution.

Another useful algebraic manipulation is to multiply through an entire equation by some nonzero constant. The corresponding row operation is:

Operation II

Any row of a matrix may be multiplied by a nonzero constant.

Notation: $R_a \leftarrow cR_a$.

Example:

$$\left[\begin{array}{ccc|c} 1 & 2 & 3 & 4 \\ 5 & 6 & 7 & 8 \\ 9 & 0 & -1 & 0.1 \end{array}\right] \rightarrow \left[\begin{array}{ccc|c} 1 & 2 & 3 & 4 \\ 5 & 6 & 7 & 8 \\ -90 & 0 & 10 & -1 \end{array}\right] \quad R_3 \leftarrow -10R_3$$

Since division is simply multiplication by a reciprocal, this operation also allows a row to be divided by any nonzero constant. Again, be sure to operate on the entire row.

The last manipulation to be discussed is that of adding one equation to another. Its row operation is:

Operation III

Any row of a matrix may be replaced by the sum or difference of that row and any other row.

Notation: $R_a \leftarrow R_a + R_b$.

Example:

$$\left[\begin{array}{ccc|c} 1 & 2 & 3 & 4 \\ 5 & 6 & 7 & 8 \\ 9 & 0 & -1 & 0.1 \end{array}\right] \rightarrow \left[\begin{array}{ccc|c} 10 & 2 & 2 & 4.1 \\ 5 & 6 & 7 & 8 \\ 9 & 0 & -1 & 0.1 \end{array}\right] \quad R_1 \leftarrow R_1 + R_3$$

When using Operation III, remember that the addition or subtraction is done column by column, and is done for every column in the row. When doing row subtraction, pay close attention to signs. Sign mistakes are the single most common error in row operations, and are almost invariably fatal. For convenience, Operations II and III are often combined, so that a multiple of one row is added to (or subtracted from) another. The obvious notation is used:

$$R_a \leftarrow R_a \pm cR_b$$

Example:

$$\begin{array}{ccc|c} 1 & 2 & 3 & 4 \\ 5 & 6 & 7 & 8 \\ 9 & 0 & -1 & 0.1 \end{array} \rightarrow \begin{array}{ccc|c} -17 & 2 & 5 & 3.8 \\ 5 & 6 & 7 & 8 \\ 9 & 0 & -1 & 0.1 \end{array} \qquad R_1 \leftarrow R_1 - 2R_3$$

Example A.1

Problem:

Solve the system of equations

$$\begin{aligned} x - 2y - 3z &= 1 \\ 2x - 3y - 5z &= 2.25 \\ x + 4y + 4z &= 2 \end{aligned}$$

Solution:

The augmented matrix of the system is

$$\begin{array}{ccc|c} 1 & -2 & -3 & 1 \\ 2 & -3 & -5 & 2.25 \\ 1 & 4 & 4 & 2 \end{array}$$

Perform the row operation $R_2 \leftarrow R_2 - 2R_1$:

$$\begin{array}{ccc|c} 1 & -2 & -3 & 1 \\ 0 & 1 & 1 & 0.25 \\ 1 & 4 & 4 & 2 \end{array}$$

Perform the row operation $R_3 \leftarrow R_3 - R_1$:

$$\begin{array}{ccc|c} 1 & -2 & -3 & 1 \\ 0 & 1 & 1 & 0.25 \\ 0 & 6 & 7 & 1 \end{array}$$

Finally, perform the row operation $R_3 \leftarrow R_3 - 6R_2$:

$$\begin{array}{ccc|c} 1 & -2 & -3 & 1 \\ 0 & 1 & 1 & 0.025 \\ 0 & 0 & 1 & -0.5 \end{array}$$

Now that the augmented matrix contains the desired triangle of zeros, the solution is easily calculated. From Row 3,

$$z = -0.5$$

Using this result in Row 2,

$$y = 0.75$$

Using both these results in Row 1,

$$x = 1$$

The solution is $(1, 0.75, -0.5)$.

Example A.2

Problem:

A practical example from circuit analysis is the system of equations resulting from mesh analysis of the circuit in Figure A.1. The augmented matrix of the system is

$$\left[\begin{array}{ccc|c} 10.2 & -3.3 & -2.2 & 6 \\ -3.3 & 10.5 & -2.9 & 9 \\ -2.2 & -2.9 & 8.3 & -12 \end{array}\right]$$

Note that the units (kilohms, milliamperes, volts) have been entirely suppressed in the augmented matrix. This point will be addressed later; it is not particularly relevant to the mechanics of Gaussian elimination.

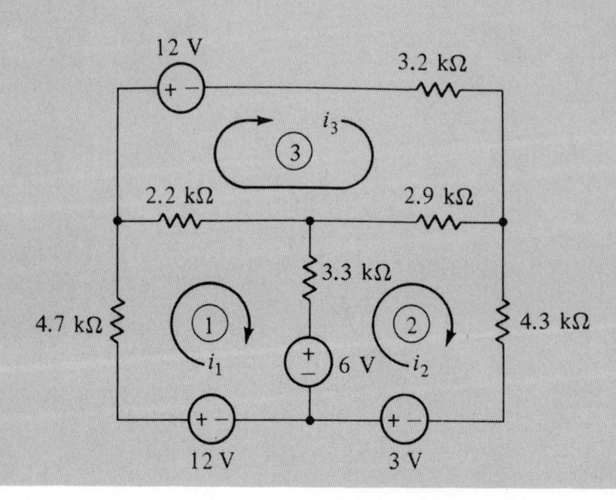

Figure A.1
Example A.2.

Solution:

The numbers in this matrix are not very convenient for calculations, which is frequently the case in circuit analysis problems. It is useful to **normalize** certain entries in the matrix to alleviate the computational difficulties.[1]

From previous examples, the reader can probably guess that the next step will be to introduce zeros in the (2, 1) and (3, 1) positions of the matrix. To facilitate this, we first normalize the (1, 1) entry by performing the following row operation:

$$R_1 \leftarrow R_1/10.2 \qquad \text{(division by 10.2)}$$

$$\left[\begin{array}{ccc|c} 1 & -0.3235 & -0.2157 & 0.5882 \\ -3.3 & 10.5 & -2.9 & 9 \\ -2.2 & -2.9 & 8.3 & -12 \end{array}\right]$$

With the (1, 1) position normalized, it is clear that the two operations to be performed are

$$R_2 \leftarrow R_2 + 3.3R_1$$
$$R_3 \leftarrow R_3 + 2.2R_1$$

$$\left[\begin{array}{ccc|c} 1 & -0.3235 & -0.2157 & 0.5882 \\ 0 & 9.4325 & -3.6118 & 10.9411 \\ 0 & -3.6117 & 7.8255 & -10.7060 \end{array}\right]$$

Next, introduce a zero into the (3, 2) position. Normalizing the (2, 2) entry helps:

$$R_2 \leftarrow R_2/9.4325$$

$$\left[\begin{array}{ccc|c} 1 & -0.3235 & -0.2157 & 0.5882 \\ 0 & 1 & -0.3829 & 1.1599 \\ 0 & -3.6117 & 7.8255 & -10.7060 \end{array}\right]$$

Now the necessary operation is seen to be

$$R_3 \leftarrow R_3 + 3.6117R_2$$

$$\left[\begin{array}{ccc|c} 1 & -0.3235 & -0.2157 & 0.5882 \\ 0 & 1 & -0.3829 & 1.1599 \\ 0 & 0 & 6.4426 & -6.5168 \end{array}\right]$$

The matrix now contains the desired triangle of zeros. Note that solving the

[1]Normalize, in this sense, means to divide a quantity by itself, leaving a quotient of one. In matrix manipulation, normalization is accomplished by a row operation.

last equation is simply normalizing the (3, 3) entry of the matrix:

$$R_3 \leftarrow R_3/6.4426$$

$$\left[\begin{array}{ccc|c} 1 & -0.3235 & -0.2157 & 0.5882 \\ 0 & 1 & -0.3829 & 1.1599 \\ 0 & 0 & 1 & -1.0115 \end{array}\right]$$

From the modified augmented matrix, $i_3 = -1.0115$ mA. Substitution of i_3 into row 2 yields $i_2 = 0.7726$ mA, which in turn yields $i_1 = 0.6200$ mA. The reader may verify that this solution is correct.

As noted early in the example, units were suppressed throughout the augmented matrix as a matter of convenience. The problem was formulated as "resistance in kilohms times current equals voltage in volts"; clearly the unit of current must be milliamperes to make the problem dimensionally correct. No purpose is served by continually rewriting unit symbols throughout the elimination process. All coefficients must be expressed in the same unit, and the right-hand sides of all equations in the system must be expressed in the same unit. The unit of the solution can then be determined by dimensional analysis, as in this example.

The process of applying Gaussian elimination to circuit analysis problems may be summarized as follows:

1. Form the system's augmented matrix.
2. Normalize the (1, 1) entry of the augmented matrix, then perform the appropriate row operations to introduce zeros in each position below the (1, 1) entry.
3. Normalize the (2, 2) entry and introduce zeros below it.
4. Repeat this process, working diagonally downward, until the matrix contains the desired triangle of zeros.
5. Use back-substitution to calculate the variable values.

It is possible to encounter a diagonal entry of the augmented matrix which is zero, and thus cannot be normalized. In such a case, simply swap the row with a lower row whose appropriate entry is nonzero, and proceed normally. Judicious rearrangement of rows can often simplify the solution process, as the next sample illustrates.

Example A.3

Nodal analysis of the circuit in Figure 1.29 (Example 1.10) yields a system whose augmented matrix is

$$\left[\begin{array}{cccc|c} 3 & -1 & -1 & 0 & 0 \\ -1 & 3 & 0 & -1 & 50 \\ -1 & 0 & 3 & -1 & -100 \\ 0 & -1 & -1 & 3 & 100 \end{array}\right]$$

Normalizing the (1, 1) element immediately introduces inconvenient fractions. Instead, consider the result of

$$R_1 \leftrightarrow R_3$$
$$R_2 \leftrightarrow R_4$$

$$\begin{array}{rrrr|r} -1 & 0 & 3 & -1 & -100 \\ 0 & -1 & -1 & 3 & 100 \\ 3 & -1 & -1 & 0 & 0 \\ -1 & 3 & 0 & -1 & 50 \end{array}$$

Now a simple sign change normalizes the first row:

$$R_1 \leftarrow -R_1$$

$$\begin{array}{rrrr|r} 1 & 0 & -3 & 1 & 100 \\ 0 & -1 & -1 & 3 & 100 \\ 3 & -1 & -1 & 0 & 0 \\ -1 & 3 & 0 & -1 & 50 \end{array}$$

The indicated operations are

$$R_3 \leftarrow R_3 - 3R_1$$
$$R_4 \leftarrow R_4 + R_1$$

$$\begin{array}{rrrr|r} 1 & 0 & -3 & 1 & 100 \\ 0 & -1 & -1 & 3 & 100 \\ 0 & -1 & 8 & -3 & -300 \\ 0 & 3 & -3 & 0 & 150 \end{array}$$

A sign change also normalizes the second row:

$$R_2 \leftarrow -R_2$$

$$\begin{array}{rrrr|r} 1 & 0 & -3 & 1 & 100 \\ 0 & 1 & 1 & -3 & -100 \\ 0 & -1 & 8 & -3 & -300 \\ 0 & 3 & -3 & 0 & 150 \end{array}$$

Perform the operations

$$R_3 \leftarrow R_3 + R_2$$
$$R_4 \leftarrow R_4 - 3R_2$$

$$\begin{array}{rrrr|r} 1 & 0 & -3 & 1 & 100 \\ 0 & 1 & 1 & -3 & -100 \\ 0 & 0 & 9 & -6 & -400 \\ 0 & 0 & -6 & 9 & 450 \end{array}$$

Normalize the third row:

$$R_3 \leftarrow R_3/9$$

$$\left[\begin{array}{cccc|c} 1 & 0 & -3 & 1 & 100 \\ 0 & 1 & 1 & -3 & -100 \\ 0 & 0 & 1 & -0.6667 & -44.4444 \\ 0 & 0 & -6 & 9 & 450 \end{array}\right]$$

The next operation is

$$R_4 \leftarrow R_4 + 6R_3$$

$$\left[\begin{array}{cccc|c} 1 & 0 & -3 & 1 & 100 \\ 0 & 1 & 1 & -3 & -100 \\ 0 & 0 & 1 & -0.6667 & -44.4444 \\ 0 & 0 & 0 & 5 & 183.3336 \end{array}\right]$$

Finally,

$$R_4 \leftarrow R_4/5$$

$$\left[\begin{array}{cccc|c} 1 & 0 & -3 & 1 & 100 \\ 0 & 1 & 1 & -3 & -100 \\ 0 & 0 & 1 & -0.6667 & -44.4444 \\ 0 & 0 & 0 & 1 & 36.6667 \end{array}\right]$$

The reader may verify that the solution is

$$v_1 = 3.3333 \text{ V}, \qquad v_2 = 30 \text{ V}$$
$$v_3 = -20 \text{ V}, \quad \text{and} \quad v_4 = 36.6667 \text{ V}$$

Note that the original problem was formulated in millisiemens and milliamperes, so the solution is in volts. The reader may also verify that working with the original matrix involves many more fractional numbers, though the answer is the same (within roundoff error). In general, anything that reduces the computational load will result in a more accurate solution in a shorter time.

Some final comments about Gaussian elimination conclude this discussion. Interested readers may consult one of the numerous introductory texts in linear algebra for more information on these and related topics.

The algorithm summarized following Example A.2 introduces the triangle of zeros in the lower left corner of the matrix. As was shown earlier in the discussion, back-substitution will work regardless of where the triangle occurs. In many cases, introducing the triangle in another position may require much less work. The elimination algorithm handles these cases easily—simply start at a different corner of the *coefficient portion* and work along a different diagonal direction. For instance, in Example A.2, start with the (1, 3) entry and work diagonally down and left, introducing zeros above the diagonal. This leads to zeros in the upper left triangle, and the back-substitution process proceeds normally. This modification can be particularly useful if the original augmented matrix has many zeros—a head start, so to speak.

Though circuit analysis problems generally lead to systems having square coefficient matrices, the technique of Gaussian elimination is applicable to systems of any dimensions. (Since only square matrices have determinants, Cramer's rule does not apply to systems which are not square.) The solutions of such systems are not unique (and may not even exist), so this discussion has been restricted to the square systems found in circuit analysis.

It is possible that during the reduction process a row having all zero coefficients may occur. In this case, the left-hand side of the corresponding equation will be zero regardless of the values of the variables. If the rightmost element of the row is nonzero, the corresponding equation has no solution, and the system is termed **inconsistent**. If the rightmost element of the row is also zero (the entire row consists of zeros), then the corresponding equation is always true, and has an infinite number of solutions, and the system is termed **dependent**. Since circuit analysis problems come from physical circuits, which have one and only one response, these results of inconsistency and dependency usually indicate an error in computation, or in derivation of the circuit model.

A powerful feature of Gaussian elimination is its ability to solve several related systems in a single operation. Suppose a designer wishes to compute the response of a certain network to two different sets of sources. Consider the systems of equations which result from two circuits which differ only in source values. The systems will have the *same* coefficient matrices, but different source vectors. Since the row operations of Gaussian elimination depend *only* on the coefficient matrix, the same row operations would be used to reduce the matrix of either system. This being the case, there is no need to repeat the entire elimination process. Simply append the second source vector to the original augmented matrix (the matrix now has a coefficient portion and *two* source vectors) and proceed with the elimination process, remembering to include the second source column in the row operations. When the matrix is reduced to triangular form, use back-substitution on the first source vector (ignoring the second), then again on the second source vector (ignoring the first). In this manner, a second system is solved for only the small amount of work added by the extra column. Any number of source vectors can be appended to a single coefficient matrix.

A development of Gaussian elimination, **Gauss–Jordan elimination**, continues the process of row reduction to introduce two triangles of zeros in the augmented matrix. In this method, the coefficient portion of the augmented matrix is reduced to a diagonal of ones, with zeros elsewhere; from such a matrix the solution is found by inspection, without the need for back-substitution. The drawback of this method is that it requires twice as many row operations as Gaussian elimination. Most engineers seem to prefer back-substitution to the additional row operations, and so use Gaussian elimination.

B Existence and Properties of the Laplace Transform

This appendix is presented for those wishing to gain a fuller understanding of the Laplace transform and its properties. The first aspect of the Laplace transform to be addressed is its very existence—that is, given an $f(t)$, is it possible to find an $F(s)$? Following that, we will prove the properties of the Laplace transform which were presented in Chapters 4 and 6 without proof.

Since the Laplace transform is defined by an improper integral, the question "Is some particular function Laplace transformable?" is in fact the question "Does some particular improper integral exist?" In general, improper integrals can be quite intractable, and so it is often difficult to determine whether a given function is Laplace transformable. Fortunately, it is not hard to set a condition which guarantees the existence of the Laplace transform for many functions of practical interest.

Existence Theorem

If $f(t)$ is piecewise continuous for $t \geq 0$, and there exist constants M and α such that

$$|f(t)| < M \cdot \exp[\alpha t] \tag{B.1}$$

then $F(s)$ exists.

Proof. Since $f(t)$ is piecewise continuous, $\exp[-st] \cdot f(t)$ is piecewise continuous, and hence integrable on any finite interval; thus the (improper) transform integral may exist. By definition

$$|F(s)| = \left| \int_0^\infty f(t) \cdot \exp[-st] \, dt \right|$$

The Laplace variable is a complex variable, $s = \sigma + j\omega$, so

$$\exp[-st] = \exp[-(\sigma + j\omega)t] = \exp[-\sigma t] \cdot \exp[-j\omega t]$$

and

$$|\exp[-st]| = |\exp[-\sigma t]| \cdot |\exp[-j\omega t]|$$

Using this fact and some properties of integrals,

$$\left|\int_0^\infty f(t) \cdot \exp[-st]\, dt\right| \le \int_0^\infty |f(t)| \cdot |\exp[-st]|\, dt$$

$$\le M \int_0^\infty \exp[\alpha t] \cdot \exp[-\sigma t]\, dt$$

$$M \int_0^\infty \exp[\alpha t] \cdot \exp[-\sigma t]\, dt = M \int_0^\infty \exp[-(\sigma - \alpha)t]\, dt$$

Provided Real$\{s\} > \alpha$, the last integral exists and is equal to $M/(\sigma - \alpha)$, so $F(s)$ exists. The requirement that $\sigma = \text{Real}\{s\} > \alpha$ need not concern us here.

Observe that if Equation (B.1) holds for $\alpha \le 0$, it certainly holds for any $\alpha > 0$, and it is usual to assume that $\alpha > 0$. The condition expressed in Equation (B.1) is often referred to as **exponential order**, i.e., the expression "$f(t)$ is of exponential order" means that there exist M and α for which Equation (B.1) holds.

It must be emphasized that this theorem is only a sufficient condition for Laplace transformability. There are many functions which do not satisfy the hypotheses of the theorem, but are still Laplace-transformable. The value of the theorem is its applicability to many functions of interest in engineering and technology; as an example, $\sinh 3t < \frac{1}{2} \cdot \exp[3t]$, so $\sinh 3t$ is Laplace-transformable.

Proofs of Laplace Transform Properties

Linearity Property

$$\mathscr{L}\{k_1 f_1(t) \pm k_2 f_2(t)\} = k_1 F_1(s) \pm k_2 F_2(s) \qquad \text{(B.2)}$$

Proof. This property follows directly from the linearity properties of the integral:

$$\begin{aligned}
&\mathscr{L}\{k_1 f_1(t) \pm k_2 f_2(t)\} \\
&\quad = \int_0^\infty [k_1 f_1(t) \pm k_2 f_2(t)] \exp[-st]\, dt \\
&\quad = k_1 \int_0^\infty f_1(t) \cdot \exp[-st]\, dt \pm k_2 \int_0^\infty f_2(t) \cdot \exp[-st]\, dt \\
&\quad = k_1 F_1(s) \pm k_2 F_2(s)
\end{aligned}$$

s-Domain Shifting Property

$$\mathscr{L}\{\exp[-at] \cdot f(t)\} = F(s + a) \tag{B.3}$$

Proof. This property follows from properties of the exponential function:

$$\begin{aligned}\mathscr{L}\{\exp[-at] \cdot f(t)\} &= \int_0^\infty \exp[-at] \cdot f(t) \cdot \exp[-st]\, dt \\ &= \int_0^\infty f(t) \cdot \exp[-(s + a)t]\, dt\end{aligned}$$

The last integral may be recognized as the Laplace transform integral with s replaced by $s + a$. Since the variable of integration is t, not s, the substitution simply carries over into the result, yielding the desired conclusion,

$$\mathscr{L}\{\exp[-at] \cdot f(t)\} = F(s + a)$$

t-Domain Shifting Property

$$\mathscr{L}\{f(t - t_0)\} = \exp[-t_0 s] \cdot F(s) \tag{B.4}$$

Proof. This property is established by a change of variables in the transform integral. By definition

$$\mathscr{L}\{f(t - t_0)\} = \int_0^\infty f(t - t_0) \cdot \exp[-st]\, dt$$

Making use of the fact that $f(t) = 0$ for $0 \le t < t_0$, we obtain

$$\mathscr{L}\{f(t - t_0)\} = \int_{t_0}^\infty f(t - t_0)\exp[-st]\, dt$$

Now make the change of variables $\tau = t - t_0$. Note that

$$t = \tau + t_0 \quad \text{and} \quad dt = d\tau$$

Also,

$$t = t_0 \;\Rightarrow\; \tau = 0, \quad \text{and} \quad t \to \infty \;\Rightarrow\; \tau \to \infty$$

Incorporating these observations yields

$$\begin{aligned}\mathscr{L}\{f(t-t_0)\} &= \int_0^\infty f(\tau)\cdot \exp[-s(\tau+t_0)]\,dt\\ &= \exp[-t_0 s]\cdot \int_0^\infty f(\tau)\cdot \exp \tau\,dt\\ &= \exp[-t_0]\cdot F(s)\end{aligned}$$

The last integral is the definition of the Laplace transform with τ as variable of integration; the change of symbol has no significance in the definition.

Periodicity Property

If $f_P(t)$ is a periodic function with period T, whose first period is represented by $f_1(t)$ [having Laplace transform $F_1(s)$], then

$$\mathscr{L}\{f_P(t)\} = \frac{F_1(s)}{1-\exp[-sT]} \tag{B.5}$$

Proof. This property follows from the linearity property, the t-domain shifting property, and the properties of the geometric series. First represent the periodic function as

$$f_P(t) = f_1(t) + f_1(t-T) + f_1(t-2T)\cdots = \sum_{n=0}^{\infty} f_1(t-nT)$$

Now

$$\begin{aligned}\mathscr{L}\{f_P(t)\} &= \int_0^\infty\left(\sum_{n=0}^{\infty} f_1(t-nT)\right)\cdot \exp[-st]\,dt\\ &= \sum_{n=0}^{\infty}\left(\int_0^\infty f_1(t-nT\cdot \exp[-st]\,dt)\right) && \text{(linearity)}\\ &= \sum_{n=0}^{\infty}\exp[-nTs]\cdot F_1(s) && \text{(time-domain shift)}\\ &= F_1(s)\sum_{n=0}^{\infty}\exp[-nTs]\end{aligned}$$

The last manipulation is possible because the summation is independent of s. Now observe that the summation

$$\sum_{n=0}^{\infty}\exp[-nTs]$$

is in the form of a geometric series with ratio $\exp[-Ts]$. From physical considerations, $T > 0$ and Real$\{s\} \geq 0$, so that $|\exp[-Ts]| < 1$, allowing the series to be summed. Using the formula for the sum of a geometric series,

$$\sum_{n=0}^{\infty} \exp[-nTs] = \frac{1}{1 - \exp[-Ts]}$$

so

$$\mathscr{L}\{f_P(t)\} = \frac{1}{1 - \exp[-Ts]} \cdot F_1(s)$$

as desired.

Differentiation Property

$$\mathscr{L}\{f'(t)\} = sF(s) - f(0) \tag{B.6}$$

Proof. This property is established via integration by parts. We assume that $f(t)$ satisfies the hypotheses of the existence theorem. By definition

$$\mathscr{L}\{f'(t)\} = \int_0^{\infty} f'(t) \cdot \exp[-st]\, dt$$

We then let

$$u = \exp[-st] \qquad \text{and} \qquad v = f(t)$$
$$du = -s \cdot \exp[-st]\, dt \quad \text{and} \quad dv = f'(t)\, dt$$

so that

$$\begin{aligned}
\mathscr{L}\{f'(t)\} &= f(t) \cdot \exp[-st]\big|_0^{\infty} - \int_0^{\infty} f(t)(-s \cdot \exp[-st])\, dt \\
&= f(t) \cdot \exp[-st]\big|_0^{\infty} + s\int_0^{\infty} f(t) \cdot \exp[-st]\, dt \\
&= f(t \to \infty) \cdot \exp[-s(t \to \infty)] - f(0) \cdot \exp[-s(0)] + s \cdot F(s)
\end{aligned}$$

The factor $\exp[-st]$ in the first term on the right causes that term to vanish as $t \to \infty$, yielding

$$\mathscr{L}\{f'(t)\} = 0 - f(0) + s \cdot F(s) = s \cdot F(s) - f(0)$$

Results for higher derivatives may be established similarly through repeated integration by parts; the general formula may be established by induction.

There are some subtle points in this proof. We tacitly assumed that both $f(t)$ and its derivative were continuous functions. If the derivative is not continuous, the integration by parts used in the proof may not be valid. The case of most interest is that in which the derivative is piecewise continuous. In this case, we break the transform integral into parts so that the derivative is continuous on each part, then the proof proceeds as before. An example of this case is the full-wave rectified sinusoid, a continuous function with a discontinuous derivative. If $f(t)$ is discontinuous, it is difficult to work with its derivative. If $f(t)$ is piecewise continuous, we consider the derivative on each interval, and derive a related result. While not difficult, this general case is somewhat cumbersome, and is not of concern here. Interested readers may consult a more advanced text.

Integration Property

$$\mathscr{L}\left\{\int_0^t f(\tau)\,d\tau\right\} = \frac{1}{s}F(s) \tag{B.7}$$

Proof. Assume again that $f(t)$ satisfies the hypotheses of the existence theorem. Since $f(t)$ is piecewise continuous, the function

$$g(t) = \int_0^t f(\tau)\,d\tau$$

is continuous. From properties of the integral,

$$|g(t)| = \left|\int_0^t f(\tau)\,d\tau\right| \le \int_0^t |f(\tau)|\,d\tau$$

and since $|f(t)| \le M \cdot \exp[\alpha t]$,

$$|g(t)| \le M\int_0^t \exp[\alpha\tau]\,d\tau = \frac{M}{\alpha}(\exp[\alpha t] - 1)$$

so that $g(t)$ satisfies the hypotheses of the existence theorem.

From the definition of $g(t)$,

$$g'(t) = f(t)$$

[except at discontinuities of $f(t)$], so that $g'(t)$ is piecewise continuous. Now the differentiation property may be applied to $g(t)$, yielding

$$F(s) = \mathscr{L}\{f(t) = \mathscr{L}\{g'(t)\} = s \cdot \mathscr{L}\{g(t)\} - g(0)$$

Note from the definition of $g(t)$ that $g(0) = 0$; thus,

$$\mathscr{L}\{g(t)\} = \frac{1}{s} \cdot F(s)$$

and by definition of $g(t)$,

$$\mathscr{L}\left\{\int_0^t f(\tau)\, d\tau\right\} = \frac{1}{s} \cdot F(s)$$

Initial Value Theorem

$$f(0) = \lim_{s\to\infty} s \cdot F(s) \tag{B.8}$$

Proof. The theorem follows from the differentiation property, which states that

$$\mathscr{L}\{f'(t)\} = \int_0^\infty f'(t) \cdot \exp[-st]\, dt = s \cdot F(s) - f(0)$$

or

$$f(0) = s \cdot F(s) - \int_0^\infty f'(t) \cdot \exp[-st]\, dt$$

Taking the limit as $s \to \infty$,

$$\begin{aligned} f(0) &= \lim_{s\to\infty}\left(s \cdot F(s) - \int_0^\infty f'(t) \cdot \exp[-st]\, dt\right) \\ &= \lim_{s\to\infty} s \cdot F(s) - \lim_{s\to\infty} \int_0^\infty f'(t) \cdot \exp[-st]\, dt \end{aligned}$$

because $f(0)$ does not depend on s. The second limit on the right is independent of t, the variable of integration, so the limit may be taken inside the integral. Since

$$\lim_{s\to\infty} \exp[-st] = 0$$

the integrand and integral are zero, hence

$$f(0) = \lim_{s\to\infty} s \cdot F(s)$$

Final Value Theorem

$$\lim_{t\to\infty} f(t) = \lim_{s\to 0} sF(s) \tag{B.9}$$

Proof. This theorem also follows from the differentiation property, which states that

$$\mathscr{L}\{f'(t)\} = \int_0^{\infty} f'(t) \cdot \exp[-st]\, dt = s \cdot F(s) - f(0)$$

Taking the limit as $s \to 0$,

$$\lim_{s \to 0} \int_0^{\infty} f'(t) \cdot \exp[-st]\, dt = \lim_{s \to 0} [s \cdot F(s) - f(0)]$$

As in the previous proof, $f(0)$ is not affected by the limit process, so

$$f(0) + \lim_{s \to 0} \int_0^{\infty} f'(t) \cdot \exp[-st]\, dt = \lim_{s \to 0} s \cdot F(s)$$

As before, we take the limit operation inside the integral. Since

$$\lim_{s \to 0} \exp[-st] = 1$$

this yields

$$f(0) + \int_0^{\infty} f'(t)\, dt = \lim_{s \to 0} s \cdot F(s)$$

By definition of an improper integral,

$$\int_0^{\infty} f'(t)\, dt = \lim_{t \to \infty} \int_0^t f'(\tau)\, d\tau = \lim_{t \to \infty} [f(t) - f(0)]$$
$$= \left(\lim_{t \to \infty} f(t)\right) - f(0)$$

yielding the desired result

$$\lim_{t \to \infty} f(t) = \lim_{s \to 0} s \cdot F(s)$$

Note that for the final-value theorem to have any validity, it is necessary to know that the limit of $f(t)$ as $t \to \infty$ actually exists. For the purpose of this discussion, an equivalent condition is that all poles of $s \cdot F(s)$ lie in the left half of the complex plane. If $F(s)$ is a proper polynomial fraction, this condition means that all its poles must have a *negative* real part, except that $F(s)$ may have a simple pole at $s = 0$. The condition is intuitively plausible, since poles with negative real part correspond to terms containing a decaying exponential factor, while the simple pole at zero corresponds to a constant term which is the final value. Note that the periodicity factor $1/(1 - \exp[-sT])$ gives rise to a nonsimple pole at zero, precluding use of the final value theorem; again this is plausible, since a periodic quantity has no "final value."

The reader is cautioned to remember the condition on the poles of $s \cdot F(s)$; improper use of the final value theorem is a common mistake. As an example of this error, suppose $F(s) = s/(s^2 + \omega^2)$. The final value theorem does not apply, because of the poles on the $j\omega$-axis (outside the left half-plane). Careless use of the theorem would lead one to believe that the limit of $f(t)$ as $t \to \infty$ is zero, while in fact $f(t) = \cos \omega t$, and the limit of $\cos \omega t$ as $t \to \infty$ does not exist.

C Laplace Transform Fundamentals

C.1 Definition

$$F(s) = \int_0^{\infty} \exp[-st] \cdot f(t)\, dt$$

where s is a *complex variable* ($s = \sigma + j\omega$).

C.2 Laplace Operator ($\mathscr{L}\{\ \}$)

$$F(s) = \mathscr{L}\{f(t)\}$$

C.3 Properties

1. Linearity property:

$$\mathscr{L}\{k_1 \cdot f_1(t) \pm k_2 \cdot f_2(t)\} = k_1 \cdot F_1(s) \pm k_2 \cdot F_2(s)$$

2. s-domain shifting property:

$$\mathscr{L}\{\exp[-at] \cdot f(t)\} = F(s + a)$$

where $F(s)$ is the Laplace transform of $f(t)$.

3. Time-domain shifting property:

$$\mathscr{L}\{f(t-t_0)\} = \exp[-t_0 s] \cdot F(s)$$

where $F(s)$ is the Laplace transform of $f(t)$.

4. Periodicity property:

$$\mathscr{L}\{f_P(t)\} = \frac{1}{1-\exp[-Ts]} F_1(s)$$

where $F_1(s)$ is the Laplace transform of $f_1(t)$, the generating function for the first period of periodic function $f_P(t)$.

5. Differentiation property:

a. First derivative:

$$\mathscr{L}\left\{\frac{d}{dt}f(t)\right\} = s \cdot F(s) - f(0)$$

where $F(s) = \mathscr{L}\{f(t)\}$ and $f(0)$ is the initial value of $f(t)$

b. Second derivative:

$$\mathscr{L}\left\{\frac{d^2}{dt^2}f(t)\right\} = s^2 \cdot F(s) - s \cdot f(0) - \left.\frac{d}{dt}f(t)\right|_{t=0}$$

c. Derivative of any order:

$$\mathscr{L}\left\{\frac{d^n}{dt^n}f(t)\right\} = s^n \cdot F(s) - s^{n-1} \cdot f(0) - s^{n-2} \cdot \left.\frac{d}{dt}f(t)\right|_{t=0}$$

$$- \cdots - s\left.\frac{d^{n-2}}{dt^{n-2}}f(t)\right|_{t=0} - \left.\frac{d^{n-1}}{dt^{n-1}}f(t)\right|_{t=0}$$

6. Integration property:

$$\mathscr{L}\left\{\int_0^{t_1} f(t)\,dt\right\} = \frac{1}{s} \cdot F(s)$$

where $F(s) = \mathscr{L}\{(f(t)\}$.

C.4 Laplace Transform Pairs

Pair	$f(t)$	$F(s)$
I.	$\delta(t)$	1
II.	$u(t)$ or 1	$1/s$
IIIa.	$t \cdot u(t)$	$1/s^2$
IIIb.	$t^n \cdot u(t)$	$n!/s^{n+1}$
IIIc.	$\dfrac{t^{n-1}}{(n-1)!} \cdot u(t)$	$1/s^n$
IV.	$\exp[-at] \cdot u(t)$	$\dfrac{1}{s+a}$
Va.	$(1 - \exp[-at])u(t)$	$\dfrac{a}{s(s+a)}$
Vb.	$\dfrac{1}{a}(1 - \exp[-at])u(t)$	$\dfrac{1}{s(s+a)}$
VIa.	$t \cdot \exp[-at] \cdot u(t)$	$\dfrac{1}{(s+a)^2}$
VIb.	$\dfrac{t^{n-1}}{(n-1)!} \cdot \exp[-at] \cdot u(t)$	$\dfrac{1}{(s+a)^n}$
VII.	$\sin \omega t \cdot u(t)$	$\dfrac{\omega}{s^2 + \omega^2}$
VIII.	$\cos \omega t \cdot u(t)$	$\dfrac{s}{s^2 + \omega^2}$
IX.	$\exp[-at] \cdot \sin \omega t \cdot u(t)$	$\dfrac{\omega}{(s+a)^2 + \omega^2}$
X.	$\exp[-at] \cdot \cos \omega t \cdot u(t)$	$\dfrac{s+a}{(s+a)^2 + \omega^2}$
XI.	$\dfrac{\exp[-at] - \exp[-bt]}{b-a} \cdot u(t)$	$\dfrac{1}{(s+a)(s+b)}$
XII.	$\dfrac{1}{\omega_d} \cdot \exp[-\alpha t] \cdot \sin \omega_d t \cdot u(t)$ where $\omega_d = \sqrt{\omega_0^2 - \alpha^2}$	$\dfrac{1}{s^2 + 2\alpha s + \omega_0^2}$
XIII.	$\dfrac{\omega_0}{\omega_d} \cdot \exp[-\alpha t] \cdot \cos[\omega_d t + \phi] \cdot u(t)$ where $\omega_d = \sqrt{\omega_0^2 - \alpha^2}$ and $\phi = \tan^{-1}(\alpha/\omega_d)$	$\dfrac{s}{s^2 + 2\alpha s + \omega_0^2}$

XIV.	$\frac{1}{2\omega_d^2}\left(\frac{1}{\omega_d} \cdot \sin \omega_d t - t \cdot \cos \omega_d t\right) \cdot \exp[-\alpha t] \cdot u(t)$ where $\omega_d = \sqrt{\omega_0^2 - \alpha^2}$	$\frac{1}{\left(s^2 + 2\alpha s + \omega_0^2\right)^2}$
XV.	$\frac{1}{2\omega_d^2}\left(\frac{\alpha}{\omega_d} \cdot \sin \omega_d t - \omega_0 t \cdot \cos[\omega_d t + \phi]\right) \cdot \exp[-\alpha t] \cdot u(t)$ where $\omega_d = \sqrt{\omega_0^2 - \alpha^2}$ and $\phi = \tan^{-1}(\omega_d/\alpha)$	$\frac{s}{\left(s^2 + 2\alpha s + \omega_0^2\right)^2}$
XVI.	$\delta(t) - a \cdot \exp[-at] \cdot u(t)$	$\frac{s}{s+a}$
XVIIa.	$(1 - at) \cdot \exp[-at] \cdot u(t)$	$\frac{s}{(s+a)^2}$
XVIIb.	$\frac{t^{n-2}}{(n-2)!}\left(1 - \frac{a}{n-1} t\right)\exp[-at] \cdot u(t)$	$\frac{s}{(s+a)^n}$, $n = 2, 3, \ldots$

D Use of SPICE in Network Analysis

D.1 INTRODUCTION

The use of SPICE (*S*imulation *P*rogram with *I*ntegrated *C*ircuit *E*mphasis) for analysis of passive linear networks and networks containing ideal OPAMPS is described in this appendix. For a more extensive treatment, the reader is directed to the *SPICE User's Guide*, which is part of the software installation for the public-domain versions; the user manuals provided with the commercial versions; or the several commercially published texts and manuals.

A general procedure for using SPICE is

1. Sketch the circuit, and assign a number to each node.
2. Assign a name to each element, and prepare element records according to the rules in this appendix.
3. Specify the type(s) of analysis.
4. Specify the output variables and the method (table or graph) of output.
5. Execute SPICE.

Usually step 5 requires an execution command suitable for the computer in use, together with the name of the input file. The input file consists of the assignments and specifications from steps 2, 3, and 4.

D.2 Types of Analysis

DC Analysis

SPICE computes the dc operating point with inductors shorted and capacitors opened. Swept dc analysis can be specified.

Small Signal AC Analysis

SPICE analyzes the circuit in the frequency domain and computes specified output variables vs. frequency. Noise and distortion analysis can be specified in conjunction with ac analysis.

Transient Analysis

SPICE simulates the circuit in the time domain and computes specified output variables vs. time. A spectral (Fourier) analysis can be specified for the transient output variables.

D.3 Input Data

SPICE input consists of the circuit description and the control directives. The method of presenting the input data to SPICE varies with the version. Most versions accept, and many versions require, the name of a separate file which contains the input data. Some versions are interactive and cue the user for data entries. More sophisticated versions are capable of capturing the data from a schematic prepared with a computer-aided design (CAD) program.

The circuit description section of the data file consists of one record (line) for each circuit element, plus one mutual inductance record for each pair of coupled inductors. Element records have the general form

ELEMENT NAME NODE CONNECTIONS PARAMETERS

The control directive section may contain as many control directives as required to complete the desired analysis. Control directives all have the general form

.DIRECTIVE PARAMETERS

TITLE Record

The first record of the data file must be a TITLE record, which is a free-field record of up to 80 characters. The characters entered into the TITLE record are printed at the beginning of each output section.

Node Numbering

The number 0 is assigned to the ground or reference node. All other nodes are assigned positive, nonzero integers in any order.

Numeric Parameters (Values)

Numeric parameters may be specified in integer, floating-point, or scientific notation, and may be followed immediately by one of the following scale factors:

G	1E+9	MIL	25.4E-4	N	1E-9
MEG	1E+6	M	1E-3	P	1E-12
K	1E+3	U	1E-6	F	1E-15

Minus signs must be used to indicate negative values. No blanks may be included within the numeric parameter.

Examples

```
1000 = 1E3 = 1K = .001MEG = 1E6M = 1E-6G
-3 = -3000E-3 = -.003K = -3E3M = -3E-6MEG
```

The .END Control Directive

The final record in the input file must be the `.END` control directive.

D.4 Passive Elements[2]

Resistors[3]

R*XXXXXXXX* *NI* *NF* *RESISTANCE* (Ω)

Order of initial node (*NI*) and final node (*NF*) is arbitrary.

Examples:

```
RLOAD 2 6 18K
R17 8 3 1.2MEG
```

Capacitors

C*XXXXXXXX* *NI* *NF* *CAPACITANCE* (*F*) ⟨IC = *INITIAL VOLTAGE* (*V*)⟩

Initial conditions are optional (see `.TRAN` directive); default is 0 V.

---

[2] Lossless transmission lines and semiconductor devices are not described here.

[3] Optional temperature coefficient parameters are not discussed here; the defaults are zero.

Examples:

```
C2 12 4 3U
CIN 2 5 400N IC=.95
```

Inductors

L*XXXXXXXX* *NI* *NF* *INDUCTANCE* (*H*) ⟨IC = *INITIAL CURRENT* (*A*)⟩

Initial conditions are optional (see .TRAN directive); default is 0 A.

The initial node must be the dotted end of coil for coupled inductors (transformers).

Examples:

```
LPRIMARY 4 0 8
L3 2 6 5M IC=10 M
```

Mutually Coupled Inductors

K*XXXXXXXX* L*AAAAAAAA* L*BBBBBBB* *COEFFICIENT OF COUPLING*

Coefficient of coupling is k; $0 < k \leq 1$.

Examples:

```
KTFR LPRI LSEC .98
KIF1 L11 L21 .6
```

D.5 Independent Sources

Voltage Sources

V*XXXXXXX* *N*+ *N*− *TYPE* *PARAMETERS*

Voltage drop within the source is from $N+$ to $N-$.

Current Sources

I*XXXXXXX NFROM NTO TYPE PARAMETERS*

Current flow within the source is from *NFROM* to *NTO*.

Type and Parameter for Independent dc Sources

⟨DC ⟨*VALUE*⟩⟩

Default type is DC; default for *VALUE* is 0 V (short circuit) or 0 A (open circuit).

Examples:

```
VBB 8 7 DC 9
ISOURCE 2 3 DC 100M
IC 2 4 12M
VAM 5 8
```

Type and Parameters for Independent ac Sources

AC ⟨*AMP* ⟨*PHASE*⟩⟩

The default for *AMP* (amplitude) is 1 (volt or ampere); the default for *PHASE* is 0°.

Examples:

```
VGEN 3 4 AC 120 30
ISRC 6 3 AC 12M
VIN 5 6 AC
```

Types and Parameters for Independent Transient Sources

If an optional parameter is omitted or set to zero, the default value will be used. Also refer to the .TRAN record.

Rectangular Pulse Waveforms

PULSE (*VI VP* ⟨*DELAY* ⟨*RISE* ⟨*FALL* ⟨*WIDTH PERIOD*⟩⟩⟩⟩)

See Figure D.1.

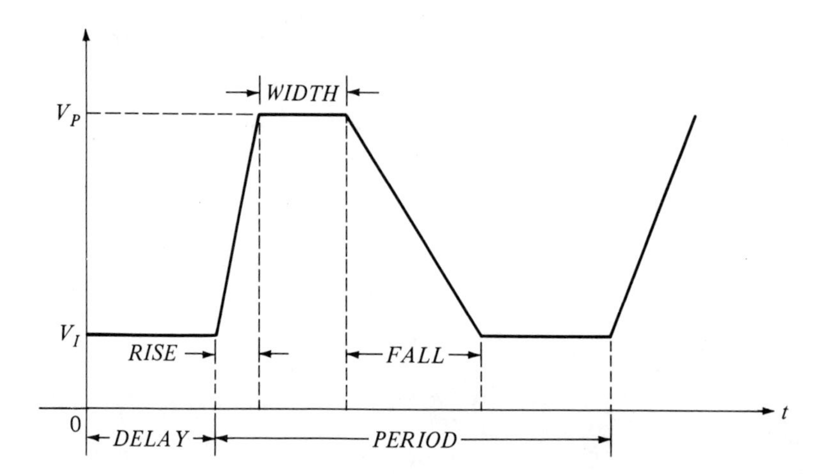

Figure D.1
PULSE specifications.

If *PERIOD* < *TSTOP*, waveform will be repeated.

	Parameter	Default	Units
VI	Initial value		V or A
VP	Pulsed value		V or A
DELAY	Delay time	0	sec
RISE	Rise time	*TSTEP*	sec
FALL	Fall time	*TSTEP*	sec
WIDTH	Pulse width	*TSTOP*	sec
PERIOD	Period	*TSTOP*	sec

Examples:

```
VTRIG 5 6 PULSE(0 4.55)
IIN 2 1 PULSE(1 3 1.2M 1U)
V12 5 8 PULSE(0 9 0 1U 1.8U .2M 1M)
```

Sinusoidal Waveforms

SIN (*VO AMP* ⟨*FREQ* ⟨*DELAY* ⟨*DAMPING*⟩⟩⟩)

See Figure D.2.

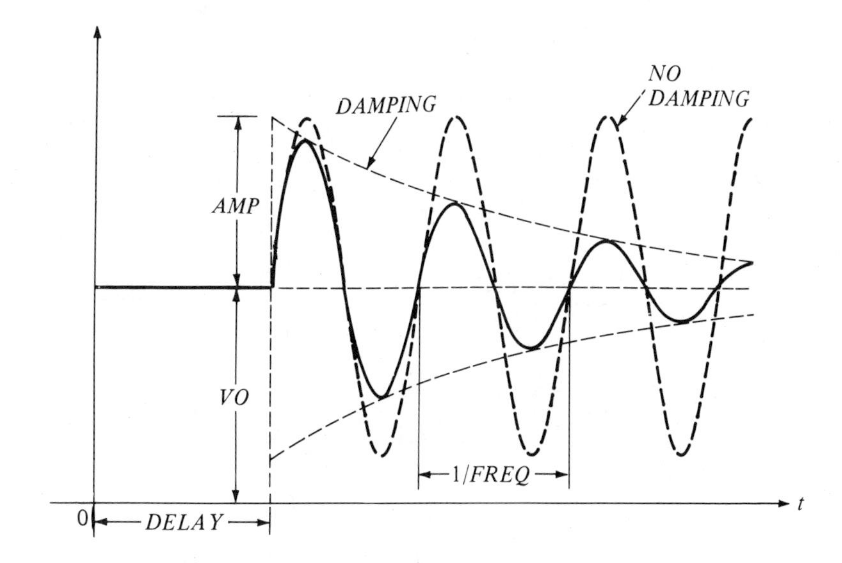

Figure D.2
`SIN` specifications.

The waveform is generated by the function

$$VO + AMP \cdot \exp[-(t - DELAY) \cdot DAMPING] \times \sin[2\pi \cdot FREQ(t - DELAY)] \cdot u(t - DELAY)$$

	Parameter	Default	Units
VO	Offset		V or A
AMP	Amplitude		V or A
FREQ	Frequency	1/*TSTOP*	Hz
DELAY	Delay time	0	sec
DAMPING	Damping factor	0	1/sec

Examples:

```
IS1 8 9 SIN(0 7 3K)
VA 2 5 SIN(1.5 1)
IOSC 3 6 SIN(0 3 120 0 6.5)
```

Exponential Waveforms

`EXP (` *VI VF* ⟨*DELAY1* ⟨*TAU1* ⟨*DELAY2* ⟨*TAU2*⟩⟩⟩⟩`)`

See Figure D.3.

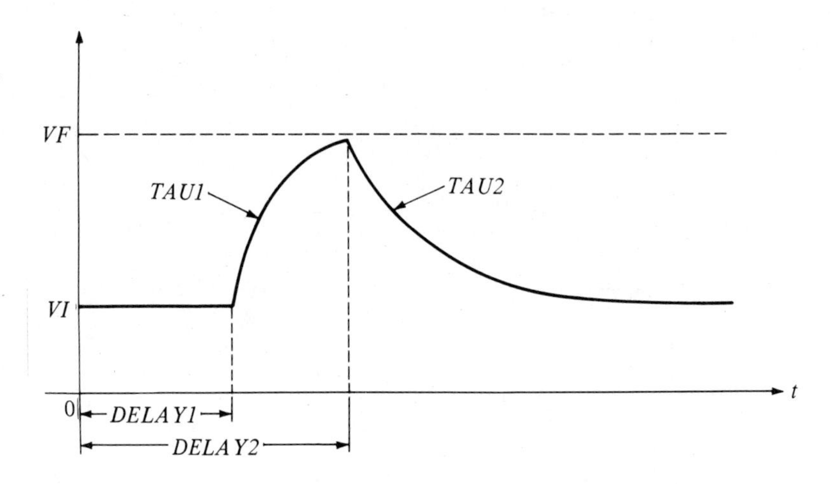

Figure D.3
EXP specifications.

The waveform is generated by the function

$$VI + (VF - VI)\left(1 - \exp\left[-\frac{t - DELAY1}{TAU1}\right]\right) \cdot u(t - DELAY1)$$

$$+(VI - VF)\left(1 - \exp\left[-\frac{t - DELAY2}{TAU2}\right]\right) \cdot u(t - DELAY2)$$

	Parameter	Default	Units
VI	Initial/final value for rise/fall		V or A
VF	Final/initial value value for rise/fall		V or A
DELAY1	Start time for rise	0	sec
TAU1	Rise time constant	*TSTEP*	sec
DELAY2	Start time for fall	*DELAY1* + *TSTEP*	sec
TAU2	Fall time constant	*TSTEP*	sec

Examples:

```
VCHARGE 1 2 EXP(0 5 0 50U 1M 50U)
IS2 3 7 EXP(6M 12M 1M .5M .1)
VSPIKE 3 5 EXP(0 6 0 0 0.001 .2M)
```

Piecewise Linear Waveforms

PWL ($T0$ $V0$ $\langle T1$ $V1$ $\langle T2$ $V2$ $\langle$ $\cdots$ $\rangle\rangle\rangle$)

See Figure D.4.

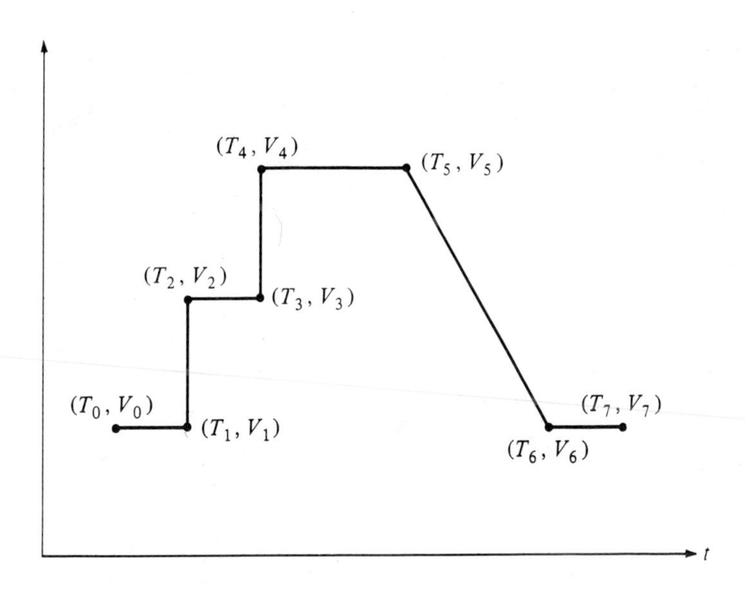

Figure D.4
PWL specifications ($T_0 < T_1 < T_2 < \cdots$).

The source value is determined by linear interpolation.

	Parameter	Units
T0	Initial time	sec
V0	Initial value	V or A
T1	Next time	sec
V1	Next value	V or A
⋮	⋮	⋮

Examples:

```
ITRIANGL 0 1 PWL(0 0 3M 100M 6M 0 10M 0)
VSQUARE 3 6 PWL(0 0 .1 5 1.9 5 2.1 -5 3.9 -5 4 0)
```

D.6 Linear Controlled (Dependent) Sources[4]

Voltage-Controlled Voltage Sources (VCVS)

E*XXXXXXX* *N* + *N* − *NC* + *NC* − *GAIN*

NC + and *NC* − are the positive and negative nodes of the controlling voltage. *GAIN* is dimensionless.

Examples:

```
EOPAMP 4 0 2 0 1E6
E1 22 9 12 13 .88
```

Current-Controlled Current Sources (CCCS)

F*XXXXXXX* *NFROM* *NTO* *CONTROL NAME* *GAIN*

CONTROL NAME is the name of a voltage source in which the controlling current flows; positive controlling current must enter the positive node of *CONTROL NAME*.

Examples:

```
FBJT 3 0 VBASE 88
F8 12 9 VAM .2
```

Voltage-Controlled Current Sources (VCCS)

G*XXXXXXX* *NFROM* *NTO* *NC* + *NC* − *TRANSCONDUCTANCE*

NC + and *NC* − are the positive and negative nodes of the controlling voltage. *TRANSCONDUCTANCE* is in siemens.

Examples:

```
GFWD 6 2 5 0 2.1E-3
G11 2 19 7 1 .08
```

[4] Nonlinear controlled sources are not described here.

Current-Controlled Voltage Sources (CCVS)

H*XXXXXXX N + N − CONTROL NAME TRANSRESISTANCE*

CONTROL NAME is the name of a voltage source in which the controlling current flows; positive controlling current must enter the positive node of *CONTROL NAME*. *TRANSRESISTANCE* is in ohms.

Examples:

```
HOUT 2 7 VAM1 40
H2211 12 19 VGEN 500
```

D.7 Analysis Directives

DC Analysis[5]

A dc analysis is performed automatically if no other type of analysis is specified, or prior to a specified transient or ac small-signal analysis. The operating-point (`.OP`) control directive can be used to force a dc analysis at a particular point in the output:

`.OP` (no other parameters)

The `.DC` control directive causes a specified dc source to be stepped or incremented, and a dc analysis is performed at each step:

`.DC` *DC SOURCE NAME START VALUE STOP VALUE INCREMENT*

DC SOURCE NAME is the name of the dc source, described in an element record, which is to be swept from *START VALUE* to *STOP VALUE*, in steps specified by *INCREMENT*. The output is available as a table or a graph.

Examples:

```
.DC VCC 3 18 3
.DC ISRC 0 20M 2M
```

[5]The small-signal dc transfer (`.TF`), and small-signal dc sensitivity (`.SENS`) directives are not described here.

Small-Signal AC Analysis[6]

The ac control directive (`.AC`) causes a small-signal ac analysis over a specified frequency range, provided the circuit contains at least one independent ac source:

`.AC` *SCALE NP FSTART FSTOP*

SCALE specifies the frequency variation, and may be `LIN` (linear), `DEC` (decade), or `OCT` (octave).

NP specifies the number of frequency points at which the analysis is to be performed. If `LIN` has been specified, *NP* is the total number of frequency points; if `DEC` or `OCT` has been specified, *NP* is the number of points per decade or per octave.

FSTART specifies the beginning frequency and *FSTOP* the ending frequency, in hertz, for the analysis.

Transient Analysis

The transient (`.TRAN`) control directive causes a transient analysis over a specified time interval:

`.TRAN` *TSTEP TSTOP* ⟨*TSTART* ⟨*TMAX*⟩⟩ ⟨*UIC*⟩

	Parameter	Default	Units
TSTEP	Output interval		sec
TSTOP	End of analysis		sec
TSTART	Start of output	0	sec
TMAX	Maximum increment for analysis	Smaller of *TSTEP* and (*TSTOP-TSTART*)/50	sec

`UIC` (for user initial conditions) is a keyword which, if present, causes SPICE to look to the capacitor and inductor element records for initial conditions (`IC =`). If `UIC` is omitted, or if `UIC` is present and `IC =` has been omitted from element records, the default (0 V or 0 A) is used.

Fourier Analysis

The Fourier analysis directive (`.FOUR`) may be used following a transient analysis to determine the amplitude and phase spectra of an output variable:

`.FOUR` *FUNDAMENTAL FREQ OV1* ⟨*OV2* ⟨*OV3* ⟨ ··· ⟩⟩⟩

Fourier analysis will be conducted for the interval *TSTOP-T* to *TSTOP*, where *T* = 1/(*FUNDAMENTAL FREQ*).

[6]The noise-analysis (`.NOISE`) and distortion-analysis (`.DISTO`) directives are not described here.

OV1, *OV2*, *OV3*,... are output variables for which the amplitude and phase spectra are required; the dc component and the first nine harmonics are computed.

D.8 Output Directives

Output Tables

The print directive causes the output variables to be printed as columns of a table. The first column of the table is the independent variable—time for transient analysis or frequency for ac analysis:

.PRINT *ANALYSIS TYPE* *OV1* ⟨*OV2* ··· ⟨*OV8*⟩⟩

ANALYSIS TYPE can be **DC**, **AC**, **TRAN**, **DISTO**, or **NOISE**, provided a directive for that type of analysis has been included in the input file.

A single print directive can specify one to eight output variables, which may be any combination of node-to-node voltages or branch currents. Additional print directives may be included, as required.

The output voltage variable

V(*N1* ⟨*N2*⟩**)**

specifies the voltage drop from *N1* to *N2*; if *N2* is omitted, the default is 0, or the ground node.

The output current variable

I(*VNAME***)**

specifies the current entering the positive node of *VNAME*, which is a named voltage source for which an element record exists in the input file.[7]

If a branch current is required, and no voltage source exists in that branch, a 0-V dc source may be inserted without altering circuit performance.

Two output variables are required to convey the results of ac analysis which are phasor quantities with amplitude and phase angle, or real and imaginary parts.

Output Variable	Quantity	Units
VM(*N1* ⟨*N2*⟩**)** or **IM(***NAME***)**	Amplitude	V or A
VP(*N1* ⟨*N2*⟩**)** or **IP(***VNAME***)**	Phase angle	deg
VR(*N1* ⟨*N2*⟩**)** or **IR(***VNAME***)**	Real part	V or A
VI(*N1* ⟨*N2*⟩**)** or **II(***VNAME***)**	Imag. part	V or A
VDB(*N1* ⟨*N2*⟩**)** or **IDB(***VNAME***)**	20 · log[Amplitude]	dB

[7]Some commercial versions will provide the current entering any element as an output variable.

Examples:

```
.PRINT TRAN V(2 5) V(1) I(VAM1)
.PRINT AC VM(4) VP(4) IR(VAM2) II(VAM2) VDB(1)
```

Output Graphs

The plot directive causes the output variables to be plotted vs. the independent variable. The format of the plot directive is essentially the same as that of the print directive, except for the option to specify plotting limits within the list of output variables:

.PLOT *ANALYSIS TYPE* *OV1* ⟨(*LO1*,*HI1*)⟩ ⟨*OV2*⟨(*LO2*,*HI2*)⟩⟩ ··· ⟨*OV8*⟩⟩

Automatic scaling to determine the minimum and maximum of all output variables is obtained by default when no plotting limits are specified.

If a plotting limits are specified, all output variables to the *left* are plotted within those limits.

Examples:

```
.PLOT TRAN V(2 3) V(4) (-12,12)
.PLOT AC VDB(6) VP(6)
```

D.9 Other Directives

Temperature Directive. Causes simulation to be performed at each specified temperature. Default temperature is 27°C.

Input File Record Width Directive. Specifies the number of columns in the input file. Default value for *COLUMNS* is 80.

Options Directive. Permits user to change a number of run-time parameters and output specifications.

Model Directive. Sets model parameters for semiconductor devices.

Subcircuit Directive. Permits use of subcircuit libraries.

Bibliography

Adby, P. R. *Applied Circuit Theory: Matrix and Computer Methods*. Ellis Horwood Limited, Chichester (John Wiley & Sons, New York), 1980.

Bell, David A. *Fundamental Electric Circuits*, second edition. Reston Publishing Company, Reston (Virginia), 1981.

Belove, Charles. *A First Circuits Course for Engineering Technology*, Holt, Rinehart and Winston, Inc., New York, 1982.

Beyer, William H., ed. *CRC Standard Mathematical Tables*. CRC Press, Boca Raton (Florida), 1978.

Bishop, Albert B. *Introduction to Discrete Linear Controls: Theory and Application*. Academic Press, New York, 1975.

Bobrow, Leonard S. *Elementary Linear Circuit Analysis*, second edition. Holt, Rinehart and Winston, New York, 1987.

Boylestad, Robert L. *Introductory Circuit Analysis*, fifth edition. Merrill Publishing Company, Columbus (Ohio), 1987.

Chirlian, Paul M. *Basic Network Theory*. McGraw-Hill Book Company, New York, 1969.

Chirlian, Paul M. *Analysis and Design of Integrated Electronic Circuits*. Harper & Row Publishers, New York, 1981.

Ciletti, Michael D. *Introduction to Circuit Analysis and Design*. Holt, Rinehart and Winston, New York, 1988.

Delagrange, Arthur D. *An Operational Amplifier Primer*. NOLTR-72-166. Naval Ordnance Laboratory, White Oak (Maryland), 1972.

Dorf, Richard C. *Modern Control Systems*, fourth edition. Addison-Wesley Publishing Co., Reading (Massachusetts), 1986.

Finkbeiner, Daniel T. *Elements of Linear Algebra*. W. H. Freeman and Company, San Francisco, 1972.

Ghausi, Mohammed S. *Electronic Circuits*. D. Van Nostrand Company, 1971.

Graeme, Jerald G. *Designing with Operational Amplifiers*. McGraw-Hill Book Company, New York, 1977.

Grove, E. A., and G. Ladas. *Introduction to Complex Variables*. Houghton Mifflin Company, Boston, 1974.

Hayt, William H., and Jack E. Kemmerly. *Engineering Circuit Analysis*, third edition. McGraw-Hill Book Company, New York, 1978.

Hubert, Charles I. *Electric Circuits AC/DC*. McGraw-Hill Book Company, New York, 1982.

Johnson, Curtis, D. *Network Analysis for Technology*. Macmillan Publishing Company, New York, 1984.

Johnson, David E. *Introduction to Filter Theory*. Prentice-Hall, Englewood Cliffs (New Jersey), 1976.

Johnson, David E., et al. *Basic Electric Circuit Analysis*. Prentice-Hall, Englewood Cliffs (New Jersey), 1978.

Kreyzig, Erwin. *Advanced Engineering Mathematics*, fourth edition. John Wiley & Sons, New York, 1979.

Kulathinal, Joseph. *Transform Analysis and Electronic Networks with Applications*. Merrill Publishing Co., Columbus (Ohio), 1988.

Kuo, Benjamin C., *Automatic Control Systems*, second edition. Prentice-Hall, Englewood Cliffs (New Jersey), 1967.

Mitra, Snajit K. *An Introduction to Digital and Analog Integrated Circuits and Applications*. Harper and Row Publishers, New York, 1980.

Muth, Eginhard J. *Transform Methods with Applications to Engineering and Operations Research*. Prentice-Hall, Englewood Cliffs (New Jersey), 1977.

Nilsson, James W. *Electric Circuits*, second edition. Addison-Wesley Publishing Co., Reading (Massachusetts), 1986.

Raven, Francis H. *Automatic Control Engineering*, second edition. McGraw-Hill Book Company, New York, 1968.

Reynolds, James A. *Applied Transformed Circuit Theory for Technology*. John Wiley & Sons, New York, 1985.

Siebert, William McC. *Circuits, Signals, and Systems*. MIT Press, Cambridge (Massachusetts), 1986.

Skilling, Hugh H. *Electrical Engineering Circuits*. John Wiley & Sons, 1957.

Smyth, Michael P. *Linear Engineering Systems: Tools and Techniques*. Pergamon Press, New York, 1972.

Stanley, William. *Operational Amplifiers with Linear Integrated Circuits*. Charles E. Merrill Publishing Co., Columbus (Ohio), 1984.

Stanley, William D. *Network Analysis with Applications*. Reston Publishing Company, Reston (Virginia), 1985.

Stanley, William D. *Transform Circuit Analysis for Engineering and Technology*, second edition. Prentice-Hall, Englewood Cliffs (New Jersey), 1989.

Temes, Gabor C., and Sanjit K. Mitra. *Modern Filter Theory and Design*. John Wiley & Sons, New York, 1973.

Tuttle, David F., Jr. *Electric Networks*. McGraw-Hill Book Company, 1965.

Van Valkenburg, M. E., and B. K. Kinariwala. *Linear Circuits*. Prentice-Hall, Englewood Cliffs (New Jersey), 1982.

Answers to Selected Problems

Chapter 1

1-1. a.

Element 1:	Active	+ 24 W
Element 2:	Passive	− 12 W
Element 3:	Passive	− 6 W
Element 4:	Passive	− 9 W
Element 5:	Active	+ 6 W
Element 6:	Passive	− 3 W

1-3. a.

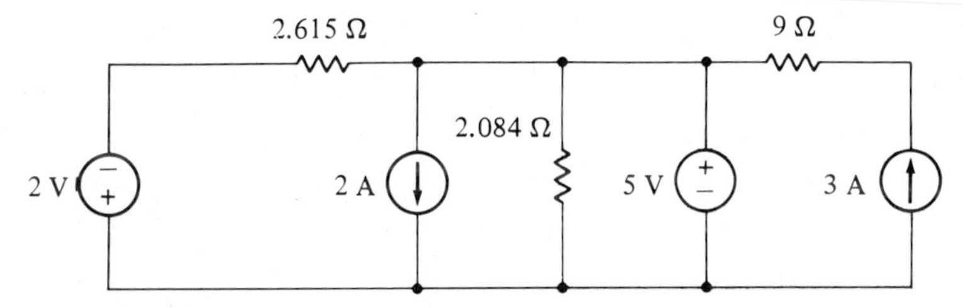

Figure 1.3

1-4. b. 0.3871 V
1-5. a. −0.26 V
1-6. b. −0.2135 A
1-7. a. 42.5 mA
1-9. a. 8 V; 2.664 A; 3 Ω; 0.3333 S
1-11. a. 6.098 V; 1.220 V; 7.317 V; 2.683 V; 122.0 mA; 146.3 mA; 268.3 mA
1-12. b. −3.246 V; −11.68 V
1-13. a. −9.760 V; −3.361 V
1-14. b. 4 V; 8.525 V; −5.791 mA
1-15. a. 4.725 V; 8.759 V; 5.794 V
1-16. b. −123.9 μA; −816.5 μA
1-17. a. −0.7299 mA; 2.176 mA
1-18. b. 3 mA; −4.962 mA; 7.962 V
1-19. a. 1.060 mA; −0.3448 mA; −1.657 mA
1-20. b. 1 V; 2 V; 3 V; 4 V
1-21. a. 5.164 V
1-22. b. 4.725 V
1-23. a. 13 V; 2 kΩ; 6.5 mA; 0.5 mS
1-24. b. 0.8274 V; 310.4 Ω; 2.666 mA; 3.222 mS
1-25. a. 6.222 V; 1.422 kΩ; 4.375 mA; 0.7032 mS
1-26. b. 0 V; 1 kΩ; 0 A; 1 mS
1-27. 0.3086 mS
1-28. b. 200 Ω
1-29. a. 125 Ω

Chapter 2

2-1. a.

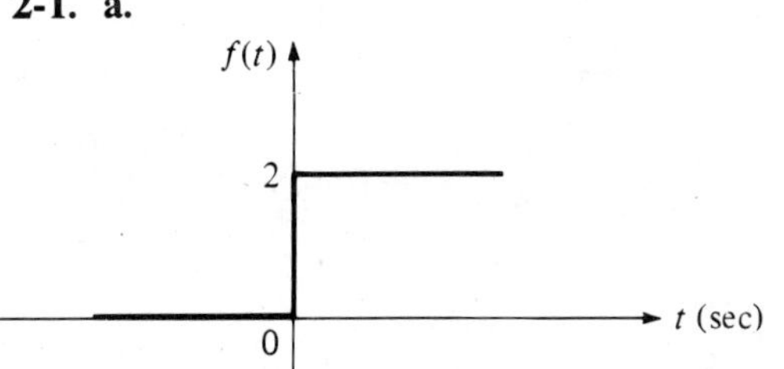

Figure 2.1a

c.

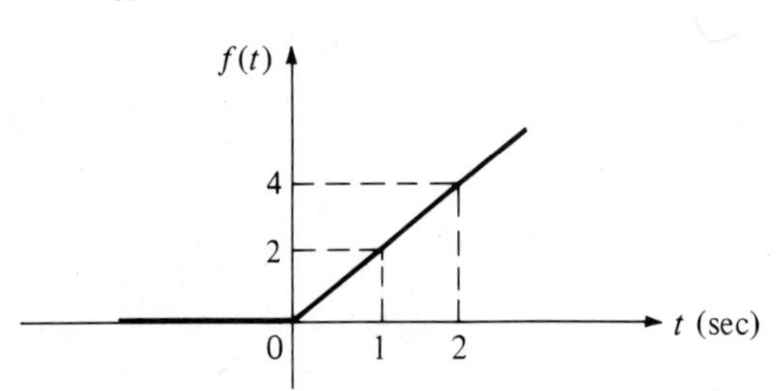

Figure 2.1c

e.

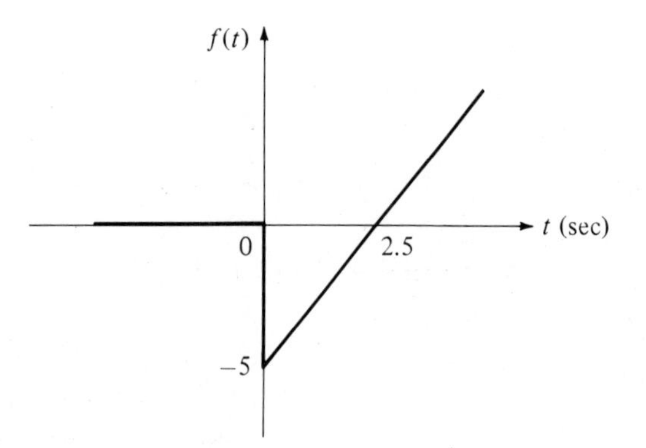

Figure 2.1e

2-2. **b.** $-2.5 \cdot u(t-4)$
d. $-2.5(t-6) \cdot u(t-5)$
f. $-2.5t \cdot u(t-5)$

2-3. **a.**

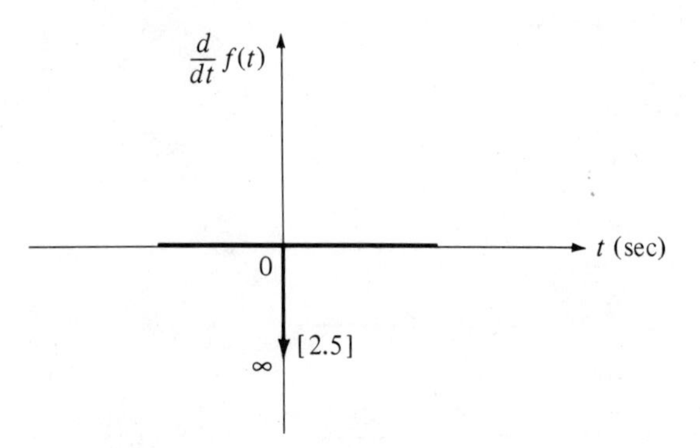

Figure 2.3a

c.

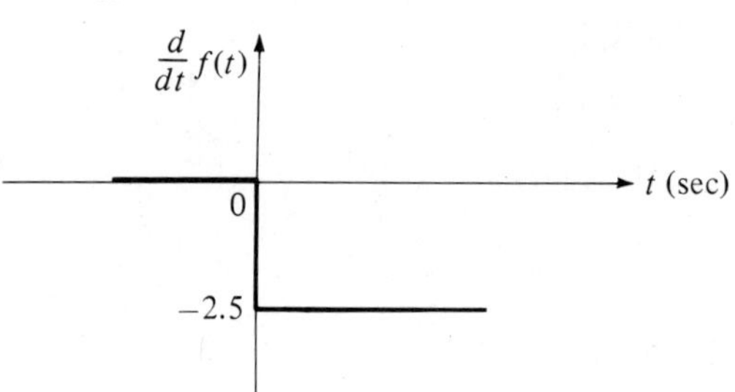

Figure 2.3c

e.

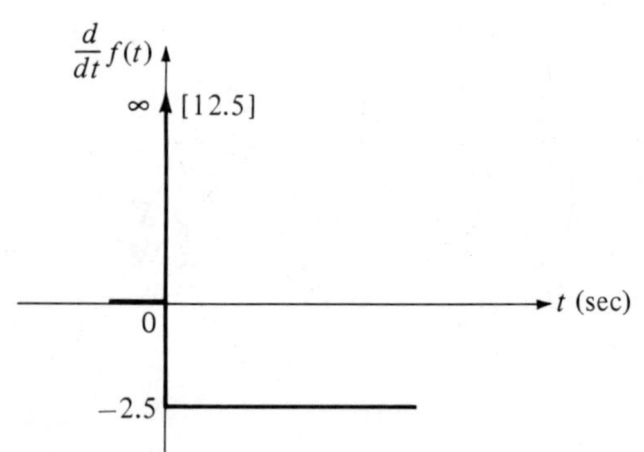

Figure 2.3e

2-4. **b.**

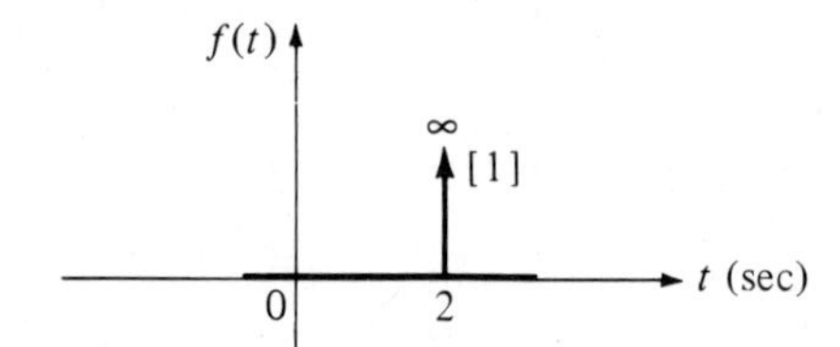

Figure 2.4b

d.

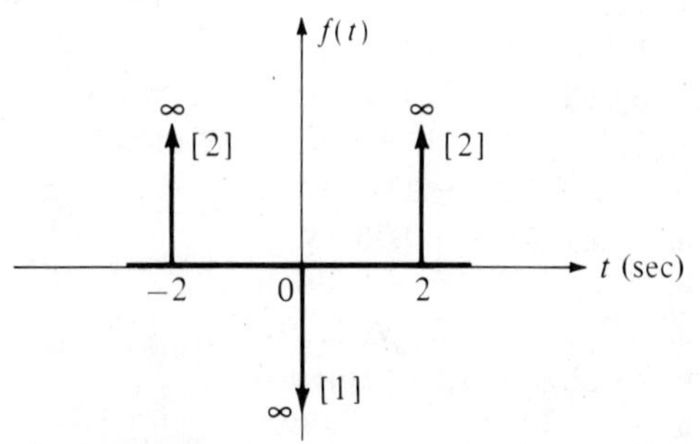

Figure 2.4d

2-5. **a.** $-3 \cdot \delta(t)$
c. $9 \cdot \delta(t-2)$

2-6. **b.**

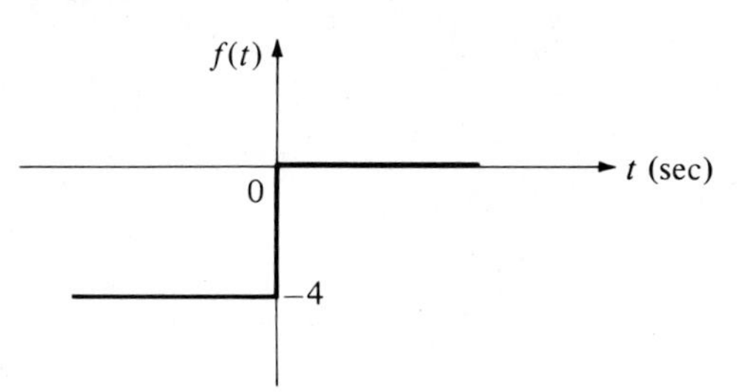

Figure 2.6

2-7. **a.** $5 \cdot u(-t-7)$
c. $5 \cdot u(7-t)$

2-8. **b.**

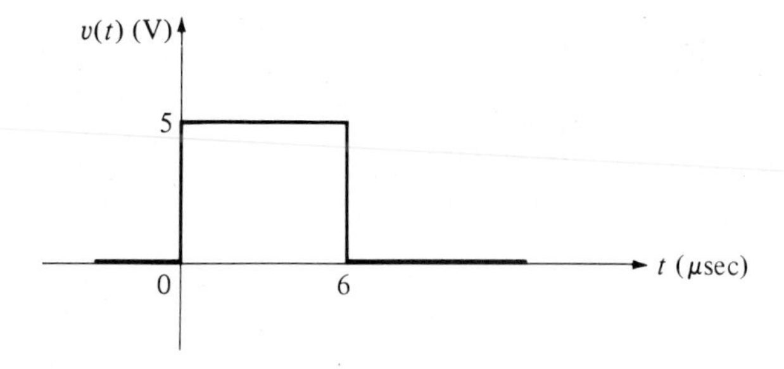

Figure 2.8

2-9. **a.** $-100 \cdot u(t-4) + 130 \cdot u(t-7) - 30 \cdot u(t-11)$
c. $20[u(t) - u(t - 40 \times 10^{-6}) + u(t - 90 \times 10^{-6}) - u(t - 120 \times 10^{-6})]$ mA

2-10. **b.**

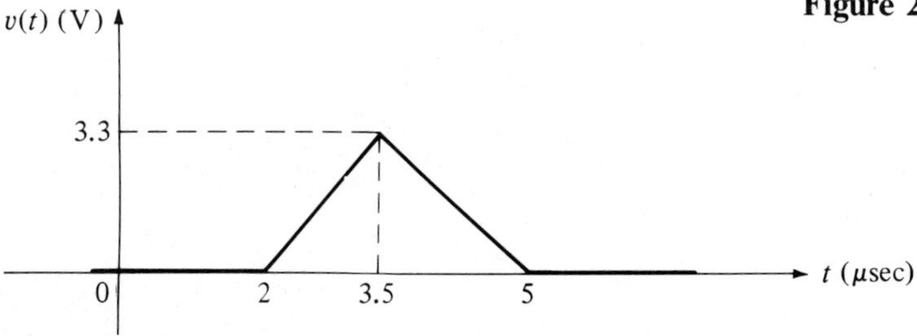

Figure 2.10

2-11. **a.** $-4000t \cdot u(t) + 6667(t-0.02)u(t-0.02) - 2667(t-0.05)u(t-0.05)$ mA
c. $125 \times 10^3(t-0.001)u(t-0.001) - 500 \cdot u(t-0.005) - 125 \times 10^3(t-0.005)u(t-0.005)$

2-12. **b.**

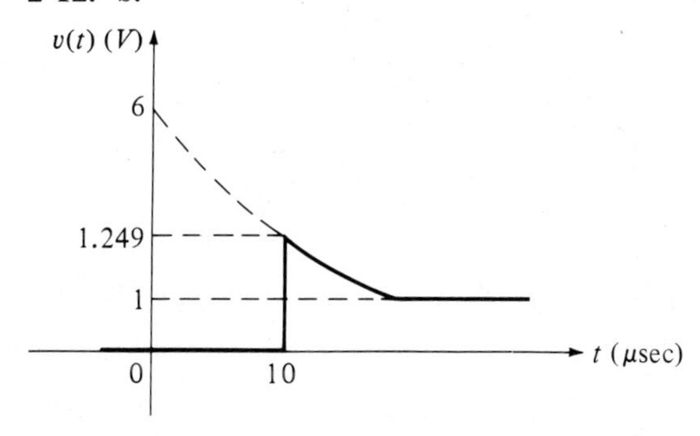

Figure 2.12

2-13. **a.** $300 \cdot \exp[-(t-40)/\tau] \cdot u(t-40)$; $\tau = 54.61$ sec
c. $18\{1 - \exp[-(t-0.01)/\tau]\} \cdot u(t-0.01)$ V; $\tau = 2.885$ msec

2-14. **b.**

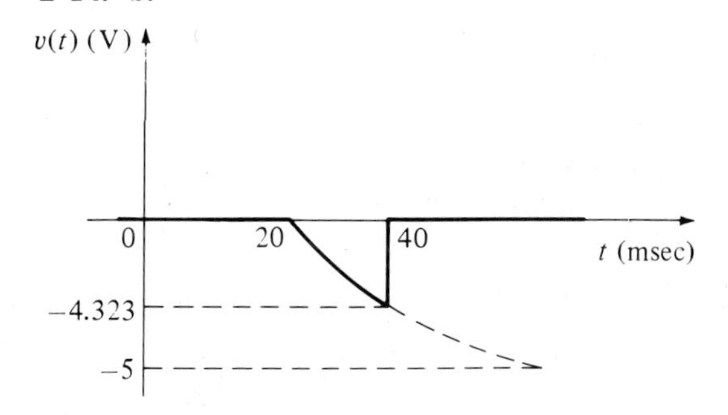

Figure 2.14

2-15. a. $1 + 2\{1 - \exp[-(t-1)/\tau]\}u(t-1)$
$- 2 \cdot u(t-3) - 0.5 \cdot \exp[-(t-3)/\tau]$
$\cdot u(t-3);\ \tau = 1.443$ sec

2-16. b.

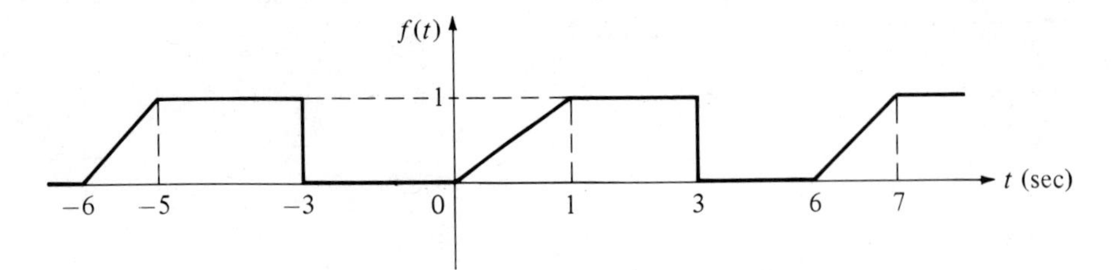

Figure 2.16

2-17. a. $0.25 \times 10^6 t \cdot u(t)$
$- 10\Sigma_{n=0}^{\infty} u(t - 20 \times 10^{-6}$
$- 40 \times 10^{-6} n)$

c. $100\Sigma_{n=0}^{\infty}\{\exp[-(t - 0.02n)/\tau]$
$\cdot u(t - 0.02n)$
$- \exp[-(t - 0.01 - 0.02n)/\tau]$
$\cdot u(t - 0.01 - 0.02n)\}$ mA;
$\tau = 1.443$ msec

2-18. b.

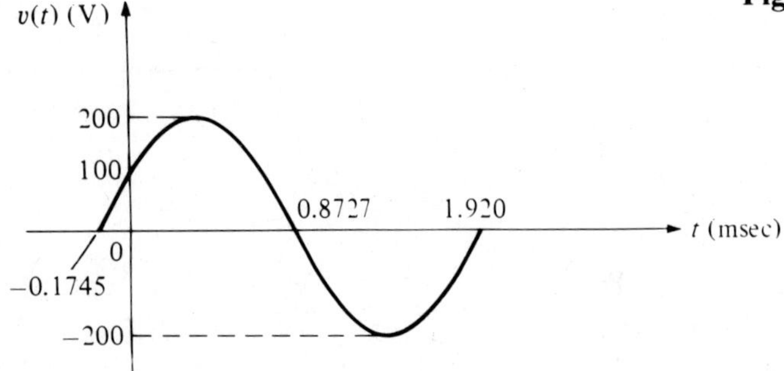

Figure 2.18

2-19. a. $10 \cdot \sin[500\pi t + 45°]$ V

c. $1 + 0.5 \cdot$
$\sin[200 \times 10^3 \pi(t - 5 \times 10^{-6})] \cdot$
$u(t - 5 \times 10^{-6})$ A

2-20. $20\Sigma_{n=0}^{\infty}(\cos[120\pi(t - T/3 - nT) + 120°]$
$\cdot u(t - T/3 - nT)$
$- \cos[120\pi(t - T - nT)]$
$\cdot u(t - T - nT))$ V; $T = \frac{1}{60}$ sec

Chapter 3

3-1. a. $0.3 \cdot \delta(t)$ μA; 0

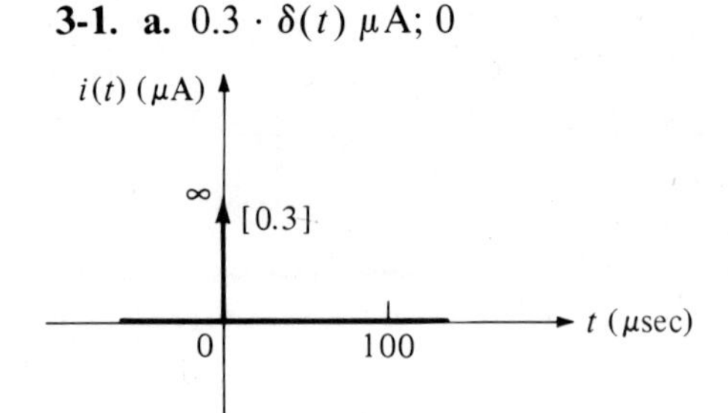

Figure 3.1a

c. $0.02 \cdot \delta(t) - 60 \cdot \exp[-3000t] \cdot u(t)$
mA; −44.45 mA

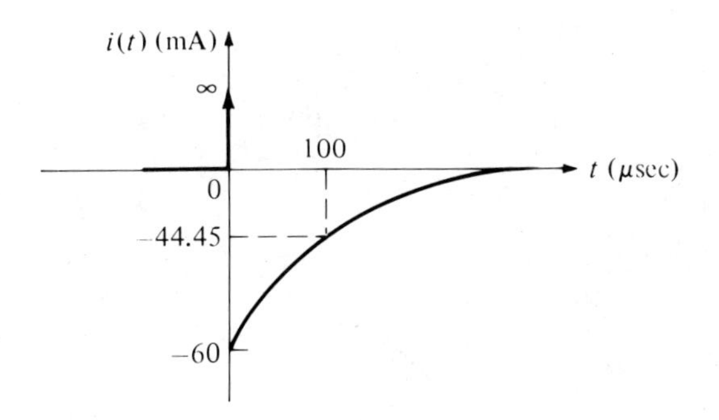

Figure 3.1c

3-2. b. $5 + 2.273 \times 10^{12}t^2 \cdot u(t)$ V; 7.273 V

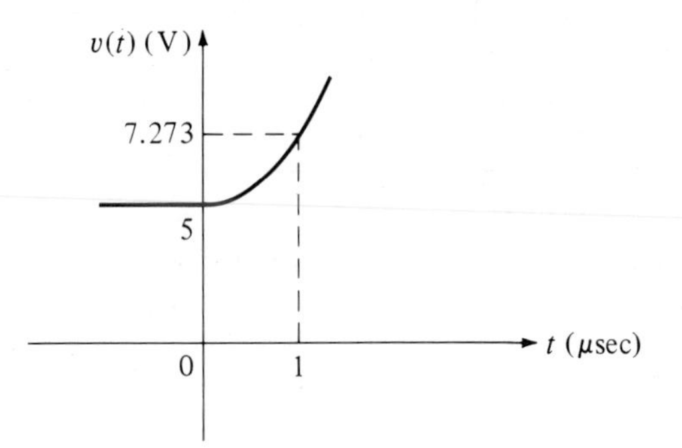

Figure 3.2b

d. $5 + 2.273\Sigma_{n=0}^{\infty}[(-1)^n \cdot u(t - 10 \times 10^{-6}n)]$ V; 7.273 V

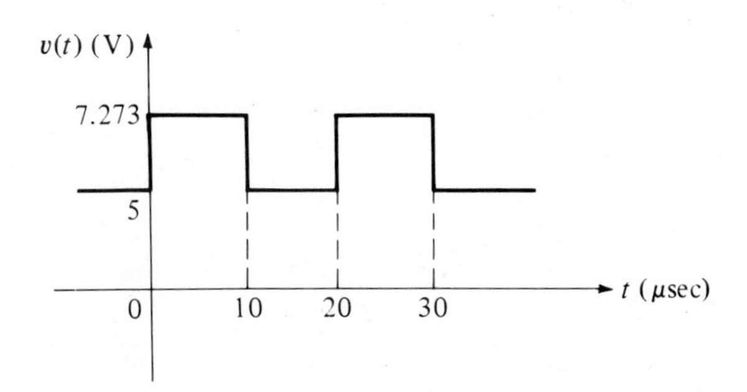

Figure 3.2d

3-3. a. 100 μF
c. 22 μF

3-4. b. $-100 \cdot u(t)$ V; -100 V

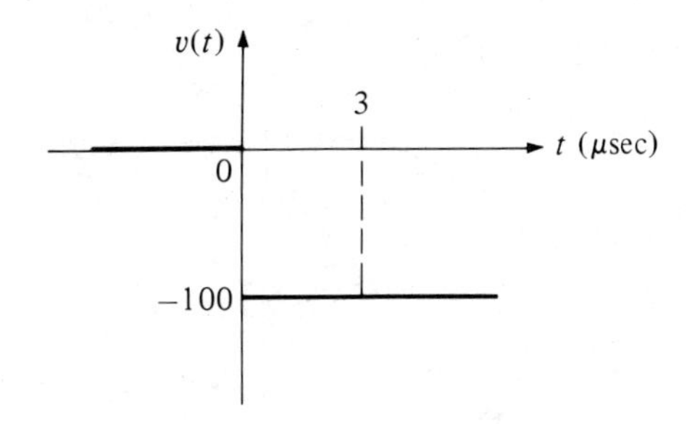

Figure 3.4b

d. $3 \cdot u(t) - 3 \times 10^{-6}\Sigma_{n=0}^{\infty}\delta(t - (n + 1) \times 10^{-6})$ V; negative impulse

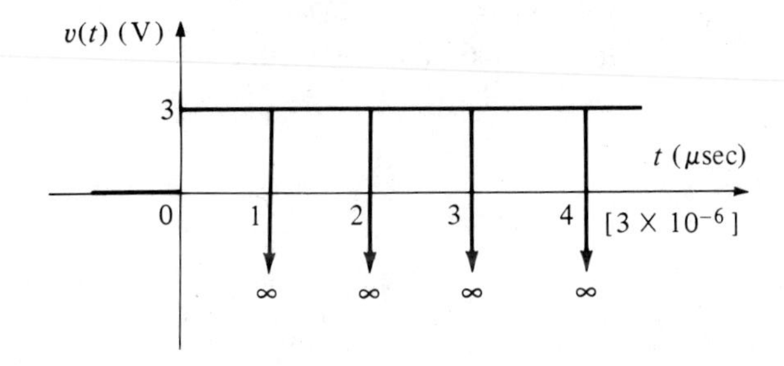

Figure 3.4d

3-5. a. $1 + 50 \times 10^3 t \cdot u(t)$ mA; 2.5 mA

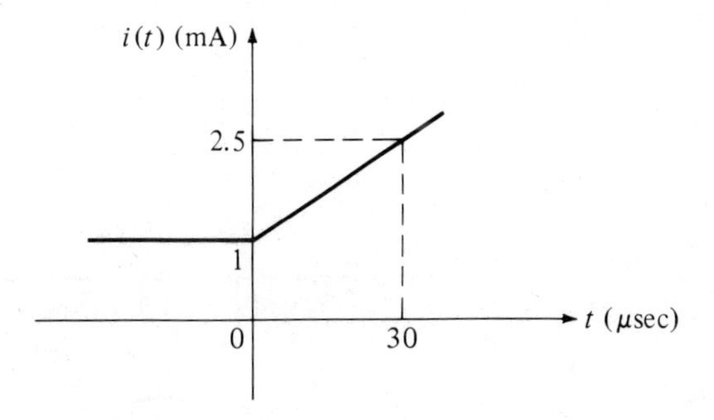

Figure 3.5a

c. $1 + 100(1 - \exp[-200t])u(t)$ mA; 6.824 mA

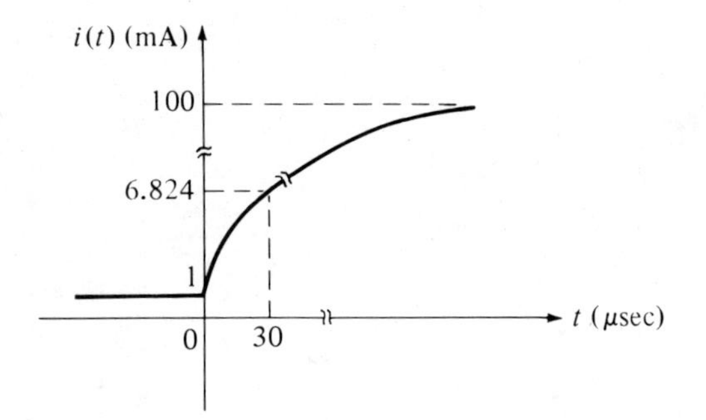

Figure 3.5c

3-6. **b.** 30 mH

3-8. $117.5[u(t) - u(t - 0.002) - u(t - 0.006) + u(t - 0.008)]$ mA

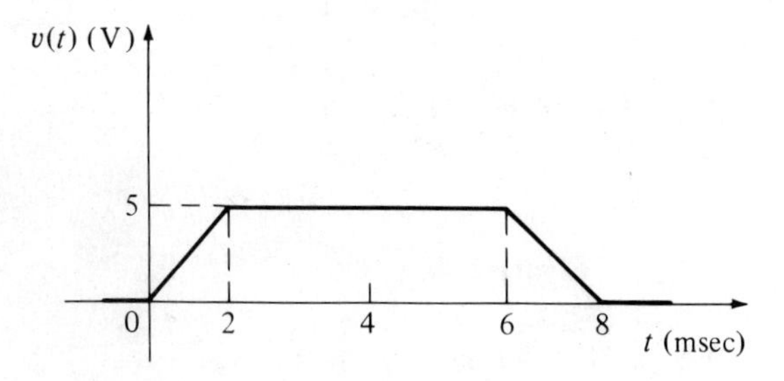

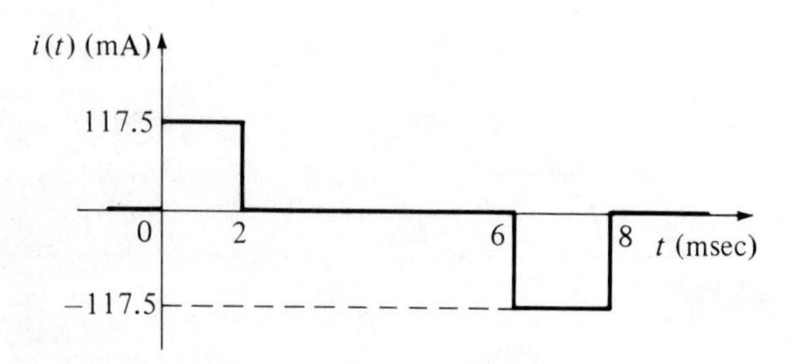

Figure 3.8

3-9. $6[u(t - 2 \times 10^{-6}) - 2 \cdot u(t - 7 \times 10^{-6}) + u(t - 12 \times 10^{-6})]$ V

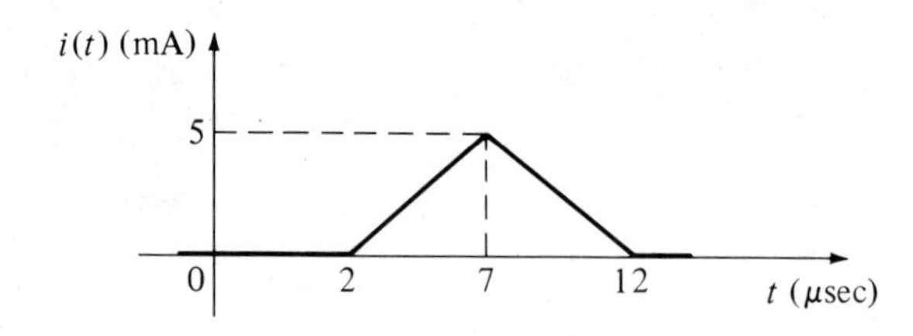

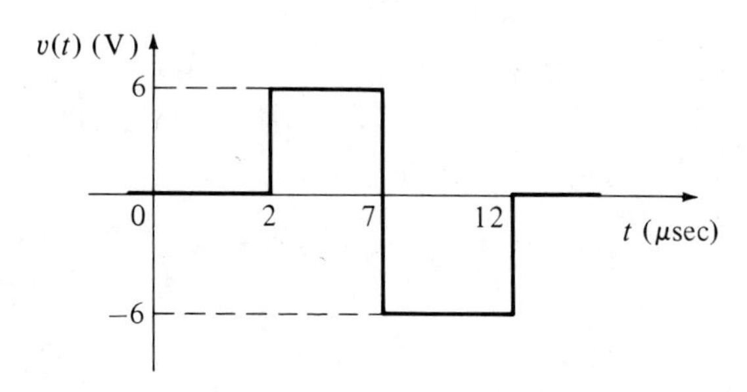

Figure 3.9

3-10. **b.** 19.36 mA; 12.90 mA
d. 36 mA; 1.8 V

3-11. **a.** $28 - 28(1 - \exp[-t/\tau])u(t)$ mA; $\tau = 320$ nsec

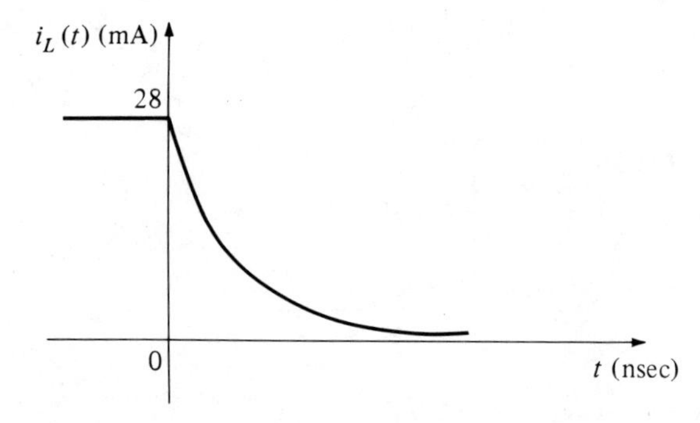

Figure 3.11a

c. $(13.33 + 24 \cdot \exp[-t/\tau])u(t)$ mA;
$\tau = 110\ \mu$sec; 140.8 μsec

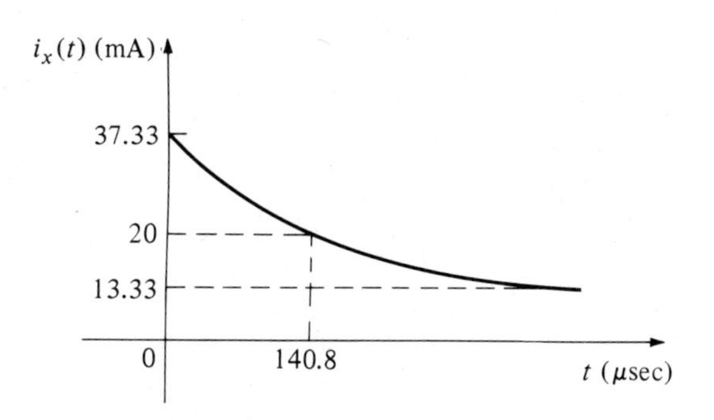

Figure 3.11c

e. $9 - (9 + 6 \cdot \exp[-t/\tau])u(t)$ V;
$\tau = 9.6$ msec; 17.2 msec

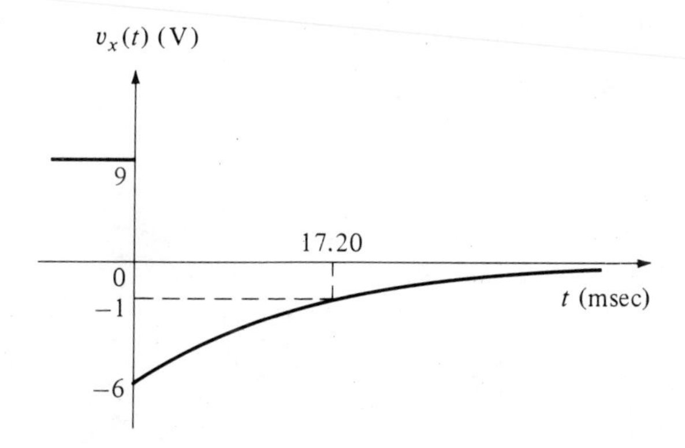

Figure 3.11e

3-12. b. $25 - 15.28(1 - \exp[-t/\tau])u(t)$ mA;
$\tau = 277.8\ \mu$sec; 12.25 mA

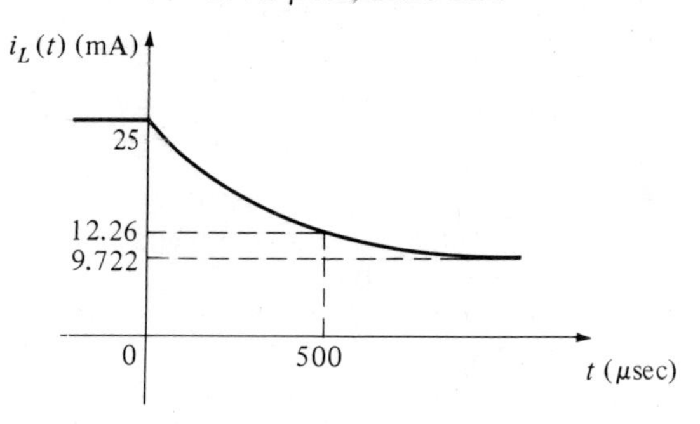

Figure 3.12b

d. $-2 \cdot \exp[-t/\tau] \cdot u(t)$ V;
$\tau = 15$ msec; 4.315 msec

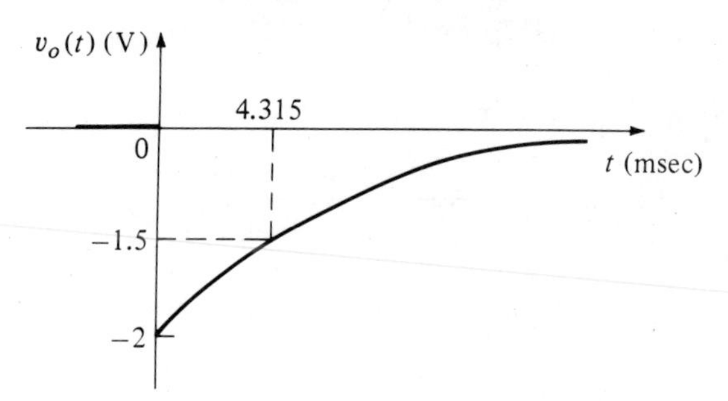

Figure 3.12d

3-13. $-3.6\sum_{n=0,2,4,\ldots\ (\text{even})}^{\infty}(1 - \exp[-\{t - (n-1)400 \times 10^{-6}\}/\tau])$
$u(t - (n-1)400 \times 10^{-6})$ V;
$-3.6\sum_{n=1,3,5,\ldots(\text{odd})}^{\infty}\exp[-\{t - (n-1)400 \times 10^{-6}\}/\tau]$
$\cdot u(t - (n-1) \times 400 \times 10^{-6})$ V;
$\tau = 10\ \mu$sec

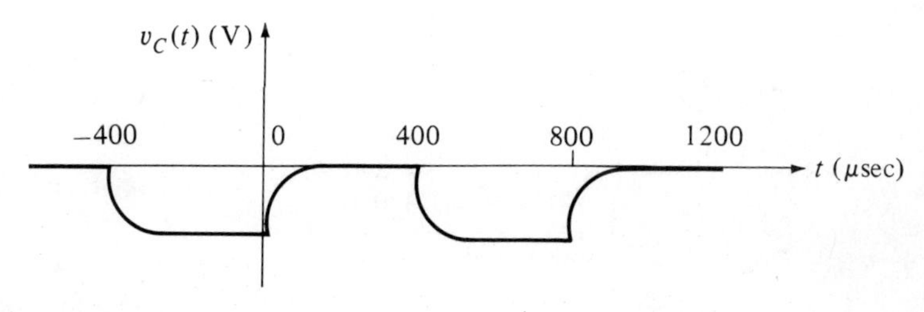

Figure 3.13

Chapter 4

4-1. **a.** $7/s^2$
c. $41(1 + 1/s + 1/s^2)$
e. $67(29 + 23/s^2)$ μA · sec

4-2. **b.** $12/(s + 83)$

4-7. **a.** $3589/[s(s + 97)]$ mA · sec
c. $-42/s + 35/(s + 5)$ V · sec
e. $696/(s + 49)^5$

4-9. **a.** $190 \times 10^3/(s^2 + 100 \times 10^6)$ V · sec
c. $942.5/(s^2 + 377^2) + 4.330s/(s^2 + 377^2)$

4-10. **b.** $31(s + 7)/[(s + 7)^2 + 300^2]$ V · sec

4-11. **a.** $2 \cdot \exp[-7s]$
c. $31(\exp[-19s] + 1/s + \exp[-31s]/s^2)$ V · sec
e. $5917 \cdot \exp[-83s]/[s(s + 97)]$ mA · sec
g. $43s \cdot \exp[-0.001s]/(s^2 + 10^6)$

4-12. **b.** $$\frac{73}{s^2} - \frac{219}{s} \cdot \frac{\exp[-3s]}{1 - \exp[-3s]}$$
c. $$\left[\frac{17}{s} + \left(\frac{11}{s^2} - \frac{39}{s}\right) \cdot \exp[-3s] + \left(\frac{11}{s^2} - \frac{1}{s}\right) \cdot \exp[-2s]\right] \cdot \frac{1}{1 - \exp[-3s]} \text{ mA} \cdot \text{sec}$$

4-13. **a.** $$F(s) = \frac{3}{s + 2}$$
c. $$F(s) = \frac{222}{s^3(s^2 + 19s + 23)} + \frac{31s}{s^2 + 19s + 23}$$
e. $$V(s) = \frac{11s^2 + 17.39s - 0.7320}{s^3 + 0.9381s^2 + 0.9588s + 0.8144} \text{ V} \cdot \text{sec}$$

4-14. **b.** $1.167t^3 \cdot \exp[-11t] \cdot u(t)$ V
d. $\exp[-7t] \cdot \cos 80t \cdot u(t)$
f. $6.539(\exp[-2t] - \exp[-15t])u(t)$ V
h. $7\{\delta(t) - 11 \cdot \exp[-11t] \cdot u(t)\}$
i. $31t \cdot \exp[-13t] \cdot u(t)$

4-15. **a.** $2 \cdot \delta(t) + 3 \cdot u(t)$ V
c. $2(t^4 + 1 - \exp[-79t])u(t)$
e. $(2 \cdot \sin 30t + 5 \cdot \cos 100t)u(t)$ mA

4-16. **b.** $17(0.06667t - 0.03556 + 0.05556 \cdot \exp[-3t] - 0.02 \cdot \exp[-5t])u(t)$
d. $9.455 \cdot \exp[-t/2] \cdot \sin 3.279t \cdot u(t)$ A
f. $\{0.6875 - 6 \cdot \exp[-t] + (9.75t + 2.75)\exp[-2t] + 2.562 \cdot \exp[-4t]\}u(t)$ V
h. $([0.01176 \cdot \cos[2.646t - 8.950°] + 0.01218 \cdot \cos[9.206t + 162.4°])\, u(t)$ mA
j. $\{(6.84t - 4.286)\exp[-3t] + 4.736 \cdot \exp[-2.5t] \cdot \cos[4.975t - 25.10°]\}u(t)$

4-25. **a.** $7 \cdot \delta(t) - (0.2 - 8.214 \cdot \exp[-t/2] \cdot \cos[2.179t - 28.73°])u(t)$
c. $3\{1 - \exp[-(t - 2)]\}u(t - 2) + 2\{(t - 17) - 1 + \exp[-(t - 17)]\} \cdot u(t - 17)$
e. $3.5(t - 97)\exp[-2(t - 97)] \cdot u(t - 97)$
f. $7 \cdot \Sigma_{n=0}^{\infty}[12.5(t - 1 - 10n)^2 - 10(t - 1 - 10n) + 1] \cdot \exp[-5(t - 1 - 10n)] \cdot u(t - 1 - 10n)$

Chapter 5

5-1. **a.** $20/s$ kΩ and $50/s$ V · sec; $50s$ μS and 2.5 mA · sec
c. $0.04s$ Ω and 12 mV · sec; $25/s$ S and $300/s$ mA · sec

5-2. b.

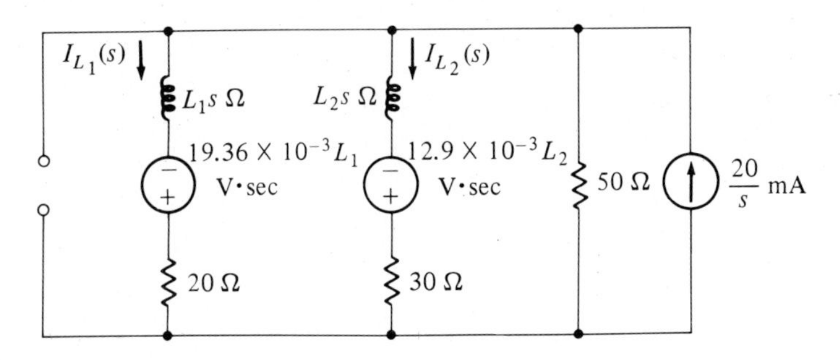

Figure 5.2b

d.

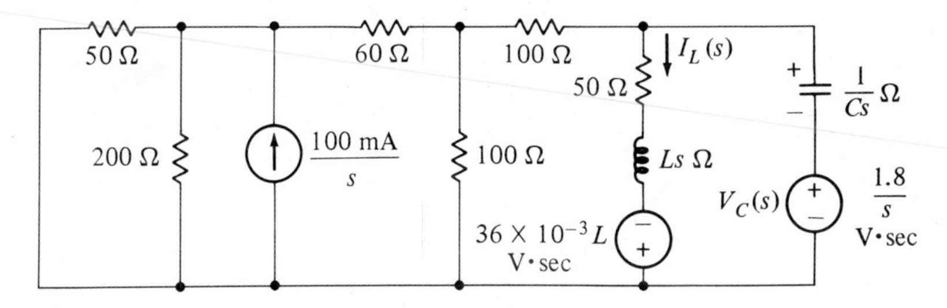

Figure 5.2d

5-3. **a.** $0.02(s^2 + 4500s + 12.5 \times 10^6)/s\ \ \Omega$; $50s/(s^2 + 4500s + 12.5 \times 10^6)$ S

c. $62.5(s^2 + 240 \times 10^3 s + 1601)/(s + 40 \times 10^3)$ pS; $16(s + 40 \times 10^3)/(s^2 + 240 \times 10^3 s + 1601)$ GΩ

5-4. **b.** $-0.05(1 - \exp[-20t])u(t)$ V

d. $0.7059(1 + 7.5 \cdot \exp[-340 \times 10^3 t]) \cdot u(t)$ V

5-5. **a.** $-10 \cdot \exp[-5t] \cdot u(t)$ mA

c. $0.1584(1 - 21.2 \cdot \exp[-1.402 \times 10^6 t]) \cdot u(t)$ mA

5-8. $0.1(1 + 11.5 \cdot \exp[-200t])u(t)$ V

5-9. 4.887 msec; 3.324 msec

Chapter 6

6-1. **a.** $100/[s^2 + 33.33(40 + R)s + 333.3 \times 10^6]$ A · sec

6-2. $(12 - 11.95 \cdot \exp[475.1t] + 11.94 \cdot \exp[-10520t])u(t)$ V

6-3. $(50 + 166.6t)\exp[-333t] \cdot u(t)$ mA

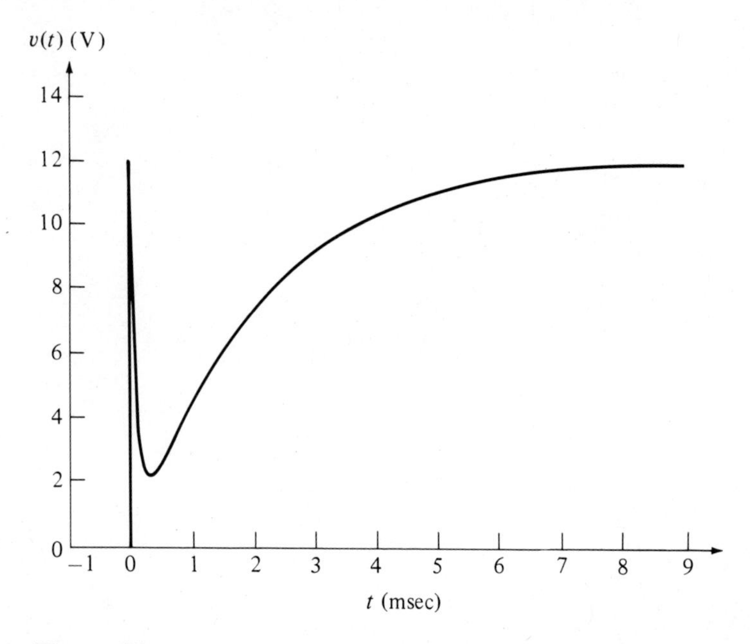

Figure 6.2

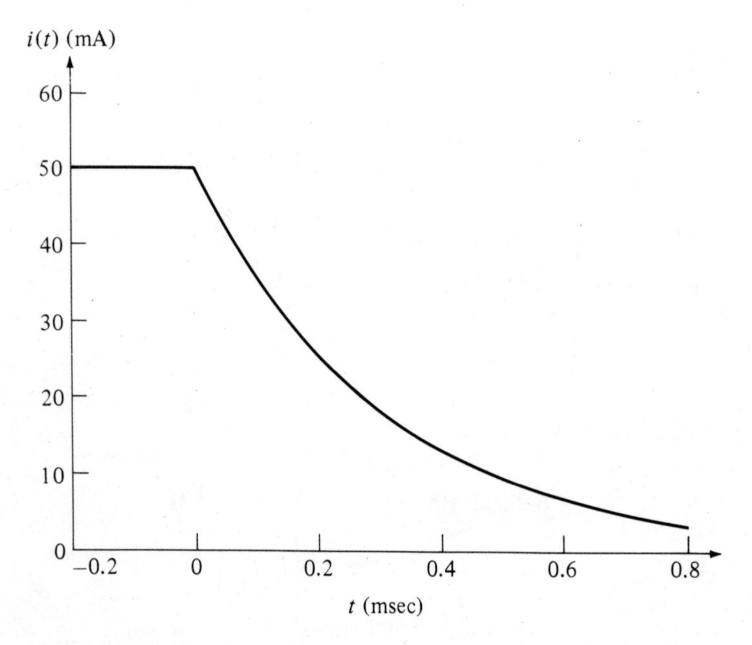

Figure 6.3

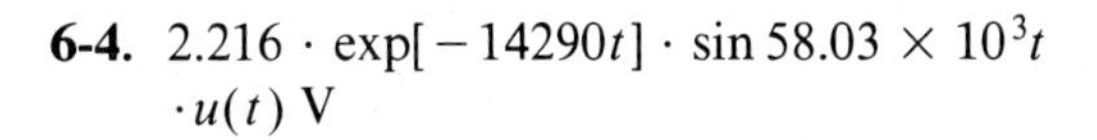

6-4. $2.216 \cdot \exp[-14290t] \cdot \sin 58.03 \times 10^3 t \cdot u(t)$ V

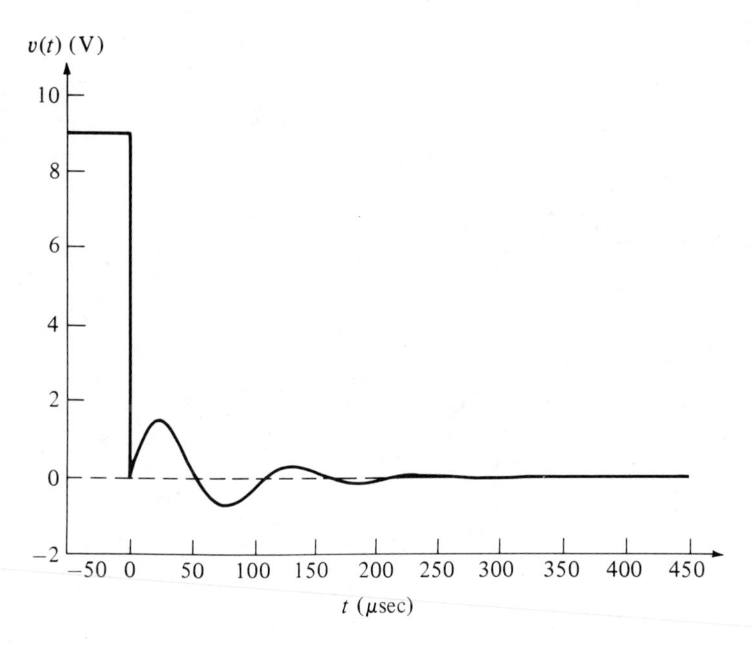

Figure 6.4

6-5. $3.203 \cdot \exp[-5000t] \cdot \sin 31.22 \times 10^3 t \cdot u(t)$ V

6-6. b. $-1139 \cdot \exp[-277.8t] \cdot \cos[52700t - 90.60°] \cdot u(t)$ V

6-7. a. 0.1 A; 0 A
c. 8 V; 12 V

6-8. $(-2.648 \cdot \exp[-400t] + 3.639 \cdot \cos[377t - 43.30°]) \cdot u(t)$ mA

6-9. a. $\dfrac{100s}{(s + 5 \times 10^6/R)(s^2 + 10^6)}$ A · sec

6-10. $(4.292 \cdot \exp[-833300t] + 2.900 \cdot \cos[10^6 t - 142.2°])u(t)$ V

6-11. $\dfrac{V_B}{s + 5000} + \dfrac{V_A - V_B}{s + 5000} \cdot \left(\dfrac{1 - \exp[-T_1 s]}{1 - \exp[-Ts]}\right)$

6-12. Case 2:
$15\Sigma_{n=0}^{\infty}\{\exp[-5000(t - 0.0015n)] \cdot u(t - 0.0015n) - \exp[-5000(t - 0.0003 - 0.0015n)] \cdot u(t - 0.0003 - 0.0015n)\}$ V
Case 4:
$10\Sigma_{n=0}^{\infty}\{\exp[-5000(t - 0.0015n)] \cdot u(t - 0.0015n) - \exp[-5000(t - 0.0003 - 0.0015n)] \cdot u(t - 0.0003 - 0.0015n)\}$ V

6-13. $0.75\Sigma_{n=0}^{\infty}(\{1 - \exp[-72730(t - 0.0002n)]\} \cdot u(t - 0.0002n) - 2\{1 - \exp[-72730(t - 0.0001 - 0.0002n)]\} u(t - 0.0001 - 0.0002n) + \{1 - \exp[-72730(t - 0.0002 - 0.0002n)]\} \cdot u(t - 0.0002 - 0.0002n))$

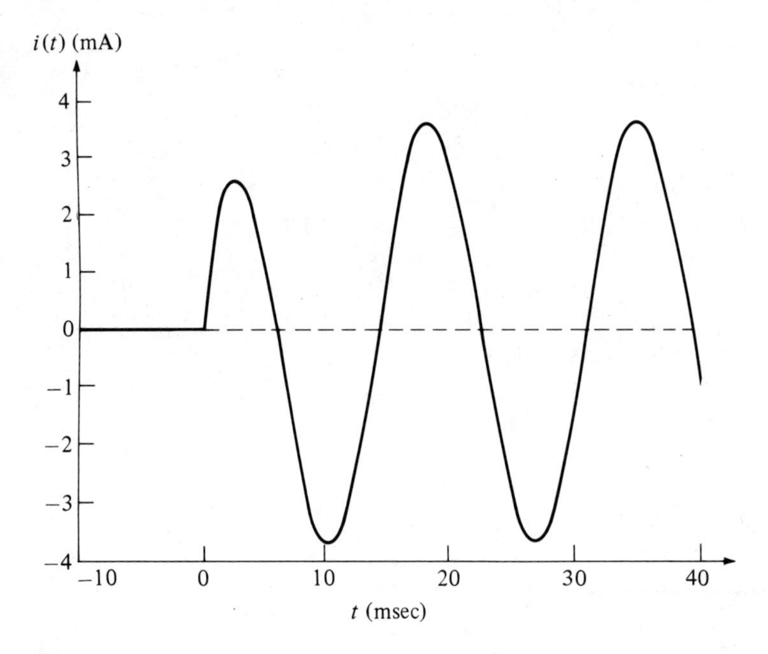

Figure 6.8

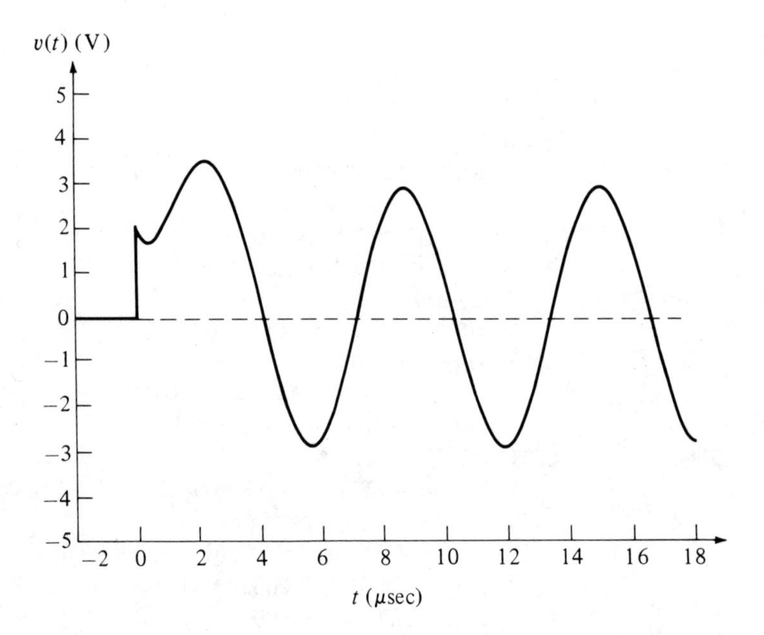

Figure 6.10

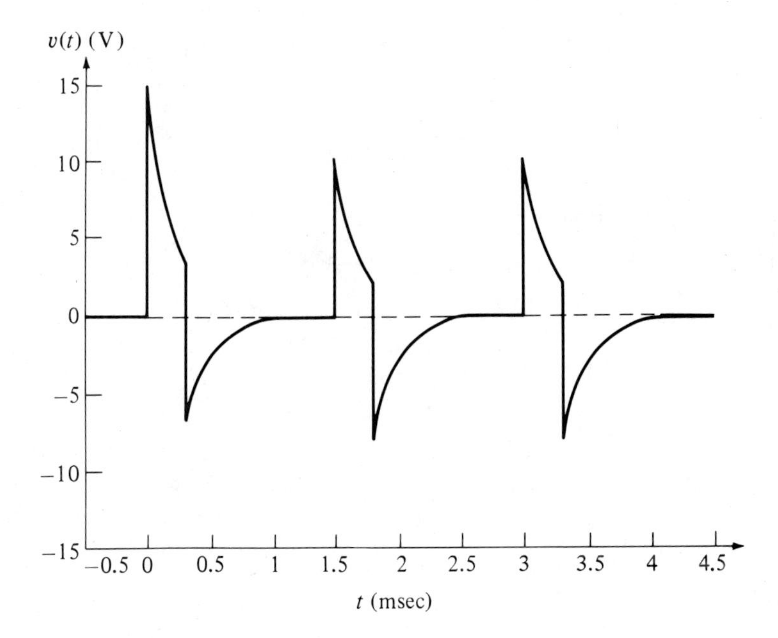

Figure 6.12a

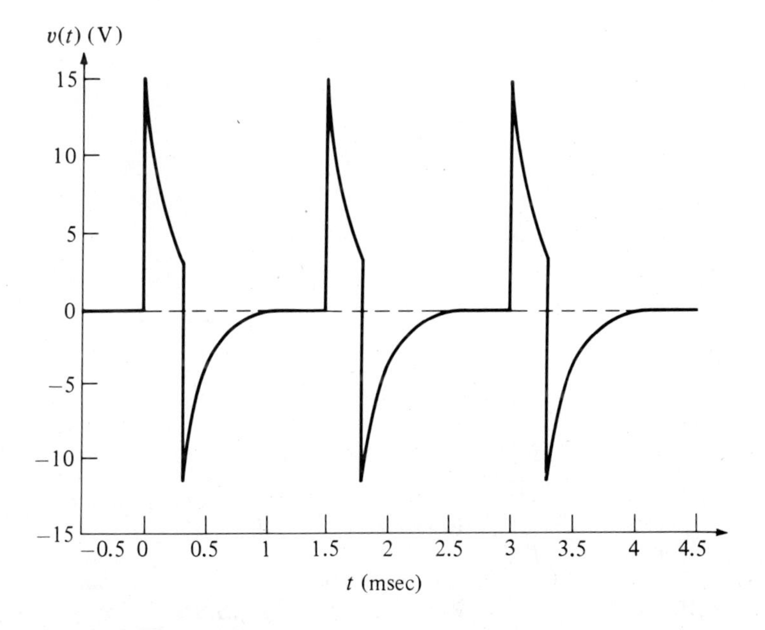

Figure 6.12b

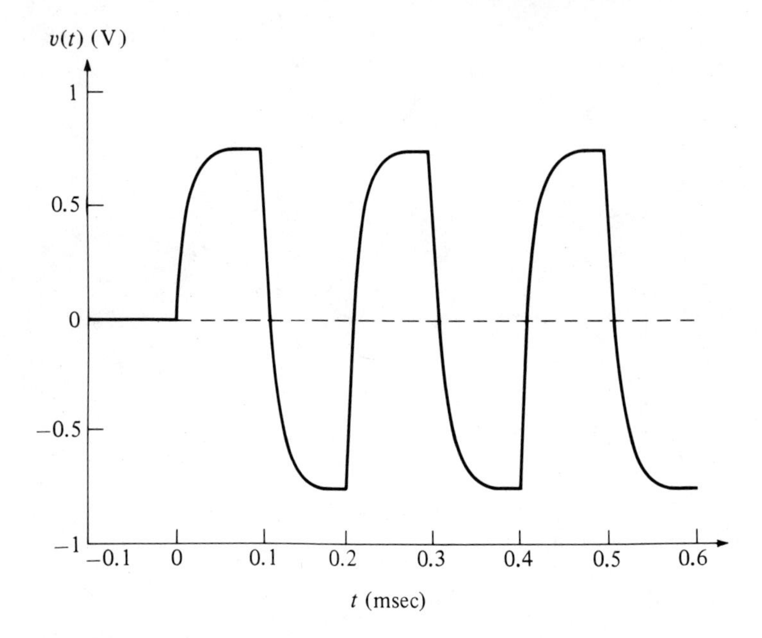

Figure 6.13

Chapter 7

7-1. a.

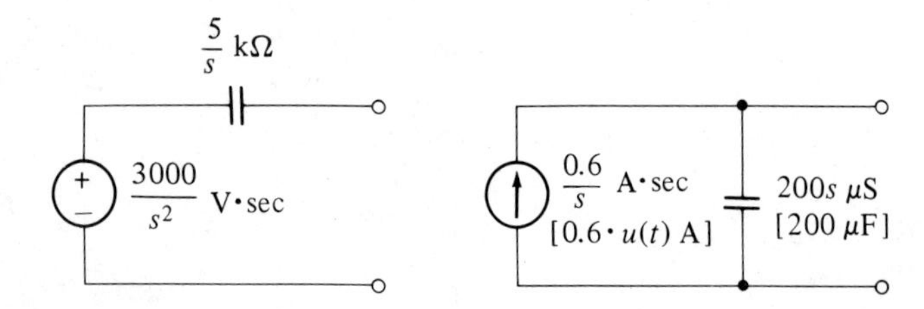

Figure 7.1

7-2.

$$
\begin{aligned}
(Y_{1A} + Y_{1B} + Y_2)V_A \quad & - Y_2 V_B & &= I_{S1} \\
-(100Y_{1B} + Y_2)V_A & + (Y_2 + Y_3 + Y_4)V_B & &= 0 \\
& - Y_4 V_B & + (Y_4 + Y_5)V_C &= 0
\end{aligned}
$$

$Y_{1A} = s/(s + 10 \times 10^3)$ mS; $Y_4 = 0.1s$ μS;
$Y_{1B} = 100(s + 10 \times 10^6)$ pS; $Y_5 = 0.1(s + 1000)$ μS;
$Y_2 = 100s$ pS; $I_{S1} = 3/(s + 10 \times 10^3)$ mA · sec
$Y_3 = 100$ μS;

7-3. $(1000t + 0.9 - 0.5 \cdot \exp[-10 \times 10^3 t] + 0.8944 \cdot \cos[5000t + 116.6°)u(t)$ V

7-4. $(2.997t - 3000 + 3000 \cdot \exp[-999.0t])u(t)$ V

7-5.

$$\begin{aligned}
(Z_1 + Z_2)I_1 \quad & -Z_2 I_2 & & = V_S \\
-(Z_2 - 100Z_4)I_1 & +(Z_2 + Z_3 - 99Z_4)I_2 & -Z_4 I_3 & = 0 \\
-100Z_4 I_1 & +99Z_4 I_2 & +(Z_4 + Z_5 + Z_6)I_3 & = 0
\end{aligned}$$

$Z_1 = (s + 10 \times 10^3)/s$ kΩ; $Z_5 = 10/s$ MΩ;
$Z_2 = 10 \times 10^3/(s + 10 \times 10^6)$ MΩ; $Z_6 = 10/(s + 1000)$ MΩ;
$Z_3 = 10 \times 10^3/s$ MΩ; $Z_4 = 10$ kΩ; $V_S = 3/s$ V · sec

7-6. $(0.4508 \cdot \exp[-11 \times 10^3 t] + 4.549 \cdot \cos[1000t + 0.5162°])u(t)$ mA

7-7. $30.24(s^2 + 9.902 \times 10^6 s + 9.901 \times 10^9)/[s^2(s + 999.0)]$ μA · sec

7-8. $-(5 - 9.572 \cdot \exp[-58580t] + 4.572 \cdot \exp[-341400t])u(t)$ V

7-9. $1200s^2/[(s^2 + 10^6)(s^2 + 730 \times 10^3 s + 1.6 \cdot 10^9)]$ V · sec

7-10. $500 \times 10^3/s^2 + 1/s$ V · sec; $0.05s + 100 + 5 \times 10^6/s$ Ω

7-12. $-3.52/(s + 14.08)$ mA · sec; 0.05 mS

7-14. $50 + 50s/(s + 500) + 150s/(s + 500)$ Ω

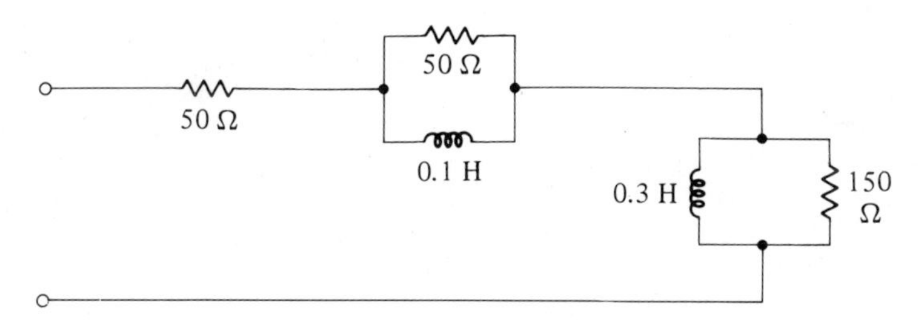

Figure 7.14

8-2.

$$\frac{V_2(s)}{V_1(s)} = \frac{s(s + 100 \times 10^3)}{s^2 + 100 \times 10^3 s + 100 \times 10^6}$$

$$\frac{V_2(s)}{I_1(s)} = 0.01(s + 1000)\ \Omega$$

$$\frac{I_2(s)}{V_1(s)} = \frac{-666.7s^2(s + 100 \times 10^3)}{s^2 + 100 \times 10^3 s + 33.33 \times 10^6}\ \text{nS}$$

$$\frac{I_2(s)}{I_1(s)} = -\frac{s(s + 100 \times 10^3)}{s^2 + 100 \times 10^3 s + 50 \times 10^6}$$

Chapter 8

8-1. a. $z_{11} = 0.01(s^2 + 100000s + 100 \times 10^6)/s$ Ω;
$z_{12} = z_{21} = 0.01(s + 100000)$ Ω;
$z_{22} = 0.01(s^2 + 100000s + 50 \times 10^6)/s$ Ω

8-3. a. $h_{11} = 500$ Ω; $h_{12} = 0.2$; $h_{21} = 80$; $h_{22} = -0.2$ mS

8-4. a. 80

8-5. a. $y_{11} = 1.111(s + 62.45 \times 10^3)/(s + 69.39 \times 10^3)$ mS; $y_{12} = 0$;
$y_{21} = 555.2/(s + 69.39 \times 10^3)$ S;
$y_{22} = 5$ μS

8-6. **b.** $5.556 \times 10^{-3}(s + 62.45 \times 10^3)/[(s + 5)(s + 69.39 \times 10^3)]$

8-8. **b.** $-1000t \cdot u(t)$ V

8-9. **a.** $-10s/[(s + 10)(s + 20)]$

c. $-2400/[s(s + 10)(s + 20)]$ V · sec

8-10. **b.** $10 \cdot u(t)$ V

8-11. $-(s + 100)$ Ω

8-12. $z_{11} = 0.1(s + 2000)$ Ω; $z_{12} = z_{21} = 0.06124s$ Ω; $z_{22} = 0.6(s + 1000)$ Ω

8-15.

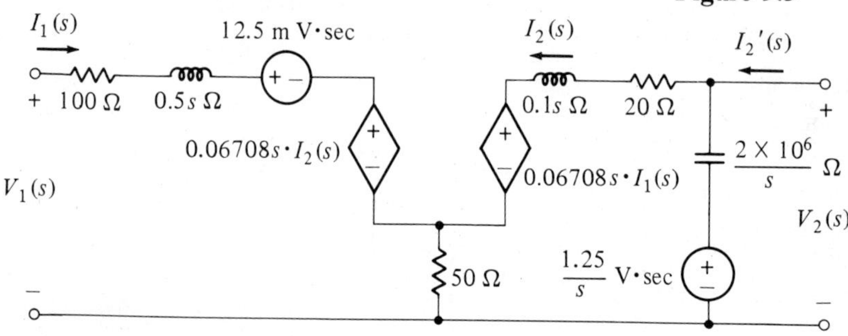

Figure 8.15

8-16.

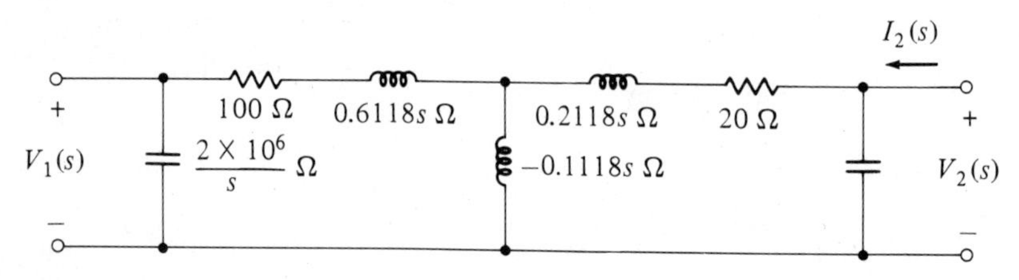

Figure 8.16

8-17. $-\dfrac{0.1225(s^2 - 408.2 \times 10^6)}{s(s^2 + 5000s + 50 \times 10^6)}$

8-18. 0.8422 H; 0.8422 H; 1.558 H

Chapter 9

9-1. **a.** $32\angle -138°$ V

c. $2.7\angle -64°$ mA

e. $636.4\angle 107°$ μA

9-2. **b.** $12\cos[4021t - 42°]$ V

d. $8 \cdot \cos[10^6 t + 36°]$ μA

9-3. **a.** $11.18\angle 66.57°$ A

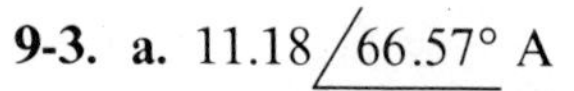

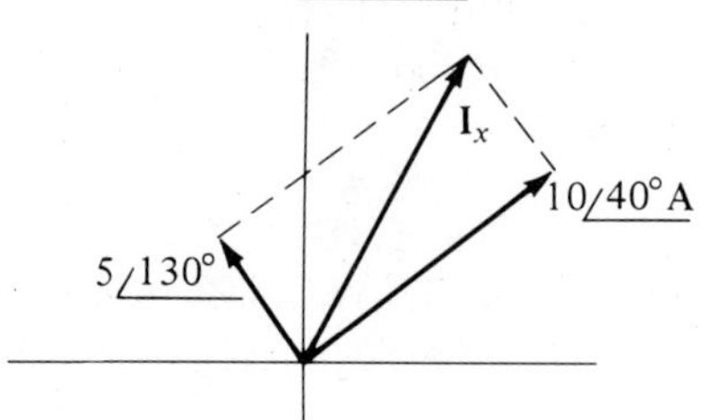

Figure 9.3

9-4. **b.** $1.202 - j0.2981$ mS

d. $73.53 - j405.9$ Ω

9-5. **a.** $232.6 \cdot \cos[600t + 114.6°]$ mA

9-6. **b.** $4.685\cos[10 \times 10^3 t - 141.3°]$ mV

9-7. **a.** $3.124\angle -8.660°$ V; $609.8 - j987.8$ Ω; $2.691\angle 49.65°$ mA; $0.4525 + j0.7330$ mS

9-8. **b.** $R_S = G_P/(G_P^2 + B_P^2)$; $X_S = -B_P/(G_P^2 + B_P^2)$; $G_P = R_S/(R_S^2 + X_S^2)$; $B_P = -X_S/(R_S^2 + X_S^2)$

9-9. **a.** $13.45\cos[1000t - 33.97°]$ V

9-10. b. 707.1 rad/sec; 707.1 rad/sec;
20 rad/sec; 35.36

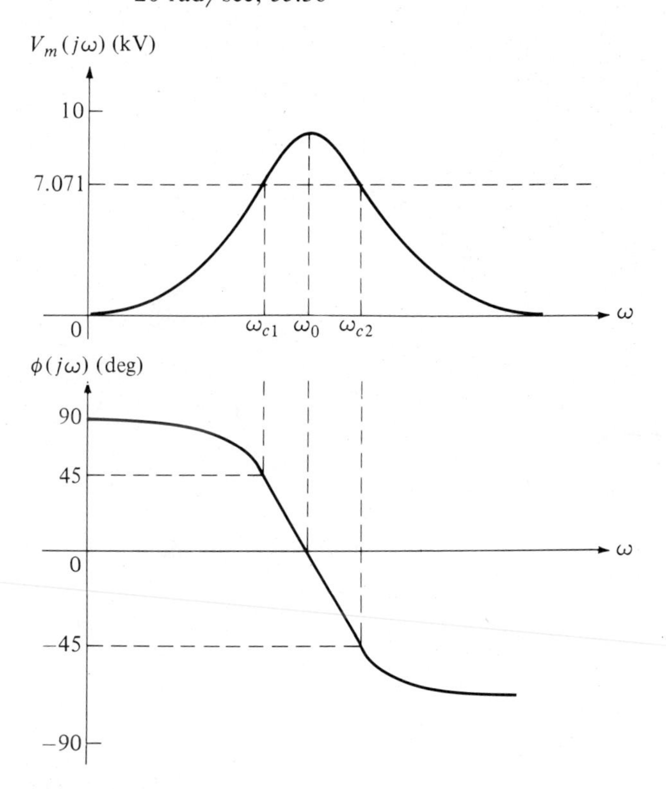

Figure 9.10b

d. 3000 rad/sec; 3169 rad/sec;
1100 rad/sec; 2.727

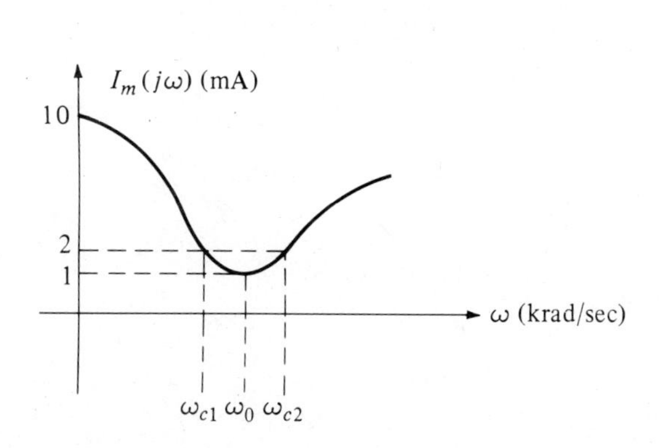

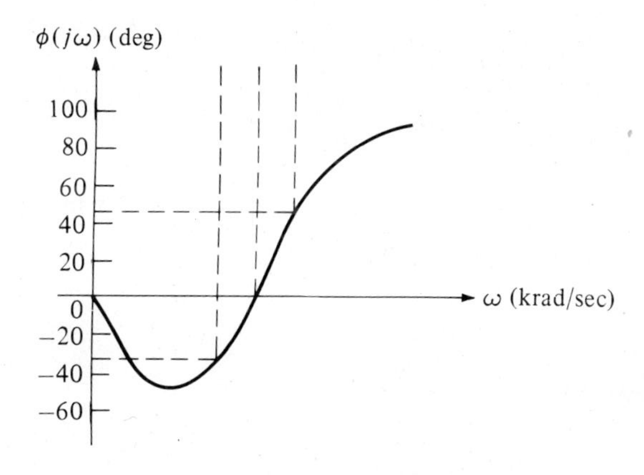

Figure 9.10d

Chapter 10

10-1. **a.** $10^9/[(10^9 - \omega^2) + j4000\omega]$

c. $-j3333\omega/[(333.3 + j\omega)(200 + j\omega)]$

10-2. **b.**

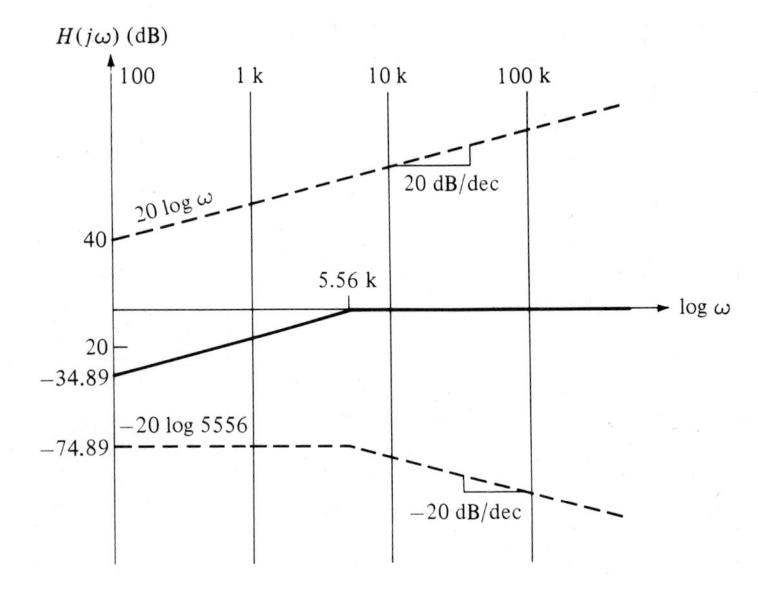

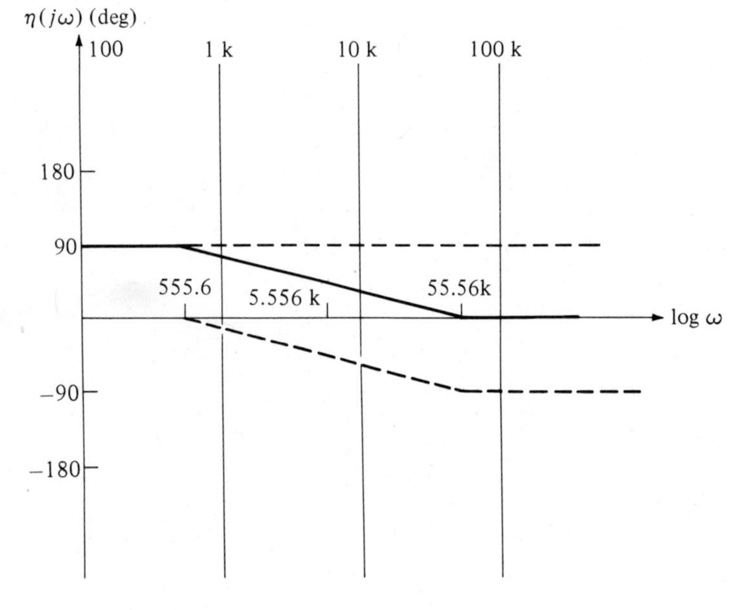

Figure 10.2

10-4. a.

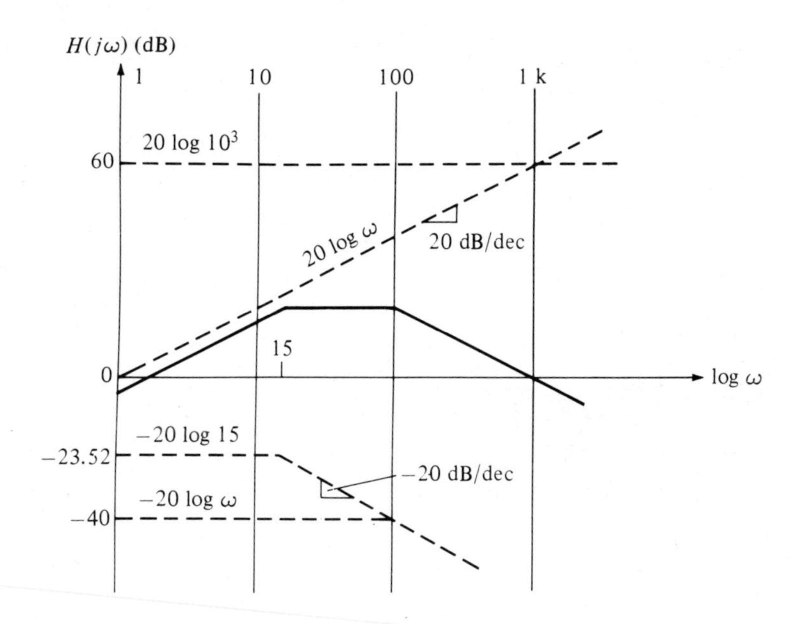

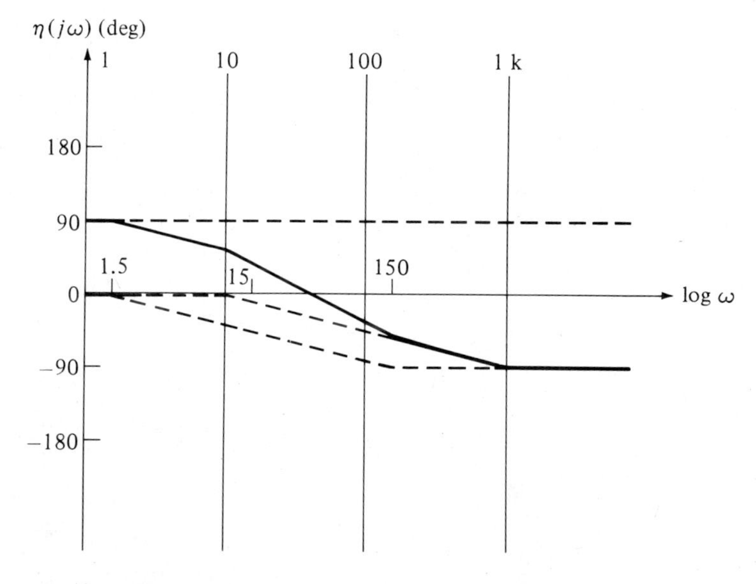

Figure 10.4a

c.

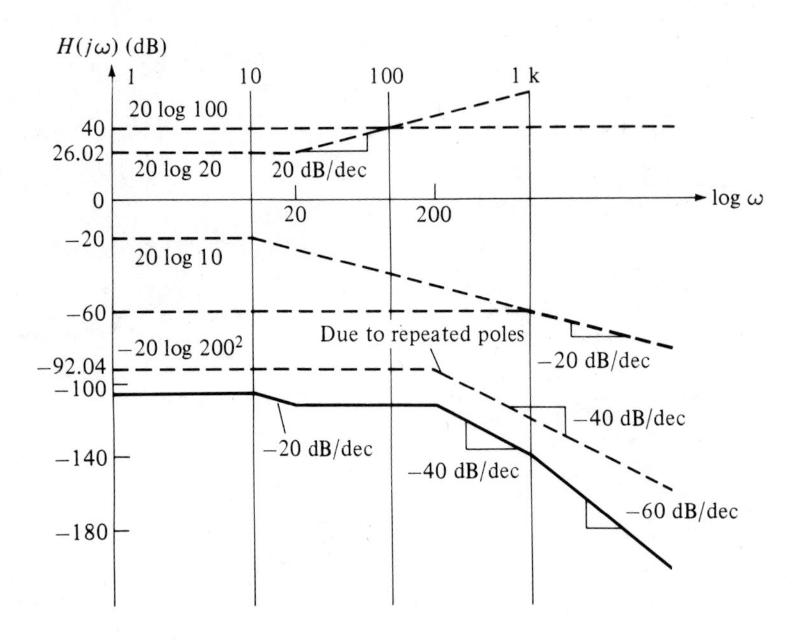

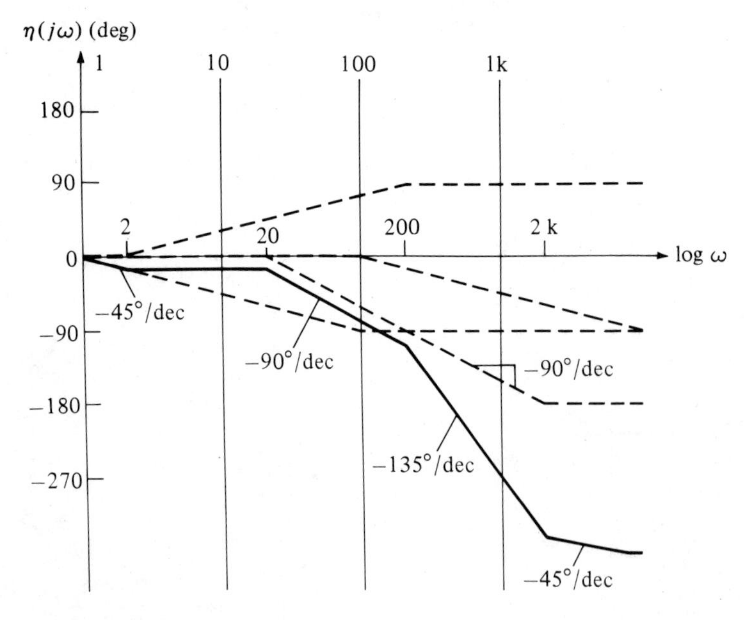

Figure 10.4c

e.

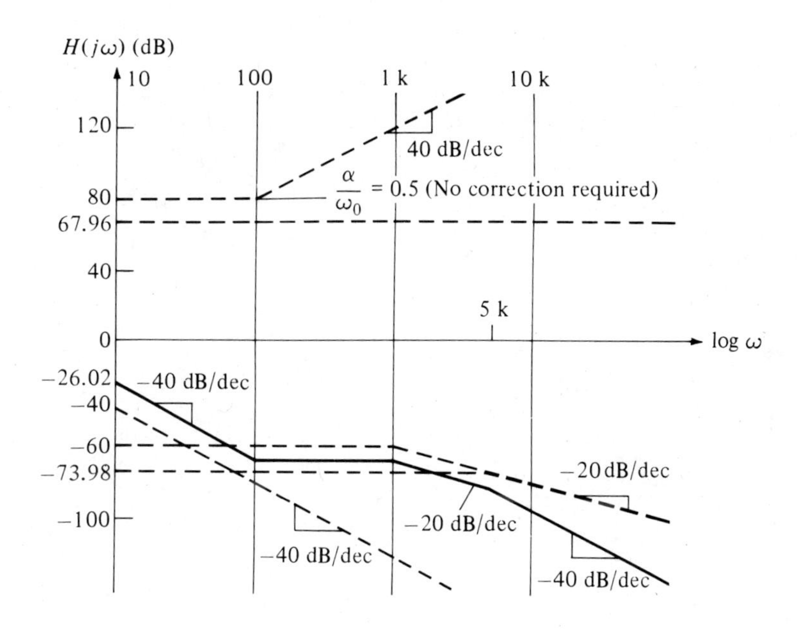

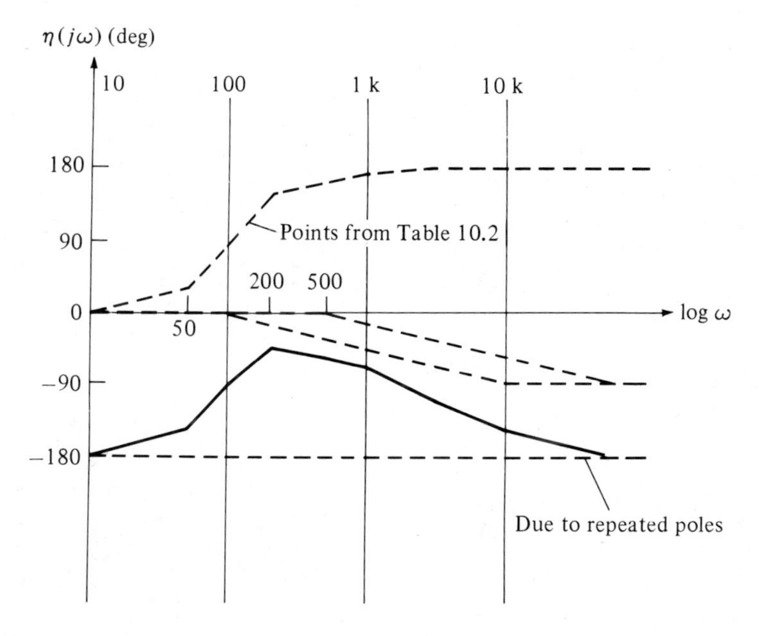

Figure 10.4e

10-6.

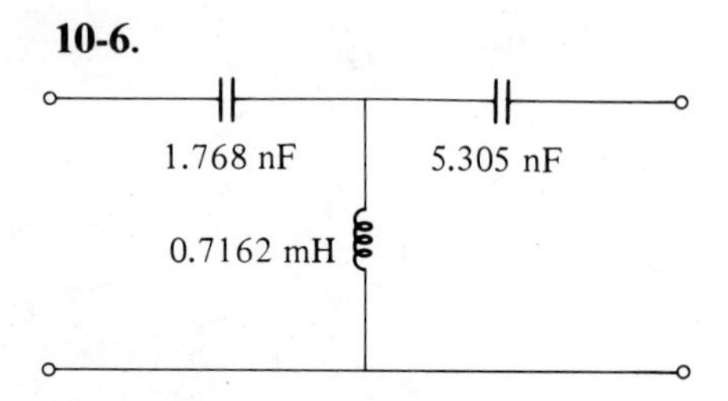

Figure 10.6

10-7.

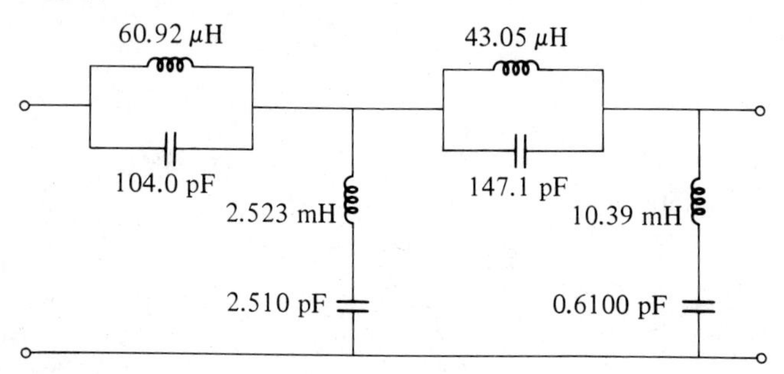

Figure 10.7

10-8.

$$\frac{(j\omega)^3}{(628300 + j\omega)\big([394.8 \times 10^9 - \omega^2] + j628.3 \times 10^3\omega\big)}$$

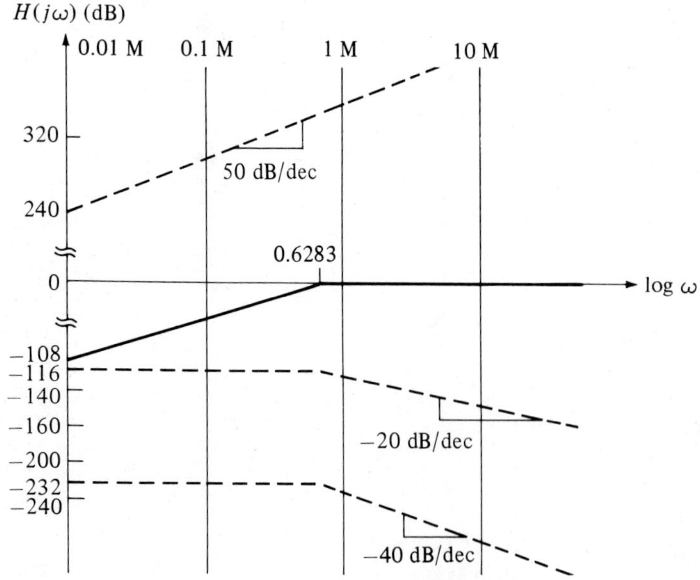

$H(j\omega)$ (dB)
0.01 M
0.1 M
1 M
10 M
320
240
50 dB/dec
0.6283
0
log ω
−108
−116
−140
−160
−200
−232
−240
−20 dB/dec
−40 dB/dec

Figure 10.8

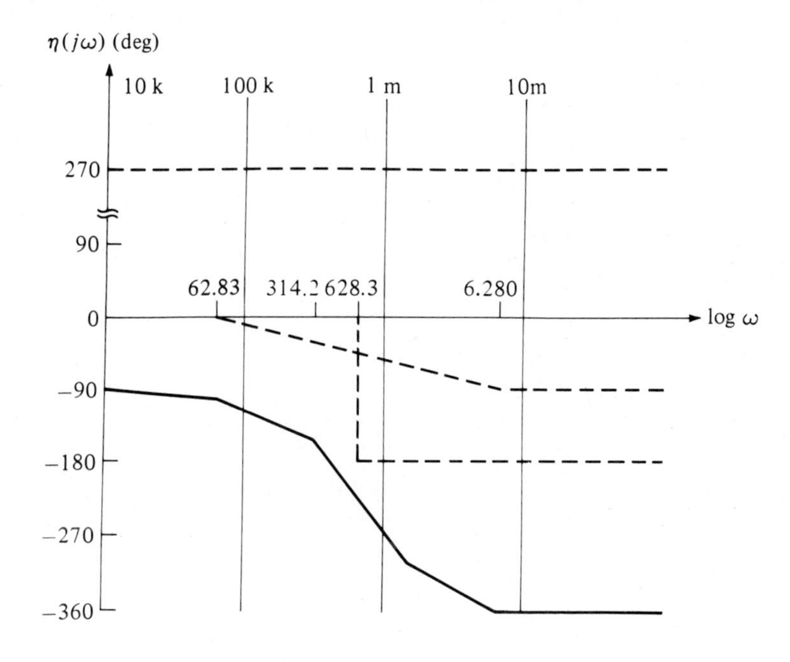

10-11. $j0.6123\omega/(2000 + j\omega)$

10-12. $\dfrac{0.1225(\omega^2 + 408.2 \times 10^6)}{j\omega\left[(50 \times 10^6 - \omega^2) + j5000\omega\right]}$

10-13.

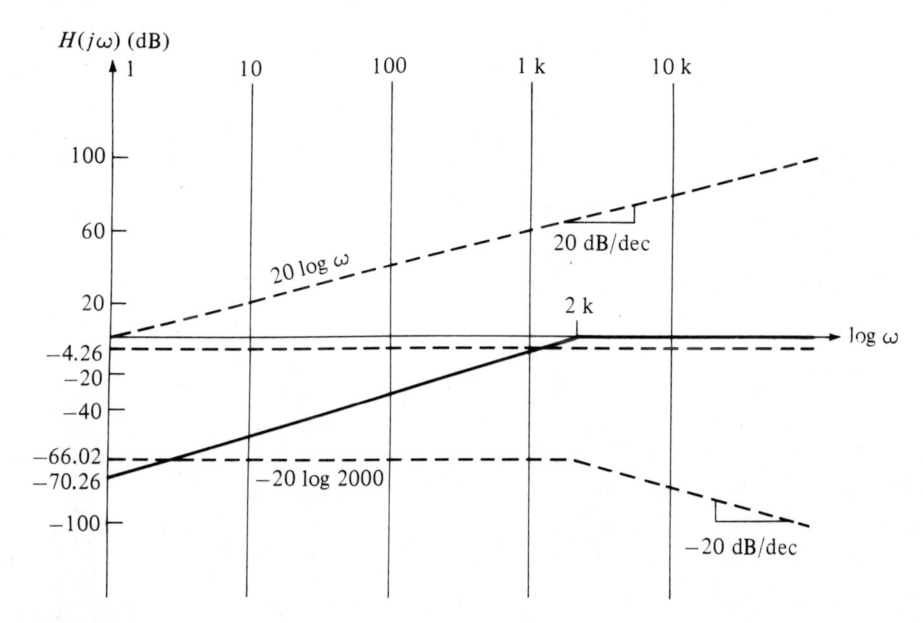

Figure 10.13

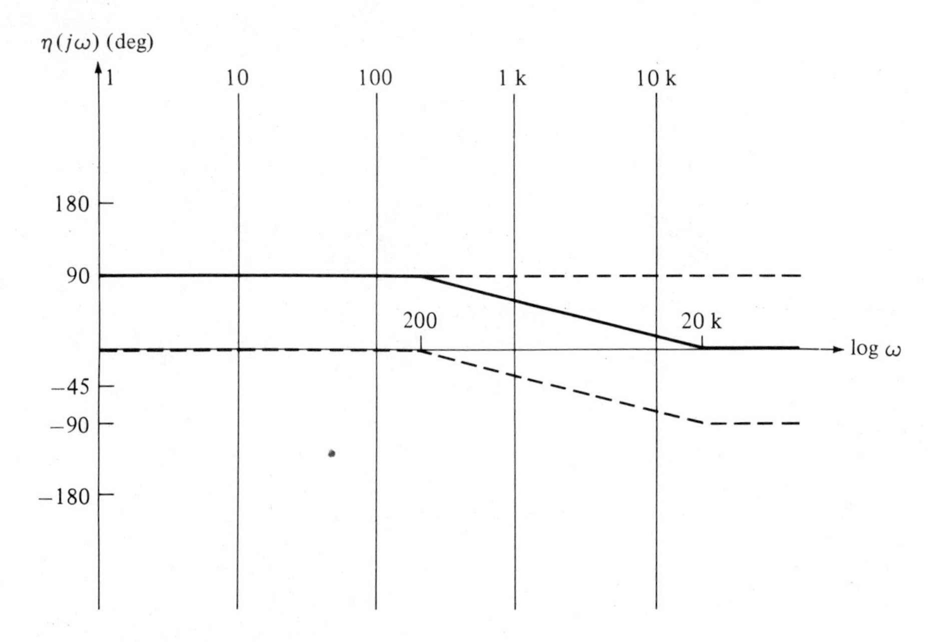

Figure 10.8 continued

10-14.

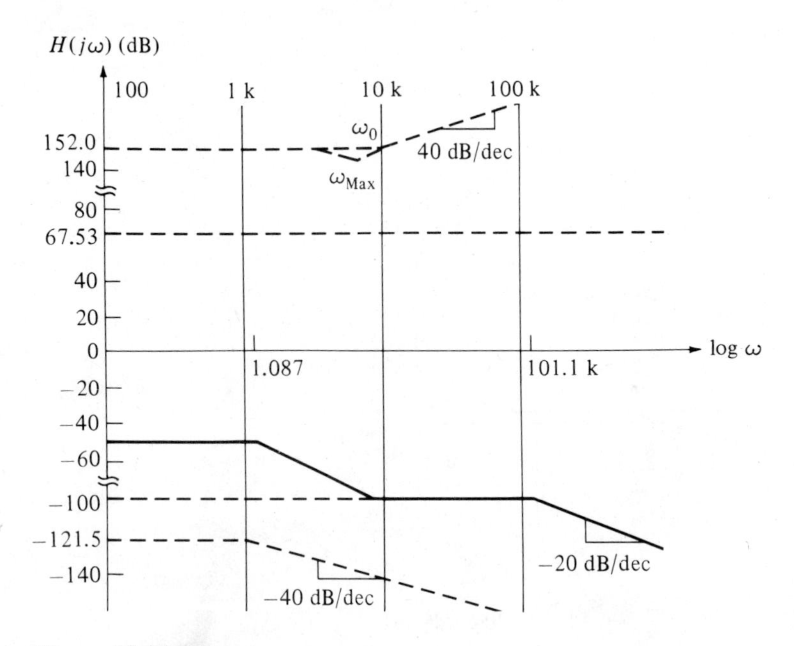

Figure 10.14

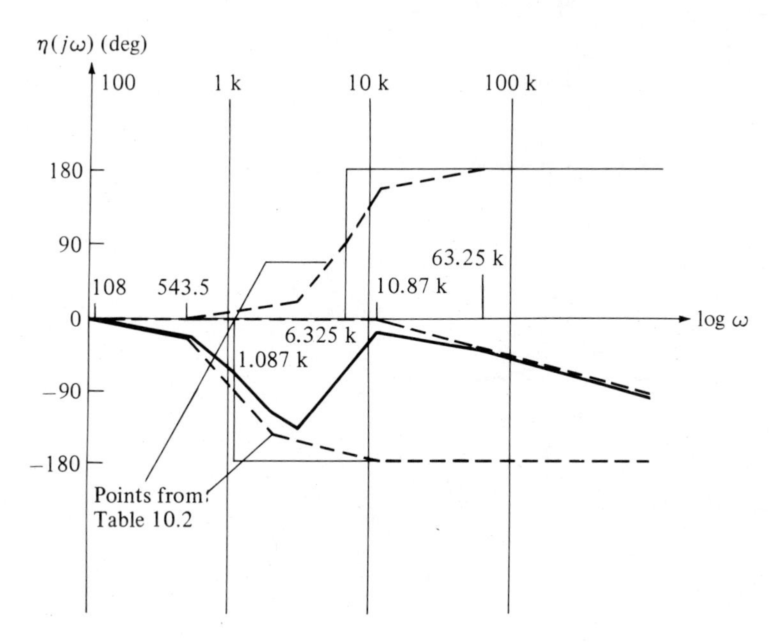

Figure 10.14 continued

10-15.

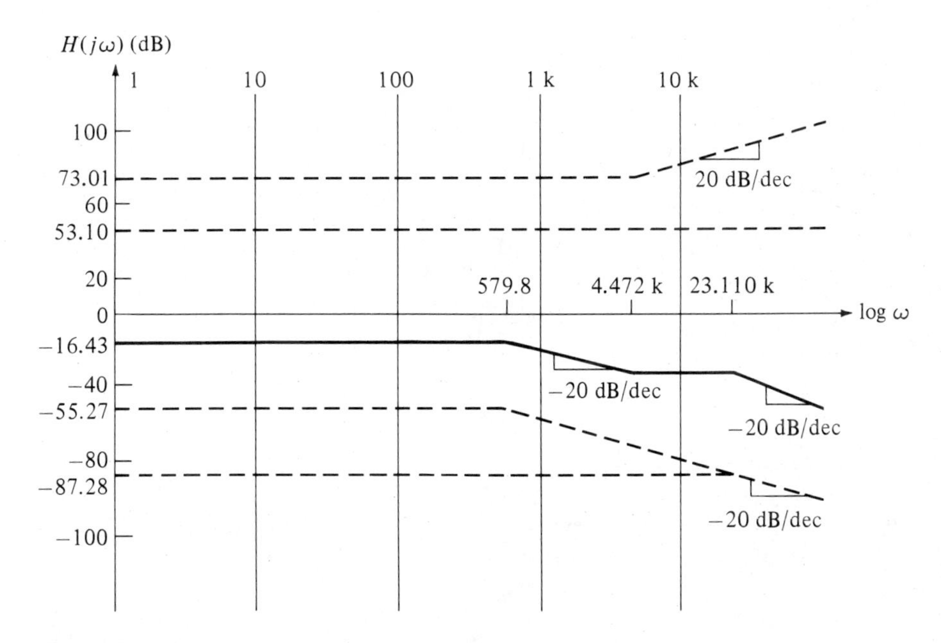

Figure 10.15

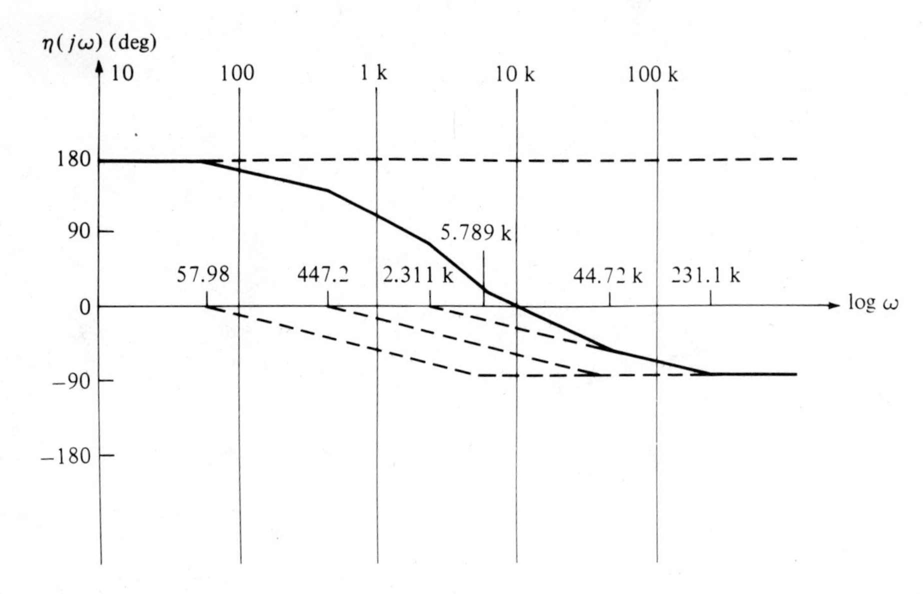

Figure 10.15 continued

10-16. 0.1367 A; 356.4 V; 3.969 A; 1.800 A; 43.46 kΩ

10-17. **a.** 2.354 A

10-18.

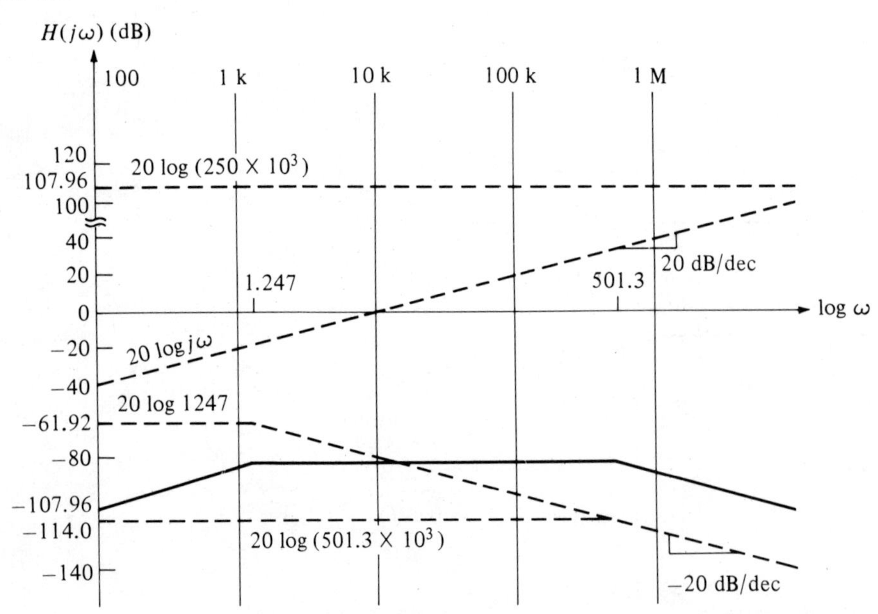

Figure 10.18

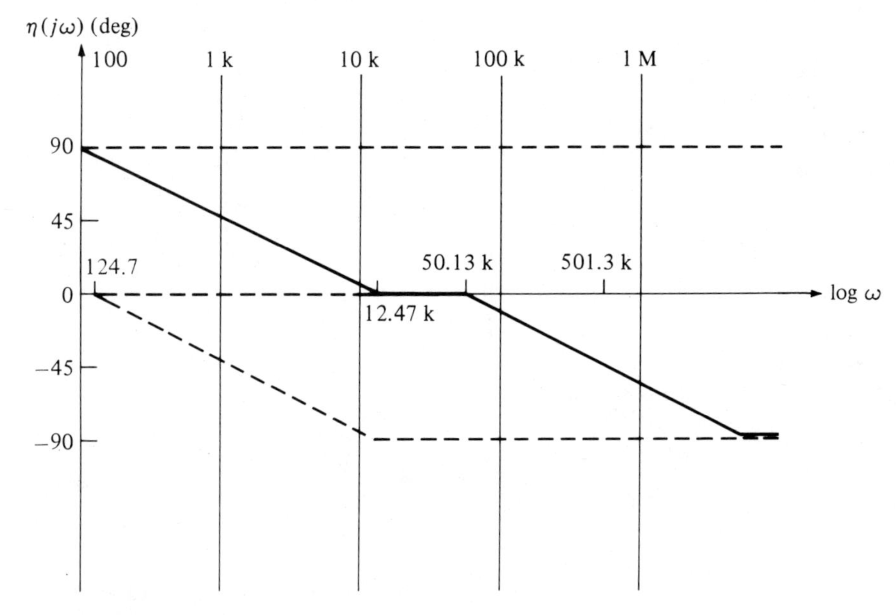

Figure 10.18 continued

10-19. 0.3062

Chapter 11

11-1. a. $a_0 = a_n = 0$; $b_n = 5/(n\pi)$

c. $a_0 = 0$; $a_n = [0.8/(n\pi)]\sin[3n\pi/4]$

11-2. b. $I_0 = 1.192$ mA; $(I_m)_n = \sqrt{568.3 + 90.92(n\pi)^2/[10 + 1.6(n\pi)^2]}$

$\theta_n = -\tan^{-1}[0.4n\pi]$

d. $V_0 = 5.955$ V; $(V_m)_1 = 1.5$ V; $(V_m)_n = 0$, n odd $(n > 1)$; $(V_m) = 6/[(n^2 - 1)\pi]$ V, n even; $\theta_1 = 90°$; θ not defined, n odd $(n > 1)$; $\theta_n = 180°$, n even

11-3. $(V_m)_4 = 0$; $(V_m)_5 = 0.4502$ V; $(V_m)_6 = 0.5305$ V; θ_4 not defined; $\theta_5 = -45°$; $\theta_5 = -90°$

11-4. $A_0 = A/2$; $(A_m)_n = 2A/(n\pi)$, n odd; $(A_m)_n = 0$, n even; $\theta_n = -90°$, n odd; θ_n not defined, n even

11-8. b. $\omega_1 = 125.6637$ krad/sec

	n	$(I_m)_n$ mA	DEG
DC	0	1.1919	UNDEF.
FUND.	1	1.4844	−51.49
2ND HARM.	2	0.8813	−68.30
3RD HARM.	3	0.6112	−75.14
4TH HARM.	4	0.4651	−78.75
5TH HARM.	5	0.3747	−80.96
6TH HARM.	6	0.3134	−82.45
7TH HARM.	7	0.2693	−83.51
8TH HARM.	8	0.2360	−84.32
9TH HARM.	9	0.2100	−84.95

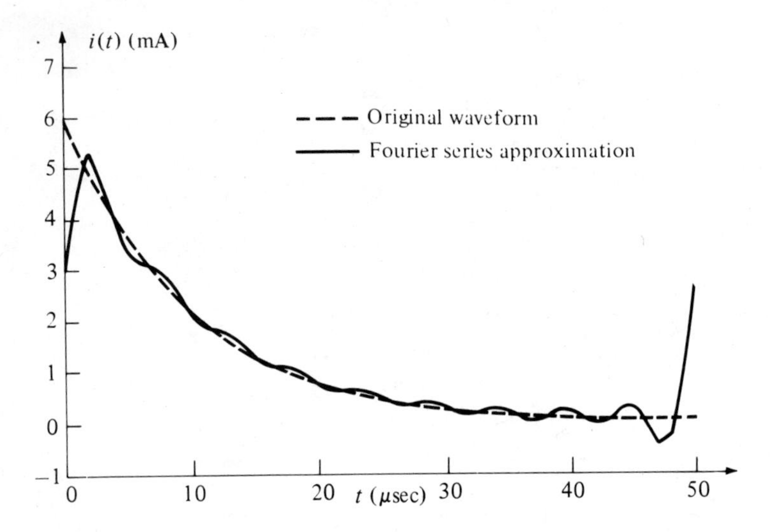

Figure 11.8b

11-8. **d.** $\omega_1 = 1570.80$ rad/sec

	n	$(I_m)_n$ A	DEG
DC	0	5.9549	UNDEF.
FUND	1	1.5000	90.00
2ND HARM.	2	0.6366	180.00
3RD HARM.	3	0.0000	UNDEF.
4TH HARM.	4	0.1273	180.00
5TH HARM.	5	0.0000	UNDEF.
6TH HARM.	6	0.0546	180.00
7TH HARM.	7	0.0000	UNDEF.
8TH HARM.	8	0.0303	180.00
9TH HARM.	9	0.0000	UNDEF.

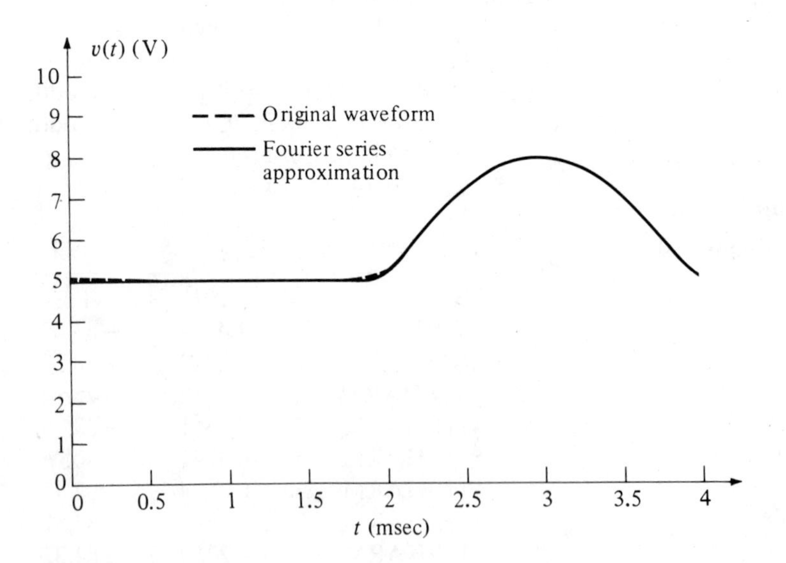

Figure 11.8d

11-9. a. $\omega_1 = 83.7758$ krad/sec

	n	$(V_m)_n$ V	DEG	$Z(jn\omega_1)$ kΩ	DEG	$(I_m)_n$ mA	DEG
DC	0	0.9000	UNDEF.	∞	UNDEF.	0.0000	UNDEF.
FUND.	1	1.6839	−36.00	41.0261	−75.8922	0.0410	39.89
2ND HARM.	2	1.3623	−72.00	22.2662	−63.3134	0.0612	−8.69
3RD HARM.	3	0.9082	−108.00	16.6104	−52.9844	0.0547	−55.02
4TH HARM.	4	0.4210	−144.00	14.1048	−44.8483	0.0298	−99.15
5TH HARM.	5	0.0000	UNDEF.	12.7799	−38.5119	0.0000	UNDEF.

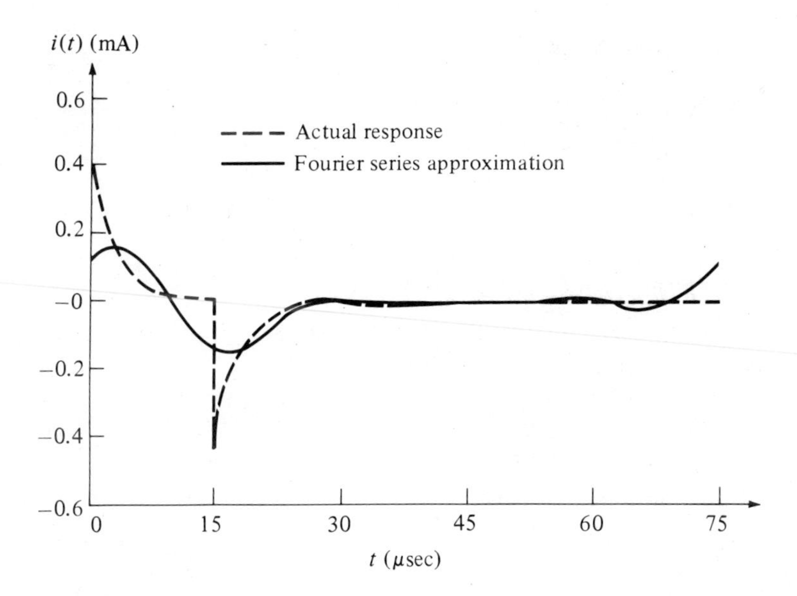

Figure 11.9a

11-9. c. $\omega_1 = 83.7758$ krad/sec

	n	$(V_m)_n$ V	DEG	$Y(jn\omega_1)$ μS	DEG	$(I_m)_n$ μA	DEG
DC	0	0.9000	UNDEF.	142.8571	UNDEF.	128.5714	UNDEF.
FUND.	1	1.6839	−36.00	138.9326	−13.4610	233.9460	−49.46
2ND HARM.	2	1.3623	−72.00	128.8533	−25.5813	175.5354	−97.58
3RD HARM.	3	0.9082	−108.00	116.0391	−35.6813	105.3858	−143.68
4TH HARM.	4	0.4210	−144.00	103.1873	−43.7544	43.4388	−187.75
5TH HARM.	5	0.0000	UNDEF.	91.5991	−50.1191	0.0000	UNDEF.

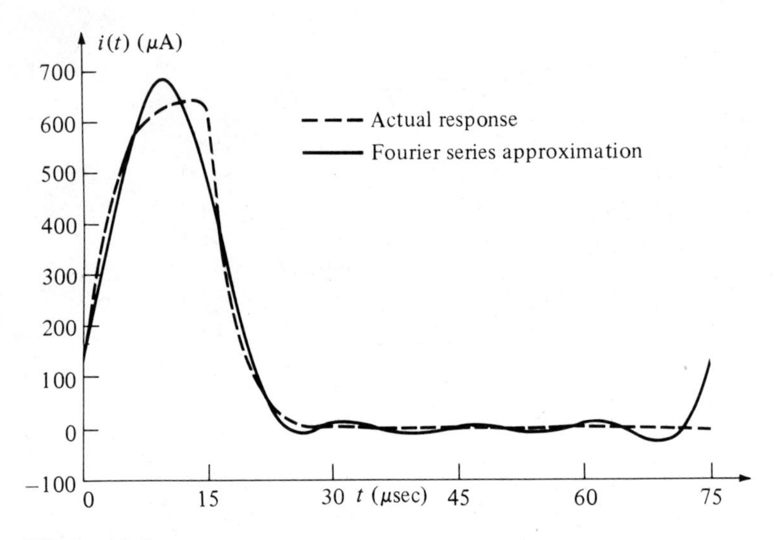

Figure 11.9c

11-10. b. $\omega_1 = 1571$ rad/sec

	n	$(I_m)_n$ μA	DEG	$Y(jn\omega_1)$ mS	DEG	$(V_m)_n$ mV	DEG
DC	0	0.0000	UNDEF.	10.0000	0.00	0.0000	UNDEF.
FUND.	1	63.6620	−90.00	11.0164	30.68	5.7788	59.32
2ND HARM.	2	0.0000	UNDEF.	14.4023	55.38	0.0000	UNDEF.
3RD HARM.	3	21.2207	−90.00	19.9932	70.52	1.0614	19.48
4TH HARM.	4	0.0000	UNDEF.	26.9501	78.67	0.0000	UNDEF.
5TH HARM.	5	12.7324	−90.00	34.5908	83.05	0.3681	6.95

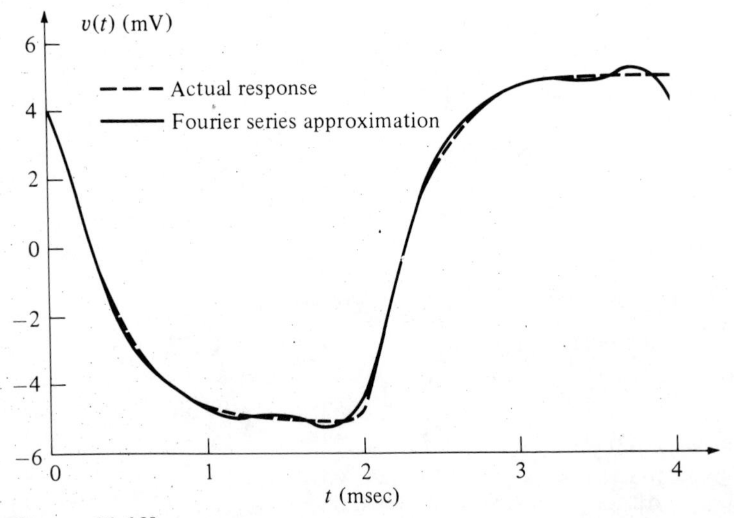

Figure 11.10b

11-10. d. ω_1 = 1571 rad/sec

	n	$(I_m)_n$ μA	DEG	$Y(jn\omega_1)$ mS	DEG	$(V_m)_n$ mV	DEG
DC	0	0.0000	UNDEF.	0.2000	0.00	0.0000	UNDEF.
FUND	1	63.6620	−90.00	0.2149	21.44	296.2833	−111.439
2ND HARM.	2	0.0000	UNDEF.	0.2543	38.15	0.0000	UNDEF.
3RD HARM.	3	21.2207	−90.00	0.3091	49.67	68.6625	−139.674
4TH HARM.	4	0.0000	UNDEF.	0.3724	57.52	0.0000	UNDEF.
5TH HARM.	5	12.7324	−90.00	0.4407	63.01	28.8916	−153.010

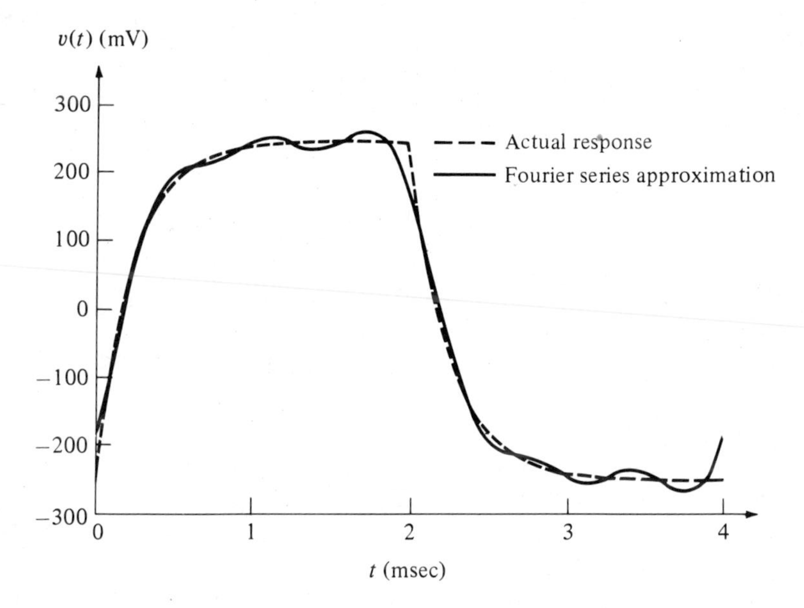

Figure 11.10d

11-11. a. ω_1 = 78.54 krad/sec

	n	INPUT $(V_m)_n$ V	DEG	$H(jn\omega_1)$	DEG	OUTPUT $(V_m)_n$ V	DEG
DC	0	−6.0000	UNDEF.	0.0000	90.00	0.0000	UNDEF.
FUND.	1	−7.6394	−90.00	0.2293	76.74	−1.7520	−13.2581
2ND HARM.	2	0.0000	UNDEF.	0.4263	64.77	0.0000	UNDEF.
3RD HARM.	3	−2.5465	−90.00	0.5772	54.75	−1.4699	−35.2548
4TH HARM.	4	0.0000	UNDEF.	0.6859	46.70	0.0000	UNDEF.
5TH HARM.	5	−1.5279	−90.00	0.7624	40.33	−1.1648	−49.6745

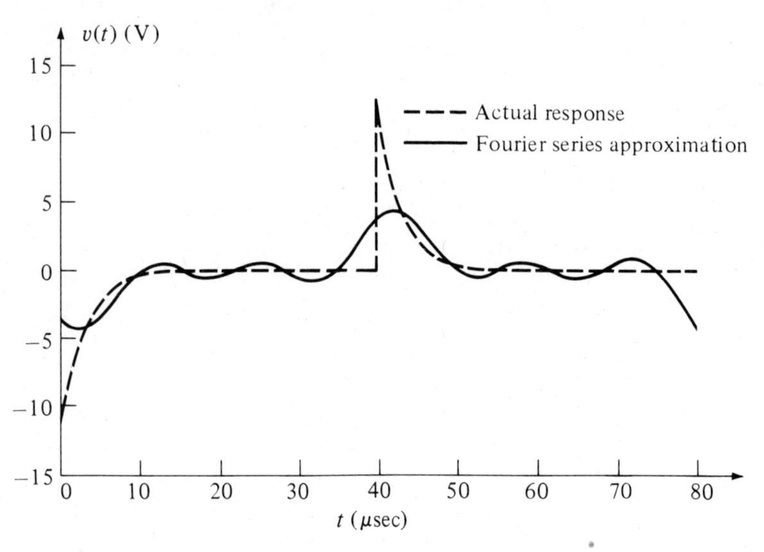

Figure 11.11a

11-11. c. $\omega_1 = 78.54$ krad/sec

	n	INPUT $(V_m)_n$ V	DEG	$H(jn\omega_1)$	DEG	OUTPUT $(V_m)_n$ V	DEG
DC	0	−6.0000	UNDEF.	0.2857	0.00	−1.7143	UNDEF.
FUND.	1	−7.6394	−90.00	0.2708	−18.60	−2.0687	−108.603
2ND HARM.	2	0.0000	UNDEF.	0.2370	−33.95	0.0000	UNDEF.
3RD HARM.	3	−2.5465	−90.00	0.2010	−45.28	−0.5120	−135.279
4TH HARM.	4	0.0000	UNDEF.	0.1704	−53.40	0.0000	UNDEF.
5TH HARM.	5	−1.5279	−90.00	0.1459	−59.28	−0.2230	−149.282

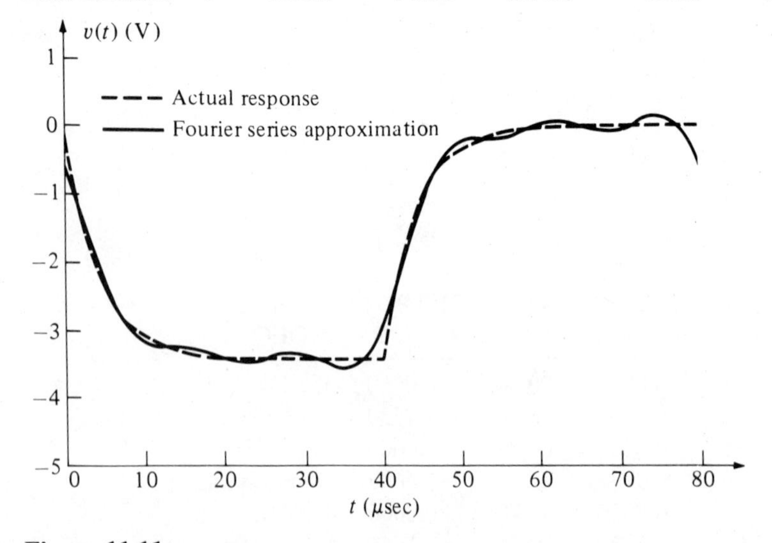

Figure 11.11c

11-12. $(V_m I_m/2) \cdot \cos[\theta - \phi]$

11-13. a.

	n	$(V_m)_n$ V	$Z(jn\omega_1)$ kΩ	DEG	$(I_m)_n$ μA	$(P)_n$ μW	ΣP μW
DC	0	0.9000	∞	UNDEF.	0.0000	0.0000	0.0000
FUND.	1	1.6839	41.0261	−75.89	41.0441	8.4231	8.4231
2ND HARM.	2	1.3623	22.2662	−63.31	61.1817	18.7160	27.1391
3RD HARM.	3	0.9082	16.6104	−52.98	54.6762	14.9474	42.0865
4TH HARM.	4	0.4210	14.1048	−44.85	29.8458	4.4539	46.5404
5TH HARM.	5	0.0000	12.7799	−38.51	0.0000	0.0000	46.5404

11-13. c.

	n	$(V_m)_n$ V	$Z(jn\omega_1)$ kΩ	DEG	$(I_m)_n$ mA	$(P)_n$ mW	ΣP mW
DC	0	0.9000	7.0000	UNDEF.	0.1286	0.1157	0.1157
FUND.	1	1.6839	6.8247	−5.07	0.2467	0.2069	0.3226
2ND HARM.	2	1.3623	6.4468	−8.25	0.2113	0.1424	0.4651
3RD HARM.	3	0.9082	6.0775	−9.47	0.1494	0.0669	0.5320
4TH HARM.	4	0.4210	5.7950	−9.52	0.0726	0.0151	0.5471
5TH HARM.	5	0.0000	5.5950	−9.05	0.0000	0.0000	0.5471

11-14. b.

	n	$(I_m)_n$ μA	$Y(jn\omega_1)$ mS	DEG	$(V_m)_n$ mV	P_n nW	ΣP nW
DC	0	0.0000	10.0000	0.00	0.0000	0.0000	0.0000
FUND.	1	63.6620	11.0164	30.68	5.7788	158.1926	158.1926
2ND HARM.	2	0.0000	14.4023	55.38	0.0000	0.0000	158.1926
3RD HARM.	3	21.2207	19.9932	70.52	1.0614	3.7561	161.9486
4TH HARM.	4	0.0000	26.9501	78.67	0.0000	0.0000	161.9486
5TH HARM.	5	12.7324	34.5908	83.05	0.3681	0.2837	162.2323

11-14. d.

	n	$(I_m)_n$ μA	$Z(jn\omega_1)$ kΩ	DEG	$(V_m)_n$ V	P_n μW	ΣP μW
DC	0	0.0000	∞	0.00	0.0000	0.0000	0.0000
FUND.	1	63.6620	15.0696	18.21	0.9594	29.0073	29.0073
2ND HARM.	2	0.0000	9.3228	8.91	0.0000	0.0000	29.0073
3RD HARM.	3	21.2207	7.0300	−13.69	0.1492	1.5379	30.5452
4TH HARM.	4	0.0000	5.6358	−19.33	0.0000	0.0000	30.5452
5TH HARM.	5	12.7324	4.6831	−18.85	0.0596	0.3592	30.9045

11-16. b. 1.897 mA
d. 6.066 V

11-17. a. $\omega_1 = 62.8319$ rad/sec

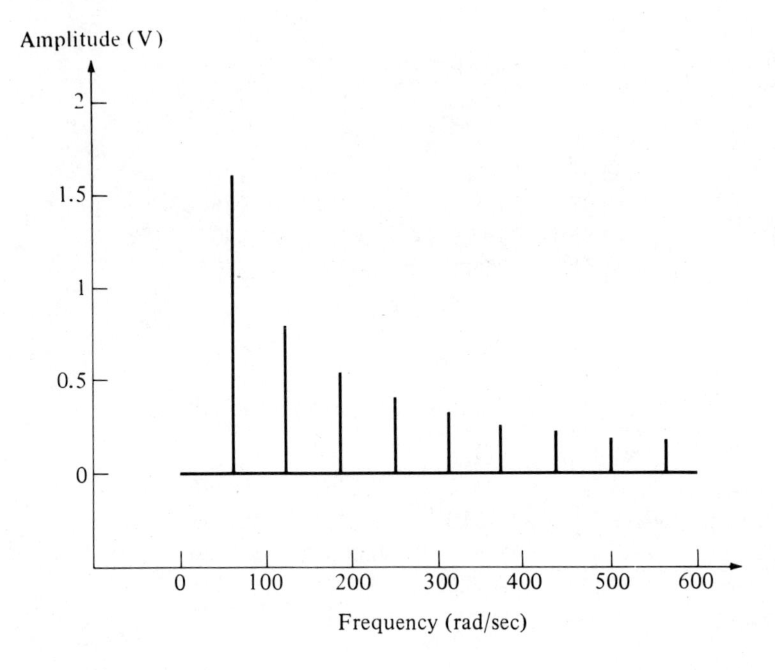

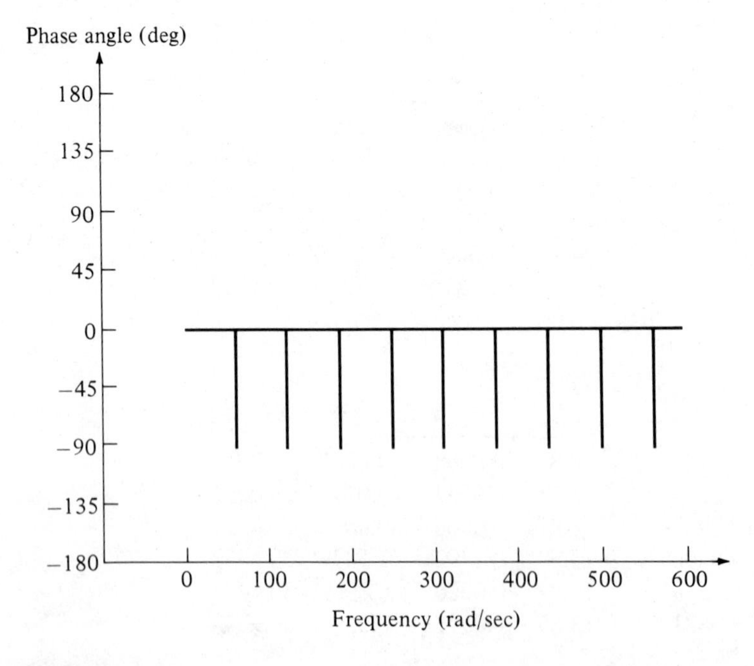

Figure 11.17a

c. $\omega_1 = 0.7854$ rad/sec

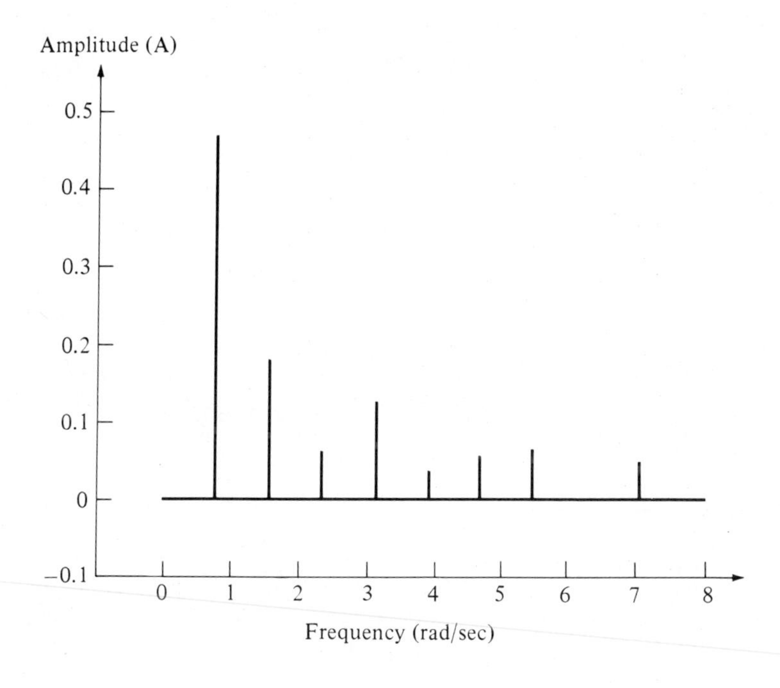

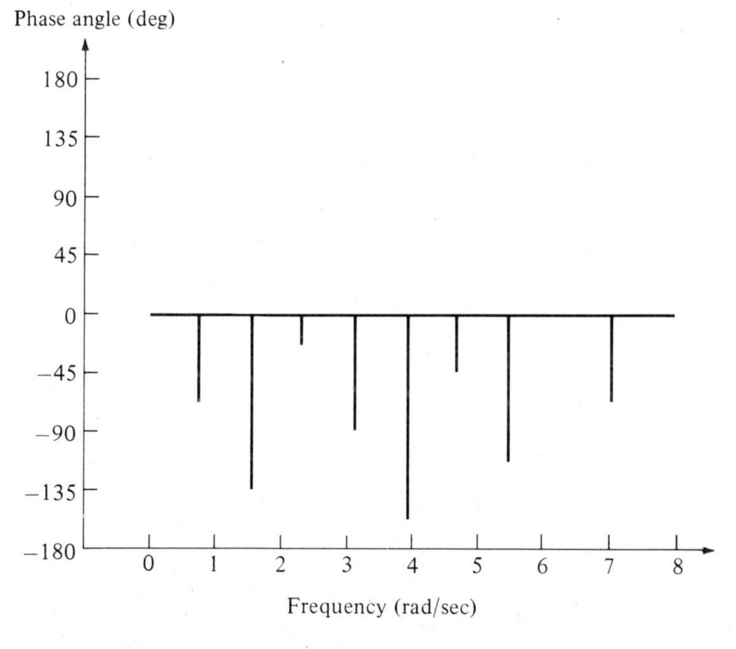

Figure 11.17c

11-18. a. Waveform b: $\omega_1 = 125.6637$ krad/sec

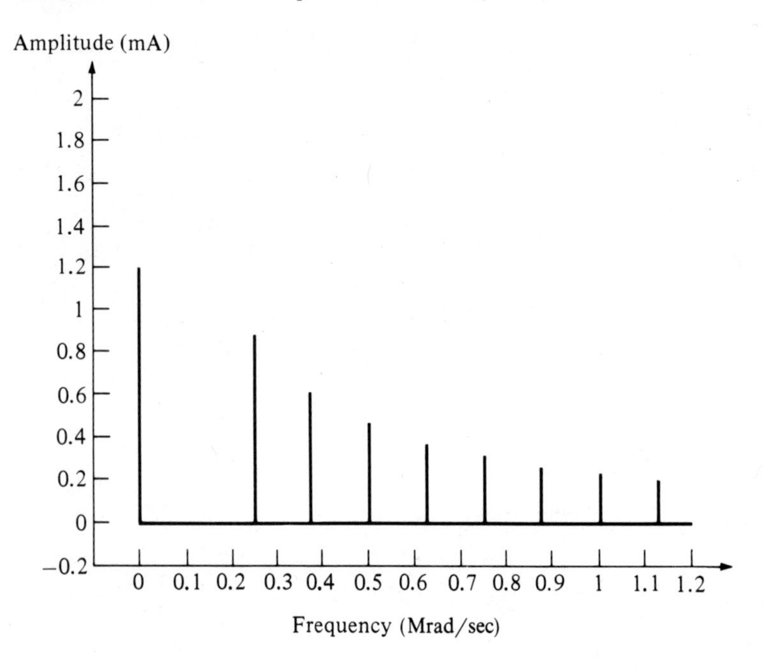

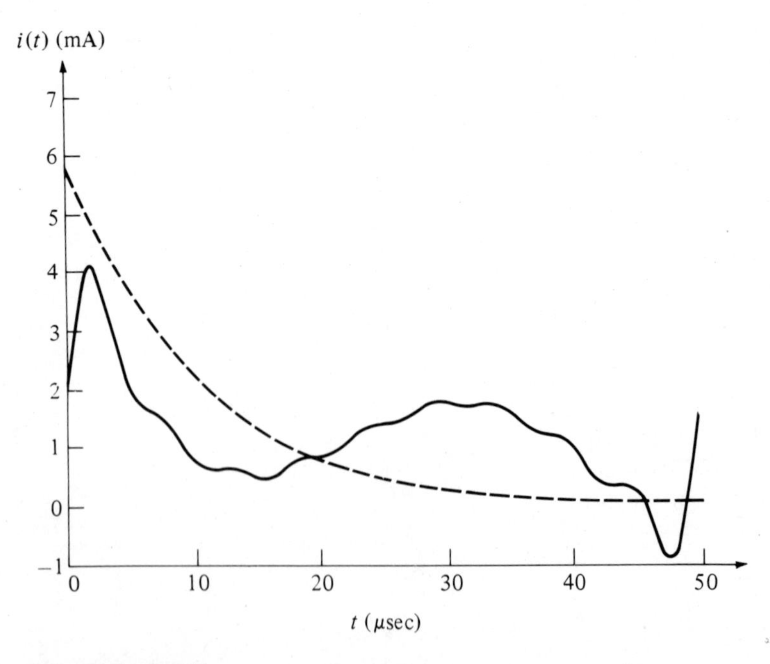

Figure 11.18a, b

Waveform d: $\omega_1 = 1570.80$ rad/sec

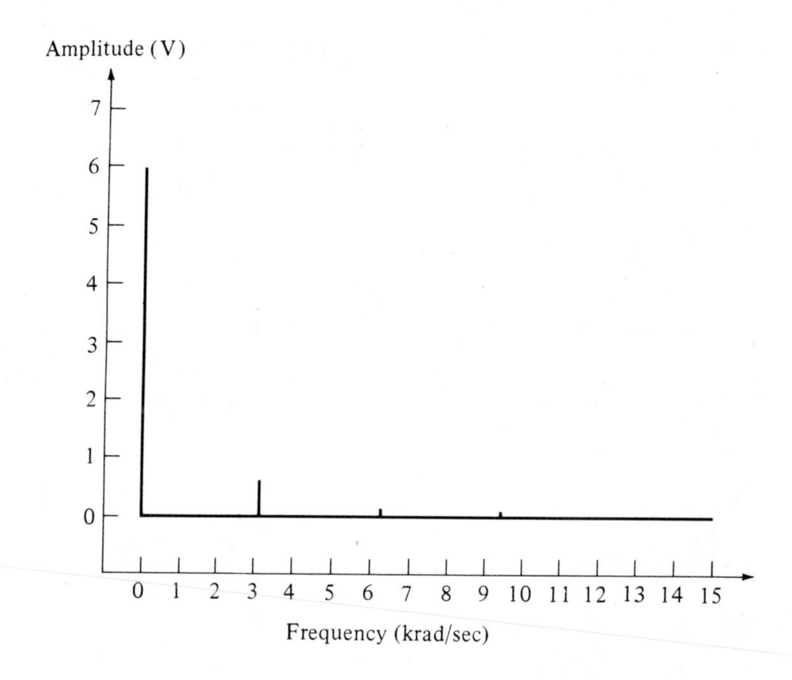

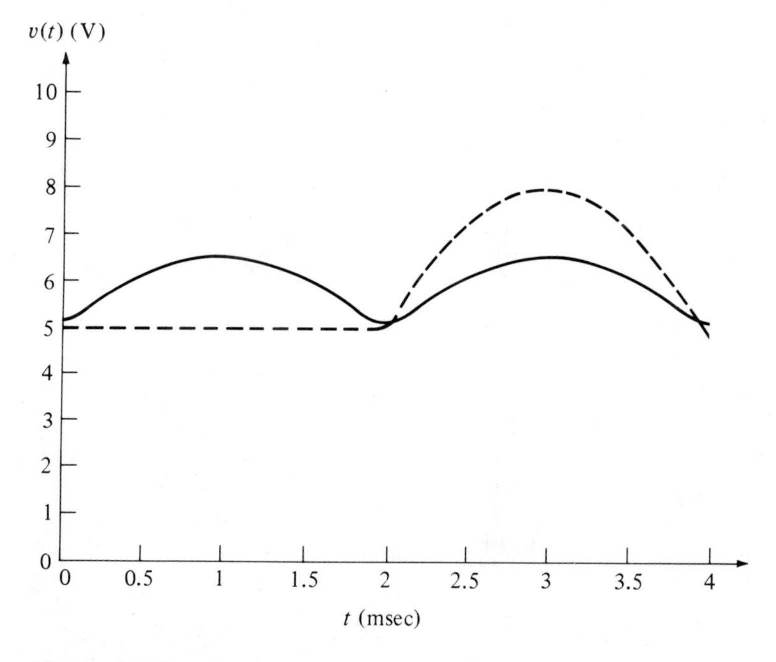

Figure 11.18a, d

b. Waveform b: $\omega_1 = 125.6637$ krad/sec

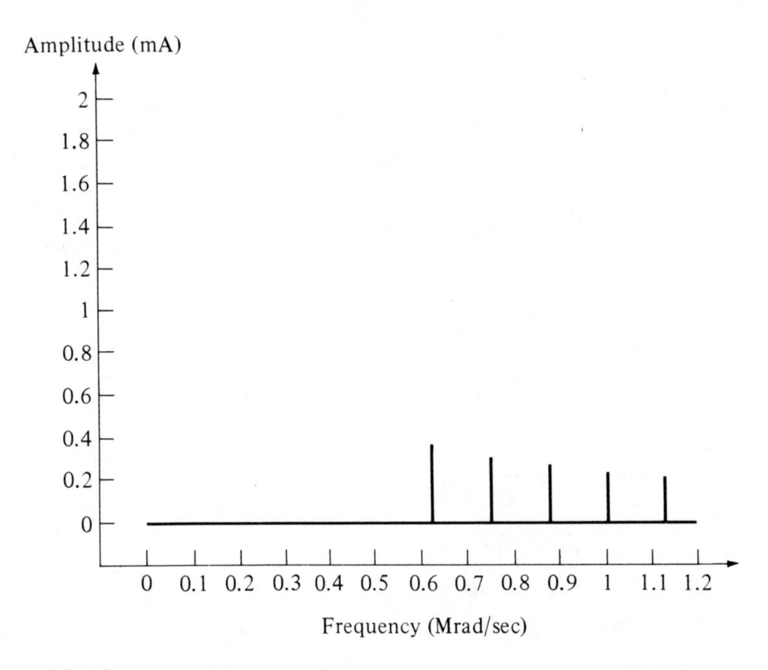

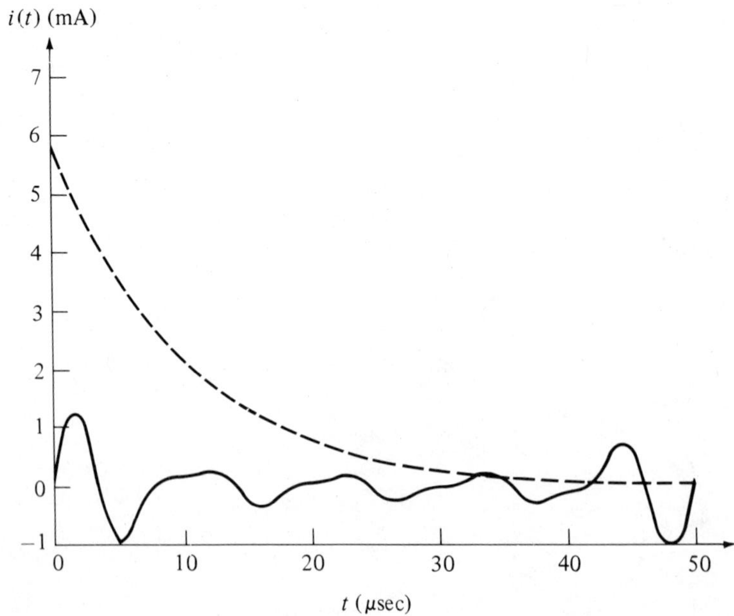

Figure 11.18b, d

Waveform d: $\omega_1 = 1570.80$ rad/sec

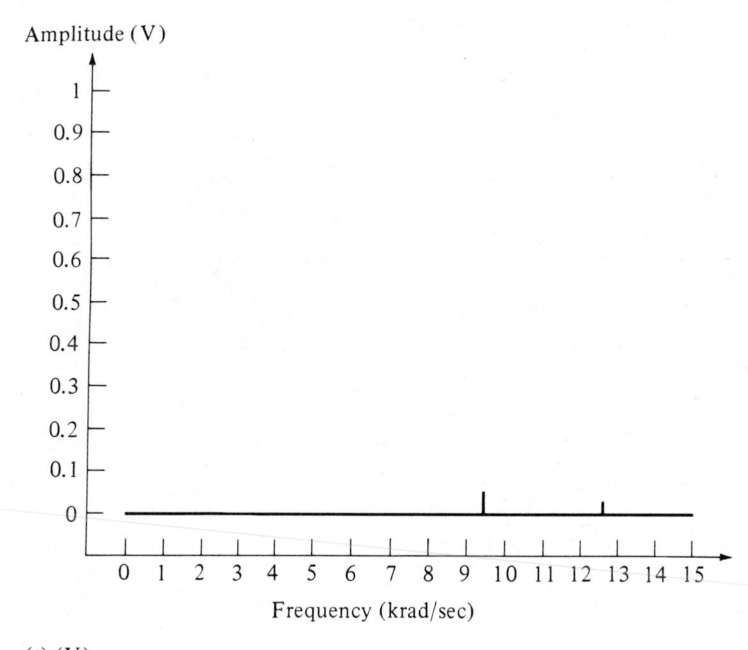

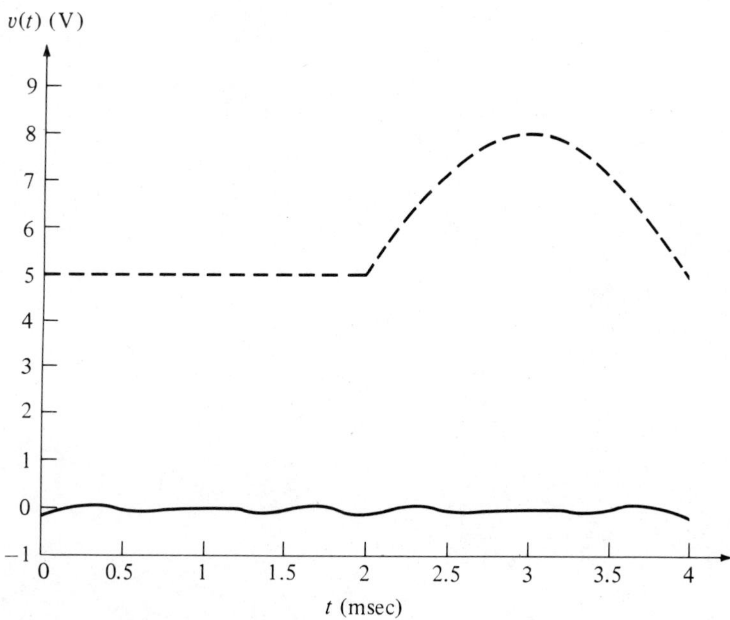

Figure 11.18b, d

c. Waveform b: $\omega_1 = 125.6637$ krad/sec

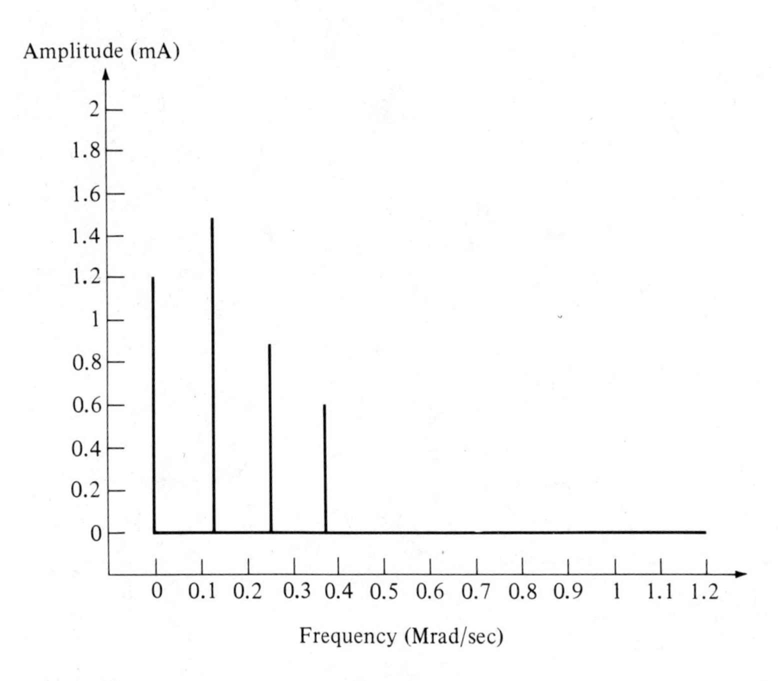

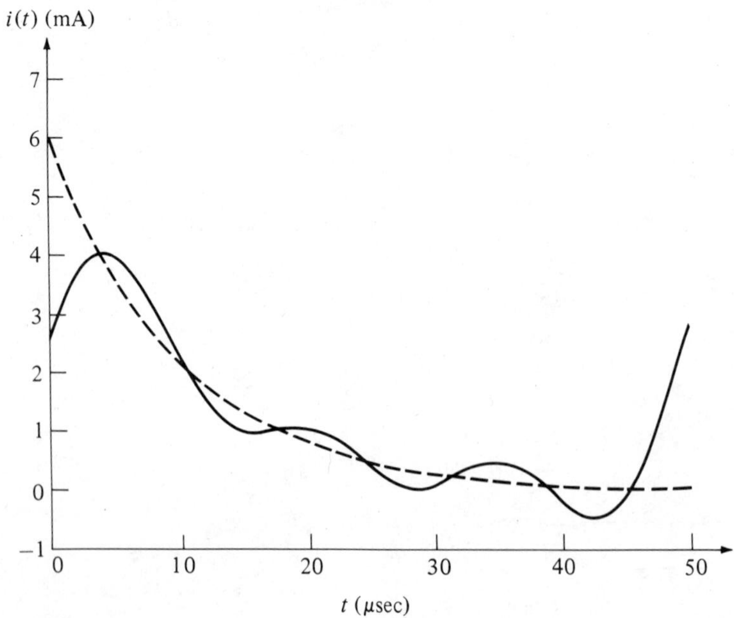

Figure 11.18c, d

Waveform d: $\omega_1 = 1570.80$ rad/sec

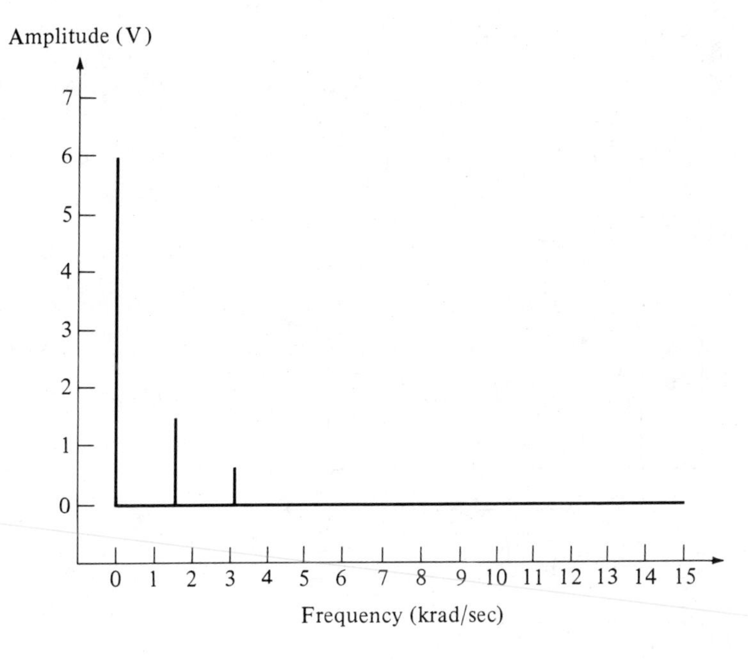

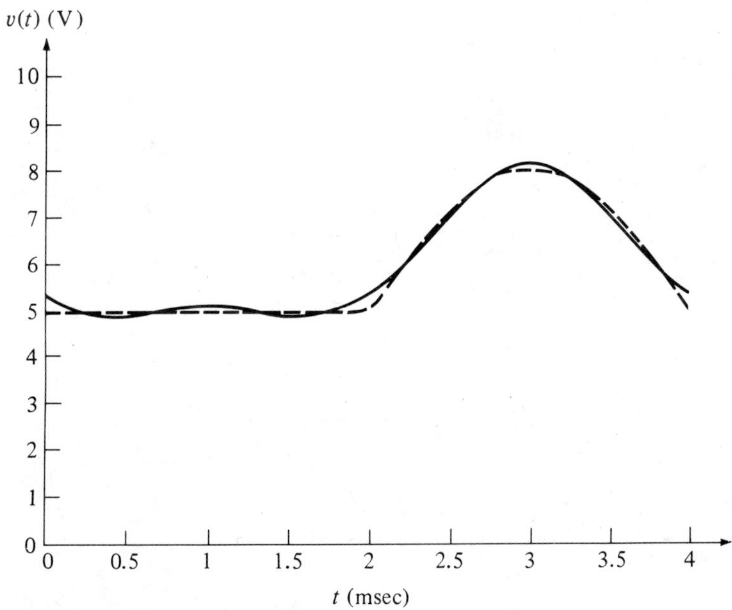

Figure 11.18c, b

d. Waveform b: $\omega_1 = 125.6637$ krad/sec

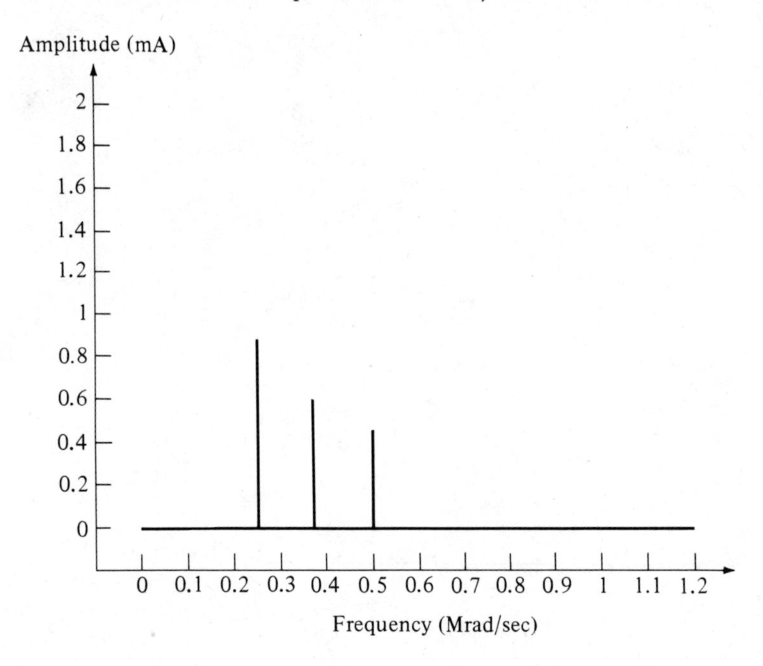

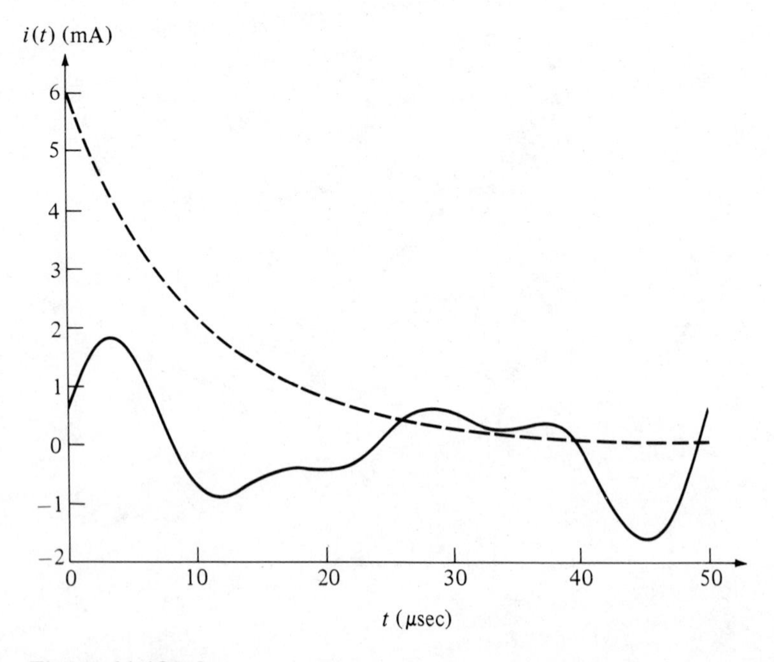

Figure 11.18d, b

Waveform d: $\omega_1 = 1570.80$ rad/sec

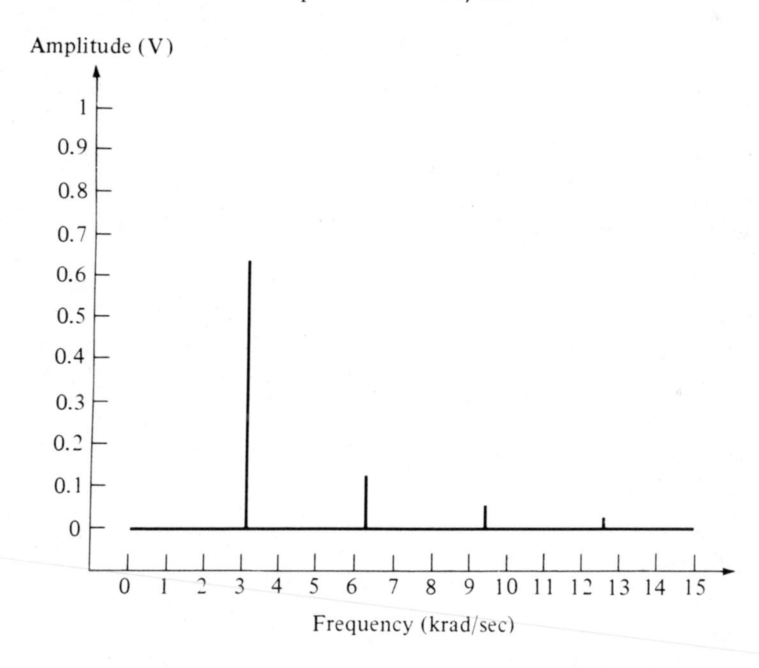

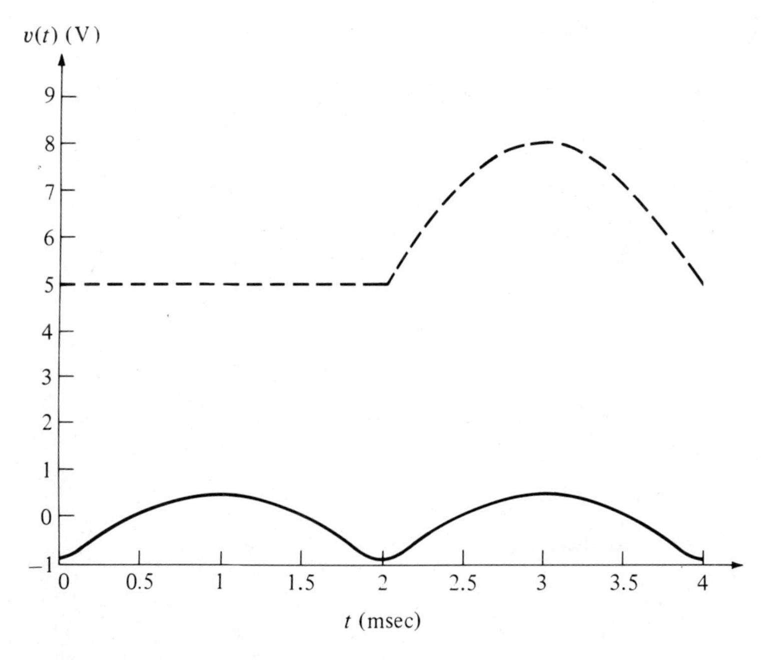

Figure 11.18d, d

11-21. 6.25 W
11-22. $36/\pi$ mW
11-23. 3.466 mW

Chapter 12

12-6. b. $4.7216z/(z^2 - 1.6065z + 0.6065)$
d. $z(z - 0.8654)/(z^2 - 1.7128z + 0.7408)$
12-7. a. $(z^5 - 5z + 4)/[z^4(z-1)^2]$
c. $2.5(z^{10} - z^8 - 0.5z^4 + 0.5)/z^9[z-1]^2$
12-8. b. $0.5\{1 - 200z/(z - \exp[-200T])\}$
d. $1 - \dfrac{3172 \cdot \exp[-250T] \cdot \sin 3152T}{z^2 - 2z \cdot \exp[-250T] \cdot \cos 3152T + \exp[-500T]}$
12-9. a. $0.5(1 - \exp[-500T])/(z - \exp[-500T])$
c. $1 - \dfrac{(z-1)(z - \exp[-250T] \cdot \cos 3152T)}{z^2 - 2z \cdot \exp[-250T] \cdot \cos 3152T + \exp[-500T]} - \dfrac{\frac{250}{3152}(z-1)\exp[-250T] \cdot \sin 3152T}{z^2 - 2z \cdot \exp[-250T] \cdot \cos 3152T + \exp[-500T]}$
12-10. b. $1000t \cdot \exp[-1000t] \cdot u(t)$
12-11. a. $(4 \cdot \exp[-400t] - 1)u(t)$ V

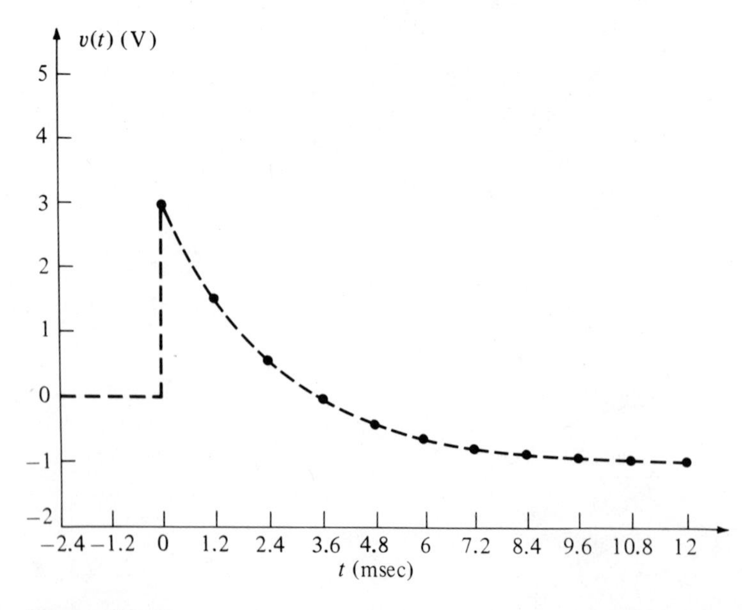

Figure 12.11a

b. $\{\exp[-2500t] + 1.2(\exp[-1250t] - \exp[-3750t])\}u(t)$

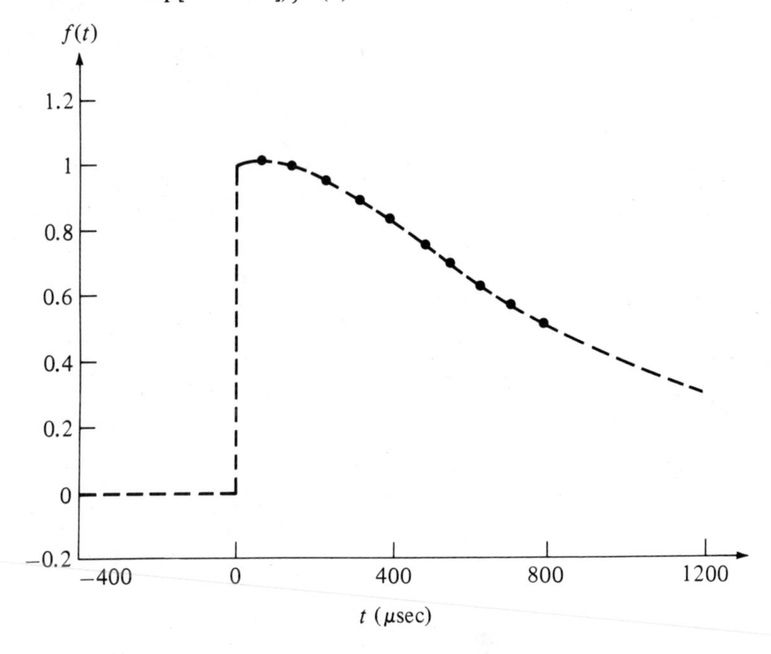

Figure 12.11b

12-13. a. $5.9442 \cdot \delta(t - T)$
$+12.1157 \cdot \delta(t - 2T)$
$+17.9606 \cdot \delta(t - 3T)$
$+23.2278 \cdot \delta(t - 4T)$
$+27.8343 \cdot \delta(t - 5T)$
$+31.7866 \cdot \delta(t - 6T)$
$+35.1351 \cdot \delta(t - 7T)$
$+37.9481 \cdot \delta(t - 8T)$
$+40.2981 \cdot \delta(t - 9T)$
$+42.2544 \cdot \delta(t - 10T)$
$+ \cdots$

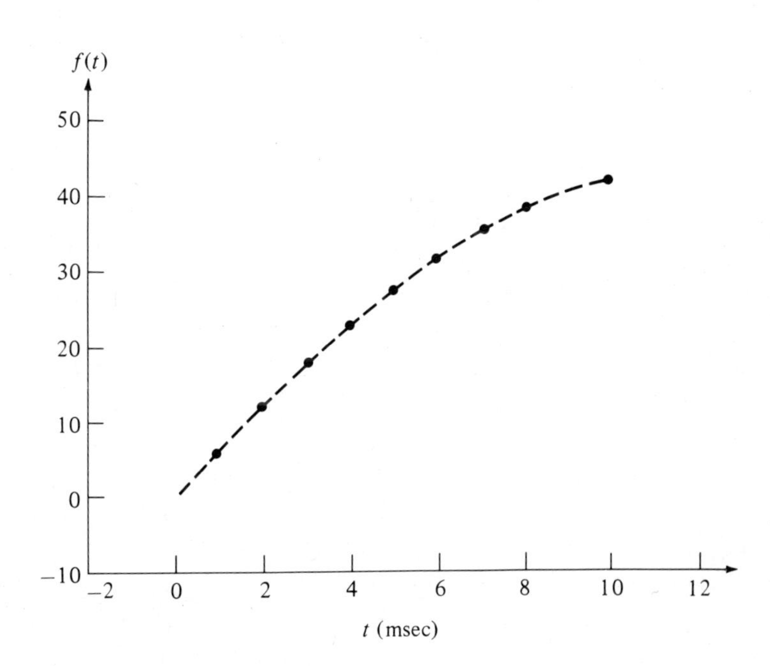

Figure 12.13a

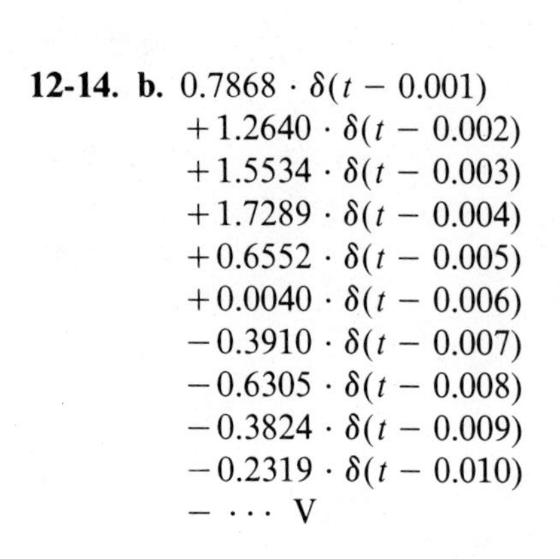

12-14. **b.** $0.7868 \cdot \delta(t - 0.001)$
$+1.2640 \cdot \delta(t - 0.002)$
$+1.5534 \cdot \delta(t - 0.003)$
$+1.7289 \cdot \delta(t - 0.004)$
$+0.6552 \cdot \delta(t - 0.005)$
$+0.0040 \cdot \delta(t - 0.006)$
$-0.3910 \cdot \delta(t - 0.007)$
$-0.6305 \cdot \delta(t - 0.008)$
$-0.3824 \cdot \delta(t - 0.009)$
$-0.2319 \cdot \delta(t - 0.010)$
$- \cdots$ V

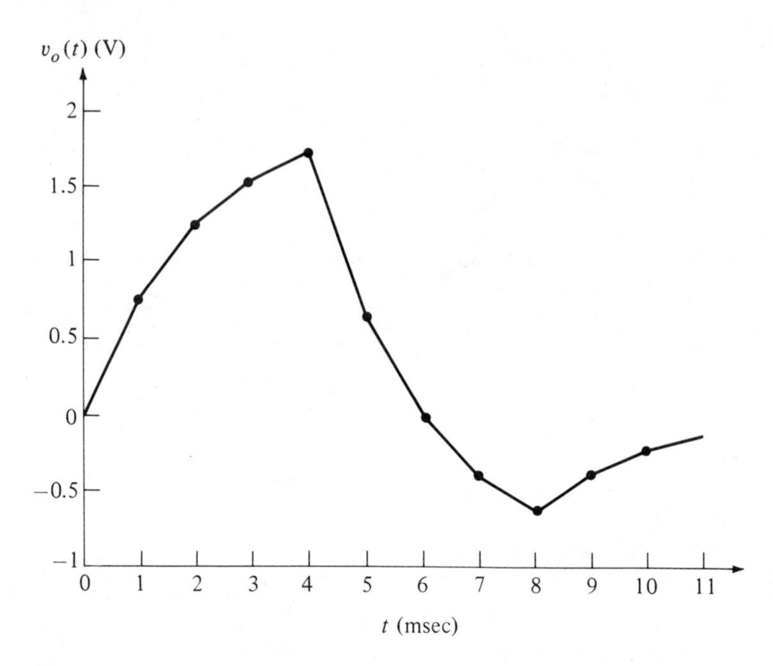

Figure 12.14b

d. $0.3469 \cdot \delta(t - 0.005)$
$+0.7717 \cdot \delta(t - 0.006)$
$+1.0293 \cdot \delta(t - 0.007)$
$+0.9711 \cdot \delta(t - 0.008)$
$+0.5889 \cdot \delta(t - 0.009)$
$+0.3572 \cdot \delta(t - 0.010)$
$+0.2167 \cdot \delta(t - 0.011)$
$+0.1315 \cdot \delta(t - 0.012)$
$+0.0799 \cdot \delta(t - 0.013)$
$+0.0485 \cdot \delta(t - 0.014)$
$+ \cdots$ V

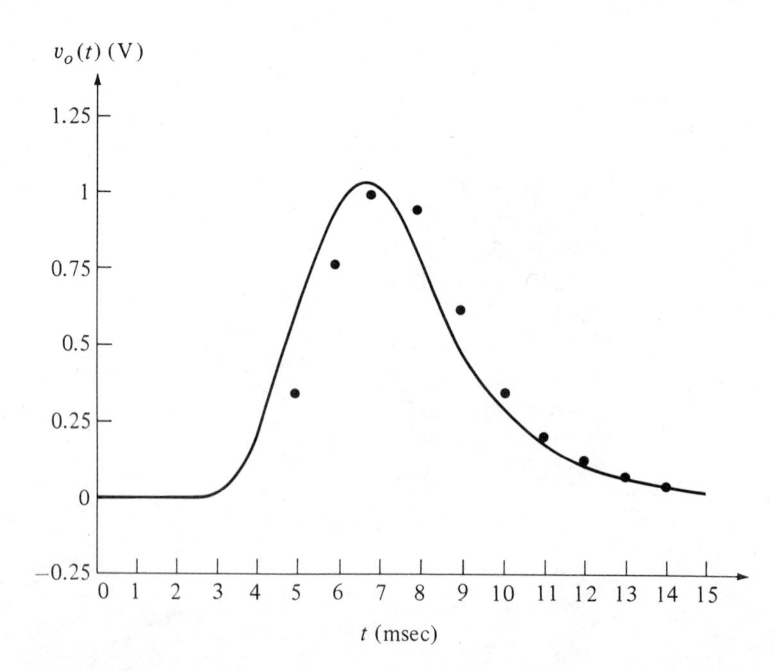

Figure 12.14d

12-15. **a.** $0.5 \cdot \delta(t - 0.001)$
$+0.9094 \cdot \delta(t - 0.002)$
$+1.2445 \cdot \delta(t - 0.003)$
$+1.5189 \cdot \delta(t - 0.004)$
$-0.7565 \cdot \delta(t - 0.005)$
$-0.6194 \cdot \delta(t - 0.006)$
$-0.5071 \cdot \delta(t - 0.007)$
$-0.4151 \cdot \delta(t - 0.008)$
$-0.3399 \cdot \delta(t - 0.009)$
$-0.2783 \cdot \delta(t - 0.010)$
$- \cdots$ V

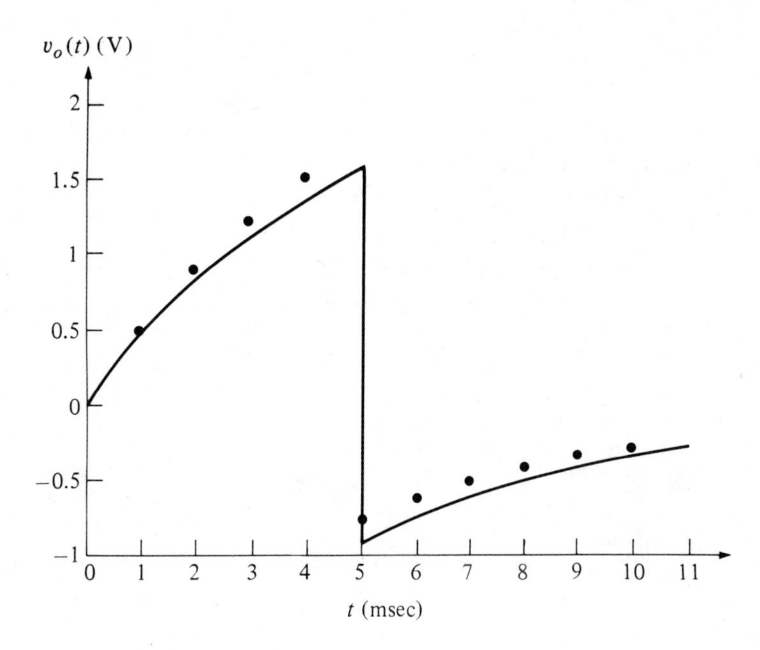

Figure 12.15a

c. $1.25 \cdot \delta(t - 0.001)$
$+2.2734 \cdot \delta(t - 0.002)$
$+1.8612 \cdot \delta(t - 0.003)$
$+1.5238 \cdot \delta(t - 0.004)$
$+1.2475 \cdot \delta(t - 0.005)$
$+1.0213 \cdot \delta(t - 0.006)$
$+0.2112 \cdot \delta(t - 0.007)$
$-0.4512 \cdot \delta(t - 0.008)$
$-0.9951 \cdot \delta(t - 0.009)$
$-1.4397 \cdot \delta(t - 0.010)$
$- \cdots$ V

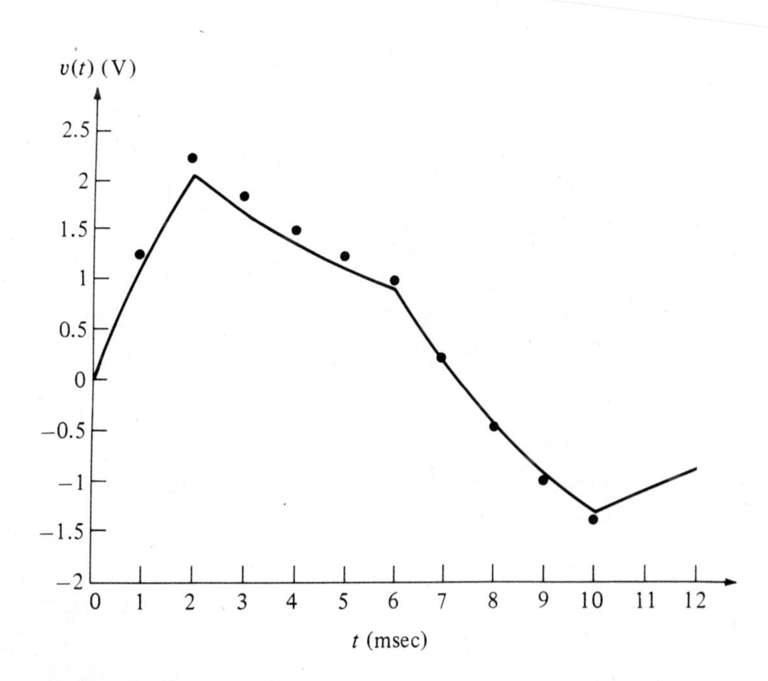

Figure 12.15c

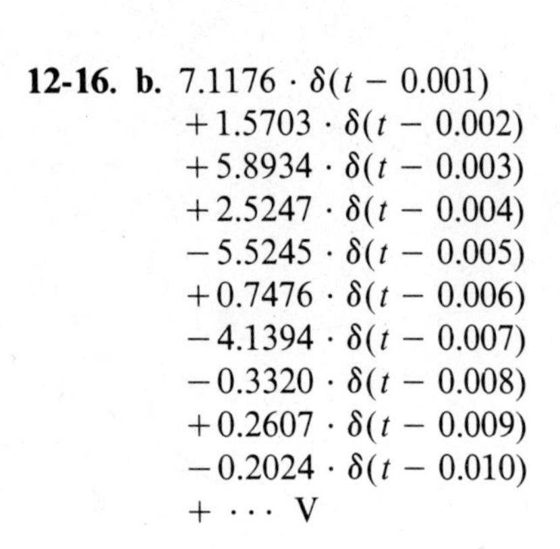

12-16. b. $7.1176 \cdot \delta(t - 0.001)$
$+1.5703 \cdot \delta(t - 0.002)$
$+5.8934 \cdot \delta(t - 0.003)$
$+2.5247 \cdot \delta(t - 0.004)$
$-5.5245 \cdot \delta(t - 0.005)$
$+0.7476 \cdot \delta(t - 0.006)$
$-4.1394 \cdot \delta(t - 0.007)$
$-0.3320 \cdot \delta(t - 0.008)$
$+0.2607 \cdot \delta(t - 0.009)$
$-0.2024 \cdot \delta(t - 0.010)$
$+ \cdots$ V

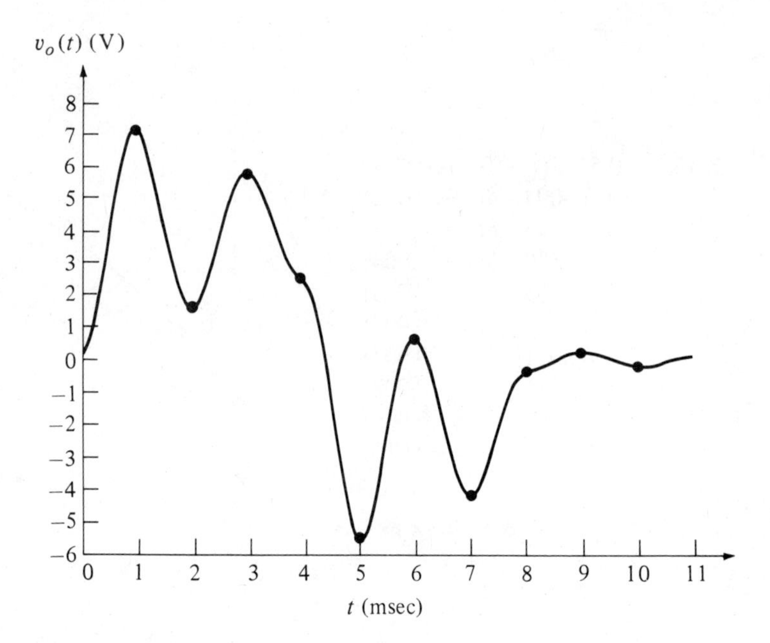

Figure 12.16b

d. $3.1378 \cdot \delta(t - 0.005)$
$+2.6314 \cdot \delta(t - 0.006)$
$+3.0256 \cdot \delta(t - 0.007)$
$+0.7789 \cdot \delta(t - 0.008)$
$-0.6083 \cdot \delta(t - 0.009)$
$+0.4732 \cdot \delta(t - 0.010)$
$-0.3693 \cdot \delta(t - 0.011)$
$+0.2880 \cdot \delta(t - 0.012)$
$-0.2236 \cdot \delta(t - 0.013)$
$+0.1754 \cdot \delta(t - 0.014)$
$+ \cdots$ V

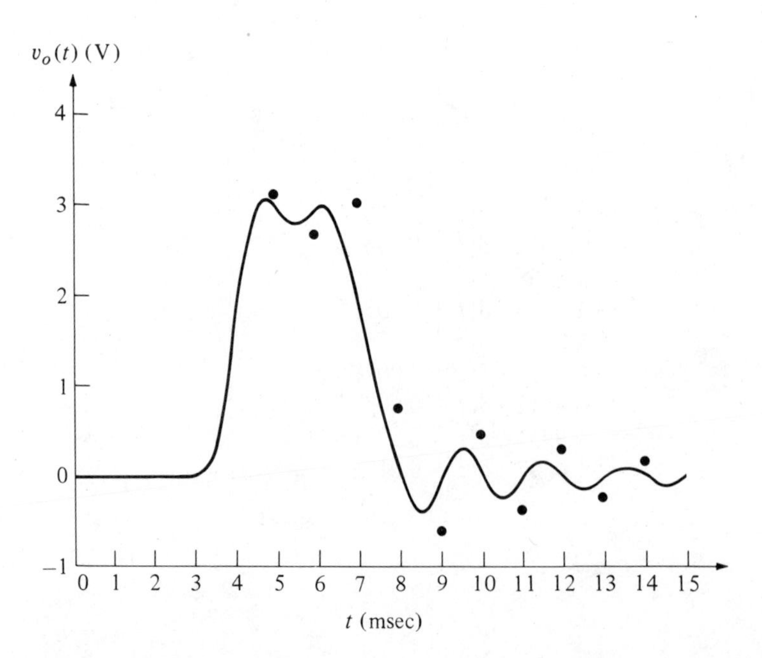

Figure 12.16d

12-17. **a.** $\delta(t - 0.001)$
$+0.2206 \cdot \delta(t - 0.002)$
$+0.8280 \cdot \delta(t - 0.003)$
$+0.3547 \cdot \delta(t - 0.004)$
$-4.2765 \cdot \delta(t - 0.005)$
$+3.3331 \cdot \delta(t - 0.006)$
$-2.5977 \cdot \delta(t - 0.007)$
$+2.0243 \cdot \delta(t - 0.008)$
$-1.5774 \cdot \delta(t - 0.009)$
$+1.2290 \cdot \delta(t - 0.010)$
$\pm \cdots$ V

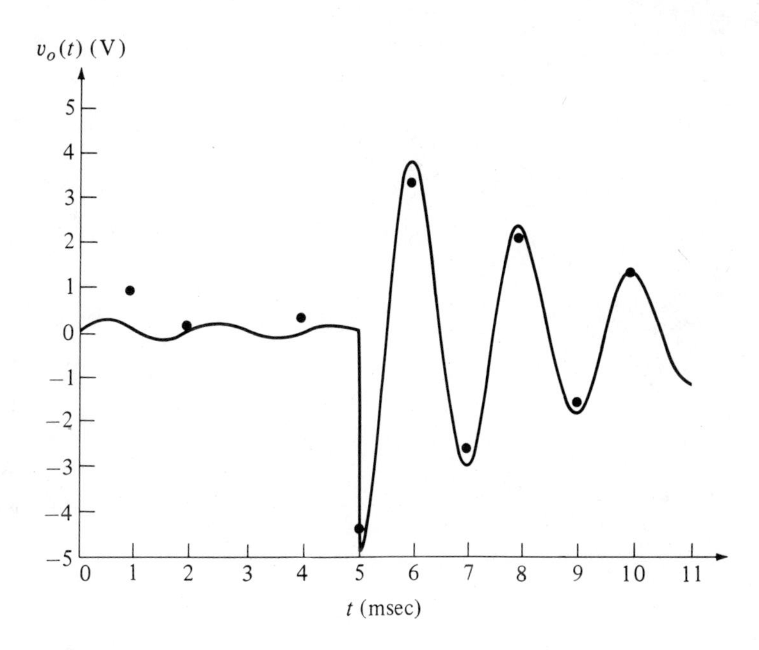

Figure 12.17a

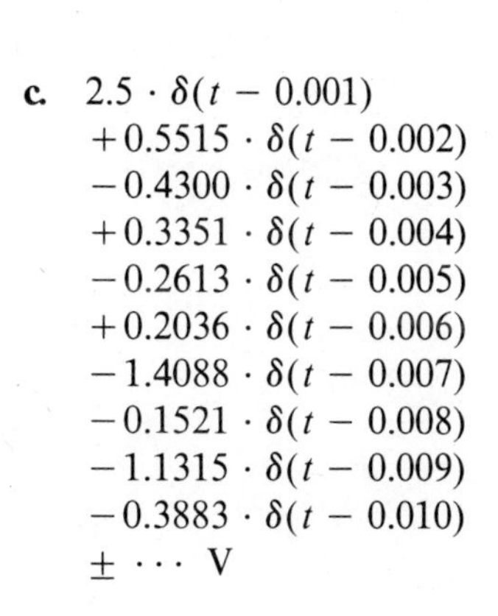

c. $2.5 \cdot \delta(t - 0.001)$
$+0.5515 \cdot \delta(t - 0.002)$
$-0.4300 \cdot \delta(t - 0.003)$
$+0.3351 \cdot \delta(t - 0.004)$
$-0.2613 \cdot \delta(t - 0.005)$
$+0.2036 \cdot \delta(t - 0.006)$
$-1.4088 \cdot \delta(t - 0.007)$
$-0.1521 \cdot \delta(t - 0.008)$
$-1.1315 \cdot \delta(t - 0.009)$
$-0.3883 \cdot \delta(t - 0.010)$
$\pm \cdots$ V

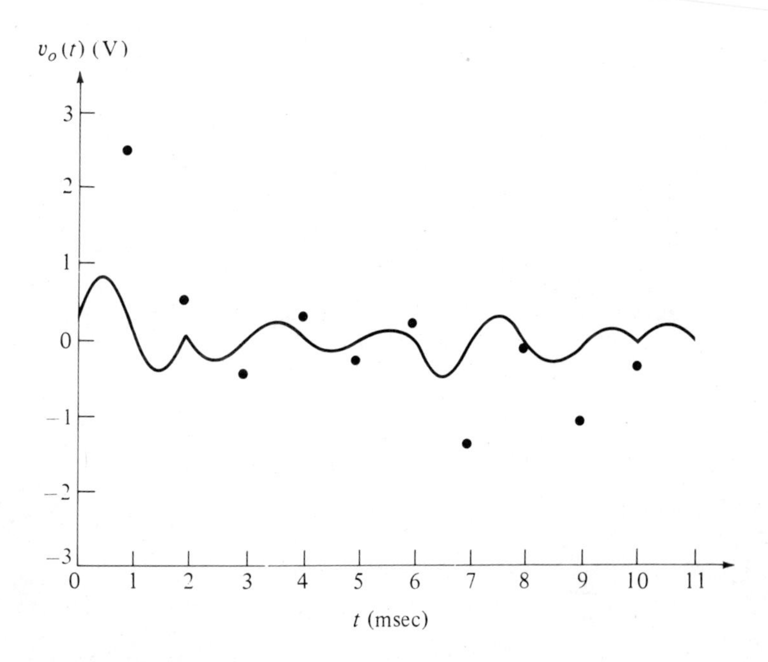

Figure 12.17c

INDEX

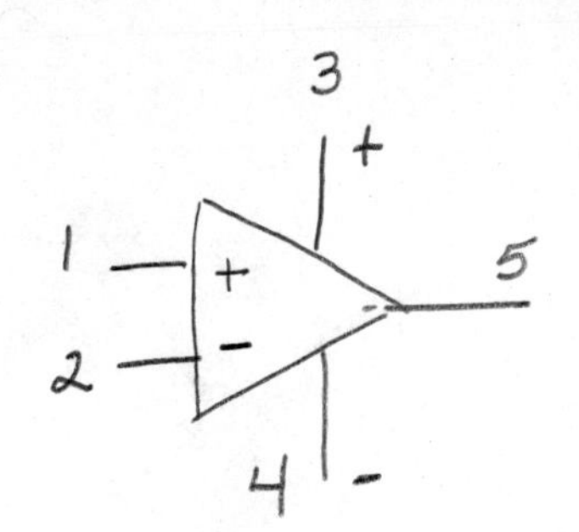

X741 1 2 3 4 5 ua741

.LIB EVAL.LIB

$s^2 + 2\zeta\omega_n s + \omega_n^2$

$0 < \zeta < 1$ underdamped

$\zeta > 1$ overdamped

$\zeta = 1$ critically damped